E. J. LONG

MOLECULAR SPECTRA
and
MOLECULAR STRUCTURE

I. SPECTRA OF DIATOMIC MOLECULES

BY

GERHARD HERZBERG, F.R.S.
National Research Council of Canada

With the co-operation, in the first edition, of
J. W. T. SPINKS, F.R.S.C.

SECOND EDITION

VAN NOSTRAND REINHOLD COMPANY

 New York   Cincinnati   Toronto   London   Melbourne

Dedicated

to the Memory of

WALTER CHARLES MURRAY

First President

of the

University of Saskatchewan

VAN NOSTRAND REINHOLD COMPANY REGIONAL OFFICES:
New York   Cincinnati   Chicago   Millbrae   Dallas

VAN NOSTRAND REINHOLD INTERNATIONAL OFFICES:
London   Toronto   Melbourne

Copyright © 1950 by LITTON EDUCATIONAL PUBLISHING, INC.

ISBN: 0-442-03385-0

First Edition Copyright 1939 by Prentice-Hall, Inc.

MANUFACTURED IN THE UNITED STATES OF AMERICA.

Published by VAN NOSTRAND REINHOLD COMPANY
450 West 33rd Street, New York, N.Y. 10001

Published simultaneously in Canada by
VAN NOSTRAND REINHOLD LTD.

21   20   19

PREFACE

Eleven years ago I published a volume entitled *Molecular Spectra and Molecular Structure I. Diatomic Molecules* which was followed in 1945 by a second volume *Infrared and Raman Spectra of Polyatomic Molecules.* The first volume has been out of print for a number of years but the demand for it seemed to justify a new edition. Although the book has been completely revised and brought up to date, its general plan has remained substantially unchanged. Concerning this plan it seems therefore appropriate to quote from the preface of the first edition:

"I have endeavored to give a presentation which is readable by the beginner in the field and also will be useful to those who do or want to do research work in this field. In order to assist the former, I have frequently made use of small type for those sections that are not necessary for an understanding of the fundamentals. For the benefit of those working in the field, numerous references to original papers have been included.

"A satisfactory presentation of molecular spectra and molecular structure is nowadays not possible without treating thoroughly, apart from the empirical results, the theoretical background also. Therefore I have included as much of the theory of molecular spectra as is possible without going into the more difficult mathematical details. A large number of diagrams, graphical representations of eigenfunctions and potential curves, as well as energy level diagrams, serve to illustrate and to explain the theory. On the other hand, I have added numerous carefully selected spectrograms of bands and band systems (some of which have been taken specially for this purpose) in order to give an accurate idea of the experimental material that forms the basis of the developments.

"While of course most of the material presented is not new, it seems that the actual procedure followed in analyzing a band spectrum has not previously been given as specifically in a book of this kind. The same holds for the applications of band spectra to other parts of physics, to chemistry, and to astrophysics given in the last chapter. I hope that both these features will be found useful."

In the eleven years since the publication of the first edition the subject *Spectra of Diatomic Molecules* has developed vigorously even though not as rapidly as in the preceding two decades. Most of the progress made has been consolidation and slow evolution rather than revolution. Exceptions to this statement are the amazing advances made by applying the new tools of molecular beams and microwaves to diatomic molecular problems.

iii

Naturally I have incorporated these advances of recent years in the present new edition. In addition I have amplified and extended the theoretical sections of the book in response to suggestions of several critics of the first edition. I have added new sections on Radiofrequency (Microwave) Spectra (in Chapter II), Hyperfine Structure (in Chapter V) and Intensities of Electronic Transitions (in Chapter VI) and I have completely rewritten many other sections. In order to improve the presentation and to bring it up to date I have scrutinized every sentence and every illustration. Thus the reader familiar with the old edition will find much revision and modification in detail. Twenty-six new illustrations have been added of which some, like Fig. 57 showing the symmetric top eigenfunctions, have not previously been given in the literature of the subject.

The table of molecular constants of the first edition included only the ground states of all diatomic molecules then known. Following the suggestions of many spectroscopists and in view of the fact that the available tables that include the excited states are now out-dated I decided to undertake the rather arduous task of preparing a comprehensive table of all known states of all known diatomic molecules. This table is presented in the Appendix (Table 39) and now covers 80 pages. I have made every effort to make this table as comprehensive and consistent as possible (see the introduction to it, p. 501). The literature up to September 1949 and in some cases up to February 1950 has been included in the table.

The manuscript of the text matter of the book was completed in August 1948, and only in a few cases was it possible to include later developments in the proofs.

As in the first edition the detailed subject index at the end of the book includes all the more important symbols and quantum numbers used as well as all the molecules treated. Part II of the bibliography of the first edition was taken over and a new Part III added giving all the references to work published since then or to earlier work not used in the first edition.

G. HERZBERG

OTTAWA, ONT.
February, 1950

ACKNOWLEDGMENTS

It gives me great pleasure to acknowledge the help and cooperation I have received from many persons during the preparation of this book. My sincere thanks are due to Dean J. W. T. Spinks who translated the original version into English and to Professor R. N. H. Haslam who contributed many stylistic improvements to the first edition. I am very grateful to Professor R. S. Mulliken for suggesting several important improvements. I should like to express my thanks to Professor S. Chandrasekhar for making many valuable suggestions and for reading parts of the manuscript and the proofs. Many colleagues helped by supplying advance information on molecular constants for Table 39. Of these I should like to mention particularly Dr. R. F. Barrow (Oxford) and Professor E. Miescher (Basel). The help given by Dr. J. G. Phillips (Williams Bay) and Dr. D. Andrychuk (Ottawa) in especially taking a number of new spectrograms for the present edition is much appreciated. Thanks are also due to the following spectroscopists for supplying the originals of various spectrograms of the new and the old edition: Dr. E. Bengtsson-Knave (Stockholm), Professor R. T. Birge (Berkeley), Dr. B. A. Brice (Philadelphia), Professor W. G. Brown (Chicago), Professor F. A. Jenkins (Berkeley), Professor R. Mecke (Freiburg), Professor R. S. Mulliken (Chicago), Dr. R. W. B. Pearse (London), Professor F. Rasetti (Baltimore), the late Lord Rayleigh, and Professor R. W. Wood (Baltimore). Finally I am greatly indebted to my wife who prepared all the illustrations of the first edition and made numerous suggestions for improving the text of the new edition.

G. HERZBERG

CONTENTS

vii

INTRODUCTION

General remarks. In the course of the last three decades, very considerable progress has been made in the investigation and theoretical interpretation of molecular spectra. As a result the study of molecular spectra has become one of the most important, perhaps the most important, means for investigating molecular structure.

From the spectra the various discrete energy levels of a molecule can be derived directly. From these again, we can obtain detailed information about the motion of the electrons (electronic structure) and the vibration and rotation of the nuclei in the molecule. The study of electronic motions has led to a theoretical understanding of chemical valence. From the vibrational frequencies the forces between the atoms in the molecule, as well as the heats of dissociation of molecules, can be calculated with great accuracy. From the rotational frequencies we obtain accurate information about the geometrical arrangement of the nuclei in the molecule—in particular, extremely accurate values of the internuclear distances.

The knowledge of the various properties of the individual molecules so obtained allows us to understand many of the physical and chemical properties of the gases under consideration and, in fact, in many instances allows us to predict these properties—for example, the specific heat and the paramagnetic susceptibility. Also, on the basis of this knowledge, chemical equilibria can be predicted theoretically with great accuracy and chemical elementary processes can be elucidated.

A further important result of the investigation of molecular spectra is that proof has been obtained of the existence of a large number of molecules which were previously unknown in chemistry or were not thought capable of free existence. Among these are CH, OH, C_2, He_2, Na_2, CP, and many others. The structures of these molecules have also been determined.

The investigation of the spectra of diatomic molecules is also of great significance for the physics of atomic nuclei, since certain nuclear properties influence these spectra in a characteristic manner and can therefore be determined from them, and since, furthermore, in some cases rare isotopes may be detected by means of these spectra.

Finally, in recent years, a new field of application of the knowledge of molecular spectra has opened up in astrophysics. Not only can the presence of various molecules on fixed stars, planets, and comets, in the upper atmosphere, and in interstellar space be detected on the basis of their spectra, but also from a more detailed analysis definite conclusions can be drawn concerning the physical conditions in these objects.

This first volume deals with the spectra of *diatomic molecules* and the con-

1

clusions that can be drawn from them concerning the structure of these molecules. The spectra of polyatomic molecules will be taken up in the second and third volumes of this series. The former, already published (28),[1] deals with the infrared and Raman spectra; the latter will deal with the electronic spectra of polyatomic molecules. Since a thorough treatment of molecular spectra presumes a certain knowledge of atomic spectra and atomic structure, a short survey of the fundamentals of atomic theory is given in the first chapter of this volume.

The experimental methods for the production and investigation of spectra will not be dealt with here. Reference should be made to Sawyer (4b), Harrison, Lord, and Loofbourow (4c), Meissner (4a), Baly (2), Brode (3), Jenkins and White (6b), and Kayser-Konen (1).

The nomenclature of the various quantities concerning the molecule—energy levels, quantum numbers, and so on—was made subject to international regulation in 1930 [Mulliken (515)]. We shall use this *international nomenclature* throughout.

TABLE 1.   PHYSICAL CONSTANTS

Velocity of light in vacuum	c	2.99776×10^{10} cm/sec
Planck's constant[1a]	h	$6.623_4 \times 10^{-27}$ erg sec
Electronic charge	e	4.8024×10^{-10} abs. e.s.u.
Electronic mass (rest mass)	m	9.1055×10^{-28} gm
$\frac{1}{16}$ mass of the O^{16} atom	M_1	$1.6597_2 \times 10^{-24}$ gm
Number of atoms in a mol:		
Referred to Aston's atomic weight scale ($O^{16} = 16$)	N_A	6.0251×10^{23}
Referred to the chemical atomic weight scale	N_{ch}	6.0235×10^{23}
Boltzmann's constant	k	1.38033×10^{-16} ergs/degree
Gas constant per mol[2] (chemical)	R	$\begin{cases} 8.31439 \times 10^7 \text{ ergs/degree/mol} \\ 1.98719 \text{ cals/degree/mol} \end{cases}$
1 thermochemical calorie (defined)[3]	cal	4.18400 abs. joules

Physical constants. The best values of various fundamental physical constants have been the subject of much discussion. We accept here the values recently recommended by DuMond and Cohen (914) which are based on and

---

[1] The numbers in parentheses refer to the bibliography, p. 583.

[1a] In an erratum received too late for inclusion in this book, DuMond and Cohen give $6.623_{73} \times 10^{-27}$ for h. The difference is smaller than the probable error.

[2] The gas constant is referred to the standard atmosphere 1013250 dynes/cm^2 as adopted by the International Commission on Weights and Measures [see Rossini (50b)] not 1013246 dynes/cm^2 as used by Birge and DuMond.

[3] This *defined calorie* should be distinguished from the calorie of the international steam tables which is used by engineers and defined by 1 I.T. steam calorie = 4.18674 abs. joules. The 15° calorie is given by 1 cal_{15} = 4.1855 abs. joules.

substantially identical with the values derived by Birge (809–811), but we use the defined (thermochemical) calorie recommended by the Bureau of Standards [Mueller and Rossini (1252)] instead of the 15° calorie used by Birge. In Table 1 the values most frequently used in this book are collected. These values differ only very slightly from those used in the first edition of this volume which were taken from an earlier version of Birge's list (107–110) and which are also used in Volume II.

Units and conversion factors. The wave lengths, λ, of the spectral lines in the infrared are measured in μ ($1\mu = 1/1000$ mm) and in the visible and ultra-violet spectral regions in Ångström units (1 Å $= 10^{-7}$ mm $= 10^{-8}$ cm). In theoretical discussions the frequency $\nu' = c/\lambda$ is much more important than the wave length. Instead of it usually the wave number $\nu = \nu'/c = 1/\lambda$ is used which is proportional to the frequency and is measured in cm^{-1} (number of waves per cm). The wave numbers can be given to the same accuracy as the wave lengths and are not limited to the accuracy with which the velocity of light is known. However, in order to have a constant proportionality factor ($1/c_{vac.}$) it is necessary to use the *wave number in vacuo*, which is the reciprocal of the wave length in vacuo. Since the wave lengths of spectral lines above 2000 Å are usually given as measured in air, they have to be converted to vacuum [by adding $(n-1)\lambda_{air}$ where n is the index of refraction of air] before the reciprocal is taken. For the region 2000–10,000 Å this calculation can be avoided by using Kayser's Tabelle der Schwingungszahlen (45).[4] According to Babcock (762) the use of these tables can be extended to 5 μ by inverting them and adding a very small correction. For still longer wave lengths $n-1$ is very nearly constant ($=2.726 \times 10^{-4}$ at 15° C and 760 mm pressure) so that the vacuum correction is easily made. In infrared work carried out before 1940 the vacuum correction has usually been neglected. This must be kept in mind when molecular constants are calculated from infrared data.

The frequency ν', and the wave number, ν, depend on the energy, E, according to the fundamental Planck relation $E = h\nu' = h\nu c$, where h is Planck's constant. We can therefore take the *wave number as a measure of the energy*, as is very frequently done in spectroscopy. Using the values in Table 1, we find that 1 cm^{-1} corresponds to $1.9855_4 \times 10^{-16}$ erg/molecule. If we wish to refer the energy to 1 mol instead of to a single atom or molecule, we have to multiply by the number of molecules N_{ch} in 1 mol. Therefore, 1 cm^{-1} corresponds to $11.959_9 \times 10^7$ ergs/mol (referred to the chemical atomic-weight scale) or, converting to calories, 2.8584_8 cal/mol.

Another unit used in measuring energy in atomic and molecular physics is the (absolute) *electron-volt* (e.v.). By V e.v. is meant an energy equal to the

---

[4] In doing so it must be remembered that Kayser's tables were calculated with values for the index of refraction of air which are now outdated. For very precise calculations the more recent values of Barrell and Sears (770) should be used [see Birge (811)].

kinetic energy of an electron (or singly charged ion) that has been accelerated through a potential difference of V volts. This energy is $eV/299.776$ ergs (the factor $1/299.776$ is introduced in converting to electrostatic units). Substituting the numerical value for e gives 1 e.v. equivalent to $1.60199_6 \times 10^{-12}$ erg/molecule, or 8068.3_2 cm^{-1}, or $23,063._2$ cal/mol. These and some other derived conversion factors for the energy units are collected in Table 2.

TABLE 2.   CONVERSION FACTORS FOR ENERGY UNITS.

Unit	cm^{-1}	ergs/molecule	cal/mol$_{chem}$	electron-volts
1 cm^{-1}	1	$1.9855_4 \times 10^{-16}$	2.8584_8	$1.23941_6 \times 10^{-4}$
1 erg/molecule	$5.0364_1 \times 10^{15}$	1	$1.43965_1 \times 10^{16}$	$6.2422_1 \times 10^{11}$
1 cal/mol$_{chem}$	0.349836_6	6.94612×10^{-17}	1	$4.3359_2 \times 10^{-5}$
1 electron-volt	8068.3_2	$1.60199_6 \times 10^{-12}$	$23063._2$	1

CHAPTER I

RÉSUMÉ OF THE ELEMENTS OF ATOMIC STRUCTURE[1]

1. Bohr Theory

Stationary energy states. Niels Bohr was the first to suggest that an atom or molecule cannot exist in states having any arbitrary energy, but only in certain discrete energy states called *stationary states*. According to the Bohr theory, these stationary states are obtained when one selects, by means of certain *quantum conditions*, some of the states from the continuous range of classically possible states. For example, in the case of a single electron moving in the Coulomb field of an atomic nucleus, we obtain, classically, elliptical orbits of any half-axes. According to the Bohr theory, only certain of these, with quite definite values of the axes, are possible. For these selected orbits, the major axis is proportional to the square of a whole number n, called the *principal quantum number* while the minor axis is proportional (with the same proportionality factor) to the product of n and a second whole number $l + 1$, where l is the so-called *azimuthal quantum number*. While the principal quantum number can take any integral value greater than 0,

$$n = 1, 2, 3, \cdots, \qquad (I, 1)$$

the azimuthal quantum number l, for a given value of n, can take only the values

$$l = 0, 1, 2, \cdots, n - 1. \qquad (I, 2)$$

The energy E of the atom contains an arbitrary additive constant. If this constant is chosen so that $E = 0$ when the electron is completely separated from the nucleus and when at the same time there is no relative kinetic energy, the following expression for the energy of the possible elliptical orbits (Bohr orbits) in the hydrogen atom or hydrogen-like ions is obtained (neglecting a very small relativity correction):

$$E_n = -\frac{2\pi^2\mu e^4}{h^2}\frac{Z^2}{n^2} = -\frac{R'Z^2}{n^2}. \qquad (I, 3)$$

In this expression, Z is the atomic number (for hydrogen, $Z = 1$) and μ is the reduced mass, $\mu = mM/(m + M)$, where m is the mass of the electron and M that of the nucleus. Thus, neglecting relativity effects, the energy depends

---

[1] A more detailed presentation has been given in the author's Atomic Spectra and Atomic Structure (Dover Publications, New York, 1944, in the following referred to as A.A.). Several other presentations are listed in the bibliography at the end of the book (9a–10).

only on the principal quantum number n.   It is negative for all stationary states. The state with lowest energy is that with $n = 1$, corresponding to the innermost orbit.   It is the ground state of the atom or ion.   With increasing n, the energy increases and eventually approaches the limit $E = 0$.

In the more general case of the motion of a single electron about an atomic core which is not a point charge, the orbits are no longer simple ellipses but, to a first approximation, ellipses whose axes rotate uniformly (*rosette motion*).   As before, they can be characterized by the two quantum numbers n and l.   But now the energy is no longer independent of l (even when relativity effects are disregarded).   To a first approximation it is given by

$$E_{n,l} = - \frac{R'Z^2}{(n + a)^2},$$ (I, 4)

where a, the so-called *Rydberg correction*, depends on the azimuthal quantum number l and rapidly approaches zero with increasing l. Z is now the charge on the core—that is, the nuclear charge (atomic number) minus the number of core electrons.

As an example, Fig. 1 gives a diagrammatic representation of the observed energy levels of the lithium atom, which has a single electron revolving about an atomic core consisting of the nucleus and two electrons.   The energy levels are indicated by horizontal lines whose vertical height corresponds to the energy of the atom as given by the energy scales to the left and right.   For each value of l there is a series of energy levels (each drawn in a separate column) which correspond to the different values of n and approach one another as n increases, finally going to the limit $E = 0$.   In agreement with the relation (I, 2) the smallest value of n for a given l is $l + 1$ except for $l = 0$ where it is 2.   The reason for this exception will be explained on p. 22.   The energy level diagram of the hydrogen atom is quite similar, except that, according to (I, 3), all levels with the same n have the same height, and that the level with $l = 0$, $n = 1$ does occur.

Radiation.   According to Bohr, electromagnetic radiation is not emitted while an electron moves in its orbit (as it should be according to classical electrodynamics), but only when the electron goes from one quantum orbit of energy E_1 to another of energy E_2.   This process is called a *quantum jump*.   The liberated energy $E_1 - E_2$ is emitted as a light quantum (*photon*) of energy $h\nu' = hc\nu$ (ν' = frequency, ν = wave number of the light) so that

$$h\nu' = hc\nu = E_1 - E_2.$$ (I, 5)

This is *Bohr's frequency condition*.   It holds for absorption as well as for emission.   That is to say, a light quantum is absorbed by an atom only when the energy of the light quantum is precisely equal to the energy which is necessary to bring the atom from the state E_2 to the state E_1.

From (I, 5) it follows that the *wave number of the emitted or absorbed light* is

$$\nu = \frac{E_1}{hc} - \frac{E_2}{hc}.\qquad\qquad\text{(I, 6)}$$

If we substitute the energy values (I, 3) of the hydrogen atom or of the hydrogen-like ions (He^+, Li^{++}, $\cdots$) into this expression, we obtain

$$\nu = RZ^2\left(\frac{1}{n_2{}^2} - \frac{1}{n_1{}^2}\right),\qquad n_1 > n_2,\qquad\text{(I, 7)}$$

where $R = R'/hc = 2\pi^2\mu e^4/ch^3$ is the *Rydberg constant* and n_2 and n_1 are the principal quantum numbers of the two states concerned. The whole spectrum

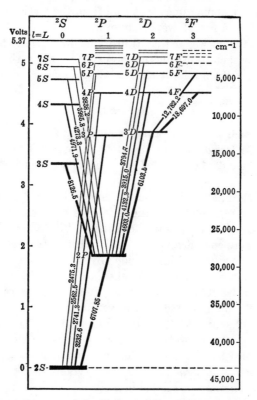

FIG. 1. **Energy Level Diagram of the Li Atom (after Grotrian).** The wave lengths of the spectral lines are written on the sloping connecting lines, which represent the transitions (see below). The symbols S, P, $\cdots$ are explained on p. 27. The numbers in front of these symbols are the principal quantum numbers n of the outer electron.

of the hydrogen atom ($Z = 1$) and of the hydrogen-like ions is represented with great accuracy by the simple equation (I, 7). When $n_2 = 2$ and $Z = 1$, the formula for the well-known *Balmer series of the hydrogen atom* results. A spectrogram of this series is given in Fig. 6, p. 30. As n_1 increases, the lines of this series draw rapidly together and go to the limit $R/4$ for $n_1 \to \infty$. The other series represented by equation (I, 7) behave correspondingly.

In the more general case, using (I, 4), we obtain

$$\nu = RZ^2 \left[\frac{1}{(n_2 + a_2)^2} - \frac{1}{(n_1 + a_1)^2} \right], \tag{I, 8}$$

where the first and second terms now depend on l as well as n. Therefore a greater number of transitions results. For fixed values of n_2, a_2, and a_1 and variable n_1 we obtain a series of lines similar to the Balmer series illustrated in Fig. 6. Such series are sometimes called *Rydberg series*. In Fig. 1, in which the transitions are indicated by sloping lines joining the levels in question, a Rydberg series corresponds to sloping lines that lead from the levels of a series, having a definite value of l, to a fixed lower-lying level.

It can be seen from Fig. 1 that transitions do not take place between *any* two levels, but only such transitions are observed for which the quantum number l changes by unity. Such a limitation is called a *selection rule*. In the present case it is written

$$\Delta l = \pm 1. \tag{I, 9}$$

Transitions for which l does not alter or alters by 2, 3, $\cdots$ units do not occur and are said to be *forbidden*. In the Bohr theory, this selection rule can be derived with the help of the so-called correspondence principle.

Terms. According to (I, 6), (I, 7), and (I, 8), the wave number ν of a spectral line can be represented as the difference between two quantities called *terms*. More precisely, the positive quantities RZ^2/n^2 or $RZ^2/(n + a)^2$ and similar, possibly more complicated, expressions are designated as terms. Terms are thus the negative energy values divided by hc. A series of terms of the form $RZ^2/(n + a)^2$ with $n = 1, 2, 3, \cdots$ is called a *Rydberg series of terms*.

The terms can be obtained empirically from the observed spectrum. The first term in (I, 8), for example, is always given by the wave number of the limit of the series of spectral lines in question, while, for the second term, a has to be chosen so that all the observed lines of the series are accurately represented.

Often, the word *term* is used synonymously with *energy state* or *quantum state*, and the expression *term value* is used synonymously with *energy value*. This practice is even more justifiable in molecular than in atomic spectroscopy, since here a zero of energy differing from that for atoms is used almost exclusively —namely, the ground state of the molecule. Hence term value and energy value differ only by the factor $1/hc$ (without the minus sign). In the following we shall employ frequently this somewhat looser usage of the concept "term."

Continuous spectra. All stable, discrete quantum states of an atom with one outer electron, according to (I, 3) or (I, 4), have negative energy values if the usual zero-point of the energy scale is used. A positive value of E corresponds to an energy greater than that of a system consisting of an electron and an ion infinitely separated and at rest. That is to say, the two particles approach or separate from each other with a kinetic energy that is not zero even at infinity. The electron then moves in a *hyperbolic orbit*, through which it goes, of course, only once. Such aperiodic motions are not quantized. The kinetic energy of the two particles can take any value; that is to say, all positive values of E are possible. Hence, extending from the limit $E = 0$ of the series of discrete energy values of an atom to higher energy values, there is a continuous region of possible energy values (*continuous term spectrum*).

Corresponding to the continuous region of energy values, there is a continuous absorption or emission spectrum, which arises when the electron jumps from a discrete state into the continuous region or from the latter into the former. The continuous absorption spectrum therefore corresponds to the *removal of the electron* (photoeffect) with more or less kinetic energy; the continuous emission spectrum corresponds to the *capture of the electron*. The continuum begins at the limit of the line series having the same principal quantum number n_2 of the lower state and extends toward shorter wave lengths. The *series limit*, which is at the same time the starting point of the continuum, corresponds to the removal or capture of the electron with zero velocity; that is to say, it gives directly the energy required to separate the electron from the atom in the lower state of the line series (*ionization potential*).

One can say quite generally that continuous spectra of gases always correspond to such splitting or recombination processes, since of all forms of molecular energy only the kinetic energy of translation is not quantized.

2. Wave Mechanics (Quantum Mechanics)

Fundamental equations. The essential point in the Bohr theory, the idea of *discrete stationary states*, is retained in wave mechanics, but the derivation of these stationary states is altered. Moreover the wave-mechanical energy values for non-hydrogen-like systems agree quantitatively with experiment, while the values calculated according to the Bohr theory do not.

Whereas the Bohr theory starts out from the classical laws of motion, and subsequently, by means of certain quantum conditions, selects as possible only a few of the orbits so obtained, wave mechanics puts in the place of the classical laws of motion another law, which is, however, so constituted that for large masses it goes over asymptotically into the classical laws of motion. Wave mechanics is therefore more comprehensive than classical mechanics, which fails for atomic dimensions.

The first step toward wave mechanics was De Broglie's fundamental idea that the motion of any corpuscle of matter is associated with a wave motion of wave length

$$\lambda = \frac{h}{mv},$$

(I, 10)

where m is the mass and v the velocity of the corpuscle. This idea was later on brilliantly confirmed by the discovery of the diffraction of corpuscular rays (electrons as well as atoms), so that the existence of waves of matter (*De Broglie waves*) is now an established fact.

Accordingly, when we want to investigate the motions in an atomic system we must investigate the wave motions associated with them. Let Ψ be the *wave function*, whose physical meaning will be discussed later. Since we are dealing with a wave motion, Ψ must vary periodically with time at every point in space. We can therefore write

$$\Psi = \psi \sin 2\pi\nu't \quad \text{or} \quad \Psi = \psi \cos 2\pi\nu't,$$

or, combining both together,

$$\Psi = \psi e^{-2\pi i\nu't}. \tag{I, 11}$$

Here ν' is the frequency of the vibrations, and ψ is the amplitude of the wave motion.

In the case of the motion of a single mass point of mass m, the amplitude ψ depends only on the coordinates x, y, and z. For this case Schrödinger put forward the equation

$$\frac{\partial^2\psi}{\partial x^2} + \frac{\partial^2\psi}{\partial y^2} + \frac{\partial^2\psi}{\partial z^2} + \frac{8\pi^2 m}{h^2}(E - V)\psi = 0, \tag{I, 12}$$

where V is the potential energy and $E = h\nu'$ is the total energy of the masspoint. He added the rather natural restriction that the solutions ψ of this equation be everywhere single-valued, finite, and continuous and vanish at infinity. The *Schrödinger equation* (I, 12) together with this restriction replaces for atomic systems the fundamental equations of motion of classical mechanics. For the potential energy V the same expression is used as in classical theory, that is, the laws of force are unchanged.

The mathematical investigation of equation (I, 12) shows that under the restrictions given, it is in general soluble only for certain definite values of E, the so-called *eigenvalues*, just as the differential equation of a vibrating string can be solved only for certain definite frequencies, those of the fundamental and the overtones. Thus the discrete energy values of an atomic system which are experimentally observed in the spectrum appear, according to Schrödinger, as the eigenvalues of the wave equation of the system.

For the case of *several (N) particles*, (I, 12) must be replaced by the equation

$$\sum_k \frac{1}{m_k}\left(\frac{\partial^2\psi}{\partial x_k^2} + \frac{\partial^2\psi}{\partial y_k^2} + \frac{\partial^2\psi}{\partial z_k^2}\right) + \frac{8\pi^2}{h^2}(E - V)\psi = 0, \tag{I, 13}$$

where ψ now depends on the $3N$ co-ordinates, x_k, y_k, z_k, and where m_k is the mass of the kth particle.

If, in (I, 12), we substitute for V the potential energy $(-Ze^2/r)$ of an electron in the field of a fixed nucleus with a charge Ze, the solution of the wave equation leads to the energy values (I, 3) with m in place of μ [see texts on wave mechanics (11a–16)]. If we drop the assumption that the nucleus is fixed, equation (I, 13) has to be used. The resulting energy values are then exactly those of (I, 3). On the other hand, when we are not dealing with the field of a point charge but with that of an atomic core, energy values of the form (I, 4) result. In both cases it is found that in addition to the discrete negative energy values (I, 3) or (I, 4) all positive values of E are possible; that is to say, a continuous region ($E > 0$) adjoins the series of discrete energy levels ($E < 0$) just as in the old Bohr theory.

Even for systems with several electrons,—for example, for He and Li$^+$—the solution of the wave equation leads to energy values that are not only qualitatively but also quantitatively in agreement with the empirical values. This is in contrast to the Bohr theory, which leads in these cases to values for the ionization potential and other term values that are entirely at variance with experiment.

Instead of starting out from the amplitude equation (I, 12) or (I, 13) one may also use, and sometimes has to use, a differential equation for the time dependent wave function Ψ. This wave equation for the non-relativistic case of several particles is, according to Schrödinger,

$$-\frac{h^2}{8\pi^2} \sum \frac{1}{m_k}\left(\frac{\partial^2\Psi}{\partial x_k{}^2} + \frac{\partial^2\Psi}{\partial y_k{}^2} + \frac{\partial^2\Psi}{\partial z_k{}^2}\right) + V\Psi = i\,\frac{h}{2\pi}\,\frac{\partial\Psi}{\partial t}. \tag{I, 14}$$

By substituting for Ψ a product of a function ψ of the coordinates alone and a function of the time alone, one is immediately led to the relation (I, 11) and, with $\nu' = E/h$, to the amplitude equation (I, 13) [see for example Pauling and Wilson (13)].

In more complicated problems it is often convenient to employ a form of the Schrödinger equation that makes use of a close formal analogy between classical and quantum-theoretical quantities. In the classical treatment of a conservative system the sum H of kinetic energy T and potential energy V of a dynamical system is a constant E:

$$H = T + V = E. \tag{I, 15}$$

H, expressed in terms of momenta and coordinates (or generally in terms of canonically conjugate quantities) is called the *Hamiltonian function*. For example, for the case of several particles, using Cartesian coordinates we have

$$H = \sum \frac{1}{2m_k}\,(p_{xk}{}^2 + p_{yk}{}^2 + p_{zk}{}^2) + V(x_1, y_1, z_1, \cdots x_k, y_k, z_k, \cdots) \tag{I, 16}$$

where p_{xk}, p_{yk}, p_{zk} are the components of the momentum and x_k, y_k, z_k the coordinates of particle k. If now the momenta in H are replaced by differential operators[2]

$$p_{xk} \to \frac{h}{2\pi i}\,\frac{\partial}{\partial x_k},\ p_{yk} \to \frac{h}{2\pi i}\,\frac{\partial}{\partial y_k},\ p_{zk} \to \frac{h}{2\pi i}\,\frac{\partial}{\partial z_k} \tag{I, 17}$$

and the whole equation is multiplied by ψ (where multiplication of an operator by ψ means that the differentiations implied in the operator are carried out on ψ) the Schrödinger equation (I, 13) for the amplitude ψ is obtained.

Symbolically we can therefore write for the Schrödinger equation

$$H\psi = E\psi \tag{I, 18}$$

[2] The symbol $\hbar$ is often used for $h/2\pi$ in this and other relations.

where H is the *Hamiltonian operator*, obtained from H of (I, 15) by the substitutions (I, 17). The usefulness of the form (I, 18) of the Schrödinger equation rests on the fact that it allows one to set up in a simple way the Schrödinger equation in terms of coordinates other than Cartesian coordinates. This is important since as a rule the potential energy can be expressed much more simply in terms of coordinates other than Cartesian coordinates. In such cases the kinetic energy has to be expressed in terms of generalized momenta p_k which are canonically conjugate to the coordinates q_k used, that is, so that

$$p_k = \frac{\partial T}{\partial \dot{q}_k} \qquad \text{(I, 19)}$$

where as usual $\dot{q}_k = dq_k/dt$.

If in (I, 18) the operator $-\dfrac{h}{2\pi i}\dfrac{\partial}{\partial t}$ is substituted for the factor E the time dependent Schrödinger equation [(I, 14) in the case of Cartesian coordinates] is obtained.

Physical interpretation of the Ψ function. The complete solution of the Schrödinger equation [(I, 12) or (I, 13)] yields not only the eigenvalues—that is, the energy values of the stationary states—but also the corresponding functions ψ (or $\Psi = \psi e^{-2\pi i(E/h)t}$), the so-called *eigenfunctions*. Just as in the case of vibrating strings or membranes these eigenfunctions have a certain integral number of nodes or nodal surfaces. These integral numbers are the quantum numbers that occur in the energy formulae.

While in the case of mechanical vibrations the wave function simply gives the displacement from the equilibrium position for every point of the system the physical interpretation of Schrödinger's Ψ function is less obvious: As was first postulated by Born $\Psi\Psi^* \, d\tau = \left|\Psi\right|^2 d\tau = \left|\psi\right|^2 d\tau$ (where Ψ^* is the complex conjugate of Ψ) gives the *probability that the particle under consideration will be found in the volume element $d\tau$ at the position given by the co-ordinates*. Consequently $\left|\psi\right|^2$ is the *probability density*. Its dependence on the co-ordinates —that is, the *probability density distribution* of the electron or electrons in an atom—is the essential thing that wave mechanics has to tell us about the motions in an atom. (See A.A., Fig. 21, p. 44). However, *it tells us nothing about the orbits of the electrons*. The eigenfunction ψ itself is sometimes referred to as the *probability amplitude*.

The electron may be observed at any point in space with a certain probability; it is, so to speak, smeared out over the whole of space. However, the ψ function decreases exponentially with increasing distance from the nucleus, and the probability of finding the electron appreciably outside the region of the old Bohr orbits is extremely small.

On the other hand, in the states of positive energy, the eigenfunction is an outgoing or incoming spherical wave corresponding to the removal or capture of the electron. Here the decrease in $\left|\psi\right|^2$ is only inversely proportional to the square of the distance from the center and not exponential.

Frequently there are several different eigenfunctions associated with one and the same eigenvalue. In that case we have a *degenerate state*. The degeneracy is d-fold if there are d linearly independent eigenfunctions for a given eigenvalue. These d eigenfunctions are not uniquely determined by the wave equation since, on account of the linearity of the latter, any linear combination of the eigenfunctions belongs also to the same eigenvalue.

The time dependent Schrödinger equation (I, 14) is fulfilled also by general solutions of the form

$$\Psi = \sum_n c_n \Psi_n = \sum_n c_n \psi_n e^{-2\pi i (E_n/h)t} \tag{I, 20}$$

where the ψ_n are eigenfunctions of the amplitude equation and where the c_n are constant coefficients. Such a wave function according to the statistical interpretation tells us that the system under consideration may be in any of the states of energy E_n, and that the probability of finding it in the state E_n is $c_n c_n{}^*$.

Orthogonality and normalization of eigenfunctions. It can be shown [see, for example, Pauling and Wilson (13)] that any two eigenfunctions of an atomic system belonging to different eigenvalues are *orthogonal* to each other, that is,

$$\int \psi_n \psi_m{}^* \, d\tau = 0 \qquad \text{for } n \neq m, \tag{I, 21}$$

where n and m represent the quantum numbers of two different states, and where the integration extends over the whole configuration space.

While degenerate eigenfunctions belonging to a given eigenvalue are not necessarily orthogonal to one another it is always possible to find a set of linearly independent functions belonging to this eigenvalue which are mutually orthogonal.

If an eigenfunction is multiplied by a constant factor the function obtained is still an eigenfunction belonging to the same eigenvalue. For many purposes it is convenient to multiply the eigenfunctions by such factors that for all of them

$$\int \psi_n \psi_n{}^* \, d\tau = 1. \tag{I, 22}$$

The eigenfunctions are then said to be *normalized to unity.*

If, in addition, the wave function Ψ of (I, 20) is normalized to unity it is easily seen that for the coefficients c_n

$$\sum c_n c_n{}^* = 1. \tag{I, 23}$$

Perturbation theory. Only in very few cases can the wave equation be solved rigorously. Usually, particularly for molecular problems, we must start out from a wave equation in which certain interactions have been neglected. Let (I, 13) with a certain potential energy V be the approximate ("unperturbed") Schrödinger equation and let

$$\sum \frac{1}{m_k} \left(\frac{\partial^2 \psi}{\partial x_k{}^2} + \frac{\partial^2 \psi}{\partial y_k{}^2} + \frac{\partial^2 \psi}{\partial z_k{}^2} \right) + \frac{8\pi^2}{h^2} (E - V)\psi + W\psi = 0 \tag{I, 24}$$

be the Schrödinger equation of the "perturbed" system where $W\psi$ is the *perturbation term* representing the originally neglected interactions. W may be an additional term in the potential energy or it may be an operator arising from an additional term in the kinetic energy (see p. 11).

Considering first the case of non-degenerate states we shall let the eigenvalues and eigenfunctions of the unperturbed Schrödinger equation (I, 13) be

$$E_1{}^0, E_2{}^0, E_3{}^0, \cdots, \psi_1{}^0, \psi_2{}^0, \psi_3{}^0, \cdots \tag{I, 25}$$

and those of the perturbed equation (I, 24)

$$E_1, E_2, E_3, \cdots, \psi_1, \psi_2, \psi_3, \cdots \tag{I, 26}$$

According to wave-mechanical perturbation theory these "true" eigenvalues and eigenfunctions can be expressed in terms of the unperturbed eigenvalues and eigenfunctions and the perturbation function. It is found that, up to the second approximation, the eigenvalues are [see, for example, Pauling and Wilson (13)].

$$E_n = E_n{}^0 + W_{nn} + \sum_{\substack{i=1 \\ i \neq n}}^{\infty} \frac{|W_{ni}|^2}{E_n{}^0 - E_i{}^0}. \tag{I, 27}$$

In this equation

$$W_{ni} = \int \psi_n{}^0 {}^* W \psi_i{}^0 \, d\tau \tag{I, 28}$$

are the so-called *matrix elements of the perturbation function* W. To a *first* approximation, perturbation theory gives for the true *eigenfunctions*[3]

$$\psi_n = \psi_n{}^0 + \sum_{\substack{i=1 \\ i \neq n}}^{\infty} \frac{W_{in}}{E_n{}^0 - E_i{}^0} \psi_i{}^0. \tag{I, 29}$$

It can be seen immediately from (I, 27 and 29) (and it is also plausible) that for a sufficiently small perturbation function W the true eigenvalues and eigenfunctions lie very close to the approximate eigenvalues and eigenfunctions. The deviations depend on the eigenfunctions and eigenvalues of all the other states of the system as well as on the magnitude of the perturbation function. This state of affairs may also be expressed by saying that the perturbation causes a mutual "interaction" of the unperturbed energy levels: Each level $(E_i{}^0)$ produces a shift of any other $(E_n{}^0)$ and at the same time contributes to the eigenfunction of the other; both effects increase with decreasing separation of the two unperturbed levels, and of course, the larger the perturbation function W, that is, the poorer the zero-order approximation, the larger will be these effects. It is easily seen that the shift of $E_n{}^0$ produced by $E_i{}^0$ is opposite and equal to that of $E_i{}^0$ produced by $E_n{}^0$, and that the direction of the shift is always such that the separation of the two levels is increased. In other words there is an *apparent repulsion* of any two states which is stronger the closer they are together. However, it must be realized that this repulsion affects only the unperturbed energy levels. The actual energy levels do not, of course, affect one another.

The formulae (I, 27) and (I, 29) hold only if the separation of $E_n{}^0$ and $E_i{}^0$ is not too small or zero. If it is zero we have degeneracy and the introduction of the perturbation will cause a *splitting of this degeneracy*. We consider only the case of two-fold degeneracy. Let E^0 be the zero order energy value and $\psi_1{}^0$ and $\psi_2{}^0$ two linearly independent, mutually orthogonal, and normalized eigenfunctions. With decreasing strength of the perturbation the true eigenfunctions will not in general approach $\psi_1{}^0$ and $\psi_2{}^0$ but will approach two linear combinations of these:

$$\chi_a{}^0 = c_{11}\psi_1{}^0 + c_{12}\psi_2{}^0$$
$$\chi_b{}^0 = c_{21}\psi_1{}^0 + c_{22}\psi_2{}^0 \tag{I, 30}$$

These zero-order eigenfunctions are often sufficiently good approximations to the true eigenfunctions. The coefficients c_{ik} are found to be determined by the equations

$$(W_{11} - \epsilon_a)c_{11} + W_{12}c_{12} = 0$$
$$W_{21}c_{11} + (W_{22} - \epsilon_a)c_{12} = 0 \tag{I, 31}$$

---

[3] Quite generally, $\psi_n = \sum_i a_{in}\psi_i{}^0$, since the functions $\psi_i{}^0$ form a complete orthogonal system and any function can be developed into a series of functions of such a complete orthogonal system.

and

$$(W_{11} - \epsilon_b)c_{21} + W_{12}c_{22} = 0$$
$$W_{21}c_{21} + (W_{22} - \epsilon_b)c_{22} = 0 \tag{I, 32}$$

where the W_{ik} are given as before by (I, 28) and where $\epsilon_a = E_a - E^0$ and $\epsilon_b = E_b - E^0$ are the shifts of the perturbed energy levels from the unperturbed position (in first-order approximation). The two equations (I, 31) give non-zero solutions for c_{11} and c_{12}, and similarly (I, 32) for c_{21} and c_{22}, only if

$$\begin{vmatrix} W_{11} - \epsilon & W_{12} \\ W_{21} & W_{22} - \epsilon \end{vmatrix} = 0 \tag{I, 33}$$

The two roots ϵ_a and ϵ_b of this secular equation added to E^0 yield the first order energy values E_a and E_b which are thus immediately given as soon as the matrix elements W_{ik} have been evaluated. Substituting ϵ_a and ϵ_b into (I, 31) and I, 32) gives relative values of the coefficients c_{ik}. The absolute values are obtained by normalizing the functions $\chi_a{}^0$ and $\chi_b{}^0$.

Heisenberg's uncertainty principle. The fact that, in wave mechanics, the orbits of the old Bohr theory have lost their meaning is closely connected with the Heisenberg uncertainty principle. According to it, the *position and momentum* (or velocity) *of a particle cannot be simultaneously measured with any desired accuracy.* Rather, if Δx is the uncertainty of the x co-ordinate and Δp_x is the uncertainty of the x component of the momentum ($m\Delta v_x$), always

$$\Delta x \cdot \Delta p_x \geqq \frac{h}{2\pi}, \tag{I, 34}$$

where h is Planck's constant. Analogous relations hold for the other co-ordinates, and also, for example, for energy and time:

$$\Delta E \cdot \Delta t \geqq \frac{h}{2\pi}. \tag{I, 35}$$

These uncertainty relations are a direct result of the wave properties of matter (see A.A., Chapter I) and represent *a limitation in principle* of the accuracy of measurements in atomic systems.

Momentum and angular momentum. Even though in wave mechanics the orbits of the electron cannot be given—that is to say, the momentum is not defined for every point—yet a probability density distribution of the momentum can be given (see A.A., p. 47). Quite definite statements may be made regarding the *angular momentum* of an atom. For any mechanical system the angular momentum is given by $\sum m_i v_i \rho_i$, where ρ_i is the perpendicular distance of the momentum $m_i v_i$ of the ith particle from a fixed point, usually the center of mass. In wave mechanics, as in classical mechanics, the angular momentum of an isolated system is constant.[4] According to wave mechanics, it can take only discrete values as can the energy; in the case of one electron, it

---

[4] For a closed system (on which no external forces act) the angular momentum is independent of the point of reference. Usually the center of mass is chosen as this point.

can have the values $\sqrt{l(l+1)}\,(h/2\pi)$, or, approximately $l(h/2\pi)$. The azimuthal quantum number l gives us therefore the orbital angular momentum of the electron in units $h/2\pi$. Consequently, the various term series of lithium, for example (see Fig. 1), differ in the angular momentum of the electron.

The fact that even in wave mechanics the angular momentum has a perfectly definite value shows that we can still speak of an electron revolving about an atomic nucleus, provided that we are quite clear that there is no point in speaking of definite orbits. Consequently, we can in many instances use the angular momentum in the same way as in the Bohr theory; for example, under certain conditions, angular momenta can be added vectorially (see section 3).

Throughout this book we shall use bold face letters for angular momentum vectors and thus distinguish them from the corresponding quantum numbers, printed in regular type. Thus l means an angular momentum vector of magnitude $\sqrt{l(l+1)}\,(h/2\pi) \approx l(h/2\pi)$.

It should be noticed that even in the continuous states of positive energy the angular momentum l of the electron can have only the discrete values given above.

In wave mechanics the classical dynamical variables are replaced by operators. Classically, the components of the angular momentum in the case of a single particle are

$$P_x = yp_z - zp_y, \qquad P_y = zp_x - xp_z, \qquad P_z = xp_y - yp_x \qquad (\text{I, 36})$$

which upon substitution of (I, 17) yields the corresponding operators

$$P_x = \frac{h}{2\pi i}\left(y\frac{\partial}{\partial z} - z\frac{\partial}{\partial y}\right), \quad P_y = \frac{h}{2\pi i}\left(z\frac{\partial}{\partial x} - x\frac{\partial}{\partial z}\right), \quad P_z = \frac{h}{2\pi i}\left(x\frac{\partial}{\partial y} - y\frac{\partial}{\partial x}\right) \quad (\text{I, 37})$$

or, introducing spherical polar coordinates, r, ϑ, φ, [see Rojansky (15b)]

$$P_x = -\frac{h}{2\pi i}\left(\sin\varphi\,\frac{\partial}{\partial\vartheta} + \cot\vartheta\cos\varphi\,\frac{\partial}{\partial\varphi}\right)$$

$$P_y = \frac{h}{2\pi i}\left(\cos\varphi\,\frac{\partial}{\partial\vartheta} - \cot\vartheta\sin\varphi\,\frac{\partial}{\partial\varphi}\right) \qquad (\text{I, 38})$$

$$P_z = \frac{h}{2\pi i}\frac{\partial}{\partial\varphi}$$

The square of the magnitude of the angular momentum is therefore represented by the operator

$$P^2 = P_x^2 + P_y^2 + P_z^2 = -\frac{h^2}{4\pi^2}\left(\frac{\partial^2}{\partial\vartheta^2} + \cot\vartheta\,\frac{\partial}{\partial\vartheta} + \frac{1}{\sin^2\vartheta}\frac{\partial^2}{\partial\varphi^2}\right) \qquad (\text{I, 39})$$

In quantum mechanics the possible values of a dynamical variable are the eigenvalues of the corresponding operator; that is, the possible values of the square of the magnitude of the angular momentum are the eigenvalues a of the equation

$$P^2\psi = a\psi. \qquad (\text{I, 40})$$

It is found that these eigenvalues are

$$a = l(l+1)\frac{h^2}{4\pi^2} \qquad (\text{I, 41})$$

which leads to the value for the magnitude of the angular momentum $|P|$ given above. The eigenfunctions of the Hamiltonian operator [equation (I, 18)] are also eigenfunctions of the angular momentum operator. That is why in every stationary state (characterized by a certain eigenfunction) the atomic system considered has not only a definite energy but also a definite angular momentum.

Space quantization. According to classical theory an atom brought into a magnetic or electric field F will carry out a precession such that the angular momentum l describes a cone with the field direction as axis and with a constant component m_l of the angular momentum as shown diagrammatically in Fig. 2.

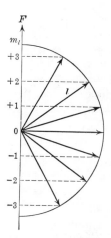

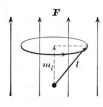

Fig. 2. Precession of the Angular Momentum l of an Electron in a Magnetic or an Electric Field F.

Fig. 3. Space Quantization of l in a Field F for $l = 3$. The figure also holds generally for other angular momentum vectors (see p. 26). The length of the vector l is $\sqrt{3 \times 4} = 3.464$ units.

In quantum theory, too, the component of the angular momentum in the field direction is constant. However, while classically m_l can have any value between $+|l|$ and $-|l|$ (that is, any angle between l and the field is possible) in quantum theory only the discrete values $m_l(h/2\pi)$ are possible where

$$m_l = l, (l-1), (l-2), \cdots, -l. \tag{I, 42}$$

that is, only certain angles of l with respect to the field direction are possible. This space quantization is shown in Fig. 3 for $l = 3$. The number m_l is called the *magnetic quantum number* of the electron. Since the magnitude of the vector l is $\sqrt{l(l+1)}\,(h/2\pi)$ which is larger than the largest value of $m_l(h/2\pi)$ it is clear that l can never point exactly in the field direction as it could in the old Bohr theory, a fact that is intimately connected with the Heisenberg uncertainty relation (see A.A. p. 102). The space quantization persists and the magnetic quantum number remains defined even for vanishingly small field. For zero field the $2l + 1$ values of m_l in (I, 42) for a given n and l correspond to $2l + 1$ different modes of motion (eigenfunctions) of the same energy. We have a $(2l + 1)$-fold degeneracy.

The component of the angular momentum in the field direction corresponds to the operator P_z of (I, 38). It can be shown that the eigenfunctions of the Hamiltonian operator are simultaneously eigenfunctions not only of the angular momentum operator P^2 but also of P_z. Instead of P_z any other component in a fixed direction may be chosen but not more than one. The eigenvalues of P_z are $m_l(h/2\pi)$ with m_l given by (I, 42).

Electron spin. Investigations of the multiplet structure of spectral lines (for example, the alkali doublets) and of the anomalous Zeeman effect have led, at first quite independently of quantum mechanics, to the assumption (Goudsmit and Uhlenbeck) that the electron has an *angular momentum of its own,* called *electron spin;* that is to say, the electron rotates about its own axis. The corresponding quantum number is $s = \frac{1}{2}$, and the angular momentum itself, s, has the magnitude $\sqrt{s(s + 1)}(h/2\pi) = \frac{1}{2}\sqrt{3}(h/2\pi)$. The electron spin appears as a necessary result of Dirac's relativistic wave mechanics. [See texts on wave mechanics (11a–16).]

Dirac showed that in a magnetic field, the spin of the electron can set itself only so that its component in the field direction has the value $m_s(h/2\pi)$, where m_s can be only $+\frac{1}{2}$ or $-\frac{1}{2}$. Without a field there is a two-fold degeneracy. But whenever there is a non-zero orbital angular momentum, an internal magnetic field arises which causes a splitting of the double degeneracy even without an external magnetic field (see also below).

Emission and absorption of radiation. If the interaction with an electromagnetic field is introduced into the Schrödinger equation of an atomic system, it is found that a non-zero probability arises of finding the system in a state E_n if originally it was in the state E_m, and if radiation of wave number $\nu = (E_n - E_m)/hc$ is present. If $E_n < E_m$, radiation of this wave number is emitted by the atomic system. Thus Bohr's frequency condition (I, 6) holds also in wave mechanics. The frequencies of the emitted or absorbed spectral lines are determined in this simple fashion by the eigenvalues of the Schrödinger equation—that is, by the energy values of the system.

The probabilities of the transitions under the influence of radiation are determined by the eigenfunctions of the states involved. Thus the *intensities* of the emitted or absorbed spectral lines can be obtained. In particular, when the eigenfunctions are known, we can calculate whether or not two states can combine with each other; that is to say, we obtain the *selection rules.*

The interaction of an electromagnetic wave having an electric vector E, with an atomic system is in a first approximation the interaction with the (variable) electric dipole moment $\boldsymbol{M}$ of the system whose components are

$$M_x = \sum e_k x_k, \quad M_y = \sum e_k y_k, \quad M_z = \sum e_k z_k \tag{I, 43}$$

where the e_k are the charges on the N particles of coordinates x_k, y_k, z_k. If the interaction energy $\boldsymbol{M} \cdot \boldsymbol{E}$ is introduced into the wave equation [see for example Pauling and Wilson (13)] it is found that the probability of a transition between two states n and m produced by the interaction is proportional to the square of the magnitude of certain vector quantities $\boldsymbol{R}^{nm}$—the *matrix elements of the electric dipole moment*—whose components depend in the following manner on the eigenfunctions ψ_n and ψ_m of the two states:

$$\boldsymbol{R}_x{}^{nm} = \int \psi_n{}^* M_x \psi_m \, d\tau, \quad \boldsymbol{R}_y{}^{nm} = \int \psi_n{}^* M_y \psi_m \, d\tau,$$

$$\boldsymbol{R}_z{}^{nm} = \int \psi_n{}^* M_z \psi_m \, d\tau \tag{I, 44}$$

The integrals are to be taken over the whole configuration space of the $3N$ coordinates. If the matrix element $\boldsymbol{R}^{nm}$ (*transition moment*) differs from zero for two states n and m, the two states combine with each other with a certain probability with emission or absorption of radiation; if it is zero the transition under consideration is forbidden as a dipole transition.

For example in the case of a one-electron system it is found that all $\boldsymbol{R}^{nm}$ vanish with the exception of those for which the quantum number l for the two states differs by only 1; that is to say, we have the selection rule

$$\Delta l = \pm 1. \tag{I, 45}$$

This is the same selection rule as that already given above in the Bohr theory [equation (I, 9)] and found to agree with experiment.

If one puts $n = m$ in (I, 44) one obtains the permanent dipole moment of the system, which of course equals zero in the case of an atom, but may be different from zero for a molecule.

A non-zero transition probability may also be produced by the interaction of an electromagnetic wave with the magnetic dipole moment, the quadrupole moment, or higher moments of the atomic system; or in the process of scattering of light a transition may be produced on account of the induced dipole moment; or a transition may be caused by collisions with electrons, atoms, or ions. In these cases the corresponding quantity—magnetic dipole moment, quadrupole moment, induced dipole moment, interaction with colliding electron, atom, or ion—has to be substituted in (I, 44) in place of M_x, M_y, M_z.

If, for a given transition, the matrix elements (I, 44) of the electric dipole moment are zero, the corresponding spectral line may still appear in absorption or emission if the matrix elements of the *magnetic dipole or quadrupole moment* are different from zero. However, calculation shows that magnetic dipole transition probabilities are only 10^{-5}, quadrupole transition probabilities only 10^{-8} of dipole transition probabilities. Therefore transitions that cannot occur with dipole radiation are always considered as *forbidden transitions*. There are of course also transitions that are forbidden for any kind of radiation.

The *intensity* of a spectral line in *emission* $I_{\text{em.}}{}^{nm}$ is defined as the energy emitted by the source per second. If there are N_n atoms in the initial state and if A_{nm} is the fraction of atoms in the initial state carrying out the transition to m per second then

$$I_{\text{em.}}{}^{nm} = N_n h c \nu_{nm} A_{nm} \qquad (I, 46)$$

where $hc\nu_{nm}$ is the energy of each light quantum of wave number ν_{nm} emitted in the transition. A_{nm} is the *Einstein transition probability of spontaneous emission* which, according to wave mechanics in the case of dipole radiation, is related to the matrix element of the transition as follows:

$$A_{nm} = \frac{64\pi^4 \nu_{nm}{}^3}{3h} \left| R^{nm} \right|^2 \qquad (I, 47)$$

Thus we have for the intensity of an emission line

$$I_{\text{em.}}{}^{nm} \sim \nu_{nm}{}^4 \left| R^{nm} \right|^2 \qquad (I, 48)$$

The order of magnitude of A_{nm} for strong dipole transitions is 10^8 sec^{-1}. Similar formulae for A_{nm} hold for magnetic dipole and quadrupole radiation; the order of magnitude of A_{nm} for these radiations is 10^3 and 1 sec^{-1}.

The definition of the intensity of an *absorption* line is complicated by the effects of natural line width. These effects can be disregarded for sufficiently thin absorbing layers. Let Δx be the thickness of the layer and ρ_{nm} the density of radiation of the incident beam of wave number ν_{nm}. Then the intensity of absorption, that is, the energy absorbed from the incident beam of 1 cm^2 cross section is given by

$$I_{\text{abs.}}{}^{nm} = \rho_{nm} N_m B_{mn} \Delta x \, hc\nu_{nm} \qquad (I, 49)$$

where N_m is the number of atoms per cm^3 in the initial (lower) state m and B_{mn} is the *Einstein transition probability of absorption*. The factor $\rho_{nm} N_m B_{mn}$ represents the number of transitions per cm^3 per second produced by the incident radiation.

If $I_0{}^{nm} = c\rho_{nm}$ is the intensity of the incident radiation (that is, the energy falling on 1 cm^2 per second) equation (I, 49) may also be written

$$I_{\text{abs.}}{}^{nm} = I_0{}^{nm} N_m B_{mn} h\nu_{nm} \cdot \Delta x \qquad (I, 50)$$

It is assumed here that the incident radiation has a constant intensity for a wave number interval about ν_{nm} sufficient to cover the whole line width. Both ρ_{nm} and $I_0{}^{nm}$ are expressed in terms of energy per unit wave number interval. For B_{mn} wave mechanics supplies the relation

$$B_{mn} = \frac{8\pi^3}{3h^2 c} \left| R^{nm} \right|^2 \qquad (I, 51)$$

which implies

$$B_{mn} = \frac{1}{8\pi h c \nu_{nm}{}^3} A_{nm} \tag{I, 52}$$

Thus we have for the intensity of an absorption line for thin absorbing layers

$$I_{\text{abs.}}{}^{nm} \sim \nu_{nm} \left| R^{nm} \right|^2 \tag{I, 53}$$

The different dependence of absorption and emission intensities on ν_{nm} should be noted.

The preceding formulae apply to the case of transitions between non-degenerate levels only. In the case of a transition between two degenerate levels of degeneracy d_n and d_m the formulae (I, 47) and (I, 51) should be replaced by

$$A_{nm} = \frac{64\pi^4 \nu_{nm}{}^3}{3h} \sum \frac{\left| R^{n_i m_k} \right|^2}{d_n} \tag{I, 54}$$

$$B_{mn} = \frac{8\pi^3}{3h^2 c} \sum \frac{\left| R^{n_i m_k} \right|^2}{d_m} \tag{I, 55}$$

where i and k number the degenerate sublevels of the upper state n, and the lower state m; and where the summation is over all possible combinations of the sublevels of the upper with those of the lower state. The relation (I, 52) is correspondingly changed to

$$B_{mn} = \frac{1}{8\pi h c \nu_{nm}{}^3} \frac{d_n}{d_m} A_{nm} \tag{I, 56}$$

If initially in an experiment there are $N_n{}^0$ atoms (or molecules) in the state n and if no new atoms in this state are produced, it follows immediately from the definition of the transition probability that there will be an exponential decrease of the number of atoms in state n; thus

$$N_n = N_n{}^0 e^{-A_{nm}t} \tag{I, 57}$$

Here it has been assumed that the transition from n to m is the only transition possible in the state n. If several transitions are possible (I, 57) has to be replaced by

$$N_n = N_n{}^0 e^{-(\sum_m A_{nm})t} \tag{I, 58}$$

After a time $\tau = 1/A_{nm}$ or $1/\sum_m A_{nm}$ the number of atoms left in the state n is $1/e$ of the initial number. This time is the *mean life* of the state n. The above relation between mean life and transition probability follows also immediately if it is realized that A_{nm} in (I, 46) is the average number of transitions per atom per second (assuming that it is immediately brought back to n once it has made the transition to m). For allowed (electric dipole) transitions τ is of the order 10^{-8} sec. If no allowed transitions can occur from the state n to any lower state, the mean life is much larger; of the order 10^{-3} sec if magnetic dipole transitions are possible, and of the order 1 sec if only quadrupole transitions are possible. Such states are called *metastable*.

3. Atoms with Several Electrons; Vector Model

Quantum numbers of the individual electrons. To a first approximation, even when a number of electrons are present, we can ascribe to each individual electron the same quantum numbers as are ascribed to a single electron moving in the field of an atomic core. This is true because, to a first approximation, we can replace the actions of the remaining electrons—on the one selected for consideration—by a mean centrally symmetric field. Thus, when several electrons are present in an atom, we have the following quantum numbers for each electron: n, the principal quantum number, which is a measure of the extent of the corresponding "electron cloud"; l, the azimuthal quantum number, which gives the orbital angular momentum of the electron; m_l, the magnetic quantum number, which gives the component of l in a given direction (for example, that of a magnetic field); and m_s, the spin quantum number, which gives the component of the spin s in a given direction.

In Table 3 the possible states of an electron are given up to $n = 4$, the restrictions for l, m_l, and m_s being taken into account. The value of m_s is indicated by an arrow directed up $(+\frac{1}{2})$ or down $(-\frac{1}{2})$. Following Mulliken (517) the states of an electron (the orbits of the Bohr theory) which are characterized by certain values of n and l and certain orbital wave functions are called *orbitals*. Electronic orbitals with different n have widely differing energies. Such groups of orbitals are called K, L, M, $\cdots$ shells, corresponding to the values of $n = 1$, $2, 3, \cdots$. The energy difference between orbitals with the same n but with different l is smaller. According as $l = 0, 1, 2, 3, \cdots$, the orbitals are designated s, p, d, f, $\cdots$ orbitals and the electrons in them s, p, d, f, $\cdots$ electrons (see the table). The value of the principal quantum number is usually placed in front of the symbol for l so that one speaks, for example, of a $2p$ orbital, or $2p$ electron. In the absence of a magnetic field, states with the same n and l but with different m_l and m_s have the same energy—that is, are degenerate with one another.

The Pauli principle. In order to represent the observed spectra of the atoms, it is necessary to assume the following principle, which was first enunciated by Pauli: *In one and the same atom, no two electrons can have the same set of values for the four quantum numbers n, l, m_l, and m_s.* This principle, also called the Pauli exclusion principle, does not result from the fundamentals of quantum mechanics but is an additional assumption.

It follows directly from Pauli's principle that, in each "cell" (orbital) with a given n and l in Table 3, there can be only as many electrons as there are arrows between the corresponding vertical lines. In the K shell $(n = 1, l = 0)$, for example, there can be at the most two electrons, since a third electron would necessarily have exactly the same four quantum numbers as one of those already present, and this, according to the above principle, is impossible. The K shell is thus *closed* with two electrons. That is why for Li the lowest state for the third electron has $n = 2$ (see p. 6). Similarly, in the L shell, there can be

TABLE 3. POSSIBLE STATES OF AN ELECTRON IN AN ATOM.

	K	L			M						N									
n	1	2			3						4									
l	0	0	1		0	1		2			0	1		2			3			
	s	s	p		s	p		d			s	p		d			f			
m_l	0	0	−1 0 +1		0	−1 0 +1		−2 −1 0 +1 +2			0	−1 0 +1		−2 −1 0 +1 +2			−3 −2 −1 0 +1 +2 +3			
m_s	⇄	⇄	⇄ ⇄ ⇄		⇄	⇄ ⇄ ⇄		⇄ ⇄ ⇄ ⇄ ⇄			⇄	⇄ ⇄ ⇄		⇄ ⇄ ⇄ ⇄ ⇄			⇄ ⇄ ⇄ ⇄ ⇄ ⇄ ⇄			

only eight electrons, two of them in the 2s subshell (orbital) and six of them in the 2p subshell. Other shells are analogous. (See also A.A., Chapter III.)

In a system containing several electrons, if any two electrons are exchanged, that is, if the coordinates of the one (say, x_i, y_i, z_i) are replaced by those of the other (say, x_j, y_j, z_j) the Schrödinger equation (I, 13) of the transformed system is

$$\sum_k \frac{1}{m_k} \left(\frac{\partial^2 \psi_{tr.}}{\partial x_k^2} + \frac{\partial^2 \psi_{tr.}}{\partial y_k^2} + \frac{\partial^2 \psi_{tr.}}{\partial z_k^2} \right) + \frac{8\pi^2}{h^2} (E_{tr.} - V_{tr.}) \psi_{tr.} = 0 \quad (I, 59)$$

where the subscript tr. indicates the transformed quantities. Because of the indistinguishability of the electrons $V_{tr.} \equiv V$ and therefore the eigenvalues $E_{tr.}$ are the same as those of the original system since the Schrödinger equation has the same form. Also the eigenfunctions are the same; but since $-\psi$ is also a solution, the transformed eigenfunction $\psi_{tr.}$ equals either $+\psi$ or $-\psi$ (the probability density distribution $|\psi|^2$ remains unaltered, as was to be expected). In other words the eigenfunctions of an atomic system that result from the Schrödinger equation are either *symmetric or antisymmetric with respect to an exchange of any two electrons*. While this has been shown here only for the case of non-degenerate states and without taking account of the electron spin, it does hold also for degenerate states and if the spin is taken into account.

It can be shown [see for example, Condon and Shortley (10)] that the states with antisymmetric eigenfunctions are those in which all electrons of the system have different sets of the four quantum numbers n, l, m_l, and m_s, whereas, in the states with symmetric eigenfunctions at least two electrons have the same set. Therefore the Pauli principle may also be stated thus: *Only those states of an atom occur in which the eigenfunctions are antisymmetric with respect to an exchange of any two electrons*. It is seen that the Pauli principle, while it represents a new assumption, can be fitted into wave mechanics in a satisfactory manner.

There are other operations which, like the exchange of two electrons, transform the system into one with the same potential energy; for example, a *reflection at the origin* (replacement of all x_k, y_k, z_k by $-x_k$, $-y_k$, $-z_k$). In a way entirely similar to the above it is found that the eigenfunctions of the system either remain unchanged or change sign if the particular operation is carried out on them. In the example (reflection at the origin) it leads to the distinction of even and odd terms of an atom (see p. 28). However, not for all symmetry operations is the situation as simple.

Quantum theoretical addition of angular momentum vectors. As we have seen previously, each electron in an atom has an orbital angular momentum l and a spin s. These individual angular momenta can be added in different ways to form a resultant. However, before we go into the particular case of the resultant angular momentum of an atom, let us briefly discuss quite generally the quantum-theoretical addition of angular momentum vectors, since such additions are also of great importance for molecules.

Let A and B be two angular momentum vectors. According to quantum theory, their magnitudes are given by $\sqrt{A(A+1)}\,(h/2\pi)$ and $\sqrt{B(B+1)}\,(h/2\pi)$, respectively [or approximately by $A(h/2\pi)$ and $B(h/2\pi)$] where A and B are the quantum numbers belonging to the angular momenta and can be integral, $0, 1, 2, \cdots$, or half integral, $\frac{1}{2}, \frac{3}{2}, \frac{5}{2}, \cdots$. The resultant C, of A and B, is obtained in the usual way from the parallelogram of vectors. However, we must remember that C is again quantized—that is to say, can take only the values $\sqrt{C(C+1)}\,(h/2\pi)$ [or approximately $C(h/2\pi)$] where C is integral when A and B are both integral or both half integral, but is half integral when only A or only B is half integral. Therefore, in order to obtain the quantized values of C, we can add A and B only in certain directions relative to one another. The possible values of C are

$$C = (A + B),\ (A + B - 1),\ (A + B - 2),\ \cdots,\ |A - B|. \qquad (\text{I, } 60)$$

A rigorous derivation of this rule may be found in Condon and Shortley (10).

Fig. 4 illustrates this vector addition for the case $A = 2$, $B = \frac{3}{2}$. For simplicity, the lengths of the vectors have been taken as proportional to the corresponding quantum numbers, which gives only an approximate picture. We see that, to this approximation, the extreme C values, $(A + B)(h/2\pi)$ and $|A - B|(h/2\pi)$, agree with the maximum and minimum classical resultants, respectively. However, not all intermediate values (as on the classical theory) but only the discrete values (I, 60) are possible. If $\sqrt{A(A+1)}\,(h/2\pi)$, $\sqrt{B(B+1)}$-$(h/2\pi)$, and $\sqrt{C(C+1)}\,(h/2\pi)$ are used for the magnitudes of the vectors it is seen that the three vectors cannot point in exactly the same direction even in the case of the extreme values of C.

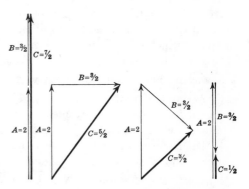

Fig. 4. **Quantum Theoretical Addition of Two Angular Momentum Vectors A and B to Form a Resultant C for the Case $A = 2$, $B = \frac{3}{2}$.** The figure gives all the possible mutual orientations for the example. The lengths of the vectors have been taken proportional to the quantum numbers (see text).

If a coupling exists between the vectors A and B, they precess about the resultant C as axis, the precession being the more rapid the stronger the coupling. The states with the different C values (I, 60) then have different energies. At the same time, A and B lose more and more their meaning as angular momenta (their direction being no longer constant). However, the quantum numbers A and B retain their significance for ascertaining the number of possible C values.

If there are several angular momenta, it is best first to add the strongly coupled vectors to form partial resultants and then to add these more weakly coupled partial resultants to give a total resultant.

With respect to any fixed direction, for example, that of an external field, the angular momentum vector A is space quantized in such a way that its component in the field direction is $M_A \cdot (h/2\pi)$, where

$$M_A = A, (A - 1), (A - 2), \cdots, - A. \tag{I, 61}$$

M_A is therefore integral or half integral according as A is integral or half integral. The space quantization corresponds exactly to that for l shown in Fig. 3. Accordingly, in a field, a state with angular momentum A has in general $(2A + 1)$ components of slightly different energy. In the absence of a field, there is a $(2A + 1)$-fold degeneracy.

Quantum numbers and angular momenta of the whole atom; term symbols. In general, in an atom containing several electrons, the orbital angular momenta $l_1, l_2, l_3, \cdots$ of the individual electrons are strongly coupled among themselves, and the spins $s_1, s_2, s_3, \cdots$ may be regarded as strongly coupled among themselves even though the strength of this latter coupling is only apparent (see A.A., p. 129). The l_i added together in the above-mentioned manner give a resultant which is the *resultant orbital angular momentum* and is designated by L. The s_i add up correspondingly to a *resultant spin* S. From our previous discussion the magnitudes of these vectors are given by $\sqrt{L(L + 1)}(h/2\pi) \approx L(h/2\pi)$ and $\sqrt{S(S + 1)}(h/2\pi) \approx S(h/2\pi)$, respectively. Since all the l_i are integral, L is also integral, whereas, since all the $s_i = \frac{1}{2}$, S is integral for an even number of electrons and half integral for an odd number of electrons.

The resultants L and S are then added together (again in just the same way as for A and B above) to give the *total angular momentum* J of the electrons of the atom. Its magnitude is $\sqrt{J(J + 1)}(h/2\pi) \approx J(h/2\pi)$, where J is integral or half integral according as S is integral or half integral. According to (I, 60), the quantum number J is given by

$$J = (L + S), (L + S - 1), (L + S - 2), \cdots, |L - S|. \tag{I, 62}$$

Thus we have the following scheme:

$$\underbrace{l_1, l_2, l_3, \cdots}_{L} \qquad \underbrace{s_1, s_2, s_3, \cdots}_{S} \tag{I, 63}$$
$$\underbrace{}_{J}$$

As an example the vector diagram for a simple case with two electrons is given in Fig. 5. The coupling assumed in the preceding discussion and in Fig. 5 is usually referred to as *Russell-Saunders coupling* or (L, S) coupling. Other types of coupling such as the so-called (j, j) coupling (see A.A., p. 174) are not of great importance for the purposes of this book.

According as $L = 0, 1, 2, 3, \cdots$, the energy levels of an atom are called $S, P, D, F, \cdots$ terms, respectively (compare the energy level diagram of Li in Fig. 1 where $L = l$, since there is only one outer electron).

Each of the possible L values can occur with each of the possible S values. Owing to the strong coupling between the l_i themselves and the s_i themselves, terms with different L or different S have very different energies, while on account of the smallness of the coupling between L and S the terms that result from one and the same L and S have only slightly different energies. The latter form together a multiplet term, which according to (I, 62) has $2S + 1$ components when $L > S$ and $2L + 1$ components when $L < S$. The number $2S + 1$ is called the multiplicity (even when $L < S$ and when therefore the actual number of components is smaller) and is added as a left superscript to the term symbol, thus: ^3P (triplet P), ^4D (quartet D), and so forth. In addition, the J value is added as a right subscript—for example, $^2P_{1/2}$, $^2P_{3/2}$. Thus the example in Fig. 5 represents a $^3G_4$ term. Sometimes the whole electron configuration (or the most important part of it) is added to the term symbol, so that we may have, for example, $1s^2 2s^2 2p^2 \, ^3P_1$. In this expression, the exponents indicate the number of electrons in the orbitals considered.

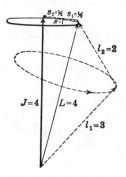

FIG. 5. **Vector Diagram for the Case of Two Electrons** with $l_1 = 3$, $l_2 = 2$, $L = 4$, $S = 1$, $J = 4$. The precessions are indicated by broken-line and solid-line ellipses. The "broken-line" precession takes place much faster than the "solid-line" precession.

As an example, let us consider an atom with an f-electron and a d-electron ($l_1 = 3$, $l_2 = 2$; see Fig. 5). According to (I, 60), L can take the values 1, 2, 3, 4, and 5, while S can take the values 0 and 1. We therefore obtain the terms P, D, F, G, and H, which can occur with $S = 0$—that is, as singlets—and with $S = 1$—that is, as triplets. Thus we have the terms $^1P_1$, $^1D_2$, $^1F_3$, $^1G_4$, $^1H_5$, $^3P_{0,1,2}$, $^3D_{1,2,3}$, $^3F_{2,3,4}$, $^3G_{3,4,5}$, and $^3H_{4,5,6}$.

When more than one electron is in one and the same orbital (same n and l)—that is, when there are *equivalent electrons*—the Pauli principle must be taken into consideration in deriving the possible terms. In the case of two equivalent p electrons, for example, it is found that only the terms ^3P, ^1D, and ^1S are possible, whereas without taking account of the Pauli principle, that is, for two non-equivalent p electrons the terms ^3P, ^1P, ^3D, ^1D, ^3S, ^1S result (see A.A., p. 129 f.) When a shell (orbital) contains the maximum number of electrons possible according to the Pauli principle (*closed shell*) it is easily seen that the resultant L and S values can only be zero and a $^1S_0$ term results. Therefore, also, in the derivation of the terms of more complicated electron configurations no account need be taken of closed shells.

A series of electron configurations differing only in the principal quantum number n of one electron (for example, $1s^2 2s \, np$) results in one or more *Rydberg series* of the corresponding terms (in the example, n^1P and n^3P).

The strength of the coupling between L and S, and therefore the magnitude of the multiplet splitting, increases rapidly with increasing atomic number. This is the reason that for heavy elements for which this coupling strength is comparable to, or larger than that between the l_i or the s_i, different coupling schemes apply, which, however, are not considered here.

Since the multiplicity is $2S + 1$ and since S is integral or half integral for an even or odd number of electrons, respectively, it follows that the multiplicity of an atomic term is odd for an even number of electrons and even for an odd number of electrons. Even and odd multiplicities therefore alternate for successive elements in the periodic system (*alternation of multiplicities*). They alternate similarly in the series of ions with single and multiple charges for a given element.

The terms of an atom are distinguished as *even* or *odd* according as $\sum l_i$, summed over all the electrons of the atom, is even or odd. It can be shown that the eigenfunctions of the even terms remain unchanged while those of the odd terms change sign for a reflection of all particles at the origin (see p. 24). In work on atomic spectra the odd terms are distinguished from the even ones by a superscript o (for example, $^2P^o$) while in work on diatomic molecules, for the sake of consistency with the corresponding molecular nomenclature, the subscripts g and u are used for even and odd terms (for example, ^2P_g, ^2P_u).

Influence of a magnetic or electric field. According to the previous discussion, when an atom with total angular momentum J is brought into a field, a space quantization takes place, so that the components of J in the field direction have the values $M_J(h/2\pi)$, where M_J can take the values

$$M_J = J, (J-1), (J-2), \cdots, -J. \tag{I, 64}$$

In a magnetic field (*Zeeman effect*), states with different M_J have different energy, but, in an electric field (*Stark effect*), only those with different $|M_J|$ do (that is, states which differ only in the sign of M_J have the same energy in an electric field.)

In a very strong field, a *Paschen-Back effect* takes place: L and S are *uncoupled* from each other by the field and are space quantized with respect to the field direction independently of each other, with components $M_L(h/2\pi)$ and $M_S(h/2\pi)$, respectively. For still stronger fields, the coupling between the various l_i and the various s_i can be broken so that an independent space quantization of all the individual angular momentum vectors takes place.

Selection rules. A determination of the selection rules according to the method previously given (p. 18 f.) leads to the following results. For dipole radiation there are two *rigorous* selection rules:

$$\Delta J = 0, \pm 1, \textit{with the restriction } J = 0 \nrightarrow J = 0 \tag{I, 65}$$

($\nrightarrow$ means "does not combine with"), and the Laporte rule, *even terms combine only with odd and odd terms combine only with even.*   (I, 66)

In addition, one obtains the following approximate selection rules. As long as the coupling between L and S is weak—that is to say, as long as the multiplet splitting is small (light elements)—to a good approximation the rules

$$\Delta L = 0, \pm 1 \tag{I, 67}$$

$$\Delta S = 0 \tag{I, 68}$$

hold. The latter rule is also known as the prohibition of intercombinations. It states that in the approximation considered terms of different multiplicities cannot combine with one another.

If the coupling between the l_i is not too strong, only one electron alters its quantum numbers in a transition and for this electron the following rule holds:

$$\Delta l_i = \pm 1. \tag{I, 69}$$

Finally, in a magnetic or electric field the selection rules for the magnetic quantum numbers are,

for a weak field:

$\Delta M_J = 0, \pm 1$; with the restriction $M_J = 0 \nrightarrow M_J = 0$ *for* $\Delta J = 0$;   (I, 70)

for a strong field:

$$\Delta M_L = 0, \pm 1; \Delta M_S = 0. \tag{I, 71}$$

Nuclear spin.   In order to explain the *hyperfine structure* in line spectra (see A.A., Chapter V) and the *intensity alternation* in band spectra (see Chapters III and V of this book) it is necessary to assume that the atomic nucleus, as well as the electron, possesses an intrinsic angular momentum (*nuclear spin*)—that is, rotates about its own axis. This nuclear angular momentum is usually designated I. Its magnitude is

$$\sqrt{I(I+1)} \; (h/2\pi) \approx I(h/2\pi),$$

where I is the corresponding quantum number which has different integral or half-integral values (including zero) for different nuclei.

According to the general rules for vector addition outlined previously, the nuclear angular momentum I added to the total angular momentum J of the extranuclear electrons gives a resultant F, the *total angular momentum of the atom, inclusive of nuclear spin.* The corresponding quantum number is given by

$$F = (J+I), (J+I-1), (J+I-2), \cdots, |J-I|. \tag{I, 72}$$

The different values of F correspond to somewhat different energies of the whole system. However, the coupling between J and I is so weak that the energy differences are extremely small. This causes a minute splitting of the spectral lines (hyperfine structure) which can be detected only by spectral apparatus of the greatest resolving power.

CHAPTER II

OBSERVED MOLECULAR SPECTRA AND THEIR REPRESENTATION BY EMPIRICAL FORMULAE

1. Spectra in the Visible and Ultraviolet Regions

Coarse structure. If the spectra of various kinds of electric discharges, of flames, or of fluorescence are investigated, it is found that in addition to the characteristic line spectra (of which Fig. 6 gives a simple example), spectra of a quite different type appear, particularly when molecular gases are used. With small dispersion, these consist not of single, sharp lines but of more or less broad wave-length regions (*bands*). We therefore speak of *band spectra*. In Figs. 7–11 some typical *emission* band spectra are reproduced. They should be compared with the typical line spectrum of the hydrogen atom in Fig. 6. In

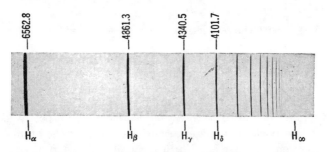

FIG. 6. **Emission Spectrum of the Hydrogen Atom in the Visible and Near Ultraviolet Region** [Balmer Series, after Herzberg (302)]. H_∞ gives the position of the series limit.

emission, bands and lines very often occur mixed together, as the figures show. In *absorption*, however, when light with a continuous spectrum (for example, the light of a filament lamp) is sent through a molecular gas, bands appear exclusively (see Figs. 13–16). Reproductions of a number of other band spectra may be found in Jevons (23) and Pearse and Gaydon (47).

As can be seen from the figures, the bands usually have at one end a sharp edge, called a *band head* (or band edge), where the intensity falls suddenly to zero, while on the other side the intensity falls off more or less slowly. According as this gradual falling off in intensity takes place toward shorter or longer wave lengths, the bands are said to be *shaded* (degraded) to the violet or the red.

Apart from this kind of bands, there are also bands occurring rather less frequently, in which the heads are not so clearly developed or not present at all. The bands of the Hg_2 molecule shown in Fig. 17 are an example of this type. In rare cases—for example, for H_2 (see Fig. 12) and the alkali hydrides—no

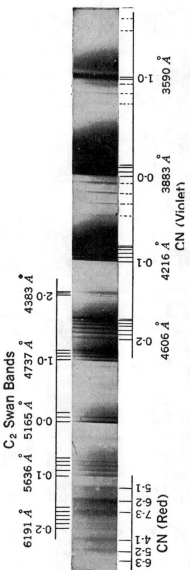

Fig. 7.   **Band Spectrum of the Carbon Arc in Air (Bands of the Molecules CN and C$_2$).**   The bands whose leading lines are connected to the same horizontal line belong to one band system.   For the violet CN bands and the C$_2$ Swan bands the numbering and wave length of only the first band of each sequence (see p. 42) is indicated.   The broken leading lines refer to the so-called tail bands (see p. 161).   The longest-wave-length group of C$_2$ is strongly overlapped and is therefore not very clear.

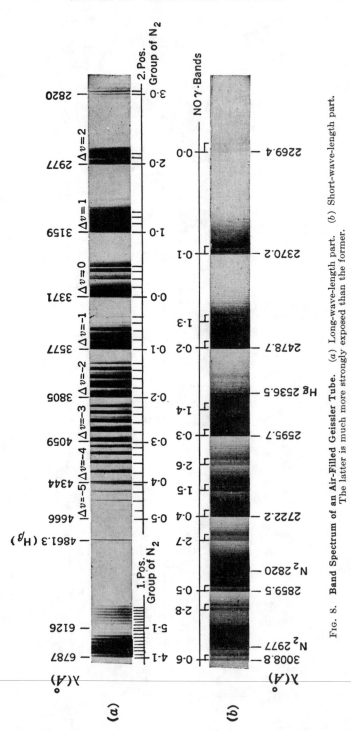

FIG. 8.   Band Spectrum of an Air-Filled Geissler Tube.   (a) Long-wave-length part.   (b) Short-wave-length part.
The latter is much more strongly exposed than the former.

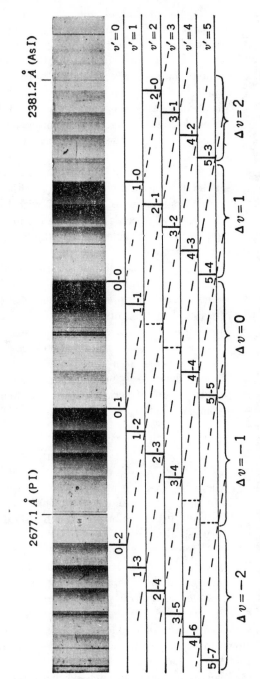

Fig. 9.   **Emission Band Spectrum of the PN Molecule [after Curry, Herzberg, and Herzberg (176)].**   The explanation of the lower part of the figure is given on p. 36.   The broken leading lines refer to unobserved bands.   Since the spectrogram is taken with a grating, the dispersion in Å/mm is approximately constant, while the wave-number dispersion increases to the left.

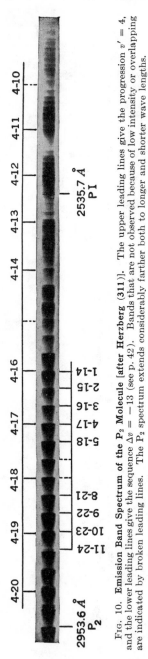

Fig. 10. Emission Band Spectrum of the P_2 Molecule [after Herzberg (311)]. The upper leading lines give the progression $v' = 4$, and the lower leading lines give the sequence $\Delta v = -13$ (see p. 42). Bands that are not observed because of low intensity or overlapping are indicated by broken leading lines. The P_2 spectrum extends considerably farther both to longer and shorter wave lengths.

well-developed bands are observed but only an enormous number of lines, a so-called *many-line spectrum* which is similar in appearance to a complicated atomic spectrum.

Finally, more extended continuous regions, so-called *continuous spectra*, or, briefly, *continua*, appear as molecular spectra. They are observed in absorption joining onto a series of bands—for example, for I_2 (Fig. 15) and other halogen molecules. However, they also appear without accompanying bands—for example, for the hydrogen halides. Continuous spectra are also observed in emission—for example, for H_2 (see Fig. 12).

A comparison of Figs. 7–17 with Fig. 6 shows that the arrangement of the bands in a molecular spectrum follows rules entirely different from those for the arrangement of lines in an atomic spectrum. In none of the cases shown is there a Rydberg series of bands, while all atomic spectra consist of such series. To be sure, such Rydberg series of bands are found in absorption in the far ultraviolet, but in the visible and near ultraviolet regions only one such case is known, the band spectrum of helium, which appears in a weakly condensed discharge through helium. The wave numbers ν of the band heads in such a series follow a Rydberg formula [compare (I, 8)],

$$\nu = A - \frac{R}{(n + a)^2} \qquad (II, 1)$$

where A, R, and a are constants and n is a whole number. The formula shows that the separation between successive bands in such a series decreases very rapidly.

However, in almost all other cases, molecular spectra consist of series of bands whose separation changes rather slowly. They are called *progressions*. In each of the absorption spectra of CO, I_2, and S_2 (Figs. 14–16), only one such progression has considerable intensity. In the emission band spectrum of PN, shown in Fig. 9, a number of progressions are present and are indicated separately below the spectrogram.

The separations between successive bands in such a progression of the PN

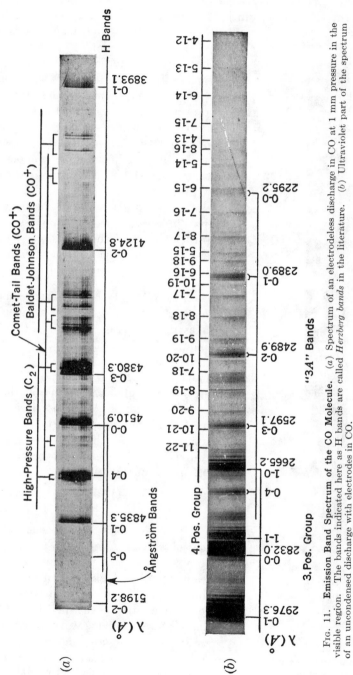

Fig. 11.   **Emission Band Spectrum of the CO Molecule.**   (*a*) Spectrum of an electrodeless discharge in CO at 1 mm pressure in the visible region.   The bands indicated here as H bands are called *Herzberg bands* in the literature.   (*b*) Ultraviolet part of the spectrum of an uncondensed discharge with electrodes in CO.

spectrum are approximately equal but decrease slowly toward longer wave lengths (to the left in the figure) if measured in cm^{-1}. On closer examination the following important regularity is found: When the progressions given in Fig. 9 are shifted relative to one another, so that the bands with the shortest wave length in each of them coincide, then all the other bands corresponding to one another in the different progressions will also coincide. However, for an exact coincidence the diagram should be drawn to a linear wave number scale, whereas the actual spectrogram has a linear wavelength scale.

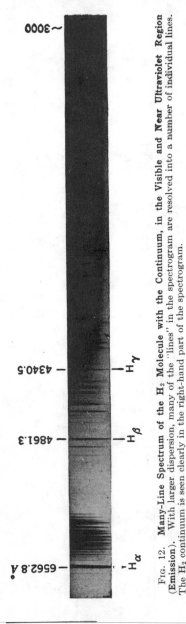

Fig. 12. Many-Line Spectrum of the H$_2$ Molecule with the Continuum, in the Visible and Near Ultraviolet Region (Emission). With larger dispersion, many of the "lines" in the spectrogram are resolved into a number of individual lines. The H$_2$ continuum is seen clearly in the right-hand part of the spectrogram.

Since the separation of the bands in a given progression decreases very slowly, the wave numbers of the bands can be represented approximately by the formula

$$\nu = \nu_{v'} - (a''v'' - b''v''^2), \quad \text{(II, 2)}$$

where $\nu_{v'}$, a'', and b'' are positive constants, ($b'' \ll a''$), and v'' is a whole number which takes the values 0, 1, 2, $\cdots$. If $b'' = 0$, equation (II, 2) would give a series of equidistant bands (separation $= a''$). If b'' is different from zero (but $\ll a''$) (II, 2) gives the observed gradual decrease in the separation of the bands. We choose the constants in (II, 2) so that for the band of highest wave number $v'' = 0$—that is, $\nu = \nu_{v'}$.[1] For example, for the second progression in the PN spectrum (Fig. 9),

$$\nu = 40{,}786.8 - (1329.38\, v'' - 6.98\, v''^2).$$
$$\text{(II, 2a)}$$

Table 4 shows the agreement between the observed wave numbers and those obtained from this formula. The fact already mentioned above that, by suitable shifting, all the band series may be made to coincide exactly, means that, in (II, 2), a'' and b'' have the same value for all the progressions and only $\nu_{v'}$ is different.

---

[1] By a suitable choice of the constants in (II, 2) we could also represent the observed bands using some other numbering (for example, $v'' = 10$ for the first band). We have chosen here the numbering which appears to be the simplest and which later proves to be of theoretical significance (see Chapter IV).

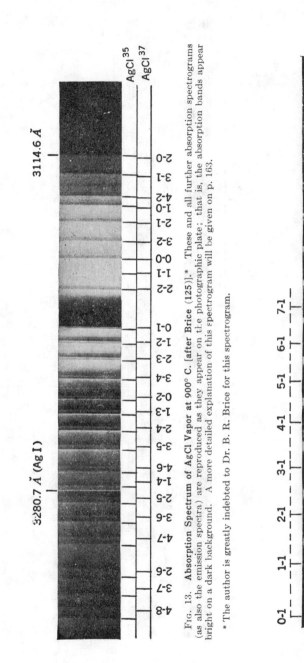

FIG. 13. **Absorption Spectrum of AgCl Vapor at 900° C.** [after Brice (125)].* These and all further absorption spectrograms (as also the emission spectra) are reproduced as they appear on the photographic plate; that is, the absorption bands appear bright on a dark background. A more detailed explanation of this spectrogram will be given on p. 163.

* The author is greatly indebted to Dr. B. R. Brice for this spectrogram.

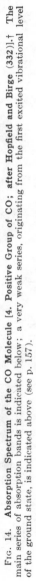

FIG. 14. **Absorption Spectrum of the CO Molecule** [4. **Positive Group of CO; after Hopfield and Birge (332)**].† The main series of absorption bands is indicated below; a very weak series, originating from the first excited vibrational level of the ground state, is indicated above (see p. 157).

† The author is greatly indebted to Professor R. T. Birge for this spectrogram.

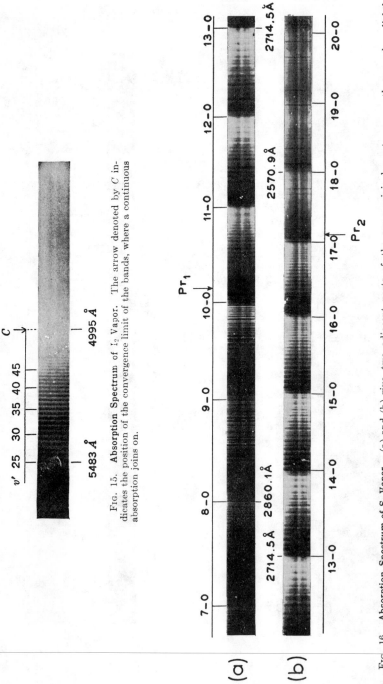

Fig. 15. **Absorption Spectrum** of I₂ Vapor. The arrow denoted by C indicates the position of the convergence limit of the bands, where a continuous absorption joins on.

Fig. 16. **Absorption Spectrum of S₂ Vapor.** (a) and (b) give two adjacent parts of the same original spectrogram; they overlap slightly (13–0 band). The arrows denoted by Pr indicate the positions at which the bands become diffuse. The emission lines at the right in (b) are lines of H₂ (first order overlapping the second order spectrum).

Table 4. v''-progression in the
PN spectrum $(v' = 1)$.

v''	$\nu_{obs.}$	$\nu_{calc.}$ according to (II, 2a)
0	40,786.2	40,786.8
1	39,467.2*	39,464.4
2	38,155.5	38,156.0
3	36,861.3	36,861.5

* This band is badly overlapped by the 0–0 band 39,698.8 cm^{-1} (see Fig. 9) and therefore not as accurately measured as the others.

It may happen that the first band in a progression—that is, the one with $v'' = 0$—is not observed. In this case a shift of this progression of bands relative to another one will result in a coincidence only when the first *observed* band of the progression under consideration is correlated with the second band of the other progression. This is, for example, the case for the fourth progression in Fig. 9.

Progressions such as those discussed here, for which the band separations decrease toward *longer* wave lengths, are called v''-*progressions*.

Taking the first band $(\nu_{v'})$ in each of the different v''-progressions (whether observed or extrapolated), we obtain again a progression, but with other constants. In particular in this progression the band separations decrease toward *shorter* wave lengths. Accordingly, they can be represented by

$$\nu_{v'} = \nu_{oo} + (a'v' - b'v'^2), \qquad \text{(II, 3)}$$

where v' takes the values 0, 1, 2, $\cdots$ and ν_{oo}, a', and b' are positive constants. In the case of the PN spectrum,

$$\nu_{v'} = 39,699.0 + (1094.80 \, v' - 7.25 \, v'^2). \quad \text{(II, 3a)}$$

Table 5 shows the agreement between the observed or extrapolated wave numbers with those calculated by this formula.

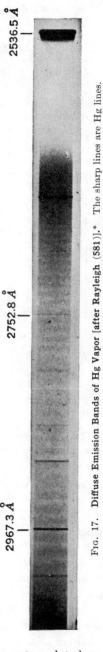

Fig. 17.   Diffuse Emission Bands of Hg Vapor [after Rayleigh (581)].*   The sharp lines are Hg lines.

* The author is greatly indebted to the late Lord Rayleigh for this spectrogram.

If we substitute $\nu_{v'}$ from (II, 3) into (II, 2), we obtain

$$\nu = \nu_{oo} + (a'v' - b'v'^2) - (a''v'' - b''v''^2);\qquad\text{(II, 4)}$$

that is, in our example,

$$\nu = 39,699.0 + (1094.80\, v' - 7.25\, v'^2) - (1329.38\, v'' - 6.98\, v''^2).\quad\text{(II, 5)}$$

TABLE 5. v'-PROGRESSION IN THE PN SPECTRUM ($v'' = 0$)

v'	$\nu_{obs.}$	$\nu_{calc.}$ according to (II, 3a)
0	39,699.1	39,699.0
1	40,786.8	40,786.6
2	41,858.9	41,859.6
3	42,919.0	42,918.2
4	43,962.0	43,962.2
5	44,991.3	44,991.8
6	46,007.3	46,006.8
7	47,005.6	47,007.4
8	47,995.0	47,993.4
9	48,964.4	48,965.0

This formula now represents all the bands. For $v' = 0$ and $v'' = 0$, we have $\nu = \nu_{oo}$, the wave number of the first band of the first series (see Fig. 9), the so-called 0–0 band (or 0,0 band), which is frequently, as in the present case, the most intense band in the system (see also Figs. 7 and 8). The totality of those bands of a molecule which can be represented by a formula of the type (II, 4), possibly with cubic or still higher terms in v' and v'', is called a band system. In many cases a number of such band systems are observed for the same molecule. In favorable cases, the different band systems of a molecule lie in different spectral regions, as, for example, for N_2 (see Fig. 8). However, they often overlap one another, as, for example, for CO (see Fig. 11).

For very extended band systems it is sometimes necessary for an accurate representation of the heads, to introduce a term $cv'v''$ in the formula (II, 4). This term arises on account of the slightly variable position of the head within a band (see Chapter IV, section 3). A fuller discussion of the empirical term $cv'v''$ is given by Jevons (23).

The wave numbers of the bands in a band system are commonly arranged in a so-called Deslandres table (or scheme of band heads) such that the bands of every v''-progression are put in a separate horizontal row. The different horizontal rows are arranged in such a way that corresponding bands of different v''-progressions are vertically below one another. As an example, the Deslandres table for the PN bands discussed above is given in Table 6. The wave-number differences for the bands in successive horizontal and vertical rows are also included. It will be noticed that these differences in a given horizontal row are very nearly constant, in agreement with what was said above—namely, that, after a suitable shift, one series comes into coincidence with the previous series.[2]

Naturally, according to (II, 4). instead of considering the band system as consisting of a number of v''-progressions (II, 2), it can also be considered to be made up of v'-progressions

$$\nu = \nu_{v''} + (a'v' - b'v'^2),\qquad\qquad\text{(II, 6)}$$

---

[2] Sometimes, particularly in older papers, wave lengths are given instead of wave numbers; however, in this case, the differences are not constant (see below).

TABLE 6. DESLANDRES TABLE OF THE PN BANDS.
(With horizontal and vertical differences)

v″ ＼ v′	0		1		2		3		4		5		6		7		8		9		10
0	39,698.8	1322.3	38,376.5	1307.8	37,068.7																
	1087.4		1090.7		1086.8																
1	40,786.2	1319.0	39,467.2	1311.7	38,155.5	1294.2	36,861.3														
	1072.9		1069.0				1071.6														
2	41,859.1	1322.9	40,536.2				37,932.9	1280.4	36,652.5	1265.3	35,387.2										
			1061.2						1060.0		1059.2										
3			41,597.4	1309.1	40,288.3				37,712.5	1266.1	36,446.4	1252.4	35,194.0								
					1042.9				1043.9				1042.6								
4					41,331.2				38,756.4				36,236.6	1238.3	34,998.3						
													1029.4								
5							41,066.1				38,519.4				36,027.7	1225.5	34,802.2				
							1015.9														
6							42,082.0		41,798.3										34,607.1	1194.5	33,412.6
																				998.7
7																					34,411.3
8											41,522.6		41,239.4								
9																					

where the $\nu_{v''}$ then form a v''-progression:

$$\nu_{v''} = \nu_{oo} - (a''v'' - b''v''^2). \tag{II, 7}$$

It follows from this that the differences between corresponding bands in two vertical rows in the Deslandres scheme must also be constant, which is likewise shown by Table 6. The progressions (II, 6) are indicated in Fig. 9 by the sloping broken lines.

In the spectrogram in Fig. 9 we notice that certain bands of different progressions form characteristic groups of bands lying relatively close together (indicated by braces). It is easily seen that in Deslandres' scheme (Table 6) the bands of such a group lie on a parallel to the diagonal from the upper left to the lower right corner. Because of this, they are sometimes referred to as diagonal groups, but usually they are called *sequences*. For the bands which lie on the diagonal itself, v' equals v''. They include the 0–0 band and in the case of PN extend from it to longer wave lengths. For the other sequences, there is a constant positive or negative difference between v' and v''. That bands of such a sequence occur close together is observed in many emission band systems (see, for example, the C_2 and CN bands in Fig. 7). However, it is not a necessary property. In the N_2 spectrum in Fig. 8 the sequences are not so clearly developed and in the P_2 band spectrum in Fig. 10 they are not conspicuous at all.

Fine structure. When spectrographs of greater resolving power are used, it is found that most of the bands (in emission or absorption) consist of a large number of individual lines, as is shown by Figs. 18–23. In general, the arrangement of the individual lines in a band is completely regular; however, the regularity is of a quite different kind from that, say, of the lines in a multiplet of an atomic spectrum. In the simplest case, which occurs fairly frequently, one finds a structure like that of the CN band shown in Fig. 18(a). This band consists of a simple series of lines which draw farther and farther apart from one another as the distance from the band head increases.

In Table 7, the wave numbers of the lines of the CN band are given, together with the differences and second differences of successive lines. As Deslandres (182) first noticed, the *separation of successive lines increases very nearly linearly*, since the second difference is a constant within the accuracy of measurement. From this it follows that the lines can be represented by a formula of the type

$$\nu = c + dm + em^2, \tag{II, 8}$$

where c, d, and e are constants and m is a whole number which numbers the successive lines.

If we allow negative values of m as well as positive, by suitable choice of c and d, we can begin the counting ($m = 0$) at any line. Now it is seen that at one point in the series a line is missing (marked ν_0 in Fig. 18). We shall see later that it is convenient to give to this missing line the value $m = 0$ (*zero*

TABLE 7. WAVE NUMBERS OF THE LINES OF THE CN BAND 3883.4 Å

[Calculated from the data of Uhler and Patterson (675)]

m	$\nu_{obs.}$ (cm^{-1})	$\Delta\nu$	$\Delta^2\nu$	$\nu_{calc.}$ according to (II, 8a)
-24	25,744.73			25,744.43
-23	45.34	0.61		45.11
-22	46.08	0.74	0.13	45.91
-21	46.99	0.91	0.17	46.85
-20	48.02	1.03	0.12	47.92
-19	49.19	1.17	0.14	49.14
-18	50.52	1.33	0.16	50.49
-17	51.98	1.46	0.13	51.98
-16	53.47	1.49	0.03	53.60
-15	55.39	1.92	0.43	55.35
-14	57.26	1.87	-0.05	57.25
-13	59.29	2.03	0.16	59.26
-12	61.42	2.18	0.15	61.42
-11	63.75	2.28	0.10	63.72
-10	66.16	2.41	0.13	66.14
-9	68.72	2.56	0.15	68.71
-8	71.35	2.63	0.07	71.41
-7	74.23	2.88	0.25	74.24
-6	77.19	2.96	0.08	77.21
-5	80.32	3.13	0.17	80.32
-4	83.53	3.21	0.08	83.55
-3	86.90	3.37	0.16	86.93
-2	90.41	3.51	0.14	90.44
-1	94.03	3.62	0.11	94.07
0				97.85
$+1$	25,801.81			25,801.77
$+2$	05.80	3.99		05.81
$+3$	10.01	4.21	0.22	10.01
$+4$	14.23	4.22	0.01	14.32
$+5$	18.77	4.52	0.30	18.77
$+6$	(23.88)*			23.37
$+7$	28.06			28.09
$+8$	33.02	4.96		32.94
$+9$	37.97	4.95	-0.01	37.94
$+10$	43.13	5.16	0.21	43.07
$+11$	48.40	5.27	0.11	48.34
$+12$	53.77	5.37	0.10	53.74
$+13$	59.28	5.51	0.14	59.26
$+14$	64.90	5.62	0.11	64.93
$+15$	70.60	5.70	0.08	70.74
$+16$	76.69	6.09	0.39	76.68
$+17$	82.73	6.04	-0.05	82.75
$+18$	88.91	6.18	0.14	88.96
$+19$	95.22	6.31	0.13	95.30
$+20$	25,901.63	6.41	0.10	25,901.78
$+21$	08.26	6.63	0.22	08.40
$+22$	14.95	6.69	0.06	15.15
$+23$	21.86	6.91	0.22	22.03
$+24$	28.83	6.97	0.06	29.05

* This line is overlapped by another band head.

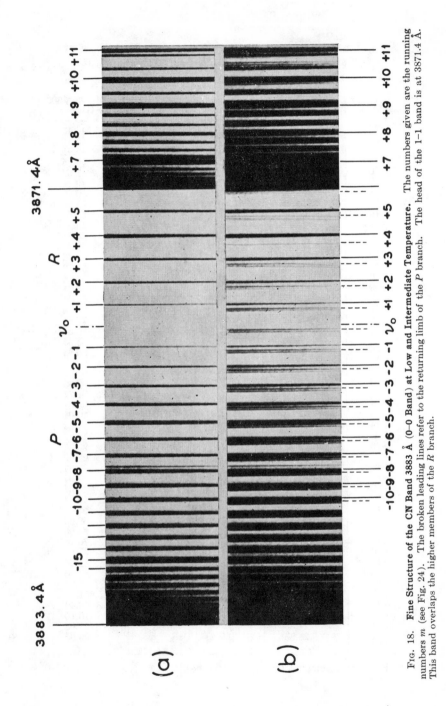

FIG. 18. Fine Structure of the CN Band 3883 Å (0–0 Band) at Low and Intermediate Temperature. The numbers given are the running numbers m (see Fig. 24). The broken leading lines refer to the returning limb of the P branch. The head of the 1–1 band is at 3871.4 Å. This band overlaps the higher members of the R branch.

line or *null line*)—that is, $c = \nu_0$. This place in the band is also called the *zero gap* (or null-line gap). In the example, using this kind of numbering (see the first column in Table 7), we obtain the formula

$$\nu = 25{,}797.85 + 3.848m + 0.0675m^2. \qquad (\text{II}, 8a)$$

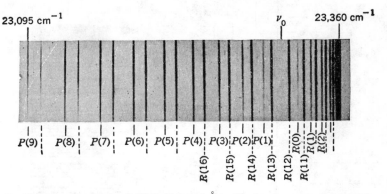

Fig. 19. **Fine Structure of the CuH Band λ4280 Å.*** The band extends much farther to the left than is shown in the spectrogram. However, it is then overlapped by another band. The numbers in parentheses after R and P are the J values (see p. 169 f.).

* The author is greatly indebted to Professor R. Mecke for this spectrogram.

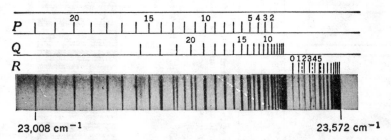

Fig. 20. **Fine Structure of the AlH Band 4241 Å [after Bengtsson-Knave (84)].*** The numbers are the J values (see p. 169). The broken leading lines refer to the returning part of the R branch.

* The author is greatly indebted to Dr. E. Knave for this spectrogram.

The values of ν calculated according to this formula are given in the last column in Table 7 and agree very well with the observed values.

The coefficient e of m^2 in the formula is obtained from the mean value of the observed second differences $\Delta^2\nu$. It is easily seen that $\Delta^2\nu = 2e$ (see also p. 113). The coefficient d of the linear term is obtained, for example, by considering that the wave-number difference between lines with $m = +1$ and $m = -1$ is equal to $2d$. One must of course allow for the fact that the wave numbers of both these lines include some error of observation. When, therefore, the ν values calculated with the coefficients thus obtained still show a systematic deviation from the observed values, one has to alter the coefficients by a small amount. The most convenient way to obtain the calculated values $\nu_{\text{calc.}}$ is by calculating the first differences

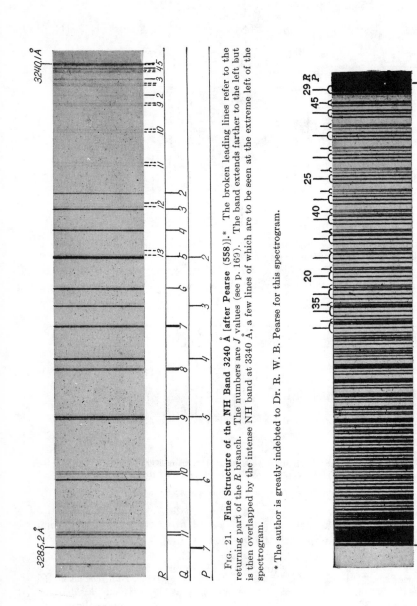

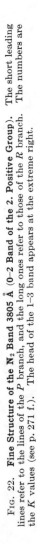

Fig. 21. **Fine Structure of the NH Band 3240 Å [after Pearse (558)].** * The broken leading lines refer to the returning part of the R branch. The numbers are J values (see p. 169). The band extends farther to the left but is then overlapped by the intense NH band at 3340 Å, a few lines of which are to be seen at the extreme left of the spectrogram.

* The author is greatly indebted to Dr. R. W. B. Pearse for this spectrogram.

Fig. 22. **Fine Structure of the N₂ Band 3805 Å (0–2 Band of the 2. Positive Group).** The short leading lines refer to the lines of the P branch, and the long ones refer to those of the R branch. The numbers are the K values (see p. 271 f.). The head of the 1–3 band appears at the extreme right.

$\Delta\nu$ from the constant second difference and by then adding or subtracting consecutively the first differences, starting with ν_0.

The series of lines corresponding to the positive values of m is called the *positive branch* or *R branch;* that corresponding to the negative values is called the *negative branch* or *P branch*. However, both together form a single simple series of lines which can be represented by one and the same formula.

When the band is shaded to the opposite side—that is to say, when the lines draw apart from one another toward the red instead of toward the violet, as they do in Fig. 18(a)—the constant e in (II, 8) is negative.

Equation (II, 8) is the equation of a parabola. It is represented graphically for the above example in Fig. 24, with ν as abscissa and m as ordinate. This representation was first used by Fortrat, and the parabola is accordingly called a *Fortrat parabola*. In the figure, the intersections of the horizontal lines, having $m = 0, \pm 1, \pm 2, \cdots$, with the parabola are indicated by small circles. The abscissae of these intersections give the wave numbers of the lines. It is seen very clearly from this method of representation how the *head of the band* is formed: The nearer one comes to the vertex of the parabola, the more the lines crowd together. The vertex itself corresponds to the head. At the head there is by no means an infinite number of lines, as for a series limit in a line spectrum, but only a finite number.

Naturally, lines are also possible which correspond to the other limb of the parabola above the vertex (broken-line part of the curve in Fig. 24). Actually, the spectrogram Fig. 18(b), for the same CN band as Fig. 18(a), (taken under conditions for which lines with higher m appear) shows such lines beyond the vertex (broken leading lines). They may be represented by the same formula as the other lines and correspond to the upper

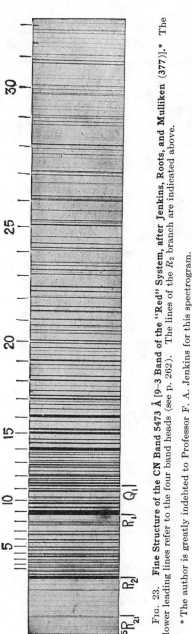

FIG. 23. Fine Structure of the CN Band 5473 Å [9–3 Band of the "Red" System, after Jenkins, Roots, and Mulliken (377)].* The lower leading lines refer to the four band heads (see p. 262). The lines of the R_2 branch are indicated above. The author is greatly indebted to Professor F. A. Jenkins for this spectrogram.

* The author is greatly indebted to Professor F. A. Jenkins for this spectrogram.

part of the Fortrat parabola in Fig. 24. The branch that forms the head (in the present case, the P branch) turns at the head and can be followed back for a certain distance. The CuH band shown in Fig. 19 gives a still better example of such a reversing (head-forming) branch, since this branch (here the R branch) is resolved almost up to the head and the head lies nearer to the zero line. As Fig. 24 shows, the lines of the head-forming branch divide the spaces between the lines of the other branch in a constant ratio whose value depends on the value of the constants d and e in (II, 8). It may happen that the lines of the reversing branch almost coincide with those of the other branch.

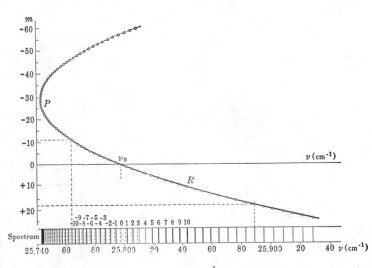

Fig. 24. **Fortrat Parabola of the CN Band 3883 Å (see Fig. 18).** The schematic spectrum below is drawn to the same scale as the Fortrat parabola above. The relation between curve and spectrum is indicated by broken lines for two points ($m = -11$ and $m = +18$). No line is observed at $m = 0$ (dotted line).

For measurements which are very accurate and extend to high m values, terms which are cubic or of even a higher power in m must be added to formula (II, 8) (see Chapter IV). They are, however, always small. The small systematic deviations of the observed from the calculated values in Table 7 are to be explained in this way.

In the bands of many band systems, in addition to the above-mentioned P and R branches, a third branch appears which is called Q *branch or zero branch.* Such a case is illustrated by the AlH band in Fig. 20. The corresponding Fortrat diagram is given in Fig. 25. The lines of the third branch lie on a parabola whose vertex lies almost on the abscissa axis. Therefore in general a second head (Q head) is present in such bands as is clearly shown in the AlH band of Fig. 20. In the formula for the Q branch, the quadratic term agrees with that for the other two branches.

There are also many bands with still more branches and correspondingly more heads. For example, in two CO^+ band systems, of which Fig. 11 shows some of the members, each band has four heads and the N_2 band in Fig. 22 has three (not very distinct) heads. Very often in such cases, two or three branches lie very close together for high m values. Fig. 22 shows this for the N_2 band. On the other hand, there are many cases where lines of a branch which are single for low values of m split into two or three components for higher values of m. This is clearly detectable in Fig. 18(b) for the lines of the reversing part of the P branch of the CN band. It is also exhibited by the NH band in Fig. 21. In spite of the more complicated appearance of these bands with more than three

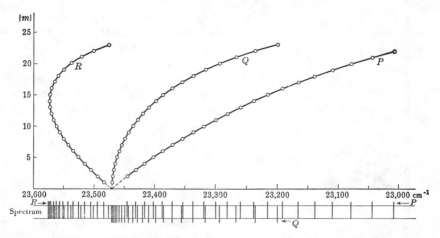

Fig. 25.   **Fortrat Diagram of the AlH Band Fig. 20.**   In contrast to Fig. 24, the Fortrat parabolae of all three branches are drawn *above* the ν axis ($|m|$ as ordinate). By reflection at the ν axis, the curve for the P branch would lie in the continuation of the R branch, as in Fig. 24. As an aid in picking out the branches in the schematic spectrogram below, the lines of the P and R branches have been extended above and those of the Q branch below, with the exception of the lines of the returning limb of the R branch.—Note that the direction of ν is reversed compared to Fig. 20.

branches, in almost all cases the individual branches may be represented by simple formulae of the type of formula (II, 8).

All bands in one and the same system have the *same number of branches*. For different bands of a system, the constants d and e in (II, 8) have only slightly different values and change regularly from band to band. Therefore, as a rule, the bands in a system are all shaded in the same direction.

A closer study of the *many-line spectra* mentioned previously shows that they arise from a superposition of bands of the type discussed above, for which, however, the line separation is so great that different bands overlap one another strongly and no definite heads appear.

The extensive *continuous* emission or absorption spectra mentioned on p. 34 show no fine structure whatsoever even when spectral apparatus of the greatest resolution is used; they are true continuous spectra.

As pointed out previously some bands do not show distinct heads. If such bands are investigated under high resolution, in some cases they are found to have a fine structure quite similar to that of bands with heads. The only distinction lies then in the fact that the constant e in (II, 8) is so small that the convergence of the lines is very slight and consequently the intensity has dropped to zero before the head is reached. Apart from these, however, bands are observed which are completely continuous even with the greatest resolution. They are called *diffuse bands*. Examples are the Hg_2 bands shown in Fig. 17. Of particular importance is the case first observed by Henri (299) (300), where, in a series of absorption bands, the bands at long wave lengths have a well-developed sharp fine structure of the above-described type, while the bands at shorter wave lengths are diffuse starting at a fairly sharply defined position. A transition case also occurs in which the fine structure of the bands can still be recognized, though the individual lines are no longer sharp, but are broadened (diffuse). Both this latter transition case and the case where the bands become completely diffuse occur in the S_2 absorption spectrum, as shown in Fig. 16 at the points marked Pr_1 and Pr_2, respectively.

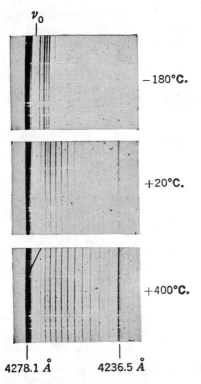

ν_0

−180°C.

+20°C.

+400°C.

4278.1 Å   4236.5 Å

FIG. 26. Intensity Distribution in the Fine Structure of the N_2^+ Band λ4278.1 Å at −180° C., 20° C., and 400° C. (Electrodeless Discharge). The P branch is not completely resolved; however, the R branch is. In the reproduction, the weak lines lying between the strong lines (intensity alternation; see text) can be seen only in the neighborhood of the intensity maximum. The head of a further band of the system lies at 4236.5 Å.

Intensity distribution. If the emission intensities of the different bands of a band system are plotted in a Deslandres table it is found that they are largest in the neighborhood of the 0–0 band and decrease more or less regularly in all directions though they may first go through one or two maxima. With increasing time of exposure, more and more bands of the system are obtained. However, it sometimes happens that, even with very much increased time of exposure, no bands appear above a definite v' (or v'') value; that is to say, only a certain number of horizontal (or vertical) rows appear in the Deslandres table. We then speak of a *breaking off* of the band system at a certain v' (or v'') value.

In *absorption* at sufficiently low temperatures (often room temperature is low enough), only one progression of a band system appears, that with $v'' = 0$ (the first vertical row in the Deslandres table). This is clearly shown by Figs. 14 and 16. All the other bands of the system in question which appear in emission have extremely small intensities in absorption at low temperatures. In the absorption spectrum of CO, Fig. 14, one progression other than that with $v'' = 0$ can be seen to appear very weakly (broken leading lines). The intensity distribution is quite different for absorption at high temperatures. As shown by the spectrum of AgCl in Fig. 13, bands with $v'' \neq 0$ do appear quite strongly.

Turning now to the intensity distribution in the *fine structure* of a band it is seen from Figs. 18–23 that in general the intensity in a branch first rises to a maximum and then gradually and regularly falls off. The position of the intensity maximum depends on the temperature. It is shifted to higher running numbers m with increasing temperature (see Chapter IV, section 4b) as is shown very clearly by Fig. 26 for the 0–1 band of the so-called negative group of nitrogen.[3] At the same time in this particular case a characteristic *alternation*

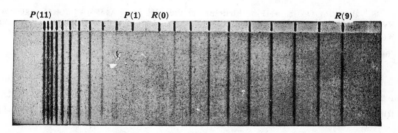

Fig. 27. **Breaking Off of the Fine Structure of the CaH Band λ3533.6 Å [after Mulliken (509)].***
The numbers following the symbols for the branches are the K values (see p. 247 f.). The R branch breaks off at $K = 9$, the P branch at $K = 11$.

* The author is greatly indebted to Professor R. S. Mulliken for this spectrogram.

of intensities can be seen—every alternate line is weak. This latter phenomenon, however, is found to occur only for molecules containing identical nuclei (homonuclear molecules).

An anomalous intensity distribution in the fine structure is often observed when we are dealing with a case of breaking off of the band system. One then finds that in the bands of the last one, two, or more observed progressions having v' constant (or sometimes v'' constant) the *branches break off* suddenly at a definite point or that a sudden decrease in intensity occurs. Two examples of this are given here—Fig. 27 (CaH band) and Fig. 28 (AlH band). A breaking off occurs in each of the two or three branches, respectively, which are present.

[3] The designations negative and positive groups (or bands) refer to the occurrence of these bands in the negative glow or the positive column, respectively, of an electric discharge. The positive groups are due to the neutral molecule, the negative groups to the singly positively charged molecular ion.

R. W. Wood (1550) (1553) was the first to study the *fluorescence* of molecular gases by illuminating them with light of a definite wave length, say one of the intense lines of a mercury lamp or cadmium lamp. He found that only a single progression with a definite v' appears in the fluorescence spectrum, and

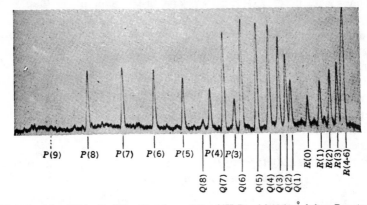

Fig. 28. **Breaking Off in the Fine Structure of the AlH Band λ4354 Å [after Bengtsson-Knave and Rydberg (87)].*** The numbers following the symbols for the branches are the J values (see p. 184). The P branch breaks off at $J = 8$, the Q branch at $J = 7$, and the R branch at $J = 6$.

* The author is greatly indebted to Dr. E. Knave for this spectrogram.

not all the bands of the system, as by excitation with white light. By excitation with another wave length, in general, another series is obtained. Investigating the fine structure of such fluorescence bands, excited by a single line, Wood found

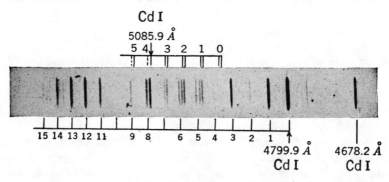
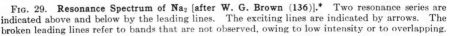

Fig. 29. **Resonance Spectrum of Na₂ [after W. G. Brown (136)].*** Two resonance series are indicated above and below by the leading lines. The exciting lines are indicated by arrows. The broken leading lines refer to bands that are not observed, owing to low intensity or to overlapping.

* The author is greatly indebted to Professor W. G. Brown for this spectrogram.

them to consist, in general, of only a few lines. In the ideal case, when the illuminating line covers only a single absorption line, the "bands" consist either of only two lines, whose separation is approximately the same for the different bands, or sometimes of only a single line. Such series of "bands" are called *resonance series*; a spectrum excited in the way described is called a *resonance spectrum*. Fig. 29 gives as an example the resonance spectrum of Na_2 excited by a Cd arc. There is a series of doublets excited by the Cd line λ5086 and a series of single lines excited by the Cd line λ4800.

2. Spectra in the Infrared Region

While a few spectra of the type described in section 1 have been observed in the very near infrared, most infrared spectra of diatomic molecular gases, particularly at longer wave lengths, are of a somewhat different kind showing very characteristic and particularly simple features. It is convenient to distinguish the spectra in the *near infrared*, with wave lengths less than about 20 μ, from the spectra in the *far infrared*, with greater wave lengths. In both cases the observations are almost always made in absorption. Below 1.3 μ specially sensitized photographic plates may be used. For longer wave lengths thermocouples, thermopiles, bolometers, radiomicrometers, or radiometers are used; [for more detailed descriptions see Lecomte (26a), Rawlins and Taylor (26b), Schaefer and Matossi (26c), Sutherland (27a), Randall (577), Sawyer (4b), and Harrison, Lord and Loofbourow (4c)]. More recently photoconductive cells have been developed and operated to 3.6 μ [Cashman (840) Sosnowski, Starkiewicz and Simpson (1431); Sutherland, Blackwell, and Fellgett (1444); Kuiper, Wilson, and Cashman (1149)] and infrared sensitive phosphors to 1.6 μ [O'Brien and collaborators (1286)]. Moreover supra-conductive metals near the critical temperature have been used

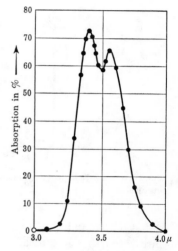

FIG. 30. **Fundamental Absorption Band of HCl in the Near Infrared [after Burmeister (148)].** With the dispersion used, the band has two maxima (Bjerrum double band). With larger dispersion, a further resolution is observed (Fig. 32).

as infrared detectors in the whole region [Andrews, Brucksch, Ziegler, and Blanchard (751) Milton (1240a) Fuson (949a)]. Attempts have also been made to "photograph" infrared spectra by evaporation of volatile solids [Czerny and Mollet (878)].

Near infrared spectra. If the absorption spectrum in the near infrared is observed with a thin layer of the absorbing gas and with small dispersion, in the whole region only a single intense absorption "line"—also called *fundamental band*—is obtained which, with somewhat greater resolution, is found to consist of two maxima close together (Bjerrum's double band). Fig. 30 shows this for HCl, for which this band lies at 3.46 μ. For HF, HBr, HI, and CO the corresponding bands lie at 2.52 μ, 3.90 μ, 4.33 μ, and 4.66 μ, respectively.   Such bands do not appear for elementary molecules such as N_2, O_2, and H_2.   For them there exists no absorption in the entire infrared region (possibly with the exception of the very near infrared, where in some cases bands appear of the type dealt with in section 1).

If the absorption is observed with thicker layers, the intensity of absorption of the fundamental band naturally increases and in addition a *second band* of similar form appears quite weakly, at approximately half the wave length or double the frequency (wave number).   If the thickness of the layer is still further increased, up to several meters at atmospheric pressure, a third and possibly

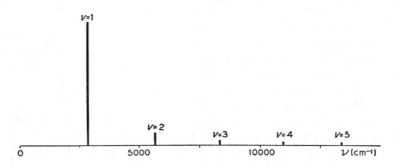

Fig. 31.   **Coarse Structure of the Infrared Spectrum of HCl (Schematic).**   The intensity actually falls off five times faster than indicated by the height of the vertical lines.

even a fourth and a fifth band appear whose wave lengths are approximately a third, a fourth, and a fifth, respectively, of that of the first band; that is to say, their frequencies are three, four, and five times as great.   Fig. 31 gives schematically the complete infrared spectrum of HCl.   In this figure the lengths of the vertical lines that represent the bands give an indication of the intensity of the bands.   However, the actual decrease in intensity is five times as fast as is indicated in the drawing.

In the second column of Table 8 the observed wave numbers of the HCl bands are given.   According to what has been said above, these wave numbers may be represented roughly by $\nu = av$, where $v = 1, 2, 3, \cdots$.   Actually the separation of successive bands (third column) is not quite constant but decreases gradually.   Therefore, for a more accurate representation, a small quadratic correction term in v must be added.   We write

$$\nu = av - bv^2 \qquad\qquad (II, 9)$$

For HCl, with $a = 2937.30$ and $b = 51.60$, we obtain the values in the last column of Table 8.   It is seen that the representation of the observed wave numbers by (II, 9) is very good.   A still better representation may be obtained if a very small cubic term is introduced in (II, 9) (see Table 39 p. 501).

TABLE 8. INFRARED BANDS OF HCl

TABLE 8. INFRARED BANDS OF HCl

The data given refer to the zero lines (see p. 112) of the bands of the isotope HCl^{35}. They have been taken from Meyer and Levin (491), Herzberg and Spinks (318), and Lindholm (1170) (1173) (partly recalculated).

v	$\nu_{obs.}$	$\Delta G_{obs.}$	$\Delta^2 G_{obs.}$	$\nu_{calc.}$
0	(0)			(0)
		2885.9$_0$		
1	2885.9$_0$		-103.7_5	2885.70
		2782.1$_5$		
2	5668.0$_5$		-103.2_2	5668.20
		2678.9$_3$		
3	8346.9$_8$		-102.8_0	8347.50
		2576.1$_3$		
4	10923.1$_1$		-102.6_9	10923.60
		2473.4$_4$		
5	13396.5$_5$			13396.50

Just as in the visible and ultraviolet spectral regions, by the use of spectrometers of sufficiently great resolution, the bands in the near infrared are resolved into a number of individual lines which are arranged in a particularly simple manner. Fig. 32 gives as an example the absorption curve of the principal band of HCl under large dispersion [Imes (354)]. Fig. 33 gives as a further example a spectrogram of the third band of HCl, which lies in the photographically accessible region. As one can see, the HCl band in Fig. 32 consists of a series of

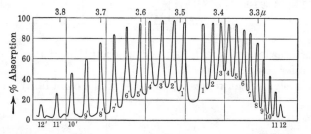

FIG. 32. **Fine Structure of the Principal Absorption Band (Fundamental) of HCl in the Near Infrared [after Imes (354)].** The ordinates give the percentage absorption, calculated from the galvanometer deflections. The numbers below the individual lines are the m values. The P branch is to the left, and the R branch is to the right. With still higher resolution [see Meyer and Levin (491) and Smith (1428)] each line is found to consist of two components on account of isotope effect (see Fig. 33 and p. 142).

almost equidistant lines; however, a line is missing in the center of the band, similar to the zero gap in the bands of the visual and ultraviolet regions. Going out from the gap there are two branches, which are again called the P branch (toward longer wave lengths) and R branch (toward shorter wave lengths). In Table 9 the measured wave numbers of the lines of the HCl band in Fig. 32, as well as their first and second differences, are given. The data of Table 9

TABLE 9. FINE STRUCTURE OF THE HCl BAND AT 3.46 μ.

The data of Meyer and Levin (491) for the isotope HCl^{35} have been used.

m	$\nu_{obs.}(m)$	$\Delta\nu(m)$	$\Delta^2\nu(m)$	$\nu_{obs.} - \nu_{calc.}$ (II, 10)	$\nu_{obs.} - \nu_{calc.}$ (II, 11)
12	3085.62			-3.52	$+0.31$
11	72.76	12.86	-0.83	-2.78	$+0.17$
10	59.07	13.69	-0.50	-2.26	-0.04
9	44.88	14.19	-0.73	-1.64	-0.02
8	29.96	14.92	-0.75	-1.14	0
7	14.29	15.67	-0.84	-0.78	-0.02
6	2997.78	16.51	-0.37	-0.66	-0.18
5	80.90	16.88	-0.78	-0.36	-0.08
4	63.24	17.66	-0.77	-0.11	-0.03
3	44.89	18.43	-0.60	-0.09	-0.03
2	25.78	19.03	-0.50	-0.06	-0.04
1	06.25	19.53		$+0.08$	$+0.08$
0					
-1	2865.09	21.53		$+0.07$	$+0.07$
-2	43.56	22.07	-0.54	$+0.03$	$+0.01$
-3	21.49	22.71	-0.64	$+0.05$	-0.01
-4	2798.78	22.99	-0.28	$+0.04$	-0.10
-5	75.79	23.76	-0.77	$+0.36$	$+0.08$
-6	52.03	24.28	-0.52	$+0.49$	$+0.03$
-7	27.75	24.69	-0.41	$+0.75$	-0.01
-8	03.06	25.33	-0.64	$+1.17$	$+0.03$
-9	2677.73	25.76	-0.43	$+1.60$	-0.02
-10	51.97	26.23	-0.47	$+2.18$	-0.04
-11	25.74	26.74	-0.51	$+2.90$	-0.05
-12	2599.00			$+3.71$	-1.12

are based on measurements with higher dispersion than that of the spectrum of Fig. 32 and refer to one isotope, HCl^{35}, only.

As previously, we can represent the lines of the band fairly well by formula (II, 8),

$$\nu = c + dm + em^2,$$

where m is again the running number, which is $+1$, $+2$, $+3$, $\cdots$ for the R branch and -1, -2, -3, $\cdots$ for the P branch. However, here the coefficient e of the quadratic term is, in comparison to d, still smaller than for CN for which it is considerably smaller than for most other bands in the visible and ultraviolet regions. The fifth column of Table 9 gives the deviations between the observed values and those calculated from

$$\nu = 2885.90 + 20.577m - 0.3034m^2. \tag{II, 10}$$

The agreement for $|m| < 7$ is clearly very good. If we wish to have a formula which also holds exactly for larger values of m, we must add a cubic term,

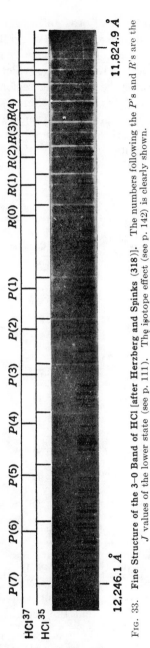

FIG. 33. Fine Structure of the 3-0 Band of HCl [after Herzberg and Spinks (318)]. The numbers following the P's and R's are the J values of the lower state (see p. 111). The isotope effect (see p. 142) is clearly shown.

gm^3, to (II, 8). The last column of the table gives the deviations from the formula

$$\nu = 2885.90 + 20.577m - 0.3034m^2 - 0.00222m^3, \qquad (II, 11)$$

which now represents all the lines very well (except $|m| = 12$).

For the third infrared band of HCl [$v = 3$ in (II, 9)] whose spectrogram is given in Fig. 33, the isotope splitting is much larger [see Chapter III,

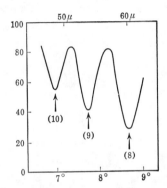

FIG. 34. Part of the Absorption Spectrum of HCl in the Far Infrared [after Czerny (178)]. The ordinates give the transmission of the gas. Thus the minima correspond to the absorption lines. The numbers in parentheses are the m values. The scale in degrees (below) refers to the spectrometer reading. The wave-length scale is given above.

section 2(g)] and it appears therefore to consist of two overlapping bands, one due to HCl^{35} and the other due to HCl^{37}. For this band the separations of successive lines vary much more rapidly than for the fundamental band and therefore a head is almost formed. Again, a formula of the form (II, 8) but with a small cubic term represents the lines very accurately.

Far infrared spectra. In the region above $30\,\mu$, for each of the hydrogen halides a simple series of absorption maxima is observed. In Fig. 34 some of these maxima are represented for the case of HCl, according to measurements by Czerny (178). The maxima, on a wave-number scale, are *very nearly equidistant* and can be represented fairly accurately by the formula

$$\nu = f_0 m, \qquad (II, 12)$$

where f_0 is a constant characteristic of the particular gas and m is an integer. The wave numbers of the maxima are thus, to a good approximation, integral multiples of the constant f_0.   In the example of HCl, the first observed maximum has $m = 4$ and the following maxima have $m = 5 \cdots 11$.   Maxima which would correspond to smaller values of m probably occur but lie outside the region investigated.   At any rate we can see that this series of absorption maxima is of remarkable simplicity.   To be sure, here also, more accurate measurements show that the maxima are not absolutely equidistant.   The change of the separation is, however, very small compared to the separation itself.

TABLE 10.   ABSORPTION SPECTRUM OF HCl IN THE FAR INFRARED.

[After Czerny (178)]

m	$\nu_{\text{obs.}}$[4] (cm^{-1})	$\Delta\nu_{\text{obs.}}$	$\nu_{\text{calc.}} = 20.68m$	$\nu_{\text{calc.}} =$ $20.79m - 0.0016m^3$
1	. . . . . .		20.68	20.79
2	. . . . . .		41.36	41.57
3	. . . . . .		62.04	62.33
4	83.03		82.72	83.06
		21.1		
5	104.1[5]		103.40	103.75
		20.2		
6	124.30		124.08	124.39
		20.73		
7	145.03		144.76	144.98
		20.48		
8	165.51		165.44	165.50
		20.35		
9	185.86		186.12	185.94
		20.52		
10	206.38		206.80	206.30
		20.12		
11	226.50		227.48	226.55

Table 10 gives for HCl the observed wave numbers and their separations, as well as the wave numbers calculated with $f_0 = 20.68$.   The slight systematic difference between observed and calculated values can be very well represented by a cubic term (a quadratic term is not present here), so that the accurate formula is

$$\nu = fm - gm^3, \qquad\qquad (\text{II, 13})$$

---

[4] These wave numbers have not been corrected for vacuum.   For $m = 11$ the correction would be -0.06 cm^{-1}, which is within the accuracy of the measurements.

[5] This value has not been measured very accurately.

where g is very small compared to f. The last column in Table 10 contains the values calculated with $f = 20.79$ and $g = 0.0016$. It is seen that these calculated values agree with the observed values to within less than 0.1 cm^{-1}.

Up to the present time, such simple spectra in the far infrared have been observed only for the hydrogen halides. Theory shows (see Chapter III) that for other molecules, such as CO, for example, they would lie at considerably longer wave lengths, where the investigation would be very difficult and thus far has not been carried out. In one instance, that of HCl, the far infrared spectrum has also been observed in emission, in a chlorine-hydrogen flame by Strong (1439). He found a considerable number of lines with high m-values ($m = 17$ to 33).

3. Radiofrequency (Microwave) Spectra

Microwave absorption. In recent years a number of molecular absorption spectra have been studied in the radiofrequency (microwave) region at wave lengths above 0.2 mm. In the short-wave part of this region from 0.2 to 2.2 mm thus far only spark-generated waves have been used [compare the recent work of Cooley and Rohrbaugh (857)] while above 2 mm various comparatively narrow regions have been investigated by radar methods using tube-generated microwaves. A large number of polyatomic molecules have been studied by these methods, but only three diatomic molecules, HI, ICl, and O_2.[5a,b] Most of these spectra represent simply the long-wave end of the far infrared spectra discussed above. Indeed according to Table 10 the first line, with $m = 1$, of HCl would occur at 0.48 mm. Although this line has not yet been observed the first three lines of HI have been observed by Cooley and Rohrbaugh (857) at 0.78, 0.39, and 0.26 mm, and the first lines of ICl^{35} and ICl^{37} at 4.30 and 4.49 cm, respectively, by Weidner (1521) and the fourth lines at 1.10 and 1.15 cm, respectively, by Townes, Merritt and Wright (1469). In the case of O_2 an exceedingly weak absorption band has been found by Beringer (795) at 0.5 cm. At lower pressure it is resolved into a number of components [see Strandberg, Meng and Ingersoll (1437a)]. This absorption, as we shall see in Chapter V, is of an entirely different type from that of the other molecular absorptions here discussed.

It must be emphasized that with increasing wave length, since $\Delta\nu = -(\Delta\lambda)/\lambda^2$, the resolving power in wave number units increases rapidly for a given resolving power in wave lengths. For this reason investigations in the radiofrequency region are of particular importance: Narrow fine structures can be resolved which in the visible and near infrared would be quite beyond the resolving power even of the best instruments. Thus the microwave "lines" of ICl show a fine structure which has led to interesting conclusions not only about the ICl molecule, but also about the nuclei involved. It is to be expected that similar fine structures will soon be reported for many other diatomic molecules studied in the same way.[5b]

[5a] Compare the comprehensive review by Gordy (994a).   [5b] For ClF see (985a).

Magnetic resonance spectra. A very different technique of studying radiofrequency spectra has been developed by Rabi and his collaborators [see the review by Kellogg and Millman (1125)]. A molecular beam is sent through two inhomogeneous magnetic fields A and B at right angles to the beam and of opposite directions. Between these two fields is a constant field C superimposed by a radiofrequency field. Molecules that absorb (or emit) quanta of radio waves change their magnetic quantum numbers, and therefore the field B will deflect them in a way different from those that go through C without change, that

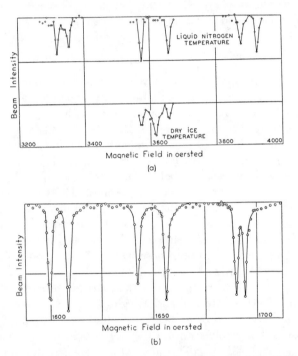

Fig. 35. **Two Sections of the Magnetic Resonance Spectrum of** H_2 [after **Kellog, Rabi, Ramsey and Zacharias (1126) (1332)**]; (a) Corresponding to $\Delta M_J = \pm 1$, $\Delta M_I = 0$, (b) **Corresponding to** $\Delta M_J = 0$, $\Delta M_I = \pm 1$. The beam intensity is plotted as a function of the magnetic field C for constant frequency of the oscillating field. For (a) this frequency was 2.4198 for (b) 6.987 megacycles/sec. Both sections were obtained at liquid nitrogen temperature except the bottom part of (a) in which the central part is repeated at dry ice temperature.

is, they will not reach the receiver. The absorption (or emission) is thus detected by the decrease in the number of molecules reaching the receiver. These absorption spectra are called *magnetic resonance spectra* since the absorbed frequencies correspond to a resonance between the radiofrequency and the magnetic field C which determines the magnitude of the splitting of the energy levels. With increasing field C, the absorbed frequency increases proportionately. Usually, instead of varying the frequency the magnetic field is varied at constant radiofrequency. As an example Fig. 35 gives two sections of the magnetic resonance spectrum of H_2 plotted as a function of the field C. For a constant field of say 10,000 oersted the centers of these two sections would occur at 0.2234×10^{-3} and 1.416×10^{-3} cm^{-1}, that is, the wave length would be 4476 and 706.2 cm, respectively. HD and D_2 are the only other diatomic molecules investigated in detail by this method. For a further discussion of these spectra see Chapter V, section 6.

4. Raman Spectra

Nature of the Raman effect. When a parallel beam of light goes through a gas, a liquid, or a transparent solid body, a small fraction of the light is *scattered* in all directions. The light beam can therefore be seen from the side (Tyndall effect). The intensity of the scattered light is inversely proportional to the fourth power of the wave length: blue light is much more strongly scattered than red (which explains the fact that the sky appears blue).

If the incident light has a discrete line spectrum and the spectrum of the scattered light is investigated, it is found that the scattered light contains exactly the same frequencies as the light producing it. This scattering is called *Rayleigh scattering*. However, if, in taking the spectrogram of the scattered light, the lines that are identical with those of the light source are strongly overexposed, some weak additional lines are found which do not appear in the spectrum of the light source. This phenomenon predicted from theory by Smekal was first discovered by Raman and his collaborators and is now commonly called the *Raman effect*.

A comparison of the wave numbers of these additional lines with those of the most intense lines of the incident light (or of the Rayleigh lines) shows that each one of the original lines is accompanied, in the *Raman spectrum*, by one or more weak lines such that the displacements (in cm^{-1}) of the *Raman lines* from the exciting lines are independent of the frequencies of the latter. If another light source with a different spectrum is used, other Raman lines are obtained for the same scattering substance. However, the displacements from the exciting lines are the same. For different scattering substances, the displacements have different magnitudes. Thus the Raman displacements are characteristic of the scattering substance under consideration. An explanation for this phenomenon will be given in Chapter III, section 1(d).

In what follows we discuss the observed phenomena only for the case of diatomic gases as scattering substances.

Large Raman displacements. When a spectrograph of small dispersion is used for the observations, for each of the diatomic gases only one relatively large Raman displacement is found; that is to say, for each exciting line only one Raman line is found and this is on the long-wave-length side. Fig. 36 gives a spectrogram for the case of HCl obtained by excitation with a mercury arc. The main Raman line is marked Q. Table 11 gives the observed Raman displacements for a number of diatomic gases. The third column of this table

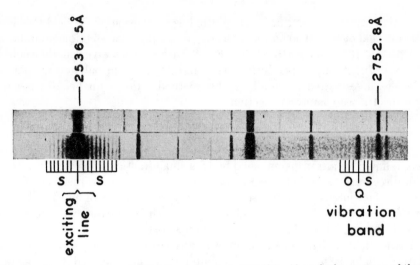

Fɪɢ. 36.   **Raman Spectrum of HCl [after Andrychuk (752a)].*** *   Above is the spectrum of the Hg arc, below the Hg spectrum scattered by HCl gas (5 atm.).   The exciting line of the Raman spectrum is the Hg line 2536.5 Å.   With lower resolution and less exposure time only the Raman line at 2737 Å (marked Q) corresponding to a shift of 2886 cm⁻¹ appears.   The spectrogram shows in addition Raman lines of small displacement on both sides of the exciting line (rotational Raman spectrum; see p. 88) and, very weakly, a fine structure accompanying the Raman line 2737 Å (rotation-vibration band; see p. 114).

* The author is greatly indebted to Dr. D. Andrychuk for this spectrogram.

contains for comparison the frequencies of the main infrared bands of the gases under consideration.   It is seen that the Raman displacements agree exactly

TABLE 11.   LARGE RAMAN DISPLACEMENTS AND INFRARED FREQUENCIES OF DIATOMIC GASES

Gas	Raman Displacement $\Delta \nu$ (cm⁻¹)	Infrared Frequency ν_0 (cm⁻¹)	References
HCl	2886.0	2885.9	(720)
HBr	2558	2559.3	(611) (563)
HI	2233	2230.1	(611) (531)
NO	1877	1875.9	(101)
CO	2145	2143.2	(68)
H₂	4160.2*		(662)
N₂	2330.7		(578)
O₂	1554.7		(578)
F₂	892		(752b)
Cl₂	556		(800)

* Refers to the transition $J = 0 \to J = 0$ (see page 114).

with the frequencies of the main bands in the near infrared, for those gases for which both have been observed.   Therefore, in these cases the Raman spectrum may be regarded as an infrared spectrum shifted into the conveniently accessible visible or ultraviolet region. It should be noticed, however, that Raman lines also appear for the elementary molecules H_2, N_2, O_2, F_2, and Cl_2, for which no infrared spectrum occurs.

Small Raman displacements. Under higher resolution, in addition to the aforementioned Raman lines, simple series of equidistant, closely spaced lines appear on either side of each exciting line. At the same time the Raman lines of large displacement are found to be somewhat broadened and, for sufficient exposure time, to be accompanied by a series of lines on either side. In the HCl spectrogram of Fig. 36 both fine structures, that near the exciting line and that near the Raman line of large displacement, are visible and just resolved, in spite of the low dispersion, since in this case the separation of successive lines is comparatively large (on the average 41.6 cm^{-1}). The high dispersion spectrograms of Fig. 37 show more clearly such Raman lines of small displacements for N_2 and O_2. Here the separation of successive lines is only 7.99 and 11.50 cm^{-1}, respectively. The displacement of these Raman lines from the exciting line is only a small fraction of the displacement of the Raman lines previously discussed.

TABLE 12. SMALL RAMAN DISPLACEMENTS FOR HCl

[After Wood and Dieke (720)]

m	$\Delta\nu_{obs.}$	$\Delta(\Delta\nu)$	$\Delta\nu_{calc.}$
2	+143.8		145.7
3	+183.3	39.5	187.4
4	+232.2	38.9	229.0
1	−101.1		104.1
2	−142.7	41.6	145.7
3	−187.5	44.8	187.4
4	−229.4	41.9	229.0
5	−271.0	41.6	270.7
6	−312.9	41.9	312.3
7	−353.0	40.0	353.9

Table 12 gives the observed displacements from the exciting line for the case of HCl. It can be seen from the table that the separations of successive lines (column 3) are very nearly constant. To a good approximation, the line displacements on both sides of the exciting line may be represented by

$$\Delta\nu = \pm(\tfrac{3}{2}p + pm), \tag{II, 14}$$

where p is the constant separation of successive lines and m takes the integral values given in the first column of Table 12.[6] The values calculated with $p = 41.64$ are shown in the last column of Table 12. The agreement is quite good. A slight residual trend can be allowed for by a cubic term, quite analogous to that for the far infrared spectra (see p. 58). It is now very striking and noteworthy that the approximately constant separation p of successive lines is very nearly twice as large as the separation of successive maxima in the far

---

[6] Naturally, one could also have chosen as the formula $\Delta\nu = \pm(\tfrac{1}{2}p + pm)$ or $\pm(\tfrac{5}{2}p + pm)$, with a corresponding alteration of the numbering. However, (II, 14) has the advantage that for N_2, for example (see Fig. 37), the first line on both sides of the exciting line has $m = 0$. For HCl, these lines are hidden by the overexposed exciting line. However, on the basis of the later theoretical discussion, there can be no doubt that they are also present for HCl.

infrared spectrum. For HCl, for example, $p = 41.64$, while f_0 in (II, 12) is equal to 20.68 cm^{-1}. This strongly suggests that the small Raman displacements correspond to the far infrared spectra, just as the large Raman displacements correspond to the near infrared spectra.

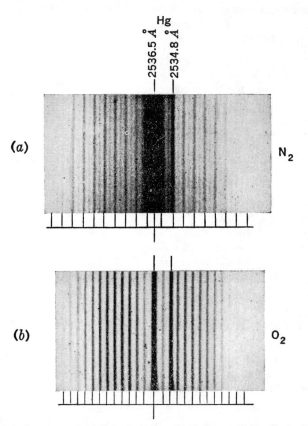

FIG. 37. **Raman Spectrum** of (*a*) N₂ and (*b*) O₂: **Small Raman Shifts, Excited by the Hg Line 2536.5 Å [after Rasetti (579)].*** The exciting line (longer leading line) is not so strongly over-exposed in (*b*), since it has been almost completely absorbed by Hg vapor before falling on the plate. The relatively strong line to the right of the exciting line, which is superimposed on a Raman line, is the Hg line 2534.8 Å. The dispersion of the enlarged spectrogram is about 0.37 Å/mm.

* The author is greatly indebted to Professor F. Rasetti for this spectrogram.

In a series of Raman lines of small displacement represented by (II, 14) the intensity first increases on going out from the exciting line and then decreases again with increasing m. For molecular nitrogen [Fig. 37(*a*)] and hydrogen there is in addition a characteristic alternation of intensities, the lines being alternately weak and strong. This is similar to the observations in the visible and ultraviolet bands (see p. 51).

The fine structure observed in the case of HCl for the Raman band of large displacement, is of the same type as that of certain bands in the visible and ultraviolet region. The only other case in which such a fine structure of a Raman band has been observed is that of H_2. A more detailed discussion of this fine structure will be given in the next chapter (p. 114 f.).

CHAPTER III

ROTATION AND VIBRATION OF DIATOMIC MOLECULES; INTERPRETATION OF INFRARED AND RAMAN SPECTRA

1. Interpretation of the Principal Features of Infrared and Raman Spectra by Means of the Models of the Rigid Rotator and of the Harmonic Oscillator

When we compare the molecular spectra observed in the infrared (see Figs. 31–34) with an atomic spectrum such as that of the hydrogen atom (Fig. 6), we see at once that even qualitatively a fundamental difference exists. To be sure, we have in both cases regular series of lines or bands. However, while for atoms the line separation in a series decreases rapidly (Rydberg series), for infrared molecular spectra it is approximately constant. Therefore we cannot hope to be able to explain these infrared molecular spectra by the same model as that used for atomic spectra—namely, by the stationary states of an electron revolving about a core—but we must search for other models.

Actually, for a diatomic molecule, two additional modes of motion, which do not occur for atoms, are possible, and these have to be considered as possible causes of the infrared spectra: First, the molecule can *rotate* as a whole about an axis passing through the center of gravity and perpendicular to the line joining the nuclei (internuclear axis), and, second, the atoms can vibrate relative to each other along the internuclear axis. We now have to investigate what type of spectrum would be expected on the basis of quantum theory for such a rotating or vibrating system, and to compare it with the observed infrared spectrum in order to find out in what way the latter is produced. Conversely, this investigation will help us to come to conclusions concerning the structure of diatomic molecules.

(a) The Rigid Rotator

The molecule as a rigid rotator. We shall begin with the simplest possible model of a rotating molecule, the so-called dumbbell model (Fig. 38). We shall consider the two atoms of masses m_1 and m_2 to be point-like and fastened at a distance r apart to the ends of a weightless, rigid rod. In so doing, we neglect, on the one hand, the finite extent of the atoms and, on the other, the fact that in reality the atoms are not rigidly bound to each other but that their distance can alter under the influence of their rotation. The neglect of the first factor is certainly well justified, since the mass of the atom is practically concentrated in the nucleus, which has a radius of only about 10^{-12} cm, while the internuclear distance in a molecule is of the order of magnitude of 10^{-8} cm (see below). The

66

effect of the neglect of the second factor is in general also very small, as we shall see later.

In classical mechanics the energy of rotation E of a rigid body is given by

$$E = \tfrac{1}{2}Iw^2 \tag{III, 1}$$

Here w is the angular velocity of the rotation and I is the moment of inertia of the system about the axis of rotation. The angular velocity is related to the number of rotations per second, ν_{rot}. (rotational frequency) by

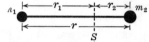

$$w = 2\pi\nu_{rot}. \tag{III, 2}$$

The moment of inertia is defined by $I = \sum m_i r_i^2$.

FIG. 38. **Dumbbell Model of a Diatomic Molecule.**

Of course, for a system freely suspended in space, the axis of rotation passes through the center of gravity C. The angular momentum of the system is given by $P = Iw$. Introducing this into (III, 1) the energy may also be expressed by

$$E = \frac{P^2}{2I} \tag{III, 1a}$$

According to (III, 1), the rotational energy depends essentially on the moment of inertia. For the dumbbell model this is

$$I = m_1 r_1^2 + m_2 r_2^2,$$

where

$$r_1 = \frac{m_2}{m_1 + m_2} r \quad \text{and} \quad r_2 = \frac{m_1}{m_1 + m_2} r \tag{III, 3}$$

are the distances of the masses m_1 and m_2 from the center of gravity C and r is the distance of the two mass points from each other (see Fig. 38). Substitution gives

$$I = \frac{m_1 m_2}{m_1 + m_2} r^2; \tag{III, 4}$$

that is, the moment of inertia is the same as that of a mass point of mass

$$\mu = \frac{m_1 m_2}{m_1 + m_2} \tag{III, 5}$$

at a distance r from the axis; μ is called the *reduced mass* of the molecule.

Thus, instead of considering the rotation of the dumbbell, we can equally well consider the rotation of a single mass point of mass μ at a fixed distance r from the axis of rotation. Such a system is called a *simple rigid rotator*.

Energy levels. In order to determine the possible energy states of such a rigid rotator according to the quantum theory, we have to solve the Schrödinger equation (I, 12) with the substitution $m = \mu$ and $V = 0$. The latter substitution is made since no potential energy is associated with the rotation as long as the rotator can be regarded as completely rigid. The Schrödinger equation appropriate to the problem is therefore

$$\frac{\partial^2 \psi}{\partial x^2} + \frac{\partial^2 \psi}{\partial y^2} + \frac{\partial^2 \psi}{\partial z^2} + \frac{8\pi^2 \mu}{h^2} E\psi = 0, \qquad \text{(III, 6)}$$

where here $x^2 + y^2 + z^2 = r^2$ is a constant. The solution of equation (III, 6) is comparatively simple [see, for example, Sommerfeld (11b) or Pauling and Wilson (13)] but will not be carried out here. The result is that solutions ψ that are single-valued, finite, and continuous (see p. 10) occur only for certain values of E—namely, the eigenvalues

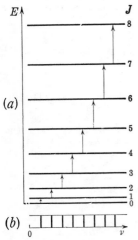

$$E = \frac{h^2 J(J+1)}{8\pi^2 \mu r^2} = \frac{h^2 J(J+1)}{8\pi^2 I}, \qquad \text{(III, 7)}$$

where the *rotational quantum number* J can take the integral values 0, 1, 2, $\cdots$. Thus we have a *series of discrete energy levels whose energy increases quadratically with increasing J* (Fig. 39).

If we compare formula (III, 7) with the classical formula (III, 1a), we see that the classical *angular momentum* of the system in the quantum state J is

Fig. 39. **Energy Levels and Infrared Transitions of a Rigid Rotator.** (a) The energy level diagram. (b) The resulting spectrum (schematic).

$$P = \frac{h}{2\pi} \sqrt{J(J+1)} \approx \frac{h}{2\pi} J. \qquad \text{(III, 8)}$$

This result agrees with the general result for the magnitude of angular momentum vectors in quantum mechanics discussed in Chapter I, p. 16. The rotational quantum number J thus gives approximately the angular momentum in units $h/2\pi$. As previously we shall use the same symbols for angular momentum vectors and corresponding quantum numbers except that the symbols referring to the former will be in heavy type. Thus $\mathbf{J}$ means an angular momentum vector of magnitude $\sqrt{J(J+1)}\,(h/2\pi) \approx J(h/2\pi)$.

Since only certain discrete energy values and angular momenta of the rigid rotator are possible, it follows that *only certain rotational frequencies* are possible. Since classically the angular momentum $P = I w$, we obtain from (III, 8) for the angular velocity

$$w = \frac{h}{2\pi I} \sqrt{J(J+1)} \approx \frac{h}{2\pi I} J.$$

Therefore, according to (III, 2), the rotational frequency is

$$\nu_{\text{rot.}} = \frac{h}{4\pi^2 I} \sqrt{J(J+1)} \approx \frac{h}{4\pi^2 I} J, \tag{III, 9}$$

that is, the rotational frequency increases approximately linearly with J.

It should be noted that the frequency of rotation can be defined only with the help of the classical formula $P = I w$. In quantum mechanics it has no exactly definable meaning except for large values of the rotational quantum number.

According to the old quantum theory, the angular momentum was exactly $J(h/2\pi)$, and therefore, according to (III, 1a), the energy of the rotator was

$$E = \frac{h^2}{8\pi^2 I} J^2 \tag{III, 10}$$

As we shall see later, experiments have decided unambiguously in favor of the wave-mechanical expression (III, 7).

Eigenfunctions.   The Schrödinger equation (III, 6) of the rigid rotator is closely related to that of the hydrogen atom, the only difference being the absence of the potential energy term.   Since this term for the hydrogen atom affects only the radial dependence of the eigenfunctions it is clear that the eigenfunctions ψ_r of the rigid rotator are identical with those obtained from the hydrogen eigenfunctions by dropping the factor depending on r; that is, they are the so-called *surface harmonics* (see text-books on wave mechanics)

$$\psi_r = N_r P_J^{|M|}(\cos \vartheta) \, e^{M\varphi} \tag{III, 11}$$

Here φ is the azimuth of the line connecting the mass point to the origin, taken about the z axis; ϑ is the angle between this line and the z axis; M is a second quantum number which takes the values

$$M = J, (J-1), (J-2), \cdots -J \tag{III, 12}$$

and which represents in units $h/2\pi$ the component of the angular momentum J in the direction of the z axis (see p. 26); $P_J^{|M|}(\cos \vartheta)$ is a function of the angle ϑ, the so-called *associated Legendre function;* and N_r is a normalization constant (see below).   The broken-line curves in Fig. 40 show the variation of the eigenfunctions (III, 11) in a plane ($\varphi = 0$ and $\varphi = 180°$, respectively) for $J = 0, 1, 2$, and 3, and the various possible $|M|$ values.   The value of the function is plotted from the origin in every direction in the plane of the figure.   The sign is indicated by $+$ or $-$ in the corresponding loop.

The probability of finding the system oriented in the direction ϑ, φ is

$$\psi_r \psi_r^* = N_r^2 [P_J^{|M|}(\cos \vartheta)]^2 \tag{III, 13}$$

Since $\psi_r \psi_r^*$ is independent of φ, the probability distribution is rotationally symmetric about the fixed axis.   The dependence of $\psi_r \psi_r^*$ on the angle ϑ is shown by the full curves in Fig. 40.   It is seen that, for $J = 0$, all directions are equally probable, while, with increasing J, more and more preferred directions exist.   For $|M| = J$, as J increases, the classical picture of a rotation about the z axis is slowly approached; that is to say, the most probable values of ϑ lie in the neighborhood of 90°.

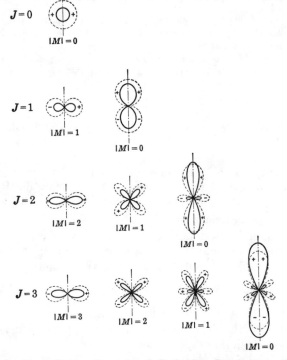

F‍IG. 40.   **Eigenfunctions (Broken Curves) and Probability Distributions (Solid Curves) for the Rotator in the Levels $J = 0, 1, 2, 3$.**   The value of the eigenfunction and the probability density distribution corresponding to a given orientation of the rotator is plotted in a polar diagram from the mid-point of each individual diagram in the direction under consideration.   The figures represent plane sections through the spatial polar diagrams.   All the figures are drawn to the same scale.

The explicit form of the normalized eigenfunctions up to $J = 2$, using the values of J and M as superscripts of ψ_r, is as follows:

$$\psi_r{}^{00} = \frac{1}{2\sqrt{\pi}}; \qquad\qquad \psi_r{}^{10} = \frac{1}{2}\sqrt{\frac{3}{\pi}}\cos\vartheta, \qquad\qquad \psi_r{}^{1\pm1} = \frac{1}{2}\sqrt{\frac{3}{2\pi}}\sin\vartheta\,e^{\pm i\varphi};$$

$$\psi_r{}^{20} = \frac{1}{4}\sqrt{\frac{5}{\pi}}\,(3\cos^2\vartheta - 1), \quad \psi_r{}^{2\pm1} = \frac{1}{2}\sqrt{\frac{15}{2\pi}}\sin\vartheta\cos\vartheta\,e^{\pm i\varphi}, \quad \psi_r{}^{2\pm2} = \frac{1}{4}\sqrt{\frac{15}{2\pi}}\sin^2\vartheta\,e^{\pm 2i\varphi}.$$

For larger J values the functions may be found in Pauling and Wilson (13).

Spectrum.   According to classical electrodynamics, an intramolecular motion leads to radiation of light only if a *changing dipole moment* is associated with it.   For a rigid rotator this can be caused by the rotating mass point possessing a charge or being associated with a *permanent dipole moment* that lies in the direction of the perpendicular from the mass point to the axis of rotation. The latter alternative applies to all diatomic molecules that consist of unlike atoms, since for them the centers of the positive and negative charges do not coincide; that is, such molecules have a permanent dipole moment which lies in the internuclear axis. During the rotation the component of the dipole in a fixed direction changes periodically with a frequency equal to the rotational frequency; that is to say, classically, light of frequency $\nu_{rot.}$ should be emitted. For molecules consisting of two like atoms, no dipole moment arises and therefore no light is emitted. Conversely, only if a permanent dipole moment is present can an infrared frequency be absorbed and thereby a rotation of the system be produced, or a rotation already present be increased. According to classical theory, the absorbed or emitted spectrum of the rotator is continuous, since $\nu_{rot.}$ can take any value.

According to quantum theory, the emission of a light quantum takes place as a result of a transition of the rotator from a higher to a lower energy level, while the absorption of a quantum of the proper frequency produces a transition from a lower to a higher level. The wave number of the emitted or absorbed quantum is, according to (I, 6),

$$\nu = \frac{E'}{hc} - \frac{E''}{hc}, \tag{III, 14}$$

where E' and E'' are the rotational energies (III, 7) in the upper and lower states, respectively. [Throughout this book we shall indicate quantities associated with the upper state by a single prime mark (') and those associated with the lower state by two prime marks ('')]. $E/hc = F(J)$ is the *rotational term*[1] (units cm^{-1}), which, according to (III, 7), is given by

$$F(J) = \frac{E}{hc} = \frac{h}{8\pi^2 cI} J(J+1) = BJ(J+1), \tag{III, 15}$$

where the constant

$$B = \frac{h}{8\pi^2 cI} \tag{III, 16}$$

is called the *rotational constant*. Apart from a factor, it is the reciprocal moment of inertia. Using (III, 15) we can now rewrite (III, 14) in the following way:

$$\nu = F(J') - F(J'') = BJ'(J'+1) - BJ''(J''+1). \tag{III, 17}$$

In order to calculate the frequencies that are actually emitted or absorbed, it is necessary to know the *selection rule* for the quantum number J. This selection rule is obtained by evaluating the matrix elements $R_x{}^{nm}$, $R_y{}^{nm}$, $R_z{}^{nm}$ of the

---

[1] It should be observed that there is no minus sign in the definition for term values here, in contrast to what is usual for atoms (see p. 8).

dipole moment (see p. 19). In the present case, if M_0 is the constant dipole moment of the rotator, its components are

$$M_x = M_0 \sin \vartheta \cos \varphi$$

$$M_y = M_0 \sin \vartheta \sin \varphi \qquad \text{(III, 18)}$$

$$M_z = M_0 \cos \vartheta$$

Substituting these expressions as well as the rotator eigenfunctions into the matrix elements (I, 44) we obtain

$$R_x^{J'M'J''M''} = M_0 \int \psi_r^{J'M'*} \sin \vartheta \cos \varphi \, \psi_r^{J''M''} \, d\tau$$

$$R_y^{J'M'J''M''} = M_0 \int \psi_r^{J'M'*} \sin \vartheta \sin \varphi \, \psi_r^{J''M''} \, d\tau \qquad \text{(III, 19)}$$

$$R_z^{J'M'J''M''} = M_0 \int \psi_r^{J'M'*} \cos \vartheta \, \psi_r^{J''M''} \, d\tau$$

It is immediately clear that these matrix elements will be different from zero—that is, emission or absorption by the rotator can occur—only when the dipole moment M_0 is different from zero, in agreement with the previous classical considerations. A more detailed discussion of the integrals in (III, 19) shows as outlined below that they vanish except when

$$J' = J'' \pm 1; \quad \text{that is,} \quad \Delta J = J' - J'' = \pm 1 \qquad \text{(III, 20)}$$

Thus the quantum number J can "jump" only by one unit.

Let us consider in a little more detail the matrix element $R_z^{J'M'J''M''}$. Substituting ψ_r from (III, 11) and $d\tau = \sin \vartheta \, d\varphi \, d\vartheta$ we have

$$R_z^{J'M'J''M''} =$$

$$M_0 N_r' N_r'' \int_0^\pi P_{J'}^{|M'|}(\cos \vartheta) \cos \vartheta \, P_{J''}^{|M''|}(\cos \vartheta) \sin \vartheta \, d\vartheta \int_0^{2\pi} e^{iM'\varphi} e^{-iM''\varphi} \, d\varphi \qquad \text{(III, 21)}$$

The second integral is zero except when $M' = M''$, in which case it has the value 2π. For the evaluation of the first integral we make use of a relation proven in books on spherical harmonics:

$$\cos \vartheta \, P_J^{|M|}(\cos \vartheta) = \frac{(J + |M|)}{(2J + 1)} P_{J-1}^{|M|}(\cos \vartheta) + \frac{(J - |M| + 1)}{(2J + 1)} P_{J+1}^{|M|}(\cos \vartheta) \qquad \text{(III, 22)}$$

and obtain with $M = M' = M''$

$$R_z^{J'M'J''M''} = 2\pi M_0 N_r' N_r'' \left[\frac{(J + |M|)}{(2J + 1)} \int P_{J'-1}^{|M|}(\cos \vartheta) P_{J''}^{|M|}(\cos \vartheta) \sin \vartheta \, d\vartheta \right.$$

$$\left. + \frac{(J - |M| + 1)}{(2J + 1)} \int P_{J'+1}^{|M|}(\cos \vartheta) P_{J''}^{|M|}(\cos \vartheta) \sin \vartheta \, d\vartheta \right] \qquad \text{(III, 23)}$$

A further theorem in spherical harmonics is that

$$\int P_{J_1}^{|M|}(\cos \vartheta) P_{J_2}^{|M|}(\cos \vartheta) \sin \vartheta \; d\vartheta = 0 \quad \text{for } J_1 \neq J_2$$

(III, 24)

$$\int P_{J_1}^{|M|}(\cos \vartheta) P_{J_2}^{|M|}(\cos \vartheta) \sin \vartheta \; d\vartheta = \frac{2}{(2J+1)} \frac{(J+|M|)!}{(J-|M|)!} \quad \text{for } J_1 = J_2$$

If this is taken into account it is immediately seen from (III, 23) that the matrix elements $R_z^{J'M'J''M''}$ vanish except when $J'' = J' - 1$ or when $J'' = J' + 1$ (and at the same time $M' = M''$). In a similar way it is found that $R_x^{J'M'J''M''}$ and $R_y^{J'M'J''M''}$ vanish except when $J'' = J' - 1$ or $J'' = J' + 1$ (and at the same time $M' = M'' \pm 1$). Thus the selection rule (III, 20) is proven. The selection rule for $M (\Delta M = 0, \pm 1)$ plays a role only in a magnetic field (see Chapter V, section 5).

With our choice of notation we always have for the rotator $J' > J''$ (since J' refers to the upper state), and therefore only $\Delta J = +1$ need be considered. Consequently we obtain for the emitted or absorbed lines of the rigid rotator

$$\nu = F(J'' + 1) - F(J'') = B(J'' + 1)(J'' + 2) - BJ''(J'' + 1)$$
$$= 2B(J'' + 1),$$

where J'' can take all integral values $0, 1, 2, \cdots$. In the future, for simplicity we shall write J for J'' when only the J value of the lower state occurs, so that

$$\nu = 2B(J + 1); \quad J = 0, 1, 2, \cdots. \tag{III, 25}$$

Thus the spectrum of the simple rigid rotator consists of a series of equidistant lines. The first of these $(J = 0)$ lies at $2B$ and the separation of successive lines is also $2B$. The corresponding transitions are indicated in Fig. 39, and the spectrum is drawn schematically at the bottom of this figure. The comparison of this theoretical spectrum with experiment will be taken up in subsection (c).

According to (III, 9) and (III, 16), the *rotational frequency* is

$$\nu_{\text{rot.}} = c \, 2B\sqrt{J(J+1)} \approx c \, 2BJ; \tag{III, 26}$$

that is, the rotational frequency in any given state of the rotator is approximately equal to the frequency of the spectral line that has this state as upper state.

When the formula (III, 10) of the *old quantum theory* for the energy of the rotator is used in place of the wave mechanical expression (III, 7), the wave numbers of the absorbed spectral lines are easily found to be

$$\nu = 2BJ + B. \tag{III, 27}$$

According to this, the first line would lie at $\nu = B$ and not at $\nu = 2B$. The line separation, however, is the same as in wave mechanics.

(b) The Harmonic Oscillator

The molecule as a harmonic oscillator. The simplest possible assumption about the form of the vibrations in a diatomic molecule is that each atom moves toward or away from the other in simple harmonic motion, that is, that the displacement from the equilibrium position is a sine function of the time.   Such a motion of the two atoms can easily be reduced to the harmonic vibration of a single mass point about an equilibrium position—that is, to the model of the harmonic oscillator.

In classical mechanics a harmonic oscillator can be defined as a mass point of mass m which is acted upon by a force F proportional to the distance x from the equilibrium position and directed toward the equilibrium position.   Therefore (since force = mass $\times$ acceleration)

$$F = -kx = m\frac{d^2x}{dt^2}. \tag{III, 28}$$

The proportionality factor k is called the *force constant*.   The well-known solution of the differential equation (III, 28) is

$$x = x_0 \sin (2\pi\nu_{\text{osc.}}.t + \varphi) \tag{III, 29}$$

Here the vibrational frequency $\nu_{\text{osc.}}$ is given by

$$\nu_{\text{osc.}} = \frac{1}{2\pi}\sqrt{\frac{k}{m}}, \tag{III, 30}$$

as may be immediately verified by substitution into (III, 28); x_0 is the amplitude of the vibration and φ is a phase constant dependent on the initial conditions.

Since the force is the negative derivative of the potential energy V, it follows from $F = -kx$ that, for the harmonic oscillator,

$$V = \tfrac{1}{2}kx^2 = 2\pi^2 m\nu_{\text{osc.}}^2\, x^2. \tag{III, 31}$$

We can therefore also define a harmonic oscillator as a system whose potential energy is proportional to the square of the distance from its equilibrium position; that is to say, the *potential energy curve is a parabola* (broken curve in Fig. 41).

The restoring force exerted by the two atoms of a molecule on each other when they are displaced from their equilibrium position is, at least approximately, proportional to the change of internuclear distance.   If we assume that this relation holds exactly, it follows immediately that the atoms in the molecule will execute harmonic vibrations when they are left to themselves after being displaced from their equilibrium positions.   For the first atom, of mass m_1,

$$m_1\frac{d^2r_1}{dt^2} = -k(r - r_e),$$

and, correspondingly for the second atom, of mass m_2,

$$m_2\frac{d^2r_2}{dt^2} = -k(r - r_e).$$

Here r_1 and r_2 are the distances of the two atoms from the center of gravity (see Fig. 38), r is the distance of the two atoms from each other, and r_e is the equilibrium distance. By substitution of (III, 3) we obtain from both equations

$$\frac{m_2 m_1}{m_1 + m_2}\frac{d^2 r}{dt^2} = -k(r - r_e),$$

or, when the reduced mass μ is introduced according to (III, 5) and when r, under the differential sign, is replaced by $(r - r_e)$ (which is possible, since r_e is constant),

$$\mu \frac{d^2(r - r_e)}{dt^2} = -k(r - r_e). \tag{III, 32}$$

This equation is identical with the general equation (III, 28) of the harmonic oscillator, except that x is replaced by $(r - r_e)$, the change of internuclear distance from its equilibrium value. Thus we have reduced the vibrations of the two atoms of a molecule to the vibration of a single mass point of mass μ, whose amplitude equals the amplitude of the change of internuclear distance in the molecule.

From (III, 32) in combination with (III, 30) it follows immediately that the classical *vibrational frequency* of the molecule is

$$\nu_{\text{osc.}} = \frac{1}{2\pi}\sqrt{\frac{k}{\mu}}. \tag{III, 33}$$

Whereas, for the rotator, classically all rotational frequencies can occur, here even on the classical theory only one vibrational frequency is possible whose magnitude depends on the two atomic masses and the force constant. However, classically the amplitude, and therefore the energy, of this vibration can assume any desired value.

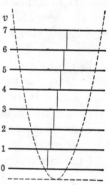

Fig. 41.   Potential Curve, Energy Levels, and Infrared Transitions of the Harmonic Oscillator. The short vertical lines representing the transitions are spread apart from one another in a horizontal direction for the sake of clarity only. The abscissa for the broken curve is the displacement from the equilibrium position (minimum).

Energy levels. Also in wave mechanics the vibrations of the two nuclei in a diatomic molecule may be reduced to the motion of a single particle of mass μ whose displacement x from its equilibrium position equals the change $r - r_e$ of the internuclear distance [see for example Weizel (22)]. If we assume the potential energy of the two nuclei to be given by (III, 31) we obtain from (I, 12) for the wave equation describing the motion of the representative particle (the harmonic oscillator)

$$\frac{d^2 \psi}{dx^2} + \frac{8\pi^2 \mu}{h^2}(E - \tfrac{1}{2}kx^2)\psi = 0 \tag{III, 34}$$

The mathematical discussion of this equation [see Pauling and Wilson (13)] shows that solutions that are single-valued, finite, and continuous and vanish at infinity do not exist for all values of E but only for the E values

$$E(v) = \frac{h}{2\pi}\sqrt{\frac{k}{\mu}}\,(v + \tfrac{1}{2}) = h\nu_{\text{osc.}}(v + \tfrac{1}{2}),\qquad\text{(III, 35)}$$

where the *vibrational quantum number* v can take only integral values, $0, 1, 2, \cdots$. The values (III, 35) are the only energy values allowed by quantum theory for the harmonic oscillator and therefore also for the harmonically vibrating molecule. They are half-integral multiples of $h\nu_{\text{osc.}}$, where $\nu_{\text{osc.}}$ is the vibrational frequency of the oscillator calculated from (III, 33) (that is, classically). Sometimes one speaks of the oscillator as having half-integral vibrational quanta. The energy level diagram of the harmonic oscillator is represented in Fig. 41. It consists of a series of *equidistant levels*.

It should be particularly noted that (in contrast to the rotator) the state of lowest energy, $v = 0$, does not have $E = 0$ but $E(0) = \frac{1}{2}h\nu_{\text{osc.}}$. Thus, even in the lowest vibrational state, vibrational energy is present, which is called *zero-point energy*. In spite of this zero-point vibration the state $v = 0$ is sometimes called, for the sake of brevity, the vibrationless state.

According to the old quantum theory, the energy of the harmonic oscillator was found to be $h\nu_{\text{osc.}}v$ (*integral* vibrational quanta). Thus, apart from the constant zero-point energy, the energy levels according to quantum mechanics are the same as in the old quantum theory.

If we transform energy values to *term values* (by dividing by hc), we obtain for the vibrational terms

$$G(v) = \frac{E(v)}{hc} = \frac{\nu_{\text{osc.}}}{c}\,(v + \tfrac{1}{2}).\qquad\text{(III, 36)}$$

In band spectroscopy, $\nu_{\text{osc.}}/c$ is generally designated by ω. Therefore,

$$G(v) = \omega(v + \tfrac{1}{2}).\qquad\text{(III, 37)}$$

ω is the vibrational frequency measured in cm^{-1}. The actual number of vibrations per second is c times as great.

Eigenfunctions. The eigenfunctions ψ_v of the Schrödinger equation (III, 34) of the harmonic oscillator are found to be the Hermite orthogonal functions [see for example, Pauling and Wilson (13)].

$$\psi_v = N_v e^{-\frac{1}{2}\alpha x^2} H_v(\sqrt{\alpha}\,x)\qquad\text{(III, 38)}$$

Here N_v is a normalization constant, $\alpha = 4\pi^2\mu\nu_{\text{osc.}}/h = 2\pi\sqrt{\mu k}/h$ and $H_v(\sqrt{\alpha}\,x)$ is an Hermite polynomial of the vth degree (see below). The vibrational eigenfunctions ψ_v for $v = 0, 1, 2, 3, 4,$ and 10 are plotted as broken-line curves in Fig. 42. It is seen that the number of nodes of each of these functions (that is, the number of times ψ_v goes through zero) equals the vibrational quantum number v. The full-line curves in Fig. 42 give the resulting probability density distributions $\psi_v\psi_v^* = |\psi_v|^2$.

The amplitude of the corresponding classical vibrational motion is obtained from the intersection of the potential energy curve (broken-line curve in Fig. 41) with the corresponding energy level, since the turning points are the positions at which the total energy is entirely potential energy. These classical turning points are indicated in Fig. 42 by vertical lines on the abscissa axis. It can be seen that the wave-mechanical probability density has large values just in the region of the classical vibration, but that also somewhat outside this region the probability density is not negligible, although it falls off exponentially there. Thus in wave mechanics the oscillator in a given energy level can reach, with a non-zero probability, regions that a classical oscillator of the same energy and the same force constant can never reach. While classically the oscillator stays for the greater part of the time at the turning points, wave-mechanically, for $v \neq 0$ there is a broad maximum of the probability distribution in the neighborhood of each of the classical turning points, but lying somewhat more toward the center. In addition, for $v > 1$, there are other maxima of the wave-mechanical probability distribution between the two outermost maxima. No classical analogues exist for these maxima.

The curve of the probability density distribution for the lowest vibrational state is particularly important. Here the difference between the classical motion and the wave-mechanical probability density is the greatest; instead of the two turning points there is now only one maximum in the center. Of course, classically, the lowest state of the oscillator is that for which the system is at

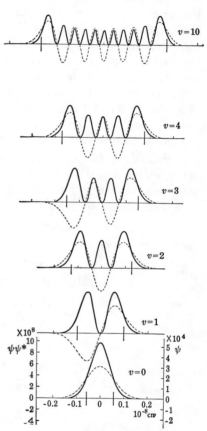

Fig. 42.   **Eigenfunctions (Broken Curves) and Probability Density Distributions (Solid Curves) of the Harmonic Oscillator for** $v = 0, 1, 2, 3, 4,$ **and 10.** The abscissae give the displacement from the equilibrium position in 10^{-8} cm. The scales for abscissae and ordinates are the same for all the figures. However, they are given explicitly only for the lowest figure ($v = 0$). The functions are plotted for $\mu_A = 10$ and $\omega = 1000$. For other values they are obtained simply by altering the scale (see text).

rest at the minimum of the potential curve (Fig. 41). According to quantum theory, such a state would contradict the Heisenberg uncertainty relation, since the position *and* velocity would then be exactly defined.

The curves in Fig. 42 are drawn for the case $\mu_A = 10$, $\omega = 1000$ cm^{-1} ($\mu_A =$

reduced mass in atomic-weight units). Corresponding curves for other cases may be obtained by changing the scale on the abscissa axis in inverse proportion to $\sqrt{\mu_A \omega}$ or, in the case of isotopic molecules, for which the force constant k is the same, in inverse proportion to $\sqrt{\mu_A}$. Applied to the molecule the abscissa in Fig. 42 gives the deviation of the internuclear distance from the equilibrium value. If one wanted to apply the curves to each nucleus separately one would have to multiply the abscissa scale by $m_2/(m_1 + m_2)$ for nucleus 1 and by $m_1/(m_1 + m_2)$ for nucleus 2 [compare formulae (III, 3)].

According to the preceding discussion, we cannot in wave mechanics speak of a proper vibrational motion with a definite frequency, $\nu_{osc.}$, any more than we can speak of electron orbits for atoms. Only the probability distribution of position (and momentum; see below) can be given. In quantum mechanics, $\nu_{osc.}$ has only a formal meaning—namely, that of the constant $(1/2\pi)\sqrt{k/\mu}$. When we speak of the vibrational frequency in the following, we shall always mean this constant, which gives the frequency that the system would have for classical motion (however, see below).

The Hermite polynomials that enter the vibrational eigenfunction ψ_v in (III, 38) are for the first few v values, with $\sqrt{\alpha}\, x = \mathsf{x}$:

$$H_0(\mathsf{x}) = 1$$

$$H_1(\mathsf{x}) = 2\mathsf{x}$$

$$H_2(\mathsf{x}) = 4\mathsf{x}^2 - 2$$

$$H_3(\mathsf{x}) = 8\mathsf{x}^3 - 12\mathsf{x} \qquad\qquad\text{(III, 39)}$$

$$H_4(\mathsf{x}) = 16\mathsf{x}^4 - 48\mathsf{x}^2 + 12$$

$$H_5(\mathsf{x}) = 32\mathsf{x}^5 - 160\mathsf{x}^3 + 120\mathsf{x}$$

The normalization factor N_v in (III, 38) is given by

$$N_v = \sqrt{\frac{1}{2^v v!}}\sqrt{\frac{\alpha}{\pi}} \qquad\qquad\text{(III, 40)}$$

It follows that for the lowest level, $v = 0$, the probability distribution is

$$\psi_0^2 = \sqrt{\frac{\alpha}{\pi}}\, e^{-\alpha x^2} \qquad\qquad\text{(III, 41)}$$

which is a simple Gauss error function (see Fig. 42).

From the eigenfunctions of the oscillator expressed in terms of linear momentum rather than of the coordinate x it is possible to derive the probability distribution of the momentum [see, for example, Rojansky (15b)]. The result for the lowest state $v = 0$ is given in Fig. 43 with an abscissa scale both for the momentum and the velocity. The curve shows that in wave mechanics a considerable range of values is possible for the momentum; therefore the energy is not zero, and hence the zero-point energy results. The two curves for the probability distributions of position and momentum for $v = 0$ give the minimum uncertainty of the position and momentum (or velocity) that is compatible with the Heisenberg uncertainty principle.

As for $v = 0$, the momentum distribution curves for $v > 0$ have the same form as the corresponding positional probability distribution curves, except that the meaning of the ordinates and abscissae is altered. It is therefore unnecessary to reproduce them here again.

Classically, the time it takes the oscillator to go from one turning point to the other is half the period—that is, $1/2\nu_{osc.}$. In quantum mechanics neither the turning points nor the velocities are sharply defined. But we can see from the range of velocities in Fig. 43 (or in corresponding diagrams for $v \neq 0$) that the average time for the oscillator to cover an appreciable fraction of the possible range of x values is of the order $1/2\nu_{osc.}$. Thus we may consider $\nu_{osc.}$ in quantum theory as a sort of *average vibration frequency*.

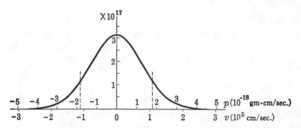

Fig. 43.   **Probability Distribution of the Momentum (Velocity) in the State $v = 0$ of the Harmonic Oscillator.** The upper abscissae scale gives the momentum, the lower the velocity. The ordinate scale gives the probability of the momentum (per unit momentum). The broken vertical lines give the maximum classical momentum (velocity) for the same energy. The curve is plotted for $\omega = 1000$ cm^{-1} and $\mu_A = 10$.

Spectrum. If the molecule in its equilibrium position has a dipole moment, as is always the case for molecules consisting of unlike atoms, this dipole moment will in general change if the internuclear distance changes. To a first approximation it may be assumed that the change of dipole moment with internuclear distance is linear. Therefore, the dipole moment changes with a frequency equal to the frequency of the mechanical vibration. On the basis of classical electrodynamics, this would lead to the emission of light of frequency $\nu_{osc.}$. Conversely, the oscillator could be set in vibration by absorption of light of frequency $\nu_{osc.}$.

Quantum-theoretically, emission of radiation takes place as a result of a transition of the oscillator from a higher to a lower state, and absorption takes place by the converse process. The wave number of the emitted or absorbed light is given by

$$\nu = \frac{E(v')}{hc} - \frac{E(v'')}{hc} = G(v') - G(v'') \qquad \text{(III, 42)}$$

where v' and v'' are the vibrational quantum numbers of the upper and the lower state, respectively. In order to determine which particular transitions can actually occur we have to evaluate the matrix elements $R_x{}^{nm}$, $R_y{}^{nm}$, $R_z{}^{nm}$ of the dipole moment. As shown below, for the oscillator these matrix elements are zero except when the permanent dipole moment is different from zero and when, in addition, v' and v'' differ by unity; that is to say, the *selection rule for the vibrational quantum number* of the harmonic oscillator is

$$\Delta v = v' - v'' = \pm 1 \qquad \text{(III, 43)}$$

Using (III, 37), we obtain

$$\nu = G(v+1) - G(v) = \omega; \tag{III, 44}$$

that is, quantum-theoretically (just as classically) the frequency of the radiated light is equal to the frequency $\nu_{\text{osc.}}$ ($= c\omega$) of the oscillator. This is true no matter what the v value of the initial state may be. The allowed transitions are indicated by vertical lines in Fig. 41. It is seen from this figure that they all give rise to the same frequency.

It must be pointed out here that for a system that is not exactly a harmonic oscillator, transitions can also appear with $\Delta v > 1$, even though very weakly [see section 2(a) of this chapter], whereas for the rotator the selection rule (III, 20) holds strictly, even if the system deviates from the model of the rigid rotator.

For a molecule consisting of two like atoms (N_2, O_2, $\cdots$), the dipole moment is zero, and therefore no transitions between the different vibrational levels occur; there is no infrared emission or absorption.

In deriving the selection rules for a non-rotating oscillator we may assume it to be oriented in the x-direction and therefore need only consider the matrix element $R_x{}^{nm}$. Since we assume (see above) the dipole moment M to vary linearly with internuclear distance we may put

$$M = M_0 + M_1 x \tag{III, 45}$$

where M_0 is the dipole moment in the equilibrium position, $x = r - r_e$ is the change of internuclear distance, and M_1 the rate of change of dipole moment with internuclear distance. Substituting (III, 45) and the vibrational eigenfunctions into the matrix element $R_x{}^{nm}$ [compare (I, 44)] we obtain

$$R_x{}^{v'v''} = M_0 \int \psi_{v'}{}^{*}\psi_v{}'' \, dx + M_1 \int x\psi_{v'}{}'^{*}\psi_v{}'' \, dx \tag{III, 46}$$

The first term on the right vanishes for $v' \neq v''$ on account of the orthogonality of the eigenfunctions (see p. 13). In order to evaluate the integral of the second term we introduce $\mathbf{x} = \sqrt{\alpha}\, x$ and have

$$x_{v'v''} = \int_{-\infty}^{+\infty} x\psi_{v'}{}'^{*}\psi_v{}'' \, dx = \frac{N_{v'}N_{v''}}{\alpha} \int_{-\infty}^{+\infty} \mathbf{x}H_{v'}(\mathbf{x})H_{v''}(\mathbf{x})e^{-\mathbf{x}^2} \, d\mathbf{x} \tag{III, 47}$$

Making use of the recursion formula for Hermite polynomials

$$\mathbf{x}H_n(\mathbf{x}) = \tfrac{1}{2}H_{n+1}(\mathbf{x}) + nH_{n-1}(\mathbf{x}) \tag{III, 48}$$

we obtain

$$x_{v'v''} = \frac{N_{v'}N_{v''}}{\alpha}\left[\tfrac{1}{2}\int H_{v'}(\mathbf{x})H_{v''+1}(\mathbf{x})e^{-\mathbf{x}^2} \, d\mathbf{x} + v''\int H_{v'}(\mathbf{x})H_{v''-1}(\mathbf{x})e^{-\mathbf{x}^2} \, d\mathbf{x}\right]$$

Since the eigenfunctions of the harmonic oscillator are orthogonal (see p. 13) it follows immediately that the first integral vanishes except when

$$v' = v'' + 1$$

while the second one vanishes except when

$$v' = v'' - 1$$

Thus the selection rule (III, 43) is proven.   At the same time it is clear from (III, 43) that an additional condition for the occurrence of a vibration spectrum is that M_1, the rate of change of the dipole moment, is different from zero.   This condition is fulfilled only for molecules with unequal nuclei.

If higher powers in the development (III, 45) are taken into account transitions with $\Delta v = \pm 2, \pm 3, \cdots$ become possible also (see p. 96).

All preceding considerations hold under the assumption of dipole radiation only.   By quadrupole radiation a transition probability may be produced even for homonuclear molecules for which the dipole moment is always zero.   However, the magnitude of the quadrupole transition probability is exceedingly small, about 10^{-8} of the dipole transition probability [see Herzberg (315) and James and Coolidge (362); see also p. 279].

(c) Comparison with the Observed Infrared Spectrum

If we compare the theoretical results so far obtained for the spectra of the rigid rotator and harmonic oscillator with the observed absorption spectra of the halogen hydrides (see Figs. 31 and 34), we are led to the following interpretation: The spectrum in the far infrared, since it consists of a series of nearly equidistant lines is a *rotation spectrum;* the molecule rotates about an axis perpendicular to the line joining the nuclei and through the center of mass; it behaves approximately like a rigid rotator and the transitions between the rotational levels give rise to the spectrum.   On the other hand, the spectrum in the near infrared, since it consists essentially of a single very intense line, is a *vibration spectrum;* the nuclei carry out approximately harmonic vibrations in the internuclear axis.   On this assumption the weak occurrence of bands with nearly two and three times the frequency would be connected with the deviations from the harmonic oscillator.

This interpretation of the infrared spectra has been confirmed by a large amount of experimental data.   For the moment we shall verify it for only one simple example.

For HCl, the separation of the lines in the far infrared is 20.68 cm^{-1} [the empirical constant f_0 in (II, 12) (see Fig. 34)].   If this spectrum is a rotation spectrum, the number 20.68 cm^{-1} must be equal to $2B$; that is, $B = 10.34$ cm^{-1} (see III, 25).   From this it follows, according to (III, 16), that the moment of inertia of HCl is $I = 2.71 \times 10^{-40}$ gm cm^2 and from this, using $\mu = 1.63 \times 10^{-24}$ gm, that the internuclear distance $r = 1.29 \times 10^{-8}$ cm.   This value is just of the order of magnitude that one would expect on the basis of values of atomic and molecular radii obtained from gas viscosity and crystal structure measurements.   We can therefore take it as proven that the far infrared spectrum is a rotation spectrum.   From the value obtained for the rotational constant B, the frequencies of rotation in the quantum states $J = 1, 2, 3,$ and 4 are, according to (III, 26), 8.7, 15.2, 21.5, and 27.8×10^{11} sec^{-1}.   The periods of rotation are the reciprocals of these values—that is, 1.15, 0.66, 0.46, and 0.36×10^{-12} sec, respectively.

In the near infrared spectrum, HCl has a single intense band at 2885.9 cm^{-1} (see p. 53 f.).   If this represents a vibration spectrum, according to (III, 44) we have $\omega = 2885.9$. Therefore the vibrational frequency is $\nu_{osc.} = \omega c = 8.65 \times 10^{13}$ sec^{-1}, which is about one hundred times greater than the rotational frequency. The period of vibration is 1.17×10^{-14} sec. From $\nu_{osc.}$ and (III, 33), the force constant $k = 4.806 \times 10^5$ dynes/cm. If we use (III, 31) to calculate the energy required to increase the nuclear separation by 1 Å, we obtain a value of about 14 volts, which is of the order of magnitude of the energy set free in chemical reactions. Therefore it is quite reasonable to interpret the near infrared spectrum as a vibration spectrum.

In explaining the infrared absorption spectrum we might also have tried to interpret the far infrared spectrum as a vibration spectrum and the near infrared spectrum as a rotation spectrum. Apart from the difficulty of explaining the structure of the spectrum on this assumption, we would have obtained internuclear distances one tenth of the known values for atomic and molecular radii (see above) and a force constant incompatible with the observed strength of chemical binding.

The above considerations have also been applied to the infrared spectra of other molecules, and lead to similar results. Also the discussion of Raman spectra (see below) confirms the explanation of the far and near infrared spectra as rotation spectra and vibration spectra, respectively.

Conversely, if we accept this explanation as correct, we can determine the *positions of the rotational and vibrational levels* with great accuracy from the observed infrared spectrum of a molecule. From the rotational levels we obtain the *moment of inertia*, the *internuclear distance*, and the *rotational frequencies* of the molecule, and from the vibrational levels we obtain the *vibrational frequency* and the *force constant*. All these quantities can be obtained more accurately from molecular spectra than by any other method.

To be sure, rotation spectra have been observed and measured for only a very few molecules. However, as we shall see later, the positions of the rotational levels (and thereby the internuclear distances) may also be obtained from the fine structure of the bands in the near infrared and in the visible and ultraviolet regions.

(d) The Raman Spectrum of the Rigid Rotator and of the Harmonic Oscillator

As we have seen (p. 61 f.), the Raman effect consists in the occurrence of (weak) displaced lines in the spectrum of the light scattered by gases (or by liquids or solids), the amount of the shift being characteristic of the substance considered.

Classical theory of light scattering and of the Raman effect. If an atom or molecule is brought into an electric field F, an electric dipole moment P is induced in the system; the center of the positive charges is moved a small distance in one direction, and that of the negative charges is moved in the opposite direction. The magnitude of the resulting dipole moment is proportional to that of the field; that is

$$|P| = \alpha |F| \tag{III, 49}$$

where α is called the *polarizability*.

Except in the case of spherical symmetry, the magnitude of the induced dipole moment depends on the orientation of the system to the field. For a diatomic molecule, for example, a field lying along the internuclear axis obviously induces a dipole moment of different magnitude from that induced by a field at right angles to the axis. In general, the direction of P does not coincide with the direction of F. However, for reasons of symmetry these two directions do coincide if F has the direction of one of the axes of symmetry of the system. Choosing these axes as coordinate axes we have

$$P_x = \alpha_{xx}F_x, \quad P_y = \alpha_{yy}F_y, \quad P_z = \alpha_{zz}F_z \tag{III, 50}$$

where α_{xx}, α_{yy}, α_{zz} are the components of the polarizability which in the most general case are all different.[2] However, for a diatomic molecule, taking the z axis in the internuclear axis, it is clear that $\alpha_{xx} = \alpha_{yy}$, since the x direction is in no way distinguished from the y direction. In any case the magnitude of P is proportional to the magnitude of F [see equation (III, 49)] which is all that matters for most of the following considerations.

From (III, 49) one obtains for the component P_F of P in the direction of the applied field:

$$P_F = \alpha_F|F| \tag{III, 51}$$

where α_F is the F-component of the polarizability (it equals α_{xx} or α_{yy} or α_{zz} if F is in the x or y or z direction, respectively, for the above choice of axes). If $1/\sqrt{\alpha_F}$ is plotted in the various directions from the origin a surface is obtained which can be shown to be an ellipsoid. It is called the *polarizability ellipsoid*. Its principal semi-axes have the lengths $1/\sqrt{\alpha_{xx}}$, $1/\sqrt{\alpha_{yy}}$, and $1/\sqrt{\alpha_{zz}}$. For diatomic molecules, since $\alpha_{xx} = \alpha_{yy}$, the polarizability ellipsoid is an ellipsoid of rotation. A somewhat more detailed discussion of the polarizability ellipsoid is given in Volume II [compare also the very detailed discussion given by Mathieu (29)[3]].

If a light wave of frequency ν' falls on an atom or molecule, there is a varying electric field

$$F = F_0 \sin 2\pi\nu't \tag{III, 52}$$

where t is the time. This field induces a varying dipole moment, which in turn causes an emission of light of the same frequency as the incident light. Thus arises the so-called *Rayleigh scattering*, which is responsible for the phenomena of refraction and of the Tyndall effect. For visible and ultraviolet incident light, essentially only the electrons move under the influence of the alternating electric field and produce the dipole moment, since the nuclei cannot follow the rapid oscillations.

If the internuclear distance changes, obviously the polarizability must also change, even if slightly. Moreover, according to the above discussion, the polarizability depends on the orientation of the molecule to the field. Thus *a change in polarizability*—that is, a change in the amplitude of the induced dipole moment—is associated with both the vibration and the rotation of the

---

[2] For the case in which there are no axes of symmetry see Volume II, p. 243.

[3] It is easily seen that Mathieu's E/A is our $1/\sqrt{\alpha_F}$.

molecule. For the vibration, to a first good approximation, we can put

$$\alpha = \alpha_{0v} + \alpha_{1v} \sin 2\pi \nu_{\text{osc.}} t, \qquad \text{(III, 53)}$$

where α_{0v} is the polarizability in the equilibrium position and α_{1v} is the amplitude of the change in polarizability during the vibration ($\alpha_{1v} \ll \alpha_{0v}$). Correspondingly, for the rotation

$$\alpha = \alpha_{0r} + \alpha_{1r} \sin 2\pi 2\nu_{\text{rot.}} t, \qquad \text{(III, 54)}$$

where α_{0r} is an average polarizability and α_{1r} is the amplitude of the change in polarizability for rotation about the rotational axis considered. The frequency with which the polarizability changes during the rotation is twice the rotational frequency, since the polarizability is the same for opposite directions of the applied field.

Substituting (III, 52) and (III, 53) into (III, 49) we obtain for the induced dipole moment in the case of a vibrating molecule

$$\boldsymbol{P}_v = \alpha_{0v} \boldsymbol{F}_0 \sin 2\pi \nu' t + \alpha_{1v} \boldsymbol{F}_0 \sin 2\pi \nu' t \sin 2\pi \nu_{\text{osc.}} t. \qquad \text{(III, 55)}$$

Similarly substituting (III, 52) and (III, 54) into (III, 49) we obtain for a rotating molecule

$$\boldsymbol{P}_r = \alpha_{0r} \boldsymbol{F}_0 \sin 2\pi \nu' t + \alpha_{1r} \boldsymbol{F}_0 \sin 2\pi \nu' t \sin 2\pi 2\nu_{\text{rot.}} t. \qquad \text{(III, 56)}$$

From this, using well-known trigonometrical formulae, we obtain

$$\boldsymbol{P}_v = \alpha_{0v} \boldsymbol{F}_0 \sin 2\pi \nu' t$$
$$+ \tfrac{1}{2}\alpha_{1v} \boldsymbol{F}_0 [\cos 2\pi (\nu' - \nu_{\text{osc.}}) t - \cos 2\pi (\nu' + \nu_{\text{osc.}}) t] \qquad \text{(III, 57)}$$

and

$$\boldsymbol{P}_r = \alpha_{0r} \boldsymbol{F}_0 \sin 2\pi \nu' t$$
$$+ \tfrac{1}{2}\alpha_{1r} \boldsymbol{F}_0 [\cos 2\pi (\nu' - 2\nu_{\text{rot.}}) t - \cos 2\pi (\nu' + 2\nu_{\text{rot.}}) t]. \qquad \text{(III, 58)}$$

Thus we see that on account of the small alteration of α during the vibration or rotation of the molecule, the induced dipole moment changes not only with the frequency ν' of the incident light, but also with the frequencies $\nu' - \nu_{\text{osc.}}$ and $\nu' + \nu_{\text{osc.}}$ or with the frequencies $\nu' - 2\nu_{\text{rot.}}$ and $\nu' + 2\nu_{\text{rot.}}$. So, according to classical theory, in the spectrum of the scattered light we have to expect *displaced lines* on both sides of the undisplaced line—in the case of an oscillator, at a distance $\nu_{\text{osc.}}$, and, in the case of a rotator, at a distance $2\nu_{\text{rot.}}$. However, while $\nu_{\text{osc.}}$ has a fixed value, $\nu_{\text{rot.}}$ can, classically, take any value. Therefore in the case of a rotator we should expect a *continuous* spectrum on either side of the undisplaced line. The intensities of the displaced lines (squares of the amplitudes), according to (III, 57) and (III, 58) are very much smaller than those of the undisplaced lines. They should be the same for the components displaced toward longer and shorter wave lengths. In the case of vibration, they should also depend on the amplitude of the vibrations since α_{1v} depends on the amplitude.

Thus, qualitatively, even classical considerations lead to the Raman effect—displaced frequencies in the spectrum of the scattered light. However, quantitatively there is no agreement. Empirically, there is no continuous Raman spectrum present for diatomic molecules, and, apart from that, in general only the long-wave components are found for the larger displacements (vibrational effect) but not the corresponding short-wave components, even though, according to (III, 57), they would be expected to have the same intensity.

Quantum theory of the Raman effect. The quantum theoretical explanation of the Raman effect is as follows: When the incident light quantum $h\nu'$ collides with a molecule, it can either be scattered elastically, in which case its energy, and therefore its frequency, remains unaltered (Rayleigh scattering), or it can be scattered inelastically, in which case it either gives up part of its energy to the scattering system or takes energy from it. Naturally, the light quantum can give to or take from the system only amounts of energy that are equal to the energy differences between the stationary states of the system. Let $\Delta E = E' - E''$ be such an energy difference. Then if the system is initially in the lower state E'', it may be brought to the upper state E' by the scattering of a light quantum, the energy ΔE being taken from the light quantum. Thus, after the scattering, the energy of the light quantum is only $h\nu' - \Delta E$. If, on the other hand, the system was initially in the state E' and is transferred to E'' by scattering, the energy of the light quantum after the scattering is equal to $h\nu' + \Delta E$. The frequency of the scattered light quantum is equal to the energy divided by h; that is, the frequencies $\nu' - (\Delta E/h)$ and $\nu' + (\Delta E/h)$ appear in the scattered light as well as the undisplaced frequency ν', that is, we have the Raman effect. In most cases ΔE may take a number of different values. If frequencies and energies are measured in wave number units, it is clear that the *Raman shifts give directly the energy differences of the system in* cm^{-1}. The Raman lines displaced toward longer wave lengths are also called *Stokes lines* and those displaced toward shorter wave lengths are called *anti-Stokes lines*.[4]

In Fig. 44, the relationships for light scattering are shown in an energy level diagram. The levels indicated by broken lines do not correspond to any possible energy states of the system but only give the energy of the light quantum above the initial state.

The elementary process underlying the Raman effect is clearly to be distinguished from the process of *fluorescence*. For the latter, the incident light quantum is completely absorbed and the system is transferred to an excited state from which it can go to various lower states only after a certain time (mean

---

[4] This nomenclature has the following origin: According to Stokes' law the frequency of fluorescent light is always smaller or at most equal to that of the exciting light. Stokes lines in fluorescence are thus such as correspond to Stokes' law, and anti-Stokes lines are such as contradict it. This nomenclature has also been adopted for the Raman effect, in spite of its difference from fluorescence.

life).   The result of both phenomena is, it is true, essentially the same: A light quantum of a frequency different from that of the incident quantum is produced, and the molecule is brought to a higher or lower energy level.   But the essential difference is that the Raman effect can take place for any frequency of the incident light (it is a light-scattering phenomenon), whereas fluorescence can occur only for the absorption frequencies.   In consequence of this also, the structure of the Raman spectrum is quite different from that of the fluorescence spectrum.

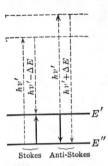

Stokes   Anti-Stokes

Fig. 44.   **Quantum Theory of the Raman Effect.**   The heavy arrows give the transitions actually taking place in the system considered.

For a mathematical formulation of the wave-mechanical theory of the Raman effect it is necessary to consider the matrix elements of the scattering moment (see p. 19),

$$[P]^{nm} = \int \Psi_n^* P \Psi_m \, d\tau \qquad \text{(III, 59)}$$

where Ψ_n and Ψ_m are the wave functions of two states of the system considered.   For P under the integral the classical expressions (III, 49) or (III, 50) are to be substituted.   Since Ψ_n^*, Ψ_m, and P have the time factors $e^{2\pi i(E_n/h)t}$, $e^{-2\pi i(E_m/h)t}$, and $e^{2\pi i\nu't}$, respectively, $[P]^{nm}$ varies with the frequency $\nu' + (E_n - E_m)/h$.   For the amplitude we obtain

$$[P^0]^{nm} = |F| \int \psi_n^* \alpha \psi_m \, d\tau \qquad \text{(III, 60)}$$

or similar expressions for the components if (III, 50) is used.   If for two states E_n and E_m the integral (III, 60) is different from zero a transition from n to m can take place under the influence of the incident light and at the same time the scattered light quantum will have the frequency $\nu' + (E_n - E_m)/h$, that is, Raman lines displaced by the amount $(E_n - E_m)/h$ appear.   The square of the integral (III, 60) is proportional to the transition probability which determines the intensity of the particular Raman lines.   For $n = m$ the undisplaced frequency ν' is obtained and (III, 60) determines the intensity of Rayleigh scattering.

If the polarizability α of a system (such as an oscillator or rotator) is constant (independent of vibration or rotation), it can be put in front of the integral sign in (III, 60).   Owing to the orthogonality of the eigenfunctions (see p. 13), the integrals are then all equal to zero, except for $m = n$; that is, when α is constant, only Rayleigh scattering and no Raman effect appears, in agreement with the results of classical theory.   A transition from one state to another, and therefore a *Raman shift, can occur only when the polarizability changes during the process under consideration* (that is, during the vibration or the rotation).

A very detailed theoretical discussion of the Raman effect has been given by Placzek (562).

Vibrational Raman spectrum.   By a study of the matrix elements of the polarizability (see below) it is found that in the case of the harmonic oscillator the same selection rule holds for the Raman effect as for the infrared spectrum; namely,

$$\Delta v = \pm 1. \qquad \text{(III, 61)}$$

A transition can take place only to the adjacent vibrational state.   Thus the Raman spectrum consists of one Stokes and one anti-Stokes line, which are shifted by an amount

$$|\Delta \nu| = G(v + 1) - G(v) = \omega \qquad (III, 62)$$

to either side of the original line. However at ordinary temperatures most of the molecules are in the lowest state ($v = 0$) and only an extremely small fraction are in the state with $v = 1$ (see p. 123) As a result of this, the *intensity of the Stokes Raman line*, which corresponds to the transition $0 \rightarrow 1$, *is very much greater than that of the anti-Stokes line*, $1 \rightarrow 0$. This agrees entirely with observation if the Raman lines of large displacement (see Fig. 36) are explained as vibrational Raman lines. In all cases of diatomic molecules thus far investigated, the corresponding line displaced toward shorter wave lengths is so weak that it has not been observed. However, anti Stokes vibrational lines have been observed for many polyatomic molecules for which smaller vibrational frequencies occur and for which therefore, the intensity ratio of Stokes and anti-Stokes lines is more favorable (see Volume II).

We have seen before that the intense bands observed in absorption in the near infrared result from the transition $v = 0 \rightarrow v = 1$, that is to say, from the same transition that takes place in the vibrational Raman effect. This is the explanation of the observed agreement between the magnitudes of the large Raman displacements and the frequencies of the intense near infrared bands (see Table 11, p. 62).

However, it must not be forgotten that the mechanism of the production of these transitions in the two cases is quite different. The occurrence of a Raman spectrum depends on the polarizability of the molecule but is entirely independent of the presence of a permanent dipole moment. Thus a Raman spectrum can also appear for molecules that have no infrared spectrum.

As we have seen (p. 86) the intensity of the Raman lines is proportional to the square of the matrix elements $[\boldsymbol{P}_0]^{nm}$ given by (III, 60). In order to evaluate them for the harmonic oscillator we have to substitute the vibrational eigenfunctions (III, 38) and an expression for the polarizability α. In a first approximation we may assume a linear variation of α with the displacement x ($= r - r_e$) from the equilibrium position:

$$\alpha = \alpha_{0v} + \alpha_v{}^1 x \qquad (III, 63)$$

Here $\alpha_v{}^1 = \left(\dfrac{\partial \alpha}{\partial x} \right)_0$ is the rate of change of the polarizability with x. The expression for α assumed here is equivalent to (III, 53) of the classical treatment (note that $\alpha_v{}^1 = \alpha_{1v}/x_0$). Substituting into (III, 60) we obtain

$$[\boldsymbol{P}^0]^{v'v''} = |F| \, \alpha_{0v} \int \psi_v{}'^* \psi_v{}'' \, dx + |F| \, \alpha_v{}^1 \int x \psi_v{}'^* \psi_v{}'' \, dx \qquad (III, 64)$$

Because of the orthogonality of the eigenfunctions, the first integral is zero except when $v' = v''$; that is, it gives the undisplaced frequency [Rayleigh scattering, corresponding to the first term in (III, 57)]. The second term contains the same integral as appears for the infrared spectrum of the oscillator [see (III, 46) on p. 80]. As we have seen, it is different from zero only when $v' = v'' \pm 1$. Thus we have the selection rule (III, 61) for the vibrational Raman spectrum.

Equation (III, 64) shows that, just as in the classical treatment, Raman lines appear only when α_v^1 is different from zero—that is, *when the polarizability changes during the vibration.* The greater α_v^1 is—that is, the more sensitive the polarizability is to changes of internuclear distance—the greater is the intensity of the vibrational Raman lines. While the dipole moment M_0 and its change M_1 are exactly zero for homonuclear diatomic molecules the polarizability and its change do not vanish exactly for any diatomic molecule.[5] However, for molecules consisting of almost undeformed ions (ionic molecules, see p. 371) α_v^1 is quite small, whereas for molecules with homopolar binding for which the electrons between the nuclei produce the binding the dependence of the polarizability on internuclear distance is considerable, that is, α_v^1 is large and therefore the Raman lines are comparatively strong. These are just the molecules for which the change of dipole moment is either small or exactly zero.

Rotational Raman spectrum. Whereas for an oscillator the same selection rule holds in the Raman effect as in the infrared spectrum, for the rotator a different selection rule is obtained—namely,

$$\Delta J = 0, \pm 2. \tag{III, 65}$$

As in the classical theory the intensity of the rotational transitions is found to depend on the change of the polarizability in a fixed direction during the rotation and is independent of the presence of a permanent dipole moment.

When, as is usually the case, molecules with different J values are present, the rotational Raman spectrum consists of a *number* of lines.[6] $\Delta J = 0$ gives the undisplaced line. For the actual transitions of the rotator ΔJ can only be positive, since it is defined as $J' - J''$, where J' refers to the upper and J'' to the lower state. Thus only $\Delta J = +2$ need be considered. Yet we obtain *two* series of lines, since the lower as well as the upper state may be the initial state in the scattering process. The first case (transition $J \to J + 2$) results in a shift to longer wave lengths (Stokes lines), and the second case (transition $J + 2 \to J$) results in a shift to shorter wave lengths (anti-Stokes lines). These transitions are indicated in the upper part of Fig. 45. Using (III, 15), we obtain for the magnitude of the frequency shift

$$|\Delta \nu| = F(J + 2) - F(J) = B(J + 2)(J + 3) - BJ(J + 1)$$

$$= 4BJ + 6B = 4B(J + \tfrac{3}{2}). \tag{III, 66}$$

The Raman spectrum to be expected is drawn schematically at the bottom of Fig. 45. According to (III, 66), we have *a series of equidistant Raman lines on either side of the undisplaced line,* as is actually observed (see Figs. 36 and 37). We can therefore identify the small Raman displacements as the rotational Raman

[5] For certain vibrations of polyatomic molecules α_v^1 may be exactly zero (see Volume II, Chapter III).

[6] For the vibrational Raman spectrum, only one Stokes and one anti-Stokes line would result, even if molecules with different v values were present, since the vibrational levels of the harmonic oscillator are equidistant [see (III, 62)].

spectrum.   The two series of rotational Raman lines both having $\Delta J = +2$ are also called *S branches*.[7]

A comparison of (III, 66) with the earlier empirical equation (II, 14) shows complete agreement.   The constant separation of the lines is $4B$ (equal to the previous p), and the separation of the first line from the undisplaced line is $\frac{3}{2}$ times as great ($6B$, corresponding to $\frac{3}{2}p$).

In the previous discussion of the observed spectra (p. 63) it was noted that the line separation in the Raman spectrum of a given molecule is twice as great as that in the far infrared spectrum of the same molecule. This observation is in striking agreement with the requirements of the theory, as is seen by comparing (III, 66) with (III, 25), and represents conversely an unambiguous proof that the spectrum discussed here really is a *rotational Raman spectrum*, and that the far infrared spectrum is a rotation spectrum.

The experimentally observed *intensities* also agree with those theoretically expected. Owing to the smallness of the rotational energy, a number of rotational states are excited at room temperature because of the thermal motion [see section 2(e)]. As a result, Stokes as well as anti-Stokes lines are present with about equal intensity—in contrast to the vibrational Raman spectrum.   The alternation of intensities observed in the rotational Raman spectrum of symmetrical molecules must be left for later discussion [see Fig. 37 and section 2(f)].

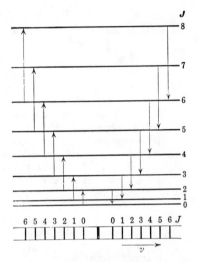

Fig. 45. **Energy Level Diagram for the Rotational Raman Spectrum.** In the schematic spectrogram below, the heavy line in the middle gives the position of the undisplaced line. To the left are the Stokes Raman lines, and to the right are the anti-Stokes lines. The numbers added to the Raman lines are the J values of the lower state.

The general proof of the selection rule (III, 65) is somewhat involved [see Placzek (562)]. We shall consider here only the case of one component of the polarizability.   Let x, y, z refer to a coordinate system fixed to the molecule and x_f, y_f, z_f to a coordinate system fixed in space. The component P_{z_f} of the induced dipole moment produced by a field $\boldsymbol{F}_{z_f}$ in the z_f direction is given by

$$P_{z_f} = \alpha_{z_f z_f} F_{z_f} \tag{III, 67}$$

On the other hand we have

$$P_{z_f} = P_x \cos\,(x, z_f) + P_y \cos\,(y, z_f) + P_z \cos\,(z, z_f)$$

If in this equation we introduce $\boldsymbol{F}_x$, $\boldsymbol{F}_y$, $\boldsymbol{F}_z$ by (III, 50), express $\boldsymbol{F}_x$, $\boldsymbol{F}_y$, $\boldsymbol{F}_z$ in terms of $\boldsymbol{F}_{z_f}$ (assum-

---

[7] According to the international nomenclature [Mulliken (515)], the symbols S, R, Q, P and O are used for branches with $\Delta J = +2$, $+1$, 0, -1, and -2, respectively, where $\Delta J = J' - J''$ is the difference between the J value in the upper and the lower state.   But, in the case of the rotational Raman spectrum, some authors use S branch for the anti-Stokes and O branch for the Stokes Raman lines, in spite of the fact that $\Delta J = +2$ for both.

ing $F_{x_f} = F_{y_f} = 0$) and compare with (III, 67) we find

$$\alpha_{z_f z_f} = \alpha_{xx} \cos^2 (x, z_f) + \alpha_{yy} \cos^2 (y, z_f) + \alpha_{zz} \cos^2 (z, z_f) \qquad \text{(III, 68)}$$

Since for diatomic molecules $\alpha_{xx} = \alpha_{yy}$, and since the angle between the axes z and z_f is the angle ϑ used in (III, 11), equation (III, 68) may also be written

$$\alpha_{z_f z_f} = \alpha_{xx} + (\alpha_{zz} - \alpha_{xx}) \cos^2 \vartheta \qquad \text{(III, 69)}$$

The corresponding matrix element is

$$\int \alpha_{z_f z_f} \psi_r'^* \psi_r'' \, d\tau = \alpha_{xx} \int \psi_r'^* \psi_r'' \, d\tau + (\alpha_{zz} - \alpha_{xx}) \int \cos^2 \vartheta \, \psi_r'^* \psi_r'' \, d\tau \quad \text{(III, 70)}$$

The first integral on the right vanishes except when $J' = J''$, that is, it gives the undisplaced line. The second integral, in a way entirely similar to the discussion of the infrared spectrum (p. 72), is easily seen to be different from zero only when either $J' = J''$ or $J' = J'' \pm 2$. The same results are obtained for the other polarizability components.

It is the occurrence of $\cos^2 \vartheta$ in (III, 70) instead of $\cos \vartheta$ as in (III, 19) that causes the selection rules for the rotational Raman spectrum to be different from those for the infrared rotation spectrum. The $\cos^2 \vartheta$ in its turn appears since the polarizability is the same in any two opposite directions, that is, for the same reason that $2\nu_{rot.}$ rather than $\nu_{rot.}$ appears in the classical treatment.

The intensity of the rotational Raman lines according to (III, 70) depends on the difference between the polarizabilities in the axis and perpendicular to the axis of the molecule in conformity with the classical treatment.

General remarks. As we have seen the study of the Raman spectrum supplies in many respects the same information about a molecule as the study of its infrared spectrum—namely, the magnitude of the vibrational quantum (that is, the constant ω) and the magnitude of the rotational quanta (that is, the constant B). From these, as previously discussed, we can derive, on the one hand, the vibrational frequency and force constant and, on the other, moment of inertia, internuclear distance, and rotational frequencies. The observed agreement between the values obtained from Raman and infrared spectra indicates that the explanation of these spectra by the models of the rotator and oscillator is a good approximation.

Even for homonuclear diatomic molecules, which have no infrared spectrum (see p. 80), the rotational and vibrational constants can be derived with the aid of the Raman spectrum. However, in general, their determination from electronic band spectra (see Chapter IV) is experimentally simpler.

2. Interpretation of the Finer Details of Infrared and Raman Spectra

The models of the rigid rotator and the harmonic oscillator explain the main characteristics of the infrared and Raman spectra. However, we have to add a number of refinements to these models if we wish to understand the finer details of these spectra.

(a) The Anharmonic Oscillator

The molecule as an anharmonic oscillator. According to equation (III, 31), a harmonic oscillator is characterized by a parabolic potential curve (broken-line curve in Fig. 46). The potential energy, and therefore the restoring force, increases indefinitely with increasing distance from the equilibrium position. It is, however, clear that in an actual molecule, when the atoms are at a great distance from one another, the attractive force is zero, and correspondingly, the potential energy has a constant value. Therefore the *potential curve of the molecule* has the form of the full curve in Fig. 46. The minimum of the curve corresponds to the equilibrium position. In its neighborhood, the curve can be represented approximately by a parabola (broken-line curve); that is why the model of the harmonic oscillator reproduces the main characteristics of the vibration spectrum quite well.

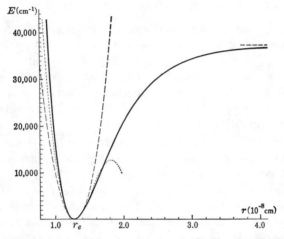

Fig. 46. **Potential Curve of the Molecule (Anharmonic Oscillator).** The full curve is drawn for the ground state of HCl. The broken-line and the dotted curves are the ordinary and the cubic parabola, respectively, that form the best approximation to the full curve at the minimum.

As a first approximation to the actual potential energy function of the molecule we may add to the quadratic function of the harmonic oscillator [equation (III, 31)] a cubic term and write

$$U = f(r - r_e)^2 - g(r - r_e)^3 \qquad \text{(III, 71)}$$

Here the coefficient g is very much smaller than f. To be sure, even this function (dotted curve in Fig. 45) does not represent the whole of the potential energy curve, but at any rate does give a much better approximation for not too great values of $(r - r_e)$. For a still better approximation quartic and higher terms in $(r - r_e)$ have to be added to (III, 71).

A mass point that moves under the influence of a potential of the type of the full or dotted curve in Fig. 46 is called an *anharmonic oscillator*.

Classical motion. The motion of a particle under the action of a potential energy of the form of the full curve in Fig. 46 may be most easily visualized if one imagines a small particle sliding under the action of gravity in a perfectly smooth tube bent to the shape of the curve and held in a vertical plane. It is quite clear that the particle, for larger amplitudes, will spend more time at the right of the equilibrium position than at the left of it. Therefore, while the motion is still strictly periodic it is no longer of the pure sine form as is the motion of the harmonic oscillator for any amplitude [see equation (III, 29)]. A more detailed calculation [see, for example, Schaefer and Matossi (26c) and Mathieu (29)] shows that the motion of the anharmonic oscillator may be represented as a *superposition of fundamental and overtone vibrations* (Fourier series) as follows:

$$x = x_{01} \sin 2\pi\nu_{osc.}t + x_{02}(3 + \cos 2\pi2\nu_{osc.}t) + x_{03} \sin 2\pi3\nu_{osc.}t + \cdots \quad \text{(III, 72)}$$

Here x_{01}, x_{02}, x_{03} are the amplitudes of the fundamental, the first, and the second overtone, respectively. As long as the anharmonicity is small (that is, $g \ll f$) we have $x_{02} \ll x_{01}$ and $x_{03} \ll x_{02}$. However, x_{02} and x_{03} are proportional to the square and cube, respectively, of x_{01}, that is, with increasing amplitude the overtones rapidly become more important. Because of the asymmetry of the potential curve the time average of the position of the oscillator is no longer at the equilibrium position ($x = 0$) but at $x = 3x_{02}$.

The frequency of the vibrations $\nu_{osc.}$ is given by $(1/2\pi)\sqrt{k/\mu}$ for very small amplitudes only; it decreases slowly as the amplitude x_{01} increases (compare the vibrations of a pendulum with large amplitude). The frequency of the overtones is two, three, $\cdots$ times this amplitude-dependent fundamental frequency.

The reduction of the motion of the two nuclei in a diatomic molecule to the anharmonic motion of a single mass point can be carried out in the same way as in the harmonic case (p. 74 f.).

Energy levels. If (III, 71) [possibly including higher powers of $(r - r_e)$] is substituted as potential energy in the wave equation (I, 12), it is found that for small anharmonicity of the oscillator ($g \ll f$) the eigenvalues of the wave equation—that is, the *energy values of the anharmonic oscillator*—are given by

$$E_v = hc\omega_e(v + \tfrac{1}{2}) - hc\omega_ex_e(v + \tfrac{1}{2})^2 + hc\omega_ey_e(v + \tfrac{1}{2})^3 + \cdots, \quad \text{(III, 73)}$$

or correspondingly the *term values* are given by

$$G(v) = \omega_e(v + \tfrac{1}{2}) - \omega_ex_e(v + \tfrac{1}{2})^2 + \omega_ey_e(v + \tfrac{1}{2})^3 + \cdots. \quad \text{(III, 74)}$$

Here v is again the vibrational quantum number and the constant $\omega_ex_e \ll \omega_e$ and $\omega_ey_e \ll \omega_ex_e$. For the given choice of signs, when g is positive in (III, 71), ω_ex_e is positive. This is practically always the case. But ω_ey_e can be positive or negative; it is often negligibly small. The reason for writing the coefficients of the quadratic and cubic terms ω_ex_e and ω_ey_e, respectively, (as is adopted internationally) is that some of the earlier authors wrote (III, 74) thus:

$$G(v) = \omega_e[(v + \tfrac{1}{2}) - x_e(v + \tfrac{1}{2})^2 + y_e(v + \tfrac{1}{2})^3 + \cdots]$$

Formula (III, 74) shows that the energy levels of the anharmonic oscillator are not equidistant like those of the harmonic oscillator, but that their separation decreases slowly with increasing v. This is shown in Fig. 47 where for the sake of clarity a faster decrease is drawn than corresponds to most observed cases.

The *zero-point energy* of the anharmonic oscillator is obtained from (III, 74) by putting $v = 0$:

$$G(0) = \tfrac{1}{2}\omega_e - \tfrac{1}{4}\omega_e x_e + \tfrac{1}{8}\omega_e y_e + \cdots. \quad (III, 75)$$

If the energy levels are referred to this lowest level as zero, we obtain

$$G_0(v) = \omega_0 v - \omega_0 x_0 v^2 + \omega_0 y_0 v^3 + \cdots, \quad (III, 76)$$

where

$$\omega_0 = \omega_e - \omega_e x_e + \tfrac{3}{4}\omega_e y_e + \cdots,$$

$$\omega_0 x_0 = \omega_e x_e - \tfrac{3}{2}\omega_e y_e + \cdots, \quad (III, 77)$$

$$\omega_0 y_0 = \omega_e y_e + \cdots.$$

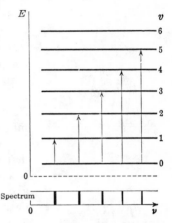

Fig. 47. **Energy Levels and Infrared Transitions of the Anharmonic Oscillator.** The absorption spectrum is given schematically below.

In order to derive the energy formula (III, 73) a perturbation calculation must be carried out (see p. 13 f.). Using the harmonic oscillator as a zero-order approximation [equation (III, 34)] the perturbation function W in (I, 24) is (including quartic terms)

$$W = gx^3 - jx^4 - \cdots \quad (III, 78)$$

The matrix elements W_{nn} in (I, 27) are easily seen to depend only on the quartic term. To the same approximation the matrix elements W_{ni} with $n \neq i$ depend only on the cubic term. The evaluation of the matrix elements consists in a repeated application of the recursion formula (III, 48). Substituting the result of this evaluation in (I, 27) the term with $(v + \tfrac{1}{2})^2$ in the energy formula of the anharmonic oscillator is obtained with

$$\omega_e x_e = \frac{3h^2}{32\pi^4\omega^2\mu^2c^2}\left(\frac{5g^2h}{8\pi^2\omega^2\mu c} - j\right) \quad (III, 79)$$

where it is assumed that the potential energy and therefore W in (III, 78) is expressed in cm^{-1} units. Substitution of the same matrix elements into (I, 29) gives the eigenfunctions of the anharmonic oscillator.

Eigenfunctions. In Fig. 48 the *eigenfunctions* (broken curves) and the *probability density distributions* (solid curves) for the levels $v = 0, 1, 2$, and 3 of a strongly anharmonic oscillator are illustrated. It can be seen that the functions are very similar to those of the harmonic oscillator (Fig. 42) except that they are not quite symmetrical. The somewhat greater heights of the right-hand maxima correspond to the fact that classically the molecule stays for a longer time on the shallow right-hand side of the potential curve (Fig. 46) than on the steeper left-hand side. In many actual cases the differences between the eigenfunctions of the harmonic and the anharmonic oscillator are much smaller than for the case of Fig. 48. For approximations to the eigenfunctions of the an-

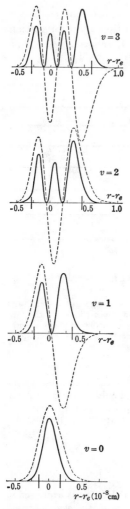

Fig. 48. **Eigenfunctions (Broken Curves) and Probability Density Distributions (Solid Curves) of an Anharmonic Oscillator for $v = 0, 1, 2, 3$.** The curves hold for the first excited triplet state of H_2 (see p. 339 f.) and are taken from Coolidge, James, and Present (162).

harmonic oscillator see Morse (504), Dunham (194a), Hutchisson (349a), Bewersdorff (99), and Norling (536).

Infrared spectrum. Just as for the harmonic oscillator, an infrared spectrum of the anharmonic oscillator arises only when a dipole moment is associated with the motion. The frequencies that are classically absorbed or emitted by the system are those with which this dipole moment changes. Therefore if the dipole moment varies linearly with the displacement from the equilibrium position (see p. 79) the classical infrared spectrum of the molecule, according to (III, 72), will consist of the fundamental frequency $\nu_{osc.}$, the first overtone $2\nu_{osc.}$ (also called second harmonic), the second overtone $3\nu_{osc.}$ (also called third harmonic), and so forth, the amplitudes (intensities) of the overtones being small compared to that of the fundamental. Qualitatively this agrees with the observed infrared spectrum of HCl (see p. 54) and of other heteronuclear molecules, since actually in addition to the intense fundamental further weak bands at about two, three, and four times its frequency are present. More accurate measurements show however (see p. 54) that the frequencies of the "overtones" are not exactly two, three, or four times the frequency of the fundamental as would be expected from the above classical consideration. This feature can only be understood on the basis of the quantum-theoretical treatment.

In considering the spectrum according to quantum theory we first have to find the *selection rules*. On account of the similarity of the eigenfunctions to those of the harmonic oscillator the selection rule $\Delta v = \pm 1$ holds at least approximately for the anharmonic oscillator, giving the most intense transitions. But now, as shown by a more detailed calculation, transitions with $\Delta v = \pm 2, \pm 3$ can also appear, even though with rapidly decreasing intensity.

The transitions that are possible in absorption when all the molecules are initially in the lowest vibrational state are indicated in Fig. 47. It can be seen immediately that the transitions with $\Delta v = 2, 3, \cdots$ have approximately, but not exactly, two, three, $\cdots$ times the frequency of the transition $\Delta v = 1$, in agreement with observation (see p. 54 f.). In spite of the (slight) deviation from the classical result, the bands 1–0, 2–0, 3–0, $\cdots$ are still frequently referred to as fundamental, first overtone (or second harmonic), second overtone (or third harmonic), and so forth. As formula for the series of absorption bands we obtain from (III, 74–76)

$$\nu_{\mathrm{abs.}} = G(v') - G(0) = G_0(v') = \omega_0 v' - \omega_0 x_0 v'^2 + \omega_0 y_0 v'^3 + \cdots. \quad \text{(III, 80)}$$

Thus, the observed absorption frequencies give directly the *positions of the vibrational levels above the lowest vibrational level*. Apart from the cubic term equation (III, 80) agrees exactly with the empirical formula (II, 9). Actually as mentioned previously, for a very accurate representation the observed wave numbers do require the introduction of a small cubic term.

Neglecting cubic terms the separation of successive absorption bands (in cm^{-1}) which is equal to the separation of successive levels, is given by

$$\Delta G_{v+\frac{1}{2}} = G(v + 1) - G(v) = G_0(v + 1) - G_0(v)$$
$$= \omega_e - 2\omega_e x_e - 2\omega_e x_e v = \omega_0 - \omega_0 x_0 - 2\omega_0 x_0 v. \quad \text{(III, 81)}$$

The observed first differences for the case of HCl are given in Table 8, p. 55. They decrease very nearly linearly with v in agreement with (III, 81). The second difference, again neglecting cubic terms in (III, 80), has the constant value

$$\Delta^2 G_{v+1} = \Delta G_{v+\frac{3}{2}} - \Delta G_{v+\frac{1}{2}} = -2\omega_e x_e = -2\omega_0 x_0. \quad \text{(III, 82)}$$

It gives directly the quantity $\omega_e x_e = \omega_0 x_0$,[7a] which is a measure of the *anharmonicity* of the oscillator. In the case of HCl, from Table 8 we obtain $\omega_0 x_0 = 51.60$ cm^{-1}. The values of ω_0 and ω_e are then obtained according to (III, 81) or (III, 80)—for example, from the frequency of the first band (1–0), since we have

$$\nu(1\text{–}0) = \Delta G_{\frac{1}{2}} = \omega_e - 2\omega_e x_e = \omega_0 - \omega_0 x_0 \quad \text{(III, 83)}$$

Thus we can *determine the vibrational constants ω_e and $\omega_e x_e$ (or ω_0 and $\omega_0 x_0$) from the observed positions of the infrared absorption bands* of a diatomic molecule. Using the figures in Table 8 for HCl, we obtain $\omega_e = 2989.10$ and $\omega_0 = 2937.50$ cm^{-1}. Somewhat improved values for ω_0 and ω_e are obtained if not only the 1–0 band but all observed bands are used; or, in other words, if ω_0 (and possibly $\omega_0 x_0$) is adjusted until the best fit of all bands is obtained. In this way, since the empirical a and b in (II, 9) are these best values of ω_0 and $\omega_0 x_0$, respectively, one obtains for HCl (still neglecting cubic terms, see below) $\omega_e = 2988.90$ and $\omega_0 = 2937.30$.

The fact that the observed bands are so well represented by the theoretical formula (III, 80) and that, in addition, the intensities agree very well with those theoretically expected (very rapid decrease in the series of bands; see Fig. 31) is further proof of the correctness of the interpretation of the near infrared spectra of diatomic molecules as vibration spectra.

[7a] This equality holds only as long as cubic and higher terms are neglected (see below)

If cubic and quartic terms $(\omega_0 y_0 v^3 + \omega_0 z_0 v^4)$ are included in (III, 80) one obtains for the separation of successive bands in place of (III, 81)

$$\Delta G_{v+\frac{1}{2}} = (\omega_e - \omega_e x_e + \omega_e y_e + \omega_e z_e) - (2\omega_e x_e - 3\omega_e y_e - 4\omega_e z_e)(v + \tfrac{1}{2})$$
$$+ (3\omega_e y_e + 6\omega_e z_e)(v + \tfrac{1}{2})^2 + 4\omega_e z_e (v + \tfrac{1}{2})^3$$
$$= (\omega_0 - \omega_0 x_0 + \omega_0 y_0 + \omega_0 z_0) - (2\omega_0 x_0 - 3\omega_0 y_0 - 4\omega_0 z_0)v$$
$$+ (3\omega_0 y_0 + 6\omega_0 z_0)v^2 + 4\omega_0 z_0 v^3 \tag{III, 81a}$$

and for the second difference, in place of (III, 82)

$$\Delta^2 G_{v+1} = -(2\omega_e x_e - 6\omega_e y_e - 14\omega_e z_e) + (6\omega_e y_e + 24\omega_e z_e)(v + \tfrac{1}{2})$$
$$+ 12\omega_e z_e (v + \tfrac{1}{2})^2$$
$$= -(2\omega_0 x_0 - 6\omega_0 y_0 - 14\omega_0 z_0) + (6\omega_0 y_0 + 24\omega_0 z_0)v$$
$$+ 12\omega_0 z_0 v^2 \tag{III, 82a}$$

For the third and fourth differences, one obtains

$$\Delta^3 G_{v+\frac{3}{2}} = \Delta^2 G_{v+2} - \Delta^2 G_{v+1} = (6\omega_e y_e + 36\omega_e z_e) + 24\omega_e z_e(v + \tfrac{1}{2})$$
$$= (6\omega_0 y_0 + 36\omega_0 z_0) + 24\omega_0 z_0 v \tag{III, 84}$$
$$\Delta^4 G_{v+2} = \Delta^3 G_{v+\frac{5}{2}} - \Delta^3 G_{v+\frac{3}{2}} = 24\omega_e z_e = 24\omega_0 z_0 \tag{III, 85}$$

In the case of HCl (Table 8) the very slight increase of the second differences indicates a non-zero third difference.   Assuming the quartic term to be zero, the constant third difference according to (III, 84) is $6\omega_0 y_0 = 6\omega_e y_e = 0.336$ cm^{-1}.   Going back to the second and first differences and adjusting in such a way that the best representation of the observed wave numbers is obtained, the following values result [Lindholm (1173)].

$$\omega_e = 2989.74, \quad \omega_e x_e = 52.05, \quad \omega_e y_e = 0.056 \text{ cm}^{-1}$$
$$\omega_0 = 2937.73, \quad \omega_0 x_0 = 51.97, \quad \omega_0 y_0 = 0.056 \text{ cm}^{-1}$$

For a more precise calculation of the *intensities of infrared bands* it is necessary to take account of the fact that the variation of the dipole moment M with internuclear distance is not strictly linear (*electrical anharmonicity*), that is (III, 45) has to be replaced by

$$M = M_0 + M_1 x + M_2 x^2 + \cdots \tag{III, 86}$$

For very large and very small internuclear distances the dipole moment must clearly approach zero except when the molecule dissociates into ions.   In almost all cases the variation of the dipole moment must therefore be of the form given in Fig. 49.   The linear variation near the equilibrium position assumed earlier corresponds to the tangent of the curve at $r = r_e$.   An electrical anharmonicity may cause the appearance of overtones even when the mechanical oscillations are harmonic.

Calculations of intensities using anharmonic oscillator eigenfunctions and a non-linear variation of the dipole moment [equation (III, 86)] have been carried out by Dunham (194a) (915) and Rosenthal (601).   According to the latter, for HCl, the observed intensity ratio of the 2–0 and 1–0 bands indicates that M_2 is nearly zero.   Scholz (625a) has made calculations on the assumption of a linear variation of the dipole moment with applications to CO.   Experimental determinations of intensities have been made by Dunham (194a), Matheson (482a), and Bartholomé (80a).   Mulliken (519a) has given a discussion of the relation of the variation of the dipole moment to the electronic structure.

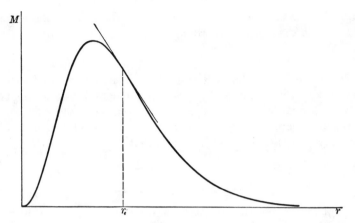

Fig. 49.   **Dipole moment as a function of internuclear distance.**   The r value of the maximum may also be larger than r_e.

Raman spectrum.   Just as in the case of the harmonic oscillator the selection rules for the Raman spectrum of an anharmonic oscillator are the same as those for the infrared spectrum.   Therefore the Raman displacements may be represented by the same formula (III, 80).   As in the infrared, the transition $1 \leftarrow 0$ would be expected to be by far the most intense; actually, no overtones have yet been observed for any diatomic molecule, since even the fundamentals are quite weak.   The observed Raman shifts give $\Delta G_{1/2}$, the first vibrational quantum, which is somewhat smaller than ω_e (or ω_0) [see (III, 81)].   So long as no overtones are observed, the vibrational constants ω_e and $\omega_e x_e$ (or ω_0 and $\omega_0 x_0$) cannot be derived from the Raman spectrum.

Just as in the case of the infrared spectrum the intensity of the overtones in the Raman spectrum depends not only on the mechanical but also on the electrical anharmonicity, that is, in the present case, the non-linear variation of the polarizability α with internuclear distance. The previous relation (III, 63) has to be replaced by

$$\alpha = \alpha_{0v} + \alpha_v{}^1 x + \alpha_v{}^2 x^2 + \cdots \qquad \text{(III, 86a)}$$

From the previous discussion of the Raman spectrum of the harmonic oscillator it is immediately clear that for negligible electrical anharmonicity ($\alpha_v{}^2 \approx 0$, $M_2 \approx 0$) the intensity ratio of fundamental and overtones of a given molecule is the same in the Raman spectrum as in the infrared. In both cases it depends on $\int x \psi_v{}'^* \psi_v{}'' \, dx$ where the same anharmonic oscillator eigenfunctions have to be substituted [see also Bhagavantam (803)].

Vibrational frequency and force constant.   For the harmonic oscillator the separation of successive vibrational levels is equal to the classical vibrational frequency (all in cm^{-1} units).   It may be shown that this also holds for the anharmonic oscillator.   To the decrease of the classical vibrational frequency with increasing amplitude of vibration (see p. 92) corresponds the decrease of the vibrational quantum with increasing vibrational quantum number.   The exact expression for the *vibrational frequency* $\nu_{osc.}(v)$ that the anharmonic oscillator would have in the state v, according to classical theory, turns out to be

$$\nu_{osc.}(v) = c\,\Delta G_v. \tag{III, 87}$$

Here it should be noted that ΔG_v not $\Delta G_{v+\frac{1}{2}}$ has to be used; that is, the vibrational frequency in the state v is intermediate between the two adjacent vibrational quanta $\Delta G_{v+\frac{1}{2}}$ and $\Delta G_{v-\frac{1}{2}}$. Because of the relation (III, 87) the vibrational quanta ΔG are sometimes referred to as vibrational frequencies. From (III, 87) and (III, 81) it follows that

$$\nu_{osc.}(v) = c[(\omega_e - 2\omega_e x_e) - 2\omega_e x_e(v - \tfrac{1}{2})]$$

$$= c[(\omega_e - \omega_e x_e) - 2\omega_e x_e v]. \tag{III, 88}$$

It is seen that for the unrealizable state with $v = -\frac{1}{2}$, for which the vibrational energy (III, 73) is zero,

$$\nu_{osc.}(-\tfrac{1}{2}) = c\omega_e. \tag{III, 89}$$

Thus ω_e gives the vibrational frequency that the anharmonic oscillator would have classically for an infinitesimal amplitude. The anharmonicity is perceptible even for the lowest vibrational state in that the corresponding classical vibrational frequency is somewhat smaller than ω_e; namely,

$$\nu_{osc.}(0) = c(\omega_e - \omega_e x_e) = c\omega_0. \tag{III, 90}$$

From the vibrational frequency ω_e for infinitesimal amplitudes the force constant of the anharmonic oscillator for infinitesimal displacements may be determined. From (III, 33) we obtain

$$k_e = 4\pi^2\mu c^2\omega_e{}^2 = 5.8883 \times 10^{-2}\mu_A\omega_e{}^2 \text{ dyne/cm}, \tag{III, 91}$$

where $\mu_A = \mu N$ (N = Avogadro number) is the reduced mass in atomic-weight units (Aston scale). For HCl, for example, from $\omega_e = 2989.74$ cm^{-1} as given above, it follows that $k_e = 5.1574 \times 10^5$ dynes/cm, which is somewhat larger than the k value previously derived from $\Delta G_{\frac{1}{2}}$ (see p. 82).

For infinitesimal amplitudes cubic and higher powers of $(r - r_e)$ can be neglected in the potential energy, which is then, comparing (III, 31) and (III, 71)

$$V_{r \to r_e} = f(r - r_e)^2 = \tfrac{1}{2}k_e(r - r_e)^2 \tag{III, 92}$$

Thus k_e determines the parabola which gives the best fit to the actual potential curve in the neighborhood of the minimum (see Fig. 46).

Continuous term spectrum and dissociation. When an anharmonic oscillator with a potential curve of the type given in Fig. 46 receives more energy than corresponds to the horizontal asymptote, the mass point will be completely removed from its equilibrium position and will not return to it. This is perhaps most clearly seen if we use again as an illustration (see p. 92) the case of a small particle sliding in a tube of the shape of the potential curve. If the particle is placed at the left limb of the tube above the level of the asymptote and then released it will leave the tube at the right after sliding through the potential minimum (assuming that there is no friction). This motion of the particle corresponds in the molecule to the atoms flying completely away from each other ($r \to \infty$); that is, the molecule dissociates. If the energy of the system just corresponds to the asymptote, the atoms at a great distance from each other will

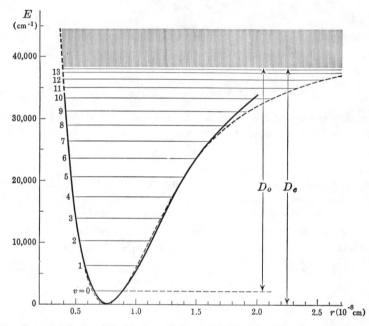

Fig. 50.   **Potential Curve of the H_2 Ground State with Vibrational Levels and Continuous Term Spectrum.** The full curve is drawn according to Rydberg's data (610). The broken curve is a Morse curve. The continuous term spectrum, above $v = 14$, is indicated by vertical hatching. The vibrational levels are drawn up to the potential curve, that is, their end points correspond to the classical turning points of the vibration. It must be remembered that in quantum theory these sharp turning points are replaced by broad maxima of the probability amplitude ψ.

have zero velocity. With increasing energy above that of the asymptote, the atoms at a great distance apart have increasing relative kinetic energy. This kinetic energy is not quantized. Therefore, above the asymptote a *continuous term spectrum, corresponding to dissociation,* joins onto the discrete vibrational term series. This is quite similar to the continuum that adjoins the atomic term series and corresponds to ionization. It is indeed a general result of quantum mechanics [see for example Pauling and Wilson (13)] that for any system for which the potential energy at infinity has a finite value $V(\infty)$ there is a continuous range of energy values above $E = V(\infty)$.

As an example, in Fig. 50 the potential curve is drawn for the special case of H_2 in its electronic ground state, together with the series of observed discrete vibrational levels and the adjoining continuum as determined from the ultraviolet spectrum (see Chapter IV). The height of the asymptote (that is, the beginning of the continuum) above the lowest vibrational level is equal to the work that must be done in order to dissociate the molecule—the so-called *heat of dissociation* or *dissociation energy*. It is designated D_0. It is immediately seen from Fig. 50 that D_0 (in cm^{-1}) is equal to the sum of all the vibrational quanta:

$$D_0 = \sum_v \Delta G_{v+\frac{1}{2}} \qquad \text{(III, 93)}$$

The energy difference, D_e, between the minimum of potential energy and the asymptote is a little greater than D_0—namely, by an amount equal to the zero-point energy:

$$D_e = D_0 + G(0) = D_0 + \tfrac{1}{2}\omega_e - \tfrac{1}{4}\omega_e x_e + \tfrac{1}{8}\omega_e y_e + \cdots \qquad \text{(III, 94)}$$

Since no discrete vibrational levels lie above the asymptote, D_e must be the maximum value of $G(v)$; that is,

$$D_e = G_{\max.}(v) \qquad \text{(III, 95)}$$

Therefore, at the limit, $\Delta G = 0$; that is, according to (III, 87) the vibrational frequency becomes zero and the period of vibration becomes infinite. This is to be expected, since the process of dissociation is, of course, aperiodic.

When the quadratic expression $\omega_0 v - \omega_0 x_0 v^2$ represents all the vibrational levels correctly, or, in other words, when ΔG is a linear function of v [see (III, 81)], ΔG_v becomes zero[7b] for

$$v_D = \frac{\omega_0}{2\omega_0 x_0}. \qquad \text{(III, 96)}$$

The next smaller integral value of v corresponds to the last discrete vibrational level before dissociation. Thus there is only a finite number of vibrational levels present, in contrast to the case of ionization, where there is an infinite number of discrete levels lying below the limit. The dissociation energy D_0 or D_e (in cm^{-1}) is given in this case by

$$D_0 = \omega_0 v_D - \omega_0 x_0 v_D{}^2 = \frac{\omega_0{}^2}{4\omega_0 x_0} \quad \text{or} \quad D_e = \frac{\omega_e{}^2}{4\omega_e x_e}. \qquad \text{(III, 97)}$$

However, in most cases ΔG is not simply a linear function of v. For the ground state of hydrogen, for example, the ΔG curve is found to be that given in Fig. 51. Here also only a finite number of vibrational levels is present.[8] The dissociation energy in this case can be derived only by means of formula (III, 93) or (III, 95). It is seen immediately that the sum of the vibrational

---

[7b] Note that we are using ΔG_v, not $\Delta G_{v+\frac{1}{2}}$ [compare equations (III, 87) and (III, 88)].

[8] An infinite number of vibrational levels below the limit is present for molecular states derived from ions [see Chapter VI, section 4(b)]—that is, when the potential energy for larger r values is the same as for an electron and an atomic ion.

quanta (III, 93) is given very closely by the *area under the ΔG curve*. For a linear ΔG curve this relation holds exactly and, as can easily be seen, leads also to equation (III, 97).

The foregoing considerations regarding the limit of the vibrational terms are not of great importance for infrared spectra, since the intensity in a series of bands falls off so rapidly that high v values are never observed in practice. However, these considerations will prove to be very important in the discussion of electronic band spectra and of the band-spectroscopic determination of heats of dissociation (Chapter VII).

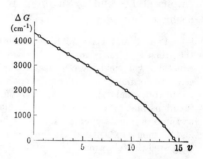

Fig. 51. **ΔG Curve for the Ground State of the H₂ Molecule [after the Data Given by Beutler (91)].** The observed $\Delta G_{v+\frac{1}{2}}$ values are plotted with the abscissae $v + \frac{1}{2}$.

Mathematical representation of the potential curves. As previously mentioned an expression with quadratic and cubic terms in $(r - r_e)$ [equation (III, 71)] represents the potential energy of a diatomic molecule near the equilibrium position only. A mathematic l expression that actually represents a potential curve of the form of the solid curve in Fig. 46, even for large values of r, has been proposed by Morse (504). It is

$$U(r - r_e) = D_e(1 - e^{-\beta(r-r_e)})^2 \qquad (III, 98)$$

Here D_e is the dissociation energy, referred to the minimum (see Fig. 50), and β is a constant whose value will be derived. It can be seen that the Morse function gives a curve of the form shown in Fig. 46, since, for $r \to \infty$, U approaches D_e and for $r = r_e$, U is a minimum—namely, $U = 0$. On the other hand, for $r = 0$, U does not approach ∞, as it must do for a correct potential energy function. However, the part of the curve in the neighborhood of $r = 0$ is of no practical importance.

Morse has shown that when (III, 98) is substituted for V, the wave equation (I, 12) can be solved rigorously [see, however, ter Haar (1014)]. The term values are found to be

$$G(v) = \beta \sqrt{\frac{D_e h}{2\pi^2 c\mu}} \left(v + \tfrac{1}{2}\right) - \frac{h\beta^2}{c8\pi^2\mu} \left(v + \tfrac{1}{2}\right)^2 \qquad (III, 99)$$

without any higher powers of $(v + \tfrac{1}{2})$. According to (III, 74), the coefficient of $(v + \tfrac{1}{2})$ is ω_e. Therefore

$$\beta = \sqrt{\frac{2\pi^2 c\mu}{D_e h}} \, \omega_e = 1.2177 \times 10^7 \, \omega_e \sqrt{\frac{\mu_A}{D_e}}, \qquad (III, 100)$$

where μ_A is the reduced mass in atomic-weight units (Aston scale) and D_e is in

cm^{-1} units. The same formula results from the coefficient of $(v + \frac{1}{2})^2$ in (III, 99) when (III, 97) is substituted.

The Morse function (III, 98) is frequently used for the representation of potential curves, since it is very convenient. In cases in which the vibrational levels cannot be represented by a two-constant formula, it is best to take the empirical D_e and ω_e in order to calculate β from (III, 100). However, the coefficient of $(v + \frac{1}{2})^2$ in (III, 99) will then not agree with the observed $\omega_e x_e$.

Poeschl and Teller (568) have shown that for a given observed set of vibrational levels the Morse function is not the only possible potential function that, on substitution in the wave equation, will yield these vibrational levels, even if the levels can be represented by a two-constant formula. In order to avoid ambiguity in the choice of the potential function, data on the rotational constants of the molecule must be used. Various authors have given potential functions that take account of the values of the rotational constants. Critical summaries of and improvements upon all this work have been given by Hylleraas (351), Coolidge, James, and Vernon (162a), and Hulburt and Hirschfelder (1078). The last-named authors have suggested the following modification of the Morse function:

$$U(x) = D_e[(1 - e^{-\beta x})^2 + c\beta^3 x^3 e^{-2\beta x}(1 + b\beta x)] \tag{III, 101}$$

In this equation $x = r - r_e$ and β is given by (III, 100) (our x and β are defined differently from those of Hulburt and Hirschfelder); c [not to be confused with the velocity of light occurring in (III, 100)] and b are constants depending on the vibrational and rotational constants in the following way:

$$c = 1 - \frac{1}{\beta r_e}\left(1 + \frac{\alpha_e \omega_e}{6B_e^2}\right)$$

$$b = 2 + \frac{1}{c}\left[\frac{7}{12} - \frac{1}{\beta^2 r_e^2}\left(\frac{5}{4} + \frac{5\alpha_e \omega_e}{12B_e^2} + \frac{5\alpha_e^2 \omega_e^2}{144B_e^4} - \frac{2\omega_e x_e}{3B_e}\right)\right] \tag{III, 102}$$

where the rotational constants B_e and α_e will be defined in subsection (c) below. The advantage of the function (III, 101) is that its five parameters can be obtained fairly readily from just those five spectroscopic constants ω_e, $\omega_e x_e$, D_e, B_e and α_e that are the only ones experimentally determined for most electronic states. Hulburt and Hirschfelder have given the values of the constants c and b for the ground states of a considerable number of molecules.

Klein (404) and Rydberg (608) (609) have given a method for constructing the potential curve point for point from the observed vibrational and rotational levels without assuming an analytical expression for the potential function. The exact curves obtained in this way are generally fairly closely approximated by the simple Morse curve (which is much simpler to calculate). The potential curve for the ground state of H_2 in Fig. 50 (full curve) has been derived by the Klein-Rydberg method. The corresponding Morse function is given as a broken-line curve [see also Hylleraas (351)]. In many cases the agreement is not quite as good. Nevertheless for many considerations the Morse function is a fairly satisfactory approximation. A very much closer approximation to the "exact" curves is provided by the modified Morse function (III, 101) at least in all cases in which a comparison has been made. For a discussion of the limitations of the Klein-Rydberg method as well as other important remarks on potential functions reference should be made to the paper by Coolidge, James, and Vernon (162a) which contains also another modification of the Morse function which has considerable merit. Rees (1343) has given an analytical formulation of the Klein-Rydberg method.

Linnett (1175) has suggested a potential function

$$U(r) = \frac{a}{r^m} - be^{-nr}$$

where the first term represents the repulsion of the atomic cores and the constants a and m (≈ 3) in it are the same for all states of a given molecule, while the constants b and n of the second term may be adjusted for each individual state. Linnett has shown that this function gives a better representation of the interrelation of the constants k_e, r_e, D, and $\omega_e x_e$ than does the Morse function. However he has not compared this function with "true" potential functions obtained, for example, by the Klein-Rydberg method.

(b) The Nonrigid Rotator

Energy levels. Thus far we have used the models of the rigid rotator and the harmonic (or anharmonic) oscillator independently of each other. However, it is quite obvious that the molecule cannot be a strictly rigid rotator when it is also able to carry out vibrations in the direction of the line joining the two nuclei. Therefore a better model for representing the rotations of the molecule is given by the *nonrigid rotator*—that is, a rotating system consisting of two mass points which are not connected by a massless rigid bar but by a massless spring.

In such a system as a result of the action of centrifugal force, the internuclear distance, and consequently the moment of inertia, increases with increasing rotation. Therefore, in expression (III, 15) for the rotational term values of a rigid rotator the factor $h/8\pi^2 cI$ depends on the rotational energy (that is, on the rotational quantum number), decreasing with increasing J. A more detailed calculation shows as indicated below that, to a very good approximation, the rotational terms of the nonrigid rotator are given by

$$F(J) = \frac{E_r}{hc} = B[1 - uJ(J + 1)]J(J + 1). \tag{III, 103}$$

That is to say, $B[1 - uJ(J + 1)]$ appears in the place of B in (III, 15). The constant B in (III, 103) is given by the same formula (III, 16) as previously but substituting for the moment of inertia I its value for zero rotational energy; u is very small compared to 1. Equation (III, 103) is usually written

$$F(J) = BJ(J + 1) - DJ^2(J + 1)^2. \tag{III, 104}$$

It should be noted that some authors use a positive sign in (III, 104) instead of the negative one used here. With our choice of sign D always has a positive value.[9]

The rotational constant D depends on the vibrational frequency ω of the molecule, since the smaller ω is, the flatter will be the potential curve [according to (III, 31)] in the neighborhood of the minimum and therefore the greater will be the influence of centrifugal force—that is, the greater will be D. It will be shown below that D is given by

$$D = \frac{4B^3}{\omega^2} \tag{III, 105}$$

if, for the vibrations, the model of the harmonic oscillator is used. As we have

[9] This D, of course, has nothing to do with the dissociation energy D.

seen, ω is always very much greater than B and therefore, according to (III, 105), D will be *very much smaller than B*.

In order to illustrate the effect of D, the rotational levels of a nonrigid rotator have been drawn in Fig. 52 (full lines) with an exaggerated value of D and are compared with the levels of the rigid rotator (broken lines). It can be seen that the effect of D is appreciable mainly for the higher rotational levels. However, in most cases occurring in practice it is considerably smaller than has been represented in Fig. 52.

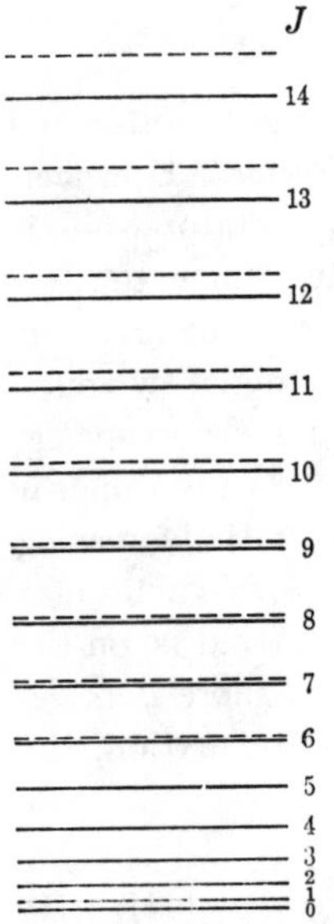

Fig. 52. **Energy Levels of the Nonrigid Rotator.** For comparison, the energy levels of the corresponding rigid rotator are indicated by broken lines (for $J < 6$, they cannot be drawn separately).

In the rotating molecule the internuclear distance assumes such a value, r_c, that the centrifugal force is just balanced by the restoring force $k(r_c - r_e)$ that arises on account of the small displacement $r_c - r_e$ from the equilibrium position (r_e). The centrifugal force F_c is given by

$$F_c = \mu w^2 r_c = \frac{P^2}{\mu r_c{}^3} \tag{III, 106}$$

where w is the angular velocity and $P = Iw = \mu r_c{}^2 w$ is the angular momentum. Equating to the restoring force we obtain for the change of internuclear distance

$$r_c - r_e = \frac{P^2}{\mu r_c{}^3 k} \approx \frac{P^2}{\mu r_e{}^3 k} \tag{III, 107}$$

The kinetic energy of rotation is $P^2/2I_c$ [see equation (III, 1a)]. In addition, for the nonrigid rotator there is the potential energy $\frac{1}{2}k(r_c - r_e)^2$ so that the total energy of rotation is

$$E = \frac{P^2}{2\mu r_c{}^2} + \frac{1}{2}k(r_c - r_e)^2 \tag{III, 108}$$

Substituting r_c from (III, 107) and neglecting higher powers of $r_c - r_e$, we obtain

$$E = \frac{P^2}{2\mu r_e{}^2} - \frac{P^4}{2\mu^2 r_e{}^6 k} + \cdots \tag{III, 109}$$

In quantum theory, as we have seen previously, the angular momentum is $\sqrt{J(J+1)}\, h/2\pi$. Therefore we obtain from (III, 109) as the energy formula for the nonrigid rotator

$$E = \frac{h^2}{8\pi^2\mu r_e{}^2} J(J+1) - \frac{h^4}{32\pi^4\mu^2 r_e{}^6 k} J^2(J+1)^2 + \cdots \tag{III, 110}$$

which is of the form (III, 104). Indeed transforming to term values and substituting $k = 4\pi^2\omega^2 c^2\mu$ the relation (III, 105) for the rotational constant D is immediately obtained. For a more rigorous proof as well as the discussion of small correction terms the reader is referred to Dunham (194) and Pekeris (1310) [see also Pauling and Wilson (13)].

If cubic and higher terms in the potential energy are taken into account and the development carried to higher powers of P^2 the rotational term values are found to be

$$F(J) = BJ(J+1) - DJ^2(J+1)^2 + HJ^3(J+1)^3 + \cdots \tag{III, 111}$$

It is only in very rare cases that the experimental data warrant the introduction of the term with H.[10]

Spectrum. The selection rule for the infrared spectrum of the rotator ($\Delta J = \pm 1$) previously derived is valid whether or not the rotator is rigid. With the term formula (III, 104) we obtain therefore for the wave numbers of the lines in the *infrared rotation spectrum*

$$\nu = F(J + 1) - F(J) = 2B(J + 1) - 4D(J + 1)^3 \qquad \text{(III, 112)}$$

That is to say, the lines are no longer exactly equidistant, as for a rigid rotator, but their separation decreases slightly with increasing J. However, this effect is very small, since $D \ll B$. Nevertheless, it has actually been found for the halogen hydrides (see Chapter II, p. 58 f.). The empirical formula (II, 13) agrees exactly with the theoretical formula (III, 112). For HCl it is found in this way that $B = 10.395$ and $D = 0.0004$ cm^{-1} (see p. 59).

The *Raman spectrum* of the nonrigid rotator also deviates in a corresponding manner from that of the rigid rotator [see (III, 66)]. We have

$$|\Delta \nu| = F(J + 2) - F(J) = (4B - 6D)(J + \tfrac{3}{2}) - 8D(J + \tfrac{3}{2})^3. \qquad \text{(III, 113)}$$

This formula shows that here also, in agreement with observations, there is a slight deviation from equidistance.

Thus we see that the model of the nonrigid rotator allows us to explain the observed far infrared spectrum and the observed rotational Raman spectrum in all details. From the observed spectrum we obtain not only the rotational constant B and with it the moment of inertia and internuclear distance in the molecule, but also the rotational constant D, which gives a measure of the *influence of centrifugal force.*

If the vibrational frequency ω is not known from the vibrational spectrum (or electronic band spectrum; see Chapter IV), an approximate value for it can be calculated from D by using the relation (III, 105). Of course, this value will not be very accurate, since D, being a correction term, cannot be determined very accurately. For HCl, for example, using the above B and D values, we obtain $\omega = 3350$ cm^{-1}, whereas the vibration spectrum yields $\omega_e = 2989.74$ cm^{-1}. The fact that in this case ω as determined from the rotational constants comes out to the right order of magnitude shows clearly that it actually is one and the same system that is carrying out the rotations as well as the vibrations.

Owing to the smallness of the rotational constant D (in practically all cases, $D < 10^{-4}B$), we may in future considerations often neglect the departure of the molecule from the model of a rigid rotator. The deviations from the model of the harmonic oscillator are, on the contrary, by no means negligible.

---

[10] In the older literature F is used in place of H which is taken as the coefficient of $J^4(J + 1)^4$. We follow the practi e of Crawford and Jorgensen (172) and avoid the use of F in two different meanings in the same equation.

(c) The Vibrating Rotator

So far we have regarded the rotation and the vibration of the molecule quite separately. However, it seems natural to assume that rotation and vibration can take place simultaneously, and, in fact, the observed fine structure of the rotation bands suggests very strongly that such a *simultaneous rotation and vibration* occurs. For this reason, we shall now consider a model in which simultaneous rotation and vibration take place—the *vibrating rotator* (or rotating oscillator).

Energy levels. If we could neglect the interaction of vibration and rotation, the energy of the vibrating rotator would be given simply by the sum of the vibrational energy of the anharmonic oscillator (III, 74) and the rotational energy of the nonrigid rotator (III, 104). Therefore a series of rotational levels of the type of Fig. 52 would exist for each of the vibrational levels of Fig. 47. However, in a more accurate treatment we must take into consideration the fact that during the vibration the internuclear distance and consequently the moment of inertia and the rotational constant B are changing.

Since the period of vibration is very small compared to the period of rotation (see p. 82), it seems plausible to use a *mean B value* for the rotational constant in the vibrational state considered; namely,

$$B_v = \frac{h}{8\pi^2 c\mu}\left[\overline{\frac{1}{r^2}}\right], \qquad (III, 114)$$

where $[\overline{1/r^2}]$ is the mean value of $1/r^2$ during the vibration. A more rigorous calculation shows that this procedure is justified [see, for example, Hylleraas (351)]. It is to be expected that B_v will be somewhat smaller than the constant B_e, which corresponds to the equilibrium separation r_e, since with increasing vibration, because of the anharmonicity, the mean nuclear separation will be greater. The value of B_e is given by

$$B_e = \frac{h}{8\pi^2 c\mu r_e^2} = \frac{h}{8\pi^2 c I_e} = \frac{27.98_{30} \times 10^{-40}}{I_e}. \qquad (III, 115)$$

According to rather involved wave-mechanical calculations [see, for example, Pauling and Wilson (13)] to a first (usually satisfactory) approximation, the rotational constant B_v in the vibrational state v is given by

$$B_v = B_e - \alpha_e(v + \tfrac{1}{2}) + \cdots \qquad (III, 116)$$

Here α_e is a constant which is small compared to B_e, since the change in internuclear distance by the vibration is small compared to the internuclear distance itself. Empirically it has been found (Birge) that α_e/B_e is only slightly larger than $\omega_e x_e/\omega_e$, a relation that is useful to remember in practical analyses of band spectra.

In a similar manner, a mean rotational constant D_v, representing the influence of centrifugal force, must be used for the vibrational state v, where, analogous to B_v,

$$D_v = D_e + \beta_e(v + \tfrac{1}{2}) + \cdots \tag{III, 117}$$

Here β_e is small compared to

$$D_e = \frac{4B_e^{\,3}}{\omega_e^{\,2}}, \tag{III, 118}$$

which refers to the completely vibrationless state [see (III, 105)].

We obtain, accordingly, for the *rotational terms in a given vibrational level*

$$F_v(J) = B_v J(J + 1) - D_v J^2(J + 1)^2 + \cdots \tag{III, 119}$$

where the second term, with D_v, is very small compared to the first and can often be neglected.

Some authors prefer to use, instead of equation (III, 119), the equation

$$F_v(J) = B_v(J + \tfrac{1}{2})^2 - D_v(J + \tfrac{1}{2})^4 + \cdots \tag{III, 120}$$

which differs from (III, 119) only by a small additive constant and by a very slight alteration in the meaning of B_v (the new B_v being the old $B_v - \tfrac{1}{2}D_v$). For the interpretation of the spectra this difference is quite unimportant, but it has to be kept in mind in comparing term values taken from different papers.

By taking into consideration the interaction of vibration and rotation, in the way described above, we obtain for the *term values of a vibrating rotator*

$$\begin{aligned}
T = G(v) + F_v(J) &= \omega_e(v + \tfrac{1}{2}) - \\
\omega_e x_e(v + \tfrac{1}{2})^2 + \cdots &+ B_v J(J + 1) - \\
D_v J^2(J + 1)^2 &+ \cdots .
\end{aligned} \tag{III, 121}$$

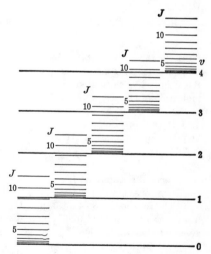

The corresponding energy level diagram is given in Fig. 53. In this figure, for clarity of representation, the individual rotational levels are represented by shorter horizontal lines than the "pure" vibrational levels $(J = 0)$.

For the *lowest vibrational state* $(v = 0)$, the rotational constant B_0 has to be used in (III, 121). According to (III, 116), B_0 is somewhat smaller than the constant B_e, which corresponds to the unrealizable

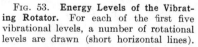

Fig. 53. **Energy Levels of the Vibrating Rotator.** For each of the first five vibrational levels, a number of rotational levels are drawn (short horizontal lines).

completely vibrationless state. It may be noted here that the rotational constant B obtained from the pure rotation spectrum (in the infrared or the Raman effect) is, of course, B_0. Correspondingly, the r value obtained (p. 81) is not r_e, corresponding to the minimum of the potential curve, but r_0, a mean value for the lowest vibrational state. r_0 is somewhat, though only very slightly, greater than r_e.

If the form of the potential curve is known the averaging of the rotational constant in a given vibrational state [equation (III, 114)] can be carried out and thus a theoretical value for α_e obtained. The result of such a calculation in the case of a potential function containing cubic and quartic terms in addition to the quadratic term [see equation (III, 71)] is according to Teller (664)

$$\alpha_e = 24 \frac{B_e^3 r_e^3 g}{\omega_e^3} - 6 \frac{B_e^2}{\omega_e}, \qquad \text{(III, 122)}$$

where g is the coefficient of the cubic term in the potential function (measured in cm^{-1})—that is, the term which determines the asymmetry of the potential curve. If α_e is obtained from the observed spectrum the coefficient g can conversely be derived from (III, 122). It is then possible to obtain the coefficient j of the quartic term from the observed value of $\omega_e x_e$ according to (III, 79).

In practically all cases, the first term in (III, 122) outweighs the second, so that α is almost always positive. However, it can be seen from (III, 122) that even for a completely symmetrical curve (parabola) the constant α is not equal to zero (as might perhaps have been expected), but has a negative value. This is due to the fact that, for a vibrating rotator, it is not the mean value of r but the mean value of $1/r^2$ (or $1/I$) that has to be used in forming the mean value of B [see equation (III, 114)].

If instead of a potential function with cubic and quartic terms in $(r - r_e)$ the Morse function (III, 98) is used, the following expression is obtained for α_e, as shown by Pekeris (1310)

$$\alpha_e = \frac{6 \sqrt{\omega_e x_e B_e^3}}{\omega_e} - \frac{6 B_e^2}{\omega_e} \qquad \text{(III, 123)}$$

Since the three constants of the Morse function are determined by ω_e, $\omega_e x_e$, and B_e (see p. 101) no new potential constants are obtained from this expression for α_e. Rather the deviation of the value of α_e calculated by means of (III, 123) from the observed value may serve as an indication of the deviation of the Morse function from the actual potential function (see also p. 102).

If very precise measurements are available sometimes higher powers of $(v + \frac{1}{2})$ have to be taken into account in (III, 116). One writes

$$B_v = B_e - \alpha_e(v + \tfrac{1}{2}) + \gamma_e(v + \tfrac{1}{2})^2 + \cdots \qquad \text{(III, 124)}$$

A theoretical formula for γ_e may be found in Dunham (194). It may serve to determine the coefficient of the fifth power term in the potential function.

The rotational constant β_e in (III, 117) can be expressed in terms of B_e, α_e, ω_e, and $\omega_e x_e$ if a power series is used for the potential function as before. One obtains from the formulae given by Dunham (104)[11]

$$\beta_e = D_e \left(\frac{8\omega_e x_e}{\omega_e} - \frac{5\alpha_e}{B_e} - \frac{\alpha_e^2 \omega_e}{24 B_e^3} \right) \qquad \text{(III, 125)}$$

With the aid of β_e obtained from this equation, D_v can be worked out according to (III, 117 and 118) if B_e, α_e, ω_e, and $\omega_e x_e$ are known. However, β_e is small compared to D_e (which itself is a small correction) and may therefore be neglected in many cases.

---

[11] This formula was first given by Birge [see Jevons (23)]. It appears that the equation as given by Jevons and many others, taking due account of their different definition of D [positive sign in (III, 119)] gives the wrong sign for β. It is believed that (III, 125) has the correct signs assuming the choice of signs in (III, 117) and (III, 119).

In exceptional cases higher powers of $J(J+1)$ in (III, 110) have to be taken into account; one writes [compare (III, 111)[10]]

$$F_v(J) = B_v J(J+1) - D_v J^2(J+1)^2 + H_v J^3(J+1)^3 + \cdots \qquad \text{(III, 126)}$$

where H_v is given by a formula similar to (III, 117). Neglecting the difference between H_v and H_e one finds from Dunham's formulae to a first approximation

$$H_v \approx H_e = \frac{2D_e}{3\omega_e^2}(12B_e^2 - \alpha_e\omega_e) \qquad \text{(III, 127)}$$

This formula was first given by Birge (19) who also gave an expression for the coefficient of $J^4(J+1)^4$.

A very careful study of the finer interaction of vibration and rotation has been made by Dunham (194) [see also Sandeman (1370)]. He expressed the term values of the vibrating rotator in the form of a double power series

$$T = \sum_{lj} Y_{lj}(v + \tfrac{1}{2})^l J^j(J+1)^j \qquad \text{(III, 128)}$$

where on the basis of the preceding considerations we would expect to have the relations

$$Y_{10} = \omega_e, \quad Y_{20} = \omega_e x_e, \quad Y_{01} = B_e, \quad Y_{02} = D_e, \quad Y_{11} = \alpha_e, \quad \cdots \qquad \text{(III, 129)}$$

However, it was actually found by Dunham that there are some very small deviations from the above relations. In other words the value of B_e that is found from the observed spectrum by the use of formulae (III, 121) and (III, 116) is not exactly $h/8\pi^2 c I_e$ as required by (III, 115) but is Y_{01} which differs slightly from it. Similarly the value of ω_e obtained is actually Y_{10} which deviates slightly from the classical vibrational frequency for infinitesimal amplitudes. The correction is of the order B_e^2/ω_e^2 which is largest for the H_2 molecule and hydride molecules. Even for these the corrections are almost always less than and usually much less than one part in 1000. For molecules other than H_2 and the hydrides these corrections are quite negligible. The corrections to the formulae (III, 122), (III, 118), (III, 125) and (III, 127) for α_e, D_e, β_e, and H_e, respectively, are increasingly larger. In the case of D_e it was shown by Dunham that the correction vanished if the potential function can be represented exactly by a Morse function. Conversely the comparison of observed and calculated values of D_e may be used as a test for the validity of the Morse function.

According to Dunham's formulae there is also a non-vanishing term Y_{00} which in terms of the other constants may be written

$$Y_{00} = \frac{B_e}{4} + \frac{\alpha_e\omega_e}{12B_e} + \frac{\alpha_e^2\omega_e^2}{144B_e^3} - \omega_e x_e$$

This term represents an addition to the zero-point energy above the value previously given for the anharmonic oscillator [equation (III, 75)].

Extensive examples for the application of Dunham's formulae have been given by Crawford and Jorgensen (173) and Dieke and Lewis (188).

In order to avoid the cumbersome nomenclature implied in (III, 128) we shall use in this book ω_e, B_e, $\cdots$ in place of the corresponding Y_{lj}, that is, assuming (III, 129) as correct without correction terms. This is permissible as long as we remember that in exceptional cases ω_e and B_e may not be related to force constant k_e and internuclear distance r_e in exactly the orthodox way given by (III, 91) and (III, 115), respectively (although always in a very good approximation), and that quite frequently the formulae for α_e, D_e, and β_e may not represent the observed values accurately.

Eigenfunctions. The eigenfunctions ψ of the vibrating rotator are of course functions of the internuclear distance r and the angles ϑ and φ. Just as for the hydrogen atom [see Pauling and Wilson (13)] a separation of the variables is possible and one may write

$$\psi = R(r)\Theta(\vartheta)\Phi(\varphi)$$

The equations for Θ and Φ are the same as for the hydrogen atom, and therefore as for the rigid rotator [see equation (III, 11)], that is,

$$\Theta(\vartheta)\Phi(\varphi) = \psi_r(\vartheta, \varphi)$$

The equation for $R(r)$ may be shown [see Pauling and Wilson (13)] to have in a first approximation (for small rotation) the solution

$$R(r) = \frac{1}{r}\psi_v(r - r_e) \tag{III, 130}$$

where ψ_v are the eigenfunctions of the linear anharmonic oscillator previously considered (see Fig. 48). Thus we have

$$\psi = \frac{1}{r}\psi_v(r - r_e)\psi_r(\vartheta, \varphi) \tag{III, 131}$$

that is, the eigenfunctions are essentially products of oscillator and rotator eigenfunctions.

The square of the radial function $R(r)$ gives the probability of finding the system with various values of r in a given direction. The probability of finding the system with various values of r irrespective of direction is proportional to $r^2[R(r)]^2 = \psi_v{}^2$, that is, the solid curves in Figs. 42 and 48 may be used as representing the radial distribution functions of the vibrating rotator (even though not the probability distribution in a fixed direction).

Even if the rotation is no longer slow, $\psi_r(\vartheta,\varphi)$ can be separated off the wave functions. However, $\psi_v(r - r_e)$ has then to be replaced by a function that depends slightly on J as well as on v, somewhat similar to the way in which the radial eigenfunctions of the hydrogen atom depend on the azimuthal quantum number l. For more details see Pekeris (1310).

Infrared spectrum. Since the eigenfunctions of the rotating vibrator are essentially products of rotator and oscillator eigenfunctions, it is readily seen (see below) that the same selection rules hold as for these systems individually, that is, v can change by any integral amount although $\Delta v = \pm 1$ gives by far the most intense transitions, and J can only change by unity [see (III, 20)]. Of course, $\Delta v = 0$ is also allowed; however this does not give rise to a rotation-vibration spectrum but rather to the pure rotation spectrum [compare the discussion of the non-rigid rotator in section 2(b)].

Substituting (III, 131) into the transition moment (I, 44) one obtains, for example, for the z component

$$R_z = \int \frac{1}{r^2}\psi_v{}'\psi_r{}'^*M_z\psi_v{}''\psi_r{}''\, d\tau$$

with $M_z = M\cos\vartheta$ and $d\tau = r^2\sin\vartheta\, dr\, d\vartheta\, d\varphi$ this becomes

$$R_z = \int M\psi_v{}'\psi_v{}''\, dr \int \cos\vartheta\, \psi_r{}'^*\psi_r{}''\sin\vartheta\, d\vartheta\, d\varphi$$

Here the second integral is precisely the one occurring for the rigid rotator (see p. 72), while the first integral is the same as that occurring for the linear anharmonic oscillator at least for slow rotation. Thus the same selection rules apply. Since the second integral is independent of the interaction of rotation and vibration, the rotational selection rule holds rigorously for the vibrating rotator.

If we now consider a particular vibrational transition from v' to v'', we can write, according to (III, 121), for the *wave numbers of the resulting lines* (neglecting the rotational constant D_v)

$$\nu = \nu_0 + B_v'J'(J' + 1) - B_v''J''(J'' + 1), \qquad \text{(III, 132)}$$

where $\nu_0 = G(v') - G(v'')$ is the frequency of the pure vibrational transition without taking account of rotation $(J' = J'' = 0)$. With $\Delta J = +1$ and $\Delta J = -1$, respectively, we obtain from (III, 132)

$$\nu_R = \nu_0 + 2B_v' + (3B_v' - B_v'')J + (B_v' - B_v'')J^2; \quad J = 0, 1, \cdots; \quad \text{(III, 133)}$$

$$\nu_P = \nu_0 \qquad - (B_v' + B_v'')J + (B_v' - B_v'')J^2; \quad J = 1, 2, \cdots. \quad \text{(III, 134)}$$

Here, as usual (see p. 73), J'' has been replaced by J. Since J can take a whole series of values (see p. 124), these two formulae represent two series of lines, which are called the *R and P branches*, respectively. The corresponding transitions are indicated in the energy level diagram Fig. 54. Since the lowest value of the rotational quantum number (J) in the upper and lower states, is zero it is clear from Fig. 54 that the smallest value of J'' $(\equiv J)$ in the R branch is 0 while in the P branch it is 1. The resulting spectrum is represented schematically at the bottom of Fig. 54. It can be seen that there is qualitatively complete agreement with the empirical fine structure of infrared bands (see p. 55 f.)

If we neglect, for a moment, the interaction between rotation and vibration, we have $B_v' = B_v'' = B$ and formulae (III, 133) and (III, 134) simplify to

$$\nu_R = \nu_0 + 2B + 2BJ, \qquad \nu_P = \nu_0 - 2BJ; \qquad \text{(III, 135)}$$

that is, we have two series of equidistant lines, the one, ν_R, going from ν_0 toward shorter wave lengths and the other, ν_P, going toward longer wave lengths. Owing to the above restriction for the J values, there is no line at ν_0; that is, we have a *zero gap*. The spectrum calculated according to (III, 135) is shown in strip (b) at the bottom of Fig. 54. It agrees qualitatively very well with the observed spectrum (Fig. 32, p. 55). The observed slight convergence of the lines is to be traced back to the *influence of the interaction of rotation and vibration*. In consequence of this interaction, B_v' differs from B_v'', giving rise to the quadratic terms in (III, 133) and (III, 134). This has the effect that, when $B_v' < B_v''$, the lines in the R branch draw closer together and those in the P branch draw farther apart [see strip (a) at the bottom of Fig. 54].

As may easily be verified, the two branches (III, 133) and (III, 134) can also be represented by a single formula—namely,

$$\nu = \nu_0 + (B_v' + B_v'')m + (B_v' - B_v'')m^2, \qquad \text{(III, 136)}$$

where m is an integral running number which takes the values 1, 2, $\cdots$ for the R branch (that is, $m = J + 1$) and the values -1, -2, $\cdots$ for the P branch (that is, $m = -J$). Thus we can also say that we have *a single series of lines for which a line is missing at $m = 0$*. The missing line, at $\nu = \nu_0$, is called the

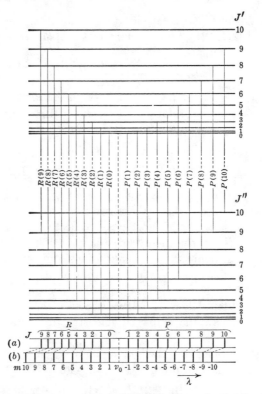

Fig. 54. **Energy Level Diagram explaining the Fine Structure of a Rotation-Vibration Band.** In general, the separation of the two vibrational levels is considerably larger compared to the spacing of the rotational levels than shown in the figure (indicated by the broken parts of the vertical lines representing the transitions). The schematic spectrograms (a) and (b) give the resulting spectrum with and without allowance for the interaction between rotation and vibration. In these spectrograms, unlike most of the others, short wave lengths are at the left.

zero line (*null line*). It would correspond to the forbidden transition between the two rotationless states $J = 0$ (see Fig. 54). ν_0 is also called the *band origin*.

Formula (III, 136) has exactly the same form as the empirical formula (II, 8). We conclude from this that the molecule actually is to be regarded as a vibrating rotator and that the spectra in the near infrared are *rotation-vibration spectra*. Furthermore we conclude that a band represents all the possible rotational transitions for a particular vibrational transition.[12]

[12] It should be remarked here that also from the old quantum theory using the expression BJ^2 for the rotational energy a formula of the form (II, 8) is obtained, but no missing line results, in disagreement with experiment.

Since α_e in (III, 116) is small, the difference between B_v' and B_v'' is, in general, small and the formulae (III, 135) give a fairly good representation of the observed spectra, particularly in the neighborhood of the zero gap. The separation of successive lines in the neighborhood of the zero gap is therefore approximately $2B$. The larger Δv is, the larger is the difference between B_v' and B_v'' [see (III, 116)]. Consequently the convergence of the lines in the fine structure is more rapid for higher overtones, as a comparison of Fig. 32 and Fig. 33 (1–0 and 3–0 bands of HCl) shows. For sufficiently high J, a *reversal of the R branch* occurs, corresponding to the vertex of the parabola represented by (III, 136). This reversal (*band-head formation*) will be discussed at greater length for electronic band spectra, since it is much more often observed there.

The *rotational constants* B_v' and B_v'' of the upper and lower st ites can be determined directly by comparing (III, 136) with the empirical formula (II, 8): we have $d = B_v' + B_v''$ and $e = B_v' - B_v''$. Thus, in the example of the 1–0 band of HCl, it follows from (II, 10) that $B_1 + B_0 = 20.577$ cm^{-1}, $B_1 - B_0 = -0.3034$ cm^{-1}, and therefore $B_1 = 10.137$ cm^{-1} and $B_0 = 10.440$ cm^{-1}.

An alternative method of determining the rotational constants is by the use of the separation $\Delta\nu(m)$ of the successive lines and the increase of this separation, the second difference $\Delta^2\nu(m)$. According to (III, 136):

$$\Delta\nu(m) = \nu(m + 1) - \nu(m) = 2B_v' + 2(B_v' - B_v'')m, \quad \text{(III, 137)}$$

and

$$\Delta^2\nu(m) = \Delta\nu(m + 1) - \Delta\nu(m) = 2(B_v' - B_v''). \quad \text{(III, 138)}$$

Thus the second difference is constant in this approximation in agreement with the empirical findings already mentioned in Chapter II. Its mean value gives a fairly accurate value for $2(B_v' - B_v'')$. A further equation for the constants B_v' and B_v'' is obtained from the first differences: for example, the intersection of the straight line representing (III, 137) with the axis $m = 0$ gives the value $2B_v'$. Other more accurate methods for the determination of B values will be discussed later (Chapter IV, p. 175 f.).

The B_v values of HCl, obtained from the 1–0, 2–0, 3–0, 4–0, and 5–0 bands by means of the more accurate methods, taking account of the rotational constant D, are summarized in Table 13. The difference ΔB_v between successive values is very nearly constant, as is to be expected from (III, 116). The mean value of this difference is the constant α_e, which is a measure of the interaction between vibration and rotation. Finally, the rotational constant B_e for the completely vibrationless state is obtained by adding $\frac{1}{2}\alpha_e$ to B_0.[13] For HCl, α_e is found to be 0.3019 cm^{-1} and B_e is found to be 10.5909 cm^{-1}; from this, according to (III, 115) we obtain for the internuclear distance in the equilibrium position

---

[13] For an accurate evaluation of α_e and B_e it is necessary to calculate the B_v values according to (III, 116) with preliminary B_e and α_e values (obtained as described above) and then when necessary, to alter α_e and B_e so that the deviations, B_v (observed) $- B_v$ (calculated), are as small as possible. This can be done graphically or by the method of least squares.

TABLE 13. ROTATIONAL CONSTANTS OF HCl IN THE DIFFERENT VIBRATIONAL
LEVELS OF THE ELECTRONIC GROUND STATE

From the data of Meyer and Levin (491), Herzberg and Spinks (318)
and Lindholm (1173).

v	$B_v (\text{cm}^{-1})$	$\Delta B_v (\text{cm}^{-1})$
0	10.4400	
		0.3034
1	10.1366	
		0.3037
2	9.8329	
		0.2986
3	9.5343	
		0.302$_3$
4	9.232	
		0.299
5	8.933	

$r_e = 1.2746$ Å, while the effective internuclear distance in the lowest vibrational level [obtained by substituting B_0 in (III, 115) in place of B_e] is $r_0 = 1.2838$ Å.

It may be noted that the B value obtained from the pure rotation spectrum, $B = 10.395$ (see p. 105), agrees closely with B_0 in Table 13, as is to be expected. The difference is within the accuracy of the measurements in the far infrared.

As can be seen from Table 9 there is a very small systematic trend in the second differences of the lines. It is due to the *influence of the rotational constant D*. If D_v in (III, 121) is not neglected one obtains in place of (III, 136)

$$\nu = \nu_0 + (B_v' + B_v'')m + (B_v' - B_v'' - D_v' + D_v'')m^2$$
$$- 2(D_v' + D_v'')m^3 - (D_v' - D_v'')m^4 \qquad \text{(III, 139)}$$

Here it is to be noted that taking account of the rotational constant D does not alter the earlier conclusion that both branches can be represented by a single formula. Since D_v' is very nearly equal to D_v'' (see p. 107 f.), the cubic term in (III, 139) suffices in most cases for a complete representation of the two branches of a rotation-vibration band even for very accurate measurements [see the empirical formula (II, 11)]. It is obvious that conversely the rotational constant D can be derived from a determination of the cubic and possibly the quartic term in the empirical formula (see also Chapter IV, p. 180 f.).

Raman spectrum.   Just as in the case of the infrared spectrum the selection rules for the Raman spectrum of the vibrating rotator are those of the anharmonic oscillator and the rotator.   The selection rule for the former is the same as for the infrared spectrum (predominant intensity of $\Delta v = \pm 1$)[13a] while that for the latter is $\Delta J = 0, \pm 2$ [compare (III, 65)].   Accordingly for a given vibrational transition, that is, for a given *Raman vibration band*, there are *three branches*. Their formulae are readily obtained from

$$\Delta \nu = \Delta \nu_0 + B_v'J'(J' + 1) - B_v''J''(J'' + 1) \qquad \text{(III, 140)}$$

by substituting $J' = J'' + 2$ (*S* branch), $J' = J'' - 2$ (*O* branch), and $J' = J''$ (*Q* branch) and using $J'' \equiv J$ [compare (III, 132)].   One finds

$$(\Delta \nu)_S = \Delta \nu_0 + 6B_v' + (5B_v' - B_v'')J + (B_v' - B_v'')J^2; \quad J = 0, 1, \cdots \qquad \text{(III, 141)}$$

[13a] This is of course in addition to the rotational Raman spectrum for which $\Delta v = 0$.

$$(\Delta\nu)_O = \Delta\nu_0 + 2B_v' - (3B_v' + B_v'')J + (B_v' - B_v'')J^2; \quad J = 2, 3, \cdots \quad \text{(III, 142)}$$

$$(\Delta\nu)_Q = \Delta\nu_0 \qquad + (B_v' - B_v'')J + (B_v' - B_v'')J^2; \quad J = 0, 1, \cdots \quad \text{(III, 143)}$$

For the $1 \leftarrow 0$ vibrational transition (the only one thus far observed in the Raman effect) the difference between B_v' and B_v'' is very small; therefore all the lines of the Q branch are very close to one another and are usually not resolved. This gives rise to an intense "line" at $\Delta\nu_0 (= \Delta G_{1/2})$ possibly somewhat diffuse at the short-wave-length side (toward smaller $\Delta\nu$), in agreement with observations (see p. 61 f.). The S and O branches are very much weaker since their lines are not superimposed. They form a series somewhat similar to the R and P branches of the infrared bands except that the line separation is twice as large. Thus far these branches have been observed only for H_2 [Rasetti (578)] and HCl [Andrychuk (752a)]. The hydrogen molecule is the only one for which the Q branch has been resolved into individual lines. In all other cases of diatomic molecules the Raman lines of large displacement represent the *unresolved Q branches* of the *Raman rotation-vibration bands*.

(d) The Symmetric Top

The diatomic molecule as a symmetric top. Thus far in our considerations we have taken as a model for the rotations of the molecule a simple rotator, with the tacit assumption that the moment of inertia about the line joining the nuclei is zero. Actually, however, there are a number of electrons revolving about the two nuclei, and, as a result, the *moment of inertia about the line joining the nuclei is not exactly zero*—though it is very small, owing to the smallness of the mass of the electron. Thus a better model for the diatomic molecule than the simple rotator would be a system somewhat like a dumbbell carrying a flywheel on its axis. Such a system is an example of the more general case of a *symmetric top*, which is defined as follows:

If for a rigid body the moment of inertia is calculated for various axes passing through the center of mass it is found according to a well-known theorem of classical mechanics that for three mutually perpendicular directions the moment of inertia is a maximum or a minimum. These directions are called the *principal axes* and the corresponding moments of inertia the *principal moments of inertia*. If the body has axes of symmetry they coincide with the principal axes. In general the three principal moments of inertia are different. If two of them are equal the rigid body is called a *symmetric top*.[14] If we disregard the finer details of the motions of the electrons but consider the nuclei as simply surrounded by a "rigid" electron cloud, the diatomic molecule is clearly a symmetric top, since the moments of inertia I_B about all the axes at right angles to the internuclear axis and passing through the center of gravity are equal to one another. The moment of inertia I_A about the internuclear axis is very much smaller than I_B. In spite of this, the corresponding *angular momenta*

---

[14] For a more detailed discussion see Volume II, Chapter I.

are of the same order of magnitude, since the electrons rotate much more rapidly than the heavy nuclei.

Angular momenta. According to classical mechanics, the total angular momentum of a symmetric top is in general not perpendicular to the figure axis (internuclear axis), as is the case for the simple rotator, since a rotation about the figure axis can now take place in addition to the rotation about an axis perpendicular to it. The two corresponding angular momenta are added together to give the *total angular momentum* **P**. Only the total angular momentum is constant in magnitude as well as direction. Its *component in the direction of the figure axis* (that is, the angular momentum about the figure axis) is *constant in magnitude* but not in direction. In fact, the figure axis rotates at a constant angle about **P** with a frequency $(1/2\pi)(P/I_B)$, which corresponds to the rotational frequency of the simple rotator (see p. 69). This motion of the figure axis is called nutation.[15] The constant component of the total angular momentum in the direction of the figure axis in the present case is due to the revolution of the electrons.

According to quantum mechanics, neglecting electron spin (see Chapter V), the component of the total angular momentum in the direction of the figure axis (that is, in the present case, the electronic angular momentum) can be only an integral multiple of $h/2\pi$. Thus, designating this angular momentum by Λ, we have

$$|\Lambda| = \Lambda \frac{h}{2\pi},$$ (III, 144)

where Λ is the *quantum number of the angular momentum of the electrons about the internuclear axis*. As previously, the *total angular momentum*, designated by J, can take only the values

$$|J| = \sqrt{J(J+1)} \frac{h}{2\pi} \approx J \frac{h}{2\pi}.$$ (III, 145)

Fig. 55 gives a diagram of the angular momentum vectors of the symmetric top. The component of the total angular momentum at right angles to the internuclear axis is designated N. It represents essentially the rotation of the nuclei alone.[16] Since Λ and J have integral values, N obviously cannot have integral values (except by accident). The magnitude of N is completely determined by Λ and J and there is no quantum number associated with N. The total angular momentum J is perpendicular to the axis only when $\Lambda = 0$ (in which case $N \equiv J$). For $\Lambda \neq 0$, the total angular momentum J subtends an angle smaller than 90° with the internuclear axis. In any case the whole system

---

[15] Frequently it is referred to as *precession*, contrary to the usage in the theory of gyroscopic motion. Precession is the motion of an angular momentum vector under the action of a force.

[16] This vector, of course, has nothing to do with the nuclear spin (see p. 29).

rotates about J with a frequency that is the same as
that of a simple rotator having the same I_B and the
same J [see (III, 26)].

It is clear from Fig. 55 that J is always larger
than Λ.   Therefore, for a given Λ, the possible values
of the quantum number J are

$$J = \Lambda, \quad \Lambda + 1, \quad \Lambda + 2, \quad \Lambda + 3, \cdots \quad \text{(III, 146)}$$

Even for the smallest value, $J = \Lambda$, the angular momentum vector J does not point exactly in the direction of the internuclear axis since the magnitude of J
is given by (III, 145), the magnitude of Λ by (III,
144).

If the sense of the rotation of the electrons
about the internuclear axis is reversed, Λ has the

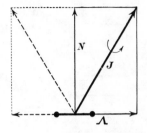

Fig. 55.   **Vector Diagram
for the Symmetric Top.**   The
curved arrow indicates the
rotation of the whole diagram
about J.   The dotted part of
the figure gives the vector
diagram when the sense of the
direction of Λ is reversed.

opposite direction and the vector diagram drawn in broken lines in Fig. 55 is
obtained.   For a given N, the same J value is obtained, or, in other words, for
each value of J there are two modes of motion of the system corresponding to
the two directions of Λ.

Energy levels.   The wave equation of the symmetric top can be solved
rigorously as was first done by Reiche and Rademacher (1345) and Kronig and
Rabi (1145)[17] [see also Pauling and Wilson (13)].   The energy levels that result
are given (in cm^{-1}) by

$$F(J) = BJ(J + 1) + (A - B)\Lambda^2, \quad \text{(III, 147)}$$

where

$$B = \frac{h}{8\pi^2 c I_B}, \qquad A = \frac{h}{8\pi^2 c I_A}. \quad \text{(III, 148)}$$

$I_B = \mu r^2$ is the "ordinary" moment of inertia of the molecule about an axis
perpendicular to the internuclear axis, as used previously, and I_A is the moment
of inertia of the electrons about the internuclear axis.   Owing to the smallness
of I_A, the constant A is very much larger than B.   For a given electronic state,
Λ is constant and has in general only a small (integral) value (see Chapter V).
As a result, *the rotational levels of the symmetrical top are the same as those of the
simple rotator except that there is a shift of magnitude* $(A - B)\Lambda^2$, which is constant
for a given electronic state, *and except that* according to (III, 146) *levels with
$J < \Lambda$ are absent.*   As an example an energy level diagram for $\Lambda = 2$ is given
in Fig. 56.   The rotational levels that do not occur $(J < \Lambda)$ are indicated by
broken lines.

The sense of direction of Λ does not enter equation (III, 147); that is, to
say, the two states represented by the full-line and broken-line vector diagrams

---

[17] Dennison (887) obtained the energy levels from matrix mechanics.

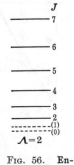

J
——— 7
——— 6
——— 5
——— 4
——— 3
——— 2
--------(1)
--------(0)
$\Lambda=2$

Fig. 56. Energy Level Diagram of a Symmetric Top with $\Lambda = 2$. The dotted levels do not occur.

in Fig. 55 have exactly the same energy (are *degenerate* with respect to each other). Thus each of the levels in Fig. 56 consists actually of two levels which coincide with each other.

The formula (III, 147) for the energy levels may be easily derived: In classical mechanics the kinetic energy of rotation of a rigid body is given by

$$E = \tfrac{1}{2}I_x w_x^2 + \tfrac{1}{2}I_y w_y^2 + \tfrac{1}{2}I_z w_z^2 = \frac{P_x^2}{2I_x} + \frac{P_y^2}{2I_y} + \frac{P_z^2}{2I_z} \quad \text{(III, 149)}$$

Here x, y, z are the directions of the principal axes; I_x, w_x, P_x are moment of inertia, angular velocity, and angular momentum, respectively, about the x axis and similarly for the other axes. In the present case $P_z^2 = \Lambda^2$, $P_x^2 + P_y^2 = N^2 = J^2 - \Lambda^2$ (see Fig. 55), $I_z = I_A, I_y = I_x = I_B$. Therefore

$$E = \frac{J^2}{2I_B} - \frac{\Lambda^2}{2I_B} + \frac{\Lambda^2}{2I_A}$$

Substituting the magnitudes of J and Λ from (III, 144) and (III, 145) and transforming to term values (cm^{-1}) the energy formula (III, 147) is obtained immediately.

Thus far we have assumed the symmetric top to be rigid. For the non-rigid symmetric top, in order to take account of the influence of centrifugal force, a term $-DJ^2(J + 1)^2$ has to be added to the energy formula (III, 147). For the vibrating symmetric top, just as previously for the rotator, effective values B_v and D_v of the rotational constants have to be used which differ slightly for different vibrational levels. Thus we have

$$F_v(J) = B_v J(J + 1) + (A - B_v)\Lambda^2 - D_v J^2(J + 1)^2 + \cdots \quad \text{(III, 150)}$$

where B_v and D_v are given by the previous formulae (III, 116) and (III, 117).

Eigenfunctions. The eigenfunctions of the symmetric top which result from a solution of the wave equation are given by [see, for example, Pauling and Wilson (13)].

$$\psi_r = \Theta_{J\bar{\Lambda}M}(\vartheta)e^{i\bar{\Lambda}\chi}e^{iM\varphi} \quad \text{(III, 151)}$$

Here ϑ, φ, and χ are the so-called Eulerian angles: ϑ is the angle of the figure axis of the top with the fixed z-axis, φ is its azimuthal angle about the z-axis, and χ is the azimuthal angle measuring the rotation about the figure axis.[18] In place of Λ, equation (III, 151) contains $\bar{\Lambda}$ which we define in the same way as Λ except that it includes the sign (compare Fig. 55), that is, for a given value of Λ we have $\bar{\Lambda} = \pm\Lambda$. M is the magnetic quantum number (see p. 69) which gives the component of J in the direction of the z-axis in units $h/2\pi$ and can have the values $J, J - 1, \cdots, -J$. $\Theta_{J\bar{\Lambda}M}(\vartheta)$ is a somewhat complicated function of the angle ϑ containing the so-called Jacoby (hypergeometric) polynomials [see Pauling and Wilson (13)]. For $\Lambda = 0$ the functions ψ_r obtained from (III, 151) are identical with those of the rigid rotator given on p. 70 except for the constant factor $1/\sqrt{2\pi}$. For $\Lambda = 1$ and $J = 1$ and 2 the factor

---

[18] The meaning of χ and φ has been interchanged here as compared to formulae (I, 26) and (IV, 51) of Volume II for the sake of consistency with the previous equation (III, 11).

$\Theta_{J\bar{\Lambda}M}(\vartheta)$ of ψ_r is found to be (including the normalization factor)

$$\Theta_{1+10}(\vartheta) = \Theta_{1-10}(\vartheta) = \frac{1}{4\pi}\sqrt{3}\,\sin\vartheta,$$

$$\Theta_{1+1\pm1}(\vartheta) = \Theta_{1-1\mp1}(\vartheta) = \frac{1}{8\pi}\sqrt{6}\,(1\pm\cos\vartheta),$$

$$\Theta_{2+10}(\vartheta) = \Theta_{2-10}(\vartheta) = \frac{1}{4\pi}\sqrt{15}\,\sin\vartheta\,\cos\vartheta, \tag{III, 152}$$

$$\Theta_{2+1\pm1}(\vartheta) = \Theta_{2-1\mp1}(\vartheta) = \frac{1}{8\pi}\sqrt{10}\,(\mp1+\cos\vartheta\pm2\cos^2\vartheta),$$

$$\Theta_{2+1\pm2}(\vartheta) = \Theta_{2-1\mp2}(\vartheta) = \frac{1}{8\pi}\sqrt{10}\,\sin\vartheta\,(1\pm\cos\vartheta).$$

It may be noted that $\Theta_{J\bar{\Lambda}M} = \Theta_{J-\bar{\Lambda}-M}$; but for a given J, Λ, and $M \neq 0$ the two functions corresponding to the two orientations of $\Lambda(\bar{\Lambda} = \pm\Lambda)$ are not the same. In Fig. 57 the functions (III, 152) and similar functions for $J = 3$ are represented graphically. They should be compared with the broken-line curves in Fig. 40 which hold for $\Lambda = 0$.[19]

Infrared spectrum. The derivation of the *selection rules* from the eigenfunctions of the symmetric top given above is rather involved [see Dennison (887), Reiche and Rademacher (1345) and Kronig and Rabi (1145)]. Different results are obtained depending on whether $\Lambda = 0$ or $\Lambda \neq 0$. For the present discussion we assume that Λ does not alter in the transition; that is, we exclude electronic transitions, which will be dealt with in the following chapters. Then

$$\text{for } \Lambda = 0, \qquad \Delta J = \pm1, \tag{III, 153}$$

and

$$\text{for } \Lambda \neq 0, \qquad \Delta J = 0, \pm1 \tag{III, 154}$$

Thus, in the first case ($\Lambda = 0$), everything is just as it was previously, when we were considering the simple rotator, since the term $(A - B)\Lambda^2$ in (III, 147) disappears and the selection rule is the same as before. We obtain exactly the same two branches of rotation-vibration bands and thus we can treat this case *as though the moment of inertia about the line joining the nuclei were zero*.

In the second case ($\Lambda \neq 0$), the energy relations are the same, except for a constant shift. However, in addition to the transitions with $\Delta J = \pm1$, there also appear transitions with $\Delta J = 0$—that is, a series of lines given by

$$\nu = \nu_0 + F'(J) - F''(J) \tag{III, 155}$$

and called a Q *branch*. Substitution of (III, 147), with B_v instead of B, gives for this branch

$$\nu_Q = \nu_0 + (B_v'' - B_v')\Lambda^2 + (B_v' - B_v'')J + (B_v' - B_v'')J^2. \tag{III, 156}$$

---

[19] Unlike the case of the simple rotator (Fig. 40) there would be no meaning here in plotting to the left of the z-axis the curves for $\varphi = \pi$ (for $\vartheta \to 0$ the values $\varphi = 0$ and $\varphi = \pi$ give in general values of the eigenfunction of opposite sign).

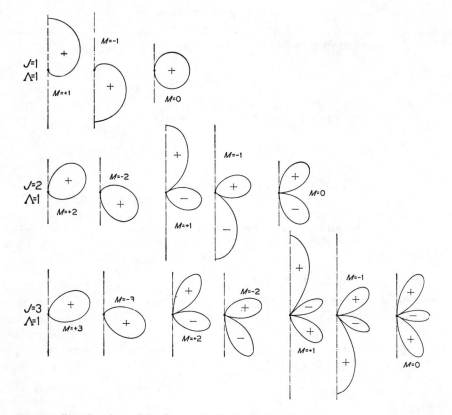

Fig. 57.   **Eigenfunctions of the Symmetric Top in the Levels $\Lambda = 1$, $J = 1, 2, 3$.**   The eigenfunctions are plotted in polar diagrams (compare Fig. 40).   The polar angle is ϑ, the angle of the figure axis of the top with the z axis, which is assumed to be vertical.   The dependence on the azimuthal angle φ and the angle of rotation χ about the figure axis is given by $e^{iM\varphi}$ and $e^{i\Lambda\chi}$ respectively but is not shown in the figure.   For a 180° rotation about the z axis the same function is obtained except that for odd M the sign (indicated in each loop) is reversed.   Similarly for a 180° rotation about the figure axis for $\Lambda = 1$ the same function is obtained with opposite sign.

The term $(B_v'' - B_v')\Lambda^2$ must be added also to the formulae for R and P branches which are otherwise the same as previously [equations (III, 133), (III, 134) and (III, 136)].   All the lines of the Q branch fall very near to the position ν_0, since B_v' is very nearly equal to B_v'' in infrared spectra.   Thus a relatively intense *"line"* would be expected at the position where, for $\Lambda = 0$, the zero gap appears.   Further, owing to the restriction of the J values by (III, 146), one or more of the lines at the beginning of the P and the R branches are missing (for further details see pp. 170, 256, 266, and 273).

A comparison of the observed infrared spectra with the theoretical spectrum of the symmetric top shows that in almost all cases there is complete agreement with the first case, $\Lambda = 0$.   That is to say, the model of the ordinary simple rotator reproduces these spectra in all details.   From this we can conclude conversely that, for the molecules for which such simple infrared spectra are observed, we have $\Lambda = 0$ in the electronic ground state; that is, the motion of the electrons is such that no angular momentum about the internuclear axis results.

The only case, thus far, of a diatomic molecule for which a Q branch (that is, a central "line") has been found in the infrared bands is that of nitric oxide (NO) [see Gillette and Eyster (986)]. In this case, therefore, we must conclude that there is an angular momentum of the electrons about the internuclear axis ($\Lambda \neq 0$), a conclusion that is confirmed by the investigation of the electronic band spectra of this molecule. However, on account of electron spin, a doubling arises which will be taken up in more detail in Chapter V.

As can be seen immediately from the selection rules the pure *rotation spectrum* of the symmetric top is the same as that of the simple rotator independent of whether $\Lambda = 0$ or $\neq 0$; but no case with $\Lambda \neq 0$ has as yet been investigated.

Raman spectrum. The selection rules for the Raman spectrum of the symmetric top are found to be the following [see Placzek (562)]:

$$\text{for } \Lambda = 0: \qquad \Delta J = 0, \pm 2, \tag{III, 157}$$

$$\text{for } \Lambda \neq 0: \qquad \Delta J = 0, \pm 1, \pm 2. \tag{III, 158}$$

Thus again in the first case ($\Lambda = 0$) everything is the same as for the simple rotator (or vibrating rotator). The Raman spectrum consists of S, O, and Q branches. However, in the second case ($\Lambda \neq 0$), in addition, an R branch ($\Delta J = +1$) and a P branch ($\Delta J = -1$) appear. As mentioned previously the only cases in which Raman vibration bands have been resolved are those of H_2 and HCl where no R and P branches appear, that is, for which $\Lambda = 0$.

In the case of the rotational Raman spectrum, where only positive values of ΔJ are possible [see section 1(d) of this chapter], we expect, in addition to the Stokes and anti-Stokes S branches ($\Delta J = +2$), a Stokes and an anti-Stokes R branch ($\Delta J = +1$). These R branches have half the spacing of the S branches (that is, $2B$ rather than $4B$). Every alternate line of the R branches coincides with a line of the S branches. Thus far such a case has not been observed for diatomic molecules. In the case of NO which does have $\Lambda \neq 0$, as shown by the infrared spectrum (see above), the rotational Raman spectrum is more complicated because of the spin doubling. Since the J values are half-integral (see Chapter V) alternate lines of the R branches do not coincide with the lines of the S branches giving a more complex appearance of the Raman spectrum. In addition transitions from one doublet component to the other occur. A partial resolution of this spectrum has been obtained by Rasetti (1341).

(e) Thermal Distribution of Quantum States; Intensities in Rotation-Vibration Spectra

As we have seen in Chapter I the intensity of a spectral line depends not only on the transition probability and the frequency ν but also on the number of molecules in the initial state. Thus for a theoretical prediction of intensities, a knowledge of the numbers of molecules in the various initial states is necessary, in addition to a knowledge of the transition probabilities. Since almost all infrared and Raman spectra are observed under conditions of *thermal equilibrium* we need only consider the distribution of the molecules over the different quantum states in thermal equilibrium.   This distribution is of importance also for an understanding of the thermodynamic properties of diatomic gases (see Chapter VIII, section 2).

Vibration.   According to the *Maxwell-Boltzmann distribution law*, the number of molecules dN_E that have a classical vibrational energy between E and $E + dE$ is proportional to $e^{-E/kT}dE$, where k is the gas constant per molecule (that is, Boltzmann's constant) and T is the absolute temperature.   The

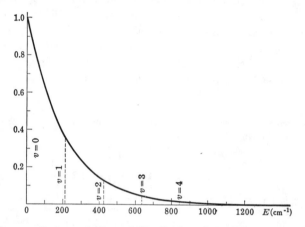

Fig. 58.   **Boltzmann Factor and Thermal Distribution of the Vibrational Levels.**   The curve gives the function $e^{-E/kT}$ for $T = 300°$ K. with E in cm^{-1}.   The broken-line ordinates correspond to the vibrational levels of the I_2 molecule.

function $e^{-E/kT}(= e^{-E/0.6952T}$, when E is expressed in cm^{-1}) is represented graphically in Fig. 58 for $T = 300°$ K. (room temperature).   Classically, there is no restriction for the E values, and therefore, according to Fig. 58, at 300° K most molecules are vibrating, even though with small energy (amplitude).

In quantum theory, only the discrete values (III, 76) are possible for the vibrational energy.   The number of molecules in each of the vibrational states is again proportional to the Boltzmann factor $e^{-E/kT} = e^{-G_0(v)hc/kT} = e^{-G_0(v)/0.6952T}$.   The zero-point energy can be left out, since to add this to the exponent would mean only adding a factor that is constant for all the vibrational levels (including the zero level).

The ordinates corresponding to the discrete values of the vibrational energy for the case of I_2 are indicated by broken lines in Fig. 58 [$G_0(v) = 213.76v - 0.596v^2$].   It is seen from this figure that the number of molecules in the higher vibrational levels falls off very rapidly.   The falling off is even more pronounced for molecules such as H_2, N_2, O_2, and so forth, which have larger vibrational quanta (see abscissa scale in Fig. 58).   In order to give some idea of the quantitative relationships, the factor $e^{-\Delta G_{1/2}hc/kT}$—that is, the ratio of the number of molecules in the first to that in the zeroth vibrational state—is given in Table 14 for a number of gases for 300° K. and 1000° K.   It is seen that this fraction is very small for gases such as H_2, HCl, CO, and N_2, even at 1000° K.   Accordingly practically all the transitions observed in infrared absorption or in the Raman effect have $v = 0$ in the initial state.   The absence of anti-Stokes vibrational lines in the observed Raman spectra of diatomic gases (see p. 87) is thus immediately understood.   It will be remembered that according to classical theory, they should have the same intensity as the corresponding Stokes lines.

TABLE 14.   RATIO OF THE NUMBER OF MOLECULES IN THE FIRST TO THAT IN THE ZEROTH VIBRATIONAL LEVEL FOR 300° K. AND 1000° K.

Gas	$\Delta G_{\frac{1}{2}}(\text{cm}^{-1})$	$e^{-\Delta G_{\frac{1}{2}}hc/kT}$	
		For 300° K.	For 1000° K.
H_2	4160.2	2.16×10^{-9}	2.51×10^{-3}
HCl	2885.9	9.77×10^{-7}	1.57×10^{-2}
N_2	2330.7	1.40×10^{-5}	3.50×10^{-2}
CO	2143.2	3.43×10^{-5}	4.58×10^{-2}
O_2	1556.4	5.74×10^{-4}	1.07×10^{-1}
S_2	721.6	3.14×10^{-2}	3.54×10^{-1}
Cl_2	556.9	6.92×10^{-2}	4.49×10^{-1}
I_2	213.2	3.60×10^{-1}	7.36×10^{-1}

A similar striking difference between the results of classical and quantum theory is found for the *vibrational part of the heat content of a diatomic gas*.   Classically all molecules would have a small amount of vibrational energy resulting in a large contribution to the heat content (corresponding to the area under the curve in Fig. 58).   Table 14 shows clearly that for most molecules the contribution of the higher vibrational levels to the heat content at room temperature is quite negligible, a result that is completely verified by experiment (see Chapter VIII, section 2).   It is only for heavy molecules, such as Cl_2 and I_2 that an appreciable fraction of the molecules is in the first vibrational level at room temperature and consequently there is an appreciable vibrational contribution to the heat content.

The quantities $e^{-G_0(v)hc/kT}$ give the relative numbers of molecules in the different vibrational levels referred to the number of molecules in the lowest vibrational level.   If we wish to refer to the total number of molecules N we have to consider that N is proportional, with the same factor of proportionality as before, to the sum of the Boltzmann factors over all states, the so-called *state sum* (or *partition function*):

$$Q_v = 1 + e^{-G_0(1)hc/kT} + e^{-G_0(2)hc/kT} + \cdots \qquad \text{(III, 159)}$$

Therefore the number of molecules in the state v is

$$N_v = \frac{N}{Q_v} e^{-G_0(v)hc/kT} \qquad \text{(III, 160)}$$

Successive terms in (III, 159) decrease very rapidly. If the second term is small compared to 1, the number of molecules, N_v, for most practical purposes, can be put equal to $Ne^{-G_0(v)hc/kT}$. Thus those figures in Table 14 that are much smaller than 1 give directly the fraction of the molecules in the first vibrational state.

Rotation. The thermal distribution of the rotational levels (unlike that of the vibrational levels) is not simply given by the Boltzmann factor $e^{-E/kT}$; we have to allow for the fact that, according to quantum theory, each state of an atomic system with total angular momentum J consists of $2J + 1$ levels which coincide in the absence of an external field; that is, the state has a $(2J + 1)$-*fold degeneracy* (see p. 26). The frequency of its occurrence (its *statistical weight*) is therefore $(2J + 1)$ times that of a state with $J = 0$.

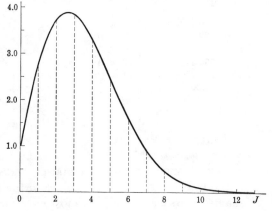

Fig. 59.   **Thermal Distribution of the Rotational Levels for T = 300° K. and B = 10.44 cm^{-1}** (That is, for HCl in the Ground State).   The curve represents the function $(2J + 1)e^{-BJ(J+1)hc/kT}$ as a function of J. The broken-line ordinates give the relative populations of the corresponding rotational levels.

The *number of molecules N_J in the rotational level J of the lowest vibrational state* at the temperature T is thus proportional to

$$(2J + 1)e^{-F(J)hc/kT}. \tag{III, 161}$$

For most practical cases ($\Lambda = 0$, rigid rotator)

$$N_J \propto (2J + 1)e^{-BJ(J+1)hc/kT}. \tag{III, 162}$$

This function is represented in Fig. 59 for $B = 10.44$ cm^{-1} (that is, for HCl) and $T = 300°$ K. Since the factor $2J + 1$ increases linearly with J, the number of molecules in the different rotational levels does not from the beginning decrease with the rotational quantum number but first goes through a *maximum*. It is easily seen that this maximum lies at

$$J_{\max.} = \sqrt{\frac{kT}{2Bhc}} - \frac{1}{2} = 0.5896\sqrt{\frac{T}{B}} - \frac{1}{2} \tag{III, 163}$$

--that is, at a value of J that increases with decreasing B and increasing T. It should be noted that the number of molecules in the lowest rotational level, $J = 0$, is not zero.

Just as for the vibration, the actual number of molecules in the rotational states is obtained by multiplying by N and dividing by the rotational state sum

$$Q_r = 1 + 3e^{-2Bhc/kT} + 5e^{-6Bhc/kT} + \cdots ; \qquad \text{(III, 164)}$$

that is,

$$N_J = \frac{N}{Q_r} (2J + 1)e^{-BJ(J+1)hc/kT} \qquad \text{(III, 165)}$$

For sufficiently large T or small B, the sum in (III, 164) can be replaced by an integral—namely,

$$Q_r \approx \int_0^\infty (2J + 1)e^{-hcBJ(J+1)/kT}\, dJ = \frac{kT}{hcB} \qquad \text{(III, 166)}$$

—so that we have

$$N_J = N \frac{hcB}{kT} (2J + 1)e^{-BJ(J+1)hc/kT}. \qquad \text{(III, 167)}$$

In the case of HCl at $300°$ K. (Fig. 59) kT/hcB is 19.98 while the exact value of the state sum is 20.39. If the ordinates in Fig. 59 are divided by this number they give directly the fractional number of molecules in the particular rotational levels. The difference between approximate and exact value of the state sum decreases rapidly for larger T or smaller B values.

For the higher vibrational levels we have

$$N_J \propto (2J + 1)e^{-(G+F)hc/kT} \qquad \text{(III, 168)}$$

instead of (III, 161). However, the factor $e^{-Ghc/kT}$ can be separated off; that is, the distribution over the rotational levels is the same, but the absolute population of all the levels is considerably smaller than for the lowest vibrational level, corresponding to the factor $e^{-Ghc/kT}$.

The *variation of the intensity of the lines* in a rotation-vibration band as a function of J is given essentially by the thermal distribution of the rotational levels; that is, to a first approximation, the intensity is proportional to the expression (III, 165). In this approximation it is assumed that the transition probability is the same for all lines of a band. Actually there is a slight dependence on J and ΔJ (see below) which, in the case of $\Lambda = 0$ (that is, when only P and R branch appear) can be taken into account by using in (III, 165) $J' + J'' + 1$ in place of $2J + 1$; that is, the intensity depends on the mean value of $2J + 1$ for the upper and lower states. Furthermore it should be noted that the J value of the initial state must be used in the exponential term—that is, in absorption J'', in emission J'. Including the dependence on ν according to (I, 53) and (I, 48) one obtains in this way for the intensities of the lines of rotation or rotation-vibration bands *in absorption*

$$I_{\text{abs.}} = \frac{C_{\text{abs.}}\nu}{Q_r}(J' + J'' + 1)e^{-B''J''(J''+1)hc/kT} \tag{III, 169}$$

and *in emission*

$$I_{\text{em.}} = \frac{C_{\text{em.}}\nu^4}{Q_r}(J' + J'' + 1)\,e^{-B'J'(J'+1)hc/kT} \tag{III, 170}$$

Here $C_{\text{abs.}}$ and $C_{\text{em.}}$ are constants depending on the change of dipole moment and the total number of molecules in the initial vibrational level.

For a given rotation-vibration band at a given temperature the factors $C_{\text{abs.}}\nu/Q_r$ and $C_{\text{em.}}\nu^4/Q_r$ in (III, 169) and (III, 170) are very nearly constants.

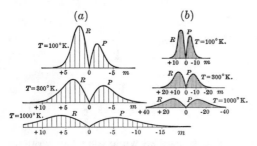

Therefore the intensity distribution resembles closely the distribution of the rotational levels (Fig. 59), a maximum of intensity occurring in each branch at about the same J value [compare (III, 163)].

To illustrate, in Fig. 60(*a*) the theoretical intensity distribution is represented for a rotation-vibration band of HCl in absorption at 100° K., 300° K., and 1000°K. The abscissae give the positions of the lines assuming that $B' = B''$; the ordinates are proportional to the intensities with the same proportionality factor

FIG. 60. **Intensity Distribution in Rotation-Vibration Bands in Absorption at 100° K., 300° K., and 1000° K.** (*a*) For $B = 10.44$ cm^{-1} (HCl). (*b*) For $B = 2$ cm^{-1}. The wave number scale of the abscissae, which is not explicitly given, is the same for all the diagrams. The lines are drawn with the separation that they would have if the constant B were the same in the upper and lower states. m is the running number of the lines (see p. 111). Note that longer wave lengths are to the right in this figure.

throughout. It is seen from Fig. 60(*a*) that with increasing temperature the band extends farther and that the intensity maxima of the two branches move outward and at the same time become flatter. The decrease of the height of the maxima with increasing temperature (for a given number of absorbing molecules) is due to the increase of the state sum Q_r in the denominator of the expression (III, 169). Naturally the total intensity of the band remains constant as long as the temperature is not so high that the number of molecules in the lower vibrational level ($v = 0$) is appreciably reduced. A comparison of the curve for 300°K. with the observed spectrum (Fig. 32 or 33) shows that the agreement between theory and experiment is very good.

Fig. 60(*b*) illustrates the same relations for the case of a molecule with $B = 2$ cm^{-1}. The maxima lie here at greater J values, but in spite of this the separation of the maxima in cm^{-1} is less than for HCl with $B = 10.44$ cm^{-1}.

If in an *unresolved band* the maxima of P and R branch can still be recognized and their separation measured, a rough value for the rotational constant B may be obtained if the temperature is known. From equation (III, 163) [or similar equations for P and R branch obtained from (III, 169)] one finds easily that the separation of the two maxima $\Delta\nu_{PR}{}^{\text{max.}}$ in cm^{-1} is given by

$$\Delta\nu_{PR}{}^{\text{max.}} = \sqrt{8BkT/hc} = 2.3583\sqrt{BT} \qquad \text{(III, 171)}$$

In the early work of Burmeister (148) the separations for HCl (compare Fig. 30) and CO at room temperature were found to be 124 and 55 cm^{-1}, respectively. From these one obtains using (III, 171) $B_{\text{HCl}} = 9.2$ and $B_{\text{CO}} = 1.81$ which may be compared with the more accurate values 10.44 and 1.922 cm^{-1} obtained from rotational analyses. The moments of inertia obtained from the former B values were among the first that were determined spectroscopically.

The formulae (III, 169) and (III, 170) apply also to the *pure rotation spectrum*. In this case, however, the factors ν and ν^4, respectively, have a much more pronounced effect since they change appreciably from line to line. The intensity maximum is shifted to higher J values. One finds easily that in a first approximation $J_{\text{max.}}$ is obtained from (III, 163) if T/B under the square root is replaced by $2T/B$ for absorption and $5T/B$ for emission. As an example, for HCl at room temperature one finds (using the rigorous formulae) in absorption $J_{\text{max.}}{}^{\text{abs.}} = 3.73$, in emission $J_{\text{max.}}{}^{\text{em.}} = 6.32$ while the maximum of the thermal distribution (Fig. 59) is at $J_{\text{max.}} = 2.66$. It may be recalled (p. 59) that in emission the HCl rotation spectrum has been observed to rather high J values.

The intensity distribution in rotation and rotation-vibration *Raman bands* is very similar to that for infrared rotation-vibration bands [Fig. 60 and formulae (III, 169) and (III, 170)] although there are certain slight differences (see below).

According to the discussion in Chapter I [equations (I, 46) to (I, 56)] the intensity of an emission line in the infrared rotation or rotation-vibration spectrum may be written, taking account of (III, 165) and remembering that $d_n = 2J' + 1$ and $d_m = 2J'' + 1$,

$$I_{\text{em.}} = \frac{64\pi^4\nu^4cN}{3Q_r}\left(\sum |R^{n_im_k}|^2\right)e^{\frac{-B'J'(J'+1)hc}{kT}}$$

$$= \frac{2C_{\text{em.}}\nu^4}{Q_r}S_J\,e^{\frac{-B'J'(J'+1)hc}{kT}} \qquad \text{(III, 172)}$$

and similarly for absorption:

$$I_{\text{abs.}} = I_0\frac{8\pi^3\nu N}{3hcQ_r}\left(\sum |R^{n_im_k}|^2\right)e^{\frac{-B''J''(J''+1)hc}{kT}}$$

$$= \frac{2C_{\text{abs.}}\nu}{Q_r}S_J\,e^{\frac{-B''J''(J''+1)hc}{kT}} \qquad \text{(III, 173)}$$

Here, the *line strength* S_J is that part of $\sum |R^{n_im_k}|^2$ that depends on J [see Condon and Shortley (10)]. A constant factor corresponding to the magnitude of the dipole moment or its change is included in $C_{\text{em.}}$ or $C_{\text{abs.}}$, respectively. By evaluating the matrix elements $R^{n_im_k}$ one obtains [see for example Sommerfeld (11b)] in the case of the rotator $(\Lambda = 0)$

$$\text{for } \Delta J = +1: \quad S_J = J + 1 \qquad \text{(III, 174)}$$

$$\text{for } \Delta J = -1: \quad S_J = J \qquad \text{(III, 175)}$$

If these expressions are substituted in (III, 172) and (III, 173) the previous formulae (III, 169) and (III, 170) are immediately obtained. If $\Lambda \neq 0$, other formulae hold which will be given later in the discussion of intensities in electronic bands (p. 208, Hönl-London formulae).

Apart from a constant factor the formula (III, 172) holds also for the Raman spectrum. However, the intensity factors are different. According to Placzek and Teller (1319) one has for $\Lambda = 0$

$$\text{for } \Delta J = +2: \quad S_J = \frac{3(J+1)(J+2)}{2(2J+3)} \tag{III, 176}$$

$$\text{for } \Delta J = -2: \quad S_J = \frac{3(J-1)J}{2(2J-1)} \tag{III, 177}$$

$$\text{for } \Delta J = \quad 0: \quad S_J = a_0(2J+1) + \frac{J(J+1)(2J+1)}{(2J-1)(2J+3)} \tag{III, 178}$$

where a_0 is a constant corresponding to the so-called trace-scattering.

(f) Symmetry Properties of the Rotational Levels

Positive and negative rotational levels. An important property of the eigenfunctions of any atomic system is their behavior with respect to a *reflection at the origin* (replacement of all x_k, y_k, z_k by $-x_k$, $-y_k$, $-z_k$). As was shown previously (p. 24) the eigenfunctions either remain unchanged or just change sign for such an *inversion*. In the case of the rigid rotator a reflection at the origin is obtained by replacing ϑ by $\pi - \vartheta$ and φ by $\pi + \varphi$. A closer investigation of the rotator functions (III, 11) shows that for such a reflection they remain unchanged for even values of J but change to $-\psi_r$ for odd values of J. This is easily verified for the functions given explicitly on p. 70 and illustrated in Fig. 40 (broken curves). In this figure in two opposing directions, the magnitude of the eigenfunction (given by the distance of the curve from the origin) is the same. However, the sign is the same only for even J values; for odd J values it is opposite.

In the case of the symmetric top, when $\Lambda \neq 0$, each rotational level is doubly degenerate [in addition to the $(2J + 1)$-fold space degeneracy], that is, has two linearly independent eigenfunctions. It can be shown (see below) that these two functions can always be chosen so that one remains unchanged while the other changes sign for a reflection at the origin.

Reflection at the origin is obtained in the case of the symmetric top by replacing ϑ by $\pi - \vartheta$, φ by $\pi + \varphi$, and χ by $-\chi$. For such a substitution the eigenfunctions (III, 151) do not remain unchanged or only change sign. However, it is easily seen that one of the linear combinations

$$\begin{aligned}
\psi_r^+ &= \Theta_{J\Lambda M}(\vartheta)e^{i\Lambda\chi}e^{iM\varphi} + \Theta_{J-\Lambda M}(\vartheta)e^{-i\Lambda\chi}e^{iM\varphi} \\
\psi_r^- &= \Theta_{J\Lambda M}(\vartheta)e^{i\Lambda\chi}e^{iM\varphi} - \Theta_{J-\Lambda M}(\vartheta)e^{-i\Lambda\chi}e^{iM\varphi}
\end{aligned} \tag{III, 179}$$

remains unchanged while the other changes sign upon reflection at the origin. It will be seen that each one of these functions contains contributions of both directions of Λ.

The rotational levels of a diatomic molecule are classified according to the behavior of the total eigenfunction (not of the rotational eigenfunction alone) with respect to reflection at the origin. *A rotational level is called positive or negative depending on whether the total eigenfunction remains unchanged or changes sign for such a reflection.* The property positive or negative is also called *parity* and corresponds to the property even or odd of atomic energy levels.

As will be shown later in more detail (p. 149) the complete eigenfunction of a molecule, to a first approximation, is a product of an electronic, a vibrational, and a rotational contribution

$$\psi = \psi_e \frac{1}{r} \psi_v \psi_r. \tag{III, 180}$$

The vibrational contribution $(1/r)\psi_v$ always remains unchanged by reflection at the origin since it depends only on the magnitude of the internuclear distance. If $\Lambda = 0$ and ψ_e remains unchanged by a reflection at the origin the parity of the rotational levels depends on the rotational eigenfunction ψ_r only; that is, *the rotational levels are positive or negative according as J is even or odd.* This is illustrated in Fig. 61(a). However, it may happen, as will be further discussed in Chapter V, that ψ_e changes sign by reflection at the origin and that, as a result, the total eigenfunction (III, 180) remains unaltered for odd J and changes sign for even J. In this case the $+$ and $-$ have to be exchanged in Fig. 61(a).

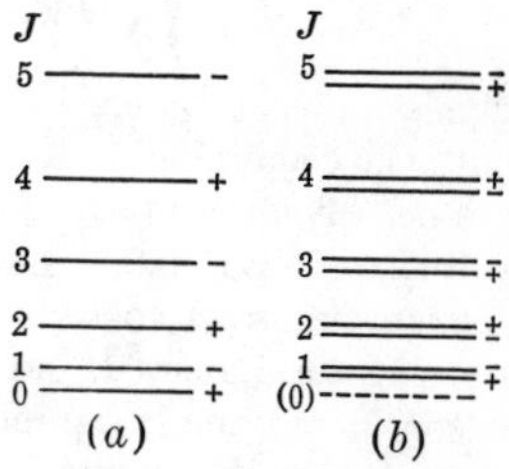

If $\Lambda \neq 0$ we have a symmetric top and therefore according to the above *for each value of J there is a positive and a negative rotational level* of equal energy. Fig. 61(b) shows an energy level diagram for $\Lambda = 1$ in which the designations $+$ and $-$ are added. The degenerate levels are drawn slightly separated. Actually, a splitting of the degenerate levels of the type indicated in the figure does appear when account is taken of the interaction of electronic motion and rotation (see Chapter V).

Fig. 61. **Symmetry Properties of the Levels** (a) **of the Rotator and** (b) **of the Symmetric Top.** For the symmetric top, $\Lambda = 1$ is assumed. The dotted level with $J = 0$ therefore does not occur.

It must be emphasized that the distinction of positive and negative rotational levels is quite independent of how closely (III, 180) approximates the complete eigenfunction of the molecule since this distinction is introduced solely by the equivalence of left and right-handed coordinate systems. The approximation (III, 180) is used only for the purpose of determining which levels are positive and which negative.

As will be shown below the following important *selection rule* holds for transitions accompanied by dipole radiation: *Positive levels combine only with negative, and vice versa.* Transitions between two positive levels or between two negative levels are forbidden. This selection rule may be written symbolically:

$$+ \rightarrow -, \quad - \rightarrow +, \quad + \nrightarrow +, \quad - \nrightarrow -. \tag{III, 181}$$

It can be seen immediately from Fig. 61(a) and (b) that this selection rule does not contradict the previous selection rules $\Delta J = \pm 1$ for the simple rotator, or $\Delta J = 0, \pm 1$ for the symmetric top with $\Lambda \neq 0$.

The opposite selection rule holds for transitions taking place in the *Raman effect*; namely, positive terms combine only with positive and negative only with negative, or symbolically:

$$+ \rightarrow +, \quad - \rightarrow -, \quad + \nrightarrow -, \quad - \nrightarrow +. \qquad (\text{III, 182})$$

Again it is easily verified that this rule does not contradict the previous selection rules $\Delta J = 0, \pm 2$ for the Raman spectrum of the simple rotator or $\Delta J = 0, \pm 1, \pm 2$ for that of a symmetric top with $\Lambda \neq 0$.

The selection rules for dipole radiation are obtained by calculating the matrix elements $\int \psi_n{}^* M_x \psi_m \, d\tau$, $\int \psi_n{}^* M_y \psi_m \, d\tau$, and $\int \psi_n{}^* M_z \psi_m \, d\tau$. The components M_x, M_y, M_z of the dipole moment [given by (I, 43)] change sign upon reflection at the origin. Therefore, if both the combining levels (with eigenfunctions ψ_n and ψ_m, respectively) are positive or if both are negative, the integrands will change sign for a reflection at the origin; that is, the integrals will also change sign. Since, however, the value of a definite integral is independent of any transformation of coordinates, it follows that the above integrals are zero; that is, such transitions are forbidden. If, on the other hand, a transition between a positive and a negative level is considered, the integrands remain unchanged for a reflection at the origin and consequently the values of the integrals may well be different from zero; that is, transitions between a positive and a negative level are allowed. This selection rule is completely analogous to the Laporte rule for atoms (see A.A., p. 154).

In the Raman effect, the transition probability depends on $\int \psi_n{}^* \, \alpha \, \psi_m \, d\tau$, where α is the polarizability (see p. 86), which has the same value for two opposite directions. Consequently, in this case the integrand remains unchanged for a reflection at the origin if ψ_n and ψ_m have the same symmetry—that is, are both positive or both negative. Thus the selection rule (III, 182) is obtained. The same rule also holds for quadrupole and magnetic dipole radiation (see p. 277).

Symmetric and antisymmetric rotational levels for homonuclear molecules.

When two identical nuclei are present in a molecule—that is, if we have a homonuclear molecule ($H_2{}^1$, $C_2{}^{12}$, $O_2{}^{16}$, and so forth) the wave equation of the system remains unchanged if the two nuclei are exchanged, that is, if the coordinates of the first nucleus x_1, y_1, z_1 are exchanged with those of the second nucleus, x_2, y_2, z_2 wherever they occur. Therefore it follows in the same way as previously shown for the exchange of two electrons (p. 24) that *for an exchange of the nuclei the total eigenfunction either remains unchanged or only changes its sign.* According as the former or the latter is the case, the state under consideration is said to be *symmetric or antisymmetric in the nuclei.*

It will be shown later (see p. 238 f.) that in a given electronic state of a molecule (in our present considerations we are always dealing with the electronic ground state) *either the positive rotational levels are symmetric and the negative are antisymmetric throughout, or, the positive are antisymmetric and the negative are symmetric throughout.* We shall limit ourselves in this chapter to the consideration of the case $\Lambda = 0$, since it is practically the only one occurring in rotation and rotation-vibration spectra. (For other cases, see Chapter V.) For $\Lambda = 0$, according to the above, either the even-numbered levels are symmetric (s) and the odd are antisymmetric (a), or the reverse is the case. These two alternatives are shown in Fig. 62(a) and (b) for electronic states whose ψ_e remain unchanged upon reflection at a plane through the internuclear axis. [In Chapter V, section 2(c) these states will be classified as Σ_g^+ and Σ_u^+.] For electronic states whose ψ_e change sign upon such a reflection (Σ_g^- and Σ_u^-) both designations $+$, $-$ and s, a have to be reversed in Fig. 62.

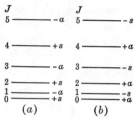

If the identical nuclei have zero nuclear spin (or if the interaction of the nuclear spin with the rest of the molecule is neglected) a simple consideration shows (see below) that there is a perfectly rigorous *prohibition of intercombinations* between the symmetric and antisymmetric states:

$$antisym. \leftrightarrow sym. \qquad (III, 183)$$

Fig. 62. **Symmetry Properties of the Rotational Levels of Homonuclear Molecules with $\Lambda = 0$.** In (a) the positive levels are symmetric (s) and the negative antisymmetric (a); in (b) the reverse is the case.

This prohibition holds *not only for transitions with radiation but also for transitions brought about in any other way* (collisions, Raman effect, and so forth).

The proof of the selection rule (III, 183) is similar to that of the rules (III, 181) and (III, 182). The components of the dipole moment $M_x = \sum e_i x_i$, $M_y = \sum e_i y_i$, $M_z = \sum e_i z_i$ remain unchanged if the two identical nuclei are exchanged. Therefore if ψ_n refers to a symmetric, ψ_m to an antisymmetric state the integrals $\int \psi_n^* M_x \psi_m \, d\tau, \cdots$ which determine the transition probability for dipole radiation, change sign for an exchange of the nuclei. Since the values of the integrals cannot depend on the mode of designating the nuclei, it follows that they are zero—that is, that the transition is forbidden. The same result is obtained for any other radiation (quadrupole radiation, magnetic dipole radiation, and so forth), since the expressions that then appear in place of $\sum e_i x_i$ are also symmetric in the two nuclei. This holds even for transitions brought about by collisions, since the interaction between the molecule and any other particle is necessarily symmetric in the two nuclei. Thus, the selection rule (III, 183) holds perfectly rigorously.

An immediate consequence of the selection rule (III, 183) is the absence of ordinary infrared rotation or rotation-vibration spectra of homonuclear molecules (which was previously derived from the vanishing dipole moment): Any two rotational levels for which the selection rule $\Delta J = \pm 1$ is fulfilled have opposite symmetry in the nuclei (see Fig. 62) and therefore cannot combine. On the other hand, in the Raman effect transitions do take place for homonuclear molecules since levels for which $\Delta J = 0, \pm 2$ is fulfilled have the same symmetry in the nuclei.

For quadrupole radiation the selection rule $\Delta J = 0, \pm 1, \pm 2$ holds [see Chapter V, section 3(c)]. Therefore a *quadrupole rotation or rotation-vibration spectrum* with $\Delta J = 0, \pm 2$ may occur without violation of the selection rule (III, 183). However, the intensity of such infrared rotation or rotation-vibration spectra of homonuclear molecules is only 10^{-9} of ordinary dipole infrared spectra of heteronuclear molecules [see Herzberg (315) and James and Coolidge (362)]. One such case has recently been observed [Herzberg (1045)].

On the basis of the selection rule (III, 183) if all the molecules of one kind (for example, all the O_2 molecules) were in symmetric states at the beginning of time (or all in antisymmetric states), they would still be in the same kind of state at the present time.   That is, it would be as if the antisymmetric (or symmetric) states had never existed.   For $\Lambda = 0$, these are either all the even-numbered or all the odd-numbered rotational levels (see Fig. 62).   If this were the case, *every second line should be missing* in the rotational Raman bands (and also in the electronic bands; see Chapter V).   This is actually observed in the Raman spectrum of O_2, as well as in the electronic band spectra of O_2, C_2, He_2, S_2, and Se_2.

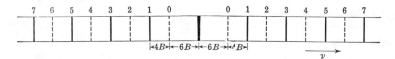

Fig. 63.   **Rotational Raman Spectrum of a Molecule with Identical Nuclei (Schematic).**   The heavy line in the center gives the undisplaced. exciting line.

It might at first be thought difficult to decide whether every second line were missing or whether the rotational lines could be represented, by a suitable choice of the rotational constant B, without the assumption of missing lines.   In order to see that an unambiguous decision is always possible, let us consider Fig. 63, in which a rotational Raman spectrum is schematically represented, the Raman lines with even-numbered lower levels (and therefore also even-numbered upper levels) being indicated by broken lines.   If no line is missing, according to (III, 66) the separation between the first line of the long-wave-length branch and the first line of the short-wave-length branch is $12B$, while the separation of successive lines in both branches is $4B$.   If the even-numbered rotational levels are missing, all the lines with even-numbered J (broken lines in Fig. 63) must be missing.   As a result, the separation of the first line of the long-wave-length branch from that of the short-wave-length branch would be $2(6 + 4)B = 20B$, while the line separation would be $8B$.   If the odd-numbered lines were missing, the separation of the first line of the long-wave-length branch from the first line of the short-wave-length branch would be $12B$, while the line separation would be again $8B$.   Thus the separation of the first long-wave-length from the first short-wave-length Raman line is to the separations of successive lines in the ratios $6:2$, $5:2$, and $3:2$, respectively, in the three cases.   Therefore, even without any knowledge of the constant B it is possible to decide unambiguously with which case we are dealing.   A measurement of the O_2 Raman spectrogram in Fig. 37(b) shows that the above-mentioned ratio is $5:2$.   It follows that the even-numbered levels are missing.   As will be shown in more detail in Chapter V (p. 251) these missing levels are the antisymmetric ones as in Fig. 62(b) (except that the parity is reversed).

Influence of nuclear spin. Apart from homonuclear molecules with bands in which every second line is missing, others are known for which the intensity of every second line is only decreased. Such an *alternation of intensities* is clearly shown, for example, by the Raman spectrum of N_2 in Fig. 37 (*a*). It is also found for H_2, Li_2, and many other molecules in their electronic band spectra.[20] It was first shown by Hund (346) that this phenomenon can be accounted for completely if it is assumed that the nuclei in molecules showing intensity alternation possess a non-zero angular momentum *I* (nuclear spin) while for the molecules with alternate missing lines the nuclear spin is zero (see also Chapter I, p. 29). A theoretical investigation shows (see below) that *in the presence of nuclear spin the selection rule sym. ↔ antisym. no longer holds absolutely*, although it is still very strict. As a result, both term systems, the symmetric and the antisymmetric, do occur, but with different statistical weights. Qualitatively the different behavior of homonuclear molecules with and without nuclear spin is connected with the fact that an exchange of nuclei when a nuclear spin is present does not necessarily lead to a completely identical state, since the nuclei may still differ by the orientation of their spins.

The spin vectors *I* of the two nuclei of a diatomic molecule form a resultant *T*, the *total nuclear spin of the molecule*. According to the rules of addition of angular momentum vectors, for two identical nuclei ($I_1 = I_2 = I$) the quantum number of the total nuclear spin is given by

$$T = 2I, \quad 2I - 1, \cdots, 0. \tag{III, 184}$$

Although the nuclear spin makes possible the occurrence of both symmetric and antisymmetric levels (even if at first only one kind occurs) a given level cannot occur with all the *T* values of (III, 184). It will be shown below that for the symmetric levels only the even *T* values and for the antisymmetric levels only the odd *T* values are possible.

According to the general discussion of angular momentum vectors (p. 26) a state with a given *T* has a statistical weight $2T + 1$. In a magnetic field a splitting into $2T + 1$ levels of slightly different energies occurs, which are distinguished by different values of M_T, the quantum number of the component of *T* in the field direction. We have

$$M_T = T, (T - 1), (T - 2), \cdots, -T. \tag{III, 185}$$

In the simplest case, when $I = \frac{1}{2}$, we have from (III, 184) the possible *T* values, $T = 1$ (parallel nuclear spins, ↑↑) and $T = 0$ (antiparallel spins, ↑↓). The first value is associated with the antisymmetric, the second with the symmetric levels. The statistical weight is 3, and 1, respectively. Therefore *for $I = \frac{1}{2}$ the antisymmetric rotational levels occur three times as frequently as the*

---

[20] Mecke (486) was the first to recognize that intensity alternation and alternate missing lines are characteristic of molecules with identical nuclei.

symmetric; that is, the total statistical weights are $3 \times (2J + 1)$ and $1 \times (2J + 1)$, respectively [see section 2(e) of this chapter]. Correspondingly instead of the one distribution curve in Fig. 59 we obtain two curves with the ratio $1 : 3$ of the ordinates, one curve for the even and the other for the odd levels. Since, according to (III, 183) the symmetric levels always combine with the symmetric and the antisymmetric always combine with the antisymmetric, and, since the symmetric and antisymmetric levels alternate (see Fig. 62) we would expect an intensity alternation $1 : 3$ to appear in the Raman bands [and similarly in the electronic bands; see Chapter IV, section 4(b) and Chapter V, section 3(b)]. This is actually observed for H_2. Conversely, from the occurrence and magnitude of the intensity alternation we may conclude that the H nucleus (the proton) has a spin of magnitude $I = \frac{1}{2}$.

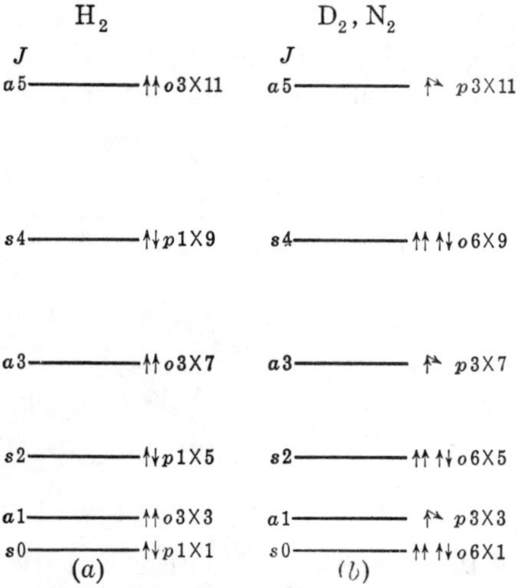

Theory shows (see Chapter VI) that the electronic eigenfunction of the ground state of H_2 is symmetric; that is, the even-numbered rotational levels are symmetric, and the odd are antisymmetric [compare Fig. 62 (a)]. In agreement with this the odd lines are observed to be strong, the even lines weak; that is, in the odd rotational levels the nuclear spins are parallel, in the even levels antiparallel. Fig. 64(a) shows the symmetries, the nuclear spin orientations, and the statistical weights for the rotational levels of H_2 in its electronic ground state.

$$H_2 \qquad\qquad D_2, N_2$$

FIG. 64. **Symmetries, Nuclear Spin Orientations, and Statistical Weights of the First Rotational Levels of H_2, D_2, and N_2, in the Ground State.** The energy scale is different for H_2, D_2, and N_2. The arrows in (a) represent $I = \frac{1}{2}$, in (b) $I = 1$. The statistical weights are written at the right of each level as products of the parts due to nuclear spin and to J.

If the nuclear spin $I = 1$, there are three possible values for the resultant nuclear spin of the molecule—namely, $T = 2$, 1, and 0. The corresponding statistical weights are 5, 3, and 1. Since, as already mentioned and as will be discussed in more detail below, the symmetric rotational levels occur with even T values, the antisymmetric levels with odd T values, their statistical weights are $(5 + 1) = 6$ and 3, respectively, and their total statistical weights are $6 \times (2J + 1)$ and $3 \times (2J + 1)$, respectively. In other words, we expect an *intensity alternation in the ratio $2 : 1$ in the bands of molecules having like nuclei whose spin $I = 1$*. Actually, just this value has been observed for N_2 [see Fig. 37(a)] and also for heavy hydrogen, D_2. Thus, conversely, we can conclude that the N and the D nuclei have a spin $I = 1$.

The symmetry of the electronic eigenfunction of D_2 is of course the same as that of H_2. This holds also for N_2 as will be shown in Chapter VI. Therefore, according to the above, the alternation of intensities is expected[21] to be opposite to that for H_2 as is indeed observed: The even lines rather than the odd ones are strong for D_2 and N_2. Fig. 64(b) serves to clarify this point by giving the symmetries, nuclear spin orientations and statistical weights for the lowest rotational levels of D_2 and N_2. It must be emphasized that for D_2 and N_2, in contrast to H_2, the "strong" and "weak" levels are not differentiated by the parallel or antiparallel orientation of the nuclear spins, but by the fact that for the weak levels the nuclear spins are at $60°$ to each other, so that their resultant $T = 1$, while for the strong levels the nuclear spins may be parallel as well as antiparallel.

In the general case, $I > 1$, the statistical weights of the symmetric and antisymmetric rotational levels are obtained by adding separately the quantities $2T + 1$ for even and odd T. One obtains $(2I + 1)(I + 1)$ and $(2I + 1)I$, respectively, for integral I, and the reverse for half-integral I. The ratio of these two numbers

$$R = \frac{I + 1}{I} \qquad \text{(III, 186)}$$

is the intensity ratio of strong and weak lines if the two nuclei of a homonuclear molecule each have spin I. Conversely, *the magnitude of the nuclear spin may be deduced from the observed magnitude of the alternation of intensities.* Actually a number of nuclear spins have been determined in this way (see Table 38), including some that could not be determined from the hyperfine structure of atomic spectra (see A.A., Chapter V). When $I = 0$, we obtain from (III, 186) $R = \infty$, that is, every second line is missing as we have already seen above. Conversely, when every alternate line is missing, as for O_2 and certain other homonuclear molecules, it follows unambiguously that the nuclear spin $I = 0$.

Influence of nuclear statistics. Let us consider a box containing a gas consisting of identical nuclei of zero spin (for example He nuclei).[22] Because of the indistinguishability of the nuclei the eigenfunctions of this system can only be symmetric or antisymmetric with respect to an exchange of any two nuclei (see p. 24). As previously, the selection rule sym. ↔ antisym. holds perfectly rigorously. Therefore if all the (He) nuclei in the box are at one time in symmetric states, they will remain permanently in symmetric states. Under these circumstances, a particular statistics holds for them—the *Bose-Einstein statistics*, or, for short, Bose statistics—which deviates in a characteristic manner from the classical Maxwell-Boltzmann statistics (different type of velocity distribution). If, on the contrary, the (He) nuclei are all initially in antisymmetric states, they must always remain in antisymmetric states. In that case, yet another statistics would hold for them—the *Fermi-Dirac statistics*, or, for

---

[21] This expectation corresponds to the present state of knowledge. At the time when the observation was first made the result was contrary to expectation. [See Rasetti (579) and Heitler and Herzberg (294).]

[22] The presence of an equivalent number of electrons to make the whole system electrically neutral would not alter the conclusions.

short, Fermi statistics[23] (with a velocity distribution different from that of Bose statistics and that of classical statistics).

The symmetry character, "symmetric or antisymmetric with respect to an exchange of any two nuclei," is not altered even when the He nuclei, each with two electrons, form He atoms and these, in pairs, form He_2 molecules. The He nuclei that follow Bose statistics in the free state would give rise to the symmetric rotational levels of the molecule, and the He nuclei that follow Fermi statistics in the free state would give rise to the antisymmetric rotational levels.

Now it would appear to be extremely peculiar if a gas of identical particles—for example, He nuclei—consisted of two modifications which could not, under any circumstances, be converted into one another and which followed quite different statistics. Actually, the band spectrum of He_2 (see above) shows that the antisymmetric levels do not occur at all; thus only those He nuclei occur in nature that follow Bose statistics and none that follow Fermi statistics,[24] or, in other words, the He nuclei always follow the Bose statistics. This has to be considered as a characteristic and important property of the He nuclei. The same has been found to hold for all other nuclei of spin $I = 0$ for which investigations of the corresponding band spectra have been made (see Table 38). Thus we can regard it as a *characteristic property of nuclei with spin $I = 0$ that they follow Bose statistics.*

According to what has been said, it is to be expected that atomic nuclei with a non-zero spin would also follow either the one statistics or the other but not both. The question is then how to explain the fact that, for homonuclear molecules with $I \neq 0$, both term systems actually appear, the symmetric and the antisymmetric, only an alternation of intensities and not a suppression of every second line being observed in their band spectra. In order to understand this fact we have to consider that the complete prohibition of intercombinations holds only between states whose *total* eigenfunctions have different symmetries and that it depends on this total eigenfunction whether the nuclei follow Bose or Fermi statistics. If the nuclei have a spin $I \neq 0$, the total eigenfunction depends also on the orientation of these spins. To zero approximation, which is a very good approximation here, the new total eigenfunction ψ' is the product of the function ψ, which we considered earlier [equation (III, 180)] and which we may call the *coordinate function*,[25] and a *nuclear spin function* β, which depends on the orientations of the spins:

$$\psi' = \psi \cdot \beta \qquad \qquad (III, 187)$$

The behavior of ψ for an exchange of the identical nuclei has been discussed above. We shall consider the behavior of β only for the case $I = 1$. Let us apply a magnetic field that is sufficiently strong[26] to uncouple the two spins. The possible spin components in the field direction are then (in units $h/2\pi$) $M_{I_1} = +1, 0, -1$ and $M_{I_2} = +1, 0, -1$. There are therefore nine different spin configurations. The corresponding spin-functions β may be ex-

[23] The differences between the three statistics first become appreciable at extremely low temperatures. The fact that a gas then no longer follows the classical statistics is called *gas degeneracy.*

[24] It should be remarked that initially the conclusion about the statistics of the He nuclei from band spectra was not quite unambiguous. By altering the explanation of the electronic structure of the He_2 molecule, Fermi-Dirac statistics would have been obtained for the He nuclei. However, experiments on the scattering of α particles in He have shown conclusively that the Bose-Einstein statistics holds for the He nuclei [see Blackett and Champion (813)].

[25] It includes *electron* spin.

[26] Actually, because of the smallness of the magnetic moment associated with the nuclear spin the uncoupling may be produced by a very weak external (or internal) field.

pressed as products of two functions β_1 and β_2 of the two spins. The nine spin functions β are

$$\beta_1{}^{+1}\beta_2{}^{+1}, \quad \beta_1{}^0\beta_2{}^0, \quad \beta_1{}^{-1}\beta_2{}^{-1}$$

$$\beta_1{}^{+1}\beta_2{}^0, \quad \beta_1{}^{+1}\beta_2{}^{-1}, \quad \beta_1{}^{-1}\beta_2{}^0 \tag{III, 188}$$

$$\beta_1{}^0\beta_2{}^{+1}, \quad \beta_1{}^{-1}\beta_2{}^{+1}, \quad \beta_1{}^0\beta_2{}^{-1}$$

where the superscripts indicate the values of M_{I_1} and M_{I_2}. The first three functions are obviously symmetric with respect to an exchange of the two nuclei. The remaining six functions occur in degenerate pairs (written one above the other). The degeneracy is removed when the interaction of the spins is taken into account and the eigenfunctions are in zero-order approximation

$$\beta_1{}^{+1}\beta_2{}^0 + \beta_1{}^0\beta_2{}^{+1} \quad \text{and} \quad \beta_1{}^{+1}\beta_2{}^0 - \beta_1{}^0\beta_2{}^{+1} \tag{III, 189}$$

and similarly for the other pairs. The first of these eigenfunctions is symmetric, the second antisymmetric. Thus of the nine nuclear spin functions six are symmetric and three are antisymmetric.

It is clear from the above that in the general case there are $2I + 1$ more symmetric than antisymmetric spin functions, that is, since the total number is $(2I + 1)^2$ there are $\frac{1}{2}[(2I + 1)^2 + (2I + 1)] = (2I + 1)(I + 1)$ symmetric spin functions and $\frac{1}{2}[(2I + 1)^2 - (2I + 1)] = (2I + 1)I$ antisymmetric spin functions.

It is interesting to see to which total spin values T the spin functions belong even though the conclusions below are independent of this correlation. For a sufficiently weak field, $\boldsymbol{T}$ is space quantized with components $M_T h/2\pi$ [see (III, 185)]. For the larger field used before (in which $\boldsymbol{T}$ is not defined) we have $M_T = M_{I_1} + M_{I_2}$. According to the adiabatic law there must be a one-to-one correspondence between the M_T values of small and large fields.[27] In our example we have $T = 2, 1, 0$. For $T = 2$ we have $M_T = +2, +1, 0, -1, -2$ of which the first and last, for large field can only correspond to the symmetric spin functions $\beta_1{}^{+1}\beta_2{}^{+1}$ and $\beta_1{}^{-1}\beta_2{}^{-1}$. Therefore the spin functions for the other M_T values of $T = 2$ must also be symmetric. Of the remaining spin functions the three antisymmetric ones account for $T = 1$ and $M_T = +1, 0, -1$ and the one symmetric function for $T = 0$. It is easily seen that in the general case[28] the spin functions are *symmetric* for $T = 2I, 2I - 2, \cdots, 1$, or 0 and *antisymmetric* for $T = 2I - 1, 2I - 3, \cdots, 0$, or 1, the number of spin functions for each T value being of course $2T + 1$.

Since according to the preceding discussion there are always symmetric as well as antisymmetric spin functions β for $I \neq 0$, by a suitable choice of β (and T), the total function ψ' can be made symmetric for a symmetric as well as an antisymmetric ψ or can be made antisymmetric for a correspondingly different choice of β (and T). This means that for nuclei following Bose statistics as well as those following Fermi statistics, if $I \neq 0$, *all the rotational levels* (even as well as odd) *can appear*—however, only with certain statistical weights (and corresponding total nuclear spin values). These relationships are collected in Table 15. We can see from this table that for Bose statistics the states with symmetric coordinate functions have the greater statistical weight, while for Fermi statistics the states with antisymmetric coordinate functions have the greater statistical weight. For brevity we shall call the states with symmetric coordinate functions *symmetric states* and those with antisymmetric coordinate functions *antisymmetric states*.

For H_2, with nuclear spin $I = \frac{1}{2}$ (see above), the antisymmetric levels (which in the

---

[27] It should be emphasized that the individual spin orientations at small and large fields are in general different even for the same M_T. Thus in the present case $(I = 1)$ the three configurations with $M_T = 0$ at small field are $\rightarrow \rightarrow$, $\nearrow \searrow$ and $\uparrow \downarrow$, at large field $\rightarrow \rightarrow$, $\uparrow \downarrow$, and $\downarrow \uparrow$.

[28] The special case $I = \frac{1}{2}$ is entirely similar to the case of two electron spins discussed at length in A.A., p. 124.

TABLE 15. SYMMETRY OF THE EIGENFUNCTIONS FOR $I \neq 0$

Coordinate Function ψ	Nuclear Spin Function β	Statistical Weight	Nuclear Spin T	Total Function ψ'	For Bose Statistics of the Nuclei	For Fermi Statistics of the Nuclei
sym.	sym.	$(2I+1)(I+1)$	$2I, 2I-2, \cdots$	sym.	occurs	
	antisym.	$(2I+1)I$	$2I-1, 2I-3, \cdots$	antisym.		occurs
antisym.	sym.	$(2I+1)(I+1)$	$2I, 2I-2, \cdots$	antisym.		occurs
	antisym.	$(2I+1)I$	$2I-1, 2I-3, \cdots$	sym.	occurs	

ground state are the odd-numbered ones) have the greater statistical weight [see Fig. 64(a)]. It follows that the H *nuclei* (protons) *follow Fermi statistics*. On the other hand, the spectrum shows that for D_2 the symmetric (even-numbered) levels have the greater statistical weight [see Fig. 64(b)]. Therefore, *Bose statistics apply to the* D *nuclei* (deuterons). The same follows for the N nuclei from the Raman spectrum of N_2 [see Fig. 37(a)]. The importance of these results for the theory of nuclear structure will be discussed briefly in Chapter VIII.

It has been found (first by the study of band spectra) that nuclei following Fermi statistics have half-integral spins, nuclei following Bose statistics have integral spins. This empirical result follows theoretically from the fact that each constituent particle of the nucleus has a spin of $\frac{1}{2}$ and follows Fermi statistics [see Rasetti (18a)]. As a consequence (see Table 15) *the symmetric states always occur with even T values* ($2I, 2I-2, \cdots$ for Bose and $2I-1$, $2I-3, \cdots$ for Fermi statistics) *the antisymmetric states occur with odd T values* ($2I-1$, $2I-3, \cdots$ for Bose and $2I, 2I-2, \cdots$ for Fermi statistics). This is the rule used in the simplified treatment above (p. 133).

The transition probability between two states with different symmetries of the *total* eigenfunctions ψ' is exactly zero. That is why the nuclei of a given species follow only one statistics (see above). The transition probability between two states of different symmetry of the *coordinate* function but the same symmetry of the total eigenfunction (which is possible only for $I \neq 0$) is not exactly zero although very close to it. Indeed as long as ψ' can be written as a product $\psi\beta$ according to (III, 187) the latter transition probability is exactly zero, since then for example, for dipole radiation (see p. 131),

$$R_x^{nm} = \int \psi_n'^* \left(\sum e_i x_i\right)\psi_m' d\tau' = \int \psi_n^* \left(\sum e_i x_i\right)\psi_m d\tau \int \beta_n^* \beta_m d\sigma, \qquad \text{(III, 190)}$$

where $d\tau'$, $d\tau$, and $d\sigma$ are the volume elements of the whole configuration space, of the coordinate space alone, and of the nuclear spin space alone, respectively. The first integral at the right and therefore R_x^{nm} (and similarly R_y^{nm}, R_z^{nm}) vanishes if the two coordinate functions have opposite symmetry in the nuclei (see p. 131). The same holds for transitions brought about in other ways. Thus in this approximation the selection rule sym. $\leftrightarrow\!\!\!\!|$ antisym. holds even if the nuclear spin is different from zero.

However, if the interaction of the nuclear spin with the rest of the molecule is taken into account, (III, 187) must be replaced by

$$\psi' = \psi\beta + \varphi \qquad \text{(III, 191)}$$

where φ depends on *all* coordinates including nuclear spin coordinates and cannot be resolved into a product.[29] Substitution of (III, 191) into R_x^{nm} (compare III, 190) shows that now

---

[29] The introduction of the term φ has no effect on the symmetry considerations since φ has necessarily the same symmetry as $\psi\beta$.

the transition probability may be different from zero even if $\int \psi_n{}^*(\sum e_i x_i)\psi_m d\tau = 0$. There is thus a small probability for transitions between symmetric and antisymmetric states if $I \neq 0$; that is, for example, transitions from the even-numbered to the odd-numbered levels of H_2, D_2, N_2, $\cdots$ can take place (see Fig. 64). Owing to the smallness of the magnetic moment associated with the nuclear spin, the interaction with the rest of the molecule is very small indeed, and, as a result, the *transition probability between symmetric and antisymmetric states is extremely small*. It is so small that its reciprocal, the mean life, is of the order of years [see Farkas (214)].

As in the case of atoms (see p. 29) the coupling of the nuclear spin with the other angular momenta in the molecule leads to a very slight splitting of otherwise single energy levels. This hyperfine structure will be discussed in section 6 of Chapter V.

Ortho and para modifications. It was shown in the preceding discussion that for homonuclear molecules with nonzero nuclear spin the selection rule (III, 183) (antisym. $\leftrightarrow$ sym.) even though no longer rigorous still holds sufficiently strictly that it may take many months, if not years, before a molecule goes from an even-numbered to an odd-numbered rotational level. As a result, as first pointed out by Dennison (181), gases such as H_2, D_2, N_2, and so forth, may be regarded as *mixtures of two modifications*—a *symmetrical* with only even-numbered rotational levels and an *antisymmetrical* with only odd-numbered rotational levels. For gases with $I = \frac{1}{2}$, such as hydrogen, the one modification has antiparallel nuclear spins and the other has parallel nuclear spins [see Fig. 64(a)].

Actually, in the case of hydrogen, it is possible to obtain these two modifications separately. When the gas is cooled to a very low temperature (temperature of liquid hydrogen), without taking into consideration the rule sym. $\leftrightarrow$ antisym., one would expect practically all the molecules to go over to the lowest rotational level, $J = 0$, whereas, considering the selection rule, only those molecules that were originally in the higher even-numbered rotational levels go to the lowest even-numbered state, $J = 0$, while all the molecules originally in the higher odd-numbered rotational levels go to the lowest odd-numbered state, $J = 1$ [see Fig. 64(a)]. This means that the distribution of rotational levels at the low temperature does not correspond to thermal equilibrium. However, when the gas is kept for a sufficiently long time (many weeks or months) at the low temperature, even the molecules that were at first in the state $J = 1$ will eventually go into the state $J = 0$ belonging to the symmetric system; that is, thermal equilibrium should eventually be obtained. If the gas is now allowed to warm up again to normal temperatures, the molecules at first can only go from $J = 0$ to the higher *even-numbered* (symmetric) levels, and not to the odd-numbered (antisymmetric) levels (owing to the rule sym. $\leftrightarrow$ antisym.). This means that for some time *only the one modification should be present*.

Bonhoeffer and Harteck (118) and, almost simultaneously, Eucken and Hiller (206) were the first to carry out experiments on hydrogen which strikingly confirmed the above theoretical conclusions and thus incidentally demonstrated the validity of wave mechanics for molecules. The former authors made the important discovery that the addition of active charcoal to the hydrogen at the low temperature greatly accelerates the establishment of thermal equilibrium, so that it was not necessary to keep the hydrogen at the low temperature for weeks or months.   That actually only the one modification was present after the temperature had been raised again was shown in the first place by observations of the heat conductivity and heat capacity, which for the separated modification differ by a predictable amount from those of the mixture (see p. 470); and, in the second place, by investigations of the spectrum, which for the gas prepared in the above-mentioned manner does not contain the lines that are strongest in the spectrum of the mixture (absence of every alternate line, just as for $I = 0$).   Thus, after such a conversion, all the molecules are in the even-numbered rotational levels, while, in normal hydrogen, three fourths of the molecules are in the odd levels.

The modification with the greater statistical weight is usually called the *ortho modification* and that with the smaller weight the *para modification*.   We differentiate correspondingly between ortho H_2 and para H_2, ortho D_2 and para D_2, ortho N_2 and para N_2, and so forth.   The designations o (ortho) and p (para) are added to the rotational levels in Fig. 64.   Thus far, apart from H_2, a separation has been effected only for D_2 [A. and L. Farkas and Harteck (211) and Clusius and Bartholomé (159)].   For N_2, owing to the smallness of the rotational quanta, a temperature of less than 1° K. would be necessary for a noticeable enrichment of one modification.   It should be noted that, while the even-numbered levels constitute the para modification of H_2 the odd-numbered levels constitute the para modification of N_2 and D_2.   Therefore ortho D_2 is produced when the method described above for the production of pure para hydrogen is applied to deuterium.

The transformation from the separated modifications back to the normal mixture of both H_2 and D_2 has been investigated under various conditions.   With increasing temperature the rate of conversion is found to increase considerably.   This thermal conversion has been shown [A. Farkas (928)] to be due to the presence of H atoms in consequence of thermal dissociation, leading to the *exchange reaction*

$$H + p\text{-}H_2 \underset{\longleftarrow}{\overset{\longrightarrow}{\rule{0pt}{0pt}}} o\text{-}H_2 + H \qquad\qquad (III, 192)$$

This reaction does not contradict the rule antisym. $\leftrightarrow$ sym. since a new molecule is formed from the incident H atom and an H atom of the p-H_2 molecule.

Another mechanism of conversion was discovered by L. Farkas and Sachsse (929) who found that paramagnetic substances such as O_2, NO, and paramagnetic ions in solutions accelerate the reconversion to the equilibrium mixture considerably.   As was shown by Wigner (1539) this effect is due to a slight breakdown of the selection rule antisym. $\leftrightarrow$ sym. in the strongly inhomogeneous magnetic field in which the p-H_2 molecule finds itself during a collision with a paramagnetic molecule or ion.   This breakdown is connected with the difference in magnetic field intensity acting on the two nuclei.   The observed magnitude of the effect (conversion probability about 5×10^{-13} per collision) was found to be in satisfactory agreement with the theoretical prediction of Wigner.   According to experiments by Bonhoeffer, A. Farkas, and Rummel (821) it is very probable that the catalytic action of active charcoal (see above) is due to paramagnetic carbon atoms at the surface [see also L. Farkas and Sandler (930)].

Isotopic molecules. In the case of isotopic molecules of homonuclear molecules such as $O^{16}O^{18}$, $Cl^{35}Cl^{37}$, HD, $N^{14}N^{15}$, the cause for the division of the rotational states into symmetric and antisymmetric ones disappears, since by exchange of the nuclei the molecule goes over into a configuration that is different from the original one. As a result, there should be *no alternation of intensities* or *no missing of every alternate line* in these isotopic molecules. This is actually found to be the case (in electronic band spectra), and this observation must be regarded as a particularly impressive confirmation of the predictions of quantum mechanics; for example, in the bands of the $O^{16}O^{18}$ molecule, all the lines appear, whereas every second line is missing for $O^{16}O^{16}$; similarly, there is no alternation of intensities in the HD bands. Correspondingly, experiments show that there are not two modifications of HD, as for H_2 or D_2 [Farkas and Harteck (211) and Clusius and Bartholomé (159)]. An HD molecule can go over without restriction from an even-numbered level to an odd-numbered level—for example, by collision. The distinction between symmetric and antisymmetric rotational levels would be absent even in molecules containing two nuclei of the same charge and mass if these nuclei are in different quantum states (isomeric nuclei). However no observations on such molecules have yet been made.

(g) Isotope Effect

Vibration. For isotopic molecules—that is, molecules that differ only by the mass of one or both of the nuclei but not by their atomic number (for example HCl^{35}, HCl^{37}, $B^{10}O$, $B^{11}O$)—the vibrational frequencies are obviously different. Assuming harmonic vibrations the (classical) vibrational frequency $\nu_{osc.}$ is given by $\nu_{osc.} = (1/2\pi)\sqrt{k/\mu}$ (see p. 75) where the force constant k, since it is determined by the electronic motion only, is exactly the same for different isotopic molecules, whereas the reduced mass μ is different. Therefore if we let the superscript i distinguish an isotopic molecule from the "ordinary" molecule we have

$$\frac{\nu_{osc.}{}^{i}}{\nu_{osc.}} = \sqrt{\frac{\mu}{\mu^i}} = \rho \qquad (III, 193)$$

The heavier isotopic molecule has the smaller frequency. This is easily visualized when we remember that the larger a mass hanging on a spring is the smaller is its vibrational frequency. If the superscript i refers to the heavier isotope the constant ρ will be smaller than 1. The values of ρ for the pairs H^1Cl^{35}, H^1Cl^{37}; H^1Cl^{35}, H^2Cl^{35}; $B^{10}O$, $B^{11}O$; $O^{16}O^{16}$, $O^{16}O^{18}$ are 0.99924, 0.71720, 0.97177, 0.97176, respectively.

By substituting (III, 193) into the term formula (III, 37) one finds for the *vibrational levels of two isotopic molecules* (still assuming harmonic oscillations)

$$G = \omega(v + \tfrac{1}{2}), \qquad G^i = \omega^i(v + \tfrac{1}{2}) = \rho\omega(v + \tfrac{1}{2}). \qquad (III, 194)$$

If anharmonicity is taken into account the calculations become rather more

involved and will not be reproduced here [see Dunham (194)]. The formulae
for the energy levels are found to be in a very good approximation

$$G(v) = \omega_e(v + \tfrac{1}{2}) - \omega_e x_e(v + \tfrac{1}{2})^2 + \omega_e y_e(v + \tfrac{1}{2})^3 + \cdots$$
$$G^i(v) = \rho\omega_e(v + \tfrac{1}{2}) - \rho^2\omega_e x_e(v + \tfrac{1}{2})^2 + \rho^3\omega_e y_e(v + \tfrac{1}{2})^3 + \cdots$$

(III, 195)

or in other words

$$\omega_e{}^i = \rho\omega_e, \quad \omega_e{}^i x_e{}^i = \rho^2\omega_e x_e, \quad \omega_e{}^i y_e{}^i = \rho^3\omega_e y_e, \cdots$$

(III, 196)

Fig. 65 shows the relative positions of the vibrational levels of two isotopic
molecules on the basis of these formulae.   The separation of corresponding levels
increases with increasing v.   The levels of the lighter isotope always lie higher
than those of the heavier.

   This vibrational isotope effect, expected according to theory, was first noticed
in the rotation-vibration spectrum of HCl by Loomis (456a) and independently
by Kratzer (420b).   On the basis of Fig. 65, since Cl has two isotopes of masses
35 and 37, instead of every single band we should expect two bands, one corresponding to HCl^{35} and the other to HCl^{37}.   The band belonging to HCl^{37}

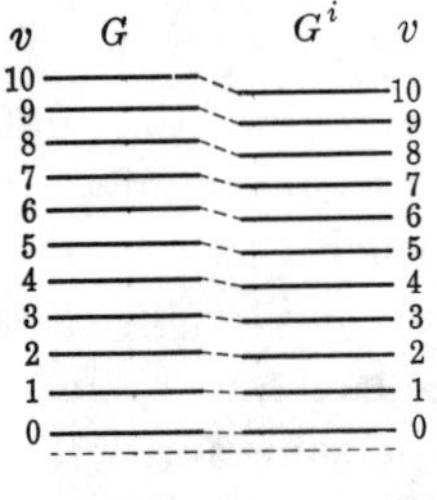

FIG. 65. **Vibrational
Levels of Two Iso-
topic Molecules (Sche-
matic)**. To the left
for the lighter isotope,
to the right for the
heavier isotope;
$\rho = 0.95$.

TABLE 16.   OBSERVED AND CALCU-
LATED VIBRATIONAL ISOTOPE DIS-
PLACEMENTS FOR THE INFRARED
HCl BANDS

[According to the data of Meyer
and Levin (491) and Herzberg and
Spinks (318)]

Band	$\Delta\nu_{obs.}$	$\Delta\nu_{calc.}$
1–0	2.01	2.105
2–0	4.00	4.053
3–0	5.834	5.845
4–0		7.481

should be shifted by a small amount toward longer wave lengths with respect to
the band belonging to HCl^{35}.   Actually, with sufficient resolution, the rotation-
vibration bands of HCl show a *doubling of all the lines*; every line has a com-
panion of smaller intensity at a constant separation to the long-wave-length
side, as is shown for the 3–0 band in Fig. 33.[30]   In Table 16 the measured shifts
(mean for all the lines) for the three bands 1–0, 2–0, and 3–0 are compared with

---

[30] In the absorption curve of the 1–0 band given in Fig. 32 the dispersion is too small for
a resolution of the doublets, particularly since the separation is only about a third of that in
the 3–0 band (Fig. 33).   The doublets were resolved in the high-dispersion work of Meyer and
Levin (491).

the theoretical shifts calculated on the basis of (III, 195).   It can be seen that the agreement is very satisfactory.

Shortly after the discovery of heavy hydrogen by Urey, Brickwedde, and Murphy (676), the infrared spectrum of the isotopic molecule H^2Cl^{35} (and H^2Cl^{37}), present in minute concentration in ordinary HCl, was investigated by Hardy, Barker, and Dennison (280).   In this case ρ^2 is nearly $\frac{1}{2}$, and, as a result, the isotope shift between H^1Cl^{35} and H^2Cl^{35} (or between H^1Cl^{37} and H^2Cl^{37}) is so great that the bands lie in quite different spectral regions.   While the fundamental band of ordinary HCl lies at 3.46 μ, that of H^2Cl lies at 4.78 μ.   It was found that the shift corresponds exactly to the above formulae (III, 195).

In deriving a formula for the actual *isotope shift* it is important to realize that the formulae (III, 196) cannot be applied to the $\omega_0{}^i$, $\omega_0{}^i x_0{}^i$, $\cdots$.   Substituting therefore ω_e, $\omega_e x_e$, from (III, 77) in (III, 80) we obtain for the frequencies of the absorption bands

$$\nu = (\omega_e - \omega_e x_e)v - \omega_e x_e v^2,$$
$$\nu^i = (\rho\omega_e - \rho^2\omega_e x_e)v - \rho^2\omega_e x_e v^2,$$
(III, 197)

where terms higher than quadratic in v have been neglected.   The isotope shift is thus

$$\Delta\nu = \nu - \nu^i = \omega_e(1 - \rho)v - \omega_e x_e(1 - \rho^2)v - \omega_e x_e(1 - \rho^2)v^2$$
$$= (1 - \rho)\{[\omega_e - \omega_e x_e(1 + \rho)]v - \omega_e x_e(1 + \rho)v^2\}.$$
(III, 198)

As long as ρ is only slightly different from 1, we can also write [according to (III, 81)]

$$\Delta\nu = (1 - \rho)v\Delta G_{v+\frac{1}{2}}.$$
(III, 199)

Rotation.   The internuclear distances in diatomic (and polyatomic) molecules are entirely determined by the electronic structure as are the force constants.   They are therefore exactly equal in isotopic molecules as long as no vibration occurs.   However, since the reduced mass μ and therefore the rotational constant $B = h/8\pi^2 c\mu r^2$ is different the rotational terms of two isotopic molecules will be different.   With $B^i = h/8\pi^2 c\mu^i r^2$ and ρ from (III, 193) we obtain [see (III, 15)]

$$F = BJ(J + 1), \qquad F^i = B^iJ(J + 1) = \rho^2 BJ(J + 1). \quad \text{(III, 200)}$$

The *rotational levels of two isotopic molecules* obtained from these formulae are shown to scale in Fig. 66.

For simultaneous vibration and rotation, in a first approximation, we simply have to add the vibrational and rotational isotope effects.   As a result, the lines of a rotation-vibration band of an isotopic molecule do not have exactly the same separations as the lines of the corresponding band of the "normal" molecule; or, in other words, the isotope displacement between corresponding lines of the two bands is not independent of J but varies slightly.   This is shown schematically in Fig. 67 where for simplicity the vibrational shift has been assumed to be zero which, of course, it actually is not.   Such a rotational effect has been observed, for example, for the 3–0 band of HCl [see Herzberg and Spinks (318)].

Since the rotational term F is small compared to the vibrational term G, the rotational isotope shift is, in general, small compared to the vibrational isotope shift, in spite of the fact that the former depends on ρ^2 rather than ρ as does the latter. In the 3–0 band of HCl, the rotational shift for the observed lines is < 0.2 cm^{-1}, while the vibrational shift amounts to 5.8 cm^{-1}.

For more precise calculations it is necessary to take account of the interaction of vibration and rotation, that is, of the rotational constants α, D, and possibly higher ones. The above relation $B^i = \rho^2 B$ holds rigorously only for

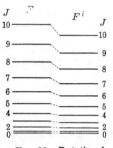

FIG. 66. **Rotational Levels of Two Isotopic Molecules.** To the left for the lighter isotope, to the right for the heavier isotope; $\rho = 0.95$. The lowest levels, $J = 0$, are drawn at the same height. Actually they have different energies owing to the vibrational effect which is always present.

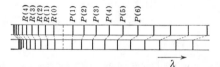

FIG. 67. **Rotational Isotope Effect in a Rotation-Vibration Band (Schematic).** The upper schematic spectrogram gives the appearance of the band of the lighter isotope, the lower that of the heavier isotope. In order to show the pure rotational effect, the zero lines of the two bands have been made to coincide, whereas actually they are separated by a certain amount (the vibrational shift). The magnitude of the rotational isotope effect is usually considerably smaller than that drawn.

B_e since the equilibrium internuclear distances are the same in the two isotopic molecules but not the effective internuclear distances for a given v, that is

$$B_e^i = \rho^2 B_e \qquad (\text{III, 201})$$

For α_e^i, D_e^i, and β^i the calculations of Dunham (194) give to a very good approximation

$$\alpha_e^i = \rho^3 \alpha_e, \qquad D_e^i = \rho^4 D_e, \qquad \beta^i = \rho^5 \beta \qquad (\text{III, 202})$$

Dunham has also given similar formulae for still higher terms in the energy formula of the vibrating rotator as well as small correction terms to the above equations. The latter can almost always be neglected except for very precise data on isotope effects involving light and heavy hydrogen (see p. 109).

Let $\nu_r = F' - F''$ be the rotational part of the wave number of a band line where F is given by (III, 119) and (III, 116). The rotational isotope shift of the band lines is then according to (III, 201) and (III, 202), but neglecting the rotational constant β,

$$\Delta \nu_r = \nu_r - \nu_r{}^t = (1 - \rho^2)[B_e{}'J'(J' + 1) - B_e{}''J''(J'' + 1)]$$
$$- (1 - \rho^3)[\alpha_e{}'(v' + \tfrac{1}{2})J'(J' + 1) - \alpha_e{}''(v'' + \tfrac{1}{2})J''(J'' + 1)]$$
$$- (1 - \rho^4)[D'J'^2(J' + 1)^2 - D''J''^2(J'' + 1)^2]. \tag{III, 203}$$

To this the vibrational shift from (III, 198) has to be added to get the total shift. The relation (III, 203) holds quite generally, even for the case of an electronic band spectrum. In the case of a rotation-vibration spectrum it is greatly simplified since $B_e{}' = B_e{}''$, $\alpha_e{}' = \alpha_e{}''$, and $D' \approx D''$.

The second and third terms in (III, 203) are, in general, very small compared to the first. We can therefore put, to a first, fairly good, approximation,

$$\Delta \nu_r = (1 - \rho^2)\nu_r \tag{III, 204}$$

In this formula, the second and third terms in (III, 203) are at least partly taken into account. The *rotational isotope shift* is thus, to a good approximation, *proportional to the distance from the zero line* (see Fig. 67).

An accurate measurement of the isotope effect can be used to obtain a precise value for the *ratio of the masses* of the two kinds of isotopic atoms concerned (from $\rho = \sqrt{\mu/\mu^i}$). Under favorable conditions, the accuracy of the ratio of the masses so obtained is comparable with the accuracy of mass-spectrographic values.[31] Apart from that, the study of the isotope effect in electronic band spectra (see Chapter IV) has led to the discovery of new isotopes and to an unambiguous confirmation of the quantum mechanical formula for the energy levels of the oscillator.

[31] See, for example, the recent work of Townes, Merritt and Wright (1439) on ICl.

CHAPTER IV

ELEMENTARY DISCUSSION OF ELECTRONIC STATES AND ELECTRONIC TRANSITIONS

The band spectra observed in the visible and ultra-violet regions of the spectrum (see Chapter II) obviously cannot be interpreted as simple rotation or rotation-vibration spectra, since their structure is generally more complicated than would be expected according to the theory of such spectra (Chapter III). Also, the frequencies in the visible and ultraviolet regions are much too great for rotational or vibrational frequencies. However, it appears highly plausible to explain these visible and ultraviolet band spectra as due to *electronic transitions in molecules* (that is, as *electronic band spectra*), in conformity with the explanation of the visible and ultraviolet line spectra as due to electronic transitions in atoms (Chapter I). In this and the following chapter we shall see that this interpretation does indeed account for all the observed features of these spectra.

1. Electronic Energy and Total Energy

The atomic nuclei in a molecule are held together by the electrons; the nuclei alone would of course repel each other. Just as for atoms, we expect different electronic states of the molecule, depending on the orbitals in which the electrons are. The energy differences between these states are of the same order as for atoms (1 to 20 volts). The electronic states are designated Σ, Π, Δ, $\cdots$ states similar to S, P, D, $\cdots$ states of atoms. A detailed discussion of the different kinds of electronic states and the exact meaning of the term symbols will be given in Chapters V and VI.

Electronic energy and potential curves; stable and unstable molecular states. The total energy of the molecule (neglecting spin and magnetic interactions) consists of the potential and kinetic energies of the electrons and the potential and kinetic energies of the nuclei. If we consider for a moment the electronic motions for fixed nuclei it is clear that the electronic energy (potential + kinetic energy) will depend on the internuclear distance r. This dependence will be different for different electronic states as is immediately clear if one compares qualitatively, for example, the interaction of two normal hydrogen atoms with the interaction of one normal and one highly excited hydrogen atom.

On account of the smallness of the mass of the electrons compared to that of the nuclei the electrons move much more rapidly than the nuclei; and therefore the electronic energy, when the nuclei are no longer fixed, takes up the

146

value corresponding to the momentary positions of the nuclei. Thus in order to change the position of the nuclei, not only must work be done against the Coulomb repulsion of the nuclei, but also work must be supplied for the necessary change of electronic energy. In other words, *the sum of electronic energy and Coulomb potential of the nuclei acts as the potential energy under whose influence the nuclei carry out their vibrations.* Only if this potential energy, in its dependence on the internuclear distance, has a minimum is the electronic state in question a *stable* state of the molecule. If there is no minimum the electronic state is *unstable*, that is, the two atoms repel each other for any value of the internuclear distance.

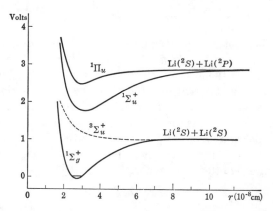

Fig. 68. **Potential Curves of the Different Electronic States of the Li₂ Molecule [after Mulliken (514)].** The repulsive state with the broken potential curve has not been directly observed.

The curves representing the variation of the effective potential energy of the nuclei (electronic energy + Coulomb potential of the nuclei) are usually referred to as *potential curves.* Each electronic state is characterized by a definite potential curve which may have a more or less deep minimum (curve of attraction, stable molecular state) or may have no minimum (curve of repulsion, unstable molecular state).

The potential curves of the Li₂ molecule, as derived from the analysis of band spectra, are represented in Fig. 68 and serve to illustrate the preceding considerations. (For the nomenclature, see Chapters V and VI.) The curve designated by $^3\Sigma_u^+$ has no minimum; this electronic state is therefore unstable.

Resolution of the total eigenfunction. The Schrödinger equation of a diatomic molecule may be written [compare (I, 13)]

$$\frac{1}{m}\sum_i\left(\frac{\partial^2\psi}{\partial x_i^2} + \frac{\partial^2\psi}{\partial y_i^2} + \frac{\partial^2\psi}{\partial z_i^2}\right) + \sum_k \frac{1}{M_k}\left(\frac{\partial^2\psi}{\partial x_k^2} + \frac{\partial^2\psi}{\partial y_k^2} + \frac{\partial^2\psi}{\partial z_k^2}\right)$$

$$+ \frac{8\pi^2}{h^2}(E - V)\psi = 0 \quad \text{(IV, 1)}$$

where x_i, y_i, z_i are the coordinates of the electrons (mass m) and x_k, y_k, z_k those of the nuclei (mass M_k). We try the approximate solution

$$\psi = \psi_e(\cdots, x_i, y_i, z_i, \cdots) \, \psi_{vr}(\cdots, x_k, y_k, z_k, \cdots) \qquad (\text{IV}, 2)$$

where ψ_e and ψ_{vr} are solutions of the equations

$$\sum_i \left(\frac{\partial^2 \psi_e}{\partial x_i^2} + \frac{\partial^2 \psi_e}{\partial y_i^2} + \frac{\partial^2 \psi_e}{\partial z_i^2} \right) + \frac{8\pi^2 m}{h^2} (E^{el} - V_e) \psi_e = 0 \qquad (\text{IV}, 3)$$

and

$$\sum_k \frac{1}{M_k} \left(\frac{\partial^2 \psi_{vr}}{\partial x_k^2} + \frac{\partial^2 \psi_{vr}}{\partial y_k^2} + \frac{\partial^2 \psi_{vr}}{\partial z_k^2} \right) + \frac{8\pi^2}{h^2} (E - E^{el} - V_n) \psi_{vr} = 0 \qquad (\text{IV}, 4)$$

respectively. The first of these equations is the *Schrödinger equation of the electrons moving in the field of the fixed nuclei* and having a potential energy V_e (which is a function of the electronic coordinates x_i, y_i, z_i). For different internuclear distances V_e is different and therefore the eigenfunctions ψ_e and eigenvalues E^{el} of this equation depend on the internuclear distance as parameter. The second equation (IV, 4) is the *Schrödinger equation of the nuclei moving under the action of the potential $E^{el} + V_n$* where V_n is the Coulomb potential of the nuclei. For a diatomic molecule whose nuclei have charges $Z_1 e$ and $Z_2 e$ and a distance r from each other we have

$$V_n = \frac{Z_1 Z_2 e^2}{r} \qquad (\text{IV}, 5)$$

If the expression (IV, 2) is substituted into the original wave equation (IV, 1) and account is taken of (IV, 3) and (IV, 4) it is readily seen that (IV, 1) is satisfied only if

$$\sum_k \frac{2}{M_k} \left[\frac{\partial \psi_e}{\partial x_k} \cdot \frac{\partial \psi_{vr}}{\partial x_k} + \frac{\partial \psi_e}{\partial y_k} \frac{\partial \psi_{vr}}{\partial y_k} + \frac{\partial \psi_e}{\partial z_k} \frac{\partial \psi_{vr}}{\partial z_k} + \psi_{vr} \left(\frac{\partial^2 \psi_e}{\partial x_k^2} + \frac{\partial^2 \psi_e}{\partial y_k^2} + \frac{\partial^2 \psi_e}{\partial z_k^2} \right) \right] \qquad (\text{IV}, 6)$$

can be neglected, that is, that the variation of ψ_e with internuclear distance is sufficiently slow so that its first and second derivatives $\dfrac{\partial \psi_e}{\partial x_k}, \cdots \dfrac{\partial^2 \psi_e}{\partial x_k^2}, \cdots$ can be neglected. That this condition is usually fulfilled to a satisfactory approximation has been shown in detail by Born and Oppenheimer (122). Thus we are justified in using $E^{el} + V_n$ as the potential energy for the motion of the nuclei and at the same time in resolving ψ into a product of ψ_e and ψ_{vr} according to (IV, 2).

As we have seen previously (p. 110) the eigenfunction ψ_{vr} of the vibrating rotator, in a first approximation, can be expressed as the product $(1/r)\psi_v \psi_r$ where ψ_v is the vibrational eigenfunction of a linear oscillator, depending only on the change of internuclear distance $(r - r_e)$, and ψ_r is the rotational eigenfunction, depending only on the orientation of the molecule in space. Thus we have in a first approximation for the *total eigenfunction*

$$\psi = \psi_e \cdot \frac{1}{r} \cdot \psi_v \cdot \psi_r \tag{IV, 7}$$

Kronig (21a) has shown that this resolution holds to a good approximation also when electron spin and magnetic interactions of angular momenta are included. As mentioned above the electronic eigenfunction depends on the internuclear distance as a parameter. Since the variation is slow (see above) it is frequently disregarded altogether.

For a given electronic state characterized by a certain electronic eigen-function and several electronic quantum numbers (see Chapter V and VI) we may have a variety of functions for ψ_v and ψ_r depending on the particular values of the vibrational and rotational quantum numbers, respectively (see Figs. 40, 42, and 57).

Resolution of the total energy. In the discussion of electronic transitions of diatomic molecules one usually considers the minimum value of $E^{el} + V_n$, that is, of the potential energy function of a given stable electronic state, as *the electronic energy* of the state, and uses for it the symbol E_e. The minimum of the lowest electronic state (the electronic ground state) of the molecule is almost always chosen as the zero point of the energy scale. This choice is different from that for atoms.

As we have seen above $E^{el} + V_n$ is to a good approximation the potential energy $U(r - r_e)$ for the (vibrational) motion of the nuclei. Therefore the excess of energy of the non-rotating molecule over the minimum electronic energy E_e must be considered as *vibrational energy* E_v (though it should be realized that the part E^{el} of the potential energy of the nuclear vibration is electronic in origin). In addition the molecule may have *rotational energy* E_r. Thus the *total energy* E of the molecule is, to a very good approximation, the sum of three component parts,

$$E = E_e + E_v + E_r, \tag{IV, 8}$$

or, if we write the equation in wave-number units (term values),

$$T = T_e + G + F. \tag{IV, 9}$$

For the vibrations and rotations of the molecule in the different electronic states, we use the model of the *vibrating rotator* [Chapter III, section 2(c)] and put

$$G = \omega_e(v + \tfrac{1}{2}) - \omega_e x_e(v + \tfrac{1}{2})^2 + \omega_e y_e(v + \tfrac{1}{2})^3 + \cdots, \tag{IV, 10}$$

and

$$F = B_v J(J + 1) - D_v J^2(J + 1)^2 + \cdots. \tag{IV, 11}$$

In general F is small compared to G and in F the second term is very small compared to the first and can be neglected entirely in many cases. The vibrational and rotational constants occurring in G and F have of course different values for the different electronic states of a molecule.

By using in equation (IV, 11) the quantities B_v and D_v which depend on v and on the vibrational constants (see p. 106 and 107 f.) we are taking into account the main part of the interaction between vibration and rotation. Similarly, if $\omega_e, \omega_e x_e, \omega_e y_e, \cdots$ are so chosen that they correspond to the potential energy

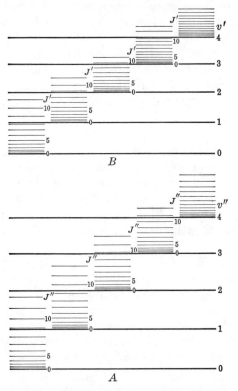

Fɪɢ. 69. **Vibrational and Rotational Levels of Two Electronic States A and B of a Molecule (Schematic).** Only the first few rotational and vibrational levels are drawn in each case.

function $U(r - r_e) = E^{el} + V_n$, we are including the main part of the interaction between vibration and electronic motion. For the present we are neglecting the interaction between electronic motion and rotation of the molecule. This interaction will be discussed in detail in the next chapter. In general, as we shall see, it produces only a small correction.

As an illustration, two different electronic states with their vibrational and rotational levels are represented graphically in Fig. 69. In the following sections we shall discuss the spectra that arise from transitions between two such sets of energy levels and we shall see that they represent in a very satisfactory way the main features of the band systems observed in the visible and ultraviolet regions (see Chapter II). In this chapter we shall consider only *singlet* electronic states.

The discussion of *multiplet* electronic states in the following chapter will lead to an explanation of the finer details of the observed band structures.

2. Vibrational Structure of Electronic Transitions

General formulae.   Using the resolution (IV, 9) for the term values of a molecule we find that the *wave numbers of the spectral lines* corresponding to the transitions between two electronic states (in emission or absorption) are given by

$$\nu = T' - T'' = (T_e' - T_e'') + (G' - G'') + (F' - F''), \quad \text{(IV, 12)}$$

where the single-primed letters refer to the upper state and the double-primed letters refer to the lower state.   According to equation (IV, 12) the emitted or absorbed frequencies may be regarded as the sums of three constituent parts (not all of which need be positive):

$$\nu = \nu_e + \nu_v + \nu_r. \quad \text{(IV, 13)}$$

For a given electronic transition, $\nu_e = T_e' - T_e''$ is a constant.   According to (IV, 12), the variable part, $\nu_v + \nu_r$, has a form similar to that for the rotation-vibration spectrum.   The essential difference is that G' and G'' now belong to different vibrational term series with different ω_e and $\omega_e x_e$, and that G' may now also be smaller than G''.   Similarly, F' and F'' belong to two quite different rotational term series with different B_e and α_e.

Since, in general, F is small compared to G, we may neglect ν_r ($= F' - F''$) for the time being in order to get a general picture.   In other words, we consider transitions between the rotationless states $F' = F'' = 0$ and obtain in this way the *coarse structure* of the electronic transition.   This structure is also called the *vibrational structure*, since only ν_v is variable.

Using (IV, 10) for G and putting $F' - F'' = 0$ in (IV, 12) we obtain the following formula for the vibrational structure [in which, for brevity, some authors use u in place of $(v + \frac{1}{2})$]

$$\nu = \nu_e + \omega_e'(v' + \tfrac{1}{2}) - \omega_e'x_e'(v' + \tfrac{1}{2})^2 + \omega_e'y_e'(v' + \tfrac{1}{2})^3 + \cdots$$
$$- [\omega_e''(v'' + \tfrac{1}{2}) - \omega_e''x_e''(v'' + \tfrac{1}{2})^2 + \omega_e''y_e''(v'' + \tfrac{1}{2})^3 + \cdots]. \quad \text{(IV, 14)}$$

This equation represents *all possible transitions between the different vibrational levels of the two participating electronic states* (see Fig. 69).   An investigation of the selection rules [see section 4(a) of this chapter] shows that for electronic transitions there is no strict selection rule for the vibrational quantum number v.   In principle, each vibrational state of the upper electronic state can combine with each vibrational state of the lower electronic state.   Thus, from (IV, 14), we can expect a large number of "lines."

The formula (IV, 14) may also be written in the following simpler way:

$$\nu = \nu_{00} + \omega_0'v' - \omega_0'x_0'v'^2 + \omega_0'y_0'v'^3 + \cdots$$
$$- (\omega_0''v'' - \omega_0''x_0''v''^2 + \omega_0''y_0''v''^3 + \cdots). \quad \text{(IV, 15)}$$

Here ν_{00} is the term independent of v' and v'' in (IV, 14); that is, it is the frequency of the transition $v' = 0 \to v'' = 0$ (the 0–0 band). By comparison of (IV, 14) and (IV, 15) it is seen that

$$\nu_{00} = \nu_e + (\tfrac{1}{2}\omega_e' - \tfrac{1}{4}\omega_e'x_e' + \tfrac{1}{8}\omega_e'y_e' + \cdots)$$
$$- (\tfrac{1}{2}\omega_e'' - \tfrac{1}{4}\omega_e''x_e'' + \tfrac{1}{8}\omega_e''y_e'' + \cdots), \quad \text{(IV, 16)}$$

while ω_0, $\omega_0 x_0$, and $\omega_0 y_0$ are given by (III, 77), p. 93. The two brackets at the right in (IV, 16) represent the zero-point vibrational energies in the upper and lower states, respectively [see equation (III, 75)]. The coefficient $\omega_e y_e = \omega_0 y_0$ is usually very small and can often be neglected. If this is done, we see that the theoretical equation (IV, 15) agrees completely with the empirical equation (II, 4) for the bands of a band system. As a matter of fact, as we have seen in Chapter II, cubic terms have to be added to equation (II, 4) in certain cases, corresponding to the cubic terms in (IV, 15).

Thus we come to the conclusion that a *band system* of the type described in Chapter II represents the *totality of the transitions between two different electronic states of a molecule* (that is, corresponds to a single line or a single multiplet of an atomic spectrum). Comparing (II, 4) with (IV, 15) we see that the empirical constants a' and a'' are to be identified with the vibrational frequencies ω_0' and ω_0'', and the empirical constants b' and b'' with the anharmonicities $\omega_0' x_0'$ and $\omega_0'' x_0''$. From these ω_e and $\omega_e x_e$ can be derived immediately by use of (III, 77) (see also below). The running numbers at first introduced empirically in the progressions are the vibrational quantum numbers in the upper and lower electronic states. Thus, on the basis of the analysis of the coarse structure of a band system in the visible or ultraviolet region of the spectrum, we can calculate the *position of the vibrational levels*, the *vibrational frequencies*, and the *anharmonicities*, as well as the *force constants* of the molecule in the two participating electronic states. Finally, from the empirical constant ν_{00} (that is, the wave number of the 0–0 band) and relation (IV, 16) we obtain ν_e, the difference in electronic energy of the two states. This ν_e is also called the *origin of the band system*.

The fact that not simply single lines appear, at the wave numbers given by (IV, 14), but whole *bands*, is due to the rotation of the molecule, as we shall see in greater detail in the following section. Formulae (IV, 14) and (IV, 15) refer to the so-called *origins* (*zero lines*) of the bands ($J' = 0 \to J'' = 0$). In contrast to this, the empirical formulae for the coarse structure of the bands usually refer to the *band heads*. Small differences between the theoretical and empirical relationships are thereby introduced. These differences will also be discussed in the following section.

For the convenience of the reader we give the formulae for ν_e, ω_e, $\omega_e x_e$, $\omega_e y_e$, $\omega_e z_e$ expressed in terms of ν_{00}, ω_0, $\omega_0 x_0$, $\omega_0 y_0$, $\omega_0 z_0$ [that is, the inverse of the formulae (III, 77) and (IV, 16)] since these formulae are always needed when the vibrational constants are to be determined from the empirical coefficients in the formulae for the band systems:

$$\omega_e z_e = \omega_0 z_0, \qquad \omega_e y_e = \omega_0 y_0 - 2\omega_0 z_0$$

$$\omega_e x_e = \omega_0 x_0 + \tfrac{3}{2}\omega_0 y_0 - \tfrac{3}{2}\omega_0 z_0 \tag{IV, 17}$$

$$\omega_e = \omega_0 + \omega_0 x_0 + \tfrac{3}{4}\omega_0 y_0 - \tfrac{1}{2}\omega_0 z_0$$

$$\nu_e = \nu_{00} - \tfrac{1}{2}(\omega_0' - \omega_0'') - \tfrac{1}{4}(\omega_0' x_0' - \omega_0'' x_0'')$$
$$\qquad - \tfrac{1}{8}(\omega_0' y_0' - \omega_0'' y_0'') + \tfrac{1}{16}(\omega_0' z_0' - \omega_0'' z_0'')$$

Examples; graphical representation. All the different bands in one and the same horizontal row in a Deslandres table (for example that of PN in Table 6, p. 41) have the same v'—that is, the same upper vibrational state—while the lower vibrational state v'' is different. These v''-*progressions* extend from the first band ($v'' = 0$) toward longer wave lengths. Such progressions are represented graphically in the energy level diagram Fig. 70(a).

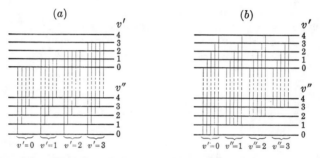

Fig. 70. **Energy Level Diagrams Representing Progressions of Bands.** (a) v''-progressions. (b) v'-progressions. The individual progressions are indicated by braces. In each case, only the first five members are drawn.

However, a band system might equally well be regarded as a series of progressions with constant v''—that is, the same lower state v'' and variable upper state v' (vertical rows in the Deslandres table). These v'-*progressions* extend from the first band, $v' = 0$, toward shorter wave lengths. They are represented graphically in the energy level diagram Fig. 70(b).

A comparison of the two diagrams in Fig. 70 with the Deslandres table shows that the constant separation of the two first v''-progressions (with $v' = 0$ and 1) gives the first vibrational quantum, $\Delta G_{\frac{1}{2}}'$, of the upper state; the constant separation of the next two v''-progressions ($v' = 1$ and 2) gives the second vibrational quantum, $\Delta G_{\frac{3}{2}}'$, of the upper state, and so on. In the example of PN (Table 6) the $\Delta G'$ values thus obtained are 1088.3, 1071.3, 1060.1, 1043.2 cm^{-1}, and so forth. Similarly, the separation of the first two v'-progressions gives the first vibrational quantum, $\Delta G_{\frac{1}{2}}''$ of the lower state, and similarly the higher vibrational quanta are obtained. In the case of PN the $\Delta G''$ values are thus found to be 1321.5, 1309.5, 1294.2, 1280.4 cm^{-1}, and so forth. From the preceding discussion it is clear that the equality of the separations of correspond ing members of two progressions is an unambiguous check for the correctness of the arrangement of the bands in the Deslandres table.

According to (III, 82), the approximately constant difference $\Delta^2 G$ of successive vibrational quanta gives the coefficients $2\omega_0 x_0 = 2\omega_e x_e$ (if we neglect $\omega_0 y_0$). For PN we obtain $\omega_0' x_0' = 7.5$ and $\omega_0'' x_0'' = 6.9$ cm^{-1}. The first vibrational quantum is $\Delta G_{1/2} = \omega_0 - \omega_0 x_0 = \omega_e - 2\omega_e x_e$. From this we obtain, in the present case, $\omega_0' = 1095.8$, $\omega_e' = 1103.3$, $\omega_0'' = 1328.4$, and $\omega_e'' = 1335.3$ cm^{-1}. More accurate values for all these constants are obtained when the calculated positions of the bands, using the above preliminary constants and equation (IV, 15), are compared with the observed positions and small adjustments of the constants are made in order to remove all systematic differences between observed and calculated values. In the case of PN, the final formula, (II, 5), was obtained in this way. The vibrational constants derived from this equation are given in Table 17, together with the force constants calculated from them [see formula (III, 91)].

TABLE 17. VIBRATIONAL CONSTANTS AND FORCE CONSTANTS IN THE UPPER AND LOWER STATES OF THE PN BANDS AND THE FOURTH POSITIVE GROUP OF CO

		PN	CO
Lower state	ω_e''	1336.36 cm^{-1}	2169.32 cm^{-1}
	ω_0''	1329.38	2156.05
	$\omega_e'' x_e''$	6.98	13.278
	k_e''	1.0144×10^6 dynes/cm	1.9005×10^6 dynes/cm
Upper state	ν_e	39,816.2 cm^{-1}	65,074.3 cm^{-1}
	ν_{00}	39,699.0	64,746.5
	ω_e'	1102.05	1515.61
	ω_0'	1094.80	1498.36
	$\omega_e' x_e'$	7.25	17.2505
	k_e'	0.6899×10^6 dynes/cm	0.9277×10^6 dynes/cm

The lower state is the ground state of the respective molecules.

The data for PN have been derived from band-head measurements; hence the true values (see Table 39) are very slightly different from those given here.

As a second example, we consider the so-called fourth positive group of CO, which forms a very extensive system of bands between 1300 and 2700 Å [see the spectrogram, Fig. 11(b), for part of the system] and often appears as an impurity in the spectra of discharges. The Deslandres table of this band system is given in Table 18. Unlike Table 6 for PN, here the band origins (corresponding to the transition $J' = 0 \to J'' = 0$; see p. 152) are given and not the band heads. In order to save space, the values of the wave-number differences between neighboring bands have been omitted. However, it can easily be seen that the differences between any two horizontal or any two vertical rows really are constant within the accuracy of the measurements (2 to 3 cm^{-1}). In the manner described above, the following formula is obtained [Read (582)]:

$$\nu = 64{,}746.5 + (1498.36v' - 17.2505v'^2)$$
$$- (2156.05v'' - 13.2600v''^2 + 0.012v''^3). \quad \text{(IV, 18)}$$

The vibrational constants derived from this formula are included in Table 17. For PN as well as CO the lower state is the ground state of the molecule (see also the energy level diagram of CO in Fig. 197, p. 452).[1]

It is particularly noteworthy that the mean difference 2143.2 cm^{-1} between the first two vertical columns of Table 18, which is, according to the above discussion, the first vibrational quantum of CO in the lower state, agrees within the accuracy of the measurements with the wave numbers of the fundamental observed in the infrared and Raman spectrum (see Table 11, p. 62). It was shown previously (Chapter III, section 2) that these fundamentals correspond to the first vibrational quanta $\Delta G_{\frac{1}{2}}$ in the ground states of the respective molecules. The second vibrational quantum obtained from the difference of the second and third column, that is, 2117.0 cm^{-1}, agrees with the difference of the first overtone and the fundamental observed in the infrared (which is 2116.7 cm^{-1}). This agreement not only confirms the assumption that the lower state of the fourth positive group is the ground state of the CO molecule (which also follows from the fact that the bands appear in absorption; see below) but in addition supplies a striking support of the interpretation of electronic band spectra in general. A similar agreement has been obtained between the infrared and ultraviolet spectrum of NO [see Gillette and Eyster (986)]. However, a complication arises here on account of the fact that the ground state is a doublet state (see Chapter V).

Absorption. As we have already emphasized in Chapter II, when a band system is observed in absorption at low temperatures, in general, *only a single v'-progression* appears—namely, that with $v'' = 0$. Thus for the CO band system just discussed, which in emission consists of the numerous bands given in Table 18, in absorption only the bands in the first column of this table appear with appreciable intensity. This is shown by the spectrogram Fig. 14. We can now readily account for this experimental result.

At room temperature practically all the molecules are in the lowest vibrational level of the electronic ground state (see Table 14, p. 123), and only extremely few are in higher vibrational levels. As a result, in absorption we have practically only transitions from the lowest vibrational level to the different vibrational levels of the upper electronic state (see Fig. 71), so that we obtain only a single progression of bands (full lines in lower part of Fig. 71). The separations of the bands decrease toward shorter wave-lengths and give directly

---

[1] The constants given in Table 39 for the ground state of CO have been obtained from a different band system; they are very slightly different from those of Table 17, but believed to be more accurate.

TABLE 18. DESLANDRES TABLE FOR THE BAND

The data are taken from Birge (103), Estey (203), Headrick and Fox (283), from the band heads, Schmid and Gerö's (621) rotational constants being used the others. In addition to those given, the bands 9–22, 11–22,

v' \ v''	0	1	2	3	4	5	6	7	8	9	10
0	(64,703)	62,601.8	60,484.7	58,393.2	56,329.4	54,291.8	(52,266)				
1	66,231.3	64,087.6	....	59,881.6	57,818.3	55,780.6	53,768.5	51,782.4	49,823.4	(47,887)	
2	67,674.8	65,533.1	63,416.1	61,325.2	....	57,224.2	55,212.9	53,227.7	51,268.2	49,334.5	47,427.3
3	69,087.8	66,944.3	64,828.1	....	60,674.5	58,636.3	....	54,640.3	52,680.6	50,747.2	48,839.7
4	70,469.5	68,323.4	(66,199)	64,116.5	62,055.3	(59,990)	58,002.9	(56,017)	(54,041)	52,125.2	50,218.2
5	71,807.2	69,666.1	(67,550)	65,458.2	....	61,356.6	(59,342)	57,360.0	55,400.9	(53,457)	51,560.0
6	(73,093)	70,973.2	68,855.9	(66,754)	(64,714)	62,665.1	....	58,667.1	....	54,774.5	(52,866)
7	....	72,248.3	70,131.1	....	(65,971)	....	(61,908)	....	....	....	(54,142)
8	....	(73,453)	71,370.8	(69,260)	(67,198)	(65,174)	....	....	....	....	....
9	....	....	72,576.9	(70,466)	....	(66,360)	....	....	....	....	....
10	....	....	73,750.0	(71,644)	....						
11	....	....	(74,902)	(72,791)	(70,717)	....					
12	....	....	....	(73,873)	(71,824)	....	(67,773)	(65,765)	....		
13	....	....	....	....	(72,893)	(70,848)	(68,846)	....			
14	....	....	....	....	....	....	....	(67,884)			

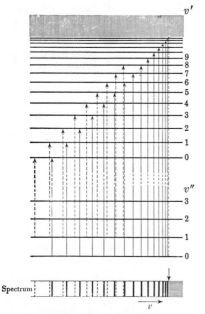

FIG. 71. **Energy Level Diagram for an Absorption Band Series with Convergence and Continuum (Schematic).** The resulting spectrum is drawn below. The vertical arrow in the schematic spectrogram gives the position of the convergence limit. The transitions starting from $v'' = 1$ at higher temperature are indicated by broken lines.

the *vibrational quanta of the upper state.* According to (IV, 15), with $v'' = 0$, the formula for this band series is

$$\nu = \nu_{00} + \omega_0'v' - \omega_0'x_0'v'^2 + \cdots .$$
(IV, 19)

With increasing v' the separations between successive vibrational levels approach the value zero, and then a *continuous* term spectrum joins onto the series of discrete vibrational levels [see Chapter III, section 2(a)]. Corresponding to this, we have to expect a *continuous absorption spectrum* joining onto the series of discrete absorption bands at the point where the separation of the bands becomes zero (the *convergence limit*). This continuous absorption spectrum corresponds to the dissociation of the molecule. It is also indicated in Fig. 71. A continuum of this nature has actually been observed in a number of cases, such as I_2, for which it is shown in Fig. 15, p. 38. In other cases, such as, for example, CO (Fig. 14), it is not observed, since the intensity of the bands has dropped to zero long before

ORIGINS OF THE FOURTH POSITIVE GROUP OF CO

Read (582), and Gerö (250). Where necessary, the origins have been calculated (see p. 171). The figures in parentheses are considerably less accurate than 12–22, 11–23, 12–23, 14–23, and 13–24 have been measured.

11	12	13	14	15	16	17	18	19	20	21
(45,529)										
46,959.2	45,101.9	(43,249)								
48,335.4	46,480.7	44,649.9	(42,845)			(37,567)	(35,866)			
49,678.2	47,821.9	45,992.8	44,187.4	42,408.2	40,655.3		(37,228)			
	49,130.0	47,300.3	45,496.1	43,716.6	(41,962)	(40,234)		(36,842)		
			46,770.2	44,990.5	43,238.1	41,509.0	(39,817)		(36,469)	
				(46,214)	44,476.5	42,748.6	41,045.5	39,367.7		
				(45,657)	43,953.9	42,251.3	40,574.0	38,920.2	(37,278)	
						(43,403)	41,744.6	40,092.2	38,464.0	
								41,229.0	(39,580)	
							(43,986)			
										(41,766)

the continuum is reached. In yet other cases, as, for example, F_2, *only* a continuum appears. These continua will be dealt with in detail in Chapter VII.

If the vibrational quanta in the ground state are small (as, for example, for the I_2 molecule), or generally when the temperature is high, the number of molecules in the first or even higher vibrational levels of the ground state is no longer negligibly small (see Table 14, p. 123). As a result, the progression $v'' = 1$ and even higher progressions appear in absorption in addition to the progression $v'' = 0$, although with smaller intensity. The separations of the bands in these progressions are exactly the same as in the progression $v'' = 0$ (see Chapter II, section 1). For an analytical representation the general formula (IV, 15) has to be used. The progression $v'' = 1$ is indicated by broken lines in Fig. 71. The absorption spectrum of AgCl reproduced in Fig. 13 and obtained at a temperature of 900° C supplies a good example for the occurrence of progressions with $v'' > 0$. In this case, on account of the high temperature and the comparatively low value of $\Delta G''$ these progressions have an intensity of the same order as that of the progression $v'' = 0$. Bands with $v'' = 1$ can also be observed in absorption at room temperature for molecules with large vibrational quanta if sufficiently thick absorbing layers or sufficiently high pressures are used. Thus some bands with $v'' = 1$ are just detectable in the CO absorption spectrum shown in Fig. 14.

It is clear that a band system can be observed in absorption only when its lower electronic state is the *electronic ground state* of the molecule unless the lower state is very close to the ground state or the temperature is extremely high. Many emission band systems have excited electronic states as the lower states. Therefore they are not observed in absorption.

Excitation of single progressions in emission.   If a molecular gas is irradiated with light having the wave length of a single absorption band, the absorbing molecules will be brought into the upper state of this particular absorption band only.   The excited molecules can then go over to the different vibrational levels of the ground state with emission of radiation.   We should thus expect, in *fluorescence*, a progression $v' = $ constant, which extends from the wave length of the exciting light toward *longer* wave lengths with *decreasing* band separation. This is in exact agreement with the observations of *resonance fluorescence* (see Chapter II).   Fig. 72(a) shows by an energy level diagram how such a *resonance series* originates.   It can be seen that the separations of the fluorescence bands are equal to the vibrational quanta in the ground state.

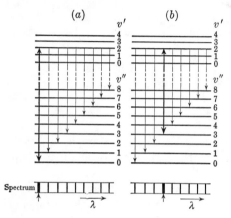

When the absorption band leading to the excitation of the fluorescence is a band with $v'' \neq 0$, as is possible for heavy molecules having small vibrational quanta (for example, I_2, Na_2, and so forth; see above) some of the fluorescence bands may have v'' smaller than that of the exciting band and, as shown by Fig. 72(b), these fluorescence bands lie on the *short-wave-length* side of the exciting band. They are called *anti-Stokes members* of the resonance series.   In the fluorescence spectrum of Na_2, which was given earlier as an example (Fig. 29), a resonance series with anti-Stokes lines (upper leading lines) is present, as well as one without such lines (lower leading lines).

Fig. 72.   **Energy Level Diagram for Resonance Series** (a) **without and** (b) **with anti-Stokes Members.**   The line exciting the fluorescence is indicated in the energy level diagrams by a double arrow, and in the spectra below by a small arrow.

In principle it is possible for the resonance series to be observed up to the point of convergence, where $\Delta G'' = 0$.   At this point, then, a continuous spectrum extending to *longer* wave lengths should join on (somewhat similar to the continuum going from the point of convergence of absorption bands to *shorter* wave lengths).   Actually, such a case of a continuous emission spectrum with preceding band series has not yet been observed, though cases of emission continua without preceding bands are known (see Chapter VII).

A selective excitation of a single v''-progression can occur in emission not only in fluorescence but under certain circumstances also in electrical discharges, and may be due to an *instability of higher vibrational levels*, to *resonance in collisions of the second kind* (see A.A., p. 231) or to an *inverse predissociation* (see Chapter VII).   As an example of the first case, we refer to the H bands of CO, for which only the v''-progression with $v' = 0$ appears, as Fig. 11(a), p. 35 clearly shows.   An example of the second case is supplied by H_2 when it is electrically excited in the presence of a large excess of argon.   Without the argon an extensive and complicated band system (many-line spectrum) is observed in the vacuum ultraviolet, which corresponds to the transitions from the first excited electronic state B of H_2 to the ground state (see Fig. 160, p. 340). However, with the argon only one simple series of bands appears, which converges toward longer wave-lengths and whose upper state has $v' = 3$ (the so-called Lyman bands). It is found that the excitation energy of this upper state is very nearly equal to the excitation energy of the metastable ^3P state of argon. By collisions of these metastable argon atoms (which are fairly abundant in a discharge) with normal H_2 molecules, according to the principle of resonance in collisions of the second kind, the $v = 3$ level of the B state of H_2 is preferentially excited [for more details see Beutler (89)]. The separations of the Lyman bands give directly the vibrational quanta for the ground state of the H_2 molecule [Witmer (718)]. A probable example of the third cause of selective excitation is the production of the $v' = 6$ bands of the Swan system of C_2 in discharges in CO at high pressure. This will be discussed in more detail in Chapter VII.

Sequences (diagonal groups). As we have already seen in Chapter II, if the constants a' and a'' in (II, 4) are not very different—that is, if the vibrational quanta in the upper and lower states have similar magnitudes—the bands that lie on the diagonal in the Deslandres table (from the upper left to the lower right), or those that lie on parallels to it, form characteristic groups in the spectrum. These groups are called *sequences* (or *diagonal groups*). For each of them $\Delta v = v' - v''$ is a constant. The reason for the formation of these groups is made clear by the energy level diagram Fig. 73 which applies to PN (see Fig. 9). Unlike Fig. 70, this figure, like the majority of the energy level diagrams in this book, is drawn so that the horizontal separation of the vertical transition arrows is proportional to the difference in length of the arrows—that is, is proportional to the separation of the bands in question. Thus the projection of the arrows on a horizontal line gives the spectrum, which is represented in the lower part of the figure. The sequences are indicated by braces. It is immediately clear from Fig. 73 that the bands of a sequence draw closer together the more nearly alike are the vibrational quanta in the upper and lower states. In addition to the PN spectrum (Fig. 9) the band systems of C_2, CN,

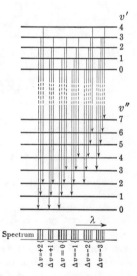

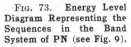

FIG. 73. Energy Level Diagram Representing the Sequences in the Band System of PN (see Fig. 9).

and N_2 shown in Fig. 7 and 8 provide examples of sequences of various degrees of compactness. If the difference of the ΔG values of the upper and lower states is large, different sequences overlap one another and thus do not stand out as particularly pronounced groups. This is the case in the P_2 spectrum (Fig. 10).

A formula representing the bands of a sequence is easily obtained by putting $v' = v'' + \Delta v$ in the general formula (IV, 15) and regarding Δv as constant. Neglecting the cubic terms, we obtain

$$\nu = \nu_{00} + \omega_0'\Delta v - \omega_0'x_0'(\Delta v)^2 - (\omega_0'' - \omega_0' + 2\omega_0'x_0'\Delta v)v''$$
$$- (\omega_0'x_0' - \omega_0''x_0'')v''^2 \tag{IV, 20}$$

In this formula the quantum number v'' takes the values 0, 1, 2, $\cdots$ for the sequences with $\Delta v \geqq 0$ (that is, for bands on the diagonal of the Deslandres table and below it), whereas it takes the values $|\Delta v|$, $|\Delta v| + 1$, $\cdots$ for the sequences with $\Delta v < 0$ (that is, for bands above the diagonal).

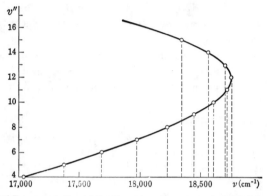

FIG. 74.   **Graphical Representation of the Sequence $\Delta v = -4$ of the N_2^+ Bands [after Herzberg (303)].**   The abscissae of the small circles give the positions of the bands in the spectrum.

It can be seen from (IV, 20) that the bands in a sequence, just as those in a progression, follow a quadratic formula except when $\omega_0'x_0'$ happens to be equal to $\omega_0''x_0''$. As long as the linear term is not too small, the sequences extend from an initial band (with the lowest v'') to longer or shorter wave lengths, according as ω_0'' is larger or smaller than $\omega_0' - 2\omega_0'x_0'\Delta v$. In the example of the PN bands considered above (Figs. 73 and 9), the sequences extend toward longer wave lengths and the bands in the sequences are approximately equidistant, since $\omega_0'x_0'$ is not very different from $\omega_0''x_0''$. Since, in general, the individual bands are shaded to the red or the violet according as ω_0'' is larger or smaller than ω_0' (see p. 174), a sequence generally begins with a band head and the shading of the individual bands is in the direction toward which the sequences extend, as is the case for PN (Fig. 9).

If in (IV, 20) the term linear in v'' is small and has the opposite sign to the quadratic term, a *reversal* can take place in the succession of the bands in a sequence. This may be seen from the graphical representation in Fig. 74, in which, similar to a Fortrat diagram of an individual band (see p. 48) the bands in the sequence $\Delta v = -4$ of N_2^+ are plotted with v'' as ordinate and ν as abscissa. If no terms higher than the quadratic occur in (IV, 20) the curve obtained is a parabola. It can be seen that with increasing v'' the bands draw together, eventually coming quite close to one another, and then draw apart again in the opposite direction. The turning point may be called a *head of heads*, since it forms a head in a series of band heads.

The value of v'' at which the *turning point* lies is obtained from (IV, 20) by putting $\dfrac{d\nu}{dv''} = 0$. We find

$$v_t'' = \frac{\omega_0' - 2\omega_0'x_0'\Delta v - \omega_0''}{2(\omega_0'x_0' - \omega_0''x_0'')}, \tag{IV, 21}$$

where the whole number nearest to v_t'' gives the v'' value of the turning point. If v_t'' comes out negative no head of heads appears.

If higher powers of v' and v'' have to be taken into consideration, the analytical expression for the turning point becomes rather more complicated. In this case, a graphical procedure for determining the turning point is much simpler [see Herzberg (303)] and, at the same time, makes even clearer the formation of the head of heads. The turning point evidently occurs when, for two successive bands of the diagonal, or a parallel to it, the vertical difference in the Deslandres table is equal to the horizontal difference. Therefore, if we plot the $\Delta G'$ values and $\Delta G''$ values against v in the same diagram (solid curves in Fig. 75, which refers to N_2^+), the intersection of the two curves gives the v value of the turning point for the sequence

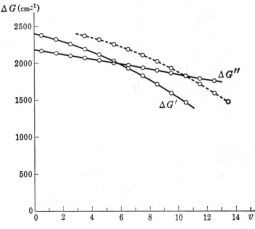

FIG. 75. Δ*G* Curves for N_2^+ in the Upper and Lower State of the Violet Bands. The broken-line curve is the $\Delta G'$ curve shifted by 3 units to the right.

$\Delta v = 0$ (in the example, at $v = 6$). The turning points in the other sequences ($\Delta v \neq 0$) are obtained, as may easily be seen, by shifting the $\Delta G'$ curve by an amount $|\Delta v|$ to the left or the right according as Δv is positive or negative, and determining the intersection with the undisplaced $\Delta G''$ curve. In Fig. 75 the broken-line curve is the $\Delta G'$ curve displaced three units. Its intersection with $\Delta G''$ gives the turning point in the sequence $\Delta v = -3$, which lies at $v'' = 10$. From this mode of presentation we can see very clearly how the turning point shifts in the different sequences.

Since the intensity of the bands in a sequence always decreases with increasing v'' (possibly after passing through an intensity maximum first), the head of heads mentioned above can be observed only in very special cases, where the ΔG curves of the two participating electronic states intersect at fairly low v values. This is true for the violet bands of CN [Jenkins (366)] and N_2^+ [Herzberg (303)] as well as for some other band systems. To be sure, for CN the turning point itself is not actually observed. However, bands are observed which lie on the returning limb of the parabola in Fig. 74 beyond the first band of the sequence considered (see the spectrogram Fig. 7). These bands, also called *tail bands*, are generally shaded in a direction opposite to that of the bands at the beginning of the sequence as will be discussed in more detail in section 3.

Vibrational analysis. When the sequences in a band system are well developed, the vibrational analysis is very simple. Starting from longer wave lengths, the separations of the sequences (that is, of their first bands) at first gradually increase up to a certain point, at which they change suddenly and then slowly decrease from the new value (see Fig. 73). This variation results from the fact that the first bands of the long-wave-length sequences have $v' = 0$, and thus form a v'' progression, while the first bands of the short-wave-length sequences have $v'' = 0$ and thus form a v' progression. The sequence for which the sudden change in the separations occurs is the main sequence with $\Delta v = 0$. In this way the correct arrangement of the bands in a Deslandres table is immediately obtained, and the vibrational quanta in the upper and lower states can be determined in the manner already described.

If the sequences are not well developed, as, for example, in the fourth positive group of CO (p. 155 f.) and in the P_2 spectrum (Fig. 10), we have to try to find series of approximately equidistant bands whose separations decrease slowly either toward long wave lengths (v''-progressions) or toward short wave lengths (v'-progressions). It must be realized of course that also in sequences the separations of successive bands may decrease slowly. Whether the progressions found are genuine may be unambiguously tested by the constancy of the separations of corresponding members of two such series. This constancy holds rigorously for progressions only, not for sequences or any other series of bands (except accidentally). Conversely, we may also adopt the procedure of looking for pairs of bands with equal separations. When a series of such pairs has been found, all having the same separation, they must represent either two v''- or two v'-progressions. If we have once found two such progressions, it is relatively easy to make further progress, since we need only to draw one of the progressions on a strip of paper and to shift it relative to the band system drawn on another strip of paper until a coincidence with another progression is found. In this way the whole Deslandres table, and from it the vibrational constants, can be obtained.

It may also happen, particularly if ω' is very close to ω'' that only one sequence is observed —namely, the principal sequence ($\Delta v = 0$). In such a case, the determination of the vibrational constants is clearly not possible, since no two bands of a sequence have the same upper or lower state. At least two sequences must be observed in order to make such a determination possible. On the other hand, when a single *progression* is observed, the vibrational constants of one of the states can be determined.

Isotope effect. The potential energy functions of two isotopic molecules, such as $B^{10}O$ and $B^{11}O$ are identical to a high degree of approximation since they depend only on the motions of the electrons and the Coulomb repulsion of the nuclei (see p. 147). The latter, of course, is entirely independent of the masses of the nuclei while the former are almost independent of these masses (see below). Not only the form of the potential curves but also the relative positions of the potential curves of different electronic states, that is, the electronic energies T_e are the same for two isotopic molecules. The mass difference affects only the vibrational and rotational energy of the molecule in each electronic state. Restricting our considerations to the non-rotating molecule we obtain from (III, 195) and (IV, 14) for the band systems of two isotopic molecules, neglecting cubic and higher terms,

$$\nu = \nu_e + \omega_e'(v' + \tfrac{1}{2}) - \omega_e'x_e'(v' + \tfrac{1}{2})^2$$
$$- [\omega_e''(v'' + \tfrac{1}{2}) - \omega_e''x_e''(v'' + \tfrac{1}{2})^2],$$

$$\nu^i = \nu_e + \rho\omega_e'(v' + \tfrac{1}{2}) - \rho^2\omega_e'x_e'(v' + \tfrac{1}{2})^2$$
$$- [\rho\omega_e''(v'' + \tfrac{1}{2}) - \rho^2\omega_e''x_e''(v'' + \tfrac{1}{2})^2].$$

(IV, 22)

Here ν_e is as previously the difference in energy of the minima of the potential curves of the two electronic states involved which is the same to a very good approximation for the two isotopic molecules.

The equations (IV, 22) can be written in the approximate form

$$\nu = \nu_e + \nu_v, \qquad \nu^i = \nu_e + \rho\nu_v,$$

if ρ^2 in the term with $(v + \frac{1}{2})^2$ is replaced by ρ. This does not introduce any great error, since this term is, in any event, small compared to the term linear in $(v + \frac{1}{2})$ and since ρ is usually quite close to unity. We can see, therefore, that the *band system of the heavier isotopic molecule is contracted compared to that of the lighter, approximately by a constant factor* $\rho = \sqrt{\mu/\mu^i}$, which is less than 1.

As an illustration, Fig. 13, p. 37, shows a spectrogram of the AgCl absorption bands after Brice (125). The bands belonging to the more abundant isotope $AgCl^{35}$ are clearly separated from those belonging to the less abundant isotope $AgCl^{37}$. It can be seen further that the isotopic displacement increases more or less linearly with increasing distance from the 0–0 band. To the right of the 0–0 band, the weaker $AgCl^{37}$ bands are overlapped by the stronger $AgCl^{35}$ bands and therefore cannot be seen separately in the reproduction. In this case of AgCl as well as in other cases, although ρ is close to 1, the isotopic displacement becomes comparable to the separation of successive bands for large ν_v values. On photographs of greater dispersion, each of the AgCl bands is once more split into two components as a result of the isotopy of silver. For the two silver isotopes, Ag^{107} and Ag^{109}, ρ is much nearer to 1 than for the Cl isotopes, giving a much smaller splitting.

In the old quantum theory the formula for the vibrational levels of the anharmonic oscillator was

$$G_0(v) = \omega_0 v - \omega_0 x_0 v^2 + \cdots$$

that is, there was no zero-point energy. This expression leads to the formula (IV, 15) for a band system which, as we have seen, is completely equivalent to (IV, 14), as far as the representation of the different bands of a band system is concerned. However, in the old quantum theory the formula for the band system of an isotopic molecule would be obtained by replacing ω_0 by $\rho\omega_0$ and $\omega_0 x_0$ by $\rho^2 \omega_0 x_0$, and therefore ν_{00}, the wave number of the 0–0 band, would remain unchanged. On the other hand the wave-mechanical formulae (IV, 22) do give a shift for the 0–0 band, since ν_v is not zero for the 0–0 band, owing to the fact that the zero-point energies in the upper and lower states, in general, have different magnitudes and are different for the two isotopic molecules. Such an *isotopic displacement for the 0–0 band* is actually observed in a number of cases, although, because of its comparative smallness, not in all the cases investigated. However, even then the existence of a displacement for the 0–0 band follows indirectly from the fact that the observed shifts of the other bands agree with formula (IV, 22) and not with the corresponding formula of the old quantum theory.

As an example of the difference between old and new quantum theory, the observed isotopic shifts in the v'-progression with $v'' = 0$, of the α bands of BO, corresponding to the two isotopes B^{10} and B^{11}, are given in Table 19. A shift

TABLE 19.   ISOTOPIC DISPLACEMENTS IN THE v'-PROGRESSION

WITH $v'' = 0$ OF THE α BANDS OF BO

[According to data of Jenkins and McKellar (373)]

Band	Observed Isotopic Displacement $B^{10}O - B^{11}O$ (cm^{-1})	Calculated from Quantum Mechanics (cm^{-1})	Calculated from Old Quantum Theory (cm^{-1})
0–0	− 8.6	− 9.08	0
1–0	+ 26.7	+ 26.29	+ 35.69
2–0	+ 60.8	+ 60.36	+ 70.09
3–0	+ 93.6	+ 93.14	+103.20
4–0	+125.2	+124.63	+135.01

of the 0–0 band is directly observed here.   The third and fourth columns give the values calculated in the two ways described above.   It can be seen that the observed values are in very satisfactory agreement with the values calculated from the wave-mechanical formula (IV, 22) (*half-integral vibrational quanta*)[1a] and deviate widely from those calculated from the old quantum theory (integral vibrational quanta).   The same has been found to be true for all other cases thus far investigated.   *Thus the existence of the zero-point vibration is proved.*

While in the great majority of cases the agreement of the observed isotope shifts with those predicted on the elementary theory presented above is exceedingly good, very slight discrepancies have been found for some spectra of deuterides for which the shift from the corresponding hydrides is particularly large [see, for example, Watson (692) and particularly Crawford and Jorgensen (173)].   A more elaborate theory of the isotope effect which accounts in a general way for these slight discrepancies has been developed by Kronig (423), Dieke (184), and Van Vleck (682).   *Four contributions to the deviations from the simple isotope theory* must be considered according to Van Vleck: (a) anharmonicity terms, (b) terms due to the interaction of electronic and nuclear motions, (c) L uncoupling terms, and (d) terms due to interaction of electronic states of the same Λ value.

The anharmonicity terms are due to the fact that the empirical constants ω_e, B_e, $\omega_e x_e$, $\cdots$ are not precisely represented by $(1/2\pi c)\sqrt{k_e/\mu}$, $h/(8\pi^2 c\mu r_e^2)$, and so forth, as was first shown by Dunham (194) (see p. 109).   Explicit formulae for these correction terms have been given by Crawford and Jorgensen (1 3).   The second contribution (b) is due to the fact that the separation of electronic and nuclear motion (see p. 148 f.) is not rigorously possible.   The motions of the electrons do not adapt themselves immediately to the varying positions of the nuclei reached during the vibration but lag behind very slightly.   In other words the potential energy acting upon the actual (moving) nuclei is not exactly the same as that obtained for infinitely slow ("fixed") nuclei.   This effect is clearly more pronounced for the lighter isotopic

---

[1a] The very small, approximately constant, difference between the observed and the calculated values is due to an *electronic* isotope effect.

molecule since the nuclei move faster than for the heavier one. In addition, the center of mass of the nuclei does not quite coincide with the center of mass of the whole molecule, but rather, on account of the motion of the electrons, the former wobbles about the latter, an effect that is, of course, largest for the lightest isotopic molecule and leads to an electronic isotope shift which is entirely analogous to the isotope effect in the spectra of light atoms (see A.A., p. 184). The contributions (c) and (d) to the deviations from the simple theory of the isotope effect are due to the interaction of different electronic states. L uncoupling will be discussed in the following chapter. It causes a deviation only for the rotational isotope effect (see p. 229 f.). The interaction of states of equal Λ affects only the vibrational isotope effect. The magnitudes of both effects are inversely proportional to the separation of the state considered from the electronic states interacting with it.

In addition to the isotope shifts caused by the difference in mass Mrozowski (1244) (1249) has discovered a *nuclear isotope shift* due to the different deviations from a Coulomb field near isotopic nuclei. Such an effect is known to be responsible for the isotope shifts in the line spectra of the heavier atoms where it produces shifts of the order of 0.2 cm^{-1} or less. For molecules such small shifts can be detected only when the ordinary mass isotope shifts are of an equally small order. This is the case for the mercury isotope effect in HgH, HgD, HgH$^+$, HgD$^+$, the zinc isotope effect in ZnH, and in certain similar cases. For HgH, in particular, isotope splittings were found by Mrozowski (1244) which are several times the shifts calculated from the simple isotope theory and are sometimes even in the opposite direction. All these deviations can be accounted for satisfactorily as nuclear isotope shifts.

Applications of the isotope effect. In several instances the study of the isotope effect has proved of great value in the solution of certain spectroscopic and other problems. One problem of this type is the *vibrational analysis of extensive band systems*. If, for example, in a series of absorption bands the intensity rises gradually from zero to a maximum, in general it is not possible to say whether or not the first observed band is the 0–0 band since further bands extending the system to longer wave lengths may be too weak for observation. Thus only a relative numbering of the individual bands can be obtained. However, when an isotope effect is observed, the absolute numbering is easily found, since the isotopic displacement must almost (though not completely) vanish for the 0–0 band, and naturally, the shifts for the other bands must also agree with the calculated values. In this way, for example, the vibrational numbering has been determined for the absorption bands of ICl [Gibson (255), Patkowski and Curtis (552) and Darbyshire (880)] and Br$_2$ [Brown (133)].

Another problem that may be solved by means of the isotope effect is that of the *carrier of* (that is, the molecule giving rise to) *an observed band system*. While in absorption experiments in pure gases there can usually be no doubt about the carrier of a given absorption band system, in emission—in electric discharges, flames, and the like—band systems often appear that belong to molecules that are not chemically stable, but which are formed in the discharge or flame and which, therefore, are not so readily identified. In addition, band systems of impurities appear frequently with a much greater relative intensity than their concentration would appear to warrant. Since the magnitude of the isotope effect depends on the masses of the two participating atoms, it can be used to decide between the different possibilities for the carrier of such band systems. For example, the above-mentioned BO bands were at first taken to be BN bands, since they appear by excitation of BCl_3 vapor with active nitrogen. Mulliken (508) then showed, however, that the magnitude of the isotope shift was not compatible with this assumption, but was compatible only with the assumption that BO is the carrier of the band system, the oxygen being due to a slight impurity. In a similar manner, for example, it has been shown definitely that the two band systems appearing in the green and the violet spectrum of the ordinary carbon arc (see Fig. 7) are due to C_2 and CN molecules, respectively, a question that was for a long time in doubt.

The investigation of the isotope effect in electronic band spectra has led, furthermore, to the *discovery of a number of new isotopes of very small abundance.* By using very long absorbing layers or, in emission, very long exposure times, it is possible to observe isotope bands that are very weak in comparison to the main bands.

The first rare isotope discovered by the use of band spectra was the *oxygen isotope* of mass 18. Already in 1927 Dieke and Babcock (186) had found in the solar spectrum an extremely weak band A' near the intense atmospheric O_2 band A at 7596 Å (0–0 band). Both bands have exactly the same structure. Fig. 76 gives a spectrogram of this region of the solar spectrum. In spite of the identical structure, the A' band does not fit into the Deslandres table of the main atmospheric oxygen bands, being much closer to the A band than the 1–1 band, which is the nearest band of the system. The explanation of this A' band remained a riddle until Giauque and Johnston (253) showed that it could be explained quantitatively if it was ascribed to a molecule $O^{16}O^{18}$, whose absorption bands could be calculated according to (IV, 22) from those of the ordinary O_2^{16} molecule. From the relative intensities of the two bands the *relative abundance* of the two isotopic oxygen atoms was estimated to be 1 : 630 [Childs and Mecke (154)]. Shortly after the discovery of O^{18}, Babcock (74) found a further band A'', five times weaker than A', which also accompanied the main atmospheric band A. This band was identified as an $O^{16}O^{17}$ band by Giauque and Johnston (254), and thereby the existence of an O^{17} isotope was proved.

Later, in addition to the 0–0 band, the 1–0 and 2–0 bands of the $O^{16}O^{18}$ molecule were also observed. Furthermore, it was established that every second line was *not* missing for the $O^{16}O^{18}$ molecule, as it is for $O^{16}O^{16}$. This difference is particularly significant since it confirms the theoretical prediction discussed earlier (p. 141). Though most of the even-numbered lines of the A' band are covered by the strong (odd-numbered) lines of the A band, nevertheless a few of them can be seen in Fig. 76. Since their discovery in the O_2 spectrum the oxygen isotopes have also been found in the spectra of other molecules containing oxygen, for example, in the absorption spectrum of NO by Naudé (527). They have also been detected by means of the mass spectrograph and in this way a more precise value for the abundance ratio, namely, $O^{16} : O^{18} : O^{17} = 506 : 1 : 0.204$, has been found [Birge (809) (810)].

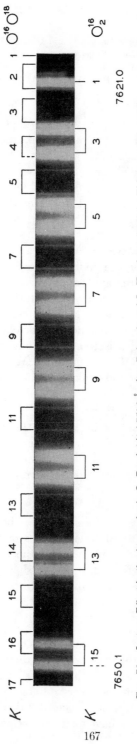

K $O^{16}O^{18}$

K O_2^{16}

7621.0

7650.1

FIG. 76. **Isotope Effect in the Atmospheric O_2 Band** (A) $\lambda 7596$ Å **after Babcock and L. Herzberg (764).*** This is part of a solar spectrum taken at low altitude of the sun. Only part of the P branch of the band is shown. The lines of O_2^{16} which appear very broad because the absorption is very intense are indicated below the spectrogram, the lines of $O^{16}O^{18}$ are indicated above it. The very weak lines inside the $O^{16}O^{18}$ doublets numbered 9 and 11 are due to $O^{16}O^{17}$.

* The author is greatly indebted to Mr. H. D. Babcock for this spectrogram.

167

As a result of the existence of the isotopes O^{17} and O^{18}, the chemical *atomic-weight scale* (based on the oxygen mixture = 16) does not agree exactly with the Aston mass-spectroscopic scale, which is based on $O^{16} = 16$ (see Table 1). From the relative abundance of the isotopes, the following conversion formula is obtained [see Birge (809) (810)]:

$$A_{\text{Aston}} = 1.000272 \, A_{\text{chem}}. \tag{IV, 23}$$

Shortly after the discovery of the oxygen isotopes the rare isotopes C^{13} of carbon and N^{15} of nitrogen were found, also by means of their band spectra, by King and Birge (400) and Naudé (527), respectively. Further work on the carbon and nitrogen isotope effects, partly with gases in which the rare isotopes had been enriched, was carried out by Jenkins and Ornstein (374), Jenkins and Wooldridge (379), Townes and Smythe (1470), Herzberg (306), Krüger (424), and Wood and Dieke (720a). Another fairly rare isotope discovered by its band spectrum is In^{113} [Wehrli (1518) (1519)].

The measurement of the relative intensities of isotopic bands may serve as a convenient means of determining the abundance ratio of isotopes, particularly in cases in which a mass spectrograph cannot be applied readily. (See also Chapter VIII, section 4). As in the case of the rotation-vibration spectra, from an accurate measurement of the isotope shifts in electronic band spectra, the ratio of the masses of the isotopes concerned can be obtained with considerable accuracy (except for the hydrogen isotopes, see above). Such determinations have been made by Birge (105) (106), Jenkins and McKellar (373), Jenkins and Wooldridge (379), and McKellar and Jenkins (1215).

3. Rotational Structure of Electronic Bands

General relations. In the foregoing section the contribution of the rotational energy to the emitted or absorbed frequencies was neglected and only the vibrational contribution considered. We shall now consider the possible changes in the rotational state for any given vibrational transition. In the expression

$$\nu = \nu_e + \nu_v + \nu_r$$

the quantity $\nu_0 = \nu_e + \nu_v$ is constant for a specific vibrational transition, while ν_r is variable and depends on the different values of the rotational quantum number in the upper and lower states. All of the possible transitions for a constant value of ν_0 taken together form, as we shall presently see in greater detail, a *single band*. As in the case of rotation-vibration spectra, we have for such a band

$$\nu = \nu_0 + F'(J') - F''(J''), \tag{IV, 24}$$

where $F'(J')$ and $F''(J'')$ are the rotational terms of the upper and lower state, respectively. ν_0 is also called the *band origin* or the *zero line*. The difference from the infrared and Raman spectra is that now F' and F'' belong to different electronic states and can therefore have very different magnitudes.

In the most general case previously considered, that is, assuming a non-rigid, vibrating symmetric top, we have for the rotational term values [compare equation (III, 150)]

$$F_v(J) = B_v J(J + 1) + (A - B_v)\Lambda^2 - D_v J^2 (J + 1)^2 + \cdots. \qquad (IV, 25)$$

Here the term containing Λ^2 is constant for a given vibrational level of a given electronic state. As a result, this term in Λ^2 can be completely neglected in the calculation of the possible rotational transitions if we choose ν_0 in (IV, 24) appropriately. Therefore, in this chapter, in order to avoid introducing yet another special symbol, we shall understand $F_v(J)$ to mean the rotational term value measured from the rotational level having $J = 0$; that is,

$$F_v(J) = B_v J(J + 1) - D_v J^2 (J + 1)^2 + \cdots. \qquad (IV, 26)$$

Then, according to (IV, 24), we have

$$\nu = \nu_0 + B_v' J'(J' + 1) - D_v' J'^2 (J' + 1)^2 + \cdots$$
$$- [B_v'' J''(J'' + 1) - D_v'' J''^2 (J'' + 1)^2 \cdots]. \qquad (IV, 27)$$

The branches of a band. The selection rules to be applied in the present case are those of the symmetric top (see p. 119). Unlike the case of infrared spectra now the upper and lower states may have different electronic angular momenta Λ. If at least one of the two states has $\Lambda \neq 0$ the selection rule for J is [compare (III, 154)]

$$\Delta J = J' - J'' = 0, \pm 1. \qquad (IV, 28)$$

If, however, $\Lambda = 0$ in both electronic states ($^1\Sigma - {}^1\Sigma$ transition; see Chapter V), the transition with $\Delta J = 0$ is forbidden and only $\Delta J = \pm 1$ appears, as for most infrared bands. Thus we have to expect *three or two series of lines* (*branches*), respectively, whose wave numbers are given by the following formulae (just as for the rotation-vibration spectra):

$$R \text{ branch: } \nu = \nu_0 + F_v'(J + 1) - F_v''(J) = R(J), \qquad (IV, 29)$$

$$Q \text{ branch: } \nu = \nu_0 + F_v'(J) \quad - F_v''(J) = Q(J), \qquad (IV, 30)$$

$$P \text{ branch: } \nu = \nu_0 + F_v'(J - 1) - F_v''(J) = P(J). \qquad (IV, 31)$$

Here the J's are the rotational quantum numbers in the lower state ($= J''$).

If we substitute (IV, 26) into (IV, 29–31), neglecting the small correction term in D_v, we obtain the following formulae which are identical with those for the branches in the rotation-vibration spectra (see Chapter III):

$$\nu = \nu_0 + 2B_v' + (3B_v' - B_v'')J + (B_v' - B_v'')J^2 = R(J), \qquad (IV, 32)$$

$$\nu = \nu_0 \qquad + (\ B_v' - B_v'')J + (B_v' - B_v'')J^2 = Q(J), \qquad (IV, 33)$$

$$\nu = \nu_0 \qquad - (\ B_v' + B_v'')J + (B_v' - B_v'')J^2 = P(J). \qquad (IV, 34)$$

As before, the P and R branches can be represented by a single formula,

$$\nu = \nu_0 + (B_v' + B_v'')m + (B_v' - B_v'')m^2, \qquad (IV, 35)$$

where $m = -J$ for the P branch and $m = J + 1$ for the R branch. Thus, when no Q branch is present ($^1\Sigma$–$^1\Sigma$ transition), there is *one simple series of lines* whose separations change regularly. This corresponds exactly to the simplest observed fine structures of bands described in Chapter II. Formula (IV, 35) has precisely the same form as the empirical equation (II, 8). The only essential difference compared to the infrared spectra is that now B_v' and B_v'' may have very different magnitudes and that, as a result, the quadratic term in (IV, 32–35) may be much greater (that is, there may be a much more rapid alteration in the separation of successive lines), in agreement with experiment.

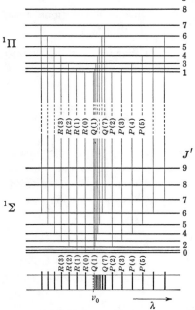

If there is no Q branch the possible transitions are the same as those represented in the previous Fig. 54 (p. 112). It is seen immediately from this figure that the smallest value of J in the R branch is $J = 0$ and in the P branch $J = 1$. As a result, according to (IV, 32 and 34), no line appears in the position $\nu = \nu_0$ (the band origin or zero line). In agreement with experiment (see Figs. 18 and 19), there is a gap in the series of lines (IV, 35), the so-called *zero gap* (or null-line gap), analogous to that in the infrared spectrum.

Fig. 77. **Energy Level Diagram for a Band with *P*, *Q*, and *R* Branches.** For the sake of clarity, in the spectrogram below, the lines of the P and R branches, which form a single series, are represented by longer lines than those of the Q branch. The separation of the lines in the Q branch has been made somewhat too large in order that the lines might be drawn separately. The convergence in the P and R branches is frequently much more rapid than shown.

The case where *all three branches* (P, Q, and R) appear is represented by an energy level diagram in Fig. 77. In this figure it is assumed that $\Lambda = 1$ in the upper state and $\Lambda = 0$ in the lower state ($^1\Pi$–$^1\Sigma$ transition; see Chapter V). As a result, the lowest level in the upper state has $J = 1$ (see p. 117). The various transitions with $\Delta J = +1, 0,$ and -1 are indicated in the figure. It can be seen that the *first lines* in the R, Q, and P branches are those having $J = 0, 1,$ and 2, respectively. As a result, there are now two lines missing in the series (IV, 35) formed by the R and P branches—namely, at $\nu = \nu_0$ and $\nu = \nu_0 - 2B_v''$. This is shown in the schematic spectrogram in the lower part of Fig. 77. However, the gap in the series formed by R and P branches is not so apparent in the present case since the Q branch begins in the neighborhood of ν_0. The first Q line ($J = 1$) lies at $\nu = \nu_0 + 2(B_v' - B_v'')$. The addition of the Q branch, which overlaps the P or R branch, gives the band a somewhat more complicated appearance than in the case considered before (compare the spectrogram Fig. 20).

Band-head formation; shading (degrading) of bands. A graphical representation of the simplest type of band with simple P and R branches by a *Fortrat diagram* has already been given in Fig. 24. In most cases, owing to the quadratic term in (IV, 35), one of the two branches "turns back" (broken part of the parabola in Fig. 24), thus forming a *band head*. This gives rise to the appearance so characteristic of most bands in the visible and ultraviolet regions. We shall now consider in somewhat greater detail the conditions under which such a band head (corresponding to the vertex of the Fortrat parabola) appears and we shall calculate its separation from the zero line (band origin).

A head is formed in the R branch if $B_v{}' - B_v{}''$ is negative, since then the linear and quadratic terms in (IV, 32) have opposite signs, and therefore a maximum of ν is reached at a certain J value (vertex of the parabola). Thus in this case the head lies on the short-wave-length side of the zero line and the band is *shaded* (degraded) *toward the red* (toward longer wave lengths). A negative value of $B_v{}' - B_v{}''$ implies that $B_v{}' < B_v{}''$, that is, that the internuclear distance in the upper state is greater than that in the lower [see (III, 115)]. Conversely, when $B_v{}' > B_v{}''$—that is, when the internuclear distance in the upper state is smaller than that in the lower—the coefficient $B_v{}' - B_v{}''$ is positive and the band head therefore lies in the P branch [compare equation (IV, 34)]; the band is *shaded* (degraded) *toward the violet* (toward shorter wave lengths), as is the case in the example given in Fig. 24.

The m value corresponding to the *vertex of the Fortrat parabola*—that is, to the band head (see Fig. 24)—is obtained by putting $d\nu/dm = 0$ in (IV, 35). This gives

$$m_{\text{vertex}} = -\frac{B_v{}' + B_v{}''}{2(B_v{}' - B_v{}'')} \tag{IV, 36}$$

If this expression is substituted in (IV, 35), we obtain for the separation between the zero line and the vertex

$$\nu_{\text{vertex}} - \nu_0 = -\frac{(B_v{}' + B_v{}'')^2}{4(B_v{}' - B_v{}'')} \tag{IV, 37}$$

Naturally, for m_{vertex} an integral value is not usually obtained. The actual head then lies at the nearest whole-numbered value of m—that is, at the nearest $\nu(m)$ in (IV, 35). However, for most practical purposes, (IV, 36 and 37) can be considered to apply to the band head. In agreement with the preceding discussion about the direction of shading, $\nu_{\text{vertex}} - \nu_0$ in (IV, 37) is positive and the band shaded to the red if $B_v{}' < B_v{}''$, while the opposite is the case for $B_v{}' > B_v{}''$. Furthermore, equations (IV, 36 and 37) show that the head lies at a greater distance from the zero line, the smaller $B_v{}' - B_v{}''$ is. If $B_v{}'$ is very nearly equal to $B_v{}''$, the head may lie at such a great distance from the band origin that it is not observed, since for the corresponding m value the intensity of the lines may have decreased to zero (see section 4). This is almost always the case for infrared rotation-vibration bands (see Chapter III). However, such cases also occur occasionally in electronic bands—for example, for certain bands of CN [Jenkins (366)] and C_2 [Herzberg and Sutton (1052) and Landsverk (1164)].

By comparing (IV, 33) with (IV, 35) it is seen immediately that the lines of the Q *branch*, if one is present, form a different parabola in the Fortrat diagram from the one corresponding to the P and R lines. Since the quadratic terms in (IV, 33) and (IV, 35) are the same, the vertices of the two parabolae point in the same direction, in agreement with the empirical results discussed in Chapter II (see Fig. 25, p. 49). If $B_v{'}$ and $B_v{''}$ are not very different, as in the example of the AlH band in Fig. 20 and 25 ($B_v{'} = 6.024$ and $B_v{''} = 6.296$ cm^{-1}), the Q-branch parabola intersects the abscissa axis almost at right angles, owing to the smallness of the linear term in (IV, 33). This means that a head is formed at the beginning of the Q branch, as is clearly shown in the spectrogram, Fig. 20. Thus, bands having a Q branch very often show *two* heads either in the P and Q branches ($B_v{'} > B_v{''}$, shading to the violet) or in the R and Q branches ($B_v{'} < B_v{''}$, shading to the red). The latter is the case for the AlH band, as well as for the previously discussed PN bands, for some of which a second head (Q head) can be seen clearly in Fig. 9, p. 33, while for others it is hidden by the R head.

As is easily seen from (IV, 33) the vertex of the Q branch parabola lies at $J = -\frac{1}{2}$ and $\nu = \nu_0 - \frac{1}{4}(B_v{'} - B_v{''})$, that is, very close to the band origin. The first line of the Q branch is the one with $J = 1$ (see above), that is, no lines very close to the vertex occur and of course there is no returning limb of the corresponding Fortrat parabola. If $B_v{'}$ is very different from $B_v{''}$ the separation of the lines of the Q branch increases rapidly and therefore the head of the Q branch is not at all pronounced, particularly, since at the same time, the separation of the R or P head from the origin is small. The bands of the fourth positive group of CO [see Fig. 11(b)] approach this type.

If $B_v{'}$ is approximately equal to $B_v{''}$, according to (IV, 33) all the lines of the Q branch nearly coincide at $\nu \approx \nu_0$. We then have an intense, almost line-like Q branch, on either side of which P and R branches extend with approximately equal separations of the lines, as in the infrared spectrum of the symmetric top. An example is the BH band at 4330 Å [see Lochte-Holtgreven and van der Vleugel (451)].

For different bands of the same band system—that is, different $v' - v''$ combinations—the pairs of values $B_v{'}$ and $B_v{''}$ are somewhat different, since, according to (III, 116), B_v depends on v. The separation of the zero line (origin) from the band head therefore changes from band to band. For example, for the PN bands discussed earlier, $\nu_{\text{head}} - \nu_0$ varies between the values 5.7 and 12.7 cm^{-1}. This variation accounts for the fact that in the Deslandres table of band heads the wave number differences between corresponding bands in two vertical or two horizontal rows are often not exactly constant: The constancy of these differences, as has been shown in the preceding section, holds rigorously for the non-rotating molecule only—that is, for the *band origins* ($J = 0$). It would hold equally for any fixed J and ΔJ value. But it does not hold rigorously for the band heads with their varying separations from the zero line and their varying J values. Nevertheless, as long as the separation $\nu_{head} - \nu_0$ is itself small—that is, as long as $B_v{}'$ and $B_v{}''$ are not too nearly the same—the deviation from constancy of the differences when band heads are used is slight and introduces only a rather small error in the vibrational constants derived, particularly since in many cases the variation of $\nu_{head} - \nu_0$ is less than in the example of PN given above.

As we have seen above, the head of the Q branch lies very close to ν_0. Thus, in a band system in which Q branches appear whose heads can be measured, the Q *heads*, and not the P or R heads, should be used to determine the vibrational constants. Naturally, still more accurate values for the vibrational constants are obtained from the *band origins* themselves (for their determination, see p. 185).

When, as is usually the case, the difference between $B_e{}'$ and $B_e{}''$ is not too small, the difference $B_v{}' - B_v{}''$ has the same sign for all the bands (owing to the smallness of the rotational constant α). Therefore all the bands of a band system are *usually shaded in the same direction*.

However, if $B_e{}'$ and $B_e{}''$ are only slightly different from each other, there may be, in one and the same band system, bands for which $B_v{}' - B_v{}''$ is positive as well as bands for which $B_v{}' - B_v{}''$ is negative—that is, bands that are shaded to the violet as well as bands that are shaded to the red. Such a case was first established by Jenkins (366) for the violet CN bands. The B_v curves for the upper and lower states of these bands are plotted in Fig. 78. It can be seen from this figure that the bands in the main sequence ($v' = v''$) are shaded to the violet as long as $v < 9$, since then $B_v{}' > B_v{}''$. On the other hand, for larger values of v, one has $B_v{}' < B_v{}''$; that is, the bands are shaded to the red. Thus a *reversal of the shading* is expected and indeed observed in the main sequence. For the bands of the other sequences the direction of shading is found by shifting the $B_v{}'$ curve by $|\Delta v|$ to the left or to the right, depending on whether $\Delta v = v' - v''$ is positive or negative. The point of intersection with the (unshifted) $B_v{}''$ curve gives the point where a reversal of the shading takes place in the particular sequence. The most intense bands of this system (with small v) are thus shaded to the violet (see the spectrogram, Fig. 7), whereas the weaker bands with larger v are shaded to the red. The latter lie for the most part on the long-wave-length side of the corresponding sequences—that is, on the returning limb of the parabola that represents the bands of a sequence (see p. 160). Some of these bands can be seen in Fig. 7 (broken lead-

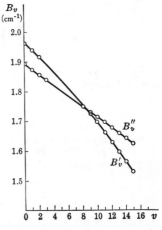

Fig. 78. B_v Curves for the **Upper and Lower States** of the **Violet CN Bands** [According to the Data of Jenkins (366) and **Jenkins, Roots, and Mulliken** (377)].

ing lines). As already remarked, they are called *tail bands*, since, before they were theoretically explained, they were thought to form the end of the preceding sequence of bands. Other cases of a reversal of shading have been observed for $N_2{}^+$ [Herzberg (303), Coster and Brons (168) (169)] and C_2 [Herzberg and Sutton (1052), Phillips (1315)]. Usually the bands in the neighborhood of the turning point of the shading ($B' \approx B''$) have no heads at all and are of the same appearance as rotation-vibration bands (see the references cited above).

The B values of the different electronic states of a molecule run fairly parallel to the ω values, that is, ω/B is nearly constant, or in other words, the smaller the internuclear distance of an electronic state, the larger is the force constant. Since, in addition, α_e/B_e roughly equals $\omega_e x_e/\omega_e$ (see p. 106) it follows, and is indeed observed, that usually when the ΔG curves of two electronic states intersect, the corresponding B_v curves also intersect, at similar v values. Thus, when a reversal of the bands is observed in the sequences (formation of heads of heads), a reversal of the shading usually occurs also (compare the above references).

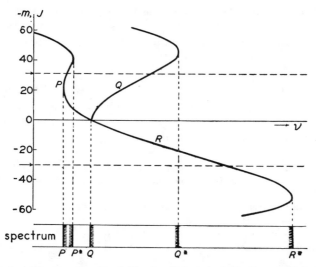

Fig. 79. **Fortrat Diagram of a Band Showing Reversal of Shading at High J Values.** The schematic spectrum below indicates the positions of the ordinary and "extra" heads formed.

Thus far, in the discussion of this section, we have neglected the *influence of the rotational constant D*, which is, of course, usually quite small. However, the introduction of this correction is necessary in the case of accurate measurements, particularly for high J values. For the P and R branches we then obtain, from (IV, 26) and (IV, 29–31), [compare equation (III, 139) for the rotation-vibration bands],

$$\nu = \nu_0 + (B_v' + B_v'')m + (B_v' - B_v'' - D_v' + D_v'')m^2$$
$$- 2(D_v' + D_v'')m^3 - (D_v' - D_v'')m^4 \qquad \text{(IV, 38)}$$

and for the Q branch

$$\nu = \nu_0 + (B_v' - B_v'')J(J + 1) - (D_v' - D_v'')J^2(J + 1)^2 \qquad \text{(IV, 39)}$$

It should be noted that now, in contrast to the case of rotation-vibration spectra, D_v' may be very different from D_v'' in which case the quartic term would be important.

While in general the cubic and quartic terms do not alter the previous considerations about the appearance of the bands, in extreme cases these terms can bring about a *change of shading in one and the same band*, particularly if the band is observed up to very high J values. Since equations of the fourth degree, such as (IV, 38) and (IV, 39), may have as many as two minima and one maximum, or one minimum and two maxima, it is clear that the series formed by P and R branch [equation (IV, 38)] may have as many as three heads, while the Q branch, since only positive J values occur, may have up to two heads. Fig. 79 gives a Fortrat diagram for a case in which all these heads occur. The heads are indicated in the schematic spectrum below. If lines only up to m or J values as indicated by the two small arrows are observed the band would have a normal appearance with a normal P and Q head. But if sufficiently high J values occur the additional heads marked P^*, Q^*, and R^* will appear with shading opposite to that of the normal heads marked P and Q. It is seen from equations (IV, 38) and (IV, 39) that such a reversal of shading will occur only if $D' - D''$ has the same sign as $B' - B''$. Even if this is the case the extra heads usually lie at such high J values that they are not observed. Rather special conditions must be met in order that the extra heads lie at sufficiently low J values to become observable.

A convenient way of ascertaining in a given case whether or not the extra heads may be observable is to plot $\Delta_1 F = F(J + 1) - F(J)$ (see below) for the upper and lower state of the band considered, and see whether, in addition to the intersection of the two curves $\Delta_1 F'$ and $\Delta_1 F''$ at $J = -1$ which corresponds to the ordinary Q head, there is another intersection for observable J values which would correspond to the Q^* head. The extra heads in P and R branch would be obtained by shifting the $\Delta_1 F'$ curve one unit to the right or left and finding the intersection with the unshifted $\Delta_1 F''$ curve (this is entirely similar to the graphical determination of the head of heads in different sequences as illustrated by Fig. 75). It is readily seen from such a representation that the extra heads will be observable, that is, that there will be a change of shading in one and the same band only when $B_v' - B_v''$ is small and $D_v' - D_v''$ is relatively large and has the same sign as $B_v' - B_v''$. Extra heads have been observed for bands of BeH [Watson (690)], of BH [Lochte-Holtgreven and van der Vleugel (451)], of the indium and gallium halides [Wehrli and Miescher (701) (495)], and of BCl [Herzberg and Hushley (1049)]. The extra head of the Q branch is usually most prominent, giving rise to a characteristic narrow "band" with a sharp long *and* short wave length edge. In some cases the maximum and minimum of the P branch (or R branch) corresponding to the two heads (Fig. 79) merge into a point of inflection which for an unresolved band gives rise to a narrow intensity maximum *without* sharp edges.

Combination relations and evaluation of the rotational constants for bands without Q branches. In principle, the rotational constants B_v' and B_v'' for electronic bands could be evaluated by comparing the empirical formula (II, 8) with the theoretical formula (IV, 35) in the manner previously described for rotation-vibration bands (see p. 113). In practice a different procedure is used to obtain the rotational constants—namely, the evaluation from the combination relations or combination differences. As we shall see, these relations permit us to obtain separately the upper and lower rotational levels from the observed band lines and then to evaluate the rotational constants separately. It must be emphasized that this procedure yields more accurate constants than that of fitting a polynomial in m or J to the observed band lines.

Let us first consider the case of a band that has only a single P and a single R branch and let us take the wave numbers of the lines as well as their correct numbering (position of the zero gap) as given. How the correct numbering can be found in cases where it is not obvious will be discussed later. As an example, the wave numbers of the lines of the 0–0 band of the green BeO bands are given in Table 20, in which the J numbering (not the m numbering) is used (see p. 169).

In such a simple band there is to every line in the P branch a corresponding line in the R branch with the same upper state. The wave number difference of these two lines, as is readily seen from Fig. 54, p. 112, is equal to the separation,

TABLE 20.   FINE STRUCTURE OF THE GREEN BeO BANDS

Wave numbers of the lines of the 0–0 band and combination differences for the bands 0–0, 0–1, and 2–1 after Rosenthal and Jenkins (602)

J	$R(J)$	$P(J)$	0–0 Band $\Delta_2 F''(J) = R(J-1)-P(J+1)$	0–0 Band $\Delta_2 F'(J) = R(J)-P(J)$	0–0 Band $\dfrac{\Delta_2 F''(J)}{J+\frac{1}{2}}$	0–0 Band $\dfrac{\Delta_2 F'(J)}{J+\frac{1}{2}}$	0–1 Band $\Delta_2 F''(J) = R(J-1)-P(J+1)$	0–1 Band $\Delta_2 F'(J) = R(J)-P(J)$	2–1 Band $\Delta_2 F''(J-1) = R(J-1)-P(J+1)$	2–1 Band $\Delta_2 F'(J) = R(J)-P(J)$
0	21,199.81									
1	202.88	21,193.25	9.84	9.63	6.560	6.420		9.44		
2	205.74	189.97	16.47	15.77	6.588	6.308	16.18	15.68		
3	208.52	186.41	23.08	22.11	6.594	6.317	22.71	21.93		
4	211.12	182.66	29.64	28.46	6.587	6.324	29.16	28.22	29.02	33.80
5	213.53	178.88	36.30	34.70	6.600	6.309	35.63	34.48		
6	215.53	174.82	42.93	40.76	6.605	6.271	42.19	40.68	42.16	46.09
7	217.71	170.65	49.23	47.06	6.564	6.275	48.59	46.94	48.65	52.19
8	219.65	166.35	55.78	53.30	6.562	6.271	55.04	53.15	55.09	58.39
9	221.43	161.93	62.37	59.50	6.565	6.263	61.49	59.48	61.57	
10	223.10	157.28	68.86	65.84	6.553	6.271	67.95	65.78	68.12	
11	224.62	152.57	75.42	72.05	6.558	6.265	74.59	71.98		
12	225.92	147.70	81.93	78.22	6.554	6.253	81.10	78.41		76.72
13	227.23	142.60	88.39	84.54	6.547	6.262	87.48	84.70	87.46	82.71
14	228.40	137.53	95.01	90.87	6.552	6.267	93.98	90.82	93.77	88.89
15	229.34	132.22	101.65	97.12	6.558	6.266	100.42	97.07	100.23	95.00
16	230.03	126.75	108.33	103.33	6.566	6.262	106.86	103.29	106.70	101.28
17	230.67	121.01	114.69	109.66	6.554	6.266	113.31	109.53	113.34	
18	230.9+	115.39	121.27	115.55	6.555	6.246	119.73	115.75		
19		109.40								

— $v'=0$ —   — $v''=1$ —

The R and P branches of the 0–0 band have been measured up to $J = 59$ and 61, respectively. For the 2–1 band a number of $\Delta_2 F$ values cannot be obtained because of overlapping of the corresponding lines by others.

$\Delta_2 F''(J)$, of one of the *lower*-state rotational levels from the next but one; for example, the difference between the lines $P(5)$ and $R(3)$ (both having $J = 4$ in the upper state) is equal to the separation of the rotational levels $J = 3$ and $J = 5$ in the lower state.   Thus

$$R(J - 1) - P(J + 1) = F_v''(J + 1) - F_v''(J - 1) = \Delta_2 F''(J) \quad \text{(IV, 40)}$$

Correspondingly, as can also be seen from Fig. 54, the difference between the wave numbers of two lines with a common lower state is equal to the separation of one of the *upper*-state rotational levels from the next but one:

$$R(J) - P(J) = F_v'(J + 1) - F_v'(J - 1) = \Delta_2 F'(J) \quad \text{(IV, 41)}$$

The correctness of the *combination relations* (IV, 40 and 41) may also be verified easily by substitution of (IV, 29 and 31).   It is important for the following discussion that the precise meaning of these relations is clearly visualized.   In order to avoid too many subscripts we are writing $\Delta_2 F(J)$ in place of $\Delta_2 F_v(J)$.

It should be noted that (IV, 40 and 41) are quite independent of the formula (IV, 26) for the rotational levels and would also hold when irregularities, so-called perturbations, occur (see Chapter V, section 4).   In the example of Table 20, $\Delta_2 F''(J)$ and $\Delta_2 F'(J)$ values derived from the observed wave numbers of the lines are given in the fourth and fifth columns.

By forming the combination differences $R(J - 1) - P(J + 1) = \Delta_2 F''(J)$ and $R(J) - P(J) = \Delta_2 F'(J)$ *the two series of rotational terms are separated from each other*, since, according to (IV, 40), the former depends only on the lower state and the latter, according to (IV, 41), depends only on the upper state.   By adding corresponding $\Delta_2 F(J)$ values, we can obtain the position of the rotational levels in the lower as well as the upper state.   However, in general it is not necessary to carry out this addition, since the rotational constants B_v and D_v, which determine the position of the rotational levels, can be obtained directly from the combination differences.   If we substitute the expression (IV, 26) for $F_v(J)$ and neglect the term in D_v we obtain

$$\Delta_2 F(J) = F_v(J+1) - F_v(J-1) = B_v(J+1)(J+2) - B_v(J-1)J$$

$$= 4B_v(J + \tfrac{1}{2}) \quad \text{(IV, 42)}$$

(About the small correction terms due to D_v, see below.)   Thus the combination differences $\Delta_2 F(J)$ form, to a first good approximation, a linear function of the rotational quantum number J going through zero for $J = -\tfrac{1}{2}$.   As an illustration, the values of $\Delta_2 F''(J)$ and $\Delta_2 F'(J)$ from Table 20 are represented graphically in Fig. 80.   The agreement with a straight line is better than can be shown in a drawing made to this scale.   From the slopes of these lines, according to (IV, 42), we obtain $4B_v''$ and $4B_v'$, respectively.

Instead of determining $4B_v$ graphically, we can also obtain it according to (IV, 42) as the mean value of $\Delta_2 F(J)/(J + \tfrac{1}{2})$.   The individual values of $\Delta_2 F(J)/(J + \tfrac{1}{2})$ are given in the sixth and seventh columns of Table 20, and

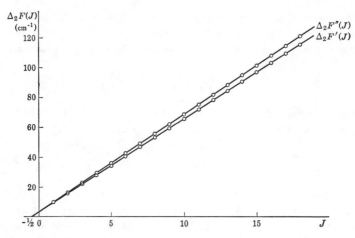

FIG. 80. $\Delta_2 F(J)$ Curves for the Upper and Lower States of the 0–0 Band of the Green BeO System.

their constancy shows the accuracy of formula (IV, 42).   From the figures in the table we obtain

$$B_0' = 1.569 \text{ cm}^{-1} \quad \text{and}$$

$$B_0'' = 1.642 \text{ cm}^{-1}.$$

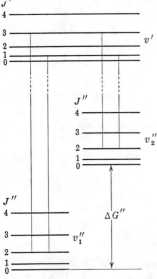

FIG. 81.   **Agreement of the Combination Differences for Two Different Bands with the Same Upper State.**   The agreement is shown for $J'' = 2$.   In both bands for the P branch $J' = 1$ and for the R branch $J' = 3$;   that is, the difference $R(2) - P(2)$ is the same for the two bands $v' \to v_1''$ and $v' \to v_2''$, and similarly for all other J values.

If two (or more) bands with the same v' are measured, the combination differences $\Delta_2 F'(J)$ for the *upper* state of the two bands *must agree exactly*, within the accuracy of measurement, for each J value and irrespective of perturbations.   This is illustrated by the energy level diagram of Fig. 81.   The combination differences $\Delta_2 F''(J)$ and $\Delta_2 F'(J)$, formed for the 0–1 band, of BeO are given in the eighth and ninth columns of Table 20.   (In order to save space, the wave numbers of the band lines are omitted.)   It can be seen that, owing to the fact that v' is the same, the $\Delta_2 F'(J)$ values for the two bands 0–0 and 0–1 agree very closely for each J value,[2] whereas the $\Delta_2 F''(J)$ values are different.

[2] The accuracy of the wave numbers of the band lines in Table 20 is about ± 0.1 cm^{-1}.   Naturally the accuracy of the differences $\Delta_2 F$ is somewhat less.   This should be taken into account when comparing the different $\Delta_2 F$ values.

A corresponding state of affairs exists, of course, for bands having a common *lower* vibrational state. For them, the $\Delta_2 F''(J)$ must agree exactly for all values of J. To illustrate this, the $\Delta_2 F''$ and $\Delta_2 F'$ values for the 2–1 band are given in the tenth and eleventh columns of Table 20. It can be seen that the $\Delta_2 F''$ values agree with those of the 0–1 band, whereas the $\Delta_2 F'$ values are different.

The agreement between corresponding combination differences for bands with the same lower or the same upper vibrational states forms an important and very sensitive *check on the correctness of a rotational analysis*. Such an agreement between the $\Delta_2 F$ values must, of course, hold also for corresponding bands of two different band systems if they have an electronic state in common and can, conversely (just as the agreement of the vibrational differences), be used for an unambiguous decision as to whether or not the electronic state is common to the two band systems.

If there are a number of bands having an upper or lower vibrational state in common, the *mean $\Delta_2 F$ values* are of course used for the determination of the corresponding B_v value. The B_v' and B_v'' values for the green BeO bands, determined in this way, are collected in Table 21. (Actually, in determining

TABLE 21.   ROTATIONAL CONSTANTS B_v OF BeO IN THE UPPER AND LOWER
STATES OF THE GREEN BANDS[3]

[After Lagerqvist and Westöö (1157) (1160)]

v	B_v''	B_v'
0	1.6415	1.5684
1	1.6223	1.5529
2	1.6035	1.5371
3	1.5846	1.5216
4	1.5656	1.5065
5	1.5465	1.4913
6	1.5273	1.4759

these B_v' and B_v'' values the correction due to the rotational constant D_v was taken into account in the manner described below). It can be seen that the B_v values decrease very nearly linearly with increasing v (see also the example of HCl, p. 114), in agreement with the earlier formula (III, 116):

$$B_v = B_e - \alpha_e(v + \tfrac{1}{2}) + \cdots \qquad (IV, 43)$$

---

[3] Note that these constants are not from Rosenthal and Jenkins (602) on whose work Table 20 is based but from the recent extensive re-investigation of the BeO spectrum by Lagerqvist and Westöö. It is for this reason that the B_0 value derived below (p. 181) does not agree exactly with the one given in this table. Unfortunately Lagerqvist and Westöö do not give the wave numbers of the band lines and it is therefore not possible to use their data exclusively.

where according to (III, 115)

$$B_e = \frac{h}{8\pi^2 c I_e} = \frac{h}{8\pi^2 c \mu r_e^2} \tag{IV, 44}$$

A graphical representation of B_v curves has been given in Fig. 78, which refers to CN. In this figure, the B_v'' curve is linear over a very wide range, whereas the B_v' curve has a marked curvature. However, the latter may also be regarded as linear for small values of v.

The *rotational constants* B_e for the unrealizable, completely vibrationless state, as well as α_e, are obtained from the observed B_v values in the manner described on p. 113. In the example of the green BeO bands (Table 21) we find that

$$B_v' = 1.5758 - 0.0154(v' + \tfrac{1}{2}),$$
$$B_v'' = 1.6510 - 0.0190(v'' + \tfrac{1}{2}),$$

where the constant terms are the B_e values. From these we can obtain the *moment of inertia* I_e and the *internuclear distance* r_e in the equilibrium position of the molecule (minimum of the potential curve) according to equation (IV, 44). Solving this equation for I_e and r_e and substituting numerical values for h, c, and N_A from Table 1, one finds

$$I_e = \mu r_e^2 = \frac{27.98_0}{B_e} \times 10^{-40} \text{ gm cm}^2 \tag{IV, 45}$$

and

$$r_e = \frac{4.10610}{\sqrt{\mu_A B_e}} \times 10^{-8} \text{ cm}, \tag{IV, 46}$$

where $\mu_A (= \mu N_A)$ is the reduced mass in Aston atomic-weight units ($O^{16} = 16$). It follows that, in the example of BeO,

$$I_e' = 17.75_8 \times 10^{-40} \text{ gm cm}^2, \quad \text{and} \quad r_e' = 1.362_2 \times 10^{-8} \text{ cm},$$
$$I_e'' = 16.94_9 \times 10^{-40} \text{ gm cm}^2, \quad \text{and} \quad r_e'' = 1.330_8 \times 10^{-8} \text{ cm}.$$

Apart from I_e and r_e, sometimes I_0 and r_0 are given. These are the values calculated from (IV, 45 and 4$_0$) when B_0 is used instead of B_e. They are mean values of the moment of inertia and internuclear distance, respectively, (obtained by averaging in a peculiar fashion, see p. 106) in the lowest vibrational level of the molecule. In the example of BeO

$$I_0' = 17.84_5 \times 10^{-40} \text{ gm cm}^2, \quad \text{and} \quad r_0' = 1.365_5 \times 10^{-8} \text{ cm},$$
$$I_0'' = 17.04_7 \times 10^{-40} \text{ gm cm}^2, \quad \text{and} \quad r_0'' = 1.334_6 \times 10^{-8} \text{ cm}.$$

In order to obtain higher accuracy in the determination of the rotational constants it is necessary to include higher rotational lines in the measurements and at the same time, in the evaluation, *take account of the non-rigidity of the rotator*, that is, of the rotational constants D_v and possibly higher correction terms. Substituting (IV, 26) with the D_v correction into

$\Delta_2 F(J) = F(J + 1) - F(J - 1)$ one obtains in place of (IV, 42)

$$\Delta_2 F(J) = (4B_v - 6D_v)(J + \tfrac{1}{2}) - 8D_v(J + \tfrac{1}{2})^3 \qquad \text{(IV, 47)}$$

If still higher correction terms are included in (IV, 26) one would obtain in (IV, 47) higher (odd) powers of $(J + \tfrac{1}{2})$. However, such terms need be taken into account in exceptional cases only [see, for example, the analysis of the spectrum of LiH by Crawford and Jorgensen (172)(173)]. In what follows we shall always neglect these higher correction terms.

Since in most cases D_v is of the order $10^{-5}B$ one can usually neglect $6D_v$ in the first bracket of (IV, 47) unless an accuracy higher than 0.001% is aimed at. For convenience we shall therefore use in the following discussion

$$\Delta_2 F(J) = 4B_v(J + \tfrac{1}{2}) - 8D_v(J + \tfrac{1}{2})^3 \qquad \text{(IV, 48)}$$

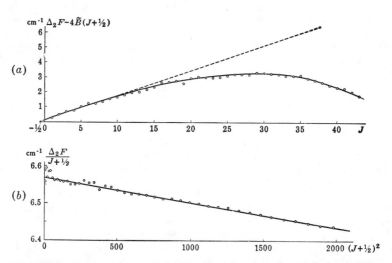

Fig. 82. Graphical Determinations of the Rotational Constants B_v and D_v for the state $v'' = 0$ of the Green BeO Bands.   (·) $\Delta_2 F(J) - 4\tilde{B}_v(J + \tfrac{1}{2})$ as a function of J with $\tilde{B}_v = 1.6$.   (b) $\Delta_2 F(J)/(J + \tfrac{1}{2})$ as a function of $(J + \tfrac{1}{2})^2$.

in place of (IV, 47). It is always easy to correct for this neglect, if it should affect the last significant figure of B_v, after the evaluations have been completed.

The formula (IV, 48) shows that for high J values the $\Delta_2 F$ curve lies somewhat below the straight line $4B_v(J + \tfrac{1}{2})$. To be sure, this is scarcely noticeable in a drawing to the scale used in Fig. 80 since D_v is so small. However, if we subtract $4\tilde{B}_v(J + \tfrac{1}{2})$ from each of the $\Delta_2 F(J)$ values, where $\tilde{B}_v$ is an approximate value for B_v, the difference $\Delta_2 F(J) - 4\tilde{B}_v(J + \tfrac{1}{2})$ can be plotted on a much larger scale and the curvature becomes noticeable. This has been done in Fig. 82(a) for the lower state with $v'' = 0$ of the green BeO bands. In drawing the figure, additional higher $\Delta_2 F''(J)$ values not given in Table 20 have been used [see Rosenthal and Jenkins (´02)]. In order to obtain an accurate value for B_v we have to draw a tangent to the curve for small J values (broken straight line). The slope of this line gives the correction $4\Delta P_v$, which has to be applied to the preliminary $4\tilde{B}_v$ in order to obtain a better value of $4B_v$. Since the figure can be drawn to a suitable scale, ΔB_v and, correspondingly, B_v itself can be very accurately determined by this graphical method. In the figure, $\tilde{B}_v$ is assumed to be 1.6. From the drawing, $\Delta B_v = 0.0423$ cm^{-1}, so that, since $v = 0$, $B_0 = 1.6423$ cm^{-1}.

The deviation of the $\Delta_2 F$ curve from the broken straight line represents the correction term $8D_v(J + \tfrac{1}{2})^3$ in (IV, 48). Thus at least a preliminary value for D_v can be obtained if the deviation for a high value of J, say $J = 40$, is divided by $8(J + \frac{1}{2})^3$. In the present case the value of D_v so obtained is 8.33×10^{-6} cm^{-1}. Since the tangent to the $\Delta_2 F$ curve cannot be drawn very accurately, it is necessary to check the correctness of the D value so obtained by calculating $8D_v(J + \frac{1}{2})^3$ for all J values and adding it to the observed $\Delta_2 F$ values. Since, according to (IV, 48),

$$\Delta_2 F(J) + 8D_v(J + \tfrac{1}{2})^3 = 4B_v(J + \tfrac{1}{2}), \tag{IV, 49}$$

if D_v is chosen correctly, the sum to the left, plotted as a function of J, must give exactly a straight line, even for high J values. This result can be tested in a drawing of the same type as Fig. 82(a), and if there is still a curvature a further correction to D_v can be found.

If the deviation of the $\Delta_2 F$ curve from a straight line cannot be determined with sufficient accuracy, the best procedure is to calculate D_v from the theoretical formulae (III, 117, 118, and 125) (but possibly neglecting the constant β_e) instead of determining it from the experimental data according to the method just given. B_v can then be accurately determined from the slope of the line $\Delta_2 F(J) + 8D_v(J + \frac{1}{2})^3$ in the manner described. In this way, at any rate, a considerably more accurate B_v value is obtained than when D_v is completely neglected.

A *second graphical method* consists of plotting $\Delta_2 F(J)/(J + \frac{1}{2})$ against $(J + \frac{1}{2})^2$. According to (IV, 48),

$$\frac{\Delta_2 F(J)}{(J + \frac{1}{2})} = 4B_v - 8D_v(J + \tfrac{1}{2})^2 \tag{IV, 50}$$

Without taking D_v into account, $\Delta_2 F(J)/(J + \frac{1}{2})$ would be a constant (see above). Owing to the D_v correction, when plotting against $(J + \frac{1}{2})^2$, we obtain a straight line slightly inclined to the abscissa axis. Fig. 82(b) shows this for the same case as Fig. 82(a) (BeO in the ground state with $v'' = 0$). The slope of the line gives $8D_v$, and the intercept on the ordinate axis gives $4B_v$ [or, more precisely, according to (IV, 47), $4B_v - 6D_v$]. In the example, the values obtained are $D_0 = 8.31 \times 10^{-6}$ and $B_0 = 1.6421$ cm^{-1}, which agree in a very satisfactory manner with the values obtained above by the first method. In some cases it will be necessary in order to obtain sufficient graphical accuracy to use an approximate $\tilde{D}_v$ and plot $\Delta_2 F(J)/(J + \frac{1}{2}) + 8\tilde{D}_v(J + \frac{1}{2})^2$. The slope of the line obtained then gives the correction $8\Delta D_v$ to be added to $8\tilde{D}_v$ in order to get $8D_v$, while the intercept as before is $4B_v$.

Instead of a graphical method we may use, of course, the *method of least squares* in order to determine B_v and D_v from the observed combination differences $\Delta_2 F(J)$. However, for a determination of both B_v and D_v this method is rather cumbersome. A somewhat simplified procedure has been given by Birge and Shea (112) and Birge (812). If D_v is calculated theoretically from (III, 117, 118 and 125), a comparatively simple formula for B_v results:

$$4B_v = \frac{\sum [\Delta_2 F(J) + 8D_v(J + \frac{1}{2})^3](J + \frac{1}{2})}{\sum (J + \frac{1}{2})^2}, \tag{IV, 51}$$

where the summations are to be taken over all the J values for which $\Delta_2 F$ is known. It appears that the graphical methods have not only the advantage of greater speed but at the same time afford the possibility of recognizing errors in the calculations and measurements much more readily and of eliminating the influence of blends and real perturbations (see Chapter V, section 4) on the final values of the constants.

From the wave numbers of the 0–0 CN band and the 1–0 rotation-vibration band of HCl in Tables 7 and 9, the reader might try to derive the B_v values of the upper and lower vibrational states according to one of the methods given. The results may be checked from Table 39 in the appendix.

The accurate B_v values obtained in one of the above described ways serve to determine precise values of the constants B_e, α_e, $\cdots$ in (IV, 43). In most cases the moments of inertia and internuclear distances obtained from the B_e values according to (IV, 45 and 46) have an equally high accuracy. However, for certain hydrides and the hydrogen molecule, particularly in some of their excited states, a number of corrections must be applied to B_e before it is substituted into (IV, 45 and 46), or in other words the value of B_e obtained from the B_v values is an *effective value*, not the true value. Three corrections must be considered (compare the discussion of the isotope effect p. 164): (a) the Dunham correction resulting from the finer interaction of vibration and rotation (see p. 109), (b) a term resulting from the finer interaction of electronic and nuclear motion including such effects as the wobbling of the nuclei on account of electronic motion and the deviation of the moment of inertia from that of two point-like atoms, (c) a term resulting from the interaction of the electronic state considered with other nearby electronic states of different Λ. A more detailed theoretical discussion of these effects including approximate formulae has been given by Van Vleck (682) while applications to H_2 and LiH have been made by Dieke and Lewis (188) and Crawford and Jorgensen (172)(173), respectively. Even in these cases which would be expected to show the largest effects, the corrections to the r_e values are less than 1% for excited states and less than 0.1% for the ground states. Rough estimates show that the corrections would be far below these limits for non-hydride molecules.

Combination relations and evaluation of rotational constants for bands with Q branches. If a Q branch is present in the bands under investigation (for example, in $^1\Pi - ^1\Sigma$ transitions: see p. 170 f.) additional combination relations arise, since now three lines have the same upper or lower state. These relations, which may be read directly from Fig. 77 and which may also be verified with the help of equations (IV, 29–31) are the following:

$$R(J) - \quad Q(J) = F_v'(J+1) - F_v'(J) = \Delta_1 F'(J) \qquad \text{(IV, 52)}$$

$$Q(J+1) - P(J+1) = F_v'(J+1) - F_v'(J) = \Delta_1 F'(J) \qquad \text{(IV, 53)}$$

$$R(J) - Q(J+1) = F_v''(J+1) - F_v''(J) = \Delta_1 F''(J) \qquad \text{(IV, 54)}$$

$$Q(J) - P(J+1) = F_v''(J+1) - F_v''(J) = \Delta_1 F''(J) \qquad \text{(IV, 55)}$$

It is seen that from these combination differences we obtain the *separation* $\Delta_1 F(J)$ *of successive rotational levels* and not, as before, the separation between one and the next but one. The first two combination differences are equal to each other and, similarly, the last two are equal to each other; that is,

$$R(J) - Q(J) = Q(J+1) - P(J+1), \qquad \text{(IV, 56)}$$

and

$$R(J) - Q(J+1) = Q(J) - P(J+1). \qquad \text{(IV, 57)}$$

Table 22 gives as an example the wave numbers of the lines of the 4–11 band of the fourth positive group of CO. The fifth to eighth columns contain the differences (IV, 52–55). It can be seen that the combination relations (IV, 56 and 57) are well satisfied. A similar agreement is found in other cases although there are quite a number for which a slight systematic discrepancy is found for high J values. This so-called *combination defect* will be explained later (see p. 252).

The agreement of the observed combination differences according to (IV, 56 and 57) provides an excellent check on the correctness of the analysis of a band. In addition, of course, if two or more bands with the same *upper*

TABLE 22.   WAVE NUMBERS OF THE LINES AND COMBINATION DIFFERENCES IN THE 4–11 BAND OF THE FOURTH POSITIVE GROUP OF CO

[After Gerö (249)]

J	$R(J)$	$Q(J)$	$P(J)$	$R(J) - Q(J) = \Delta_1 F'(J)$	$Q(J+1) - P(J+1) = \Delta_1 F'(J)$	$R(J) - Q(J+1) = \Delta_1 F''(J)$	$Q(J) - P(J+1) = \Delta_1 F''(J)$	$\dfrac{\Delta_1 F'(J)}{J+1}$	$\dfrac{\Delta_1 F''(J)}{J+1}$
0	48,338.37	.....	.....	.....	.....	3.37	.....	.....	3.37
1	340.94	48,335.00	.....	5.94	5.37	6.99	6.42	2.828	3.353
2	342.87	333.95	48,328.58	8.92	8.50	10.23	9.81	2.903	3.340
3	345.16	332.64	324.14	12.52	12.00	14.32	13.80	3.065	3.515
4	346.38	330.84	318.84	15.54	15.28	17.80	17.54	3.082	3.534
5	347.17	328.58	313.30	18.59	18.39	21.02	20.82	3.082	3.487
6	347.17	326.15	307.76	21.02	20.91	24.57	24.46	2.995	3.502
7	347.17	322.60	301.69	24.57	24.10	27.94	27.47	3.042	3.463
8	346.38	319.23	295.13	27.15	27.14	31.33	31.32	3.016	3.481
9	345.16	315.05	287.91	30.11	30.20	34.34	34.43	3.016	3.439
10	344.02	310.82	280.62	33.20	33.38	37.84	33.02	3.026	3.448
11	341.85	306.18	272.80	35.67	36.32	41.09	41.74	3.000	3.451
12	339.63	300.76	264.44	38.87	39.23	44.50	44.86	3.004	3.437
13	337.06	295.13	255.90	41.93	42.18	48.22	48.47	3.039	3.453
14	333.95	288.84	246.66	45.11	44.95	51.84	51.68	3.002	3.451
15	330.31	282.11	237.16	48.20	48.14	55.16	55.10	3.011	3.446
16	326.15	275.15	227.01	51.00	51.27	58.40	58.67	3.008	3.443
17	321.71	267.75	216.48	53.96	54.13	61.92	62.09	3.003	3.445
18	316.73	259.79	205.66	56.94	57.11	65.21	65.38	3.001	3.437
19	311.67	251.52	194.41	60.15	60.28	68.88	69.01	3.011	3.447
20	306.18	242.79	182.51						

state have been measured, the differences (IV, 52 and 53) formed for these bands must agree exactly for each J value, while for bands with the same *lower* state the differences (IV, 54 and 55) must agree exactly. At the same time the combination differences $\Delta_2 F(J)$ formed from the P and R branches must agree in the same way as for bands without Q branches.

Substituting $F(J) = B_v J(J + 1)$ in $\Delta_1 F(J)$, we obtain

$$\Delta_1 F(J) = F(J + 1) - F(J) = 2B_v(J + 1). \tag{IV, 58}$$

Thus, to a first good approximation, the $\Delta_1 F(J)$ increase linearly with J, as did the $\Delta_2 F(J)$. That such a linear variation actually obtains in the example of the CO band is shown by the constancy of the $\Delta_1 F'(J)/(J + 1)$ and $\Delta_1 F''(J)/(J + 1)$ values given in the last two columns of Table 22. These values were calculated from the means of corresponding $\Delta_1 F(J)$ values in the fifth and sixth and in the seventh and eighth columns, respectively. The value of $2B_v$ may be determined graphically from the slope of the line (IV, 58) or by calculation of the mean of the $\Delta_1 F(J)/(J + 1)$ values. The B_v values so obtained agree, of course, with those obtained from the $\Delta_2 F$ values. In the example of Table 22 we find from the $\Delta_1 F$ values

$$B_4' = 1.505, \qquad B_{11}'' = 1.723 \text{ cm}^{-1},$$

and from the $\Delta_2 F$ values (not given explicitly in the table)

$$B_4' = 1.504, \qquad B_{11}'' = 1.722 \text{ cm}^{-1}.$$

In general, the determination of the B_v values from the $\Delta_2 F$ values is to be preferred (see Chapter V). The $\Delta_1 F$ values are used for the determination of rotational constants only when, for some reason, the P or the R branch is not observed or can be measured only inaccurately.

If the non-rigidity of the rotator is taken into account, that is, if the term $-D_v J^2 (J + 1)^2$ in (IV, 26) is not neglected one obtains in place of (IV, 58)

$$\Delta_1 F(J) = F(J + 1) - F(J) = 2B_v(J + 1) - 4D_v(J + 1)^3 \tag{IV, 59}$$

For an accurate determination of the rotational constants B_v and D_v from the $\Delta_1 F$ values according to this formula one may use a graphical procedure entirely similar to that described above for the $\Delta_2 F$ values (p. 181 f.). In the example of the CO band (Table 22) the introduction of the D correction raises the B values by about 0.003 cm^{-1}.

Determination of the band origins (zero lines). As we have seen above, if we want to determine the vibrational constants of a molecule very accurately, we must use the origins (zero lines) ν_0 and not the band heads. An approximate value for ν_0 is obtained, according to (IV, 32), when we subtract $2B'$ from the measured first line $R(0)$ of the R branch or $2B' + (3B' - B'')1 + (B' - B'')1^2$ from the second line $R(1)$ of the R branch, and so on. Thus, for example, from $R(0)$ of the 0–0 band of BeO (Table 20) we obtain $\nu_0 = 21{,}196.67$ cm^{-1}. Similarly, ν_0 may be obtained from the lines of the P or the Q branch. In order to get the best possible value of ν_0 an average of all band lines should be formed which would obviously be a rather tedious procedure and would require, incidentally, a highly precise knowledge of the rotational constants.

A much more convenient method that makes use of most measured lines and does not require any knowledge of the rotational constants is the following: If only a P and an R branch are present in the band considered, $R(J-1)+P(J)$ is formed from the observed wave numbers for all J values for which both branches have been measured. According to (IV, 32 and 34)

$$R(J - 1) + P(J) = 2\nu_0 + 2(B_v' - B_v'')J^2 \qquad \text{(IV, 60)}$$

Thus,. when $R(J - 1) + P(J)$ is plotted against J^2, a straight line is obtained whose intercept with the ordinate axis gives $2\nu_0$ and whose slope gives

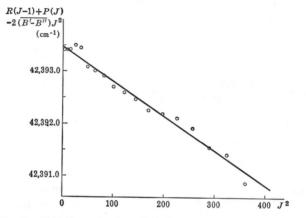

Fig. 83.   Graphical Determination of the Zero Line of the 0–0 Band of the Green BeO System from $R(J - 1) + P(J)$.

$2(B_v' - B_v'')$. In order to be able to plot the graph on a sufficiently large scale, it is advantageous to subtract $2\overline{(B_v' - B_v'')}J^2$ from $R(J - 1) + P(J)$, where $\overline{B_v' - B_v''}$ is an approximate value for $B_v' - B_v''$, and then to plot the difference $R(J - 1) + P(J) - 2\overline{(B_v' - B_v'')}J^2$ against J^2. This is done in Fig. 83 for the 0–0 band of BeO (Table 20), taking $\overline{B_0' - B_0''} = -0.07$ cm^{-1}. The straight line obtained intersects the ordinate axis at 42,393.48 cm^{-1}. Therefore it follows that $\nu_0 = 21,196.74$ cm^{-1}. The slope of the line gives a correction 0.0033_3 to the approximate $\overline{B_0' - B_0''}$ so that $B_0' - B_0'' = -0.0733_3$, a value that agrees well with the B_v values of Table 21 (see also below).

If an intense Q branch is present, for example in a $^1\Pi$—$^1\Sigma$ transition, the most convenient way of determining ν_0 is to use the Q branch. According to (IV, 33)

$$Q(J) = \nu_0 + (B_v' - B_v'')J(J + 1) \qquad \text{(IV, 61)}$$

that is, if $Q(J)$ is plotted against $J(J + 1)$, a straight line is obtained whose intersection with the ordinate axis gives ν_0 and whose slope gives $(B_v' - B_v'')$.

This last method has, for example, been used for those PN bands (Fig. 9) whose fine structure has been analyzed. The resulting zero lines are given in the scheme of band origins (Deslandres Table) in Table 23. It can be seen that the constancy of the differences between the bands in the two horizontal rows is fulfilled with much greater accuracy than in the corresponding scheme of heads (Table 6).

TABLE 23. SCHEME OF BAND ORIGINS OF THE PN BANDS

[After Curry, Herzberg, and Herzberg (176)]

v'' v'	0	1	2	3
0	39,688.57	38,365.30	37,055.97	
	1088.51		1088.52	
1	40,777.08		38,144.49	36,849.18

Usually the fine structure is analyzed only for a few bands of a band system. For the remaining bands, if Q heads are absent or not measured, approximate values for the origins can be obtained by subtracting from the wave numbers of the band heads the separations origin—band head, which may be obtained from (IV, 37) by use of approximate B_v values.

If the effect of non-rigidity (D correction) is taken into account, the term

$$-2(D_v' - D_v'')J^2(J^2 + 1)$$

has to be added at the right-hand side of (IV. 60) while

$$-(D_v' - D_v'')J^2(J + 1)^2$$

has to be added to (IV, 61). On account of these terms a curvature appears for high J values in graphs of the type of Fig. 83. The deviation from the tangent drawn at $J^2 = 0$ [or $J(J+1) = 0$] gives $D_v' - D_v''$ in much the same way as D_v was determined from the $\Delta_2 F''$ values [Fig. 82(a)]. If necessary the curve can be transformed into a straight line by adding the above corrections with opposite sign to the observed $R(J - 1) + P(J)$ or $Q(J)$ values.

For *homonuclear molecules* of zero nuclear spin it is not possible to form $R(J - 1) + P(J)$. Instead one can use $R(J) + P(J)$ in order to obtain ν_0 and $B_v' - B_v''$. It is easily seen that

$$R(J) + P(J) = 2\nu_0 + (2B_v' - 4D_v') + 2(B_v' - B_v'' - 6D_v')J(J + 1)$$

$$- 2(D_v' - D_v'')J^2(J + 1)^2 \tag{IV, 62}$$

By plotting $R(J) + P(J)$ against $J(J + 1)$ the band origin ν_0 and $B_v' - B_v''$ can be determined in substantially the same way as from $R(J - 1) + P(J)$ above. This method is useful also for homonuclear molecules with non-zero nuclear spin since it is then possible to use the strong lines only.

The $B_v' - B_v''$ values obtained in this way are frequently more accurate than those obtained from the combination differences $\Delta_2 F(J)$. In order to obtain the best possible B_v values it is therefore often advisable to determine only one B_v value from the $\Delta_2 F$ values (choosing of course the one for which this is most accurately possible) and obtaining the others from appropriate $B_v' - B_v''$ values. In this way a higher *relative* accuracy of the B_v values is obtained which implies a higher absolute accuracy of the α values.

For the determination of the *vibrational quanta*, instead of using the differences of band origins determined in the above described way one may also use [as first suggested by Jenkins and McKellar (373) the following combination differences between corresponding lines of two bands having the same *upper* state:

$$R_{v'v_1''}(J) - R_{v'v_2''}(J) = Q_{v'v_1''}(J) - Q_{v'v_2''}(J) = P_{v'v_1''}(J) - P_{v'v_2''}(J)$$

$$= G''(v_2'') - G''(v_1'') - (B_{v_1''} - B_{v_2''})J(J+1) \qquad \text{(IV, 63)}$$

and between corresponding lines of two bands having the same *lower* state:

$$R_{v_2'v''}(J-1) - R_{v_1'v''}(J-1) = Q_{v_2'v''}(J) - Q_{v_1'v''}(J)$$

$$= P_{v_2'v''}(J+1) - P_{v_1'v''}(J+1)$$

$$= G'(v_2') - G'(v_1') - (B_{v_1'} - B_{v_2'})J(J+1) \qquad \text{(IV, 64)}$$

The correctness of these relations is readily verified from Fig. 81 or similar diagrams. By plotting these combination differences against $J(J+1)$ a straight line is obtained whose intercept with the ordinate axis gives a very precise value of $G''(v_2'') - G''(v_1'')$ or $G'(v_2') -$

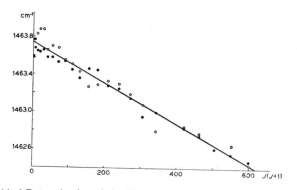

FIG. 84. **Graphical Determination of the First Vibrational Quantum** $\Delta G_{1/2}$ **of the Ground State of BeO from the 0–0 and 0–1 Green Bands [Using the Measurements of Rosenthal and Jenkins** (602)]. $R_{00}(J) - R_{01}(J) + 0.0170J(J+1)$ and $P_{00}(J) - P_{01}(J) + 0.0170J(J+1)$ are plotted as $\circ$ and $\bullet$ respectively against $J(J+1)$.

$G'(v_1')$ and whose slope gives a precise value of $B_{v_1''} - B_{v_2''}$ or $B_{v_1'} - B_{v_2'}$. As an example Fig. 84 gives the differences $R_{00}(J) - R_{01}(J)$ and $P_{00}(J) - P_{01}(J)$ for the two green BeO bands 0–0 and 0–1 (compare Table 20) plotted against $J(J+1)$. In order to be able to plot on a convenient scale $0.0170J(J+1)$ has been added to all the differences. In this case the intercept with the ordinate axis gives $\Delta G_{1/2}''$ which is thus found to be 14 3.74 cm^{-1} while the slope of the line added to 0.0170 gives $B_0'' - B_1'' = \alpha'' = 0.0190_2$ cm^{-1}, a value that is probably more accurate than the value given on p. 180 which was obtained from $\Delta_2 F$ values.

It should be noted that the D correction would introduce only a term $(D_{v_1''} - D_{v_2''})J^2 \times (J+1)^2$ in (IV, 63) and $(D_{v_1'} - D_{v_2'})J^2(J+1)^2$ in (IV, 64), which is usually negligible since the constant β is small compared to D_v. For this reason the use of the relations (IV, 63) and (IV, 64) represents the most accurate way of determining the vibrational differences (ΔG values) as well as the differences of B_v' or B_v'' values (α' and α'' values). This method has the additional advantage that it is applicable to cases of incomplete data, when, for example, only one branch has been measured and when, therefore, the $\Delta_2 F$ values or the $R(J-1) + P(J)$ values cannot be obtained.

A final check on the correctness of the rotational constants and the zero lines is supplied if the individual band lines are calculated from these constants and compared with the measured values. The best way of making these calculations is to calculate first the rotational term values of the upper and lower state and then form the appropriate differences according to (IV, 29–31). It is clear that if the constants have been correctly determined, no systematic differences between observed and calculated values should occur.

Determination of the numbering in the branches of incompletely resolved bands. Thus far we have assumed that the bands investigated are completely resolved and measured and that, in particular, the zero gap can be clearly recognized. In such cases the numbering is immediately obvious. When only P and R branches are present, the first line to the long-wave-length side of the zero gap is $P(1)$ and the first line to the short-wave-length side is $R(0)$. The other lines of the branches join on to these in the order of their J values. If a Q branch is present, its first line is $Q(1)$ (assuming a $^1\Pi$—$^1\Sigma$ or $^1\Sigma$—$^1\Pi$ transition). This line lies very near one of the "missing lines" in the series formed by the P and R branches. The first lines in the P and R branches in such a transition have $J = 2$ and $J = 0$, or both $J = 1$, depending on whether the transition is $^1\Pi$—$^1\Sigma$ or $^1\Sigma$—$^1\Pi$ (see p. 256). The first case applies when the Q head lies near the position of the first missing line in the PR series (coming from the side of the R branch); the second case applies when the Q head lies near the second missing line.

However, a complete resolution of a band is often not obtainable with the means available, especially in the case of heavy molecules. In particular, in the neighborhood of the band head and the origin, the lines are often so close together that they cannot be separated by a spectrograph of limited resolving power. In such cases the numbering of the individual lines of the branches is by no means obvious. Yet there are various ways in which it may be found.

Let us assume that, in a certain case, two or three series of lines with regularly changing separation are observed at some distance from the head. It is generally easy to decide *which of these branches is the P branch, which is the Q branch, and which is the R branch.* Since the lines of the P and R branches can be represented by one and the same quadratic formula the two branches run approximately parallel (see p. 48). The Q branch, on the other hand, which follows a different formula "intersects" the P and R branches more or less frequently; that is, if at one place in the band a Q line almost coincides with an R (or P) line a little further along Q and R (or P) lines will get out of step until finally they coincide again, but so that between these two points of coincidence there are $n + 1$ or $n - 1$ Q lines against n R (or P) lines (see for example the spectrogram Fig. 20). The Q branch is therefore, in general, easily recognizable as such. At a large distance from the band head, the weaker of the two branches that run parallel to each other is the one that forms the head, since its lines have much higher J values; that is, the weaker branch is the R branch for bands shaded to the red and the P branch for bands shaded to the violet [for an example, see Fig. 18(b)].

Clearly, if only P and R branches are present, we cannot obtain the correct numbering for an incompletely resolved band if it is the only band of the system that has been measured. This is because in the Fortrat diagram the P and R branches form a single parabola, for which the position of the abscissa axis is not known if the zero gap is not known (see Fig. 24). For the various possible positions of the zero gap we obtain different linear terms in equation (II, 8) —that is, different B_v' and B_v'' values, between which, in general, a decision is not possible. If, however, two different bands of a system are measured that are known (from the vibrational analysis) to have, for example, the same upper state, then *for the correct numbering in both bands the combination differences* $R(J) - P(J) = \Delta_2 F'$ *must agree exactly for each J value.* This is a very sensitive criterion of whether or not the correct numbering has been found. A similar reasoning applies to $\Delta_2 F''$ when the two bands have the same lower state.

Thus, in order to find the correct numbering we start out from a tentative numbering of the lines in the two bands (which is guessed and is likely to be wrong) and then vary this numbering systematically until we have found a numbering in the two bands for which the $R(J) - P(J)$ or $R(J - 1) - P(J + 1)$ values (depending on whether the bands have the upper or lower state in common) agree exactly [see, for example, Pomeroy (567)].   Even if the preliminary numbering happens to represent the correct relative numbering in each of the two bands (that is, represents the J values apart from an additive constant which is the same for both branches but not in both bands) a shift of the two sets of Δ_2F' (or Δ_2F'') values relative to each other will still be necessary in order to produce agreement and have the same relative numbering in both bands (that is, have the same additive constant in both).   In general, however, the preliminary numbering will not be the correct relative numbering and therefore one will have to change the numbering in one of the branches of each band by 1, 2, 3, $\cdots$ units until agreement of the two sets of Δ_2F values is obtained.   Provided that no chance agreement occurs and the accuracy of the measurements suffices, this procedure always leads to the desired end eventually—that is, to a determination of the correct relative numbering of the two branches in the two bands.   Other methods toward this end have been described in various papers on individual molecules.   A method described by Rosenthaler (1364) in connection with the analysis of the spectrum of BBr is particularly noteworthy.

Once the correct relative numbering has been established it is easy to find the *absolute numbering*, since according to (IV, 42) for the correct numbering

$$\Delta_2F = 4B_v(J + \tfrac{1}{2})$$

and therefore successive Δ_2F values differ by $4B_v$.   Consequently, dividing one of the Δ_2F values by this value of $4B_v$ yields immediately $J + \tfrac{1}{2}$, that is, the absolute J value.   (The data in Table 20 may serve to exemplify this.)   In other words, if the combination differences are plotted against a preliminary numbering, the correct absolute numbering is obtained by shifting the abscissa scale until the straight line formed by the Δ_2F values goes through the point $-\tfrac{1}{2}$ (see Fig. 80).   Once the correct absolute numbering has been derived the determination of the constants can be carried out in the manner described above.

When the numbering for two bands in a band system has been found, the analysis of further bands is no longer so tedious.   If a third band has a state in common with one of the first two, we know beforehand one set of the correct combination differences that must appear in the third band.   We can therefore determine the correct numbering in this band simply by shifting its relative numbering until agreement of the $R(J) - P(J)$ or $R(J - 1) - P(J + 1)$ with the known combination differences is obtained.   Even for a third band of the system that has no state in common with the two already analyzed, the numbering can easily be determined, since the combination differences must at least have similar magnitudes to those of the first two bands (since the B_v values in a band system do not vary very much).

The method described above is always applicable.   However, in some cases, the following procedures for the determination of the numbering may lead to the desired end more quickly.

If a Q *branch* is present in addition to P and R branches, the numbering can be found for a single band, even if the region of the zero gap is unresolved.   In the Fortrat diagram the Q branch forms a parabola which has its vertex almost on the ν axis (see p. 172 f.).   This point is approximately ν_0.   Therefore, if we extrapolate the Q branch, observed at larger J values, to smaller J values, where it is no longer resolved (calculating with a constant second difference; see Table 7, p. 43), ν_0 must lie approximately at the position where the calculated branch forms a head.   The *numbering of the Q branch* is thereby given (at least within a few units depending on the uncertainty of the extrapolation), since at its head $J = -\tfrac{1}{2}$ (see p. 172).   When the P and R branches are similarly extrapolated, we can also obtain their numbering, at least approximately, since ν_0 is now known.   The numbering found can be checked by the combination relations (IV, 56 and 57).   If the combination relations are not fulfilled for the approximate numbering, we have to vary systematically the relative numbering of P, Q, and R branches within the single band considered, in a way entirely similar to that described above for two different bands with P and R branches.   Once the relative numbering has been established the absolute numbering is obtained either from the relation $\Delta_1F(J) = 2B_v(J + 1)$, or, as above, from the relation $\Delta_2F(J) = 4B_v(J + \tfrac{1}{2})$.

Finding the numbering becomes particularly simple if so-called *perturbations* occur in the branches. These perturbations, which will be discussed in greater detail in Chapter V, section 4, consist in the deviation of one line or several successive lines in a branch from a smooth curve. They always appear at a corresponding position in the P and R branches—that is, at the same J' or at the same J''. They are thus perturbations of the upper or lower rotational term series. Therefore, when this phenomenon appears, we can use it to determine which lines of the P and R branches correspond to each other—that is, have the same J' or the same J''. We can decide immediately which of the two cases applies if the disturbed band can be compared with another one with the same upper or the same lower state. If the perturbation also occurs in the other band, the common state is the one that is perturbed; if it does not occur in the other band, the state not in common is the one that is perturbed. In this way we learn whether the perturbed lines have the same J' or the same J'', and thus we obtain the relative numbering. The absolute numbering is obtained in the same manner as above.

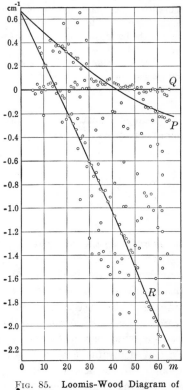

FIG. 85. **Loomis-Wood Diagram of the 1–1 Band of the Blue-Green System of Na$_2$.**

The picking out of branches. In the case of singlet bands, which are the only ones considered in this chapter, the identification of branches is not difficult as long as the individual bands of a band system do not overlap much. Series of lines that form branches usually stand out fairly clearly (see Figs. 18–20). All we have to do in order to be sure that a particular series forms a branch is to see that the separation of successive lines changes approximately linearly ($\Delta^2 \nu = $ constant; see p. 42) and that the intensity changes regularly. As explained before, it is generally not difficult to decide which is the P, which is the R, and which is the Q branch (if the latter is present).

However, it often happens that the individual bands of a band system lie so close to one another that their *fine structures overlap considerably*. The resultant fine structure may then have such a complex appearance that it is difficult to pick out a number of branches belonging to the same band. In such cases a procedure first given by Loomis and Wood (467) for the picking out of branches has proved very valuable. It is also valuable for multiplet bands (see Chapter V) with their large number of branches.

Let us assume that by inspection of the spectrogram a few lines have been found that apparently belong to one and the same branch. We then form the first and second differences for this "branch," and, keeping the second difference constant, we calculate the expected positions of further lines on both sides of the observed part of this branch. We then have a series of wave-number values that might represent a branch. For each line of this calculated branch we now form the differences with all the neighboring observed wave numbers and plot them in a diagram against the arbitrary running numbers in the branch. We obtain a diagram such as that shown in Fig. 85 for the case of an Na$_2$ band.

If, as in the example, the originally assumed branch is real, there must be points in the diagram lying in the neighborhood of the abscissa axis for every value of the running number.

Thereby this one branch is completed as far as the data permit. At the same time, however, we obtain in the diagram the other branches belonging to the same band. Since the branches of a band all have the same quadratic term $(B_v' - B_v'')J^2$, the distances between successive lines must be rather similar for large values of J, except if B_v' happens to be nearly equal to B_v''. Therefore the other branches must appear in the same diagram as series of points lying on smooth, slightly curved lines, as is the case in Fig. 85. It can be seen that, in the example, three branches are present. The two curves running at an angle to the abscissa axis correspond to the P and R branches, and the points lying in the neighborhood of the abscissa axis correspond to the Q branch (the branch from which we started). The latter intersects the P as well as the R branch. The points lying in between the three curves correspond to lines belonging to other bands. Since the line separation in these other bands is either much larger or much smaller than in the band considered, the corresponding points form curves with such a large slope that they are difficult to recognize.

If the branches of a band have been determined in this manner for at least a part of their course, the numbering can be found by the procedure previously described and the band constants can be determined.

Isotope effect. As we have seen in Chapter III, section 2(g) the rotational constants of two isotopic molecules differ; they are related by the formulae

$$B_e{}^i = \rho^2 B_e, \qquad \alpha_e{}^i = \rho^3 \alpha_e, \qquad D_e{}^i = \rho^4 D_e, \qquad \beta_e{}^i = \rho^5 \beta_e \qquad \text{(IV, 65)}$$

where

$$\rho^2 = \frac{\mu}{\mu^i} \qquad \text{(IV, 66)}$$

is the ratio of the reduced masses. The formula (III, 203) for the rotational isotope shift $\Delta\nu_r = \nu_r - \nu_r{}^i$ previously derived for rotation-vibration bands remains valid for electronic bands. But account must be taken of the fact that the rotational constants in the upper and lower states of a band may be quite different. As in the case of infrared bands, the rotational isotopic shift is approximately *proportional to the distance from the origin*, the lines of the heavier isotope lying nearer to the origin. As an illustration, Fig. 86(a) gives the Fortrat diagram of a band for two isotopic molecules. The parabola corresponding to the heavier isotope is indicated by a broken line. Fig. 86(b) and (c) give schematically the corresponding spectrograms. It can be seen that in the head-forming branch the rotational isotopic displacement becomes zero on going through the origin again.

The vibrational isotope shift (see p. 162) must be added, of course, to the rotational shift. In general, the former shift is much larger than the latter except where ν_r is comparable to or larger than ν_v, which is often the case for the 0–0 band.

The rotational isotope effect in electronic bands was first investigated in detail for BO by Jenkins and McKellar (373). Later on, the hydrogen-deuterium rotational isotope effect in a large number of hydrides was studied by various authors [see particularly Crawford and Jorgensen (172)(173) for LiH and LiD, Dieke and Lewis (188) for H_2, HD, and D_2]. More recent work on non-hydrides, CN and Li_2, has been carried out by Jenkins and Wooldridge (379) and McKellar and Jenkins (1215), respectively.

If band heads are used in the investigation of the vibrational isotope effect (see p. 162 f.), for precise evaluations it is of course necessary to take account of the rotational isotope effect since the separation of the band head from the origin is different for the different isotopes and varies from band to band.

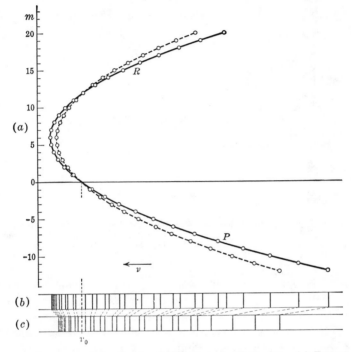

Fig. 86.   **Fine Structure of the Bands of Two Isotopic Molecules.**   (*a*) Fortrat parabolae. (*b*) and (*c*) Spectra of the lighter and the heavier isotopic molecule, respectively.   In order to show the influence of the rotational isotope effect alone, the zero lines ν_0 of the two bands have been drawn one above the other.

4. Intensities in Electronic Bands

In this section we deal only with the intensity distribution within a given band system postponing the discussion of the absolute intensities of different band systems until after the electronic structure of diatomic molecules has been more fully treated (Chapter VI).

(a) Intensity Distribution in the Vibrational Structure

Observed intensity distribution in absorption. Three typical *cases* of intensity distribution in absorption band series are represented schematically in Fig. 87.   In the first case [Fig. 87(*a*)], coming from long wave lengths, the first band, the 0–0 band of the system, is very intense.   Joining onto it, a few further bands of the $v'' = 0$ progression appear with very rapidly decreasing intensities.   Such a case is observed, for example, for the atmospheric oxygen bands which appear in the red part of the solar spectrum (see Fig. 76).   In the second case [Fig. 87(b)], the intensity of the bands in the pro gression $v'' = 0$ at first increases somewhat with decreasing wave length and then decreases slowly.   The absorption bands of CO, whose spectrogram is reproduced in Fig. 14, exemplify such a case.   In the third case [Fig. 87(c)], there is a long

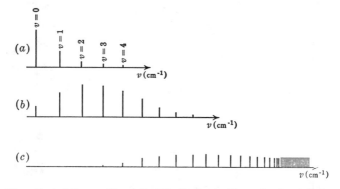

FIG. 87.   **Three Typical Cases of Intensity Distribution in Absorption Band Series** (Schematic). For the sake of simplicity, the bands are drawn with the same separations in the three cases.   Naturally, these cases would be observed in different band systems, which would in general not have the same band separations.

progression of absorption bands whose intensity rises gradually from zero at the long-wave-length end.   The first observed band is usually not the 0–0 band (as may be shown, for example, by an investigation of the isotope effect; see p. 165).   Toward shorter wave lengths the bands draw ever closer together until they come to a convergence limit, at which the previously mentioned continuum joins on.   The maximum of the intensity lies either at very high v' values or possibly even in the continuum.   An example of this kind of intensity distribution is presented by the I_2 absorption spectrum in Fig. 15.   Between these three cases there are all possible transition cases.   Thus, observation shows that for a given v'' the possible v' values vary greatly—that is, that there is *no strict selection rule for the vibrational quantum number v.*

The Franck-Condon principle: absorption.   The different cases of intensity distribution are explained in an easily visualized manner by the *Franck-Condon principle.*   Franck's (229) main idea, which was developed mathematically and later given a wave-mechanical basis by Condon (160) is the following: *The electron jump in a molecule takes place so rapidly in comparison to the vibrational motion that immediately afterwards the nuclei still have very nearly the same relative position and velocity as before the "jump."*   In order to apply this principle let us consider Fig. 88(a), (b), and (c) in which are drawn the potential curves that were assumed by Franck and Condon for the upper and lower electronic states in the three typical cases of intensity distribution.

In Fig. 88(a) the potential curves of the two electronic states have been so chosen that their *minima* lie *very nearly one above the other* (equal internuclear distance). In absorption, the molecule is initially at the minimum of the lower potential curve, if we disregard the zero-point vibration. It can be seen that for a transition to the minimum of the upper potential curve (0–0 band) the requirement of the Franck-Condon principle that the change of position and momentum be small is satisfied. On the other hand, a transition into a high vibrational state [*CD* in Fig. 88(a)] would be possible only when, at the moment of the electronic jump, either the position (transition from *A* to *C*) or the velocity

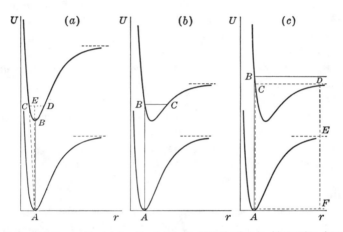

Fig. 88. **Potential Curves Explaining the Intensity Distribution in Absorption According to the Franck-Condon Principle.** In (c), *AC* gives the energy of the dissociation limit, *EF* the dissociation energy of the ground state, and *DE* the excitation energy of the dissociation products (see p. 389 f.).

(transition from *A* to *E*) or both alter to an appreciable extent. At the point *E*, of course, the molecule has the amount of kinetic energy *EB*. Only at the turning point *C* or *D* are the velocity and the kinetic energy zero, as in the initial state at *A*. Thus, on the basis of the Franck-Condon principle, a transition from $v'' = 0$ to such a high vibrational level is forbidden or at least highly improbable. For the level $v' = 1$, the necessary alteration of the position or of the velocity during the electron jump is comparatively small. Therefore the 1–0 band will still appear, though with much smaller intensity than the 0–0 band. For the 2–0, 3–0, $\cdots$ bands the necessary alteration of position and velocity increases, and consequently a rapidly decreasing intensity is to be expected. Thus we obtain an intensity distribution of the type illustrated in Fig. 87(a).

In Fig. 88(b) the *minimum of the upper potential curve lies at a somewhat greater r value than that of the lower.* Therefore the transition from minimum to minimum (0–0 band) is no longer the most probable, since the internuclear distance must alter somewhat in such a transition. The most probable of the transitions is that from *A* to *B* in Fig. 88(b) (vertically upwards). For this transition there is no change in the internuclear distance at the moment of the "jump" and no change of the velocity.   Thus, immediately after the electron jump the two nuclei still have their old distance from each other and zero relative velocity.   Since, however, the equilibrium internuclear distance has a different value in the new electronic state, the nuclei start to vibrate between B and C. The vibrational levels whose left turning points lie in the neighborhood of B are the upper levels of the most intense bands.   For still higher vibrational levels an appreciable change of the internuclear distance or velocity must take place, as a result of which the intensities of the bands decrease again with increasing v'. Thus the observed intensity distribution in the second case [Fig. 87(b)] is explained.   The same intensity distribution results when the minimum of the upper potential curve lies at a somewhat *smaller r* value than in the ground state.   However, this case is met much less frequently in absorption than that shown in Fig. 88(b).

In Fig. 88(c) the *minimum of the upper potential curve lies at a still greater internuclear distance*.   The Franck-Condon principle is strictly fulfilled for the transition AB.   However, the point B on the upper potential curve lies above the asymptote of this curve and therefore corresponds to the continuous region of the vibrational term spectrum of the upper state.   After such an electron jump the atoms will fly apart.   Transitions to points somewhat below B (that is, in the discrete region) and somewhat above B (that is, in the continuous region) are also possible.   In this way the third case of intensity distribution [Fig. 87(c)] is explained.

Summarizing, we can say that in absorption the most intense transition from $v'' = 0$ is always that corresponding to a transition from the minimum of the lower potential curve *vertically upward*.   In all cases in which a fine-structure analysis has been carried out the internuclear distances r_e have actually been found to be in agreement with the values assumed according to the above discussion (Fig. 88) for an explanation of the intensity distribution.

The meaning and mode of action of the Franck-Condon principle can be demonstrated and made clearer by a *mechanical model*.   If we bend a flexible strip of tin into the shape of the potential curve of the lower state and illustrate the motion of the molecule by the motion of a small cylindrical body rolling on this surface, the position of rest of the cylinder will be at the bottom of the channel thus formed.   If the shape of the channel is now suddenly altered (corresponding to the electron jump), the cylinder, which as a result of its inertia retains for an instant its original position (r value) will in general no longer be at the minimum.   It will therefore start vibrating about the new equilibrium position with an amplitude equal to the change of equilibrium position.   If this change is sufficiently large the cylinder will fly out at the other side of the channel after traversing the curve once.   This will be the case when the cylinder, immediately after the "electron jump," is above the asymptote of the new potential curve.   On the other hand, it is clear that the cylinder will not be set in vibration when the position of the minimum does not alter.

The Franck-Condon principle: emission (Condon parabola).   According to the Franck-Condon principle, the variation of intensity in a *band progression with* $v' = 0$ in emission corresponds exactly to that of a progression with $v'' = 0$

in absorption: There is an intensity maximum at a v'' value that is determined by the relative position of the minima of the two potential curves. The greater the difference of the r_e values the larger is the v'' value of the intensity maximum.

However, the intensity distribution is different for *band progressions in emission* having $v' \neq 0$ (the same is true for the absorption band progressions with $v'' \neq 0$ which appear at higher temperatures). In order to understand

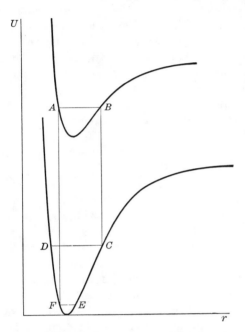

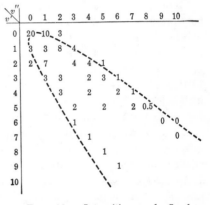

v' \ v''	0	1	2	3	4	5	6	7	8	9	10
0	20	10	3								
1	3	3	8	4							
2	2	7			4	4	1				
3		3	3		2	3	1				
4		3			2		2	1			
5			2			2		2	0.5		
6			1						0	0	
7				1						0	
8					1						
9							1				
10											

Fig. 89. Potential Curves Explaining the Intensity Distribution in Emission According to the Franck-Condon Principle.

Fig. 90. Intensities and Condon Parabola in the Band System of PN.

this, let us consider Fig. 89. During the vibration in the upper state the molecule stays preferentially at the turning points A and B of the vibrational motion, while the intermediate positions are passed through very rapidly. As a result, the electron jump takes place preferentially at the turning points. If it takes place at the turning point B and if there is to be no change in position and velocity, immediately after the jump the molecule will be at C, vertically below B, and C forms the right turning point of the new vibrational motion C–D. However, the electron jump can take place from A as well as from B. In this case, according to Franck and Condon, the transition takes place to F, and F forms the left turning point of the new vibrational motion E–F.

Thus we can see that there are *two* v'' values for which the probability of the transition from a given v' is a maximum; that is, there are *two intensity maxima* to be expected in a v''-progression ($v' = $ constant), one at small v'' and a second at large v''. In Fig. 90 the estimated intensities of the PN bands

(Table 6) are plotted in an array similar to a Deslandres table. It can be seen that there actually are two intensity maxima present in all the horizontal rows with the exception of that with $v' = 0$. With increasing v' the point C in Fig. 89 will in general move up more rapidly than the point F, that is, the two intensity maxima should separate from each other with increasing v' and, apart from that, go to higher v'' values. This also is in agreement with the observations (Fig. 90). If we join up the most intense bands in the array, we obtain a parabolic curve whose axis is the principal diagonal. It is called the *Condon parabola*. This Condon parabola can be obtained theoretically according to the above method of construction when the potential curves are known. The broken curve given in Fig. 90 is the theoretical curve, which, as we can see, reproduces the observed intensity maxima in the band progressions, $v' =$ constant, quite well. Similarly, in Table 18, the observed CO bands (which are naturally the most intense bands) are seen to lie on a parabolic curve.

On the basis of Fig. 89 it is clear that the separation of the two intensity maxima in a progression $v' =$ constant, and therefore the width of the Condon parabola, increases with increasing difference of the equilibrium internuclear distances (minima of the potential curves) of the upper and lower states. Thus in the case of the PN bands (Fig. 90) the parabola is less open than for CO (Table 18) since the internuclear distances differ less (compare the r_e values in Table 39). If the two potential minima lie at the same or approximately the same r value, the two maxima in each progression $v' =$ constant coincide and the parabola degenerates into a straight line. In such cases the most intense bands are those with $\Delta v = 0$ (principal diagonal of the Deslandres table) while the bands with $\Delta v = \pm 1$ are either much weaker or not observed at all. At the same time these are often the cases in which heads of heads and tail bands appear (see p. 160 f.). In general, the more open the Condon parabola is, the greater is the number of bands that appear in a band system (see the above examples of PN and CO). This is easily understood on the basis of the above considerations.

It is obvious that the Condon parabola will be well developed only when a large number of successive vibrational levels of the upper state are about equally populated. If, on the other hand, the level with $v' = 0$, for example, is preferentially excited, the Condon parabola derived in the above manner no longer gives the most intense bands in the system. However, it gives in every case the relative intensity maxima in the individual progressions $v' =$ constant.

The same considerations that apply to v''-progressions apply of course also to v'-progressions ($v'' =$ constant). The intensity maxima in the latter lie on the same Condon parabola. Such v'-progressions with $v'' \neq 0$ occur in absorption at elevated temperatures. However, except for molecules with very small vibrational quanta, rather high temperatures are necessary in order to bring about a clear development of the Condon parabola in the intensity distribution in absorption.

Wave-mechanical formulation of the Franck-Condon principle. As Condon (160) has shown (see below) a wave-mechanical treatment confirms in a general way the basic assumption of the Franck-Condon principle that *transitions vertically upward or downward* in the potential energy diagram correspond to the most intense bands. However, we have to remember that, according to wave mechanics, an oscillator can never be quite at rest but rather that in the lowest vibrational level we have the probability density distribution given by Fig. 42. Therefore a transition "vertically upward" starting from $v'' = 0$ may take place within a certain *range* of r values, particularly since also in the various upper vibrational levels a certain range of r values is possible (see the eigenfunctions in Fig. 91). This consideration makes it quite clear why instead of only one band a number of bands of the v'-progression with $v'' = 0$ appear in absorption with a more or less broad intensity maximum. The extent of the band series and of the continuum which may possibly join onto it is the greater the steeper the upper potential curve.

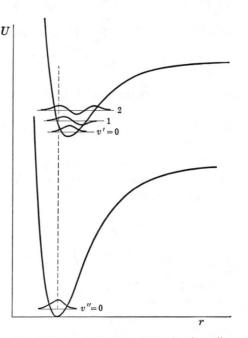

FIG. 91. **Franck-Condon Principle According to Wave Mechanics.** The potential curves are so drawn that the "best" overlapping of the eigenfunctions occurs for $v' = 2$, $v'' = 0$ (see the broken vertical line).

In a similar way the intensity distribution in other progressions is qualitatively understood.

We proceed now to a more detailed proof and a quantitative refinement of the preceding qualitative formulation. The probability of a transition between two states, which are characterized by the total eigenfunctions ψ' and ψ'' is proportional, according to (I, 47) or (I, 51), to the square of the corresponding matrix element of the electric moment (the so-called transition-moment):

$$R = \int \psi'^* M \psi'' \, d\tau \qquad (IV, 67)$$

where M is a vector with components $\sum e_i x_i$, $\sum e_i y_i$, and $\sum e_i z_i$. As will be shown below we may neglect the rotation of the molecule altogether and put

$$\psi = \psi_e \psi_v \qquad (IV, 68)$$

where ψ_e is the electronic and ψ_v the vibrational eigenfunction. Furthermore we can resolve the electric moment M into one part depending on the electrons and one depending on the nuclei:

$$M = M_e + M_n \qquad \text{(IV, 69)}$$

Substituting this in (IV, 67) and remembering that $\psi_v^* = \psi_v$ we obtain

$$R = \int M_e \psi_e'^* \psi_v' \psi_e'' \psi_v'' \, d\tau + \int M_n \psi_e'^* \psi_v' \psi_e'' \psi_v'' \, d\tau \qquad \text{(IV, 70)}$$

Since M_n does not depend on the coordinates of the electrons the second integral may be written

$$\int M_n \psi_v' \psi_v'' \, d\tau_n \int \psi_e'^* \psi_e'' \, d\tau_e,$$

where $d\tau_n$ and $d\tau_e$ are, respectively, the elements of volume of the space of the nuclear coordinates and of the space of the electronic coordinates. The electronic eigenfunctions belonging to different electronic states are orthogonal to one another; that is $\int \psi_e'^* \psi_e'' \, d\tau_e = 0$. We therefore obtain

$$R = \int \psi_v' \psi_v'' \, dr \int M_e \psi_e'^* \psi_e'' \, d\tau_e, \qquad \text{(IV, 71)}$$

where dr is substituted for $d\tau_n$, since the vibrational eigenfunctions ψ_v depend on the internuclear distance r only.

The second integral in (IV, 71)

$$R_e = \int M_e \psi_e'^* \psi_e'' \, d\tau_e \qquad \text{(IV, 72)}$$

is the *electronic transition moment*. (Its square is proportional to the *electronic transition probability*). As has been mentioned previously the electronic eigenfunctions depend slightly on the internuclear distance as a parameter and therefore, for a given electronic transition, R_e depends to some extent on the internuclear distance. The wave-mechanical formulation of the Franck-Condon principle rests on the assumption that this *variation of R_e with r is slow* and that R_e may be replaced by an average value $\bar{R}_e$. Then we have

$$R^{v'v''} = \bar{R}_e \int \psi_v' \psi_v'' \, dr. \qquad \text{(IV, 73)}$$

The transition probability and therefore the intensity (see p. 20 f.) is proportional to the square of this expression, that is, is proportional to the square of the integral over the product of the vibrational eigenfunctions of the two states involved (the so-called *overlap integral*). More specifically we have from (I, 46) and (I, 47) for the band intensities in *emission*

$$I_{\text{em.}}^{v'v''} = \tfrac{64}{3} \pi^4 c N_{v'} \nu^4 \bar{R}_e{}^2 \left[\int \psi_v' \psi_v'' \, dr \right]^2 \qquad \text{(IV, 74)}$$

and from (I, 50) and (I, 51) in *absorption*

$$I_{\text{abs.}}{}^{v'v''} = \frac{8\pi^3}{3hc} I_0 \Delta x N_{v''} \nu \overline{R}_e{}^2 \left[\int \psi_v' \psi_v'' \, dr \right]^2 \qquad \text{(IV, 75)}$$

where $N_{v'}$ and $N_{v''}$ are the numbers of molecules in the vibrational levels v' and v'' respectively.

Let us first consider the intensity of the 0–0 band given by (IV, 74 or 75). As illustrated in Fig. 91, the vibrational eigenfunctions of the upper and lower states are bell-shaped curves. Obviously, the product of the two eigenfunctions for each value of r, and therefore also the integral of this product over all r values, is greatest if the minima of the two potential curves lie exactly one above the other (unlike the drawing). The more the minima are separated from each other, the smaller is the overlap integral and therefore the smaller is the intensity of the 0–0 band, entirely in agreement with the result of the elementary form of the Franck-Condon principle.

If the minima of the two potential curves lie at equal internuclear distances the value of the overlap integral (IV, 73) while large for the 0–0 band, is obviously quite small for the 1–0 band since there are approximately as many positive as negative contributions to the integral (indeed, for the harmonic oscillator the integral is exactly zero). If we now imagine the potential curve for the upper state to be shifted to larger or smaller r values, the integral, and therefore the intensity of the 1–0 band, at first increases (whereas it decreases for the 0–0 band). The integral has a maximum value when the maximum (or the minimum) of the upper vibrational eigenfunction lies approximately vertically above the maximum of the lower eigenfunction. At the same time the overlap integral for the 0–0 band is clearly smaller. If the r values of the minima of the potential curves differ still more, the overlap integral for the 1–0 band, and therefore the intensity of the 1–0 band, decreases again. Entirely similar considerations apply to the 0–1 band.

The eigenfunctions of the higher vibrational levels have broad maxima or minima near the classical turning points of the motion. Between these terminal loops there are smaller and narrower maxima and minima (see Figs. 42 and 48). The contributions of the latter to the overlap integral usually cancel one another, at least in a rough approximation. In a series of absorption bands with $v'' = 0$ the overlap integral (that is, the intensity) has therefore a maximum value for transitions to those upper vibrational levels *whose eigenfunctions have their broad terminal maxima or minima* (roughly) *vertically above the maximum of the eigenfunction of the lower state* ($v'' = 0$). The value of the integral decreases for greater as well as smaller values of v'. Entirely similar considerations apply to emission bands with $v' = 0$. Thus we see that essentially the same results follow from wave mechanics as from the elementary conception of the Franck-Condon principle (see above). The extension of the progressions with $v'' = 0$ or $v' = 0$ on both sides of the maximum of intensity is determined mainly by the width of the eigenfunction of the state with $v'' = 0$ or $v' = 0$—that is, ultimately by the Heisenberg uncertainty principle (see p. 77 f.).

The intensity distribution in v''-progressions with $v' \neq 0$ or v'-progressions with $v'' \neq 0$ is perhaps most easily understood if we consider as an example the eigenfunctions $v = 10$ and $v = 4$ of Fig. 42 and regard them as belonging to two different electronic states with different r_e values. We can see then that the overlap integral (IV, 73) and therefore the intensity of the transition $v' = 10 \rightarrow v'' = 4$ will be a maximum if either the right terminal maximum of the upper eigenfunction lies vertically above the right terminal maximum of the lower eigenfunction or if the left terminal maximum of the upper lies vertically above that of the lower eigenfunction (for large differences of the r_e values the left upper terminal maximum may lie vertically above the right lower maximum). Therefore, for a given position of the potential curves and a given v' and varying values of v'' there is in general an intensity maximum for two different values of v'', and these v'' values are close to those obtained in semi-classical fashion from a construction of the type of Fig. 89 since the classical turning points are close to the terminal loops of the eigenfunctions.

The preceding considerations explain only the broad features of the intensity distribution since we have assumed that the effects of the intermediate maxima and minima of the eigenfunctions cancel out. Qualitatively, it is clear that for a favorable relative position of the eigenfunctions these maxima and minima may lead to a large value of the overlap integral, that is, a large intensity even for a band that does not lie near the Condon parabola; and on the other hand, for an unfavorable relative position, an anomalously small intensity of a band near the Condon parabola may result. As shown by Fig. 90 and Table 18 appreciable intensities have indeed been observed for bands that correspond to points between the two limbs of the Condon parabola. Similarly in the resonance series of I_2 and Na_2 (see Fig. 29) certain members are found to be anomalously weak, others anomalously strong.

Quantitative comparisons of theoretical and observed intensities have been carried out by a number of investigators. Hutchisson (1082) using the harmonic oscillator approximation for both electronic states derived explicit formulae for the intensities. The agreement with the observations in a number of cases was only fair [see also Wurm (727)]. In a second paper Hutchisson (349a) developed intensity formulae which take account of the anharmonicity to a certain approximation. These formulae are, however, so involved that very little use has been made of them. An alternative, suggested by Brown (136) is the substitution in (IV, 73) of harmonic oscillator eigenfunctions adapted to each vibrational level instead of using one and the same harmonic oscillator for all vibrational levels of a given electronic state. The adaptation is accomplished by using an effective $\omega = G(v)/(v + \frac{1}{2})$ and an effective r calculated from B_v. These values are then substituted in Hutchisson's original formulae. With this method Brown succeeded in representing the observed rather irregular intensity distributions in several resonance series of Na_2 (see Fig. 29) in a very satisfactory manner. Bewersdorff (99) has actually determined the anharmonic oscillator eigenfunctions and calculated the intensities by numerical integration for the case of a band progression (Lyman bands) of the HD molecule explaining some apparent anomalies.

It is readily seen from (IV, 73), as was first pointed out by Wehrli (700), that, if the equilibrium positions of upper and lower state coincide and harmonic oscillator functions are used, bands with odd Δv should have zero intensities; that is, the decrease of intensity from the main sequence ($\Delta v = 0$), which represents the Condon parabola in this case, is not uniform but there are subsidiary maxima for even Δv. Such subsidiary maxima have indeed been found by Wehrli for GaI and InI for which the conditions mentioned are approximately fulfilled.

In the opposite limiting case of a large difference in r_e, in a number of cases [for P_2 by Herzberg (311), for AgCl by Jenkins and Rochester (376), for As_2 by Almy and Kinzer (748), for RbH by Gaydon and Pearse (961)] the intense bands inside the Condon parabola have been found to form *subsidiary parabolae*. Gaydon and Pearse (961) in the case of RbH have been able to explain these subsidiary parabolae on the basis of distorted harmonic oscillator functions.[3a] Qualitatively the reason for these subsidiary parabolae is the fact that the terminal loops of the vibrational eigenfunctions of the lower state (with small r_e and large ω_e) are narrow and comparable in width to the intermediate loops of the upper state (with large r_e and small ω_e). Thus large values of the overlap integral result not only when the terminal loops of the two states occur at the same r value but also when the terminal loops of the lower state coincide with the intermediate loops of the upper state.

The proof for the legitimacy of neglecting the rotation in discussing the Franck-Condon principle is as follows: If we substitute the expression (IV, 7) for the total eigenfunction in the transition moment (IV, 67) we obtain for example for the z component

$$R_z = \int \psi_e'^* \frac{1}{r} \psi_v' \psi_r'^* M_z \psi_e'' \frac{1}{r} \psi_v'' \psi_r'' \, d\tau$$

With $M_z = M \cos \vartheta$ and $d\tau = d\tau_e \, r^2 \sin \vartheta \, dr \, d\vartheta \, d\varphi$, where $d\tau_e$ is the volume element of the configuration space of the electrons, this becomes

$$R_z = \int \psi_e'^* \psi_v' M \psi_e'' \psi_v'' \, d\tau_e \, dr \int \sin \vartheta \cos \vartheta \, \psi_r'^* \psi_r'' \, d\vartheta \, d\varphi \qquad \text{(IV, 76)}$$

The second integral is a constant for a given J', J'' combination while the first integral is identical with (IV, 70). The preceding wave-mechanical formulation of the Franck-Condon principle is therefore valid for each J', J'' combination and since the first integral to a good approximation is independent of J (see p. 110) this formulation holds also for the whole bands.

Vibrational sum rule and vibrational temperature. If the vibrational eigenfunctions are properly normalized it can be shown from elementary properties of systems of orthogonal functions that the sum of the squares of the overlap integrals summed over all values of the vibrational quantum numbers of the upper or of the lower state is equal to one:

$$\sum_{v'} \left[\int \psi_v' \psi_v'' \, dr \right]^2 = \sum_{v''} \left[\int \psi_v' \psi_v'' \, dr \right]^2 = 1 \qquad \text{(IV, 77)}$$

Combining these relations with (IV, 74 and 75) it follows immediately that

$$\sum_{v''} \frac{I_{\text{em.}}^{v'v''}}{\nu^4} \propto N_{v'}, \qquad \sum_{v'} \frac{I_{\text{abs.}}^{v'v''}}{\nu} \propto N_{v''} \qquad \text{(IV, 78)}$$

There is thus a (vibrational) sum rule for the intensities of the bands of a band system [Jabłoński (1088)]. Calling *band strength* the emission intensity divided by ν^4 or the absorption intensity divided by ν (similar to line strength, see p. 127) this sum rule can be formulated as follows: *the sums of the band strengths of all bands with the same upper or the same lower state are proportional to the number of molecules in the upper and lower state, respectively.* In absorption, in place of the intensity one may also use the integrated absorption coefficient. In emission, the intensities used in the sum rule must if necessary be corrected for self absorption.

One can readily see from the above derivation that the sum rule is valid only if the electronic transition moment R_e is a constant for all vibrational transitions that give an appreciable

---

[3a] See also the recent work of Pillow (1318b) on MgH.

contribution to the sum.   If this condition is fulfilled the sum rule may be used for the determination of the temperature of the gas emitting or absorbing the band system: Since in thermal equilibrium the population N_v of the initial state is proportional to $e^{-G_0(v)hc/kT}$ we obtain from (IV, 78)

$$\log \sum_{v''} \frac{I_{\text{em.}}{}^{v'v''}}{\nu^4} = C_1 - \frac{G'(v')hc}{kT}$$

$$\log \sum_{v'} \frac{I_{\text{abs.}}{}^{v'v''}}{\nu} = C_2 - \frac{G''(v'')hc}{kT}$$

$$(IV, 79)$$

where C_1 and C_2 are constants.   Therefore, by plotting the logarithms of the sums of the band strengths, as measured in the various v'' or v'-progressions, against the vibrational term values $G(v)$ a straight line is obtained whose slope is hc/kT and thus gives the temperature T.   The band intensities used in this method of course need only be relative intensities.   A disadvantage of this method is the necessity of measuring the intensities of all bands in the progressions used. If emission bands are employed this method will yield reliable results only if the excitation of the band system is purely thermal.   But even for non-thermal excitation often a straight line is obtained in the above-mentioned plot, and at least an *effective "vibrational" temperature* can be determined from the slope of this line.   This vibrational temperature will not necessarily agree with the rotational temperature to be discussed below except in actual thermodynamic equilibrium.

If the intensities of a sufficient number of bands cannot be measured a determination of the vibrational temperature can still be obtained if the overlap integrals are calculated for the measured bands.   Dividing the band strength of each band by the square of the overlap integral gives according to (IV, 74 and 75) again quantities that are proportional to the number of molecules in the initial state and may be plotted in the same way as above.   (For applications of this method see Chapter VIII, sections 2 and 4.)

For non-thermal excitation (for example by electron collisions) it still holds that the sums of the band strengths in a v''-progression [see equation (IV, 78)], or the individual band strengths divided by the squares of the appropriate overlap integrals, are proportional to the numbers of molecules in the vibrational levels of the upper state.   Thus, it is possible to study relative *probabilities of excitation* for the various upper vibrational levels.   For the case of excitation by electron collisions Langstroth (439a) found in this way experimentally that the relative excitation probability of different upper vibrational levels is proportional to the square of the overlap integral formed for the initial and final level of the excitation process ($v'' = 0, v'$) and independent of the electron velocity, a result that Langstroth showed to be in agreement with a wave-mechanical treatment [see also Duffendack, Revans, and Roy (191)].   Only very close to the minimum excitation potential does the relative excitation probability of different vibrational levels differ from that given by the overlap integral.   The absolute excitation probability does of course vary considerably with the electron velocity.

(b) Intensity Distribution in the Rotational Structure

The line intensities in most branches of electronic bands vary in essentially the same way as in the branches of rotation-vibration bands [see Chapter III, section 2(e)].   In each branch there is an intensity maximum which lies at a higher J value the higher the temperature and the smaller the rotational constant B.

$^1\Sigma$—$^1\Sigma$ transitions.   In the special case of $^1\Sigma$—$^1\Sigma$ transitions (see p. 169 f. and Chapter V) for which there is only a P and an R branch, the intensity relations are given quantitatively by the previous formulae (III, 169 and 170) and Fig. 60, p. 126.   In Fig. 60 we have of course to allow for the fact that in electronic bands one of the branches usually forms a head.   However, this does not alter the intensities of the lines.   The diagrams in Fig. 60 refer to absorption. For emission, as is easily seen by comparing (III, 169) and (III, 170), the same diagrams represent the intensity distribution but with P and R exchanged (and correspondingly the sign of m reversed).   The band fine structures reproduced in Figs. 18 and 19 show qualitatively such intensity distributions.

From (III, 170) one obtains immediately

$$\log \frac{I_{\text{em.}}}{J' + J'' + 1} = A - \frac{B_{v'}J'(J' + 1)hc}{kT} \tag{IV, 80}$$

where $A = \log (C_{\text{em.}}v^4/Q_r)$ may be considered as a constant since for a given band v covers only a very small range of values.   A formula similar to (IV, 80) is obtained from (III, 169) for the case of absorption.   Formula (IV, 80) shows that by plotting $\log I_{\text{em.}}/(J' + J'' + 1)$ against $J'(J' + 1)$ a straight line is obtained whose slope is $B_{v'}hc/kT$.   Thus if the line intensities have been measured and the rotational constant is known the temperature of the source may be determined.[4]   This method of determining a *"rotational" temperature* has been used in a considerable number of cases (see also Chapter VIII, section 2).

Strictly speaking the formulae (III, 170) and (IV, 80) for the intensity distribution in emission hold only in the case of purely thermal excitation, since only then is the number of molecules in the different states given by the Boltzmann factor and the statistical weight.   However, it has been found experimentally that the intensity distribution in emission bands occurring in electric discharges is of the same type.   This is easily understood, since, if a molecule is excited by electron collision, no great change in the angular momentum of the system can be produced (owing to the smallness of the electron mass), and therefore the distribution of the molecules over the different rotational levels in the upper electronic state is practically the same as in the ground state.   But in the latter, owing to the numerous molecular collisions, the distribution corresponds to thermal equilibrium at a certain effective temperature, and therefore this will also be at least approximately the case in the upper state. Thus we obtain the intensity distribution (III, 170).

However, we must be quite clear that this normal intensity distribution in electric discharges results from the circumstance that the angular momentum is not strongly altered in excitation by electron collisions.   For other modes of excitation—for instance, by collisions with metastable atoms, by chemical reactions or by dissociation of polyatomic molecules in electric discharges—large deviations from the normal thermal distribution may occur.   This has been

---

[4] If higher rotational lines are included for which the D correction becomes important, $B_v'J'(J' + 1)$ in (IV, 80) should be replaced by $F_v'(J')$ and the plotting should be against $F_v'(J')$ rather than against $J'(J' + 1)$.

observed, for example, by Bonhoeffer and Pearson (822), Oldenberg (539), and Lyman (471b) in the excitation of OH bands by discharges in H_2O vapor and by Lochte-Holtgreven (447) in the excitation of C_2 and CH bands by discharges in hydrocarbons. In both cases, much higher rotational lines appear than one would expect on the basis of the temperature. Furthermore, large deviations from the thermal intensity distribution occur when molecular gases are excited by monochromatic light (see below), as well as when the molecules can predissociate (see Chapter VII).

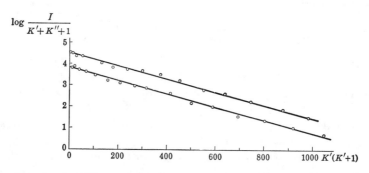

Fig. 92. Log $[I/(K' + K'' + 1)]$ for the 0–0 Band of the Negative Nitrogen Bands Plotted Against $K'(K' + 1)$ [after Ornstein and van Wijk (549)]. The log $[I/(K' + K'' + 1)]$ values for corresponding lines of the P and R branches (with the same K') agree within the accuracy of the measurements, as they should. They have therefore not been separately plotted, but rather their average has been plotted.

How well the thermal intensity distribution holds in most cases of electric discharges is illustrated by Fig. 92, where the observed values of $\log [I_{em.}/(K' + K'' + 1)]$ for the 0–0 band ($\lambda3914$) of the negative nitrogen group in an electric discharge are plotted against $K'(K' + 1)$. Here K is used instead of J, since the bands represent a $^2\Sigma$–$^2\Sigma$ transition [see Chapter V, section 3(b)]. It is seen that the points fall on two parallel straight lines indicating that a definite temperature is indeed defined. From the slope an effective (rotational) temperature of 970° K is obtained. Why two parallel straight lines are obtained instead of only one will be explained later.

The *influence of temperature on the intensity distribution* in an electronic band is shown by Fig. 26, in which photographs of the N_2^+ band $\lambda4278$ at the three temperatures $-180°$ C., $20°$ C., and $400°$ C. are reproduced. It can be seen very clearly how the maxima in the P and R branches shift and become flatter with increasing temperature and how the appearance of the band (development of the band head, distinctness of the zero gap) is thereby altered. Still more marked differences in appearance are observed for bands occurring in the electric arc as well as in weak electric discharges—for instance, the CN bands. The J value of the intensity maximum is essentially given by the previous formula (III, 163) and conversely the observed value of J_{max} may be used for a rough determination of the temperature.

If there were strictly no change of angular momentum upon excitation of a molecule by electron collisions and if no redistribution through collisions occurs in the upper state one would expect the rotational distribution to be determined by the B value of the ground state (from which the excitation takes place) rather than that of the initial state of the emission. In other words if B_v' is used in (IV, 80) the temperature obtained would be too small by a factor B_v'/B_v''. A case of this type has been found by Ginsburg and Dieke (987) for H_2 when it is excited in an electric discharge at very low pressure. The apparent rotational temperature from (IV, 80) was found to be appreciably less than room temperature since the B value of the upper state is appreciably smaller than that of the ground state. At higher pressures (and current densities) the actual temperature is raised by the discharge and in addition a redistribution of the molecules over the rotational levels of the upper state occurs. For both these reasons the apparent temperature obtained from (IV, 80) is then found to be higher.

Other transitions. If the quantum number Λ differs from zero in one or both of the participating electronic states, the intensity distribution in the P and R branches is substantially the same as for $^1\Sigma$—$^1\Sigma$ transitions (Fig. 60). However, in addition, a Q branch appears for which the intensity distribution depends on whether $\Delta\Lambda = \pm1$ or $\Delta\Lambda = 0$. Using the exact intensity formulae (see below) the two typical distributions represented in Fig. 93 are obtained assuming $B/T = 0.015$.

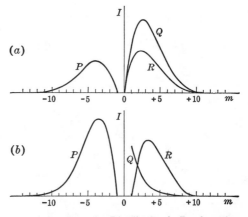

The first case [Fig. 93(a)] represents a $^1\Pi$—$^1\Sigma$ *transition* (see Chapter V). Similar curves apply to all other transitions with $\Delta\Lambda \neq 0$ except that for $\Delta\Lambda = -1$ the P branch is slightly stronger than the R branch, whereas in the example ($\Delta\Lambda = +1$) the R branch is the stronger one (both in emission and absorption). It is noteworthy that according to the theory [see Fig. 93(a)] the lines of the Q branch have approximately twice the in-

Fig. 93.   **Intensity Distribution in Bands with a Q Branch (in Emission).** (a) $^1\Pi$—$^1\Sigma$ transition. (b) $^1\Pi$—$^1\Pi$ transition. The running number m has been chosen as abscissa (see p. 111 f.). For the Q branch, $m = J$.

tensity of the corresponding lines of the P branch or the R branch. It can be seen qualitatively from the spectrograms Fig. 20 and Fig. 21 (pp. 45 and 46) that this is actually the case. Also quantitatively, the calculated intensity distribution has been confirmed for a number of transitions of this type.

The second case [Fig. 93(b)] refers to a $^1\Pi$—$^1\Pi$ *transition* for which $\Delta\Lambda = 0$ and $\Lambda = 1$. In this case the intensity of the Q branch decreases very rapidly from the beginning without going through a maximum [it is approximately proportional to $(1/J)e^{-F_v(J)hc/kT}$]. As a result, the Q branch is observed in such cases only when the region in the neighborhood of the zero line is well resolved.   The same applies to other transitions having $\Delta\Lambda = 0$ and $\Lambda \neq 0$. The intensity distribution in the P and R branches in these cases is even more similar to the case of Σ—Σ transitions, the P branch being the stronger one in emission, the R branch in absorption.

In both cases ($\Delta\Lambda = \pm 1$ and $\Delta\Lambda = 0$) the difference of intensity of P and R branches is much less pronounced for higher temperatures and non-hydrides with their smaller B values (compare Fig. 60).

As we have seen in Chapter III, p. 127, the intensities in thermal equilibrium may be expressed, apart from a constant and from the frequency factor, as a product of the line strength S_J and the Boltzmann factor.   The precise formulae for the *line strengths of the symmetric top* (which are, of course, independent of whether or not thermal equilibrium applies) were first given, on the basis of the old quantum theory, by Hönl and London (329) and were later derived on the basis of wave mechanics by Dennison (887), Reiche and Rademacher (1345), and others.

For $\Delta\Lambda = 0$ the *Hönl-London formulae* are

$$S_J{}^R = \frac{(J'' + 1 + \Lambda'')(J'' + 1 - \Lambda'')}{J'' + 1} = \frac{(J' + \Lambda')(J' - \Lambda')}{J'}$$

$$S_J{}^Q = \frac{(2J'' + 1)\Lambda''^2}{J''(J'' + 1)} = \frac{(2J' + 1)\Lambda'^2}{J'(J' + 1)} \qquad\qquad \text{(IV, 81)}$$

$$S_J{}^P = \frac{(J'' + \Lambda'')(J'' - \Lambda'')}{J''} = \frac{(J' + 1 + \Lambda')(J' + 1 - \Lambda')}{J' + 1},$$

for $\Delta\Lambda = +1$

$$S_J{}^R = \frac{(J'' + 2 + \Lambda'')(J'' + 1 + \Lambda'')}{4(J'' + 1)} = \frac{(J' + \Lambda')(J' - 1 + \Lambda')}{4J'}$$

$$S_J{}^Q = \frac{(J''+1+\Lambda'')(J''-\Lambda'')(2J''+1)}{4J''(J'' + 1)} = \frac{(J'+\Lambda')(J'+1-\Lambda')(2J'+1)}{4J'(J'+ 1)} \qquad \text{(IV, 82)}$$

$$S_J{}^P = \frac{(J'' - 1 - \Lambda'')(J'' - \Lambda'')}{4J''} = \frac{(J' + 1 - \Lambda')(J' + 2 - \Lambda')}{4(J' + 1)},$$

and for $\Delta\Lambda = -1$

$$S_J{}^R = \frac{(J'' + 2 - \Lambda'')(J'' + 1 - \Lambda'')}{4(J'' + 1)} = \frac{(J' - \Lambda')(J' - 1 - \Lambda')}{4J'}$$

$$S_J{}^Q = \frac{(J''+1-\Lambda'')(J''+\Lambda'')(2J''+1)}{4J''(J'' + 1)} = \frac{(J'-\Lambda')(J'+1+\Lambda')(2J'+1)}{4J'(J' + 1)} \qquad \text{(IV, 83)}$$

$$S_J{}^P = \frac{(J'' - 1 + \Lambda'')(J'' + \Lambda'')}{4J''} = \frac{(J' + 1 + \Lambda')(J' + 2 + \Lambda')}{4(J' + 1)}$$

Here the superscript R, Q, or P indicates the branch for which the particular expression holds ($\Delta J = +1$, 0, -1 respectively).   Of the two alternative forms given the first is more useful for absorption, the second for emission, since the Boltzmann factor contains J'' and J', respectively.   The previous qualitative statements with regard to the intensities of the different branches are easily verified from the formulae (IV, 81–83).

If the line strengths of the three transitions with the same J'' are added for a given electronic transition, apart from a constant factor, the value $2J'' + 1$ (or $2J' + 1$ for transitions with the same J') is obtained, that is, a (rotational) *sum rule* holds: the sums of the line strengths of all the transitions from or to a given rotational level are proportional to the statistical weight of that level (compare the Burger-Dorgelo-Ornstein sum rule for atoms, see A.A., p. 161).

While for most singlet band systems the Hönl-London formulae would be expected to represent the intensity distribution to a high degree of approximation, it should be emphasized that they are derived under the assumption of negligible interaction of rotation and electronic motion. Considerable deviations from these formulae apparently due to failure of this assumption (that is, due to perturbation by other electronic states), have been found, for example, for H_2 by Ginsburg and Dieke (987).

Intensity alternation. According to Chapter III, section 2(f), *for homonuclear molecules with* $\Lambda = 0$, either the even-numbered or the odd-numbered rotational levels have the greater statistical weight. The ratio of the weights of the "strong" and the "weak" levels is $(I + 1)/I$, where I is the quantum number of the nuclear spin. As we have seen in Chapter III, for a given homonuclear molecule the strong levels are symmetric and the weak levels are anti-symmetric for all electronic states, or vice versa (depending on the statistics of the nuclei). Therefore and on account of the selection rule (III, 183) (sym. $\leftrightarrow$ antisym.) even in electronic transitions strong levels combine only with strong and weak with weak. Since in addition, the selection rule $\Delta J = \pm 1$ holds, two electronic states with $\Lambda = 0$ (Σ states) can combine with each other only if in one the even-numbered levels are the "strong" levels and in the other the odd are the "strong" levels. This is shown in Fig. 94. It can be seen that, owing to the alternating statistical weights of the rotational levels in the upper as well as in the lower states, an *alternation of intensity* must occur, exactly as in the Raman spectra of homonuclear molecules. This alternation of intensities is very clearly seen in the spectrograms of the N_2^+ band given in Fig. 26. From the magnitude of the intensity alternation the nuclear spin can be derived, just as it can be from the rotational

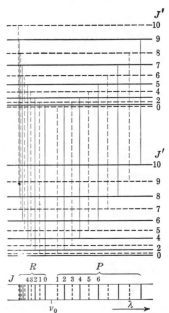

Fig. 94. **Origin of the Intensity Alternation in an Electronic Band** ($^1\Sigma$—$^1\Sigma$) **of a Homonuclear Molecule.** The "strong" and "weak" levels and transitions are indicated by full and broken lines, respectively.

Raman spectrum [see Chapter III, section 2(f)]. If every alternate line is missing, the nuclear spin is zero (see p. 135).

The intensity alternation is the reason for the occurrence of the two parallel lines in Fig. 92, which also refers to N_2^+. Their vertical separation is $0.69 = \log_e 2$. The points corresponding to successive lines in the fine structure lie alternately on one and the other straight line in the diagram; that is, an intensity alternation in the ratio 2 : 1 is present. Consequently, the spin of the N^{14} nucleus is $I = 1$. The same N_2^+ bands have also been produced with the heavy isotope N^{15} and an intensity alternation in the ratio 3 : 1 has been found yielding $I = \frac{1}{2}$ for the spin of the N^{15} nucleus [Krüger (424), Wood and Dieke (720a)]. For other nuclei see Table 38 and Chapter VIII, section 2.

The intensity alternation for bands involving at least one state with $\Lambda \neq 0$ will be taken up in the next chapter.

Wood's resonance series. The regular intensity distributions illustrated by Figs. 60 and 93 can be expected in emission only when the mechanism of excitation is non-selective and independent of the rotational quantum number J. Extreme deviations from such distributions will occur if excitation is produced by irradiation with monochromatic light (for example, the green mercury line obtained from a mercury arc through a suitable filter) which will excite only one rotational level in the upper state. The spectrum of the resulting resonance fluorescence will consist of a progression of bands each of which consists only of a few lines. For example, in a $^1\Sigma$—$^1\Sigma$ transition we expect *doublets* consisting of one line of the P branch and one line of the R branch. They are, as we can see from Fig. 54, the lines $R(J - 1)$ and $P(J + 1)$, where J is the rotational quantum number of the state excited by the irradiation. Such *resonance series* were first observed and studied by Wood and his collaborators (1550) (1553) for Na_2 and I_2 long before the theory of band spectra was developed.

An example of a doublet resonance series of Na_2 excited by the Cd line 5086 Å is shown in the previous Fig. 29. According to (IV, 40 and 42), the doublet separation is given by

$$\Delta \nu = R(J - 1) - P(J + 1) = 4B_v''(J + \tfrac{1}{2}) \qquad \text{(IV, 84)}$$

It changes from doublet to doublet in the resonance series corresponding to the change in B_v''. The rotational constant α can be determined from this change.

For $^1\Pi$—$^1\Sigma$ transitions, as before, a doublet resonance series appears when the excitation is by means of a line of the P or R branch, whereas only a series of *single* lines appears when the excitation is by means of a line of the Q branch. On the basis of Fig. 77 one might expect to obtain triplets in either case. Why only doublets and singlets appear will become clear in the next chapter (p. 254). A singlet resonance series, excited by the Cd line 4800 Å, is also present in the spectrogram in Fig. 29.

For heavier molecules, such as I_2, it may easily happen that the irradiating line, particularly when it is not very sharp, covers a number of absorption lines—for example, one of the P and one of the R branch. Naturally, if this is the case, we obtain a superposition of a number of resonance series [see, for example, Loomis (457)].

The investigation of resonance spectra is particularly useful as an aid to the fine-structure analysis of bands of heavier molecules, for which the rotational structure is not completely resolved. However, the rotational constants B_v'' can be obtained from (IV, 84) only if J is known, for instance from an analysis of one of the complete bands as obtained in absorption.

If a foreign gas that does not quench the fluorescence (for example a rare gas) is added to the fluorescing gas, *collisions* of the excited molecules with the foreign gas molecules may occur before the fluorescence radiation is emitted. These collisions will produce transitions of the excited molecules to other rotational levels and ultimately, with sufficient pressure of the foreign gas, the complete bands may appear in spite of the excitation by a single line only. In this connection a particularly interesting phenomenon was observed by Wood and Loomis (721) for I_2—namely, that only every alternate line appears in the I_2 bands produced in this way. Since the selection rule sym. $\leftrightarrow$ antisym. holds also for collisions (see p. 131), only lines with symmetrical upper levels (or only those with antisymmetrical upper levels) can appear in the fluorescence spectrum, even after collisions, if the originally excited state was a symmetrical (or antisymmetrical) state. On the other hand in absorption, or in fluorescence produced by white light, all the rotational lines appear, since the nuclear spin of iodine is not zero.

The problem of the relative intensities of different members of a resonance series has been discussed in the preceding subsection.

Further details about resonance series may be found in Pringsheim (25) (573).

CHAPTER V

FINER DETAILS ABOUT ELECTRONIC STATES AND ELECTRONIC TRANSITIONS

1. Classification of Electronic States; Multiplet Structure

In general, each molecule has a number of band systems of the type discussed in the foregoing chapter and has therefore a number of electronic states. It is to be expected that the classification of these electronic states is analogous to the classification of atomic energy states, which are, of course, "pure" electronic states. This analogy is particularly apparent in the cases in which Rydberg series of bands are observed (see Chapter II). The bands of such a series must clearly have a Rydberg series of electronic states as their upper states and the reason that only one band (the 0–0 band) of each electronic transition is observed, must be due to the Franck-Condon principle, the potential curves involved all having nearly the same r_e value.

Orbital angular momentum. The motion of the electrons in an atom takes place in a spherically symmetrical field of force. As a consequence (see Chapter I) the electronic orbital angular momentum L is a constant of the

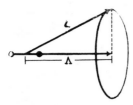

FIG. 95. Precession of the Orbital Angular Momentum L about the Internuclear Axis.

motion as long as the effect of electron spin is small or neglected. In a diatomic molecule the symmetry of the field in which the electrons move is reduced; there is only *axial symmetry* about the internuclear axis (which for the present we consider as fixed). As a consequence only the *component* of the orbital angular momentum of the electrons about the internuclear axis is a constant of the motion. The situation is essentially the same as that of an atom in a strong electric field, which is here the electrostatic field of the two nuclei. A *precession* of L takes place about the field direction (internuclear axis) with constant component $M_L(h/2\pi)$, where M_L can take only the values

$$M_L = L, L - 1, L - 2, \cdots, -L. \tag{V, 1}$$

For the case of the molecule this precession is illustrated in Fig. 95.

In an electric field, unlike a magnetic field, reversing the directions of motion of all electrons does not change the energy of the system but does change M_L into $-M_L$. Therefore, in diatomic molecules, states differing only in the sign of M_L have the same energy (are degenerate). On the other hand, states with different $|M_L|$ have in general widely different energies since the electric field which causes this splitting is very strong. As the strength of the field increases, L precesses faster and faster about the axis of the field and consequently loses more and more its meaning as angular momentum while its component M_L remains well defined. It is therefore more appropriate to classify the electronic states of diatomic molecules according to the value of $|M_L|$ than according to the value of L. Following the international nomenclature, we put

$$\Lambda = |M_L| \tag{V, 2}$$

The corresponding angular momentum vector Λ represents the *component of the electronic orbital angular momentum along the internuclear axis*. Its magnitude is $\Lambda(h/2\pi)$. The quantum number Λ is therefore identical with that previously introduced in the treatment of the symmetrical top [see Chapter III, section 2(d)].

According to (V, 1), for a given value of L, the quantum number Λ can take the values

$$\Lambda = 0, 1, 2, \cdots, L. \tag{V, 3}$$

Thus in the molecule for each value of L there are $L + 1$ distinct states with different energy. However, often the value of L cannot be given at all, since the corresponding angular momentum L is not defined.

According as $\Lambda = 0, 1, 2, 3, \cdots$, the corresponding molecular state is designated a $\Sigma, \Pi, \Delta, \Phi, \cdots$ state, analogous to the mode of designation for atoms. Greek letters are used throughout in the designation of molecular quantities referring to components of electronic angular momenta, while the corresponding italic letters used for atoms refer to the electronic angular momenta themselves.

$\Pi, \Delta, \Phi, \cdots$ states are *doubly degenerate* since M_L can have the two values $+\Lambda$ and $-\Lambda$ (see above); Σ states are non-degenerate.

Mathematically the angular momentum Λ may be introduced in the following way: We consider the nuclei as fixed in the z axis and introduce cylindrical coordinates z_i, ρ_i, φ_i for each electron, where ρ_i is the distance from the axis and φ_i the azimuth. Furthermore it is convenient to introduce relative azimuths $\varphi_i' = \varphi_i - \varphi_1$ for all but the first electron. If the wave equation is expressed in terms of z_i, ρ_i, φ_i', changing φ_1 simply means a rotation about the axis which must leave the wave equation unchanged. For this reason the electronic eigenfunction ψ_e must have the form [see Wigner and Witmer (712) and Hund (348)]

$$\psi_e = \chi e^{+i\Lambda\varphi_1} \quad \text{or} \quad \psi_e = \bar{\chi} e^{-i\Lambda\varphi_1} \tag{V, 4}$$

where χ and $\bar{\chi}$ are functions of all electronic coordinates z_i, ρ_i, φ_i', except φ_1, and where the function $\bar{\chi}$ differs from χ only in that the φ_i' are replaced by their negatives. The constant Λ is a positive integer. If Λ were not integral ψ_e would not be single-valued. It is easily verified that the form (V, 4) is the only one that leaves $\psi_e^*\psi_e$ invariant against any change of φ_1.

The possible values of the angular momentum about the z axis are the eigenvalues of the operator $(h/2\pi i)\partial/\partial\varphi_1$ (see p. 16). Operating with this operator on the two functions (V, 4) yields $(h/2\pi)\Lambda\psi_e$ and $-(h/2\pi)\Lambda\psi_e$, respectively, that is, the corresponding values of the angular momentum are $\Lambda h/2\pi$ and $-\Lambda h/2\pi$, respectively. Thus the two functions (V, 4) correspond to the rotation of the electrons in one sense or the opposite sense, respectively, about the line joining the nuclei and for a given Λ and χ belong to the same eigenvalue (twofold degeneracy). Upon reflecting at any plane through the line joining the nuclei one of the functions (V, 4) goes over into the other just as do the corresponding classical motions.

Owing to the linearity of the Schrödinger equation (I, 13) any linear combination of the two functions (V, 4),

$$\psi_e = a\chi e^{+i\Lambda\varphi_1} + b\bar{\chi}e^{-i\Lambda\varphi_1},\tag{V, 5}$$

is also a solution belonging to the same eigenvalue. If $\Lambda = 0$ (Σ state), however, there is only one solution and ψ_e does not depend on φ_1.

Spin. A closer investigation of the fine structure of electronic bands shows in many cases a *multiplet structure* either of the whole bands or of individual lines or both (see Chapter II). As in the case of atoms this multiplet structure is due to the *electron spin*.

Just as for atoms, the spins of the individual electrons form a resultant S, the corresponding quantum number S being *integral or half integral according as the total number of electrons in the molecule is even or odd*. In Σ states (similar to S states of atoms) the resultant spin S, since it is unaffected by an electric field, is fixed in space as long as the molecule does not rotate and if there is no external magnetic field. On the other hand, if $\Lambda \neq 0$, (II, Δ, $\cdots$ states), there is an internal magnetic field in the direction of the internuclear axis resulting from the orbital motion of the electrons. This magnetic field causes a *precession of S* about the field direction (that is, in this case the internuclear axis) *with a constant component $M_S(h/2\pi)$*. For molecules, M_S is denoted by Σ, in order to bring out the analogy better (this quantum number Σ must not be confused with the symbol Σ for terms with $\Lambda = 0$).[1] The values of Σ allowed by the quantum theory are (see Chapter I for M_S)

$$\Sigma = S, S - 1, S - 2, \cdots, -S.\tag{V, 6}$$

That is to say, $2S + 1$ different values are possible. In contrast to Λ, the quantum number Σ can be positive and negative. It is not defined for states with $\Lambda = 0$—that is, Σ states.

Total angular momentum of the electrons; multiplets. The total electronic angular momentum about the internuclear axis, denoted by Ω, is obtained by adding Λ and Σ, just as the total electronic angular momentum J for atoms is obtained by adding L and S. Whereas, however, a vector addition has to be carried out for atoms, for molecules an algebraic addition is sufficient, since the vectors Λ and Σ both lie along the line joining the nuclei. Thus for the *quantum number of the resultant electronic angular momentum about the internuclear axis* we have

$$\Omega = |\Lambda + \Sigma|.\tag{V, 7}$$

If Λ is not equal to zero, according to (V, 6) there are $2S + 1$ different values of $\Lambda + \Sigma$ for a given value of Λ (that is, of Ω as well, if $\Lambda \geq S$). As a result of the interaction of S with the magnetic field produced by Λ, these dif-

---

[1] It will be remembered that, correspondingly, for atoms the letter S is used both as the spin quantum number and as symbol for a term with $L = 0$.

ferent values of $\Lambda + \Sigma$ correspond to somewhat different energies of the resulting molecular states. Thus an electronic term with a given $\Lambda \neq 0$ splits into a *multiplet of $2S + 1$ components*. On the other hand, if Λ equals zero, there is no magnetic field in the direction of the internuclear axis (Σ is not defined) and consequently no splitting occurs. Σ states are single as long as the molecule does not rotate. Nevertheless $2S + 1$ is called the *multiplicity* of a state, quite independent of whether or not Λ is greater than 0—that is, regardless of whether or not (without rotation) an actual splitting is present (compare the similar convention for atoms). The relative orientations of the vectors Λ and S are illustrated in Fig. 96(a) for a term with $\Lambda = 2$ and $S = 1$. Fig. 96(b) shows the corresponding energy level diagram.

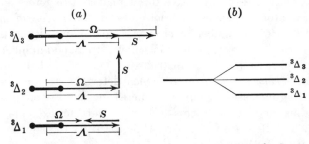

FIG. 96. (a) **Vector Diagrams and** (b) **Energy Level Diagram for a** $^3\Delta$ **State** ($\Lambda = 2$, $S = 1$). In (a) the approximate magnitude $S(h/2\pi)$ [rather than $\sqrt{S(S+1)}\,(h/2\pi)$] is used. In (b), to the left the term is drawn without taking the interaction of Λ and S into account; to the right, taking account of it.

Just as for atoms, according to the international nomenclature, the multiplicity $2S + 1$ is added to the term symbol as a left superscript. Furthermore the value of $\Lambda + \Sigma$ is added as a subscript (similar to J for atoms). Thus the example deals with a $^3\Delta$ term whose components are designated $^3\Delta_3$, $^3\Delta_2$, and $^3\Delta_1$.

The internal magnetic field H_i that causes the precession of S about the internuclear axis is proportional to Λ. The magnetic energy of the spin in this magnetic field is $\mu_{H_i} H_i$ where the component μ_{H_i} of the magnetic moment in the field direction is proportional to Σ. The *electronic energy of a multiplet term* is therefore given to a first approximation by

$$T_e = T_0 + A\Lambda\Sigma \qquad (V, 8)$$

where T_0 is the term value when the spin is neglected and A is a constant for a given multiplet term.

The relation (V, 8) shows clearly that the components of a molecular multiplet term with $\Lambda \neq 0$ are equidistant [see Fig. 96(b)] unlike the components of an atomic multiplet term. Moreover it shows that the number of multiplet components is equal to the number of Σ values, that is $2S + 1$, irrespective of whether S is smaller or larger than Λ. Thus molecular electronic states, with $\Lambda \neq 0$ unlike atomic states, always have the full multiplicity.

The last remark is irrelevant for singlet, doublet, and triplet states which are the ones mostly observed for molecules. For singlet terms ($S = 0$) we have $\Omega = \Lambda$, for example $^1\Pi_1$; for doublet terms ($S = \frac{1}{2}$) we have $\Omega = \Lambda \pm \frac{1}{2}$, for example $^2\Pi_{\frac{1}{2}}$, $^2\Pi_{\frac{3}{2}}$; for triplet terms ($S = 1$) we have $\Omega = \Lambda$ and $\Lambda \pm 1$, for example $^3\Pi_0$, $^3\Pi_1$, $^3\Pi_2$. For quartet terms ($S = \frac{3}{2}$) the possible values of $\Lambda + \Sigma$ are $\Lambda + \frac{3}{2}$, $\Lambda + \frac{1}{2}$, $\Lambda - \frac{1}{2}$ and $\Lambda - \frac{3}{2}$, which for $\Lambda = 1$ ($^4\Pi$ state) gives $\frac{5}{2}$, $\frac{3}{2}$, $\frac{1}{2}$, and $-\frac{1}{2}$. According to (V, 8), the four Σ values correspond to four different equidistant energy levels even though two of them have the same Ω value ($\Omega = |\Lambda + \Sigma| = \frac{1}{2}$). That is why $\Lambda + \Sigma$ rather than Ω is used to distinguish the multiplet components. For the $^4\Pi$ state the components are designated $^4\Pi_{\frac{5}{2}}$, $^4\Pi_{\frac{3}{2}}$ $^4\Pi_{\frac{1}{2}}$, $^4\Pi_{-\frac{1}{2}}$. For $^4\Delta$, $^4\Phi$, $\cdots$ states this slight complication does not arise, since $\Lambda + \Sigma$ is positive throughout and therefore equal to Ω.

Just as for atoms there is for molecules an *alternation of multiplicities:* Molecules with an even number of electrons have odd multiplicities (singlets, triplets, $\cdots$) since S is integral, while molecules with an odd number of electrons have even multiplicities (doublets, quartets, $\cdots$) since S is half-integral.

The *coupling constant A* in (V, 8) which determines the magnitude of the multiplet splitting increases rapidly with increasing number of electrons just as the corresponding quantity for atoms does. For BeH, for example, the splitting of the first excited $^2\Pi$ state is 2 cm^{-1}, while for HgH the splitting of the corresponding state is 3684 cm^{-1}.

Just as for atoms the coupling constant A may be positive or negative giving rise to *normal (regular) or inverted terms*, respectively—that is, terms for which the multiplet components lie in the order of their Ω values or terms for which they lie in the inverse order.

The spin does not alter the *twofold degeneracy of states with $\Lambda \neq 0$*. Each of the multiplet components is doubly degenerate, even those with $\Omega = 0$, as long as $\Lambda \neq 0$. However, for the latter case ($\Omega = 0$), in contrast to the former ($\Omega \neq 0$), a degeneracy exists only in a first approximation. If the finer interaction of the electrons is taken into account a slight splitting into two states results, while for $\Omega \neq 0$ the degeneracy persists in any approximation as long as the molecule is not rotating. Thus for a $^3\Pi_0$ state there are two sublevels of very slightly different energy which are designated $^3\Pi_{0^+}$ and $^3\Pi_{0^-}$ (see below).

In the above discussion we have assumed an interaction of the electrons in a molecule of the same type as is assumed in the *Russell-Saunders coupling* [(L, S) coupling] for atoms. Since S is now precessing about Λ the coupling in the molecule is sometimes referred to as (Λ, S) *coupling*. In either case the coupling of L and S is assumed to be small. If the coupling between L and S is large as is frequently the case for heavy molecules, the electric field in the internuclear axis may not be sufficient to cause a complete Paschen-Back effect, that is, an independent space quantization of L and S with components Λ and Σ; but rather the resultant of L and S, which we call here J_a, will precess about the axis with component ω (see Fig. 104, p. 225). In such cases Λ and Σ are no longer defined and only ω retains its meaning as electronic angular momentum about the internuclear axis. Such states cannot be classified as Σ, Π, Δ, $\cdots$ states. Following Mulliken (516) they are simply called 0, $\frac{1}{2}$, 1, $\cdots$ states according to the value of Ω. Other related types of coupling will be discussed later.

As long as the coupling of spin and orbital motion is small the electronic eigenfunction ψ_{es} may be written as a product of a coordinate function ψ_e, as in (V, 4) or (V, 5), and a spin function β, that is

$$\psi_{es} = \psi_e \beta \tag{V, 9}$$

The form of the spin function is the same as for atoms with the same number of electrons. For the case of two electrons see A.A., p. 124 and Kronig (21a), for the general case see Wigner and Witmer (712).

Symmetry properties of the electronic eigenfunctions. For the classification of molecular electronic states, in addition to the quantum numbers introduced above, the symmetry properties of the electronic eigenfunctions are of great importance. These symmetry properties depend on the symmetry properties of the field in which the electrons move (compare the general discussion in Chapter I, p. 24).

In a diatomic molecule (and similarly in a linear polyatomic molecule) any plane through the internuclear axis is a plane of symmetry. Therefore the electronic eigenfunction of a non-degenerate state (Σ state) either remains unchanged or changes sign when reflected at any plane passing through both nuclei. In the first case, the state is called a Σ^+ state, and in the second case it is called a Σ^- state.

The eigenfunctions of Σ states according to (V, 4) do not depend on φ_1 but only on the relative azimuths of the electrons. In the case of two electrons, an eigenfunction that changes sign for a reflection at any plane through the internuclear axis (Σ^-) is, for example, $\sin \varphi_2'$, if φ_2' is the azimuthal angle between electrons 1 and 2, whereas $\cos \varphi_2'$ is an example of an eigenfunction that does not change sign for such a reflection (Σ^+).

For degenerate states (Π, Δ, $\cdots$ states) the electronic eigenfunction need not remain unchanged or only change sign when a symmetry operation is carried out, but may also go over into another of the functions belonging to the same eigenvalue (other than $-\psi_e$). For example, by reflection at a plane through the internuclear axis (exchange of all the azimuthal angles with their negative values), the first of the functions (V, 4) goes over into the second, and vice versa. However, linear combinations (V, 5) can easily be found such as remain unchanged or go over into their negatives by the reflection—namely, the functions

$$\psi_e^+ = \chi e^{i\Lambda \varphi_1} + \overline{\chi} e^{-i\Lambda \varphi_1}$$
$$\psi_e^- = \chi e^{i\Lambda \varphi_1} - \overline{\chi} e^{-i\Lambda \varphi_1} \tag{V, 10}$$

It can be seen immediately that when all the φ_i are replaced by $-\varphi_i$, the function ψ_e^+ remains unaltered, whereas ψ_e^- goes over into $-\psi_e^-$. The functions ψ_e^+ and ψ_e^- are linearly independent of each other and can therefore also be chosen as *the* two eigenfunctions of the degenerate state in place of (V, 4). That is to say, all the eigenfunctions of this state can be represented by

$$\psi_e = c\psi_e^+ + d\psi_e^-, \tag{V, 11}$$

where c and d are two constants.

According as the eigenfunction is ψ_e^+ or ψ_e^-, we may therefore also distinguish Π^+, Π^-, Δ^+, Δ^-, $\cdots$ states, similar to the Σ^+ and Σ^- states but with the difference that Π^+ and Π^- (and correspondingly the other pairs) have *exactly* equal energies, so that it is usually superfluous to make this distinction. However, if we take account of the influence of the rotation of the molecule (see below), a splitting takes place into two components of slightly different energy with just the eigenfunctions ψ_e^+ and ψ_e^-, and not into two components with eigenfunctions

$\chi e^{i\Lambda\varphi_1}$ and $\bar{\chi} e^{-i\Lambda\varphi_1}$. According to (V, 10), each of the states Π^+ and Π^- (and correspondingly for larger Λ) contain, so to speak, rotations of the electrons in *both* senses, since both $e^{+i\Lambda\varphi}$ and $e^{-i\Lambda\varphi_1}$ occur in ψ_e^+ as well as in ψ_e^-.

In the case of states with $\Omega = 0$, $\Lambda \neq 0$ a slight splitting exists even for the fixed center system (see above). Again the eigenfunctions of one of the components remains unchanged, that of the other changes sign upon reflection at any plane through the two nuclei. In the case of a $^3\Pi_0$ state they are distinguished as $^3\Pi_{0^+}$ and $^3\Pi_{0^-}$. If Λ is not defined one distinguishes 0^+ and 0^- states corresponding to Σ^+ and Σ^-.

If the two nuclei in the molecule have the *same charge* (for example, $O^{16}O^{16}$, but also $O^{16}O^{18}$), the field in which the electrons move has, in addition to the symmetry axis, a *center of symmetry;* that is to say, the field remains unaltered by a reflection of the nuclei at this center of symmetry (midpoint of the internuclear axis). For brevity it is sometimes said that the *molecule* has a center of symmetry. In consequence of this symmetry *the electronic eigenfunctions remain either unchanged or only change sign when reflected at the center* (that is, when the coordinates of all the electrons x_i, y_i, z_i, are replaced by their negatives $-x_i, -y_i, -z_i$). In the first case the state to which the eigenfunction belongs is called an *even state,* in the second case an *odd state* (similar to even and odd states of atoms, see A.A., p. 153 f.). It can be shown [see Wigner and Witmer (712)] that this property applies to degenerate as well as to non-degenerate states and that for the former both components always have the same symmetry with respect to reflection at the center. The symmetry property even or odd is indicated by adding a subscript g or u, respectively, to the term symbol (from the German "gerade" and "ungerade"). Thus for molecules with like nuclei we have Σ_g, Σ_u, Π_g, Π_u, $\cdots$ states. The electron spin has no influence on the symmetry property even-odd and therefore the components of a given multiplet term are either all even (g) or all odd (u).

Even if Λ is not defined (see above) the symmetry property even-odd is defined and one has 0_g^+, 0_u^+, 0_g^-, 0_u^-, $\frac{1}{2}_g$, $\frac{1}{2}_u$, 1_g, 1_u, $\cdots$ states.

According to the above, the distinction between even and odd electronic states is independent of whether the molecule is homonuclear or not, provided the nuclei have the same charge. To a certain approximation, even the electronic eigenfunctions of molecules such as CN, for which the two nuclei differ only by one unit of charge, possess this symmetry property.

2. Coupling of Rotation and Electronic Motion

In section 1 we have considered the electronic motions in the field of two fixed nuclei (two-center system) and have disregarded the fact that in the actual molecule rotation and vibration take place simultaneously with the electronic motions. It is now necessary to consider in which way these different motions influence one another.

The mutual interaction of vibrational and electronic motions is essentially taken into account if the vibrational levels are so chosen as to fit the potential curve of the electronic state [see Chapter III, section 2(a)], since this potential curve represents the dependence of the electronic energy (including nuclear repulsion) on the internuclear distance (see p. 147 f.). The mutual interaction of rotation and vibration has already been discussed in Chapter III, section 2(c). However, we must now consider more closely the *influence of rotational and electronic motions on each other;* that is, we must find out by what quantum numbers we can describe the rotational levels in the different types of electronic states, how their energies depend on these quantum numbers, and what symmetry properties the corresponding eigenfunctions possess.

(a) Hund's Coupling Cases

The different angular momenta in the molecule—electron spin, electronic orbital angular momentum, angular momentum of nuclear rotation—form a resultant that is always designated *J*, as is the total angular momentum of an atom (in both cases disregarding nuclear spin). If the spin *S* and the orbital angular momentum Λ of the electrons are zero—that is, if we have a $^1\Sigma$ state—the angular momentum of nuclear rotation is identical with the total angular momentum *J* and we have the simple rotator previously treated (see Chapter III). In all other cases we have to distinguish different modes of coupling of the angular momenta, as was first done by Hund.

Hund's case (a). In Hund's case (a) it is assumed that the interaction of the nuclear rotation with the electronic motion (spin as well as orbital) is very weak, whereas the *electronic motion* itself is *coupled very strongly to the line joining the nuclei.* This case therefore approaches very closely the two-center system used in section 1. Even in the rotating molecule the electronic angular momentum Ω is then well defined. Ω and the angular momentum *N* of nuclear rotation form the resultant *J*. The coupling is entirely similar to that for the symmetric top previously discussed [Chapter

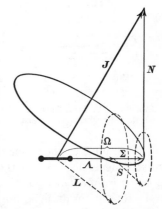

Fig. 97. **Vector Diagram for Hund's Case (a).** The nutation of the figure axis is indicated by the solid-line ellipse; the much more rapid precessions of *L* and *S* about the line joining the nuclei are indicated by the broken-line ellipses.

III, section 2(d)], except that now we have Ω in place of Λ. Fig. 97 gives the vector diagram for this case. The vector *J* is constant in magnitude and direction. Ω and *N* rotate about this vector (nutation). At the same time, the precession of *L* and *S* takes place about the internuclear axis (see p. 212 f.). In Hund's case (a) this precession is assumed to be very much faster than the nutation of the figure axis whose frequency is given by (III, 26).

The *rotational energy* in case (a) must clearly be that of a symmetric top with Ω as the angular momentum about the figure axis, that is, neglecting centrifugal stretching terms [see (III, 147)]

$$F_v(J) = B_v[J(J+1) - \Omega^2].\qquad\text{(V, 12)}$$

Here we have omitted the term $A\Omega^2$ since it is a constant for a given electronic state and can therefore be included in the electronic energy. For a given multiplet term B_v has the same value for each multiplet component except when the multiplet splitting is very large.

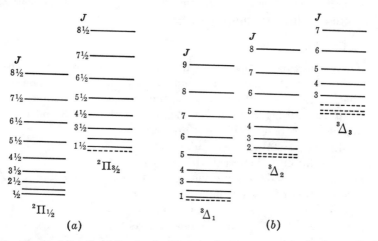

F<small>IG.</small> 98.   **The First Rotational Levels of a ²Π and a ³Δ State in Hund's Case (a).**   The dotted levels do not occur, since J must be $\geqq \Omega$. The Λ-type doubling [see section 2(b)] is ignored.

According to the discussion of the preceding section, Ω is integral or half-integral depending on whether the number of electrons is even or odd. Since Ω is the component of J it follows (see p. 26) that J is integral when Ω is integral—that is, for an even number of electrons—whereas J is half-integral when Ω is half-integral—that is, for an odd number of electrons. Naturally J cannot be smaller than its component Ω. Therefore for a given Ω we have

$$J = \Omega,\ \Omega + 1,\ \Omega + 2,\ \cdots\qquad\text{(V, 13)}$$

Levels with $J < \Omega$ do not occur. Except for these missing levels and apart from an additive constant, the rotational energy (V, 12) for a given multiplet component is the same as for the simple vibrating rotator. The electronic energy of the multiplet components in Hund's case (a) is given, to a first approximation, by (V, 8).

As an example, the rotational levels of a $^2\Pi$ and a $^3\Delta$ state in case (a) are represented in Fig. 98. The missing levels are indicated by broken lines. Fig. 99 gives a somewhat different representation of the rotational levels of a $^3\Pi$ state. If case (a) holds strictly, the three curves representing the rotational levels can be brought into coincidence simply by a vertical shift [see however, section 2(b)].

Hund's case (b). As we have seen above, when $\Lambda = 0$, and $S \neq 0$, the spin vector S is not coupled to the internuclear axis at all. This means that Ω is not defined. Therefore Hund's case (a) cannot apply here. Sometimes, particularly for light molecules, even if $\Lambda \neq 0$, S may be only *very weakly coupled to the internuclear axis*. This weak (or zero) coupling of S to the internuclear

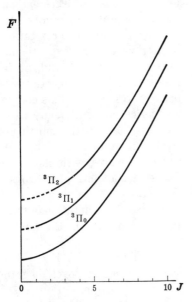

FIG. 99. **Rotational Levels of a** $^3\Pi$ **State in Case (a).** Only the solid parts of the curves correspond to actually occurring rotational levels. The Λ-type doubling is ignored (see Fig. 106).

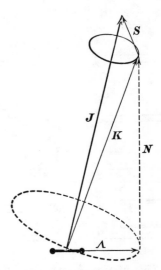

FIG. 100. **Vector Diagram for Hund's Case (b).** The nutation of the figure axis, represented by the broken-line ellipse, is much faster than the precessions of K and S about J, represented by the solid-line ellipse. For $\Lambda = 0$, K ($\equiv N$) is perpendicular to the internuclear axis.

axis is the characteristic of Hund's case (b). In this case the angular momenta Λ (when it is different from zero) and N form a resultant [in the same manner as Ω and N in case (a)] which is here designated K (see Fig. 100). The corresponding quantum number K can have the integral values

$$K = \Lambda, \Lambda + 1, \Lambda + 2, \cdots. \tag{V, 14}$$

K is the *total angular momentum apart from spin.* If $\Lambda = 0$ the angular momentum K is identical with N and is therefore perpendicular to the internuclear axis; the quantum number K can then have all integral values from 0 up.

The angular momenta K and S form a resultant J, the *total angular momentum including spin* (Fig. 100). The possible values of J for a given K are, according to the principles of vector addition (see Chapter I, p. 25), given by

$$J = (K + S), (K + S - 1), (K + S - 2), \cdots, |K - S|. \qquad (V, 15)$$

Thus, in general (except when $K < S$), each level with a given K consists of $2S+1$ components; that is, the number of components is equal to the multiplicity. Again J is half-integral for an odd number of electrons and integral for an even number of electrons. The molecular rotation produces a very slight magnetic moment in the direction of K and this in turn causes a slight magnetic coupling between S and K. This coupling as well as certain other causes (see below) produce a slight splitting of the levels with different J and equal K which increases with increasing K. The corresponding precessions of K and S about J are always very slow compared to the nutation of the figure axis (or, for $\Lambda = 0$, the nuclear rotation).

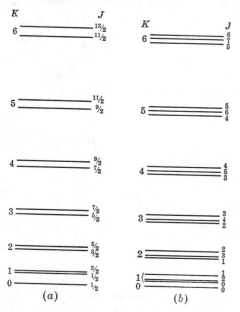

Fig. 101.   **First Rotational Levels** (a) **of a** $^2\Sigma$ **and** (b) **of a** $^3\Sigma$ **State.** It should be noted that, for $^3\Sigma$, for a given K value, the levels do not lie in the order of their J values. The doublet or triplet splitting is drawn to a much larger scale than the separation of levels with different K.

As illustrations, Fig. 101 shows schematic energy level diagrams of the two most important cases, $^2\Sigma$ and $^3\Sigma$. It should be noted that the lowest rotational level ($K = 0$) of all Σ states is single since in this case $J \equiv S$ and therefore, in agreement with (V, 15), only one J value ($J = S$) occurs. This conclusion agrees of course with the one drawn previously when the rotation was neglected altogether, namely, that in the non-rotating molecule multiplet Σ states are single like the S states of atoms.

Hund (346) and Van Vleck (678) have shown that in the case of $^2\Sigma$ states the rotational term values are given by [see Mulliken (512)]

$$F_1(K) = B_v K(K + 1) + \tfrac{1}{2}\gamma K$$
$$F_2(K) = B_v K(K + 1) - \tfrac{1}{2}\gamma(K + 1) \qquad (V, 16)$$

where $F_1(K)$ refers to the components with $J = K + \tfrac{1}{2}$ and $F_2(K)$ refers to those with $J = K - \tfrac{1}{2}$. The splitting constant γ is very small compared to B_v and is usually, although not necessarily, positive. Because of the smallness of γ the energy levels (V, 16) are nearly those of the rotator; however, each level is split into two components. The *splitting* is given by $\gamma(K + \frac{1}{2})$, that is, it *increases linearly with K*. In the earlier literature this splitting is referred to as ρ-type doubling. For a more accurate representation a term $D_vK^2(K + 1)^2$ and possibly higher terms have to be added to (V, 16) [see Chapter III, section 2(b) and (c)]. As an illustration in Fig. 102 the two term series given by (V, 16) are plotted for an anomalously large γ value.

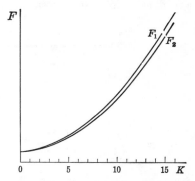

FIG. 102. **Rotational Levels of a** $^2\Sigma$ **State Plotted as a Function of K.** The doublet splitting is greatly exaggerated.

As has been shown by Van Vleck (678) the splitting of a $^2\Sigma$ state is only partly due to the magnetic moment in the direction of K produced by the motion of the nuclei. The other part is due to a magnetic moment that arises from the electronic orbital angular momentum L which in a Σ state precesses at right angles about the internuclear axis. In the rotating molecule the speed of this precession is no longer uniform (rotational distortion) and on the average a non-zero magnetic moment results in the direction of K. This magnetic moment depends inversely on the separation of Π states from the Σ state considered and in general gives the main contribution to the splitting.

In the case of $^3\Sigma$, $^4\Sigma$, $\cdots$ states an additional important contribution to the splitting is provided by the spin-spin interaction and was first recognized by Kramers (420a). He showed that this interaction is equivalent to an interaction of S with the figure axis (rather than with K) and therefore depends only slightly on K, unlike the magnetic interaction of K and S. Including both effects the rotational levels of a $^3\Sigma$ state are given by [Schlapp (617)]

$$F_1(K) = B_vK(K+1) + (2K+3)B_v - \lambda - \sqrt{(2K+3)^2B_v^2+\lambda^2-2\lambda B_v} + \gamma(K+1)$$

$$F_2(K) = B_vK(K + 1) \tag{V, 17}$$

$$F_3(K) = B_vK(K+1) - (2K-1)B_v - \lambda + \sqrt{(2K-1)^2B_v^2+\lambda^2-2\lambda B_v} - \gamma K$$

where F_1, F_2, and F_3 refer to the levels with $J = K + 1$, K, and $K - 1$, respectively, and where λ and γ are constants.[2] Both splitting constants are comparatively large when a $^3\Pi$ state is close to the $^3\Sigma$ state considered [see Hebb (284)]. For the $^3\Sigma_g^-$ ground state of O_2 the splitting constants are found to have the values $\lambda = 1.984$, $\gamma = -0.0084\,\mathrm{cm}^{-1}$ while $B_0 = 1.43777\,\mathrm{cm}^{-1}$ [Babcock and L. Herzberg (764)]. Fig. 103 shows graphically the variation of the splitting as a function of K in the ground state of O_2 up to fairly high K values. For large B/λ the formula (V, 17) can be approximated by the original formula of Kramers:

$$F_1(K) = B_vK(K + 1) - \frac{2\lambda(K + 1)}{2K + 3} + \gamma(K + 1)$$

$$F_2(K) = B_vK(K + 1) \tag{V, 18}$$

$$F_3(K) = B_vK(K + 1) - \frac{2\lambda K}{2K - 1} - \gamma K$$

---

[2] For $K = 1$, $J = 0$ the sign in front of the square root has to be inverted.

This form shows more clearly that for small K and $\gamma \ll \lambda$ (which is usually the case) the terms F_1 and F_3 are close together and below F_2 see Fig. 101(b)]. It also shows that for small γ the splitting is almost independent of K.

Similar formulae have been given by Budó and Kovács (140) (834a) for $^4\Sigma$ states. On account of the spin-spin interaction here also terms with equal $|J - K|$ are close together.

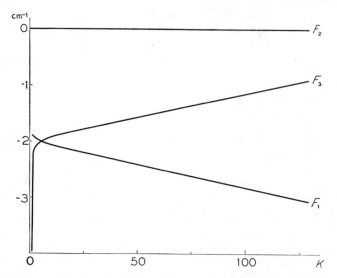

Fig. 103.   **Variation with K of the Triplet Splitting in the $^3\Sigma_g{}^-$ Ground State $(v = 0)$ of O_2.** The quantity $B_0 K(K + 1)$ has been subtracted from the term values. The splittings have been observed up to $K = 39$. Above this K value the term values have been calculated according to (V, 17).

If $\Lambda \neq 0$ an additional term, which depends on Λ and K, has to be added to the energy formulae (V, 16–18). Explicit expressions will be given below when the transition from case (a) to case (b) is considered.

In all cases the splitting constants γ (and λ) depend on v in a way similar to the dependence of B_v on v. However, since these constants are small this variation is often neglected [see however, Mulliken and Christy (523)].

Naturally, for *singlet states* the distinction between cases (a) and (b) is pointless. For them, $\Lambda = \Omega$ and $K = J$, and we have the simple symmetrical top [see Chapter III, section 2(d)].

Cases (a) and (b) are the most important of the five coupling cases distinguished by Hund.

Hund's case (c). As mentioned previously in the discussion of the two-center system, the *interaction between L and S*, in certain cases, particularly for heavy molecules, may be *stronger than the interaction with the internuclear axis*. In this case Λ and Σ are not defined; rather, L and S first form a resultant J_a which is then coupled to the internuclear axis with a component Ω. This coupling is entirely similar to that of an atom in a weak electric field when no Paschen-Back effect occurs. The electronic angular momentum Ω and the angular momentum N of nuclear rotation then form the resultant angular momentum J just as in case (a). This type of coupling is Hund's case (c). Fig. 104 gives the vector diagram for this case. The rotational energy levels and their J values are given by the same formulae (V, 12) and (V, 13) as for case (a).

A more detailed discussion of case (c) may be found in Mulliken (513) (1263).

Hund's case (d). If the *coupling between L and the internuclear axis is very weak* while that between L and the axis of rotation is strong one has Hund's case (d). In this case the angular momentum of nuclear rotation which is called R here (rather than N) is quantized, that is, has magnitude $\sqrt{R(R+1)}\, h/2\pi$ where the quantum number R can have the values

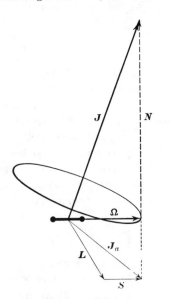

Fig. 104. **Vector Diagram for Hund's Case (c).** The precessions of L and S about J_a and of J_a about the line joining the nuclei is not indicated.

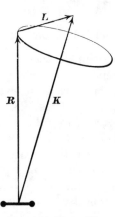

Fig. 105. **Vector Diagram for Hund's Case (d).** The addition of K and S to form J is not shown, since it is of no practical importance.

$0, 1, 2, \cdots$. The angular momenta R and L are added vectorially, as shown in Fig. 105, giving the total angular momentum apart from spin, which as previously, is designated by K. The corresponding quantum number K, for a given R, can have the values

$$K = (R + L), (R + L - 1), (R + L - 2), \cdots, |R - L| \qquad (V, 19)$$

Thus there are $2L + 1$ different K values for each R, except when $R < L$.

The angular momenta K and S form the total angular momentum J in the same way as in case (b) (see Fig. 100). Usually, however, in case (d) the coupling between K and S is so small that S and therefore also J can be disregarded and we need to use only K. Mulliken (512) distinguishes the case in which J can be disregarded, as case (d'), from the case (d) in which it cannot be disregarded.

The *rotational energy* in case (d) is given to a first approximation, by

$$F(R) = B_v R(R + 1) \qquad (V, 20)$$

Each of the terms given by this formula is split into $2L + 1$ close components [see the right-hand part of Fig. 107] each of which, strictly speaking, is further split into $2S + 1$ components.

Formulae for these splittings may be found in Weizel (22) who also considers a partial case (d) in which the orbital angular momentum of the inner electrons is strongly coupled to the internuclear axis while that of an outer electron is very weakly coupled to this axis.

Hund's case (e). In principle a case is conceivable in which *L and S are strongly coupled* but otherwise the coupling conditions are similar to case (d) (that is, weak interaction of *L* and *S* with the internuclear axis). As in case (c) *L* and *S* form a resultant J_a which is now combined with *R* to form *J* similar to the combination of *L* and *R* in case (d). No practical examples of this case (e) have as yet been observed. For more details about this case see Mulliken (512) and Weizel (22).

(b) Uncoupling Phenomena

Hund's coupling cases represent idealized limiting cases. Nevertheless they do often represent the observed spectra to a good approximation. However, not infrequently, small or even large deviations from these limiting cases are observed. These deviations have their origin in the fact that interactions which were neglected or regarded as small in the idealized coupling cases really have an appreciable magnitude, and particularly that the relative magnitude of the interactions changes with increasing rotation. Therefore, sometimes, with increasing rotation, a *transition* takes place *from one coupling case to another*. Angular momentum vectors coupled to the internuclear axis for small rotation are *uncoupled* from it with increasing rotation.

Λ-type doubling. In Hund's cases (a) and (b) the interaction between the rotation of the nuclei and *L* has been neglected. For larger speeds of rotation this interaction must be taken into account and is found to produce a *splitting into two components for each J value* in the states with $\Lambda \neq 0$ which are doubly degenerate without rotation. In general, this splitting increases with increasing rotation—that is, with increasing *J*. It is present for all states with $\Lambda \neq 0$ and is called Λ-*type doubling*. Such a splitting is shown qualitatively in the energy level diagram in Fig. 61(b), p. 129, which applies to a $^1\Pi$ state. The two component levels with somewhat different energies, into which the one level of the ordinary rotator is split, have the same value of *J*, thus differing from the two component levels into which each rotational level of a $^2\Sigma$ state is split [see Fig. 101(a)].

For *multiplet states*, the rotational levels of each component of the multiplet, including those with $\Omega = 0$, split into two Λ-type components. The splitting of the rotational levels of $^1\Pi$, $^2\Pi$, and $^3\Pi$ states is shown in (a), (b), and (c) respectively, of Fig. 106. The magnitude of the Λ-type splitting is much exaggerated in the figure. In general, it will amount to only a fraction of a cm^{-1}. Only in special cases and for large *J* values will it reach a value of a few cm^{-1}. The splitting is relatively the greatest for terms with the smallest Ω (see the figure). In particular, for $\Omega = 0$ (in the case of a $^3\Pi$ state), it is relatively large even for small *J* and is approximately independent of *J*, in contrast to the other terms. As was mentioned previously (p. 216) even in the two-center system, disregarding rotation entirely, a slight splitting of a $^3\Pi_0$ state occurs.

The Λ-type splittings of $^1\Delta$, $^2\Delta$, and $^3\Delta$ states are similar to those of the $^1\Pi$, $^2\Pi$, and $^3\Pi$ states (Fig. 106) except that they are considerably smaller.

Following Mulliken (513), the two slightly different rotational term series of a state with $\Lambda \neq 0$ are usually distinguished by subscripts c and d (in the older literature a and b). For example, for a Π term we have a Π_c and a Π_d com-

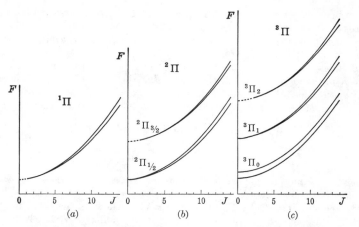

Fig. 106. **Λ-Type Doubling for the First Rotational Levels of $^1\Pi$, $^2\Pi$, $^3\Pi$ States.** The parts of the curves corresponding to non-occurring levels are dotted. The magnitude of the Λ-type doubling is exaggerated considerably.

ponent. The quantities $F(J)$ are given corresponding subscripts: $F_c(J)$ and $F_d(J)$.

It is necessary to guard against the error of supposing that the c sublevels correspond to one orientation of the electronic angular momentum along the line joining the nuclei and that the d levels correspond to the other. Rather the electronic eigenfunctions of the c levels as well as those of the d levels are combinations of the eigenfunctions corresponding to the two directions of rotation (see p. 213).

Indeed it turns out that the zero approximation eigenfunctions are the functions (V, 10) p. 218, that is, the Π_c and Π_d levels might also be called Π^+ and Π^- levels (or vice versa) as was done on p. 217.

Kronig (421) and Van Vleck (678) have shown that, if the interaction of rotation and electronic motion is taken into account in the wave equation of the molecule, the following formula for the rotational levels of a state with $\Lambda \neq 0$ results (omitting terms independent of J):

$$F_i(J) = B_v J(J+1) + D_v J^2(J+1)^2 + \cdots + \Phi_i(J), \qquad (V, 21)$$

where the subscript i stands for c or d. In by far the most cases $\Phi_i(J)$ is a very small correction which may be represented by

$$\Phi_i(J) = \kappa_i + \delta_i J(J+1) + \mu_i J^2(J+1)^2 + \cdots, \qquad (V, 22)$$

where $\mu_i \ll \delta_i$ and κ_i is of the order of magnitude of δ_i. From this it follows that, if κ_i is neglected, we can also write

$$F_i(J) = B_v{}^i J(J+1) + D_v{}^i J^2(J+1)^2 + \cdots, \qquad (V, 23)$$

where $B_v{}^i$ and $D_v{}^i$ are *effective* B_v and D_v values,[3] which are related to the true B_v and D_v values according to

$$B_v{}^i = B_v + \delta_i,$$
$$\qquad\qquad (V, 24)$$
$$D_v{}^i = D_v + \mu_i$$

Thus the rotational levels of each of the substates c or d may be represented in exactly the same manner as those of the simple vibrating rotator [formula (IV, 26)]. This is indeed confirmed by the observed spectra. However, the effective B_v and D_v values no longer have exactly the same physical meaning as before.

For $^1\Pi$ states or multiplet Π states belonging to case (b) (in the latter case J must be replaced by K in the formulae), calculation shows that the difference $\delta_c - \delta_d$ is of the same order of magnitude as δ_c (or δ_d). Therefore $(\mu_c - \mu_d)J^2(J+1)^2$ may be neglected compared to $(\delta_c - \delta_d) J(J+1)$, and, to a first good approximation, the Λ-type splitting is

$$\Delta\nu_{cd}(J) = F_c(J) - F_d(J) = qJ(J+1), \qquad (V, 25)$$

where

$$q = B_v{}^c - B_v{}^d = \delta_c - \delta_d. \qquad (V, 26)$$

Thus, in contrast to the spin splitting of $^2\Sigma$ terms, the Λ-*type splitting increases quadratically with the rotational quantum number.*

The theoretical treatment by Kronig and Van Vleck shows further that, for $^1\Delta$ states and multiplet Δ states of case (b), $\delta_d \approx \delta_c$. As a result, according to (V, 21–23), the splitting is proportional to $J^2(J+1)^2$, where the proportionality factor $(\mu_c - \mu_d)$ is very much smaller than δ_i. This means that for Δ states the Λ-type doubling is negligibly small for not too large J values.

The analysis of the empirical data concerning Π and Δ states yields the effective B values, $B_v{}^c$ and $B_v{}^d$. The *true* B_v *values*, a knowledge of which is necessary for the calculation of the moments of inertia and the internuclear distances, can be obtained from $B_v{}^c$ and $B_v{}^d$ only if δ_c and δ_d are known individually while from the empirical data only $q = \delta_c - \delta_d = B_v{}^c - B_v{}^d$ can be determined. In a few favorable cases the individual values of δ_d and δ_c can be obtained on the basis of the theoretical formulae from the positions of neighboring electronic states (see below). Fortunately it has always been found that δ_d and δ_c are very small compared to B_v and are actually of the order of magnitude of $B_v{}^c - B_v{}^d$ (that is, roughly $10^{-3} B_v$ or smaller). In practice, the *mean* of $B_v{}^c$ and $B_v{}^d$ is usually used as an approximation for the true B_v value. It should be remarked that, strictly speaking, a term $\Phi(J)$ has also to be added to the usual formula for the rotational levels of $^1\Sigma$ states. This inclusion of $\Phi(J)$ has the result that, for $^1\Sigma$ states also, the experimentally derived B_v value is only an effective B_v value which deviates by a small amount from the true B_v value. The deviation—that is, $\Phi(J)$—is greater (for Σ as well as Π and Δ states) the closer other electronic states are to the state under consideration (see p. 292). It is usually exceedingly small for the ground state of a molecule since the ground state is as a rule far from other electronic states that may perturb it. The deviations of the true from the effective B_v values on account of the interaction of rotation and electronic motion as discussed here, occur in addition to those discussed previously (p. 109), which were caused by the finer interaction of rotation and vibration.

Comparatively simple formulae are obtained for the constant q of Λ-type doubling when the Π state considered is fairly close to one particular Σ state but far away from other Σ states

---

[3] These are not to be confused with the symbols for isotopes used in equations (III, 201 and 202).

and when both Σ and Π state have the same orbital angular momentum L which precesses under two different angles to the internuclear axis in the two states. This situation has been called the case of *"pure precession"* by Van Vleck (678). In this case assuming that L is due to one electron only and is thus equal to the l of this electron Van Vleck found

$$q = \frac{2B_v{}^2 l(l+1)}{\nu(\Pi, \Sigma)} \tag{V, 27}$$

where $\nu(\Pi, \Sigma)$ is the separation of the Π from the Σ state assuming that B_v is the same in the two states. The relation (V, 27) shows that the splitting constant q depends on the vibrational quantum number in the same way as $B_v{}^2$.

For *multiplet states belonging to case* (a) (large multiplet splitting) Van Vleck's formulae show that the variation of the Λ-type doubling with J is different for the different multiplet components. For $^2\Pi$ states, the Λ-type doubling of the $^2\Pi_{1/2}$ component varies linearly with J, whereas for the $^2\Pi_{3/2}$ component it varies with the third power of J and for small J is very small compared to that of $^2\Pi_{1/2}$. As mentioned above, for $^3\Pi$ states, the Λ-type splitting of the $^3\Pi_0$ component is relatively large even for small J and is, to a first approximation, independent of J. The $^3\Pi_1$ component shows about the same variation of the Λ-type splitting as a $^1\Pi$ term; that is, it is proportional to $J(J+1)$, whereas the $^3\Pi_2$ component shows a splitting which is proportional to $J^2(J+1)^2$ and which, for not too large J, is very much smaller than that of the two other triplet components (see Fig. 106). Similar relations hold for $^4\Pi$ states [see Budó and Kovács (834)].

In *Hund's case* (c) a doubling that is entirely similar to the Λ-type doubling occurs for all electronic states with $\Omega \neq 0$. Since Λ is not defined in case (c) this doubling is called Ω-*type doubling*. States with $\Omega = 0$ (0^+ and 0^- states, see above) are non-degenerate and no Ω-type doubling occurs for them. For $\Omega \neq 0$ the dependence of the doubling on J is similar to the dependence in case (a) for the same Ω values.

A full discussion of Λ-type doubling, including intermediate cases and extensive applications is to be found in a paper by Mulliken and Christy (523). Thus far, a complete analysis of Λ-type doubling has been carried out for only relatively few molecules.

Transition from case (b) to case (d) (*L* uncoupling). The Λ-type doubling discussed in the foregoing may be regarded as the beginning of a gradual transition of the coupling in the molecule from case (a) or case (b) to case (d). We shall now consider the complete transition, but only from case (b) to case (d), since this is the only transition observed in practice—for example, for some electronic states of H_2 and He_2. For these molecules it happens fairly often that the interaction between the orbital angular momentum L and the internuclear axis is so small that for the higher rotational levels it is outweighed by the interaction with the rotational axis [case (d)] that is, with increasing rotation, L is *uncoupled* from the internuclear axis. While, without rotation, there are $L + 1$ terms with different Λ for a given value of L, of which those with $\Lambda \neq 0$ are doubly degenerate, for large rotation [in case (d)] there is a splitting into $2L + 1$ components—that is, there is no degeneracy. It is therefore clear that for an incipient L uncoupling a splitting of the levels that are degenerate with each other must take place—that is, just the Λ-type doubling.

As an example of the complete transition, in Fig. 107 the first rotational levels for the case $L = 1$ are drawn to the left for case (b) (strong coupling of L with the internuclear axis) and to the right for case (d) (vanishing coupling with the internuclear axis). If, starting out from case (b), we now imagine the coupling to be decreased successively (say by decreasing the nuclear separation in the molecule), the rotational levels to the left must eventually go over into those to the right and must do so in such a way that K, which in the present case is equal to the total angular momentum J, remains unchanged for a given level and, in addition, in such a way that levels with equal K do not cross one another (see section 4). In this way we obtain the connecting lines drawn in Fig. 107, which of course represent only schematically the behavior of the levels for intermediate coupling conditions. We can see how in the $^1\Pi$ state the levels split with increasing uncoupling (Λ-type doubling) and finally go over into two levels with different R (but, of course, with equal K).

In an actual case, of course, the coupling does not change for a particular level but rather changes from one level to the next, approaching case (d) for the higher rotational levels. This change of coupling from one level to the next is indicated by the broken-line curves in Fig. 107 for weak uncoupling and by the dotted curves for strong uncoupling. The points of

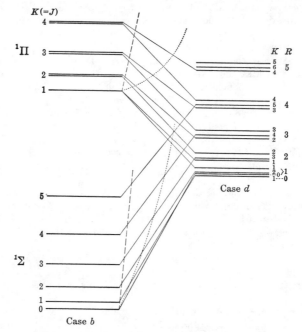

FIG. 107.   **Transition from Case (b) to Case (d) for the Lowest Rotational Levels, for $L = 1$.** The strength of the interaction between L and the internuclear axis decreases from left to right.

intersection of these curves with the connecting lines between right and left give very roughly the positions of the rotational levels. If we follow the broken-line curves, we can see immediately how the Λ-type doubling arises and how a difference between the effective and the true B value is brought about (this difference is much larger in the figure than in most practical cases). The dotted curves show that for strong uncoupling the rotational levels, particularly in the $^1\Pi$ state, follow quite an abnormal course. Such abnormal rotational term series have been observed in a number of electronic states of H_2 and He_2 and have been thoroughly discussed by Weizel, Dieke, Richardson, and others [see (22) and (32)]. Thus far, such cases have not been observed with certainty for other molecules.

In Hund's cases (a) and (b) the quantum number L is usually poorly defined (see p. 219 f.) and its value is not easily determined. However, when L uncoupling occurs for the higher rotational levels L is readily obtained since the number of component levels with a certain R is $2L + 1$.

It should be observed that, as shown in Fig. 107, in case (d) (complete uncoupling) the levels with different K of a given R do not lie in the order of the K values but the levels with equal $|K - R|$ lie near one another. The reason for this is that the coupling between L and R is due mainly to gyroscopic forces or their quantum-theoretical analogues and not, or only to quite a secondary extent, to magnetic forces (as is the coupling between L and S in an atom).

Detailed formulae for the energy levels in a general case between (b) and (d) have been given by Kovács and Budó (1139a).

Transition from case (a) to case (b) (spin uncoupling). While multiplet Σ states always belong strictly to case (b), multiplet Π, Δ, $\cdots$ states frequently belong to cases intermediate between (a) and (b). For these states usually (apart from H_2, He_2, and some hydrides) case (a) is a good approximation for no rotation or very small rotation; that is, S is so coupled with Λ that a resultant Ω is defined. However, if the multiplet splitting is not too great, as J increases, the rotational velocity of the molecule becomes comparable with the precessional velocity of S about Λ (see Fig. 97), and finally, with still further

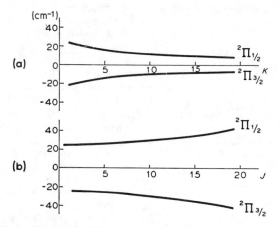

Fig. 108.   **Doublet Splitting in the** $^2\Pi$ ($v = 7$) **State of CN; (a) as a Function of** K **and (b) as a Function of** J. The quantities $B_7 K(K + 1)$ and $B_7 J(J + 1)$ respectively have been subtracted from the term values.

increasing J, the influence of the molecular rotation predominates. Consequently, S is uncoupled from the molecular axis and forms with K (the resultant of Λ and N) the total angular momentum J, in accordance with case (b). This process is called *spin uncoupling*.

For large rotation [case (b)] there is then only a small splitting of the levels with a given K, while for small rotation [case (a)] there is a much greater splitting. As an illustration, Fig. 108(a) shows the variation of the rotational term values in the $^2\Pi$ state of CN, the amount $BK(K + 1)$ being subtracted from the observed values (using a mean B). It can be seen that the terms draw together with increasing K. To be sure for small rotation the angular momentum K is not defined, since then S is coupled to the molecular axis. However, in spite of that, we can at least formally assign K values to the levels ($K = J \pm \frac{1}{2}$ for doublet terms) if we extend to small K the numbering used for large K values (where it has a definite physical meaning).

Fig. 109 illustrates in a way similar to Fig. 108(a) the behavior of the components of the $B\,^3\Pi$ state of the N_2 molecule as an example of a triplet state that changes from case (a) to case (b). Δ states behave in a similar manner; however, few have been investigated thus far.

The rotational term values of the components of doublet states have been calculated theoretically by Hill and Van Vleck (323) for any magnitude of the coupling between S and Λ, but neglecting the coupling between K and S [see Mulliken (512)]. They obtained

$$F_1(J) = B_v\left[(J+\tfrac{1}{2})^2 - \Lambda^2 - \tfrac{1}{2}\sqrt{4(J+\tfrac{1}{2})^2 + Y(Y-4)\Lambda^2}\right] - D_vJ^4$$
$$F_2(J) = B_v\left[(J+\tfrac{1}{2})^2 - \Lambda^2 + \tfrac{1}{2}\sqrt{4(J+\tfrac{1}{2})^2 + Y(Y-4)\Lambda^2}\right] - D_v(J+1)^4$$

$$(V,\,28)$$

Here $Y = A/B_v$, where the coupling constant A is a measure of the strength of the coupling between the spin S and the orbital angular momentum Λ.[4] $F_1(J)$ is the term series that

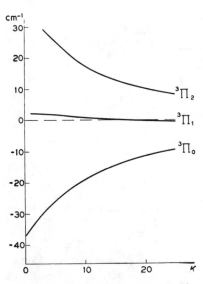

Fig. 109. **Triplet Splitting in the $B^3\Pi$**
($v = 6$) State of N_2 as a Function of K.
The deviations from $B_6K(K+1) - D_6K^2(K+1)^2$ have been plotted using the formulae (V, 31) and the observed values of B_6 and D_6.

forms for large rotation the levels with $J = K + \tfrac{1}{2}$, while $F_2(J)$ is that which forms for large rotation the levels with $J = K - \tfrac{1}{2}$. The formulae (V, 28) hold both for regular doublets $(A > 0)$ and inverted doublets $(A < 0)$. As is immediately seen from the form of (V, 28) F_1 and F_2 form $^2\Pi_{1/2}$ and $^2\Pi_{3/2}$ for a regular, but $^2\Pi_{3/2}$, $^2\Pi_{1/2}$, respectively, for an inverted $^2\Pi$ state and similarly for other values of Λ. However an exception arises for $0 < Y < 2$ when the lowest level $J = \Lambda - \tfrac{1}{2}$ belongs to F_2 even though the other rotational levels of the same doublet component ($^2\Pi_{1/2}$, $^2\Delta_{3/2}, \cdots$) belong to F_1.

In the formulae (V, 28) the influence of the centrifugal force has been assumed to be the same as in case (b), that is, to be given by $-D_v(K + \tfrac{1}{2})^4$. This approximation is usually quite satisfactory, since D_v is very small compared to B_v. A more rigorous discussion of the effect of centrifugal force has been given by Almy and Horsfall (63). Their formulae show a rather more complicated dependence on D_v than (V, 28). It should be mentioned that in both formulae (V, 28) a term independent of J has been omitted [see Mulliken (512)].

On substituting $J = K + \tfrac{1}{2}$ and $J = K - \tfrac{1}{2}$, respectively, in the two equations (V, 28) it is readily seen that F_1 equals F_2 for the same K value if $Y = 0$ or 4, and that F_1 and F_2 are nearly equal when Y is small—that is, we have case (b). Even for larger Y values (that is, larger values of the doublet splitting in the rotationless state), $F_1(K)$ and $F_2(K)$ become more and more alike with increasing K but this transition from case (a) to case (b) occurs at higher and higher J values when A increases.

When A is very small or for very large K additional terms $+\tfrac{1}{2}\gamma K$ and $-\tfrac{1}{2}\gamma(K+1)$

---

[4] That A in (V, 28) has essentially the same meaning as in (V, 8) is readily verified: for $\Lambda = 1$ the difference $F_2 - F_1$ for $J = \tfrac{1}{2}$ is $A - 2B_v$. Of course only one of the levels with $J = \tfrac{1}{2}$ actually exists.

should be added to the two formulae (V, 28). They represent the effect of the magnetic interaction of K and S mentioned previously (compare also the discussion of $^2\Sigma$ states p. 222). In most practical cases these terms can be neglected.

For *small spin uncoupling* [large A, case (a)] the formulae (V, 28) may be replaced by

$$F(J) = F_{\text{eff.}} \, (J + 1) - D_v J^2 (J + 1)^2 \qquad (V, 29)$$

where the effective rotational constant B is somewhat different for the two multiplet components and where terms independent of J have been omitted. According to Mulliken (513), for doublet states

$$B_{\text{eff.}} = B \left(1 \pm \frac{B}{A\Lambda} + \cdots \right), \qquad (V, 30)$$

For *large uncoupling* (very small Y) the square root in (V, 28) can be replaced by

$$2(J + \tfrac{1}{2}) + \frac{Y(Y - 4)\Lambda^2}{4(J + \tfrac{1}{2})} + \cdots$$

Substituting this and $J = K + \tfrac{1}{2}$ and $J = K - \tfrac{1}{2}$, respectively, into (V, 28) one obtains

$$F_1(K) = B_v \left[K(K + 1) - \Lambda^2 + \frac{Y(4 - Y)}{8(K + 1)} \Lambda^2 + \cdots \right] - D_v (K + \tfrac{1}{2})^4$$

$$F_2(K) = B_v \left[K(K + 1) - \Lambda^2 - \frac{Y(4 - Y)}{8K} \Lambda^2 + \cdots \right] - D_v (K + \tfrac{1}{2})^4$$

$$(V, 28a)$$

which for $\Lambda = 0$ would go over into the energy formula (V, 16) for $^2\Sigma$ states [except for the terms with γ which have been omitted in (V, 28a), see above].

It is seen from (V, 23) that for any uncoupling the mean of the two components with equal J follows rigorously the simple rotator formula (V, 29) with the true B_v value. This is not the case for the mean of the two components with equal K [see (V, 28a)].

In Fig. 110 the transition from case (a) to case (b) for a regular $^2\Pi$ state is illustrated by an energy level diagram similar to Fig. 107 for the L uncoupling. The figure shows how levels with equal K draw together with increasing uncoupling. On the other hand levels with equal J go over into levels with different K. Therefore their separation increases with increasing J and increasing uncoupling [see the formulae (V, 28)]. This is illustrated by Fig. 108(b) in which the splitting of the $^2\Pi$ state of CN is plotted as a function of J in contrast to Fig. 108(a) in which it is plotted as a function of K.

The effect of the uncoupling on the rotational energy levels in two typical cases is shown in Fig. 111(a) and (b), the former representing the observed levels in the regular $^2\Pi$ state of CaH ($Y = +18.5$), the latter in the inverted $^2\Pi$ ground state of OH ($Y = -7.41$). Levels with equal K are connected by oblique lines. In both cases for the same K the levels F_2 lie higher than F_1. The levels F_2 lie lower than F_1 (as is assumed in Fig. 110) only if A/B is between 0 and 4 [see equation (V, 28a)]. In conformity with the previous rules the levels F_1 form the $^2\Pi_{1/2}$ state of CaH and the $^2\Pi_{3/2}$ state of OH while the levels F_2 form $^2\Pi_{3/2}$ of CaH and $^2\Pi_{1/2}$ of OH. As mentioned previously the quantum number K is well defined only for the higher rotational levels. If in the case of CaH (and similarly for other regular $^2\Pi$ states) one continues the K numbering to the lower levels one is led to assigning $K = 1$ formally to the two lowest levels of $^2\Pi_{1/2}$ while otherwise the two levels of each K belong to different doublet components.

For *triplet states* Budó (139) on the basis of Hill and Van Vleck's general equations has

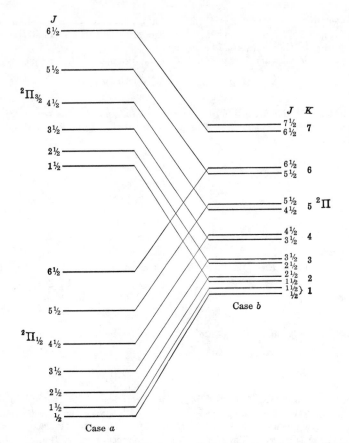

F<small>IG.</small> 110.   Transition of a regular $^2\Pi$ State from Case (a) to Case (b) (Spin Uncoupling).

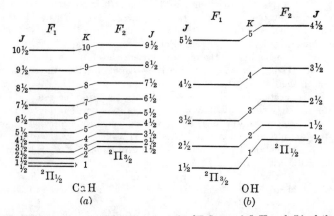

F<small>IG.</small> 111.   The Lowest Rotational Levels (a) of the $^2\Pi$ State of CaH and (b) of the $^2\Pi$ Ground State of OH.   The Λ-type doubling is not shown.   It is negligible on the scale of the diagram.

given the following formulae for the rotational term values for any degree of spin uncoupling[5] [see also Gilbert (260)]

$$F_1(J) = B_v[J(J+1) - \sqrt{Z_1} - 2Z_2] - D_v(J - \tfrac{1}{2})^4$$

$$F_2(J) = B_v[J(J+1) + 4Z_2] - D_v(J + \tfrac{1}{2})^4 \qquad \text{(V, 31)}$$

$$F_3(J) = B_v[J(J+1) + \sqrt{Z_1} - 2Z_2] - D_v(J + \tfrac{3}{2})^4$$

where

$$Z_1 = \Lambda^2 Y(Y-4) + \tfrac{4}{3} + 4J(J+1)$$

$$Z_2 = \frac{1}{3Z_1}[\Lambda^2 Y(Y-1) - \tfrac{4}{9} - 2J(J+1)] \qquad \text{(V, 32)}$$

As before $Y = A/B_v$ is a measure of the strength of the coupling of the spin to the internuclear axis. For large rotation F_1, F_2, and F_3 go over into the term series with $J = K + 1$, K, and $K - 1$, respectively, which for a given K lie close to one another and, in general, in the inverse order of the J values [see Mulliken (513) and Jevons (23)]. This holds for both regular and inverted triplet states.

If the spin uncoupling is slight, just as for doublet states the rotational levels of each triplet component may be represented by the simple rotator formula (V, 29) with an effective rotational constant which for $\Sigma = 0$ (that is, $^3\Pi_1$, $^3\Delta_2$, $\cdots$) is equal to the true B_v, but for $\Sigma = \pm 1$ is given by

$$B_{\text{eff.}} = B_v\left(1 \pm \frac{2B_v}{A\Lambda} + \cdots\right) \qquad \text{(V, 33)}$$

While for larger spin uncoupling the central triplet component no longer follows exactly the simple rotator formula (V, 29) it is easily seen from (V, 31) that the mean of the three components taken for each J value does follow the simple rotator formula with the true B_v value which can in this way be determined from the observed rotational levels.[6]

Challacombe and Almy (152) have discussed the dependence of the position of the rotational levels on Y on the basis of (V, 31) and related formulae.

Formulae for the rotational levels of *quartet states* in the general case between (a) and (b) have been given by Brandt (124) and Budó (140) [see also Nevin (530)]. They are very similar to the formulae (V, 31) for triplet states.

Thus far, in considering spin uncoupling we have neglected the Λ-*type doubling* (that is, incipient L uncoupling), which takes place simultaneously and also increases with increasing rotation. In most cases the Λ-type doubling is small compared to the spin splitting so that the relations valid for case (a) can be applied (see p. 226). However, cases do occur in which the two uncouplings are of comparable magnitude. As an example, Fig. 112 gives the rotational levels of the $^2\Pi$ state of CaH as a function of K. It is seen that in this case for large K the Λ-doubling is larger than the spin doubling.

The Λ-type doubling may be introduced into the formulae (V, 28) and (V, 31) as before by adding a function $\Phi_i(J)$ which is different for the two Λ-doubling components. The form of this function is now rather more involved than that given previously for singlet states [equation (V, 22)]. We do not give explicit formulae here but refer to Van Vleck (678), Mulliken and Christy (523), Hebb (284), Gilbert (260) and Challacombe and Almy (152). It should be noted that in a multiplet state intermediate between case (a) and (b) the Λ-type doubling is very different for the different multiplet components for small J whereas for large rotation the splitting is nearly the same (see the example Fig. 112). The mean of all component levels for each J value can again be represented by the simple rotator formula (V, 29).

---

[5] The formulae do not hold for the rotational level with $J = \Lambda - 1$.

[6] It may be noted that this conclusion does not apply to the mean taken for each K value.

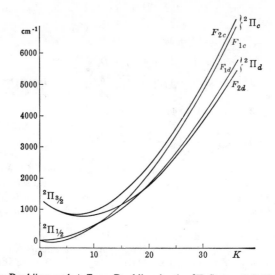

FIG. 112. Spin Doubling and Λ-Type Doubling in the $^2\Pi$ State of CaH [after Mulliken and Chᵢity (523)]. The ordinate gives the energy values for the component F_{1d}. The deviations of the other components, F_{1c}, F_{2c}, and F_{2d}, from the former are plotted on a 20-fold scale.

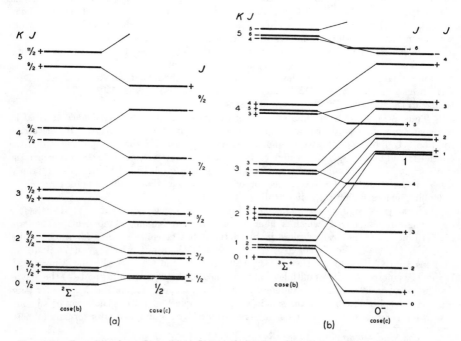

FIG. 113. Transition from Case (b) to Case (c); (a) for a $^2\Sigma^-$ State, (b) for a $^3\Sigma^+$ State. The transition for $^2\Sigma^+$ and $^3\Sigma^-$ is similar except that $+$ and $-$ are everywhere interchanged.

Transition from case (b) to case (c). Cases intermediate between case (a) or (b) and (c) occur for small internuclear distances (for example in certain hydrides) or for large internuclear distances (for example in the halogen molecules or generally for vibrational levels close to the dissociation limit) when the electric field in the internuclear axis is not sufficient to break down the atomic (L, S) coupling.

The transition from case (a) to case (c) needs no special discussion since Ω is defined in both cases and the rotational energy formulae are the same. The only point to be remembered is that $^3\Pi_0$ states (and similarly $^5\Delta_0$) split into two 0 states, a 0^+ and a 0^- state, which in case (c) are independent of one another as are the other multiplet components.

The transition from case (b) to case (c) is important only when $\Lambda = 0$, since for case (b) states with $\Lambda \neq 0$ the spin-orbit coupling is usually weak throughout and a transition to case (c) does not occur for any internuclear distance. This last remark applies also to multiplet Σ states having $L = 0$. However, Σ states having $L \neq 0$ will go over into case (c) states when the effect of the internuclear axis on L becomes small. Fig. 113 shows the correlation of the rotational levels of a $^2\Sigma^-$ and of a $^3\Sigma^+$ state in the two coupling cases. The $^2\Sigma^-$ state goes over into a $\frac{1}{2}$ state in case (c); the $^3\Sigma^+$ state into a 1 and 0^- state. That this must be so even without rotation is readily seen if the states derived from $L = 1$, $S = \frac{1}{2}$ and 1 for case (c) and case (b) are compared with each other [see also Mulliken (512)]. It should be emphasized that an actual transition from case (b) to case (c) in a given molecule cannot be brought about by increasing rotation but may occur for increasing vibration of the molecule.

(c) Symmetry Properties of the Rotational Levels

The symmetry properties of the rotational levels introduced in Chapter III, section 2(f)—positive-negative, and, for identical nuclei, symmetric-antisymmetric—are, as we have already mentioned, influenced by the symmetry of the electronic eigenfunctions. Now that we have become acquainted with the different kinds of electronic eigenfunctions, we can discuss this in greater detail.

Σ **states.** A more detailed discussion (see below) shows that *for Σ^+ states the even-numbered rotational levels are positive and the odd are negative*, whereas *for Σ^- states the even are negative and the odd are positive*. For multiplet Σ states the character positive-negative depends on whether K (not J) is even or odd [see Wigner and Witmer (712)]. The symmetry properties of the rotational levels of $^1\Sigma^+$, $^2\Sigma^+$, $^3\Sigma^+$, and $^1\Sigma^-$, $^2\Sigma^-$, $^3\Sigma^-$ states are represented schematically in Fig. 114(a). The positive and negative rotational levels are indicated by $\oplus$ or $\ominus$, respectively, on the horizontal lines. The spacing of the circles does not correspond to the separation of the rotational levels.

A molecular state is called positive or negative [see Chapter III, section 2(f)] according as the total eigenfunction remains unaltered or changes its sign by reflection of all the particles (including the nuclei) at the origin (inversion). The vibrational and rotational eigenfunctions, ψ_v and ψ_r, depend on the co-ordinates of the nuclei only. According to our previous discussion, ψ_r remains unchanged or changes its sign for an inversion depending on whether K is even or odd [see Chapter III, section 2(f)].[7] The vibrational eigenfunction ψ_v remains unaltered by the inversion, since it depends only on the magnitude of the internuclear distance. In order to see how the electronic eigenfunction ψ_e behaves for an inversion we must take account of the fact that an inversion is equivalent to a rotation of the molecule through 180° about

---

[7] In the discussion in Chapter III J was used. However, it is clear that K determines the symmetry of ψ_r if $S \neq 0$ (see also below).

an axis perpendicular to the internuclear axis, followed by a reflection at a plane perpendicular to this rotational axis and passing through the internuclear axis. The first operation does not influence ψ_e, since ψ_e depends only on the co-ordinates of the electrons *relative* to the nuclei (as well as on the internuclear distance), whereas the second operation (which affects the electrons only) leaves ψ_e unaltered for Σ^+ states and changes its sign for Σ^- states (see the definition of Σ^+ and Σ^- states, p. 217); that is, ψ_e remains unaltered or changes its sign for an inversion depending on whether the electronic state is Σ^+ or Σ^-. From these considerations it follows that the total eigenfunction $\psi = \psi_e \cdot (1/r)\psi_v \cdot \psi_r$ for a Σ^+ state remains unaltered or changes its sign for an inversion—that is, rotational level of a Σ^+ state is positive or negative—according as the rotational quantum number is even or odd, whereas for Σ^- states just the converse holds.

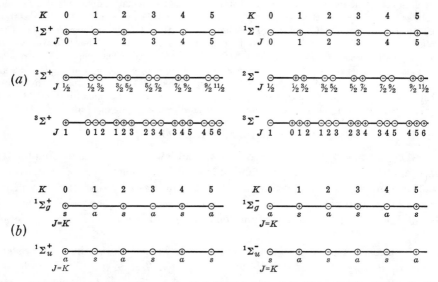

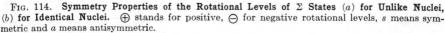

Fɪɢ. 114. **Symmetry Properties of the Rotational Levels of Σ States** (a) **for Unlike Nuclei,** (b) **for Identical Nuclei.** $\oplus$ stands for positive, $\ominus$ for negative rotational levels, s means symmetric and a means antisymmetric.

Thus far ψ_e was considered to be the electronic coordinate function. It has to be multiplied by a function representing the orientation of the electron spins. In case (b) (which always holds for Σ states) this spin function is independent of an inversion [see Kronig (21a)] and therefore levels with different J but equal K have the same $(+, -)$ symmetry [see Fig. 114(a)].

If the two nuclei of the molecule are identical we have to consider in addition the *symmetry with respect to an exchange of the nuclei.* It is easily seen (see below) that the positive rotational levels are symmetric and the negative are antisymmetric for *even* electronic states (for example, Σ_g), and the negative are symmetric and the positive are antisymmetric for *odd* states (for example, Σ_u). These properties are represented in Fig. 114(b) for $^1\Sigma_g{}^+$, $^1\Sigma_u{}^+$, $^1\Sigma_g{}^-$, and $^1\Sigma_u{}^-$. Multiplet Σ states show a corresponding behavior.

An exchange of the nuclei can be brought about by first reflecting all particles at the origin and then reflecting only the electrons at the origin. In the first reflection the eigenfunction remains unchanged for positive rotational levels and changes sign for negative levels. In the second reflection the eigenfunction remains unchanged for even electronic states and changes sign for odd electronic states (see p. 218). From this the above rule follows immediately.

The symmetry properties of 0^+ and 0^- states in case (c) are the same as those of Σ^+ and Σ^- states.

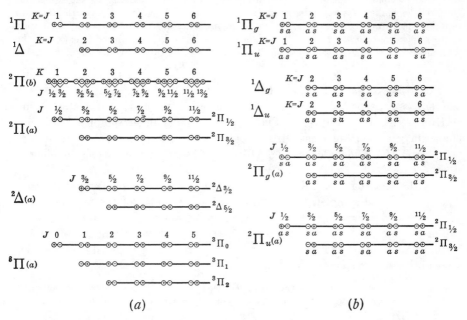

(a)   (b)

Fig. 115. Symmetry Properties of the Rotational Levels of Π and Δ States.
(a) For unlike nuclei.   (b) For identical nuclei.

$\Pi, \Delta, \cdots$ states. For $\Pi, \Delta, \cdots$ states there is a two-fold degeneracy without rotation, which, as we have seen, is removed with increasing rotation (Λ-type doubling). It can be shown (see below) that *for each J value one component level is positive and the other is negative*, and that it is alternately the upper and the lower that is positive, as is illustrated in Fig. 115(a) for $^1\Pi$ and $^1\Delta$ [see also Fig. 61(b), p. 129].

In the discussion of Λ-type doubling it was found that the upper components for each J value form one series, the lower components another with slightly different effective B values. It is seen from Fig. 115(a) that in each of these series, designated c and d, positive and negative levels alternate, as for Σ states. The one behaves like a Σ^+ state and therefore may be designated Π^+ (or Δ^+); the other behaves like Σ^- and may therefore be designated Π^- (or Δ^-).

Because of the possibility of confusion of the superscripts $+$ and $-$ with the symmetry property $+$ or $-$ of the individual rotational levels we shall use the designation c-levels for a series of rotational levels in which the even levels are $+$ and the odd $-$, as in Σ^+, Π^+, Δ^+, $\cdots$, and d-levels for a series in which the even levels are $-$ and the odd $+$ as in Σ^-, Π^-, Δ^-, $\cdots$.

Thus, as previously, the two almost coinciding rotational term series of a Π state may be distinguished as $F_c(J)$ and $F_d(J)$; but we have now specified which is c and which is d. It should be noted that our choice does not in all cases coincide with that of Mulliken (513) who calls the series c or d depending on whether the lowest level is $+$ or $-$. For Π states, for example, this is the converse of the notation used here.

The electronic eigenfunctions of the two components, $Π^+$ and $Π^-$ (or $Π_c$ and $Π_d$) of a Π state and similarly for $Δ$, $Φ$, $\cdots$ states, are given to a first approximation by the expressions (V, 10), even if the $Λ$-type splitting is taken into account. Whether the $Π^+$ or the $Π^-$ term series lies higher depends on the position of neighboring $Σ$ states. If there is only a $Σ^+$ state in the neighborhood of the Π state the $Π^+$ term series $[F_c(J)]$ is the upper (lower) one if the $Σ^+$ state is below (above) the Π state. The converse is the case if the neighboring state is $Σ^-$. For more details of these relations see Van Vleck (678) and Mulliken and Christy (523).

Multiplet states with $Λ > 0$ behave in a similar way. As examples, the symmetry properties of the rotational levels of $^2Π$, $^2Δ$, and $^3Π$ states belonging to Hund's case (a) are given in Fig. 115(a). The symmetry properties do not change in going to Hund's case (b). However, for convenience a $^2Π$ state belonging to case (b) is included in Fig. 115(a). Except for the $Λ$-type doubling it is similar to $^2Σ$ [Fig. 114(a)].

In the case of *identical nuclei* there are even and odd electronic states. The rules for the symmetry with respect to an exchange of the nuclei are the same as for $Σ$ states, that is, in even electronic states the positive levels are symmetric and the negative are antisymmetric, and conversely in odd electronic states. Fig. 115(b) shows this for ^1Π_g, ^1Π_u, ^1Δ_g, ^1Δ_u, ^2Π_g, and ^2Π_u. Other states behave correspondingly.

The symmetry properties of the rotational levels in $\frac{1}{2}$, 1, $\frac{3}{2}$, $\cdots$ states of Hund's case (c) are the same as those of the multiplet components in Hund's case (a) with the same $Ω$ value.

3. Types of Electronic Transitions

(a) Selection Rules

The quantum numbers and symmetry properties introduced in the two foregoing sections affect the spectra of diatomic molecules through the selection rules that hold for them. Conversely if these selection rules are known the quantum numbers and symmetry properties of the various energy levels of a molecule may be determined from the observed band spectra.

The selection rules are obtained by evaluating the matrix elements R^{nm} of the electric dipole moment given by (I, 44). We must distinguish between selection rules that hold quite generally, independent of the coupling case to which the electronic state under consideration belongs, and those that hold only for a definite coupling case.

General selection rules. For any atomic system the selection rule for the quantum number J of the total angular momentum is

$$\Delta J = 0, \pm 1, \textit{ with the restriction } J = 0 \nleftrightarrow J = 0, \tag{V, 34}$$

This rule holds rigorously for electric dipole radiation. Furthermore, the symmetry selection rules already mentioned in Chapter III hold quite generally: *Positive terms combine only with negative, and vice versa,* or symbolically

$$+ \leftrightarrow -, \quad + \nleftrightarrow +, \quad - \nleftrightarrow -, \tag{V, 35}$$

and, for identical nuclei, *symmetric terms combine only with symmetric and antisymmetric only with antisymmetric,* or

$$s \leftrightarrow s, \quad a \leftrightarrow a, \quad s \nleftrightarrow a. \tag{V, 36}$$

Finally, we have, in the case of a molecule with nuclei of equal charge the selection rule that *even electronic states combine only with odd;* that is,

$$g \leftrightarrow u, \quad g \nleftrightarrow g, \quad u \nleftrightarrow u. \tag{V, 37}$$

Thus, for example, a Σ_g state can combine with a Σ_u state but not with another Σ_g state.

The proof of the selection rule (V, 34) for J is somewhat involved and will not be given here. It may be found for example in Condon and Shortley (10). The proofs of the selection rules (V, 35) and (V, 36) have been given previously (p. 130 and 131). The proof for (V, 37) is similar to that for (V, 35) except that we have to use the electronic eigenfunctions instead of the total eigenfunctions. In the case that not only the charge but also all other properties of the nuclei are the same (homonuclear molecules). (V, 37) follows immediately from (V, 35 and 36) if we take account of the fact (see Figs. 114 and 115) that for identical nuclei in an even electronic state all the positive levels are symmetric and the negative antisymmetric while the opposite is the case in an odd electronic state. But the rule (V, 37) is more general, holding also for such heteronuclear molecules as $O^{16}O^{18}$, even though the property symmetric-antisymmetric is not defined for them and therefore (V, 36) does not apply. However, in such a case on account of the influence of rotation the rule (V, 37) is not perfectly rigorous, as it is for homonuclear molecules. To a certain approximation the rule (V, 37) holds even for molecules in which the nuclei have slightly different charges, such as CN (see p. 218).

Selection rules holding for case (a) as well as case (b). Apart from the preceding perfectly general selection rules, there are some that hold in Hund's case (a) as well as in case (b) but not in other coupling cases. Since these two coupling cases are by far the most frequent, these selection rules still have a fairly general significance.

In cases (a) and (b), the quantum number Λ is defined and for it there is the selection rule

$$\Lambda = 0, \pm 1. \tag{V, 38}$$

This means that Σ—Σ, Σ—Π, Π—Π, Π—Δ, $\cdots$ transitions but not Σ—Δ, Σ—Φ, Π—Φ, $\cdots$ transitions can occur. The selection rule for Λ corresponds exactly to the selection rule for M_L for atoms in an electric or magnetic field (see p. 2?).

Furthermore, Σ^+ *states cannot combine with* Σ^- *states;* that is,

$$\Sigma^+ \leftrightarrow \Sigma^+, \quad \Sigma^- \leftrightarrow \Sigma^-, \quad \Sigma^+ \nleftrightarrow \Sigma^-. \tag{V, 39}$$

However, Σ^+ as well as Σ^- states combine with Π states.

The proof of this selection rule is similar to that for (V, 35) (p. 130). Let ψ_n and ψ_m be the electronic eigenfunctions of a Σ^+ and a Σ^- state, respectively. If M_z is the component of the dipole moment in the direction of the internuclear axis it is clear that the matrix element $\int \psi_n^* M_z \psi_m d\tau_e$ vanishes, since the integrand, for reflection at any plane through the internuclear axis, changes sign. The integrands of the matrix elements $\int \psi_n^* M_x \psi_m d\tau_e$ and $\int \psi_n^* M_y \psi_m d\tau_e$ change sign for reflection at the xz and yz planes, respectively, and therefore these matrix elements are also zero, that is, a Σ^+—Σ^- transition is forbidden. If the same matrix elements are considered for a Σ^+—Σ^+ or a Σ^-—Σ^- transition it is readily seen that only the matrix element $\int \psi_n^* M_z \psi_m d\tau_e$ is different from zero; that is for Σ—Σ transitions the transition moment lies in the direction of the internuclear axis. It can be shown that this holds for all transitions with $\Delta\Lambda = 0$, whereas for transitions with $\Delta\Lambda = \pm1$ the transition moment is perpendicular to the internuclear axis.

It should be emphasized that the selection rules (V, 38) and (V, 39) have been derived by using the electronic eigenfunctions only. If the interaction of rotation and electronic motion is strong, violations of these rules may be expected for high J values since then $(1/r)\psi_e\psi_v\psi_r$ is no longer a good approximation to the total eigenfunction.

In cases (a) and (b) the resultant spin S is defined, and for the corresponding quantum number there is the selection rule (as for atoms)

$$\Delta S = 0 \qquad\qquad (V, 40)$$

This means that *only states of the same multiplicity combine with one another*. To be sure, this prohibition of intercombinations holds less and less rigorously with increasing interaction of S and Λ (increasing multiplet splitting), that is, with increasing nuclear charge.

The possible electronic transitions on the basis of the selection rules so far mentioned, are summarized in Table 24.

We now come to the special selection rules that hold only if *both* the participating electronic states belong to the same coupling case. The relative simplicity of a large number of multiplet bands (compare, for example, the spectrogram Fig. 22) results from the fact that usually some of these special selection rules are fulfilled.

Selection rules holding only in case (a). In Hund's case (a) the quantum number Σ of the component of the spin in the internuclear axis is defined. If both states belong to case (a), the following rule holds for this quantum number:

$$\Delta\Sigma = 0 \qquad\qquad (V, 41)$$

This means that in an electronic transition *the component of the spin along the internuclear axis does not alter*. Therefore, transitions such as $^2\Pi_{1/2}$—$^2\Pi_{1/2}$, $^2\Pi_{3/2}$—$^2\Pi_{3/2}$, $^2\Pi_{1/2}$—$^2\Delta_{3/2}$, $^2\Pi_{3/2}$—$^2\Delta_{5/2}$, $^3\Pi_0$—$^3\Pi_0$, $\cdots$ but not $^2\Pi_{1/2}$—$^2\Pi_{3/2}$, $^2\Pi_{1/2}$—$^2\Delta_{5/2}$, $^3\Pi_0$—$^3\Pi_1$, $\cdots$ take place. This selection rule corresponds exactly to the selection rule $\Delta M_S = 0$ for atoms in a strong electric or magnetic field (see p. 29).

Like (V, 40) it holds to a good approximation as long as the interaction between spin and orbital angular momentum is not too great.

The quantum number Ω of the total electronic angular momentum about the internuclear axis is equivalent to $|M_J|$ for atoms (see p. 28). For M_J the selection rule $\Delta M_J = 0, \pm 1$ holds (p. 29) and therefore we have for Ω

$$\Delta\Omega = 0, \pm 1 \qquad\qquad (V, 42)$$

This selection rule adds nothing to the rules (V, 38) and (V, 41) as long as these latter rules hold. However, (V, 42) holds even for strong interaction of spin and

TABLE 24. ALLOWED ELECTRONIC TRANSITIONS

Unequal Nuclear Charge	Equal Nuclear Charge
$\Sigma^+ \longleftrightarrow \Sigma^+$	$\Sigma_g^+ \longleftrightarrow \Sigma_u^+$
$\Sigma^- \longleftrightarrow \Sigma^-$	$\Sigma_g^- \longleftrightarrow \Sigma_u^-$
$\Pi \longleftrightarrow \Sigma^+$	$\Pi_g \longleftrightarrow \Sigma_u^+, \Pi_u \leftarrow\rightarrow \Sigma_g^+$
$\Pi \longleftrightarrow \Sigma^-$	$\Pi_g \longleftrightarrow \Sigma_u^-, \Pi_u \longleftrightarrow \Sigma_g^-$
$\Pi \longleftrightarrow \Pi$	$\Pi_g \longleftrightarrow \Pi_u$
$\Pi \longleftrightarrow \Delta$	$\Pi_g \longleftrightarrow \Delta_u, \Pi_u \longleftrightarrow \Delta_g$
$\Delta \longleftrightarrow \Delta$	$\Delta_g \longleftrightarrow \Delta_u$
$\ldots$	$\ldots$

All the above transitions are possible as singlets, doublets, triplets, and so on, but intercombinations—singlet-triplet and similar transitions—are forbidden.

orbital angular momentum when (V, 38) and (V, 41) lose their validity [Hund's case (c), see below].

If $\Omega = 0$ for both electronic states the following additional restriction holds

$$\Delta J = 0 \text{ is forbidden for } \Omega = 0 \rightarrow \Omega = 0; \qquad (V, 43)$$

that is, only transitions with $\Delta J = \pm 1$ occur. This restriction is entirely similar to the rule for atoms that $M_J = 0$ does not combine with $M_J = 0$ for $\Delta J = 0$ [see the rule (I, 70)]. As a consequence of (V, 43) for example, no Q branch appears for the $^3\Pi_0$—$^3\Pi_0$ component of a $^3\Pi$—$^3\Pi$ transition.

Apart from these selection rules, which state whether or not a given transition can take place at all, the *intensities of the allowed lines* can be obtained theoretically by an evaluation of the matrix elements of the transition moment. References where exact intensity formulae, obtained in this way may be found will be given later in the discussion of individual band types. However, a few general qualitative results must be mentioned here. As we have seen a case (a) multiplet band consists of several sub-bands corresponding to the various values of Ω' and Ω'' for a given Λ' and Λ''. The intensity distribution in these sub-bands is very similar to that for singlet bands discussed previously (p. 204 f.):

for $\Delta\Omega \neq 0$ (for example $^2\Pi_{3/2}$—$^2\Delta_{5/2}$) the intensity distribution is given qualitatively by Fig. 93(a), that is, an intense Q branch is present; for $\Delta\Omega = 0$ (for example, $^2\Pi_{3/2}$—$^2\Pi_{3/2}$) the intensity distribution is given by Fig. 93(b)— that is, a weak Q branch is present whose intensity decreases rapidly with increasing J. If, in addition, Ω itself is zero, no Q branch is present, as in Σ—Σ bands. The appearance or non-appearance of an intense Q branch can accordingly be used to establish whether or not $\Delta\Omega$ (and therefore $\Delta\Lambda$) differs from zero.

Selection rules holding only in case (b). In Hund's case (b) the quantum number K of the total angular momentum apart from spin is defined. If both states belong to case (b), the following selection rule holds for this quantum number

$$\Delta K = 0, \pm 1 \qquad (V, 44)$$

with the added restriction

$$\Delta K = 0 \text{ is forbidden for } \Sigma\text{—}\Sigma \text{ transitions.} \qquad (V, 45)$$

These rules are similar to the selection rules (V, 34) and (V, 43) for J and hold to the extent that the interaction of the electron spin S with K can be neglected. The restriction (V, 45) may also be considered as a consequence of (V, 35) and (V, 39) as is immediately seen by considering Fig. 114.

Furthermore some qualitative statements can be made about the dependence of the *intensity* on J or K: for $\Delta\Lambda = 0$—that is, for Π—Π, Δ—Δ, $\cdots$ transitions [for Σ—Σ transitions see (V, 45)]—the branches with $\Delta K = 0$ decrease very rapidly in intensity with increasing K [approximately proportional to $(1/K)e^{-E_r/kT}$; see Fig. 93(b)] and therefore are often not observed. Similarly, in branches for which $\Delta J \neq \Delta K$, the intensity falls off very rapidly with increasing K. These branches are often called *satellite branches*, since their lines always lie very near to those of the corresponding main branches with the same ΔK but having $\Delta J = \Delta K$. The intensity of satellite branches is always small compared to that of the main branches as long as case (b) applies to both states.

Selection rules holding only in case (c). In Hund's case (c) apart from J only the quantum number Ω of the total electronic angular momentum about the internuclear axis is well defined. For this quantum number we have the selection rule

$$\Delta\Omega = 0, \pm 1 \qquad (V, 46)$$

In addition, the analogue of (V, 39) holds, that is

$$0^+ \longleftrightarrow 0^+, \quad 0^- \longleftrightarrow 0^-, \quad 0^+ \longleftrightarrow\!\!\!| \; 0^- \qquad (V, 47)$$

Therefore we have transitions of the same kind as those given in Table 24 if we replace Λ by the quantum number Ω. For the same reasons as in case (a) the restriction (V, 43) holds, that is, if both states have $\Omega = 0$ no Q branch occurs.

Selection rules holding only in case (d). In Hund's case (d) the quantum numbers K, R, and L are defined. For K the rule (V, 44) holds, for L the same selection rule holds as for atoms, that is $\Delta L = 0, \pm 1$. Since the coupling of electronic motion and rotation of the molecule is assumed to be slight, no change of R can occur for an electronic transition, that is $\Delta R = 0$.

More general cases. The special selection rules just discussed hold only if *both* participating states belong to one and the same coupling case. If, for example, the one state belongs to case (a) and the other to case (b), these special selection rules lose their validity and there remain only the selection rules holding for both coupling cases. Therefore, the number of allowed branches in such bands is much greater.

As a result of the spin uncoupling usually present, it often happens that both states approach case (b), at least for large rotations. Then the special selection rules for case (b) hold at least for large rotation. Also it not infrequently happens that both electronic states belong to case (a) for small rotation and to case (b) for large rotation (see below).

(b) Allowed Electronic Transitions

Notation. In designating a given electronic transition, the upper state is always written first and then the lower. $^1\Pi$—$^1\Sigma$ thus means a transition for which the upper state is $^1\Pi$ and the lower $^1\Sigma$; $^1\Sigma$—$^1\Pi$ means a transition for which $^1\Sigma$ is the upper state and $^1\Pi$ is the lower state.[8] Sometimes an arrow is used to indicate whether the transition under consideration is observed in emission or absorption: $^1\Pi \rightarrow {}^1\Sigma$ means a transition from $^1\Pi$ to $^1\Sigma$ in emission, and $^1\Pi \leftarrow {}^1\Sigma$ indicates a transition in absorption from the lower $^1\Sigma$ state to the upper $^1\Pi$ state. Naturally, the validity of a selection rule is independent of which state is the upper and which is the lower. This independence is indicated in the foregoing by a double arrow $\leftrightarrow$; for example, $\Pi \leftrightarrow \Sigma^+$ in Table 24 means that Π—Σ^+, as well as Σ^+—Π, is possible.

If several electronic states of a molecule are known, some of which may be of the same type, they are distinguished by a letter $X, A, B, \cdots, a, b, \cdots$, in front of the term symbol or sometimes by one or more asterisks added to it. Thus we describe transitions by symbols such as $A\,^1\Pi$—$X\,^1\Sigma$ or $B\,^1\Sigma$—$X\,^1\Sigma$ (or Σ^{**}—Σ^*) and so on. X is frequently used for the ground state of the molecule in question. Instead of this brief characterization of the various electronic states of the molecule the electronic configurations are also used (see Chapter VI). The following discussion of the various types of electronic transitions is, of course, independent of these additional symbols.

$^1\Sigma$—$^1\Sigma$ **transitions.** For singlet transitions, there is no object in distinguishing between Hund's cases (a) and (b). They can be treated equally

---

[8] This convention is the converse of that for atomic spectra. The difference is connected with the fact that for atoms the energies of the quantum states are usually negative, since they are referred to the ionized state, whereas for molecules they are positive, since they are referred to the lowest state.

well according to the one or the other case. If we adopt case (b), in addition to the general rules (V, 34–37 and 39), we have, for $^1\Sigma$—$^1\Sigma$, the special rule (V, 45): $\Delta K = 0$ is forbidden, or, since here $J = K$, $\Delta J = 0$ is forbidden, which is the rule (V, 43) in case (a).

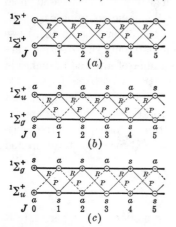

FIG. 116. $^1\Sigma$—$^1\Sigma$ **Transitions.**
(a) $^1\Sigma^+$—$^1\Sigma^+$. (b) $^1\Sigma_u{}^+$—$^1\Sigma_g{}^+$.
(c) $^1\Sigma_g{}^+$—$^1\Sigma_u{}^+$. The values of J given refer to the upper as well as the lower state. In diagrams (b) and (c), the transitions corresponding to the antisymmetrical levels are indicated by broken lines.

Thus only the transitions with $\Delta J = \Delta K = \pm 1$ appear, and we obtain the band structure already treated in detail in the foregoing chapter —namely, *a single P and a single R branch* (see Figs. 19, 24 and 54)—which is fairly often observed experimentally.

In Fig. 116(a) the branches of $^1\Sigma$—$^1\Sigma$ bands are represented in a manner somewhat different from that used previously. It can be seen that the selection rule $+ \leftrightarrow -$ is fulfilled for both branches. Fig. 116(a) holds for a $^1\Sigma^+$—$^1\Sigma^+$ transition. For a $^1\Sigma^-$—$^1\Sigma^-$ transition, $+$ and $-$ must be everywhere interchanged; this interchange, however, does not alter the transitions that occur. It is therefore not possible to determine from the fine structure of an observed $^1\Sigma$—$^1\Sigma$ band system alone which of the two cases applies.

Examples of $^1\Sigma$—$^1\Sigma$ transitions are the previously discussed BeO bands [see Rosenthal and Jenkins (602) and Lagerqvist and Westöö (1160)] and the CuH bands [see Heimer and Heimer (290)], of which Fig. 19 gives a spectrogram.

For *like nuclei*, the even-odd symmetry has to be considered. Therefore only the transitions $^1\Sigma_u{}^+$—$^1\Sigma_g{}^+$, $^1\Sigma_g{}^+$—$^1\Sigma_u{}^+$, $^1\Sigma_u{}^-$—$^1\Sigma_g{}^-$, and $^1\Sigma_g{}^-$—$^1\Sigma_u{}^-$ are possible. The first two of these are represented in Fig. 116(b) and (c). In all four cases the transitions with $\Delta K = \Delta J = \pm 1$ are in agreement with the rule sym. $\leftrightarrow$ antisym. as well as with the rule $+ \leftrightarrow -$. The origin of the intensity alternation has already been explained by Fig. 94. The alternation is indicated in Fig. 116(b) and (c) by full and broken lines. If $+$ and $-$ are everywhere interchanged, Fig. 116(b) would represent a $^1\Sigma_g{}^-$—$^1\Sigma_u{}^-$ transition and Fig. 116(c) a $^1\Sigma_u{}^-$—$^1\Sigma_g{}^-$ transition.

It can be seen from Fig. 94 that the strong lines of the P branch form the continuation of the series of the weak lines rather than of the strong lines of the R branch. Thus, if alternate lines are entirely missing (nuclear spin $I = 0$), the P and R branches (unlike the behavior for molecules with unequal nuclei) do not appear to form a single series—that is, if we take no account of the missing lines. This is an unambiguous criterion of whether or not alternate lines are missing in a band.

Examples of $^1\Sigma_u{}^+$—$^1\Sigma_g{}^+$ transitions are the longest-wave-length absorption band systems of the alkali vapors [for Li_2 see Almy and Irwin (64) and McKellar and Jenkins (1215); for Na_2 see Frederickson (233)], the ultraviolet P_2 bands [Herzberg (311), Ashley (69), Marais (1195)], and a number of band systems of H_2 [see Richardson (32)] and He_2 [see Weizel (22)]. The latter two molecules also exhibit examples of $^1\Sigma_g^+ - ^1\Sigma_u^+$ transitions. $^1\Sigma^- - ^1\Sigma^-$ transitions have not yet been observed.

In Chapter III it was shown that in all electronic states of a given homonuclear molecule (including ionized states) either the symmetric rotational levels have the greater statistical weight throughout or the antisymmetric levels have the greater weight throughout, depending on whether the nuclei follow Bose statistics or Fermi statistics. Thus, according to Fig. 116(b) and (c) for one and the same molecule the transition $^1\Sigma_u^+ - ^1\Sigma_g^+$ is distinguished from $^1\Sigma_g^+ - ^1\Sigma_u^+$ by the fact that for the one the even-numbered (or odd-numbered) rotational lines are more intense and for the other the odd (or even) are more intense. Therefore, if the statistics of the nuclei is known, we can decide between the two transitions mentioned on the basis of the observed alternation of intensities, or conversely, can draw conclusions concerning the *nuclear statistics* if the type of the electronic states is known. However, we cannot decide in this way between $^1\Sigma_u^+ - ^1\Sigma_g^+$ and $^1\Sigma_g^- - ^1\Sigma_u^-$ since in these two transitions the same lines are strong, and similarly we cannot decide between $^1\Sigma_g^+ - ^1\Sigma_u^+$ and $^1\Sigma_u^- - ^1\Sigma_g^-$.

For H_2, for example, theory shows (see Chapter VI) that the ground state is $^1\Sigma_g^+$. The Lyman bands of H_2 which occur in absorption and have P and R branches only, are therefore a $^1\Sigma_u^+ - ^1\Sigma_g^+$ transition. From the observation that in these bands the odd-numbered rotational lines—that is, those corresponding to the antisymmetric levels—are the stronger, we can conclude immediately that the H *nuclei (protons) follow Fermi statistics*. For D_2 the intensity alternation is just the opposite; therefore the D *nuclei (deuterons) follow Bose statistics* [compare also the discussion in Chapter III, section 2(f)].

$^2\Sigma - ^2\Sigma$ **transitions.** $^2\Sigma$ states always belong strictly to Hund's case (b) and therefore for $^2\Sigma - ^2\Sigma$ transitions the selection rule $\Delta K = \pm 1$ holds, $\Delta K = 0$ being forbidden [see (V, 45) above]. The separation of the two sublevels with $J = K + \frac{1}{2}$ and $J = K - \frac{1}{2}$ for a given K is, in general, very small compared to the separation of successive rotational levels. Therefore, with not too great resolution, we have exactly the same band structure as for $^1\Sigma - ^1\Sigma$ bands except that the lines are now to be numbered by K instead of J. There is a P and an R branch, for which exactly the same formulae hold as previously (see Chapter IV, section 3). Because of this fact we were able to use the violet CN bands, which represent a $^2\Sigma - ^2\Sigma$ transition, as an example of the simplest type of band in Chapter II.

However, with larger resolution, as is shown by Fig. 117, each "line" of the P and R branches, according to the rule $\Delta J = 0, \pm 1$, is split into three components. For one of these—namely, that with $\Delta J = 0$—ΔJ is unequal to ΔK. According to the previous discussion (p. 244), the intensity of these components falls off very rapidly with increasing K. (They are indicated by broken lines in Fig. 117.) Therefore, in practice, except for very small K, a splitting of each of the lines (single for small dispersion) into two components of about equal intensity and of a separation increasing with increasing K is to be expected (full lines in Fig. 117). In other words there is a *doublet* P and a *doublet* R branch.

Such bands occur for many molecules. The best known *example* is provided by the violet CN bands, one of which is shown in the spectrograms reproduced in Fig. 18. In one of these spectrograms the splitting into doublets is quite noticeable for the largest K values. Also the violet SiN and CP bands [Jenkins and Laszlo (372) and Baerwald, G. and L. Herzberg (77)] the ultraviolet bands

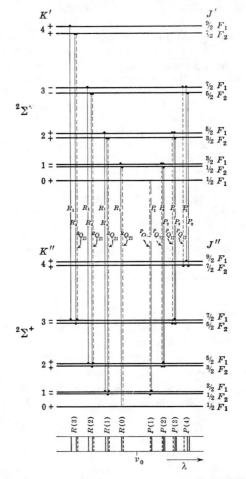

Fɪɢ. 117. **Energy Level Diagram for the First Lines of a $^2\Sigma^+$—$^2\Sigma^+$ Band, with Schematic Spectrum.** The numbers in brackets behind P and R in the schematic spectrum at the bottom give the K'' values. The full designation of the branches (see below) is written on the vertical lines representing the transitions. The doublet splitting has been much exaggerated.

of CO^+ [Coster, Brons and Bulthuis (865), Woods (1555)] and SiO^+ [Woods (1554)] and a large number of hydride bands are of this type. They are all $^2\Sigma^+$—$^2\Sigma^+$ transitions. $^2\Sigma^-$—$^2\Sigma^-$ transitions would have the same structure, but no example is known.

The two branches indicated by broken lines in Fig. 117 have $\Delta J = 0$ and are thus to be called Q branches. However, as may be seen from the figure, the lines of these Q branches lie very close to the corresponding lines (with equal K and ΔK) of the P and R branches (they form satellite branches). These Q branches therefore have the form of an R and a P branch, respectively, and not that of a Q branch whose Fortrat parabola would have its vertex on the ν axis. Such branches are consequently called *R-form Q branches* or *P-form Q branches* [abbreviated ${}^R Q$, ${}^P Q$ branches; see Mulliken (515)].

If we distinguish the term components having $J = K + \frac{1}{2}$ by the subscript 1 and those having $J = K - \frac{1}{2}$ by the subscript 2 (see p. 222), we obtain for the four *main branches*

$$R_1(K) = \nu_0 + F_1'(K+1) - F_1''(K), \tag{V, 48}$$

$$R_2(K) = \nu_0 + F_2'(K+1) - F_2''(K), \tag{V, 49}$$

$$P_1(K) = \nu_0 + F_1'(K-1) - F_1''(K), \tag{V, 50}$$

$$P_2(K) = \nu_0 + F_2'(K-1) - F_2''(K), \tag{V, 51}$$

and correspondingly for the two *satellite branches*

$$^R Q_{21}(K) = \nu_0 + F_2'(K+1) - F_1''(K), \tag{V, 52}$$

$$^P Q_{12}(K) = \nu_0 + F_1'(K-1) - F_2''(K). \tag{V, 53}$$

The subscripts 21 or 12 of ${}^R Q$ or ${}^P Q$ are to indicate that the transition takes place from a term of the series F_2 to one of the series F_1, or vice versa.

According to (V, 16) and (V, 48–51), the *line splitting* in the P and R branches is

$$\Delta \nu_{12}(P) = P_1 - P_2 = (\gamma' - \gamma'')K - \tfrac{1}{2}(\gamma' + \gamma''),$$
$$\Delta \nu_{12}(R) = R_1 - R_2 = (\gamma' - \gamma'')K + \tfrac{1}{2}(3\gamma' - \gamma''). \tag{V, 54}$$

Thus the splitting of the lines in the branches increases linearly with K (as does the splitting of the terms), the magnitude of the splitting depending essentially on the difference of the splitting factors in the upper and lower states. This difference can be derived very accurately

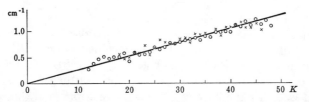

Fig. 118. **Doublet Splitting in the P and R Branches of the 3–0 Band of the Violet CP System** ($^2\Sigma$—$^2\Sigma$). The small circles refer to the P branch, and the crosses refer to the R branch.

from the observed splitting of the branches even if, as is usually the case, these values are very small and, as a result, the doublets are not resolved for very small K values. In this case, the splitting in the P and in the R branches for equal K is practically the same. Fig. 118 gives as an example the splitting in the 3–0 band of the $^2\Sigma$—$^2\Sigma$ system of CP in its dependence on K. It can be seen that the linear relations (V, 54) are well fulfilled. From the slope of the straight line we obtain $\gamma_3' - \gamma_0'' = 0.0263$ cm^{-1}.

It is seen from (V, 48–51) and also from Fig. 117 that the *combination differences* (IV, 40 and 41) must be formed either between R_1 and P_1 or between R_2 and P_2. From (V, 16) follows for the $\Delta_2 F$ values (disregarding terms with D_v)

$$\Delta_2 F_1(K) = 4B_v(K + \tfrac{1}{2}) + \gamma, \qquad \Delta_2 F_2(K) = 4B_v(K + \tfrac{1}{2}) - \gamma, \tag{V, 55}$$

which means that the $\Delta_2 F$ curves for the two components of a $^2\Sigma$ state form parallel lines whose vertical separation is 2γ. Thus, while for different bands with the same lower vibrational state the values of $R_1(K-1) - P_1(K+1)$ [and correspondingly $R_2(K-1) - P_2(K+1)$] must agree exactly even if perturbations are present, $R_1(K-1) - P_1(K+1)$ is slightly different from $R_2(K-1) - P_2(K+1)$ and may be widely different in case of perturbations. Similar considerations apply to $R_1(K) - P_1(K)$ and $R_2(K) - P_2(K)$. In principle, the values of γ for the upper and lower states can be obtained from the differences of the observed $\Delta_2 F_1$ and $\Delta_2 F_2$ values according to (V, 55). In practice, however, the magnitude of γ is often smaller than the accuracy of the $\Delta_2 F$ values. In the above example of CP (Fig. 118), from the mean of the differences of corresponding $\Delta_2 F''$ values, $2\gamma_0'' = -0.034$ is obtained for the lower state. This value is much less accurate than the difference of the γ values in the upper and lower states given above (the accuracy of the individual $\Delta_2 F$ values is only ± 0.05 cm^{-1}). It should be mentioned that the rotational constant B_0'' in this case is 0.796 cm^{-1} which, when compared to the value of γ, shows that γ represents only a small correction term.

At any rate, it can be seen from (V, 55) that, if there is no perturbation, the value of B_v for a $^2\Sigma$ state, and consequently the moment of inertia and internuclear distance, can be derived without a knowledge of γ by using the mean value of $\Delta_2 F_1$ and $\Delta_2 F_2$. We have

$$\tfrac{1}{2}[\Delta_2 F_1(K) + \Delta_2 F_2(K)] = 4B_v(K + \tfrac{1}{2}) + \cdots, \tag{V, 56}$$

from which the constant B_v can be obtained as described on p. 177 f.

The *line strengths* of the individual lines of a $^2\Sigma - ^2\Sigma$ transition according to Mulliken (513) are given by

$$S_J^R = \frac{(J''+1)^2 - \tfrac{1}{4}}{J''+1}, \quad S_J^Q = \frac{(2J''+1)}{4J''(J''+1)}, \quad S_J^P = \frac{J''^2 - \tfrac{1}{4}}{J''} \tag{V, 57}$$

These formulae show that in both P and R branches the components with $J = K + \tfrac{1}{2}$, that is P_1 and R_1, have a larger intensity than the components with $J = K - \tfrac{1}{2}$, that is, P_2 and R_2. The difference is slight for high K values but quite appreciable for low K values. This difference is important because it supplies the only way of ascertaining which branches are P_1 and R_1 and which are P_2 and R_2.

The effect of the symmetry properties on $^2\Sigma - ^2\Sigma$ transitions of *homonuclear molecules* is quite analogous to that for $^1\Sigma - ^1\Sigma$ transitions. We only have to replace J by K in the previous discussion, since now states with equal K have the same symmetry properties (see Fig. 114, p. 238). Therefore, the two components of the doublets are alternately both strong and both weak (or missing) that is, either the lines with even K are strong and those with odd K are weak, or vice versa, depending on the nuclear statistics and on whether the lower state is $^2\Sigma_g^+$, $^2\Sigma_u^-$, $^2\Sigma_u^+$, or $^2\Sigma_g^-$. Thus, when the doublets are not resolved the intensity alternation in a $^2\Sigma - ^2\Sigma$ band is the same as in the corresponding $^1\Sigma - ^1\Sigma$ band.

The so-called negative nitrogen bands due to N_2^+, represent the only example of a $^2\Sigma - ^2\Sigma$ transition of a homonuclear molecule observed up to now [see Coster and Brons (169)]. More specifically it is a $^2\Sigma_u^+ - ^2\Sigma_g^+$ transition. Fig. 26 shows one of these bands under medium dispersion (no resolution of the doublets). The intensity alternation is clearly to be seen.

$^3\Sigma - ^3\Sigma$ **transitions.** $^3\Sigma - ^3\Sigma$ transitions are very similar to $^2\Sigma - ^2\Sigma$ transitions. With small dispersion the bands consist of a single P and a single R branch, as do $^1\Sigma - ^1\Sigma$ bands. With larger dispersion, each "line" is resolved into three components of about the same intensity.   Thus there are in all *six main branches*.   Apart from that, six very weak satellite branches appear for which $\Delta J \neq \Delta K$.   The formulae for the branches are entirely similar to (V, 48–53) for $^2\Sigma - ^2\Sigma$ bands except that (V, 17) has to be substituted for $F_i(K)$ and that corresponding formulae for R_3, P_3, Q_{13}, Q_{31}, Q_{23}, Q_{32} have to be added.

It is clear from the energy formulae (V, 17) for $^3\Sigma$ states and Fig. 103 that only for rather large K values will the line splitting be similar to that in $^2\Sigma - ^2\Sigma$ bands, that is, will it increase linearly with increasing K.   For intermediate and small K values, depending on the difference of the splittings in the upper and lower states there are regions in which two of the three components of P or R branch are so close together that they may not be resolved.   The rotational constants, according to (V, 17), are best determined from $\Delta_2 F_2(K)$ although the average of $\Delta_2 F_1$ and $\Delta_2 F_3$ taken for the same J may also be used for this purpose.

The only $^3\Sigma - ^3\Sigma$ bands thus far known for unlike nuclei are the SO bands for which the triplet splitting has been completely resolved at large K values [Martin (482)].

For identical nuclei, the intensity alternation is again determined by K. The best-known example of a $^3\Sigma - ^3\Sigma$ transition for identical nuclei is the ultraviolet absorption band system of O_2 (Schumann-Runge bands), for which every alternate triplet in the P and R branches is missing, since the nuclear spin of the oxygen atom is zero [see Mulliken (510), and Lochte-Holtgreven and Dieke (449)].   For most of the branches in these bands the triplet splitting has only been resolved incompletely so far.   A more complete resolution has been obtained for the corresponding bands of S_2 [see Olsson (548), Naudé (1272)].

Since in both the O_2 and S_2 bands the even-numbered rotational lines are missing and since the O and S nuclei follow Bose statistics, the lower states, that is, the ground states of the O_2 and S_2 molecules, must be $^3\Sigma_g^-$ or $^3\Sigma_u^+$ state.   The electron configuration (see Chapter VI) shows that they can only be $^3\Sigma_g^-$ states.   In a similar way the ultraviolet bands of the B_2 molecules found by Douglas and Herzberg (900) were shown to be due to a $^3\Sigma_u^- - ^3\Sigma_g^-$ transition.

Thus far, $\Sigma - \Sigma$ transitions of still higher multiplicities have not been observed.   According to what has been said, it would be easy to predict their structure.

$^1\Pi - ^1\Sigma$ **transitions.**   For a $^1\Pi - ^1\Sigma$ transition the distinction between case (a) and case (b) is again pointless.   According to the selection rules previously discussed transitions with $\Delta J = 0$ as well as $\Delta J = \pm 1$ are possible and we obtain (according to the discussion in Chapter IV) *one P, one Q, and one R branch*.   As long as the Λ-type doubling of the $^1\Pi$ state can be neglected, the band structure is given exactly by the formulae (IV, 29–34).   Here it must be remembered that, in a $^1\Pi$ state, according to (V, 13) the smallest J value is $J = 1$.   As a result, the lines $P(1)$ and $Q(0)$ do not occur (see Fig. 77).

If we take account of the Λ-*type doubling*, we might at first think that each line of the three branches P, Q, and R would split into two components.   However, this is not observed.   The reason for this becomes clear when we remember the selection rule (V, 35): $+ \leftrightarrow -$.   On its basis it follows immediately from

Fig. 119(a) that only simple branches are to be expected: Since in the upper state one of the two components of a Λ-type doublet is + and the other is −, a given lower state can combine only with the one *or* the other, according as the lower state is − or +. Thus line doublets do not appear. In the kind of Λ-type doubling assumed in Fig. 119(a), the lines of the Q branch always have the *lower* Λ component as upper state, whereas the lines of the P and R branches have the *upper* Λ component as upper state. If the Λ-type doubling were

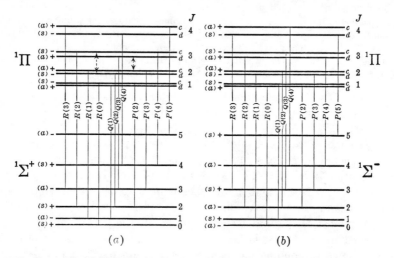

(a)   (b)

Fig. 119.   **Energy Level Diagram for the First Lines of** (a) **a** $^1\Pi$—$^1\Sigma^+$ **Transition and** (b) **a** $^1\Pi$—$^1\Sigma^-$ **Transition.** For the sake of clarity, the Λ-type doubling in the $^1\Pi$ state has been greatly exaggerated. The broken arrow to the left in (a) gives $R(2)$—$Q(2)$, and the one to the right gives $Q(3)$—$P(3)$. Their difference gives the sum of the Λ doublings for $J = 2$ and $J = 3$ in the upper state. The designations s and a, added in parentheses, hold for a $^1\Pi_u$—$^1\Sigma_g$ transition for identical nuclei. For the designations c and d see p. 239.

such that the order of + and − in the $^1\Pi$ state was reversed, the lines of the Q branch would start from the upper component and those of the P and R branches would start from the lower component.

Thus, since the lines of the P and R branches always have a somewhat different upper state from the lines of the Q branch, it follows that the combination relations (IV, 56 and 57) no longer hold exactly. A so-called *combination defect* occurs—that is to say,

$$R(J) - Q(J) = Q(J + 1) - P(J + 1) + \varepsilon \approx \Delta_1 F'(J), \qquad (V, 58)$$

and

$$R(J) - Q(J + 1) = Q(J) - P(J + 1) + \varepsilon \approx \Delta_1 F''(J). \qquad (V, 59)$$

As shown by Fig. 119(a) for $J = 2$, the combination defect ε is equal to the sum of the Λ splittings of the terms with J and $J + 1$.

The Λ-type doubling is scarcely noticeable in the $^1\Pi$—$^1\Sigma$ CO band previously given (Table 22). A better example is the AlH band 4241 Å (see Fig. 20)

TABLE 25. WAVE NUMBERS OF THE LINES AND COMBINATION DEFECTS IN THE AlH BAND 4241 Å

[0–0 band of the $^1\Pi$–$^1\Sigma$ system according to Bengtsson-Knave (84)]

J	$R(J)$	$Q(J)$	$P(J)$	$R(J)-Q(J)$	$Q(J+1)-P(J+1)$	ε	$R(J)-Q(J+1)$	$Q(J)-P(J+1)$
0	23,483.54	......	......	......	......	......	13.34	25.14
1	494.36	23,470.20	......	24.16	24.01	0.15	25.29	37.65
2	505.18	469.07	23,445.06	36.11	35.91	0.20	37.85	50.24
3	515.46	467.33	431.42	48.13	47.87	0.26	50.50	62.65
4	525.09	464.96	417.09	60.13	59.72	0.41	63.06	75.02
5	534.00	462.03	402.31	71.97	71.37	0.60	75.62	87.14
6	542.20	458.38	387.01	83.82	82.82	1.00	88.14	99.64
7	549.59	454.06	371.24	95.53	94.46	1.07	100.71	111.73
8	556.03	448.88	354.42	107.15	105.83	1.32	113.05	123.78
9	561.55	442.98	337.15	118.57	116.93	1.64	125.42	135.80
10	566.10	436.13	319.20	129.97	128.03	1.94	138.74	147.68
11	569.51	428.36	300.33	141.15	138.87	2.28	149.96	159.46
12	571.55	419.55	280.68	152.00	149.42	2.58	162.04	171.10
13	572.39	409.51	260.09	162.88	159.92	2.96	174.06	182.60
14	571.55	398.33	238.41	173.22	169.89	3.33	185.93	193.91
15	569.00	385.62	215.73	183.38	179.52	3.86	197.77	205.06
16	564.58	371.23	191.71	193.35	189.08	4.27	209.33	216.24
17	557.97	355.25	166.17	202.72	198.14	4.58	220.82	
18	548.80	537.15	139.01	211.65	......	......	232.27	
19	......	316.53						

which is the 0–0 band of a $^1\Pi-^1\Sigma$ system. The wave numbers of the observed lines of this band and the combination differences (V, 58 and 59) are given in Table 25. The Λ-type splitting is much greater here than for CO, as can be seen from the increasing difference ε between columns 5 and 6 or 8 and 9.

On account of the combination defect (ε) exact values of the rotational constants B in the upper and lower states cannot be obtained from the combination differences $\Delta_1 F(J)$ if, as for AlH, the Λ-type doubling is large. However, for most non-hydrides the defect is small. In either case, for an exact determination of the B values, the $\Delta_2 F$ values given by (IV, 40–42) must be used.

It can be seen from Fig. 119(a) that the $\Delta_1 F''$ values for different bands with the same lower state (unlike the $\Delta_1 F'$ values for different bands with the same upper state) need not agree exactly, since the Λ-type doubling in the different upper states may be of different magnitude. On the other hand the $\Delta_2 F''$ values of bands with the same lower state must always agree exactly, even if perturbations are present.

We can now also understand the fact (see p. 210 f.) that the resonance series corresponding to absorption bands with P, Q, and R branches ($^1\Pi-^1\Sigma$) do not consist of triplets but only singlets or doublets, depending on whether excitation is by a Q line or a P or R line. Since, as shown above, the upper state of a Q line is different from that of the P and R lines with the same upper J value, for excitation by a Q line only Q lines can appear in fluorescence; that is, we have a singlet resonance series. However, for excitation by a P or R line both P and R lines but no Q lines appear in fluorescence (doublet resonance series).[9]

As for $^1\Sigma-^1\Sigma$ transitions, the $\Delta_2 F''$ values of a $^1\Pi-^1\Sigma^+$ band yield directly the rotational constants B'' and D'' of the lower ($^1\Sigma^+$) state. However, only the effective B' and D' values of the one component of the upper state, designated by c (or Π^+ see p. 239), are obtained from $\Delta_2 F_c' = R(J) - P(J)$ [see Fig. 119(a)]. The effective B' and D' values of the other component of the upper state, designated by d (or Π^-) can be obtained only with the help of the Q branch—for example, by using the method described on p. 186 for obtaining $B_v' - B_v''$ from the Q branch (assuming the previously determined B_v'') or from

$$\Delta_2 F_d'(J) = Q(J+1) - Q(J-1) + R(J-1) - P(J+1), \qquad (V, 60)$$

a relation that may be easily verified by referring to Fig. 119(a).

In equations (V, 58 and 59) ε is the sum of the Λ-type splittings of two successive levels. The splitting of one level is very nearly one half of this. Therefore, according to (V, 25 and 26), we have

$$\tfrac{1}{2}\varepsilon = (B_v^c - B_v^d)J(J+1) = qJ(J+1) \qquad (V, 61)$$

The $\tfrac{1}{2}\varepsilon$ values derived from Table 25 are represented graphically in Fig. 120. The solid curve represents $qJ(J+1)$ with $q = 0.0080$ cm^{-1}. It can be seen that the observed points follow this curve closely. The value of B_v is the mean of B_v^c and B_v^d (see p. 228 f.); in the present case $B_0 = 6.020$ cm^{-1}.

---

[9] On the other hand, for a $^1\Sigma-^1\Pi$ transition (see below) triplet resonance series would indeed occur by excitation with a line of any one of the three branches; but such a case has not yet been observed.

Fig. 119(a) holds for a $^1\Pi$—$^1\Sigma^+$ transition. A $^1\Pi$—$^1\Sigma^-$ transition, represented in Fig. 119(b), is completely analogous except that the upper states of the P and the R branches on the one hand, and of the Q branch on the other, are interchanged.

If only a single $^1\Pi$—$^1\Sigma$ band system is observed for a molecule it is not possible to decide from the band structure alone (without knowledge of the electronic structure of the molecule) whether the system is a $^1\Pi$—$^1\Sigma^+$ or a $^1\Pi$—$^1\Sigma^-$ transition since the order of the c and d levels in the $^1\Pi$ state is not known a priori. However, if two different $^1\Sigma$ states combine with one and the same $^1\Pi$ state, we can decide from the band structure alone whether the two $^1\Sigma$ states have like or unlike symmetry, as may be readily seen by comparing Fig. 119(a) and (b).

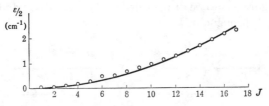

Fig. 120. **Λ-Type Doubling in the $^1\Pi$ State of AlH.** The circles represent the observed splittings (see Table 25). The curve represents $0.0080J\,(J+1)$.

The combination defect ε would be the same for like symmetry but would have opposite sign for unlike symmetry. Therefore, if the symmetry (Σ^+ or Σ^-) of the one state can be determined on the basis of electronic structure (see Chapter VI), that of the other can be determined from the band structure. At the same time one obtains the order of the $+$ and $-$ levels in the $^1\Pi$ state. Multiplet Π and Σ states can be treated in a similar manner.

The *intensity distribution* in the three branches of $^1\Pi$—$^1\Sigma$ bands has been discussed previously in detail [see p. 207 f. and Fig. 93(a)].

Numerous *examples* of $^1\Pi$—$^1\Sigma^+$ transitions have been analyzed. We mention only (in addition to the CO and AlH bands discussed above) the BH bands [see Thunberg (667) and Almy and Horsfall (63)], the red and infrared bands of BeO [L. Herzberg (320) and Lagerqvist (1156)] and the PN bands (see Chapter IV). In the papers quoted further details about the analysis and structure of $^1\Pi$—$^1\Sigma$ bands may be found. Thus far an example of a $^1\Pi$—$^1\Sigma^-$ transition is not known with certainty.

For *homonuclear molecules* every second line in the branches is weak or missing, as for $^1\Sigma$—$^1\Sigma$ transitions. This can be seen immediately from Fig. 119(a) and (b), in which the symbols s and a, which would apply to a $^1\Pi_u$—$^1\Sigma_g^+$ and a $^1\Pi_u$—$^1\Sigma_g^-$ transition, respectively, have been added in parentheses. The intensity alternation would be just the opposite for $^1\Pi_g$—$^1\Sigma_u$ transitions.

Examples of $^1\Pi_u$—$^1\Sigma_g^+$ transitions are the Werner bands of H$_2$, which appear both in emission and absorption [see Jeppesen (380)], several other band systems of H$_2$ [see Richardson (32)], the blue-green absorption systems of Li$_2$ [see Harvey and Jenkins (282)] and Na$_2$ [see Loomis and Wood (467)], and the infrared bands of C$_2$ [Phillips (1314)]. A number of $^1\Pi_g$—$^1\Sigma_u^+$ transitions have been observed for He_2 [see Weizel (22)]. Thus far $^1\Pi-^1\Sigma^-$ transitions have not been found for homonuclear molecules.

$^1\Sigma-^1\Pi$ **transitions.** The rotational structure of a $^1\Sigma-^1\Pi$ transition is very similar to that of a $^1\Pi-^1\Sigma$ transition. Three branches, P, Q, and R appear as before. The P and R branches have one component of the $^1\Pi$ state as lower state, the Q branch the other. The relations (V, 58) and (V, 59) hold as for $^1\Pi-^1\Sigma$ but the combination defect ε now gives twice the Λ-type splitting in the *lower* state. Fig. 121 gives diagrams of the transitions for $^1\Sigma_g{}^+-^1\Pi_u$ and $^1\Sigma_g{}^--^1\Pi_u$ bands. The intensity alternation is indicated by broken and full lines. For unequal nuclei the distinction between s and a and broken and full lines in the figure has to be disregarded. For $^1\Sigma_u-^1\Pi_g$ bands, s and a have to be interchanged throughout.

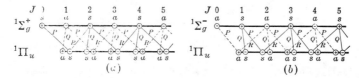

FIG. 121. $^1\Sigma-^1\Pi$ **Transitions.** (a) $^1\Sigma_g{}^+-^1\Pi_u$. (b) $^1\Sigma_g{}^--^1\Pi_u$. The broken-line transitions correspond to the antisymmetric levels.

The main *difference* between $^1\Pi-^1\Sigma$ and $^1\Sigma-^1\Pi$ bands is that for the former the lines $P(0)$, $P(1)$ and $Q(0)$ are missing, whereas, for the latter, $P(0)$, $R(0)$, and $Q(0)$ are missing (see Figs. 119 and 121). Apart from that, as we have seen previously [Chapter IV, section 4(b)] there is a slight difference in the intensity distribution: for $^1\Pi-^1\Sigma$ the R branch is more intense than the P branch, and for $^1\Sigma-^1\Pi$ the P branch is the more intense, independent of whether the transition is observed in absorption or emission. This difference is appreciable only for small J values.[10] Therefore, for an observed band with single P, Q, and R branches, we can decide whether $^1\Pi$ or $^1\Sigma$ is the lower state from the band structure alone only if the region in the neighborhood of the zero line is resolved, that is, if it can be decided which of the two lines, $R(0)$ and $P(1)$, is missing or which of the branches P and R is the more intense for low J values. Only if other band systems of the same molecule (or analogous systems of an analogous molecule) are observed is a distinction possible without resolution of the region in the neighborhood of the zero line.

The determination of the rotational constants and of the Λ-type doubling from $^1\Sigma-^1\Pi$ bands proceeds in the same way as for $^1\Pi-^1\Sigma$ bands.

The Ångstrom bands of CO in the visible region are the best known example of a $^1\Sigma^+-^1\Pi$ transition [Coster and Brons (165), Schmid and Gerö (619); see Fig. 11]. Another example is the AlH band at 4752 Å [Bengtsson and Hulthén (85)]. For equal nuclei such transitions have been observed for H_2

[10] As can be seen from the formulae (IV, 82 and 83) the difference is more pronounced in emission for $^1\Sigma-^1\Pi$ and in absorption for $^1\Pi-^1\Sigma$.

[see Richardson (32)], He$_2$ [see Weizel (22)] and N$_2$ [see Gaydon (954) and Herzberg (1041)]. A $^1\Sigma^-$—$^1\Pi$ transition has been found for AlH by Holst (326).

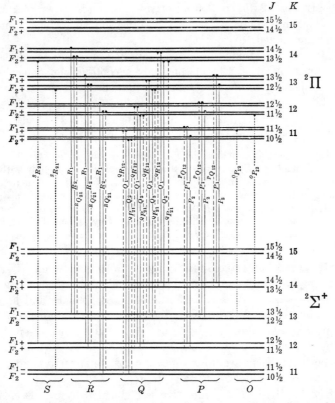

Fig. 122. **Energy Level Diagram for a** $^2\Pi\,(b)$—$^2\Sigma^+$ **Band.** The levels are drawn only for medium K values (11 to 15), since, in general, case (a) is approached for small K values. A number of consecutive lines are drawn for each branch. In the upper state, two levels occur for each J and K value, owing to the Λ-type doubling. Often in the upper state, the order of F_1 and F_2 is opposite to that drawn. The broken-line branches are satellite branches; the dotted branches (with $\Delta K = -2$) do not appear in a strict case (b). The last dotted vertical line to the right should go to the lower of the two levels with $K = 12$, $J = 12\frac{1}{2}$ of the $^2\Pi$ state (not $J = 11\frac{1}{2}$).

$^2\Pi$—$^2\Sigma$ **transitions.** While a $^2\Sigma$ state always belongs to Hund's case (b), a $^2\Pi$ state may belong either to case (a) or to case (b) or to cases intermediate between (a) and (b).

We shall consider first those $^2\Pi$—$^2\Sigma$ transitions for which the $^2\Pi$ *state belongs to case* (b). It is clear that, if the doublet separation in both states is so small that the doublet components cannot be separated in the spectrum, exactly the same band structure will be obtained as for a $^1\Pi$—$^1\Sigma$ transition—that is, one P, one Q, and one R branch, the Q branch being the most intense. For somewhat greater doublet splitting *each line of the three branches is split into two components* (similar to the fine structure of $^2\Sigma$—$^2\Sigma$ transitions). This is illustrated by the solid vertical lines in Fig. 122. Both in the upper and lower states the quantum number K is defined for which the selection rule (V, 44) holds.   The three doublet branches correspond to the three ΔK values, $+1$, 0, and -1. These branches are designated R_1, R_2, Q_1, Q_2, P_1, and P_2 in the way indicated in Fig. 122.   In Fig. 124(b), p. 260 a Fortrat diagram is shown of a MgH band at 5203 Å which represents a ${}^2\Pi(b)$—${}^2\Sigma$ transition and which shows the six branches just mentioned.

As for ${}^1\Pi$—${}^1\Sigma$ transitions, the Λ-type doubling produces no additional doubling of the lines for ${}^2\Pi$—${}^2\Sigma$ bands but only a combination defect (difference between the upper states of the Q_1 and Q_2 branches and those of the R_1, R_2 and P_1, P_2 branches).   As may easily be verified from Fig. 122, the relations (V, 58 and 59) hold individually for the branches distinguished by the subscript 1 and for those distinguished by the subscript 2.

In addition to the main branches there are also four *satellite branches* (broken lines in Fig. 122), for which $\Delta J \neq \Delta K$, and whose intensity decreases very rapidly with increasing K (see p. 244).   Their designation as ${}^R Q_{21}$, ${}^Q R_{12}$, ${}^Q P_{21}$, and ${}^P Q_{12}$ will be clear from what was previously said for ${}^2\Sigma$—${}^2\Sigma$ transitions.   They have the same form as the six main branches, and, for a small doublet splitting of the ${}^2\Sigma$ state, their lines lie very close to the corresponding lines of the main branches.   These satellite branches are very seldom observed in case (b).

General formulae for the *line strengths* in ${}^2\Pi(b)$—${}^2\Sigma$ bands have been given by Mulliken (513).

If the ${}^2\Pi$ state belongs to case (a) (large separation between ${}^2\Pi_{\frac{1}{2}}$ and ${}^2\Pi_{\frac{3}{2}}$), the selection rule $\Delta K = 0$, ± 1 no longer applies and all transitions in accord with the selection rules $\Delta J = 0$, ± 1, and $+ \leftrightarrow -$ are possible and appear with comparable intensities.   We can now divide each of the bands of a ${}^2\Pi$—${}^2\Sigma$ transition into *two sub-bands*, ${}^2\Pi_{\frac{1}{2}}$—${}^2\Sigma$ and ${}^2\Pi_{\frac{3}{2}}$—${}^2\Sigma$, which are separated from each other by the amount of the doublet splitting of the ${}^2\Pi$ state.   Correspondingly, for each band, there are *two zero lines* whose separation is approximately constant for different bands of a band system.

In Fig. 123, a ${}^2\Pi(a)$—${}^2\Sigma^+$ transition is illustrated by means of an energy level diagram.   In this diagram the splitting of the ${}^2\Pi$ state had to be drawn on a smaller scale than the separation of the rotational levels.   It is seen from the figure that for each sub-band six branches are possible, making twelve branches in all.   The branches of the first sub-band (with F_1 upper levels) are designated P_1, Q_1, R_1 (or P_{11}, Q_{11}, R_{11}) and P_{12}, Q_{12}, R_{12} depending on whether the lower levels are F_1 or F_2, and similarly for the second sub-band.   Fig 124(a) shows as an example the Fortrat diagram of the CdH band at 4400 Å for which all twelve branches have been observed.

While for ${}^2\Pi(b)$—${}^2\Sigma$ bands the branches are to a very good approximation given by the formulae for ${}^1\Pi$—${}^1\Sigma$ bands, for ${}^2\Pi(a)$—${}^2\Sigma$ bands different formulae hold.   Neglecting the Λ-type doubling in the ${}^2\Pi$ state and the spin doubling in the ${}^2\Sigma$ state, that is, using for the ${}^2\Pi$ state [with $B_{\text{eff.}}$ from (V, 30)]

$$F_1'(J) = B_{\text{eff.}}^{(1)} J(J+1), \qquad F_2'(J) = B_{\text{eff.}}^{(2)} J(J+1) \tag{V, 62}$$

and for the ${}^2\Sigma$ state

$$F_1''(J) = B'' K(K+1) = B''(J - \tfrac{1}{2})(J + \tfrac{1}{2}),$$
$$F_2''(J) = B'' K(K+1) = B''(J + \tfrac{1}{2})(J + \tfrac{3}{2}), \tag{V, 63}$$

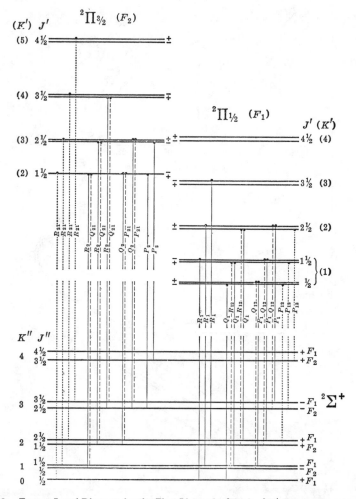

Fig. 123. **Energy Level Diagram for the First Lines of a** $^2\Pi\,(a)$—$^2\Sigma^+$ **Band.** In actual cases the spin-doublet splitting in the upper state is often much larger than shown, while, on the other hand, the spin-doublet splitting in the lower state and the Λ-doublet splitting in the upper state are in general much smaller. If the $^2\Pi$ state belongs strictly to case (a), the dotted and broken-line transitions are of the same intensity as the full-line transitions; however, in going over to case (b), they change over into the satellite branches given there.

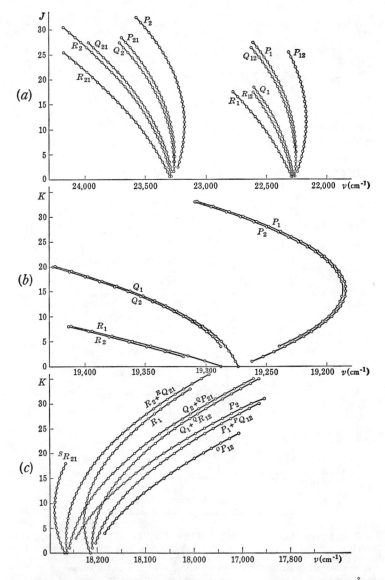

Fig. 124. **Fortrat Diagrams of Typical $^2\Pi$—$^2\Sigma$ Bands.** (a) CdH band at 4400 Å. (b) MgH band at 5203 Å. (c) CN band at 5473 Å (see Fig. 23). The ordinate is J or K, and not m, as in previous Fortrat diagrams. The upper state of the CN band is an inverted $^2\Pi$ state. For this reason the first lines of the branches do not agree with those in Fig. 123.

one obtains for the branches

$$P_1(J) = \nu_0^{(1)} + F_1'(J-1) - F_1''(J) = \nu_0^{(1)} + \tfrac{1}{4}B'' - B_{\text{eff.}}^{(1)}J + (B_{\text{eff.}}^{(1)} - B'')J^2$$

$$Q_1(J) = \nu_0^{(1)} + F_1'(J) - F_1''(J) = \nu_0^{(1)} + \tfrac{1}{4}B'' + B_{\text{eff.}}^{(1)}J + (B_{\text{eff.}}^{(1)} - B'')J^2$$

$$R_1(J) = \nu_0^{(1)} + F_1'(J+1) - F_1''(J)$$
$$= \nu_0^{(1)} + 2B_{\text{eff.}}^{(1)} + \tfrac{1}{4}B'' + 3B_{\text{eff.}}^{(1)}J + (B_{\text{eff.}}^{(1)} - B'')J^2$$

$$P_{12}(J) = \nu_0^{(1)} + F_1'(J-1) - F_2''(J)$$
$$= \nu_0^{(1)} - \tfrac{3}{4}B'' - (B_{\text{eff.}}^{(1)} + 2B'')J + (B_{\text{eff.}}^{(1)} - B'')J^2$$

$$Q_{12}(J) = \nu_0^{(1)} + F_1'(J) - F_2''(J) \approx P_1(J+1)$$

$$R_{12}(J) = \nu_0^{(1)} + F_1'(J+1) - F_2''(J) \approx Q_1(J+1)$$

$$\text{(V, 64)}$$

$$P_2(J) = \nu_0^{(2)} + F_2'(J-1) - F_2''(J)$$
$$= \nu_0^{(2)} - \tfrac{3}{4}B'' - (B_{\text{eff.}}^{(2)} + 2B'')J + (B_{\text{eff.}}^{(2)} - B'')J^2$$

$$Q_2(J) = \nu_0^{(2)} + F_2'(J) - F_2''(J)$$
$$= \nu_0^{(2)} - \tfrac{3}{4}B'' - (2B'' - B_{\text{eff.}}^{(2)})J + (B_{\text{eff.}}^{(2)} - B'')J^2$$

$$R_2(J) = \nu_0^{(2)} + F_2'(J+1) - F_2''(J)$$
$$= \nu_0^{(2)} + 2B_{\text{eff.}}^{(2)} - \tfrac{3}{4}B'' + (3B_{\text{eff.}}^{(2)} - 2B'')J + (B_{\text{eff.}}^{(2)} - B'')J^2$$

$$P_{21}(J) = \nu_0^{(2)} + F_2'(J-1) - F_1''(J) \approx Q_2(J-1)$$

$$Q_{21}(J) = \nu_0^{(2)} + F_2'(J) - F_1''(J) \approx R_2(J-1)$$

$$R_{21}(J) = \nu_0^{(2)} + F_2'(J+1) - F_1''(J)$$
$$= \nu_0^{(2)} + 2B_{\text{eff.}}^{(2)} + \tfrac{1}{4}B'' + 3B_{\text{eff.}}^{(2)}J + (B_{\text{eff.}}^{(2)} - B'')J^2$$

$$\text{(V, 65)}$$

Here $\nu_0^{(1)}$ and $\nu_0^{(2)}$ are the origins of the two sub-bands. It should be noted that in this approximation in which the spin splitting of $^2\Sigma$ is neglected [that is, $F_2''(J-1) = F_1''(J)$] the branches Q_{12}, R_{12}, P_{21}, and Q_{21} coincide with P_1, Q_1, Q_2, and R_2, respectively. In comparing the above formulae with (IV, 32–34) for $^1\Pi$—$^1\Sigma$ bands it is seen that while the quadratic terms are the same the linear terms are rather different. In the event that $B_{\text{eff.}}' \approx B''$, the separation of successive lines is not $2B$ as for the P and R branches of $^1\Pi$—$^1\Sigma$ and Σ—Σ bands but is B for P_1, Q_1, Q_{12}, R_{12}, Q_2, R_2, P_{21}, and Q_{21}, while it is $3B$ for R_1, P_{12}, P_2, and R_{21}. Extending the notation previously mentioned, Mulliken (513) describes these branches as being of $\tfrac{1}{2}P$, $\tfrac{1}{2}R$, $\tfrac{3}{2}P$ and $\tfrac{3}{2}R$ form. Mulliken has also given more accurate formulae taking account of the Λ-type doubling of the $^2\Pi$ state and the spin splitting of the $^2\Sigma$ state as well as approximate formulae similar to (IV, 35) representing the branches in pairs.

In the majority of actual cases the $^2\Pi$ state belongs neither strictly to case (a) nor strictly to case (b), but usually to a *transition case* which approximates case (a) for small rotation, while for large rotation a transition to case (b) gradually takes place. This transition takes place at a lower J value the smaller the doublet splitting is in the rotationless state (see p. 232 f.). In Fig. 124(c) a Fortrat diagram is shown for such an intermediate case, a red CN band (see the spectrogram Fig. 23). It should be compared with the Fortrat diagrams of $^2\Pi(a)$—$^2\Sigma$ and $^2\Pi(b)$—$^2\Sigma$ bands in Fig. 124(a) and (b). Since for the CN band the doublet splitting in the $^2\Sigma$ state is quite small, lines with the same K'' in the branches P_1 and Q_{12}, Q_1 and R_{12}, P_{21} and Q_2, and Q_{21} and R_2 fall so close together that they are not resolved [compare equations (V, 64 and 65)]. Therefore each sub-band shows only four branches, of which two form heads, as can be seen from the diagram. Thus these bands have *four heads*, as shown by the spectrogram Fig. 23.[11] It can be seen in Fig. 124(c) that, with increasing J, corresponding pairs of branches, P_1P_2, Q_1Q_2, R_1R_2, draw closer and closer together, although to a smaller extent than in Fig. 124(b). This drawing-together indicates clearly the transition to case (b).

As previously mentioned, the numbering by K in the $^2\Pi$ state, applicable for high rotation [since then usually case (b) is approximated] can be formally extended to small J values. The K values so obtained are given in Fig. 123. By comparison of this figure with Fig. 122 it can be seen that the branches P_1, Q_1, and R_1 and P_2, Q_2, and R_2 of case (a) correspond to the six main branches of case (b) and therefore can be designated by the same symbols. The branches Q_{12}, R_{12}, P_{21}, and Q_{21} of case (a) go into the satellite branches $^PQ_{12}$, $^QR_{12}$, $^QP_{21}$, and $^RQ_{21}$ of case (b). The branches R_{21} and P_{12} of case (a) have $\Delta K = \pm 2$ and correspond to the branches $^SR_{21}$ and $^OP_{12}$, which are forbidden in case (b) (dotted transitions in Fig. 122). These branches would have the *form of O or S branches* near case (b), since $\Delta K = \pm 2$. The formulae for these branches in case (b) would be [neglecting spin doubling in both upper and lower state, that is, putting $F_1(K) = F_2(K)$]

$$^SR_{21}(K) = \nu_0 + F_2'(K+2) - F_1''(K) = \nu_0 + 6B' + (5B' - B'')K + (B' - B'')K^2$$
$$^OP_{12}(K) = \nu_0 + F_1'(K-2) - F_2''(K) = \nu_0 + 2B' - (3B' + B'')K + (B' - B'')K^2 \qquad \text{(V, 66)}$$

Thus if B' and B'' are not too different, the separation of successive lines for these branches is twice as great (approximately $4B$) as for the P and R branches (see also p. 115). The formulae for the branches P_1P_2, Q_1Q_2, R_1R_2 for large K values in the intermediate case are the same as in case (b), that is, similar to those for the P, Q, and R branches of $^1\Pi$—$^1\Sigma$ bands. For small K values, that is, near case (a) the formulae (V, 64 and 65) hold and none of the branches have the form of regular S, R, Q, P, or O branches (see above).

Since the satellite branches in case (b) have an appreciable intensity only for small K, while the branches with $\Delta K = \pm 2$ have zero intensity, and since in most $^2\Pi$ states, at least for large rotation, a transition to case (b) takes place, it follows that in most $^2\Pi$—$^2\Sigma$ transitions for large rotation only the six main branches will appear with appreciable intensity. Accordingly, Fig. 124(c) shows that the branches $^SR_{21}$ and $^OP_{12}$ in the CN band are observed only for low K values. Explicit formulae for the intensities of the lines in the various branches for any value of the coupling constant $Y = A/B$ have been given by Earls (197).

In the *analysis* of $^2\Pi$—$^2\Sigma$ bands each individual sub-band is dealt with in the manner previously described for singlet bands (see p. 175 f.). For large J, the three most intense branches of one sub-band are the branches P_1, Q_1, and R_1 and of the other, P_2, Q_2, and R_2, whose correct numbering and identification can be tested by combination relations similar to (IV, 56 and 57). The Δ_2F'' values and from them the rotational constants of the lower state ($^2\Sigma$) are determined by (see Fig. 123)

$$\Delta_2F_1''(K) = R_1(K-1) - P_1(K+1),$$
$$\Delta_2F_2''(K) = R_2(K-1) - P_2(K+1). \qquad \text{(V, 67)}$$

As explained above for $^2\Sigma$—$^2\Sigma$ transitions, Δ_2F_1'' and Δ_2F_2'' differ in general only by a small

---

[11] Sometimes, particularly if $B_{\text{eff}}.'$ and B'' are very different from each other, only the shortest- or longest-wave-length head is prominent.

constant amount, 2γ. From the mean of the $\Delta_2 F''$ values very accurate B and D values for the lower state may be obtained.

There is an excellent check if the *satellite branches* are also observed. From Fig. 123 it can be seen that the following relations must hold:

$$R_1(K-1) - P_1(K+1) = R_{21}(K-1) - P_{21}(K+1),$$
$$R_2(K-1) - P_2(K+1) = R_{12}(K-1) - P_{12}(K+1). \tag{V, 68}$$

These combination relations hold exactly, even allowing for Λ-type doubling.

Furthermore, if the satellite branches are observed, the *doublet splitting* for each rotational level of the lower state can be determined quite independent of any assumption about its variation—in fact, from four different combinations, namely, according to Fig. 123,

$$F_1''(K) - F_2''(K) = R_{12}(K) - Q_1(K) = Q_{12}(K) - P_1(K)$$
$$= R_2(K) - Q_{21}(K) = Q_2(K) - P_{21}(K). \tag{V, 69}$$

From these individual doublet separations the spin-coupling constant γ is obtained according to (V, 16).

For the *upper state* in a case approaching pure case (a) one would obtain from

$$\Delta_2 F_i'(J) = R_i(J) - P_i(J) \tag{V, 70}$$

effective B and D values for $^2\Pi_{1/2}$ and $^2\Pi_{3/2}$. In an intermediate case the general formulae (V, 28) have to be used. Since as mentioned previously $\frac{1}{2}[F_1(J) + F_2(J)]$ follows the simple formula (V, 29) it is clear that $\frac{1}{2}[\Delta_2 F_1'(J) + \Delta_2 F_2'(J)]$ may be used for a determination of B and D by the method described in Chapter IV.[12] In both cases the B and D values obtained refer only to the one Λ doublet component of the $^2\Pi$ state. The effective B and D values for the other Λ doublet component and the magnitude of the Λ-type doubling can be obtained in a manner entirely similar to that for $^1\Pi$—$^1\Sigma$ transitions (see p. 254).

Which of the two components of the upper state is $^2\Pi_{1/2}$ and which is $^2\Pi_{3/2}$—that is, whether the $^2\Pi$ term is *normal* or *inverted*—can be decided in two different ways. First, one may use the magnitude of the Λ-type doubling, that is, the combination defect, in the two sub-bands; according to our previous discussion (p. 229), for $^2\Pi_{3/2}$ the Λ doubling is considerably smaller than for $^2\Pi_{1/2}$, at least for small J values [case (a)]. Second, one may use the fact that the missing lines in the neighborhood of the zero gap are different for the two sub-bands, as may be seen from Fig. 123.

The magnitude of the spin-doublet splitting in the $^2\Pi$ state may be derived for each value of J from relations such as

$$F_2'(J) - F_1'(J) = R_{21}(J-1) - R_1(J-1) = Q_2(J) - Q_{12}(J) \tag{V, 71}$$

and similar ones that can be read from Figs. 122 and 123. According to the general formulae (V, 28) this splitting is given by

$$F_2'(J) - F_1'(J) = B_v' \sqrt{4(J + \tfrac{1}{2})^2 + Y(Y-4)} \tag{V, 72}$$

so that from the observed values (V, 71) the spin coupling constant $A = B_v Y$ in the $^2\Pi$ state can be obtained. Near case (b) more accurate values for A are obtained from K doublets [that is, from $F_2'(K) - F_1'(K)$] rather than from J doublets. Suitable formulae have been given by Almy and Horsfall (63) who in addition have provided other important directions for the analysis of $^2\Pi$—$^2\Sigma$ bands.

Examples of $^2\Pi$—$^2\Sigma$ transitions, apart from those already mentioned, are

---

[12] The refinement of Almy and Horsfall mentioned on p. 232 has only the effect of changing $4B_v - 6D_v$ in (IV, 47) to $4(B_v - D_v)$, a change that is usually negligible.

the α bands of BO [Jenkins and McKellar (373)] and the comet-tail bands of CO^+ [Schmid and Gerö (1384), Coster, Brons, and Bulthuis (865)(143)]. Both of these are similar to the red CN bands [see Fig. 124(c)] and like them have an inverted $^2\Pi$ state as upper state. Examples of $^2\Pi$—$^2\Sigma$ bands with normal $^2\Pi$ states have been found for a considerable number of alkaline-earth hydrides and similar iso-electronic ions (see Table 39 for references on BeH, BH^+, MgH, AlH^+, CaH, SrH, BaH, ZnH, CdH, and HgH).

All these examples are $^2\Pi$—$^2\Sigma^+$ transitions. (Compare Figs. 122 and 123.) $^2\Pi$—$^2\Sigma^-$ transitions are of course quite analogous, but with an appropriate change of the $+ -$ symmetry (see above for $^1\Pi$—$^1\Sigma^-$). Thus far, examples of $^2\Pi$—$^2\Sigma^-$ transitions are not known.

For like nuclei, $^2\Pi$—$^2\Sigma$ transitions may occur only for ionized molecules. In each sub-band they would show an intensity alternation like the $^1\Pi$—$^1\Sigma$ transitions. However, no example of such a case has yet been observed.

$^2\Sigma$—$^2\Pi$ transitions. For $^2\Sigma$—$^2\Pi$ transitions exactly the same branches occur as for $^2\Pi$—$^2\Sigma$ transitions. As before, for medium doublet splitting of the $^2\Pi$ state, each band has four characteristic heads. But the relative intensities of the branches are somewhat different [see Earls (197)] and at the beginning of the branches, different lines are missing. Which lines are missing for $^2\Sigma$—$^2\Pi$ can easily be seen from an energy level diagram similar to Fig. 123.

Examples of $^2\Sigma^+$—$^2\Pi$ transitions are the ultraviolet OH bands [Tanaka and Koana (661)], the HCl^+ bands [Norling (536)], the NO γ bands [Schmid (618)], and the Baldet-Johnson bands of CO^+ [Bulthuis (143)(144); see Fig. 11(a)]. Two examples of $^2\Sigma^-$—$^2\Pi$ transitions are known—namely, the CH band at 3900 Å [see Mulliken (513) and Gerö (974)] and an MgH band at 4400 Å [Guntsch (277)].

$^3\Pi$—$^3\Sigma$ transitions. For a $^3\Pi$—$^3\Sigma$ transition, if the $^3\Pi$ state approaches case (a), the band structure is of course still more complicated than for $^2\Pi(a)$—$^2\Sigma$ transitions. Fig. 125 represents the energy level diagram for such a case assuming the $^3\Sigma$ state to be $^3\Sigma^-$. There are now three sub-bands: $^3\Pi_0$—$^3\Sigma^-$, $^3\Pi_1$—$^3\Sigma^-$, and $^3\Pi_2$—$^3\Sigma^-$. In each sub-band there are nine branches, three for each triplet component of the lower state. Thus there are in all *27 branches*. The designations of these branches are indicated in Fig. 125. They are similar to those for $^2\Pi(a)$—$^2\Sigma$ transitions (compare Fig. 123). In Fig. 125 K values are formally assigned to the levels in order to indicate the relation to a $^3\Pi(b)$—$^3\Sigma^-$ transition. In addition to branches of O, P, Q, R, S form there are also branches with $\Delta K = \pm 3$, that is, of T and N form.

The first case of a $^3\Pi(a)$—$^3\Sigma$ band that was completely analyzed and for which all 27 branches were identified was the PH band at 3400 Å [Pearse (557) (1087)]. It is a $^3\Pi(a)$—$^3\Sigma^-$ transition with an inverted $^3\Pi$ state. The well-known first positive group of N_2 (see Fig. 8, p. 32) forms an example of a $^3\Pi_g(a)$—$^3\Sigma_u^+$ transition [Naudé (528)].

If the $^3\Pi$ state is close to case (b) the selection rule $\Delta K = 0, \pm 1$ operates and transitions with $\Delta K \neq \Delta J$ are very weak (form satellite branches) except at the lowest J values. Thus only the nine branches $P_{11}, Q_{11}, R_{11}, P_{22}, Q_{22}, R_{22}, P_{33}, Q_{33}, R_{33}$ retain an appreciable intensity. An example of a $^3\Pi(b)$—$^3\Sigma^-$ transition of this type is the NH band system at 3300 Å [Funke (240)]; an example of an intermediate case is the similar band system of OH^+ [Loomis and Brandt (460)].

The analysis of $^3\Pi$—$^3\Sigma$ bands proceeds in a way very similar to that for $^2\Pi$—$^2\Sigma$ bands described above, taking account of the previously discussed energy formulae for $^3\Pi$ and $^3\Sigma$ states (p. 235 and p. 223). More details may be found in the papers quoted and particularly

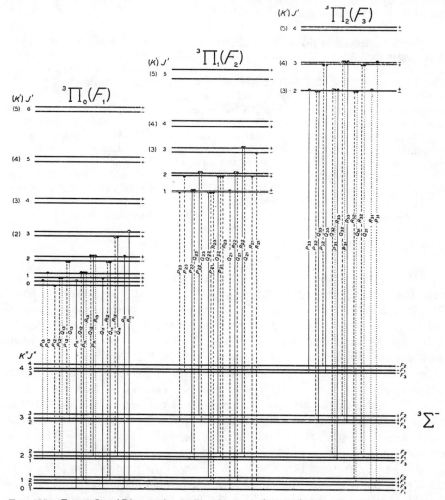

Fig. 125. **Energy Level Diagram for the First Lines of a** $^3\Pi\,(a)$—$^3\Sigma^-$ **Band.** The Λ type doubling in the upper state and the spin tripling in the lower state is much exaggerated. The triplet splitting of the upper state is usually much larger. If the upper state is strictly a case (a) state all 27 branches indicated have comparable intensities. In going over to case (b) the branches indicated by dotted lines ($\Delta K = \Delta J \pm 2$) and those indicated by broken lines ($\Delta K = \Delta J \pm 1$) become weaker or disappear altogether. Some of them go over into the satellite branches of case (b).

in papers by Budó (139) and Challacombe and Almy (152). General formulae for the intensity distribution in the branches have been given by Nolan and Jenkins (1285a) and Budó (141).

$^3\Sigma$—$^3\Pi$ **transitions.** The structure of $^3\Sigma$—$^3\Pi$ bands is closely similar to that of $^3\Pi$—$^3\Sigma$ bands except for a difference in the missing lines at the beginning of the branches and for a slight difference of the intensity distribution [see Budó (141)].

Three band systems of CO, the third positive group, the 3A bands and the Asundi bands, all with the same lower $^3\Pi$ state form examples of $^3\Sigma$—$^3\Pi\,(a)$ transitions showing most of the 27 branches [see Dieke and Mauchley (190) and Gerö (969)(975)].   Examples of $^3\Sigma$—$^3\Pi\,(b)$ transitions are the BH band at 3690 Å and the AlH band at 3800 Å [Almy and Horsfall (63), Holst (1057), Challacombe and Almy (152)].

$\Pi\longleftrightarrow\Sigma$ transitions of higher multiplicities.   Only two $\Pi\longleftrightarrow\Sigma$ transitions of a multiplicity higher than 3 have as yet been analyzed in detail, both by Nevin: The first negative bands of oxygen which represent a $^4\Sigma_g^-$—$^4\Pi_u$ transition of O_2^+ [Nevin (530)(1277)] and the MnH bands at 5677 and 6237 Å which represent a $^7\Pi$—$^7\Sigma$ transition [Nevin (1278)].   If the Π states were strictly case (a) there would be 48 and 147 branches, respectively.   For intermediate cases the branches with large ΔK values become very weak and finally in case (b) only 12 and 21 main branches, respectively, remain.   In the case of the O_2^+ bands Nevin has actually identified 40 branches while for MnH he identified 49.   By proceeding in a manner similar to that described above for $^2\Pi$—$^2\Sigma$ transitions Nevin determined the rotational constants as well as the constants describing the multiplet splitting and Λ-type doubling.

A theoretical treatment of $^4\Sigma$—$^4\Pi$ transitions including intensity formulae for the two limiting cases has been given by Budó (140).   The observed splittings of the $^4\Pi_u$ state of O_2^+ do not fit well with the theoretical formulae.

In the spectrum of FeCl a $^6\Pi$—$^6\Sigma$ and a $^4\Pi$—$^4\Sigma$ transition have been tentatively identified by Miescher (494) and Müller (1254) on the basis of the heads of the different sub-bands (without resolution of the fine structure).

$^1\Pi$—$^1\Pi$ transitions.   If for a moment we ignore Λ-type doubling it is clear from the selection rule $\Delta J = 0,\ \pm 1$ that a $^1\Pi$—$^1\Pi$ transition, like a $^1\Pi$—$^1\Sigma$ transition, has a P, a Q, and an R branch; but here the Q branch is weak and its intensity decreases rapidly with increasing J, since $\Delta\Lambda = 0$ [see Fig. 93(b) and the accompanying discussion].   If we take account of the Λ-type doubling, each line splits into two components, as Fig. 126 shows (this is different from a $^1\Pi$—$^1\Sigma$ transition, where no line splitting is produced by the Λ-type doubling of the levels; see Fig. 119).   The reason that, for a given $J' - J''$ combination, only two lines result and not four, lies in the selection rule $+ \longleftrightarrow -$, as can be seen immediately from Fig. 126.   A $^1\Pi$—$^1\Pi$ band thus has *six branches*, two P, two Q, and two R branches, of which the two Q branches are very weak.

As can be seen from Fig. 126, the *first lines* of the branches are $P(2)$, $Q(1)$, and $R(1)$.   Thus in the series (IV, 35), formed by the P and R branches, three lines are missing (see lower part of Fig. 126), and not just one line, as for Σ—Σ transitions.

If the region of the zero line and the head is not resolved and only the P and R branches are observed at large J values, a $^1\Pi$—$^1\Pi$ band has the same appearance, for a superficial examination, as a $^2\Sigma$—$^2\Sigma$ band.   However, in a $^2\Sigma$—$^2\Sigma$ transition, the splitting of the lines increases linearly with K (see p. 249), whereas, in a $^1\Pi$—$^1\Pi$ transition, in general it increases *quadratically* with J, since the Λ-type doubling of each $^1\Pi$ state increases quadratically with J (see p. 228).   In this way we can distinguish between the two types of transitions if it is not already clear, from a knowledge of the carrier of the band system or from other reasons, whether we are dealing with a doublet or a singlet transition.

The sole example of a $^1\Pi$—$^1\Pi$ transition thus far analyzed for unlike nuclei is the AlH band at 3380 Å [Holst (325)].

If the Π^+ rotational levels are distinguished from the Π^- levels by subscripts c and d in agreement with the usage previously adopted (p. 239), it is clear from Fig. 126(a) (in which the upper one of each pair of levels is a c level) that for the P and R branches $c \longleftrightarrow c$ and $d \longleftrightarrow d$ whereas for the Q branches $c \longleftrightarrow d$. Therefore the six branches may be designated P_c,

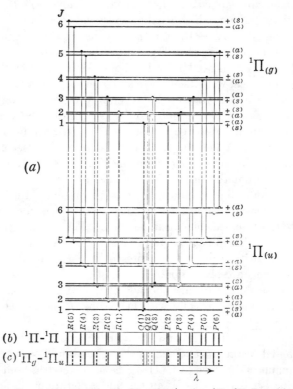

Fig. 126. **Energy Level Diagram for the First Lines of a $^1\Pi$—$^1\Pi$ Transition.** The Λ-type doubling of the levels, as well as of the lines, is much exaggerated. The B values in the upper and lower states are taken as equal. Consequently the spacing of the lines in the Q branch could not be drawn to scale. The schematic spectrogram (c) refers to the case of identical nuclei, the lines corresponding to the antisymmetrical levels being broken.

P_d, R_c, R_d, and Q_{cd}, Q_{dc}. The combination differences $\Delta_2 F(J)$ for the upper and lower states must of course be formed separately for P_c, R_c, and P_d, R_d, yielding effective rotational constants B^c, B^d, D^c, D^d of the two Λ-type components of the two states. Their means give a very good approximation to the true B and D values, respectively (see p. 228 f.).

The magnitude of the splitting of the line components in the P and R branches is equal to the difference of the Λ-type doublet separations of the upper and lower states, and that in the Q branch to their sum, if, as in Fig. 126, the order of the c and d levels is the same in the two $^1\Pi$ states. The reverse is the case if the order of the c and d levels is opposite in the two $^1\Pi$ states. Therefore only for those levels for which Q as well as P and R lines are observed and resolved, can the magnitude of the Λ-type doubling be determined directly.   For other levels one must be satisfied with an indirect determination from the effective B values according to (V, 25) and (V, 26).

The *intensity distribution* in both sets of branches P_c, R_c, Q_{cd}, and P_d, R_d, Q_{dc} is exactly the same and given by the formulae (IV, 81); that is, the components of a line doublet due to Λ-type doubling have equal intensities unlike the components of line doublets due to spin doubling (compare $^2\Sigma$—$^2\Sigma$ bands above).   This holds for molecules with unequal nuclei only and under the assumption that there are no perturbations (see below).

For *homonuclear molecules*, $^1\Pi_u$ can combine only with $^1\Pi_g$.   In Fig. 126(a) the designations s and a, which would correspond to a $^1\Pi_g$—$^1\Pi_u$ transition, are added in brackets (for $^1\Pi_u$—$^1\Pi_g$, s and a would have to be exchanged throughout). In the schematic spectrogram in Fig. 126(c) the lines corresponding to the anti-symmetric levels are indicated by broken lines.   It can be seen that, as a result of the different statistical weights of the s and a levels, here again strong and weak lines alternate in a given branch.   However, now there are always two branches, with opposite intensity alternation, lying close together.   Consequently, if the Λ-type doublets are not resolved, a resultant branch with no apparent intensity alternation appears.   When the nuclear spin is zero, every second line is missing in each individual branch [broken lines in Fig. 126(c)].   However, even then, as may be seen from the figure, for a casual inspection only a single branch without any missing lines would appear to be present.   But the figure shows that then the even-numbered lines are displaced to one side and the odd-numbered lines to the other side of a mean position.   This effect is called *"staggering."* This staggering is also to be expected to a smaller degree if no lines are missing but only an intensity alternation is present and the Λ-type doublets are un-resolved, since an unresolved doublet, consisting of a strong and a weak line, will appear to be shifted toward the side of the stronger line.   The staggering increases of course quadratically with increasing J and may not be detectable for small J.

An example of a $^1\Pi_g$—$^1\Pi_u$ transition are the violet C_2 bands discovered by Deslandres and d'Azambuja and analyzed by Dieke and Lochte-Holtgreven (189) and Kopfermann and Schweitzer (418) [see also Herzberg and Sutton (1052)]. The staggering is clearly observed in the fine structure of the individual bands. An example of a $^1\Pi_u$—$^1\Pi_g$ transition is a new ultraviolet band system of N_2 discovered by Gaydon (954) [see also Herzberg (1041)].

$^2\Pi$—$^2\Pi$ **transitions.**   When both $^2\Pi$ states of a $^2\Pi$—$^2\Pi$ transition belong to case (a), the selection rule $\Delta\Sigma = 0$ holds (see p. 242).   As a result, a $^2\Pi$—$^2\Pi$ band splits into *two sub-bands*, $^2\Pi_{1/2}$—$^2\Pi_{1/2}$ and $^2\Pi_{3/2}$—$^2\Pi_{3/2}$.   As Fig. 127(a) shows in detail, each sub-band has a structure similar to that of a $^1\Pi$—$^1\Pi$ transition; that is, each sub-band has six branches, which form three close pairs, two P, two Q, and two R branches, the two Q branches being very weak. Thus a $^2\Pi(a)$—$^2\Pi(a)$ band has twelve branches in all.   Since the Λ-type doubling is always small and since the Q branches are very weak, each sub-band has only one head and thus each band has two heads. The separation of the two heads is approximately constant for all the bands of a system.

The two sub-bands differ in the number of missing lines at the beginning of the branches and in the magnitude of the Λ-type doubling, which is appreciably greater for $^2\Pi_{1/2}$ than for $^2\Pi_{3/2}$. On this basis it is possible to ascertain in an observed $^2\Pi$—$^2\Pi$ band which one of the sub-bands is $^2\Pi_{1/2}$—$^2\Pi_{1/2}$ and which is $^2\Pi_{3/2}$—$^2\Pi_{3/2}$. In case (a) the rotational levels of each component of a $^2\Pi$ state can be represented with effective rotational constants B_v and D_v

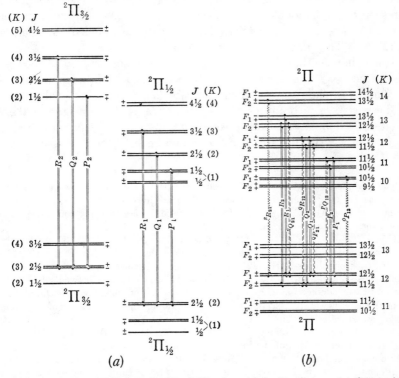

Fig. 127. **Energy Level Diagrams for $^2\Pi$—$^2\Pi$ Bands.** (a) In Hund's case (a) [$^2\Pi(a)$—$^2\Pi(a)$]. (b) In Hund's case (b) [$^2\Pi(b)$—$^2\Pi(b)$]. Only one line of each branch is given. In the designation of the branches components of the same Λ-type doublet are not distinguished. In (a) the $^2\Pi_{1/2}$ and $^2\Pi_{3/2}$ levels form the $F_1(J)$ and $F_2(J)$ series, respectively. The dotted branches in (b) do not appear when both $^2\Pi$ states belong strictly to Hund's case (b).

[according to (V, 29)]. As long as this is the case the determination of the rotational constants can be carried out separately for each sub-band in a manner corresponding exactly to that described for $^1\Pi$—$^1\Pi$ bands. The magnitude of the doublet splitting of the band (separation of the zero lines of the two sub-bands) is the difference or the sum of the doublet splittings of the two states involved, according as the two states are of the same kind (both normal or both inverted) [Fig. 127(a)] or of the opposite kind. Thus the doublet splitting of $^2\Pi(a)$ states cannot be determined directly from $^2\Pi(a)$—$^2\Pi(a)$ bands. However, if there is a slight deviation from case (a) and formulae (V, 28) have to be used for a representation of the observed $\Delta_2F_i(J)$ values, approximate values of the coupling constants $A = YB$ in the upper and lower states can be determined.

If both $^2\Pi$ states approach case (b), the selection rule $\Delta K = 0, \pm 1$ holds and, in addition, the rule that branches with $\Delta K \neq \Delta J$ are very weak. On the basis of these rules it is readily seen from Fig. 127(b) that there are again twelve main branches [full-line transitions in Fig. 127(b)], which correspond completely to those of case (a), except that now they do not form two separate sub-bands. Disregarding Λ-type doubling and neglecting the weak Q branches and the satellite branches, the band structure is quite similar to that of a $^2\Sigma$—$^2\Sigma$ transition (four strong branches). However, as a result of the Λ-type doubling each component of a spin doublet in the spectrum is once more split into two components. The satellite branches with $\Delta K \neq \Delta J$ are indicated by broken lines in Fig. 127(b). They are usually not observed.

It may happen that in both $^2\Pi$ states case (a) holds for small J and case (b) holds for large J. Then, as may easily be verified by a comparison of Figs. 127(a) and (b), the twelve branches of case (a), present for small J, go over into the twelve main branches of case (b) for large J. For large J there are characteristic doublet branches (disregarding Λ-type doubling) which separate from each other with decreasing rotation and belong to different sub-bands for small J values.

The *determination of the rotational constants* can be carried out in these cases in much the same way as for $^2\Pi$—$^2\Sigma$ transitions, that is, by using the mean of $\Delta_2 F_1(J)$ and $\Delta_2 F_2(J)$. The differences between $\Delta_2 F_1(J)$ and $\Delta_2 F_2(J)$ may be used to determine the coupling constant A. Intensity formulae for case (b) may be found in Mulliken (513).

If one $^2\Pi$ state approaches case (a) and the other approaches case (b), neither the rule $\Delta\Sigma = 0$ nor the rule $\Delta K = 0, \pm 1$ holds. Therefore, the satellite branches with $\Delta K \neq \Delta J$ [indicated by broken lines in Fig. 127(b)] as well as the branches with $\Delta K = \pm 2$ (indicated by dotted lines), appear with appreciable intensity. In all there are twenty-four branches, or rather twelve doublet branches (corresponding to Λ-type doubling), the four doublet Q branches being very weak. Except for the Λ-type doubling and the weakness of the Q branches the band structure is very similar to that of $^2\Pi(a)$—$^2\Sigma$ bands discussed earlier (p. 258 f.); the forms of the branches are given by the formulae (V, 64 and 65) [see Fig. 124(a) and (c)]. Therefore the determination of the rotational constants as well as of the spin splitting can be carried out in exactly the same way as described for $^2\Pi(a)$—$^2\Sigma$. Unlike the other types of $^2\Pi$—$^2\Pi$ bands here the individual splittings for each J value can be obtained with the help of the satellite branches [from relations like (V, 71)] and internal checks of the analysis within a given band are possible: the relations (V, 68) and similar ones for the upper state hold.

Examples of $^2\Pi$—$^2\Pi$ transitions for molecules with unequal nuclei are the band at 4550 Å of MgH [Guntsch (275)] for which both $^2\Pi$ states approach closely case (b), the β bands of NO [Jenkins, Barton and Mulliken (369)] for which both $^2\Pi$ states go from case (a) to case (b) with increasing rotation, and the α bands of SiF [Eyster (210)] for which the lower state belongs to case (a), the upper to case (b). The rotation-vibration spectrum of NO also belongs here since the electronic ground state is $^2\Pi$ [Gillette and Eyster (986)]. Since in this case the doublet splitting without rotation is the same in the upper and lower state the two sub-bands coincide except for higher K values. The analogues in PO and NS of the β bands of NO very probably represent cases in which both upper and lower state are close to Hund's case (a) but the fine structure of these bands has not yet been analyzed.

For homonuclear molecules the alternation of statistical weights leads to an intensity alternation of the same type as in $^1\Pi$—$^1\Pi$ transitions (see p. 268). An example is provided by the ultraviolet O_2^+ bands [see Stevens (653) and v. Bozóky (123)] which represent a $^2\Pi_u(b)$—$^2\Pi_g(a)$ transition.

$^3\Pi$—$^3\Pi$ **transitions.** The structure of $^3\Pi$—$^3\Pi$ bands is very similar to that of $^2\Pi$—$^2\Pi$ bands. If both $^3\Pi$ states belong to case (a), on account of the selection rule $\Delta\Sigma = 0$, there are *three sub-bands*: $^3\Pi_0$—$^3\Pi_0$, $^3\Pi_1$—$^3\Pi_1$, and $^3\Pi_2$—$^3\Pi_2$. If we disregard Λ-type doubling, each sub-band has a strong R, a strong P, and, with the exception of $^3\Pi_0$—$^3\Pi_0$, a weak Q branch. Therefore each band has three heads. The same six strong branches occur if both $^3\Pi$ states belong to case (b) or if both go over from (a) to (b) with increasing rotation. For all K values of the first case and for large K values of the second case the three P branches are close together and the three R branches are close together. These characteristic triplets are clearly seen in the spectrogram Fig. 22 of a band of the second positive group of N_2 which represents such a transition. The spectrogram also shows that with decreasing K the triplet splitting becomes larger indicating that the two $^3\Pi$ states approach case (a) for low K values. There are indeed three separate heads corresponding to the three sub-bands.

Fig. 128 shows the main branches of a $^3\Pi$—$^3\Pi$ transition in an energy level diagram. A rapid transition from case (a) to case (b) has been assumed in this diagram. Just as for $^1\Pi$—$^1\Pi$ and $^2\Pi$—$^2\Pi$ transitions each branch is split into two on account of Λ-type doubling.

In order to evaluate the *rotational constants* one has to form the combination differences $\Delta_2 F_i(J)$ for the individual multiplet components of the upper and lower states. As was pointed out before, the mean of $\Delta_2 F_i(J)$ yields the B and D values. With regard to the determination of the triplet splitting in the $^3\Pi$ states the same remarks apply as for $^2\Pi$—$^2\Pi$ transitions (see above).

The theoretical line strengths in the various branches for different coupling conditions have been given by Budó (141).

Examples of $^3\Pi$—$^3\Pi$ transitions for heteronuclear molecules are the α bands of TiO [Christy (155)], the similar bands of ZrO [Lowater (471)] and the BN bands [Douglas and Herzberg (901)].

For homonuclear molecules an intensity alternation becomes noticeable only when the Λ-type doubling is resolved, just as for $^1\Pi$—$^1\Pi$ bands. With the resolution used in the spectrogram Fig. 22 no intensity alternation is visible in the N_2 band, which represents a $^3\Pi_u$—$^3\Pi_g$ transition. With somewhat greater dispersion the Λ-type doubling in the branches has actually been resolved and at the same time the intensity alternation observed [Hulthén and Johannson (344), Lindau (446), Coster, Brons, and van der Ziel (167)]. The first lines of the weak Q branches have also been observed [Guntsch (274)].

Examples of $^3\Pi_g$—$^3\Pi_u$ transitions are the Swan bands [Johnson (390),

Shea (635), Budó (139)] and a system of ultraviolet bands of C_2 [Fox and Herzberg (227), Phillips (1316)]. Their structure is very similar to that of the N_2 bands, but on account of the zero nuclear spin of carbon there is a "stagger-

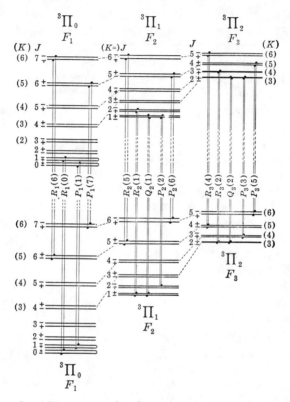

Fig. 128. **Energy Level Diagram for a $^3\Pi$—$^3\Pi$ Transition.** For the upper as well as the lower state a very rapid transition from case (a) to case (b) is assumed. The Λ-type doubling is greatly exaggerated. The first line of every branch, as well as one line with a larger J value, is shown except for the Q branches, which are not observed for higher J values. It is seen how for larger J the triplet R and triplet P branches arise. The K values are not given for the lowest levels, since they are not defined in case (a). Of course, formally it is possible to continue the K values from the higher levels to the lower, but the K values thus obtained do not always correspond to the true K values that one would obtain if a transition to case (b) were actually carried out. For example, strictly speaking, the three lowest levels of $^3\Pi_0$ belong to $K = 1$.

ing" in each of the branches (see the discussion of $^1\Pi$—$^1\Pi$ transitions for homonuclear molecules).

Π—Π transitions of higher multiplicity. On the basis of the preceding discussion it would be easy to derive qualitatively the structure of $^4\Pi$—$^4\Pi$, $^5\Pi$—$^5\Pi$, $\cdots$ transitions. No such transitions have been analyzed in detail. However, Müller (1254) has identified $^7\Pi$—$^7\Pi$ transitions of MnCl and MnBr by the observation of the heads of all the sub-bands. Here, on account of a very large value of $Y = A/B$ the selection rule $\Delta\Sigma = 0$ breaks down [transition to Hund's case (c)] and all transitions possible according to $\Delta\Omega = 0, \pm 1$ occur.

$\Pi \leftrightarrow \Delta$ **transitions.** For $^1\Pi \leftrightarrow {}^1\Delta$ *transitions* we obtain exactly the same *six branches* as for $^1\Pi - {}^1\Pi$ transitions (Fig. 126). It is therefore unnecessary to reproduce a special energy level diagram. The only differences are that now the two Q branches are very intense ($\Delta\Lambda \neq 0$; see p. 207 f.) and that more lines are missing near the zero gap. As may easily be seen, the first lines are $P(2)$, $Q(2)$, and $R(2)$ for $^1\Pi - {}^1\Delta$ and $P(3)$, $Q(2)$, and $R(1)$ for $^1\Delta - {}^1\Pi$.

The NH band at 3240 Å, reproduced in the spectrogram Fig. 21, is an example of a $^1\Pi - {}^1\Delta$ transition [see Pearse (558) and Dieke and Blue (187)]. The six branches, which form three close pairs, not resolved for small J, can be clearly seen. Furthermore it can be seen that the lines $P(1)$, $R(0)$, $R(1)$, $Q(0)$, and $Q(1)$ are indeed missing. Further bands of this kind have been observed for H_2 and He_2 [see Richardson (32) and Weizel (22)].

$^2\Pi \leftrightarrow {}^2\Delta$, $^3\Pi \leftrightarrow {}^3\Delta$, $\cdots$ transitions are similar to $^2\Pi - {}^2\Pi$, $^3\Pi - {}^3\Pi$, $\cdots$ transitions, except that the Q branches are very intense. As long as both states belong to the same coupling case, there are twelve intense main branches for $^2\Pi \leftrightarrow {}^2\Delta$ and eighteen intense main branches for $^3\Pi \leftrightarrow {}^3\Delta$ which fall together in close pairs (Λ-type doubling). In the most general case— that is, if both terms belong to different coupling cases—there are twenty-four branches for $^2\Pi \leftrightarrow {}^2\Delta$ and fifty-four branches for $^3\Pi \leftrightarrow {}^3\Delta$.

Examples of $^2\Delta - {}^2\Pi$ transitions are the CH band at 4315 Å [see Mulliken (513)] and the corresponding SiH band [see Rochester (593)]. The only examples of $^3\Delta - {}^3\Pi$ transitions thus far investigated are certain bands of He_2 [see Weizel (22)] for which the triplet splitting is not resolved and which have therefore the appearance of $^1\Delta - {}^1\Pi$ bands.

$\Delta - \Delta$ **transitions.** The structure of $^1\Delta - {}^1\Delta$ bands is almost identical with that of $^1\Pi - {}^1\Pi$ bands except that one more line is missing at the beginning of each of the branches, that there is a corresponding slight change of intensity distribution [see the formulae (IV, 81)], and that the Λ-type doubling is much smaller. Up to the present time no such transitions have been identified with certainty.

Similarly, $^2\Delta - {}^2\Delta$ and $^3\Delta - {}^3\Delta$ transitions resemble $^2\Pi - {}^2\Pi$ and $^3\Pi - {}^3\Pi$ transitions, respectively. Gaydon and Pearse (245) have found one component, $^2\Delta_{\frac{5}{2}} - {}^2\Delta_{\frac{5}{2}}$, of a $^2\Delta - {}^2\Delta$ transition in the spectrum of NiH [see also A. Heimer (288)]. Mahanti (477) has tentatively identified the VO bands occurring in an electric arc containing vanadium as a $^2\Delta - {}^2\Delta$ transition. $^3\Delta - {}^3\Delta$ transitions have not yet been observed.

Band structures in Hund's cases (c) and (d). Thus far, in our discussion of different band types, it has always been assumed that the participating electronic states belong to Hund's case (a) or case (b) or a case intermediate between them. Since case (c) may be regarded as a case (a) with very large multiplet splitting, the band structures in case (c) are quite similar to those in case (a), with only the difference that S and Σ are no longer "good" quantum numbers and, as a result, the selection rules $\Delta S = 0$ and $\Delta\Sigma = 0$ no longer apply. The electronic states combine according to the selection rule $\Delta\Omega = 0$, ± 1. The structure of a band with given Ω' and Ω'' is then of exactly the same type as that of the above-described singlet bands: Transitions with $\Delta\Omega = 0$ and $\Omega \neq 0$ have a structure similar to that of a $^1\Pi - {}^1\Pi$ transition; transitions with $\Delta\Omega = 0$ and $\Omega = 0$ are similar to $^1\Sigma - {}^1\Sigma$ transitions; transitions with $\Delta\Omega = \pm 1$ are similar to $^1\Pi \leftrightarrow {}^1\Sigma$ or $^1\Delta \leftrightarrow {}^1\Pi$ transitions, depending on whether or not Ω is zero in one of the states.

An important example is afforded by the visible absorption bands of the halogens, which have the appearance of $^1\Sigma - {}^1\Sigma$ transitions, but for which actually the upper state is the $^3\Pi_{0^+}$

component of a $^3\Pi$ term with extremely large splitting [see Mulliken (516)], while the lower state is the $^1\Sigma^+$ ground state of the molecule. The proper case (c) designation of these bands is 0^+—0^+.

In Hund's case (d), since $\Delta R = 0$ (see p. 245) and since R determines the energy according to (V, 20), the bands consist of Q form branches each line of which is split into a number of components, corresponding to the coupling of L and R. A pure case (d) for both upper and lower state is, however, quite rare. On the other hand several examples of intermediate cases between (b) and (d) have been found for some higher excited states of H_2 and He_2. In these cases the band structures are quite complicated. For extended discussions of these structures the reader is referred to Weizel (22).

General remarks on the technique of the analysis of multiplet bands.
It is not always easy to decide to which of the various types of transitions the bands of an empirically observed band system belong. The first step in the analysis is always the picking out of branches in the manner previously described (see p. 191 f.). The number (and kind) of branches found narrows down the choice of possible band types, but one must of course allow for the fact that some branches may have eluded observation. From the various band types possible on this basis the most probable is chosen and made the starting point of an attempt to find the numbering of the lines in the branches by calculation of the combination differences in the manner previously described. If the combination differences and the variation of the splitting of the levels obtained from the line splittings (if such splittings occur) agree in all details with those to be expected theoretically on the basis of the assumed band type, and if in addition the checks mentioned above for the various band types are fulfilled, then the bands under consideration must actually belong to this type, and the types of the upper and lower electronic state are thereby established. The rotational constants for the two electronic states can then be determined in the manner indicated above for the individual band types. If deviations from the theoretical expectations occur, the other possible band types must be tried out.

If, for example, we observe a band system whose bands at some distance from the band head are resolved into two doublet branches which run parallel to each other and whose splitting increases with increasing rotation, we may have a $^2\Sigma$—$^2\Sigma$ or a $^1\Pi$—$^1\Pi$ transition assuming that the Q branches are too weak to be observed for the $^1\Pi$—$^1\Pi$ transition. Which of the two cases we actually have can be decided by determining the variation of the doublet splitting in the branches. If it increases linearly with an arbitrary running number, we have a $^2\Sigma$—$^2\Sigma$ transition; if it increases quadratically, we have a $^1\Pi$—$^1\Pi$ transition.

It is clear that a reliable determination of the moments of inertia and internuclear distances in the various electronic states of a diatomic molecule is possible only when the multiplet structure of the bands has been completely analyzed. As we shall see in Chapter VII a knowledge of the multiplet structure is also required for a determination of the heat of dissociation of the molecule considered. These are just the data that are of greatest importance for many applications (see Chapter VIII) as well as for progress in the theoretical understanding of the structure of diatomic molecules. A considerable body of data in this field concerning a large number of molecules has been accumulated in the past 25 years but much work remains to be done both with regard to molecules that have not yet been investigated and with regard to new electronic states of known molecules.

(c) Forbidden Electronic Transitions

In addition to the transitions treated in the foregoing subsection, under certain conditions transitions occur that contradict the selection rules discussed in subsection (a). They are called *forbidden transitions.* They can be observed in absorption by using very long absorbing paths (considerably longer than are necessary for the ordinarily allowed transitions). In emission they appear only under quite special conditions of excitation. Apart from the weak occurrence of whole band systems in violation of some electronic selection rule, certain branches that are forbidden by the ordinary selection rules may appear very weakly in the bands of an otherwise non-forbidden band system.

The occurrence of forbidden transitions may have one of the three following reasons (see also A.A., p. 154):

1. The selection rule that is violated may hold *only to a first approximation.*

2. The selection rule may hold strictly for dipole radiation but not for *quadrupole radiation* or *magnetic dipole radiation.*

3. The selection rule may hold only for the completely free and uninfluenced molecule and may be violated in the presence of external fields, collisions with other molecules, and the like (*enforced dipole radiation*).

In the case of atomic spectra, in addition, the interaction of electronic motion and nuclear spin has been found to cause the very weak occurrence of certain forbidden lines (for example the line $^1S_0$—$^3P_0$ of Hg). No case has as yet been found for molecules in which this cause is operating.

Violation of approximate selection rules. An example of a selection rule that holds only in a certain approximation is the rule $\Delta S = 0$. It holds less and less strictly with increasing spin-orbit interaction, that is, with increasing atomic number. Actually, *singlet-triplet intercombinations* have been observed even for such a relatively light molecule as CO. In this case there is a $^3\Pi$—$^1\Sigma$ transition (Cameron bands), which can be observed by use of an absorbing path of 1 m or more at atmospheric pressure, whereas a fraction of a millimeter suffices as absorbing path for the corresponding allowed $^1\Pi$—$^1\Sigma$ transition [see Hopfield and Birge (332) and Gerö, Herzberg, and Schmid (252)]. Under certain conditions (at very low pressure of CO, or in a mixture of Ne and CO) the Cameron bands may also be observed in emission in electric discharges [see Hansche (1019), Gerö (968), Rao (1335)]. Similar intercombinations ($^3\Pi$—$^1\Sigma$) have also been observed for the Ga and In halides [Miescher and Wehrli (495) (701)], and the halogen molecules [Mulliken (516)]. For the heavier molecules they represent fairly intense absorption and emission bands [transition to Hund's case (c)]. If the $^3\Pi$ state belongs to Hund's case (b) nine branches are possible in $^3\Pi$—$^1\Sigma$ bands. However, for large multiplet splitting, as case (a) and finally case (c) is approached, on account of the selection rule $\Delta\Omega = 0, \pm1$ only the sub-bands $^3\Pi_0$—$^1\Sigma$ (with two branches) and $^3\Pi_1$—$^1\Sigma$ (with three branches) occur. Theoretical formulae for the intensity distribution in intermediate cases have been given by Budó (141).

An example of a $^3\Sigma_u^+$—$^1\Sigma_g^+$ transition are the Vegard-Kaplan bands of N_2 which occur in the luminescence of solid nitrogen [Vegard (684)] and in certain electric discharges in gaseous nitrogen [Kaplan (396), Wulf and Melvin (1564), Janin (1096)]. Even though these bands involve the ground state of the N_2 molecule they have not yet been observed in absorption. Nor has their fine structure been completely analyzed. As Fig. 129 shows there should be four branches. The intensity distribution in these branches has been derived from theory by Schlapp (1381)(617). Unlike $^2\Sigma$—$^2\Sigma$ or $^3\Sigma$—$^3\Sigma$ transitions here the branches with $\Delta J \neq \Delta K$, that is ^PQ and ^RQ, are not weak but have approximately the same intensity as the corresponding branches with $\Delta J = \Delta K$. Schlapp also showed that $^1\Sigma^- $—$^3\Sigma^+$ or $^1\Sigma^+$—$^3\Sigma^-$ transitions may occur with an intensity of the same order as $^1\Sigma^+$—$^3\Sigma^+$ or $^1\Sigma^-$—$^3\Sigma^-$ transitions but of course with a different structure (branches with $\Delta K = 0$ and $\Delta K = \pm 2$ in place of branches with $\Delta K = \pm 1$). Both types of transition are made possible because spin-orbit interaction

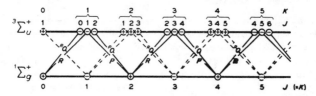

FIG. 129. **Branches of a** $^3\Sigma_u^+$—$^1\Sigma_g^+$ **Band.** The broken-line circles refer to antisymmetric rotational levels which are absent for zero nuclear spin. For a heteronuclear molecule the difference between full-line and broken-line levels and transitions should be ignored. The order of the levels of given K in $^3\Sigma$ is actually different from that shown (see Fig. 101).

causes the $^1\Sigma$ state to "borrow" some of the properties of a $^3\Pi_0$ state or because the $^3\Sigma$ state "borrows" some of the properties of a $^1\Pi$ state.

No example of *doublet-quartet transitions* has yet been found [see Bernstein and Herzberg (798)]. Budó and Kovács (833) have given formulae for the intensity distribution in such bands.

Two other examples of approximate selection rules that may be violated in a higher approximation are the rule that $\Delta K = 0$ is forbidden for Σ—Σ transitions and the rule that

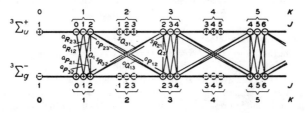

FIG. 130. **Branches of a** $^3\Sigma_u^+$—$^3\Sigma_g^-$ **Band.** Transitions between antisymmetrical levels (broken circles) have been omitted. They would be absent for zero nuclear spin as in O_2 (see also caption of Fig. 129).

Σ^+ does not combine with Σ^-. For Σ^+—Σ^+ and Σ^-—Σ^- bands transitions with $\Delta K = 0$ are also excluded by the $+ \longleftrightarrow -$ rule (V, 35) and are therefore rigorously forbidden for dipole radiation. But, for Σ^+—Σ^- bands transitions with $\Delta K = 0$ (and similarly $\Delta K = \pm 2$) would be allowed by the $+ \longleftrightarrow -$ rule while $\Delta K = \pm 1$ would be rigorously forbidden (see Fig. 114). Both the rule $\Sigma^+ \longleftrightarrow \Sigma^-$ and the rule for ΔK are strictly valid only when electronic and rotational motion can be separated according to (IV, 7). Therefore rotational distortion of the electronic motion may cause the appearance of $^1\Sigma^+$—$^1\Sigma^-$ transitions with $\Delta K = 0$. Such bands would have Q branches only. For non-zero spin, in addition, spin-orbit interaction may be responsible for the occurrence of Σ^+—Σ^- transitions. In this case, as shown by Present (569), also transitions with $\Delta K = \pm 2$ are possible. The near ultraviolet absorption bands of oxygen found by Herzberg (310) represent such a transition, more specifically $^3\Sigma_u^+ \longleftarrow ^3\Sigma_g^-$. Fig. 130 shows the origin of the various branches in this case. Under low resolution these bands appear to consist only of Q branches ($\Delta K = 0$). Under high resolution all the predicted branches have been found [Herzberg (1046)]. It should be emphasized that the absolute intensity of these bands is considerably less than that of the red atmospheric oxygen bands to be discussed below.

For transitions involving multiplet Π, Δ, $\cdots$ states of case (b) branches with $|\Delta K| > 1$ are forbidden. Nevertheless branches with $\Delta K = \pm 2$ are frequently observed (see p. 260) since case (b) represents the coupling conditions only approximately and at least for small K a tendency toward case (a) exists. Such transitions are usually not classified as forbidden transitions.

Quadrupole and magnetic dipole radiation. For quadrupole as well as magnetic dipole radiation the following selection rule holds for the symmetry property $+$ $-$ [see Van Vleck (681)]:

$$+ \longleftrightarrow +, \qquad - \longleftrightarrow -, \qquad - \longleftrightarrow\!\!\!| \; + \qquad\qquad (V, 73)$$

which is just the opposite of (V, 35). For the quantum number J the same selection rule (V, 34) holds for magnetic dipole radiation as for electric dipole radiation, and for the quantum number Λ the rule $\Delta\Lambda = 0, \pm 1$. However, unlike the case of electric dipole radiation, for magnetic dipole radiation, $\Delta\Lambda = 0$ can occur only for intersystem combinations (singlet-triplet transitions) since otherwise the magnetic dipole moment would not change in the transition. For quadrupole radiation one finds the following selection rule for J:

$$\Delta J = 0, \pm 1, \pm 2$$
$$(J = 0 \longleftrightarrow\!\!\!| \; J = 0, \quad J = \tfrac{1}{2} \longleftrightarrow\!\!\!| \; J = \tfrac{1}{2}, \quad J = 1 \longleftrightarrow\!\!\!| \; J = 0). \qquad (V, 74)$$

Thus far only one case of such forbidden transitions is known for a *heteronuclear* molecule, namely, in the $^2\Sigma$—$^2\Pi$ bands of OH. Here very weak satellite branches have been found [Dieke (889), Watson (1513), Jack (1091), Mulliken (1257)] which contradict the $+$ $-$ rule (V, 35) but are in agreement with (V, 73). They correspond to transitions to the "wrong" Λ-doublet component (see Fig. 123). Their intensity ratio to the main branches is of the correct order of magnitude for magnetic dipole radiation [Van Vleck (681)]. Unlike other satellite branches (see p. 244) they have their intensity maximum at about the same J value as the main branches.[12a]

A number of forbidden transitions of the type here considered have been found for *homonuclear* molecules. In each case they yield *forbidden band systems*, not just forbidden branches in allowed band systems. For these molecules, of course, the rule *sym.* $\longleftrightarrow\!\!\!|$ *antisym.* remains unaltered. Therefore, according to (V, 73), as can be seen from Fig. 114(b), only the transitions g—g and u—u can appear as magnetic dipole or quadrupole radiation, whereas the transitions g—u, which are the allowed transitions for ordinary dipole radiation, cannot occur.

The band structure for *magnetic dipole transitions* is the same as for similar electric dipole transitions since the selection rule for J is the same. Even the intensity distribution in the branches is the same. For example, magnetic dipole $^1\Pi_g$—$^1\Sigma_g{}^+$ bands [Fig. 131(a)] have the same structure as electric dipole $^1\Pi_u$—$^1\Sigma_g{}^+$ bands [Fig. 119(a)]. In order to distinguish them one must either know the absolute intensity of the transition—smaller by a factor 10^{-5} for magnetic than for electric dipole radiation—or know the types of the states involved from other evidence. Such other evidence may be based on electron configurations (see Chapter VI) or on the observation of other band systems combining with both the upper and lower state of the band system considered. It is immediately evident that in a cycle of three transitions between three electronic states of a homonuclear molecule at least one must be of the type g—g or u—u, that is, must be due to magnetic dipole radiation. In this way it was shown recently [Herzberg (1041)] that the *Lyman-Birge-Hopfield bands* of N_2 observed in absorption as well as emission (see Fig. 196) represent a $^1\Pi_g$—$^1\Sigma_g{}^+$ transition [see Fig. 131(a)]. This conclusion is also in much better agreement with the electron configuration and with the absolute intensity of the bands than the original assumption, based on the band structure alone, that they represent a $^1\Pi_u$—$^1\Sigma_g{}^+$ transition.

[12a] In more recent work Dieke and Crosswhite (895a) were unable to obtain these forbidden satellite branches.

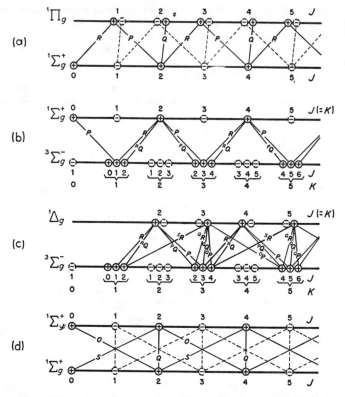

Fig. 131. **Branches of Magnetic Dipole and Quadrupole Bands** (a) $^1\Pi_g$—$^1\Sigma_g{}^+$, (b) $^1\Sigma_g{}^+$—$^3\Sigma_g{}^-$, (c) $^1\Delta_g$—$^3\Sigma_g{}^-$, (d) $^1\Sigma_g{}^+$—$^1\Sigma_g{}^+$. For the sake of clarity transitions between antisymmetric levels (broken circles) have been omitted in (b) and (c). They are absent in the only known examples of each of these (O_2) because of zero nuclear spin. Note that in (b) and (d) the lines $J = 0 \to J = 0$ are missing on account of the restriction (V, 34). (See also caption of Fig. 129.)

Similarly the structure of the well-known *atmospheric oxygen bands*[13] which appear in the red part of the solar spectrum can be equally accounted for on the assumption that they represent an electric dipole $^1\Sigma_u{}^-$—$^3\Sigma_g{}^-$ transition (see Fig. 129) or that they represent a magnetic dipole $^1\Sigma_g{}^+$—$^3\Sigma_g{}^-$ transition [see Fig. 131 (b)]. In either case there are four branches as observed. However, both the electron configuration [Mulliken (510) (514)] and the low absolute intensity of the bands [Van Vleck (681)]—requiring the whole atmosphere to give strong absorption—decide definitely in favor of the second alternative.[14] The absolute transition probability A_{nm} for these bands has been found by Childs and Mecke (154) to be about 0.14 $\sec^{-1}$ [see also van de Hulst (1487)]. This may be compared to 10^7–10^8 $\sec^{-1}$ for ordinary electric dipole radiation and 10^2–10^3 $\sec^{-1}$ for magnetic dipole radiation. The reason that the observed transition probability is only about one-thousandth of

---

[13] Recently Kaplan (1119) and Herman, Hopfield, Hornbeck, and Silverman (1035a) have obtained these bands in emission in a certain afterglow and in the CO—O_2 flame, respectively.

[14] An independent proof would be possible from an investigation of the Zeeman effect. The Zeeman effect of the atmospheric bands has been investigated by Schmid, Budó, and Zemplén (1383) but the resolution was not sufficient to obtain such a proof.

that expected for magnetic dipole radiation is the fact that the atmospheric bands represent a singlet-triplet intercombination.

A further example is provided by the infrared atmospheric oxygen bands at 1.27 μ and 1.07 μ which represent a magnetic dipole $^1\Delta_g$—$^3\Sigma_g^-$ transition [Ellis and Kneser (203), L. and G. Herzberg (312)(1053), Van Vleck (681)]. Fig. 131(c) shows the possible branches. On account of the presence of a rotational level for each J value in the $^1\Delta_g$ state, in addition to the four branches with $\Delta K = \pm 1$ which are the same as for $^1\Sigma_g^+$—$^3\Sigma_g^-$, three branches with $\Delta K = 0$ and one branch each with $\Delta K = +2$ and -2 are expected and are indeed observed. The same band structure would be obtained for an electric dipole $^1\Delta_u$—$^3\Sigma_g^-$ transition; but again the low intensity, which is even less than for the $^1\Sigma_g^+$—$^3\Sigma_g^-$ system, and the electron configuration (see Chapter VI) prove that the transition is $^1\Delta_g$—$^3\Sigma_g^-$.

Both the red and infrared oxygen bands are also allowed as quadrupole radiation. However, the contribution of quadrupole radiation to the transition probability is expected to be only 10^{-3}–10^{-4} of that of magnetic dipole radiation [see Van Vleck (681)]. This contribution could be detected only by way of the branches with $\Delta J = \pm 2$ which cannot occur for pure magnetic dipole radiation. No such branches have as yet been observed even though for the red bands absorbing paths 10^4 times as large as necessary to produce the branches with $\Delta J = 0, \pm 1$, have been used.

As a further example Fig. 131(d) shows the branches of a $^1\Sigma_g^+$—$^1\Sigma_g^+$ transition which can occur only as quadrupole radiation since a $^1\Sigma$ state has no magnetic dipole moment. Here, different from the other cases, no P and R branches are possible (on account of the symmetry selection rules) but Q, S, and O branches occur corresponding to $\Delta J = 0, +2$, and -2. No such transition has as yet been observed.

A homonuclear molecule in its ground state has in general a non-zero quadrupole moment which changes during the vibrational motion. Therefore, as mentioned previously, *a quadrupole rotation-vibration spectrum* may occur [Herzberg (315)] while the ordinary dipole rotation-vibration spectrum is, of course, rigorously forbidden. The structure of quadrupole rotation-vibration bands is exactly like that of $^1\Sigma_g^+$—$^1\Sigma_g^+$ electronic bands as shown in Fig. 131(d) assuming that the ground state of the molecule is a $^1\Sigma_g^+$ state; that is, there is one Q, one S and one O branch. Calculation shows [James and Coolidge (362)] that the intensity of this quadrupole rotation-vibration spectrum is about 10^{-9} of that of ordinary dipole rotation-vibration spectra of heteronuclear molecules and thus requires even in favorable cases an absorbing path of several kilometers. For H_2 several lines of the 2–0 and 3–0 bands of this quadrupole rotation-vibration spectrum have recently been found by Herzberg (1044)(1045) with the predicted intensity.

The transitions with $\Delta J = 0$ in Fig. 131(d) would also be in conformity with the magnetic dipole selection rules but a magnetic dipole rotation-vibration spectrum cannot arise for homonuclear molecules with $^1\Sigma_g^+$ ground states since no magnetic dipole moment is present. However, for O_2 with its $^3\Sigma_g^-$ ground state a magnetic dipole contribution to the rotation-vibration spectrum is possible. No spectra of this type have been observed.

Still another magnetic dipole absorption of the oxygen molecule has recently been recognized and observed in the *microwave* region: transitions within a triplet of levels of given K of the $^3\Sigma_g^-$ ground state [Van Vleck (1488), Beringer (795), Lamont (1163)]. Since the three components of such a triplet level have the same $+$ $-$ symmetry and since they are distinguished by the orientation of the electron spin S and its associated magnetic moment, transitions from $J = K \pm 1$ to $J = K$ are possible as magnetic dipole radiation. There is no change of the rotational or vibrational state during such transitions. They might perhaps best be described as a *spin-reorientation spectrum*. Since the splitting between the levels $J = K \pm 1$ and $J = K$ is approximately 2 cm^{-1} in O_2 (except for $K = 1$; see p. 224) all these transitions, for the different K values (except $K = 1$) almost coincide at a wave length of 0.5 cm, as has been observed.[14a]

[14a] In the recent work of Strandberg, Meng and Ingersoll (1437a) the absorption at 0.5 cm has been partially resolved into its components corresponding to different K values.

Enforced dipole radiation. Thus far, only a few examples of enforced dipole radiation are known for diatomic molecules, and they have not been explained in detail. One example is provided by the various continuous absorption spectra of bromine which are greatly enhanced (up to more than 100%) when foreign gases are added to bromine vapor [Bayliss and Rees (790)] or in going from the vapor to the liquid or solutions [Porrett (1320), Aickin, Bayliss, and Rees (745)]. Apparently for a molecule in the neighborhood of disturbing molecules the selection rules are less strict. The effect is different for different electronic transitions [see Mulliken and Rieke (1268)].

Further examples are found in the absorption spectrum of oxygen at high pressure and in the liquid and solid states: The $^1\Delta_g$—$^3\Sigma_g^-$ system is very greatly strengthened relative to the $^1\Sigma_g^+$—$^3\Sigma_g^-$ system and in the ultraviolet a new system of bands occurs which is similar to but not identical with the $^3\Sigma_u^+$—$^3\Sigma_g^-$ system. However, the situation is complicated here by the formation of O_4 molecules. For more details see Ellis and Kneser (202), Salow and Steiner (612), and Finkelnburg (219).

Condon (161) has discussed theoretically the possibility of the occurrence of rotation-vibration spectra of homonuclear molecules due to the presence of an external electric field. Recently Crawford, Welsh and Locke (873a) have observed the fundamentals of N_2, O_2, and H_2 in absorption at high pressures and found an increase of intensity proportional to the square of the pressure. Similar absorptions were found in liquid O_2 and N_2 by Oxholm and Williams (1300b). It is probable that these absorptions are induced by the collisions in a way closely related to that envisaged by Condon for an external field.

4. Perturbations

Observed phenomena. As has been mentioned on several previous occasions, in some bands perturbations are observed in the otherwise smooth course of the branches.[15] In such cases one line or several successive lines deviate more or less strongly from formula (II, 8). Sometimes even a splitting into two lines appears, or for multiplet bands the multiplet splitting may be abnormally great at a certain place in the band. A type of perturbation also appears in which the intensity is abnormally small for one or more lines in the band. *Displacement* and *weakening* in intensity may also appear simultaneously. When the perturbations appear for a number of successive J values, they usually have a resonance-like behavior; the deviation from the normal course increases rapidly to a maximum with increasing J and then decreases rapidly again to zero.

As an example, a spectrogram of the $^2\Sigma$—$^2\Sigma$ CN band $\lambda 3921$ Å is reproduced in Fig. 132(a). The deviation (in cm^{-1}) from the regular course in the P branch is represented graphically in Fig. 132(b). It can be seen clearly from the spectrogram as well as from the graphical representation that the doublet splitting, which is unnoticeable for small K, suddenly increases very rapidly at $K = 12$ and then, with a change of sign, decreases again to a small value. However, only the one doublet component is influenced.

It is clear that the perturbations in the band fine structure are due to *perturbations in the rotational term series* either of the upper or of the lower state. Therefore, if a perturbation appears, for example, at a certain point in a P

[15] Compare also the perturbations appearing in atomic spectra (see A.A., p. 170).

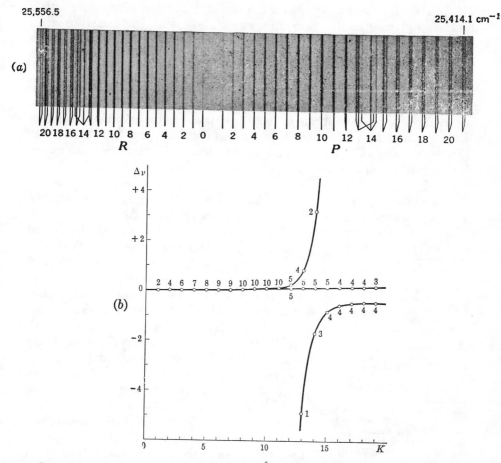

FIG. 132. Perturbation in the 11–11 Band λ3921 Å of the Violet CN System after Jenkins (366). (a) Spectrogram.* (b) Graphical representation of the perturbation in the *P* branch. The long leading lines in (a), on which the *K* values are written, give the unperturbed doublet components, and the shorter lines give the perturbed components. In each case the latter are joined to the former. In (b) the deviation $\Delta\nu$ of the lines of the *P* branch from the regular course is given. The numbers on the curve are the estimated intensities of the respective lines.

* The author is greatly indebted to Professor F. A. Jenkins for this spectrogram.

branch of a band, it will also be observed for the same J' or J'' value in the corresponding R branch of the band and will have the same type and magnitude. This is clearly shown by Fig. 132(a) for the CN band. Furthermore, the perturbation will appear also in all other bands that have the same perturbed vibrational state as upper or lower state.

The magnitude of the perturbations is sometimes considerable. Deviations up to 20 cm^{-1} from the normal position have been observed. In rare cases, so many and such strong perturbations may be present in a band that the rotational analysis presents great difficulties, since the branches are no longer easily discernible. It should be emphasized, however, that the combination relations previously given (except where otherwise stated) must hold exactly even if strong perturbations are present, and that therefore, in principle, an analysis is always possible [an example of such a complicated case has, for instance, been described by Gerö (251)].

Apart from these perturbations in the rotational structure of bands, deviations of whole bands from the regular course of a band progression have been observed; this means that one or more vibrational levels have abnormal positions. Such perturbations have been found, for example, for N_2, CS, CN, CP, S_2, P_2, and other molecules, and are called *vibrational perturbations*, as distinguished from the aforementioned *rotational perturbations*.

General considerations. As we have seen in Chapter I, the introduction of an interaction term into the wave equation produces *shifts of the zero-approximation energy levels* (that is, of the levels that result from the approximate wave equation without the interaction term). The shift of a given level depends inversely on its separations from the other energy levels. The smaller the separation between two levels the larger are their shifts on account of the interaction term; these shifts are always in the sense of a *"repulsion"*; that is, the higher level is displaced upward and the lower downward by an equal amount. In addition, each of the two states assumes properties of the other; both states form a kind of *hybrid*.

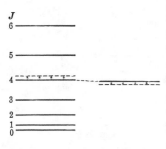

FIG. 133. **Explanation of Perturbations.** The level $J = 4$ of the left series is perturbed by the level to the right, which has nearly the same energy but belongs to a different series of levels. The direction of the shift (repulsion) is indicated by the small arrows.

The mutual repulsion of two states of approximately equal energy just mentioned immediately gives an explanation for the occurrence of perturbations in band spectra since it can easily happen that a level belonging to a different electronic state has the same energy as a level of a given series. Thus, as illustrated by Fig. 133, the latter will be displaced from its "normal" position.

General formulae for the perturbed energy levels have been given previously both for the case of several interacting non-degenerate levels [equation (I, 27)] and for the case of two levels that in zero approximation are degenerate with each other [equation (I, 33)]. It remains to give a formula for the slightly more general case, particularly appropriate to the present discussion, in which two levels in zero approximation are close together but do not coincide. The energy difference in zero order may be considered as due to a perturbation in which matrix elements of the type W_{12} in (I, 33) have been neglected but terms of the type W_{11} and W_{22} have been included, so that in the actual perturbation calculation only terms of the type W_{12}, W_{21} need be considered. It is readily seen that (I, 33) then assumes the form

$$\begin{vmatrix} E_1 - E & W_{12} \\ W_{21} & E_2 - E \end{vmatrix} = 0 \qquad (V, 75)$$

where E_1 and E_2 are the "unperturbed" energies. From this equation one obtains for the perturbed energy E (with $W_{21} = W_{12}{}^*$)

$$E = \tfrac{1}{2}(E_1 + E_2) \pm \tfrac{1}{2}\sqrt{4\,|W_{12}|^2 + \delta^2} \qquad (\mathrm{V}, 76)$$

where $\delta = E_1 - E_2$ is the separation of the unperturbed levels. According to (V, 76) the shifts of the two levels $E - E_1$ and $E - E_2$ are equal in magnitude $[\tfrac{1}{2}(\sqrt{4\,|W_{12}|^2 + \delta^2} - \delta)]$ but of opposite sign (repulsion). The larger the matrix element W_{12} the greater are these shifts. For a given W_{12} they are largest for $\delta = 0$ and decrease with increasing δ. In Fig. 134 the positions of the perturbed levels are plotted as a function of δ. The unperturbed levels are indicated by the broken lines.

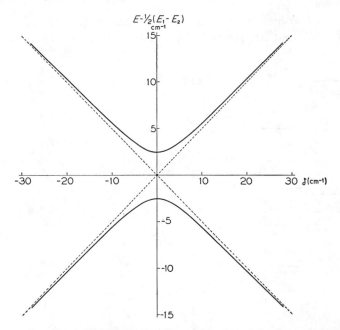

Fig. 134. **Perturbation of Two Energy Levels as a Function of the Separation of the Unperturbed Levels.** The broken lines represent the positions of the unperturbed levels, the solid curves those of the perturbed (actual) levels. A value $W_{12} = 2.5$ cm^{-1} has been assumed.

From (I, 31) and with proper normalization ($c_{11}{}^2 + c_{12}{}^2 = 1$) one finds for the eigenfunctions of the perturbed levels[16]

$$\psi_a = c\psi_1 - d\psi_2$$
$$\psi_b = d\psi_1 + c\psi_2 \qquad (\mathrm{V}, 77)$$

where

$$c = \left(\frac{\sqrt{4\,|W_{12}|^2 + \delta^2} + \delta}{2\sqrt{4\,|W_{12}|^2 + \delta^2}}\right)^{\!\tfrac{1}{2}} \qquad d = \left(\frac{\sqrt{4\,|W_{12}|^2 + \delta^2} - \delta}{2\sqrt{4\,|W_{12}|^2 + \delta^2}}\right)^{\!\tfrac{1}{2}} \qquad (\mathrm{V}, 78)$$

For $\delta = 0$ we obtain $c = d = 1/\sqrt{2}$, that is, we have a fifty-fifty mixture. If δ is large, c approaches 1, while d approaches 0, that is, $\psi_a \to \psi_1$, and $\psi_b \to \psi_2$.

---

[16] In case W_{12} is complex and has a phase factor $e^{i\alpha}$ the constant d should be multiplied by $e^{-i\alpha}$ in the first and $e^{+i\alpha}$ in the second equation [see Dieke (894)].

The case of a mutual perturbation of three nearly coinciding levels has been discussed in detail by Kovács and Singer (1139b).

Selection rules for perturbations. According to the preceding considerations the magnitude of a perturbation depends not only on the smallness of the energy difference of the unperturbed levels but also on the magnitude of the matrix element

$$W_{12} = \int \psi_1{}^* W \psi_2 \, d\tau \qquad\qquad (V, 79)$$

of the perturbation function W, that is, it depends on the eigenfunctions of the two states involved. An approximate equality of the energy arises very fre-

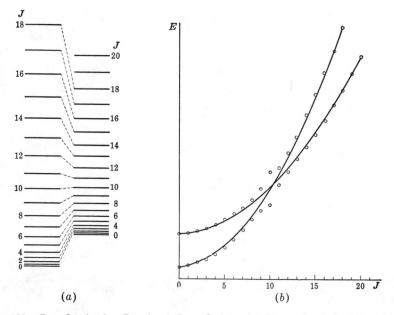

(a) (b)

Fig. 135. **Two Overlapping Rotational Term Series.** (a) Energy level diagram. (b) Term curves. In (a), only the unperturbed levels are drawn, which are represented in (b) by the full-line curves. The position of the perturbed levels is indicated in (b) by the circles. However, the deviations from the curves have been drawn to a larger scale than the curves themselves.

quently in a molecule. As shown by Figs. 69 and 135(a) there are numerous rotational levels of one set which have nearly the same energy as those of another set belonging to the same or to a different electronic state [for example in Fig. 135(a) the levels $J = 7, 8, 9, 10, 12, 14,$ and 16 of the left-hand set nearly coincide with $J = 4, 6, 8, 10, 13, 16,$ and 19, respectively, of the right-hand set]. Yet in most of these cases no perturbations occur because W_{12} vanishes. The conditions for non-vanishing W_{12}, that is, the *selection rules* for perturbations have been derived by Kronig (421) in so far as they are determined by the quantum numbers and symmetry properties of the two states considered.   These selection rules are:

(1) Both states must have the same total angular momentum J; that is, $\Delta J = 0$.

(2) Both states must have the same multiplicity; that is, $\Delta S = 0$.

(3) The Λ values of the two states may differ only by 0 or ± 1 ($\Delta \Lambda = 0$, ± 1).

(4) Both states must be positive or both must be negative; that is, $+ \leftrightarrow -$.

(5) For identical nuclei, both states must have the same symmetry in the nuclei; that is, $s \leftrightarrow a$.

The first, fourth, and fifth rules are perfectly rigorous.   The second rule ($\Delta S = 0$) holds only approximately, just as for transitions with radiation, that is, with increasing multiplet splitting perturbations between states of different multiplicity increase in magnitude.   The third rule ($\Delta \Lambda = 0, \pm 1$) holds of course only as long as Λ is defined, that is, in Hund's cases (a) and (b).   In Hund's case (c) it must be replaced by $\Delta \Omega = 0, \pm 1$.   In Hund's case (b) the quantum number K of the total angular momentum apart from spin is defined. It is clear that for it the same selection rule must hold as for J, that is, $\Delta K = 0$, provided that pure case (b) applies.   However, with increasing coupling of S with the internuclear axis this rule will break down.

Perturbations with $\Delta \Lambda = 0$ are perturbations between states of the same type.   Mulliken (522) therefore called them *homogeneous* perturbations in contrast to the perturbations with $\Delta \Lambda = \pm 1$ which he called *heterogeneous* perturbations.[17]   The homogeneous unlike the heterogeneous perturbations may occur even in an approximation in which the finer interaction of rotation and electronic motion is neglected.

The selection rules (4) and (5) follow immediately from (V, 79) and from the fact that the perturbation function W remains unchanged for an inversion and for an exchange of identical nuclei.   The proof of the other selection rules requires somewhat more elaborate derivations [see Kronig (421)].

In Kronig's selection rules the vibrational quantum number v is not considered.   As will be shown below while there is no strict selection rule for v an analogue to the Franck-Condon principle holds [see Hulthén (343), Weizel (704), Herzberg (309), and Dieke (183)]: *Two vibrational states belonging to two different electronic states and lying at approximately the same height will influence each other strongly only if classically the system could go over from the one state to the other without a large alteration of position and momentum*—that is, if the levels lie in the neighborhood of the intersection of the potential curves of the two electronic states.   This is illustrated by Fig. 136(a) and (b).   According to the

---

[17] Dieke (185) designated them as B and A perturbations, respectively, and more recently (894) as vibrational and rotational perturbations.   We are using here the latter designations in a different way (see below).

above rule, no great perturbation can take place between the levels A and B in these figures, whereas a large perturbation is to be expected between the levels C and D, provided that also the Kronig selection rules are fulfilled.

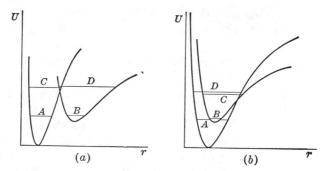

Fɪɢ. 136.   **Potential Curves Explaining the Franck-Condon Principle in Perturbations.**

The matrix element W_{12} which according to (V, 76) determines the magnitude of the perturbation can be split into two parts, one depending on the electronic and rotational eigenfunctions and the other depending on the vibrational eigenfunction (similar to the transition moment for radiation; see p. 200 f.). The latter is found to be

$$W_{12}^v = \int \psi_1^v W^v \psi_2^v \, dr, \tag{V, 80}$$

where W^v is the part of the interaction energy depending on the internuclear distance; that is, W^v represents the finer interaction between vibration and electronic motion. According to (V, 80), a *strong perturbation* will occur only *if the vibrational eigenfunctions overlap in a suitable manner.*

For the levels A and B in Fig. 136 (a), the eigenfunctions practically do not overlap at all, which means that W_{12}^v is very nearly zero and a perturbation does not occur, even if the resonance is very sharp. The same holds for the levels A and B in Fig. 136 (b), although to a smaller extent, since the vibrational eigenfunctions between the turning points are not zero. In this case, for favorable energies of the two levels, a small rotational perturbation may become noticeable. On the other hand, for the levels C and D in the neighborhood of the points of intersection of the potential curves [Fig. 136 (a) and (b)] a very favorable mutual overlapping of the eigenfunctions occurs, giving a large value to the integral (V, 80). Therefore, if the Kronig selection rules are also fulfilled, a large perturbation can appear. In such cases, the perturbation can be quite considerable, even if the energy condition $E_1 \approx E_2$ is not very well fulfilled.

Rotational perturbations. We shall first apply Kronig's selection rules to the *mutual perturbation of two rotational term series.* As may be seen from Fig. 135 (a), the first selection rule obviously imposes a considerable restriction on the possibilities of perturbation. A noticeable perturbation will take place only if, *for one and the same J in both states,* the levels have about the same energy. In the figure this is the case for $J = 10$. The separation of corresponding levels is appreciably greater for higher and lower J values, and therefore the perturbation is considerably smaller if at all noticeable. The positions of the levels expected as a result of the mutual repulsion of levels with equal J are given qualitatively in Fig. 135(b) for the two different rotational term series of Fig. 135(a). Naturally, in this figure it is assumed that the other selection rules (2)–(5) are also fulfilled (see below). The full-line curves give the regular course of the levels which would be expected without allowing for the perturbation; the small circles give the actual perturbed levels (plotted on a different scale). It is seen that in each term series the deviations from the normal course increase at first in a resonance-like manner and then jump over to the other side of the smooth curve and decrease again rapidly. This behavior is in striking agreement with observations (see Fig. 132).

The width of the perturbation region depends clearly on the relative spacing of the two sets of rotational levels, that is, on the respective B_v values. If the difference of the B_v values is large the energy difference of levels with the same J increases rapidly above and below the point of best resonance (maximum perturbation) and therefore a perturbation may be noticeable only in a very few, sometimes only one rotational level, particularly if W_{12} is small. A number of such cases have been found.

As mentioned before the shifts of two interacting levels are *equal in magnitude but opposite in sign*. This must hold for the whole course of the perturbations in two interacting term series (see Fig. 135). Experimentally, usually only one (perturbed) term series is observed; but in a few cases—for example, CN [Rosenthal and Jenkins (603)], He₂ [Dieke (183)], CuH [T. Heimer (292)], and BeO [Lagerqvist (1156)] —in addition to the perturbed levels, the perturbing levels have been found by means of other band systems. In all these cases the shifts have actually been found to be opposite and equal in the two states involved.

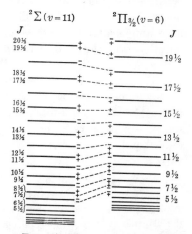

Fig. 137. **Energy Level Diagram for the Perturbation in the CN Band shown in Fig. 132.** The spin doublets of the $^2\Sigma$ state, as well as the Λ doublets of the $^2\Pi$ state, are not resolved in the drawing. However, the splitting is indicated by giving the symmetry $+ -$, as well as the J values for the spin doublets. The levels which, according to the selection rules, can perturb each other are joined by broken lines. The vibrational quantum number of the $^2\Pi_{\frac{3}{2}}$ state should be changed to 7 [see Herzberg and Phillips (1051)].

The perturbation in the CN band reproduced in Fig. 132(a) is due to a perturbation in the lower state ($^2\Sigma$) as shown by the observation of the same perturbation in other bands with the same v'' and by the absence of a corresponding perturbation in other bands with the same v'. The perturbation is caused by the $v = 7$ level of the $^2\Pi_{3\frac{1}{2}}$ component of the $^2\Pi$ upper state of the red CN bands. The corresponding perturbation in these latter bands has also been observed [Rosenthal and Jenkins (603)]. The two term series are represented graphically in Fig. 137. It can be seen that, on the basis of the selection rule

$\Delta J = 0$, only one component of the $^2\Sigma$ state will be perturbed in the neighborhood of $K = 14$, as is observed (Fig. 132). The $^2\Pi_{1/2}$ component of the $^2\Pi$ state produces perturbations of $^2\Sigma$ at different K values.

The selection rule $\Delta J = 0$ discussed so far determines the broad features of all perturbations. The other Kronig selection rules (2)–(5) (p. 285) prevent the occurrence of perturbations between certain electronic states or certain components of electronic states. Thus according to the second selection rule ($\Delta S = 0$), for example, a triplet state cannot perturb a singlet state; according to the third selection rule ($\Delta \Lambda = 0, \pm 1$), for example, a Δ state cannot perturb a Σ state (or vice versa). From the fourth selection rule it follows, for example (compare Fig. 114), that a $^1\Sigma^-$ state cannot perturb a $^1\Sigma^+$ state, since for equal J the $+ -$ symmetry is not the same. From the same selection rule it also follows that a $^1\Sigma$ state can perturb only one Λ-doublet component of a $^1\Pi$ state, since the $+ -$ symmetry for equal J is the same only for one Λ component. On the other hand, both Λ components are influenced by the mutual perturbation of two $^1\Pi$ states.

In the case of multiplet states the relations are somewhat more complicated. For example, a $^2\Sigma$ state can perturb both Λ components of a $^2\Pi$ state but not at the same J value. Thus, as shown by Fig. 137, in the example of CN near $J = 13\frac{1}{2}$ only one Λ component of the $^2\Pi_{3/2}$ state is perturbed, but for higher J values (above those shown in the figure) the other will be perturbed. The first perturbation is caused by the F_2 levels of $^2\Sigma (J = K - \frac{1}{2})$, the second by the F_1 levels $(J = K + \frac{1}{2})$. A similar pair of perturbations occurs for $^2\Pi_{1/2}$. As another example consider the mutual perturbation of two $^2\Sigma$ states. If both are $^2\Sigma^+$ there will be a pair of strong perturbations $F_1 \leftrightarrow F_1$ and $F_2 \leftrightarrow F_2$ (assuming of course that the energy condition is fulfilled); if one state is $^2\Sigma^+$ and the other $^2\Sigma^-$ no perturbation would occur if the rule $\Delta K = 0$ were rigorous (see Fig. 114). But since it is not, a pair of weak perturbations, $F_1 \rightarrow F_2$ and $F_2 \rightarrow F_1$, may occur. More detailed discussions of such cases including intercombinations, have been given by Ittmann (357), Coster and Brons (166), Brons (131), Kovács (420), and Budó and Kovács (142).

Both for singlet and multiplet states it follows from the fourth and fifth selection rule that if the nuclei are alike, the states perturbing each other must *either both be even or both be odd* (see Figs. 114 and 115). For example a Π_u state cannot perturb a Π_g state. This rule holds rigorously.

As mentioned previously homogeneous perturbations ($\Delta \Lambda = 0$) occur even in an approximation in which the finer interaction of rotation and electronic motion is neglected. Therefore, for them the matrix element W_{12} in a first approximation is independent of J. This has been assumed to be the case in Fig. 135(b). Heterogeneous perturbations ($\Delta \Lambda = \pm 1$) can arise only on account of the finer interaction of rotation and electronic motion and therefore, as has been shown by Kronig (421), the matrix element W_{12} is proportional to J. Consequently the magnitude of such a perturbation increases with increasing J and is zero for $J = 0$ even if the two levels are in very close resonance. This dependence of W_{12} on J does not change the qualitative features of a perturbation, that is, the existence of a maximum shift at the point of closest resonance. However, it has the effect that the shift does not quite go back to zero at higher

J values as it does for homogeneous perturbations [Fig. 135(b)]. This difference makes it possible to distinguish experimentally between the two cases $\Delta\Lambda = 0$ and $\Delta\Lambda = \pm 1$ [see Ittmann (357), Brons (131), and Dieke (185)]. For example in the case of the perturbation in the $^2\Sigma$ state of CN represented in Fig. 132 the shift does not go back to zero, that is, we must have $\Delta\Lambda = \pm 1$, a conclusion that has been confirmed by direct observation of the perturbing state which is $^2\Pi$ (compare Fig. 137).

The preceding considerations show that a closer study of the perturbations found in a certain electronic state may lead to rather definite conclusions with regard to the type of the perturbing state even if this state cannot be directly observed.

In favorable cases it is even possible to obtain approximate values for the *rotational and vibrational constants of the perturbing state* without the observation of bands involving this state. Three different methods have been suggested for this purpose:

(a) As has been emphasized before the shifts of corresponding levels of the perturbed and the perturbing state are opposite and equal. Therefore the mean of the actual energy levels follows a simple quadratic formula

$$\overline{T} = \tfrac{1}{2}C + \tfrac{1}{2}(B_1 + B_2)J(J + 1) \tag{V, 81}$$

where C is the energy difference of the two term series for $J = 0$ and where B_1 and B_2 are the unperturbed B values of the two series. Frequently in the perturbation region lines due to the perturbing as well as to the perturbed state are observed [see Fig. 132(b) and below]. If at least two such pairs are observed their mean wave numbers substituted in (V, 81) give values of C and B_1 assuming that B_2, the rotational constant of the perturbed state. is known. For a more detailed discussion of this method the reader is referred to Schmid and Gerö (62) who first suggested it and to Kovács (419)

(b) Coster and Brons (166) have made use of the separation of the three points of maximum perturbation produced by a $^3\Sigma$ state in a $^1\Pi$ state. It is readily seen that if J_1, J_2, J_3 are the J values in the $^1\Pi$ state where maximum perturbation occurs the corresponding K values in the $^3\Sigma$ state are $J_1 - 1$, J_2, and $J_3 + 1$. From that and the energy values of the points of perturbation the rotational constant of the $^3\Sigma$ state can be obtained. This method is applicable only for certain types of perturbations.

(c) Since both the perturbed and the perturbing electronic state has a series of vibrational levels it is to be expected that a perturbation will occur more or less periodically with increasing vibrational quantum number depending on the spacing of the vibrational levels in the two states. Such a periodicity was first observed for the $^2\Sigma$ —$^2\Sigma$ bands of the N_2^+ molecule for which characteristic perturbations occur for $v' = 0, 1, 3, 5, 9, 13$ [Herzberg (302), Coster and Brons (169), Childs (847), Parker (1304), Brons (830)(131)]. Schmid and Gerö (250) (623) have given a convenient method for representing the perturbations in the different vibrational levels. The values $T_e + G(v) + F(J)$ of the different vibrational levels of the perturbing and the perturbed electronic states are plotted against $J(J + 1)$. Two sets of intersecting lines for the two electronic states are obtained, as is illustrated by Fig. 138, which refers to CN [compare similar diagrams for BeO given by Lagerqvist and Westöö (1156)(1160)]. The points of intersection correspond to positions of equal J for the two states—that is, to positions of maximum perturbation. Actually, in the case of CN, because of the doublet character of the two states involved each point of intersection corresponds to a group of four perturbations. If in such a diagram only one set of lines corresponding to one electronic state is known but with the points of perturbation produced by another state, the set of lines corresponding to this other state can be drawn in and its rotational and vibrational constants can thus be determined approximately.

Thus far we have ignored the effects of the *change of the eigenfunctions* produced by the perturbation. As previously mentioned each of the eigenfunctions of the perturbed levels is a mixture of the eigenfunctions of the two unperturbed levels with coefficients as given in (V, 77) and (V, 78). The closer the unperturbed levels are together the more nearly a 50–50 mixture arises.

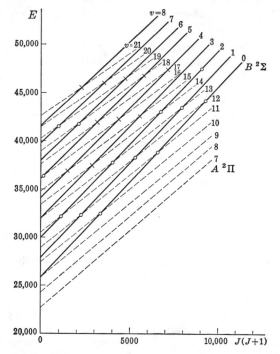

Fig. 138.   **Rotational Energy Levels and Perturbations in the $B\,^2\Sigma$ and $A\,^2\Pi$ States of the CN Molecule [Adaptation of a Figure by Schmid, Gerö, and Zemplén (624a)].**   The ordinates give the energy above the ground state of CN; the abscissae are $J(J+1)$.   The solid straight lines represent the energies of the rotational levels of the $B\,^2\Sigma$ state, and the broken straight lines represent those of the $A\,^2\Pi$ state.   The small circles indicate the points of observed perturbations; the short dashes indicate points of intersection for which no perturbations have been observed because the corresponding bands have not been analyzed.   Since this diagram was prepared an additional point of perturbation has been found by Wager (1505) in the $v = 0$ level of the $B^2\Sigma$ for low J values, in agreement with expectation from the diagram.

As a result of this mixing of the eigenfunctions, the perturbed level assumes the properties of the perturbing level, and vice versa; and for very close resonance it is no longer possible to say unambiguously which of the perturbed levels belongs to a given unperturbed level.   The most important consequence of the mixing is the appearance of *extra lines* in the perturbed branches.   The way in which these extra lines arise is illustrated by Fig. 139 in which B and A are, respectively, the upper and the lower state of a band and C the state that perturbs the upper state.   The normal lines of the R branch are indicated by full vertical lines.   The corresponding lines of the transition $C \rightarrow A$ are indicated by broken lines.   Normally they do not occur or are very much weaker than the lines of the transition $B \rightarrow A$ (for example, on account of the Franck-Condon principle).   However, in the perturbation region (near $J = 8$ in Fig. 139) state C partially assumes properties of B and therefore does combine strongly with A leading to the occurrence of a few lines of the R branch of the transition $C \rightarrow A$.   Similar considerations apply to the other branches.   Such extra lines are clearly recognizable in the example of Fig. 132(a): There are three lines each for $K = 13$ and $K = 14$ in both branches instead of the ordinary two.

A study of Fig. 134 readily shows that the extra lines must fit into smooth curves with the ordinary lines, corresponding either to the upper or the lower curve of this figure; and indeed in the example of CN this is found to be the case [see Fig. 132(b)]: In addition to the unperturbed doublet component (horizontal row of observed points) there are for $K = 13$

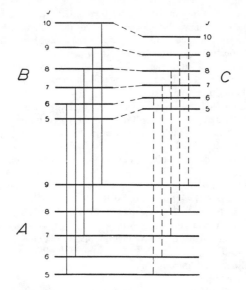

FIG. 139.   **Energy Level Diagram Explaining Extra Lines in Perturbations.**

and 14 observed points both in the upper and lower curve.   Usually one designates the weaker of the two lines as "the extra line" which in the example is in the lower curve for $K = 13$ and in the upper for $K = 14$.   But this distinction is quite formal.   The separation of the extra line from the regular line equals the separation of perturbed and perturbing level for the particular J value.

Since in the region of perturbation the eigenfunctions of the upper levels of the "normal" lines ($B \rightarrow A$ in Fig. 139) are mixed with those of the perturbing levels (C) which do not combine with the same lower state (A), the intensity of the "normal" lines is reduced compared to what it would be without the perturbation.   As has been shown by Dieke (894) the sum of the intensities of each pair of normal and extra lines is the same with and without perturbation.   Thus one may say that the extra lines *"borrow"* their intensity from the regular lines. In the example Fig. 132 it is seen that these intensity rules are qualitatively fulfilled.   In favorable cases when both δ and W_{12} in (V, 76) and (V, 78) are small a large *perturbation of intensity* may occur even for a small (and possibly not noticeable) perturbation of the position of the line.   For quantitative intensity formulae which are somewhat different for homogeneous and heterogeneous perturbations the reader is referred to Dieke (894).

The preceding discussion of intensities in perturbations applies rigorously only to absorption spectra or emission spectra produced by thermal excitation.   Large deviations may occur when the excitation mechanism favors one of the mutually perturbing states.   Suppose that the electronic state to which C in Fig. 139 belongs is n times more often excited than B.   In this case, assuming the same life time for B and C, all lines with an admixture of C, that is, all the perturbed lines in the band $B \rightarrow A$ will have an increased intensity and particularly the extra lines will be anomalously strong. The same result obtains when the excitation of the two states is the same but the life time of C is n times that of B, provided that collisions keep up a normal distribution of C during its life time. A very strong anomaly of this kind was observed in CN by Herzberg (1037) and first correctly explained by Beutler and Fred (799) [see also Wager (1505)]. Here the lines with $K' = 4$, 7, and 15 under certain conditions of excitation predominate over all other band lines and these lines are just the ones that are perturbed and accompanied by extra lines (unresolved at low dispersion). If the excitation conditions are opposite to the ones assumed above, anomalously small intensities of the perturbed lines may occur. Such cases have been found for example in the red CN bands at the places corresponding to the enhancements observed in the violet bands [see Wager (1505)].

If two mutually perturbing states are fairly widely separated the magnitude of the perturbation is of course small. The separation δ of corresponding levels with the same J is very nearly constant and therefore, if $\Delta\Lambda = 0$, the shift on account of the perturbation is practically the same for all J values: We have a (very small) vibrational perturbation (see below). On the other hand if $\Delta\Lambda = \pm 1$ the shift increases quadratically with J, that is, the rotational constants B of the two states are slightly altered by the perturbation. In particular if a $^1\Pi$ state is perturbed by a $^1\Sigma$ state (lying at some distance from it) according to the selection rule (4) only one Λ component is shifted by amounts that increase quadratically with J and thus a splitting of the Λ-type doublets results. This is an ordinary Λ-*type doubling*. Indeed all cases of Λ-type doubling are produced in this way; they are special cases of perturbations and like all other perturbations are due to the finer interaction of electronic motion and rotation in the molecule. For large δ the shift produced by a perturbation is inversely proportional to δ as can be easily seen from (V, 76) [see also (I, 27)]. That is why the magnitude of the Λ-type doubling of a $^1\Pi$ state is inversely proportional to the separation from the nearest $^1\Sigma$ state (see p. 229). The theory of the Λ-type doubling by Van Vleck and Kronig referred to earlier (p. 227) is based on a perturbation calculation. Similar considerations apply to the theory of spin uncoupling and L-uncoupling.

Finally it should be mentioned that the occurrence of certain *forbidden transitions* [see section 3(b)] is due to similar perturbations between states of fairly large separation and of the same Ω. Thus a $^1\Sigma$ state may combine with a $^3\Sigma$ state if the former is perturbed by a $^3\Pi_0$ state. On account of the mixing of the eigenfunctions the $^1\Sigma$ state assumes to a slight extent properties of the $^3\Pi_0$ state and can therefore very weakly combine with $^3\Sigma$.

Vibrational perturbations. If the Franck-Condon principle for perturbations (p. 285) is well fulfilled the perturbation between two states may be fairly strong even if the energy resonance is not very sharp. In such a case the energy discrepancy $(E_1 - E_2)$ may be about the same for all J values. (For example if the two rotational term series of Fig. 135(a) are shifted relative to each other by half the height of the figure, $E_1 - E_2$ will have about the same magnitude for all pairs of levels with equal J). If such a perturbation is homogeneous $(\Delta\Lambda = 0)$ the shifts in both levels will be nearly the same for all J values and we have what was called above empirically, a *vibrational perturbation*, that is, a displacement of a whole vibrational level. The slight dependence of $E_1 - E_2$ on J will lead to a small change of the rotational constant B compared to the unperturbed value.

Numerous cases of vibrational perturbations have been found (see the examples quoted on p. 282). However, in several cases a closer study has shown that actually a rotational perturbation at low J values produces a shift of the band head and thus leads to the appearance of a vibrational perturbation. For example, several apparent vibrational perturbations in P_2 found from band head measurements by Herzberg (311) are very probably due to rotational perturbations at low J values [see Herzberg, Herzberg, and Milne (1048) and Marais (1195)]. An example of a genuine vibrational perturbation in the above defined sense seems to be that observed for the $v = 4$ level of the upper $^3\Pi_u$ state of the second positive group of N_2 [Büttenbender and Herzberg (147)].

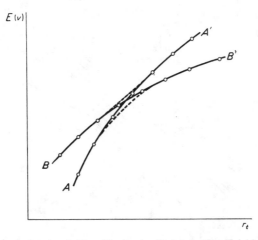

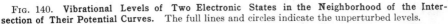

Fig. 140. **Vibrational Levels of Two Electronic States in the Neighborhood of the Intersection of Their Potential Curves.** The full lines and circles indicate the unperturbed levels.

Since there is no strict selection rule for the vibrational quantum number, vibrational perturbations do not always follow a simple resonance-like course as do rotational perturbations. In Fig. 140 the unperturbed vibrational levels of two electronic states are plotted as a function of r_t, the r values of the right-hand turning points of the classical vibrational motion (or, in other words, the right-hand parts of the potential curves are plotted). If only levels of nearly the same r_t could perturb one another the resulting perturbed levels would follow approximately the broken-line curves. However, since the Franck-Condon principle is not a sharp rule but allows the perturbation of levels with fairly different r_t, it is clear from Fig. 140 that in the neighborhood of the point of intersection of the potential curves, the vibrational levels may be very irregular so that it becomes quite ambiguous to which unperturbed level an observed level in this region belongs. For this reason it may frequently be impossible to decide on the basis of the observed levels alone whether the levels on branch A' belong to those on A or on B, and similarly for branch B'. It is only when a number of successive levels on branches A and A' can be represented by one and the same formula of the usual kind, that one would be justified in assuming that these two branches belong together, that is, that "originally" there was an intersection of the two potential curves.

The situation is usually complicated still further by the simultaneous occurrence of rotational perturbations which, as we have seen, are not fundamentally different from vibrational perturbations and occur when certain energy relations between the two states are fulfilled. An interesting case is that of the upper $^1\Sigma$ state of the AgH bands which was studied in detail by Gerö and Schmid (979) [see also Bengtsson and Olsson (86)]. Fig. 141 is a schematic diagram of the observed rotational levels (similar to Fig. 138). The probable course of the "original" unperturbed levels is indicated by broken lines. The perturbing state is called

$^1\Sigma_b$, the perturbed state $^1\Sigma_a$. Strong rotational perturbations are expected and do occur near the intersections of the broken lines. On account of these the rotational levels of $^1\Sigma_a$ ($v = 0$) with increasing J go over into those of $^1\Sigma_b$ ($v = 0$); the rotational levels of $^1\Sigma_a$ ($v = 1$) first tend to go to those of $^1\Sigma_b$ ($v = 0$), then to those of $^1\Sigma_a$ ($v = 0$), and finally go over into those of $^1\Sigma_b$ ($v = 1$); and so on. For higher v values, because of the strength of the perturbations, resulting curves are formed which no longer have a clear relation to the individual crossing points. For the non-rotating molecule ($J = 0$) the lowest vibrational levels of $^1\Sigma_a$ are close

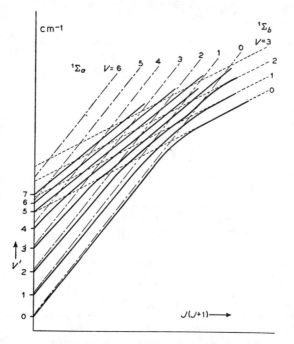

FIG. 141. **Perturbations in the Upper State of the AgH Bands, Schematic** [after Gerö and Schmid (979)]. The full lines represent the observed rotational levels plotted against $J(J + 1)$ (see Fig. 138); the broken lines represent the unperturbed rotational levels. The numbers at the left are the vibrational quantum numbers assigned without regard to the presence of the perturbing levels.

to the unperturbed levels, but where the vibrational levels of $^1\Sigma_b$ overlap those of $^1\Sigma_a$, the perturbations are so large that it is not possible to assign the vibrational levels unambiguously to one or the other electronic state. If one assumes all observed levels to belong to $^1\Sigma_a$ the ΔG curve of Fig. 142 is obtained. The bend in this ΔG curve at $v = 5$ must be taken as evidence that a point of intersection of the potential curve of the $^1\Sigma_a$ state with that of another state of the same Λ and S (that is, $^1\Sigma_b$) is in this neighborhood. It is clear that such a curve cannot be used for an extrapolation of the dissociation limit ($\Delta G = 0$, see p. 100). Similar anomalies in ΔG curves indicating strong perturbations have been observed for example for N_2 [Büttenbender and Herzberg (147)] and Br_2 [Brown (132)].

The vibrational eigenfunctions in the region of a strong (vibrational or rotational) perturbation are mixtures of the unperturbed eigenfunctions [compare equations (V, 77)]. Since each unperturbed vibrational eigenfunction has two broad maxima near the classical turning points (see Fig. 42) and since for strong perturbations the right-hand (or left-hand) turning points are close together it is clear that the perturbed eigenfunctions will have in general *three* broad maxima corresponding to a vibrational motion for which there is no classical analogue. This type of vibrational eigenfunction is simply an expression of the fact that the vibrational level belongs to both electronic states.

Intersection of potential curves and the non-crossing rule. According to the preceding discussion, in the case of strong homogeneous perturbations, the assignment of the vibrational levels to the two electronic states and the correlation of the levels below and above the point of maximum perturbation may become quite ambiguous. However, there is no ambiguity with regard to the correlation of the potential energy curves. As mentioned previously a mutual perturbation of electronic states with the same Λ and the same symmetry properties (in future called *states of the same species*) occurs even for the non-rotating and non-vibrating molecule, that is, in the two-center problem in which only the electronic motions are considered. On account of this interaction (repulsion) the following non-crossing rule holds as was first shown by von Neumann and Wigner (529): *For an infinitely slow change of internuclear distance two electronic states of the same species cannot cross each other* (or, we may also say, always "avoid" each other). Since a potential curve is defined as the curve representing the

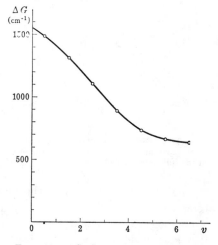

FIG. 142. ΔG Curve for the Upper $^1\Sigma$ State of AgH [after Bengtsson and Olsson (86)].

energy of an electronic state with the nuclei held fixed at each point (two-center system; see p. 148) the above rule may also be formulated thus: *The potential curves of two electronic states of the same species cannot cross each other.*

It does happen frequently that *in a certain approximation* two potential curves of states of the same species intersect; but according to the non-crossing rule in a sufficiently high approximation this intersection is avoided (see Fig. 140); that is, the upper curve on the right of the point of intersection goes over into the upper curve on the left, and the lower curve on the right goes over into the lower curve on the left. If the intersection is present only in a very low-order approximation and therefore the interactions causing the avoidance are very strong, the potential curves resulting in higher order will show very little evidence that "originally" an intersection was present. This is illustrated in Fig. 143(a). If only a fairly high-order approximation leads to avoidance the resultant potential curves will have anomalous forms which allows one to recognize the "original" intersection and for certain considerations it may be preferable to consider a lower-order approximation with intersection. Such cases are illustrated by Fig. 143(c) and (d) while Fig. 143(b) represents an intermediate case.

Fig. 143(d) illustrates the case where an "attractive" potential curve in a low approximation is crossed by a "repulsive" potential curve. On account of the finer interaction (that is, in higher approximation) the intersection is avoided leading to a *potential maximum* of the lower of the resulting potential curves if the interaction is not too strong. A few cases of such potential curves with maximum have been observed by means of predissociation phenomena (see Chapter VII), for example, for AlH and BH [Herzberg and Mundie (1050)].

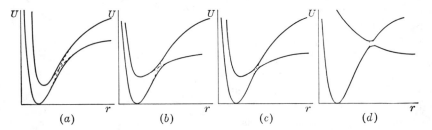

Fig. 143. **Different Cases of "Avoidance" of Potential Curves of the Same Species.** The broken parts of the curves give the course of the potential energy in a low-order approximation, and the full curves give the true course.

In other cases comparison of the molecular states arising from the separated atoms (see Chapter VI, section 1) with the observed states makes it necessary to assume potential maxima—sometimes rather high ones. Thus for NO the vibrational levels of the lowest observed $^2\Sigma^+$ state extend to 6.33 e.v. above the ground state (without by far having reached their point of convergence) while for large r there is a $^2\Sigma^+$ state resulting from normal atoms and having an energy of 5.29 e.v. (the dissociation energy). Since according to the non-crossing rule these two states must join, a potential maximum of at least 1.04 e.v. height above the asymptote is indicated.[18] In addition to the potential maxima due to intersections there are smaller maxima due to van der Waals interaction (see Chapter VI, section 4).

It must be emphasized that in the case of strong interaction of two electronic states of the same species the usefulness of the concept of potential curves becomes quite problematic. This usefulness is based on the fact that in general the electronic motions are very fast compared to the nuclear motions and therefore electronic and nuclear motions may be separated in the way outlined in Chapter IV, section 1. However, this separability no longer exists in the neighborhood of "intersections" of the type illustrated in Fig. 143 since here the electronic eigenfunctions are mixtures of those of the two original electronic states and a change over from

[18] Gaydon and Penney (962) have implicitly assumed that such high potential maxima cannot occur and consider that the values of the dissociation energies of NO, N_2, and CO must be revised in order to avoid such maxima. Their contention that the earlier values of the dissociation energies lead to contradictions to the non-crossing rule must be rejected. Even with their value for the dissociation energy of NO it would be necessary to assume a potential maximum in the $^2\Sigma^+$ state.

one to the other takes place within a time $h/(\Delta E)$ if ΔE is the shift produced by the perturbation. This time may be of the order of the time of a vibration of the nuclei $[h/(\Delta G)]$. In such a case it is clearly no longer possible to consider the vibrational motion of the nuclei as taking place in the average field produced by the electrons at various internuclear distances; that is, electronic and vibrational motion can no longer be separated. In other words, the potential curves obtained from the two-center system do not represent the situation completely: If one were to calculate separately for each resultant potential curve (full lines in Fig. 143) the vibrational levels on the basis of the wave equation, the vibrational levels obtained near the point of intersection would *not* coincide with the actual levels. The latter can only be obtained correctly if in addition the mutual interaction of the levels of the two electronic states is taken into account. Instead, it may often be more practical to start out from a lower approximation in which an intersection takes place and in which the zero-approximation vibrational levels are readily obtained. The anomalies are then due to the mutual interaction of these intersecting states rather than to the mutual interaction of states that approach one another closely and have anomalous shapes of their potential curves.

If the shifts ΔE are much larger than ΔG the going over from one to the other electronic state will be fast compared to the period of oscillation and the case of separability is approached again; the two resultant potential curves [Fig. 143(a)] give a fair representation of the situation. However if ΔE is much smaller than ΔG the going over does not happen in general during one oscillation but only after a great number of them. In such a case it is much more appropriate to describe the situation on the basis of the approximation in which the two electronic states cross and consider the anomalies in the positions of the vibrational levels as due to the interaction of these states [see Weizel (704) and Heitler (293)]. This is all the more justified since in such a case the higher vibrational levels above the point of intersection, will closely approach the unperturbed levels obtained *with* crossing. Even when ΔE near the intersection is of the same order as ΔG the vibrational levels sufficiently high above the point of intersection will approach the unperturbed levels simply because for such a level the time taken to traverse the region of large ΔE is much smaller than $h/(\Delta E)$, the time necessary to change over to the other electron configuration.

Another way of describing the situation is as follows: If AA' and BB' in Fig. 140 are the "original" potential curves with electronic eigenfunctions ψ_A and ψ_B respectively, then along the resultant curve $A'B$ the electronic eigenfunction changes from ψ_A to $a\psi_A + b\psi_B$ and then to ψ_B. If the system moves with a certain velocity starting from A' there is thus a certain probability that instead of continuing to B it will jump over to the other curve and come out at A with eigenfunction ψ_A. This probability is zero if the motion is infinitesimally slow. In that case the resulting potential curves describe the situation completely (two-center system). However, the probability of going to A rather than B increases with increasing velocity and approaches the value *one* for very large velocities; that is, for vibrational levels sufficiently high above the point of intersection, a crossing is the rule rather than the exception. These conclusions were first drawn by London (456) in whose paper a more detailed and quantitative treatment may be found [see also Stueckelberg (657) and the more recent paper of Valatin (1485)]. The same problem arises in the discussion of chemical elementary reactions [see Chapter VIII, section 3 and Glasstone, Laidler, and Eyring (40b)].

Summarizing the preceding discussion we may say that while the non-crossing rule of electronic states of the same species holds rigorously for the two-center system, practical cases may arise in which the observed vibrational levels indicate an intersection, particularly when the interaction of the two states is small. This apparent contradiction is due to the fact that the approximation in which the potential curves represent the observed levels accurately breaks down (on account of the non-zero velocity of the nuclei) when the potential curves come close together. Nevertheless the potential curves are extremely useful in correlating and systematizing the energy levels. Particularly in the discussion of the dissociation energies of diatomic molecules (Chapter VII) will the concept of potential curves and its limitations become of considerable importance.

5. Zeeman Effect and Stark Effect

The investigation of the Zeeman and Stark effects in molecular spectra by conventional means is experimentally very difficult since, even without a field, the lines in a band lie very close together, and since the individual band lines in a magnetic or electric field generally split into a larger number of closer components than do atomic lines. However, the recent introduction of Rabi's *molecular beam magnetic* (or *electric*) *resonance method* and of the use of microwaves into this field avoids some of these difficulties since far smaller term intervals can be separated. These methods will undoubtedly become very powerful tools in the investigation of the magnetic, electric, and even mechanical properties of diatomic molecules.

In the case of atomic spectra the study of the Zeeman effect is frequently an indispensable tool in the analysis of the spectrum. In the case of molecular spectra the investigation of the rotational structure usually gives all necessary information about the types of the electronic states concerned. Nevertheless the investigation of the Zeeman effect in band spectra is of considerable interest since it serves to test the theories previously developed and gives direct information about the magnetic properties of the molecules. Moreover the study of the Stark effect by the molecular beam method has led to determinations of moments of inertia of molecules which defy investigation by ordinary means.

The following discussion can only deal with the most important points. For details of the older investigations the reader is referred to the review by Crawford (171) [see also Jevons (23)].

General remarks on the splitting of molecular energy levels in a magnetic field. In a magnetic (or an electric) field, the total angular momentum J of the molecule can have only certain orientations (*space quantization*) such that the component in the field direction is $Mh/2\pi$, where

$$M = J, J - 1, \cdots, -J$$

(see Fig. 3 and p. 26). If the molecule has a magnetic moment, that is, if an interaction between the field and the molecule exists, a precession of J takes place at the corresponding angle, about the field direction as axis (see Fig. 2, p. 17). In addition, the states with different M have somewhat different energies—namely, in the case of a magnetic field (see A.A., p. 98),

$$W = W_0 - \bar{\mu}_H H \tag{V, 82}$$

where W_0 is the energy of the molecule in the absence of the field, $\bar{\mu}_H$ is the mean value of the component of the magnetic moment in the field direction, and H is the field strength.

There are three contributions to the *magnetic moment of a diatomic molecule:* (1) the magnetic moment associated with the orbital and spin angular momenta of the electrons, (2) the magnetic moment produced by the rotational motion of the molecule, and (3) the magnetic moments associated with the spins of the nuclei. The first contribution is of the order of a Bohr magneton $[\mu_0 = (e/2m_ec)(h/2\pi)]$, the second and third are of the order of a nuclear magneton $[\mu_{0n} = (e/2m_pc)(h/2\pi); \ m_p = \text{mass of proton}]$, that is, of the order of $1/1850$ of the first. Therefore, whenever the orbital or spin angular momenta of the electrons are different from zero (that is, in all but $^1\Sigma$ states) the second and third contributions may in general be neglected.

As for atoms (see p. 29) we have for dipole radiation the selection rule

$$\Delta M = 0, \pm 1 \qquad (M = 0 \nrightarrow M = 0 \text{ for } \Delta J = 0) \qquad \text{(V, 83)}$$

where $\Delta M = 0$ gives rise to lines polarized parallel to the field and $\Delta M = \pm 1$ to lines polarized perpendicular to the field assuming observation perpendicular to the field.

The Zeeman splitting of $^1\Sigma$ states. Since for $^1\Sigma$ states both Λ and Σ are zero, only the second and third of the above contributions to the magnetic moment arise, and consequently the splitting will be exceedingly small even for large magnetic field intensities. We leave a discussion of the effect of nuclear spin until section 6 and consider here only the effect of the rotation of the molecule.

If only the nuclei were present their rotational motion would give rise to a magnetic moment of the order of a nuclear magneton and proportional to the angular momentum J (in the case of identical nuclei of charge Ze it would be $Z\sqrt{J(J+1)}\,\mu_{0n}$). If we assume the electron cloud to rotate like a rigid body with the nuclei they will give rise to a magnetic moment of the same order of magnitude but opposite in sign. More detailed calculations for the case of H_2 [see Frisch and Stern (945)] show that the second contribution overcompensates the first yielding a magnetic moment of about $-3\sqrt{J(J+1)}\,\mu_{0n}$. Actually, of course, the electron cloud is not rigid. A calculation of the finer interaction of rotation and electronic motion [Wick (1531)(1532)] shows that a circulation of the electrons opposite to the direction of rotation arises whose magnetic moment almost cancels the electronic contribution just mentioned.[19] Thus at least for hydrogen the magnetic moment due to the rotation is positive. Its direction is always that of J and its magnitude μ_J is proportional to $|J|$, that is

$$\mu_J = g_r\sqrt{J(J+1)}\,\mu_{0n} \qquad \text{(V, 84)}$$

where g_r for H_2 is somewhat less than 1.

Since the angle between the direction of J and the field direction is given by (see Fig. 3)

$$\cos(J, H) = \frac{M}{\sqrt{J(J+1)}} \qquad \text{(V, 85)}$$

---

[19] This does not mean that the electron cloud remains stationary and only the nuclei rotate. Rather it means that within the oblong electron cloud which moves with the nuclei the electrons have a drift opposite to the direction of rotation but without changing the relative orientation of nuclei and cloud.

we obtain from (V, 82)

$$W = W_0 - \mu_J \frac{M}{\sqrt{J(J+1)}} H = W_0 - g_r\mu_{0n}HM \qquad \text{(V, 86)}$$

Thus we have for each J a set of $2J + 1$ equidistant levels whose spacing is $g_r\mu_{0n}H$, independent of J.

Transitions between successive levels of these sets ($\Delta M = \pm1$) are allowed as magnetic dipole radiation and occur in the radio-frequency region. According to (V, 86) combined with the Bohr frequency condition their position is given by

$$\nu = \frac{g_r\mu_{0n}H}{hc} \qquad \text{(V, 87)}$$

Such transitions have been found by Ramsey (1332) for H_2, HD, and D_2 using the molecular beam magnetic resonance method. For a field of 3600 oersted and $J = 1$ they occur at 8.043_4, 6.041_4, and $4.032_0 \times 10^{-5}$ cm^{-1}, respectively (or 2.411_2, 1.811_1, and 1.208_7 megacycles/sec.).[20] Actually, instead of single lines, groups of lines are observed on account of the nuclear spin (see the following section). From the observed positions of the "lines" and (V, 87) one finds for g_r

$$g_r^{H_2} = 0.8787, \qquad g_r^{HD} = 0.6601, \qquad g_r^{D_2} = 0.4406.$$

The decrease from H_2 to D_2 corresponds exactly to the decrease of rotational velocity. To an accuracy of better than 1 in 1000 the above g_r are in the inverse ratio of the reduced masses of the three molecules. In the case of H_2 the corresponding transitions for $J = 2$ were also found. Within the accuracy of the measurements they occur at the same frequency as for $J = 1$ (they can be separated on account of the hyperfine structure for $J = 1$). This shows that the proportionality of the magnetic moment with the angular momentum [equation (V, 84)] is accurately fulfilled.

Zeeman splittings of lines of inversion and rotation spectra of polyatomic molecules in the microwave region have been found by Coles and Good (855) and Jen (1097). They are due to magnetic moments of the same type as discussed here [formula (V, 84)] (see also p. 314).

Zeeman effect in Hund's case (a). If there is a resultant angular momentum of the electrons a magnetic moment of the order of a Bohr magneton arises leading to splittings that are 2000 times larger than those discussed in the preceding paragraphs. We are therefore justified in neglecting entirely the magnetic moment associated with the rotation of the molecule. The magnetic moment due to an orbital angular momentum of the electrons is $e/(2m_ec)$ times the magnitude of the angular momentum, the magnetic moment due to the spin is twice as large. The component μ_H of the magnetic moment in the field direction depends now on the way in which the electronic angular momenta are coupled to the rotation of the molecule. In Hund's case (a) (see p. 219) the electronic angular momenta are strongly coupled to the internuclear axis. The *magnetic moment* associated with the resultant angular momentum $\Omega = |\Lambda + \Sigma|$ in the internuclear axis is according to the above

$$\mu_\Omega = \frac{e}{2mc}(\Lambda + 2\Sigma)\frac{h}{2\pi} = (\Lambda + 2\Sigma)\mu_0 \qquad \text{(V, 88)}$$

Thus, for a $^1\Pi$ state, for example, there is a magnetic moment of exactly one Bohr magneton ($1\mu_0$) in the line joining the nuclei; for a $^2\Pi_{3/2}$ state it is $2\mu_0$; for a $^2\Pi_{1/2}$ state it is $0\mu_0$; and s) on.

As before we now have to find the time average $\bar{\mu}_H$ of the component of the magnetic moment in the field direction [see (V, 82)]. Since the nutation of Ω about J takes place very

---

[20] Actually in these experiments the magnetic field, rather than the frequency is varied.

rapidly compared to the precession of J about the field direction, we can obtain $\bar{\mu}_H$ by first forming $\bar{\mu}_J$, the time average of the component of μ in the direction of J, and then taking its component in the field direction. Fig. 144 (a) shows the vector diagram. It can be seen from this diagram that

$$\bar{\mu}_J = \mu_\Omega \cos (\Omega, J) = (\Lambda + 2\Sigma)\mu_0 \frac{\Lambda + \Sigma}{\sqrt{J(J + 1)}} \cdot \tag{V, 89}$$

Therefore we obtain for $\bar{\mu}_H$

$$\bar{\mu}_H = \bar{\mu}_J \cos (H, J) = \bar{\mu}_J \frac{M}{\sqrt{J(J + 1)}}$$

$$= \frac{(\Lambda + 2\Sigma)(\Lambda + \Sigma)M}{J(J + 1)} \mu_0. \tag{V, 90}$$

This formula, in combination with formula (V, 82), shows that with increasing J the maximum splitting in the magnetic field (for $M = \pm J$) decreases inversely proportionally to $J + 1$. This decrease is qualitatively obvious from Fig. 144 (a) since, for large J, the electronic angular momentum Ω is almost perpendicular to J and therefore $\bar{\mu}_J$ is very small compared to μ_Ω. Thus, except for the smallest J values, the total splitting $2\bar{\mu}_H H$ is considerably smaller than the

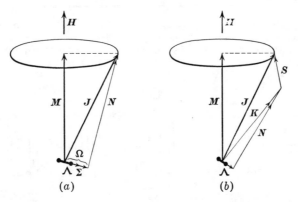

Fig. 144. **Vector Diagrams of the Molecule in a Magnetic Field.** (a) In Hund's case (a). (b) In Hund's case (b). The ellipses indicate the precession of J in the magnetic field. At the same time, the much faster precession (or nutation) of the other vectors takes place about J (see Figs. 97 and 100).

splitting $2\mu_0 H$ for the normal Zeeman effect of atoms. Fig. 145 (a) shows the splitting for the first rotational levels of a $^1\Pi$ state (the same figure also holds for a $^3\Pi_1$ state). It can be seen that the number of components $(2J + 1)$ increases with increasing J, while at the same time the total splitting decreases rapidly, so that, even for as low a value of J as 5, the components can no longer be drawn separately.

It is clear from this that well-resolved splitting patterns in bands involving case (a) states are to be expected only for lines with very low J values. The patterns to be expected, on the basis of the selection rule $\Delta M = 0, \pm 1$, for the first lines of the three branches of a $^1\Sigma$—$^1\Pi$ transition are drawn in Fig. 146. Since the splitting in the $^1\Sigma$ state is negligibly small, the line splittings give directly the term splittings in the $^1\Pi$ state. We shall not discuss here the theoretical formulae for the intensities of the Zeeman components [see Crawford (171)]. But the theoretical values are indicated in Fig. 146 by the heights of the vertical lines. For the CO Ångström bands which represent such a $^1\Sigma$—$^1\Pi$ transition, the theoretical splitting patterns

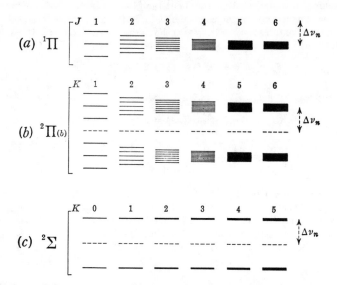

FIG. 145. **Term Splitting in a Magnetic Field.** (a) $^1\Pi$ $(^3\Pi_1)$. (b) $^2\Pi(b)$. (c) $^2\Sigma$. The position of the levels without field is indicated by broken lines. For comparison, the magnitude of the splitting in the normal Zeeman effect $\Delta\nu_n$ is given to the right by the broken-line arrow.

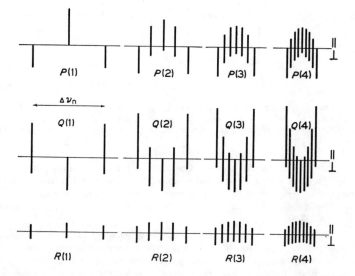

FIG. 146. **Splitting of the First Lines of a $^1\Sigma$—$^1\Pi$ Band in a Magnetic Field** [after Crawford (171)]. The length of the vertical lines gives the intensity. The part above the horizontal line is polarized parallel, the part below it perpendicular to the field direction.

have been confirmed quantitatively by experiment, at least for the first two lines of each of the three branches, and a qualitative agreement has been found for the succeeding lines whose splitting patterns cannot be fully resolved. For further examples see Crawford (171).

It should be noted that, as long as there is no uncoupling, the splitting is quantitatively the same for all like transitions, independent of the special molecular constants. This behavior is analogous to that of atoms.

Zeeman effect in Hund's case (b). In Hund's case (b), the orbital angular momentum of the electrons is coupled to the internuclear axis, whereas the spin is coupled to the rotational axis [see Fig. 144(b)]. The time average $\bar{\mu}_J$ of the magnetic moment in the direction of J is composed of the contributions due to $\mu_\Lambda = \Lambda\mu_0$ and $\mu_S = 2\sqrt{S(S+1)}\,\mu_0$. The precession of J about the field is slow compared to that of K and S about J while the latter in turn is slow compared to the nutation of Λ about K. Therefore in finding $\bar{\mu}_J$ we average first over the nutation of Λ and then over the precession of K and S about J obtaining

$$\bar{\mu}_J = \Lambda\mu_0 \cos(\Lambda, K) \cos(K, J) + 2\sqrt{S(S+1)}\,\mu_0 \cos(S, J) \qquad (V, 91)$$

where $\cos(\Lambda, K) = \Lambda/\sqrt{K(K+1)}$. Finally averaging over the precession of J about the field, that is, multiplying by $\cos(J, H) = M/\sqrt{J(J+1)}$ we obtain

$$\bar{\mu}_H = \left[\frac{\Lambda^2}{\sqrt{K(K+1)}} \cos(K, J) + 2\sqrt{S(S+1)} \cos(S, J)\right] \frac{M}{\sqrt{J(J+1)}}\mu_0, \qquad (V, 92)$$

where the two cosines are to be taken from the obtuse-angled triangle in Fig. 144(b) (see also A.A., p. 110). According to (V, 92), there is again a splitting into $2J + 1$ equidistant components. However, now the total splitting ($M = \pm J$) for large J, when the first term in the bracket is small, is *approximately independent of J* and is of the order of magnitude of the normal Zeeman splitting.

To be sure, it very often happens in case (b) that the coupling between S and K is so weak that these two angular momenta are uncoupled even by a small field. The *molecular analogue to the Paschen-Back effect* then appears (see A.A., p. 112); S and K are space quantized independently of each other in the field direction with components M_S and M_K. As a result, $\bar{\mu}_H$ is given by the simpler formula

$$\bar{\mu}_H = \frac{\Lambda^2\mu_0}{\sqrt{K(K+1)}} \cos(K, H) + 2\sqrt{S(S+1)}\mu_0 \cos(S, H)$$

or, since $\cos(K, H) = M_K/\sqrt{K(K+1)}$ and $\cos(S, H) = M_S/\sqrt{S(S+1)}$,

$$\bar{\mu}_H = \frac{\Lambda^2 M_K}{K(K+1)}\mu_0 + 2M_S\mu_0. \qquad (V, 93)$$

The first term in (V, 93) agrees exactly with $\bar{\mu}_H$ for case (a), as is seen if in (V, 90) we put $\Sigma = 0$, $J = K$, and $M = M_K$. Formula (V, 93) holds as long as the multiplet splitting is small compared to the Zeeman splitting. The first term in (V, 93) decreases rapidly with increasing K, whereas the second term is independent of K. Fig. 145(b) shows the Zeeman splitting of a $^2\Pi$ state in this coupling case. For higher K there is practically only a splitting into two components corresponding to $M_S = \pm\frac{1}{2}$, with a separation $2\mu_0 H$. Fig. 145(c) gives the splitting for a $^2\Sigma$ term. Here the first term in (V, 93) is zero, since $\Lambda = 0$, and there remains only a splitting into two components, independent of K. Only in higher approximation, as a result of the interaction between K and S, does there appear a very small splitting of each component with a given M_S into $2K + 1$ components, which is of the same order of magnitude as the doublet splitting without field. In Fig. 145(c) this is indicated by a broadening of the levels. Triplets and higher multiplets behave correspondingly.

If the doublet (multiplet) splitting without field is appreciably greater than the normal Zeeman splitting $(\Delta\nu_n)$, formula (V, 92) holds; that is, a splitting of each doublet (multiplet) component into $2J + 1$ components takes place, the total splitting being of the order of $2\Delta\nu_n$.

When the Paschen-Back effect is present in both states participating in a transition, the selection rule $\Delta M_S = 0$ holds (see p. 29); that is, for doublet states $M_S = +\frac{1}{2}$ in the upper state combines only with $M_S = +\frac{1}{2}$ in the lower state, and correspondingly for $M_S = -\frac{1}{2}$. Therefore since the Zeeman splitting due to S is the same in the upper and lower states the set of lines arising for $M_S = +\frac{1}{2}$ coincides with that for $M_S = -\frac{1}{2}$; that is, the Zeeman splittings of the lines of a doublet band of case (b) are of exactly the same type and magnitude as those of a corresponding singlet transition; and similarly for triplet (and higher multiplet) bands. In particular, for a $^2\Sigma-^2\Sigma$ transition the Zeeman splittings will be unobservably small as for $^1\Sigma-^1\Sigma$; for $^2\Sigma\longleftrightarrow^2\Pi$, only a splitting for small K values is to be expected; and so on. Actually, the lines of the violet CN bands $(^2\Sigma-^2\Sigma)$, for example, show no splitting in a magnetic field.[21] Furthermore, the triplet bands of the He_2 molecule, for which the triplet splitting is vanishingly small, show exactly the same Zeeman splittings as the corresponding singlet bands of He_2.

Other cases.   In many practical examples of multiplet states with $\Lambda \neq 0$ the coupling is intermediate between Hund's cases (a) and (b). In such transition cases the behavior in a magnetic field is much more complicated. Hill (322) has given detailed formulae for these cases [see also Crawford (171)]. The Zeeman splitting for the different J values depends strongly on the value of $Y = A/B$, where A is the spin-coupling constant (see p. 232) and B is the rotational constant. As an example, consider a $^2\Pi$ state that belongs to case (a) for small rotation but goes to case (b) for larger J. Here the Zeeman splitting of the $^2\Pi_{\frac{1}{2}}$ component unlike $^2\Pi_{\frac{3}{2}}$ is zero as long as case (a) is a good approximation, since $\Lambda + 2\Sigma = 0$; whereas if case (b) is approximated we have the splitting shown in Fig. 145(b) in which both $^2\Pi_{\frac{1}{2}}$ and $^2\Pi_{\frac{3}{2}}$ are affected.

As we have seen previously (p. 223) for $^3\Sigma$ states frequently the spin-spin interaction is large causing a noticeable and almost constant splitting even for small K. In such cases the formula (V, 92) cannot be applied except for very strong fields. Formulae for the case of weak and intermediate fields have been given by Schmid, Budó, and Zemplén (1383). The line splittings of the individual band lines would be correspondingly complicated in these cases. Such experimental data as there are, confirm the theoretical results [Crawford (171)].

For many of the excited states of the molecules H_2 and He_2, the transition to Hund's case (d) is also important. In this case the electronic orbital angular momentum L is quantized with respect to the rotational axis, and therefore the Zeeman splittings for the singlet states also (including $^1\Sigma$ if $L \neq 0$) are of the order of the normal Zeeman splitting even for large K. The behavior of the splitting in intermediate cases, which actually occur more often than the limiting cases, is very complicated. It will not be discussed here [see Kovács and Budó (1139a)].

Finally, the *Zeeman splitting of perturbed lines* must be mentioned. It has been found experimentally that perturbed lines are much more strongly influenced by a magnetic field than the neighboring normal lines of the same band. This is easily understood from the fact that only levels with equal M can perturb each other. As long as no magnetic field is present, this selection rule for perturbations plays no part, since all the levels with different M and equal J coincide, and therefore it was not mentioned in our previous discussion of perturbations. However, in a magnetic field, the splitting of the perturbing state is in general different from that of the perturbed state, and therefore the interaction of the individual component levels of the perturbed and the perturbing state is not the same for all values of M, since their separation varies. Therefore a large perturbation of certain component levels can be indirectly

---

[21] To be sure, for lines with very high K, which without field are split into doublets, a small diminution of the doublet separation in a magnetic field has been found. This behavior may be understood theoretically by taking into account the interaction of S and K.

brought about by the field, and thus a comparatively large line splitting in the magnetic field is observed. The magnitude and type of the interaction naturally depend very strongly on the kind of states perturbing each other, and therefore an investigation of the Zeeman effect of perturbed lines lends itself to determining the type of the perturbing state. Thus far, however, very few data are available for this purpose [see Crawford (171) and Schmid and Gerö (620)].

The selection rule $\Delta M_S = 0$ which holds for strong fields in case (b) breaks down not only in cases intermediate between (a) and (b) but also for *intercombinations* even if case (b) applies. Therefore the lines of $^1\Sigma$—$^3\Sigma$ bands should show for all K values a large splitting corresponding to the splitting of the $^3\Sigma$ state into three components with spacing $2\mu_0 H$ [compare the splitting of $^2\Sigma$ in Fig. 145(c)]. Such a splitting, although complicated by spin-spin interaction (see above), has been found for the atmospheric oxygen bands by Schmid, Budó, and Zemplén (1383).

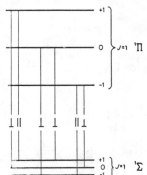

Polarization of resonance fluorescence. It was discovered by Wood (1551) that the resonance fluorescence of diatomic gases is partially polarized when excited by natural or polarized light and observed at right angles to the incident beam. This phenomenon can be understood on the basis of the preceding discussion of the splittings in a magnetic field and the polarization rules of the Zeeman components. The relations for zero field can be derived from those for a very weak field applied in the direction of the electric vector of the incident light [see Heisenberg (1024)].

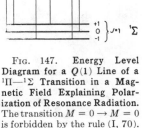

Consider, for example, the line $Q(1)$ of a $^1\Pi$—$^1\Sigma$ transition for which Fig. 147 gives an energy level diagram. According to this diagram if the incident light is polarized parallel to the field, only the levels $M = \pm 1$ of the upper state are excited. On returning to the lower state the $\perp$ components with $M = 0$ in the upper state are missing and therefore the whole line (without field) will be partially polarized in

Fig. 147. **Energy Level Diagram for a $Q(1)$ Line of a $^1\Pi$—$^1\Sigma$ Transition in a Magnetic Field Explaining Polarization of Resonance Radiation.** The transition $M = 0 \rightarrow M = 0$ is forbidden by the rule (I, 70).

a direction at right angles to the incident beam. Even for excitation by natural light the upper magnetic sublevels will not be uniformly populated (this would be the case only for isotropic excitation) and therefore there will be a deficiency of perpendicularly polarized light in the fluorescence. For other lines the situation is similar.

The degree of polarization

$$P = \frac{I_{||} - I_\perp}{I_{||} + I_\perp} \tag{V, 94}$$

can be calculated quantitatively if the intensities of the Zeeman components are known. The results for the lines of a singlet band are shown graphically in Fig. 148 for linearly polarized incident light. The polarization P_n for natural incident light may be obtained from P_l, the polarization for linearly polarized incident light by

$$P_n = \frac{P_l}{2 - P_l} \tag{V, 95}$$

The polarization for the lines of the R and P branches is different for excitation by the same line from that for excitation by the corresponding lines of the P and R branches, respectively. This is indicated by the symbols in brackets in Fig. 148. For large J values the polarization of the lines of P and R branches approaches the value 0.143 while that of the Q branch approaches 0.50. A striking confirmation of both these theoretical values has been obtained by Pringsheim (1322) for the resonance series of Na_2 [see also Mrozowski (1248)].

It must be emphasized that the theoretical values for the polarization hold only at sufficiently low pressure when the molecule is free to radiate without the interference of collisions. Conversely the study of the depolarization brought about by collisions may be used for an investigation of the mechanism of such collisions [see for example Mrozowski (1247)].

Magnetic rotation spectra. As is well known, when linearly polarized light traverses a medium in the direction of a strong magnetic field a rotation of the plane of polarization takes place (Faraday effect). For a gas this effect can occur only if the wave-length of the incident light is close to an absorption line of the gas (that is, in the region of anomalous dispersion). A further condition for the occurrence of this effect is, as shown by a more detailed theory [see, for example, Born (7)], that the absorption line must show an appreciable Zeeman effect. The effect was discovered for a monatomic gas by Macaluso and Corbino. It was first found for a diatomic gas—Na_2—by Wood (719).

In these experiments white light is sent through two crossed Nicols, between which is placed a tube filled with Na_2 vapor. As long as there is no magnetic field, no light goes through the system. However, when a longitudinal magnetic field is applied, a small part of the light does go through —namely, that part for which the plane of polarization has been rotated. An investigation of the spectrum of this transmitted light shows that it consists of individual bright lines or narrow bands whose positions agree very closely with those of the heads of the absorption bands of the vapor. However, the spectrum is much simpler than the absorption spectrum, since, instead of each band, there is only one line or at most a very narrow band. Such *magnetic rotation spectra* have also been found and investigated for a number of other molecules, including I_2, Br_2, K_2, and Bi_2. In particular, with their help, Loomis and co-workers [(463)–(465), (459), and (462)] have followed the vibrational quanta for the upper states of the alkali molecules to considerably higher v' values than was possible by the absorption spectrum.

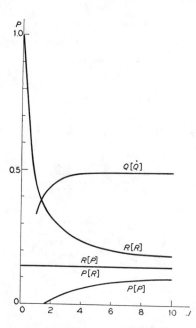

Fig. 148. **Degree of Polarization of the Lines of Singlet Bands in Resonance Fluorescence Excited by Linearly Polarized Light [after Mrozowski (1246); see also the formulae given by Placzek (562)].**

According to the previous discussion, the explanation for the magnetic rotation spectra of diatomic molecules is clear: Since a rotation of the plane of polarization takes place only in the immediate neighborhood of such absorption lines as show a noticeable Zeeman effect, in general only the lines in a band with the lowest J values cause a magnetic rotation. Therefore, only in their immediate neighborhood is light transmitted by the crossed Nicols (unless the magnetic field is extremely strong). That is why only part of each band appears in the magnetic rotation spectrum. This effect is strengthened if the head of the band lies at low J values, since then part of the light that gives strong magnetic rotation lies just outside the band, and is therefore not absorbed by the gas.

According to the preceding discussion, magnetic rotation spectra should appear only for band systems showing a noticeable Zeeman effect. Indeed, the absorption systems of the alkalis that have been particularly thoroughly investigated are $^1\Pi$—$^1\Sigma$ transitions, which show a marked Zeeman splitting for small J (see above). For $^1\Sigma$—$^1\Sigma$ transitions there should be no magnetic rotation spectrum. However, in two cases—the red Na_2 absorption bands and the infrared K_2 absorption bands, which represent such transitions—weak magnetic rotation spectra have been found. It has been shown by Carroll (150) that the explanation for this apparent contradiction to the theory lies in the fact that the upper $^1\Sigma$ state of the bands is perturbed by a $^3\Pi$ state. The perturbed levels of the $^1\Sigma$ state assume to a certain extent the properties of the $^3\Pi$ state—that is, they have a magnetic moment (whereas the nonperturbed levels have none)—and are therefore split in the magnetic field. In consequence of this, the perturbed lines (and only these) will produce a magnetic rotation. In agreement with this explanation the magnetic rotation lines observed in these band systems do not in general lie in the neighborhood of the zero lines of the bands, as is the case for $^1\Pi$—$^1\Sigma$ magnetic rotation spectra.

Stark effect. In an electric field E the space quantization of J is the same as in a magnetic field (see p. 17). However, since in an electric field (unlike a magnetic field) a reversal of the sense of rotation leaves the energy of the system unchanged, states which differ only in the sign of M have the same energy, that is, the electric field produces a splitting of a given state into $J + 1$ (or $J + \frac{1}{2}$) states of which all but the one with $M = 0$ are doubly degenerate. The energies of these states in the field are given by [similar to (V, 82)]

$$W = W_0 - \bar{\mu}_E E \tag{V, 96}$$

where $\bar{\mu}_E$ is the mean component of the *electric moment* in the field direction. (Note that the μ used here has nothing to do with the magnetic moment designated previously by the same symbol).

For *molecules having no permanent dipole moment*, just as for atoms, a small dipole moment is *induced* by the field and is proportional to the field, that is, $\bar{\mu}_E = a_{J|M|}E$ where $a_{J|M|}$ is a constant depending on J and $|M|$. Thus we have a splitting of the energy levels that increases quadratically with the electric field intensity (*quadratic Stark effect*). The values of $a_{J|M|}$ can be determined by second order perturbation theory [see Macdonald (473)] and depend on the positions of other electronic states. The ground states of most homonuclear molecules are widely separated from all other electronic states and therefore for them the $a_{J|M|}$ and consequently the Stark splittings are exceedingly small and have not yet been detected in a single instance. For the excited states the separations from other suitable states may be small and therefore appreciable Stark splittings may result. Thus far the H_2 molecule is the only one for which such splittings have been found in the spectrum [Kiuti and H sunuma (402) (402a), Macdonald (472), Snell (641), Rave (580)]. Just as for the Zeeman effect the splittings are largest and most readily understood for small J values. In addition to the splittings of normal transitions some additional lines are observed which would be forbidden without the field, similar to the Stark effect of non-hydrogen-like atoms (see A.A., p. 118).

If a *molecule with a permanent electric dipole moment* μ is in a Σ state the dipole moment is perpendicular to the total angular momentum (apart from spin); that is $\mu_J = 0$ and therefore $\bar{\mu}_E = 0$. There is no first order Stark splitting. However, there is a second order effect on account of the nonuniform rotation brought about by the action of the field on μ (or in other words, brought about by the mutual perturbation between neighboring rotational levels of the *same* electronic state).[22] A more detailed calculation gives [see Van Vleck (35)]

$$\mu_E = -\frac{4\pi^2 I_0 \mu^2}{h^2}\left[\frac{J(J+1) - 3M^2}{J(J+1)(2J-1)(2J+3)}\right]E; \tag{V, 97}$$

that is, combining with (V, 96) we see that there is a splitting proportional to E^2 whose magnitude is entirely determined by the dipole moment μ, the moment of inertia I_0 and the value of J (independent of the positions of other electronic states). The splitting decreases rapidly with increasing J. For HCl in the state $J = 1$ the splitting between $|M| = 1$ and $M = 0$

---

[22] In an electric field states differing in J by one unit can perturb each other. Moreover states of opposite $+ -$ symmetry can perturb each other.

in a field of 10000 volts/cm is (taking $\mu = 1.03 \times 10^{-18}$ e.s.u., $I_0 = 2.68 \times 10^{-40}$ gm cm²) $\Delta W = 0.00433$ cm⁻¹. No such splittings have been found by ordinary spectroscopic methods.

However, recently Hughes (1076) using the molecular beam electric resonance method has directly observed the transition $J = 1$, $M = 0 \rightarrow J = 1$, $M = \pm 1$ in the ground state of the CsF molecule at 0.001043 cm⁻¹ for $E = 265$ volts/cm. From this frequency according to (V, 97) and (V, 96) an accurate value of $I_0\mu^2$ can be deter.nined. The slight dependence of the effective moment of inertia and of the dipole moment on the vibrational quantum number produces slight differences in the position of the line for the different vibrational levels. A series of lines corresponding to $v = 0, 1, 2, 3, 4$ has been found by Trischka (1471) for CsF and by Grabner and Hughes (997) for KF.

For very precise measurements a higher order term $b\mu^4 I_0{}^3 E^3$ must be added to (V, 97). If this term can be determined experimentally it yields a second equation for μ and I_0 and thus these quantities can be evaluated separately. For CsF Hughes obtained in this way the approximate values $I_0 = 187 \times 10^{-40}$ gm cm², $\mu = 7.3 \times 10^{-18}$ e.s.u. No values obtained by the more conventional methods are available for CsF. This new method of obtaining moments of inertia and dipole moments for molecules for which they cannot readily be obtained by other methods promises to give important results. (For the results on KF see Table 39.)

In the case of a linear triatomic molecule (OCS) the Stark splitting of the transition $J = 2 \leftarrow J = 1$ has been observed in microwave absorption by Dakin, Good and Coles (879a). In this case since the moment of inertia is known from the field free position of the line the dipole moment is immediately obtained from the observed splitting according to (V, 96 and 97).

If a molecule with a permanent electric dipole moment is in a Π, Δ, ··· state, a non-zero component of the dipole moment in the direction of J arises and consequently even in first order approximation a component in the field direction results. In a way entirely similar to the magnetic case [equation (V, 90)] one obtains (assuming $S = 0$)

$$\bar{\mu}_E = \frac{\mu M \Lambda}{J(J + 1)} \tag{V, 98}$$

which together with (V, 96) gives a splitting linear in E and linear in M, but inversely proportional to $J(J + 1)$, very similar to the Zeeman splitting (see Fig. 145). There is one important difference from the magnetic case: The above formula holds only if the Λ-type doubling is very small compared to the Stark splitting. For very small fields a quadratic splitting of each Λ component arises which with increasing field intensity goes over into the above described linear effect, but so that each component level has a positive and negative M value [see Penney (560)].

According to (V, 98) and (V, 96) for the case of the level $J = 1$ of a ¹Π state assuming $\mu = 1 \times 10^{-18}$ the overall splitting would be 0.168 cm⁻¹ in a field of 10000 volts/cm. No such splittings have yet been observed even though they are within reach of ordinary spectroscopic methods. In addition to the linear effect there is also in the case $\Lambda \neq 0$ a quadratic effect similar to that for Σ states [see Van Vleck (35)].

6. Hyperfine Structure

Hyperfine structure without field. The nuclear spin is not included in the total angular momentum J of the molecule. As for atoms (see p. 29) we call $\boldsymbol{F}$ the *total angular momentum inclusive of nuclear spin*. If $\boldsymbol{T}$ is the resultant nuclear spin (see p. 133) we have for the corresponding quantum numbers, similar to (I, 72)

$$F = (J + T), (J + T - 1), \cdots, |J - T| \tag{V, 99}$$

that is, except for very small J there are $2T + 1$ component levels. These component levels will have slightly different energies if there is a coupling of the nuclear spins with the rest of the molecule. In the case of a purely magnetic coupling the component levels will follow the interval rule; that is, the separation of successive levels will be proportional to $F + 1$ This is shown qualitatively in Fig. 149 (a). The magnitude of the splitting will be of the same order as the hyperfine structure splitting in atomic spectra if the mean magnetic moment $\bar{\mu}_J$ of the outer electrons in the direction of J is of the order of a Bohr magneton. This is the case, as we have seen in the preceding section, for the low rotational levels of Π, Δ, $\cdots$ states belonging to Hund's case (a) as well as for multiplet Σ, Π, $\Delta \cdots$ states belonging to case (b). For the latter $\bar{\mu}_J$ and therefore the hyperfine structure splitting is essentially independent of J, for the former it decreases rapidly with J (see Figs. 144 and 145). For $^1\Sigma$ states the magnetic moment of the outer electrons is of the order of a nuclear magneton and therefore the hyperfine structure splitting will be only about 1/1000 of that for atoms. To be sure in this case the splitting will increase proportionally with J.

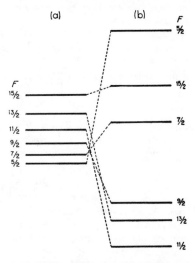

FIG. 149. **Hyperfine Structure of Molecular Levels for $J = 5$, $T = I = \frac{5}{2}$ (Schematic); (a) Produced by Nuclear Magnetic Moment, (b) by Nuclear Quadrupole Moment.**

In recent years it has been found, at first from the hyperfine structure in atomic spectra and later also from the hyperfine structure of molecular spectra, that many nuclei possess, in addition to a magnetic dipole moment, an *electric quadrupole moment*. In other words, the electric charge within the nucleus is in general not distributed in a spherically symmetric way except for nuclei with $I = 0$ and $I = \frac{1}{2}$ (for which the corresponding wave functions have spherical symmetry). Rather, the charge distribution has the shape of an oblate or prolate ellipsoid whose axis of symmetry coincides with the axis of spin. The electrostatic field produced by such a charge distribution of total charge Ze at a point x, y, z is given by [see Born (7)]

$$V = \frac{Ze}{r} - \frac{eQ}{4r^5}(x^2 + y^2 - 2z^2) \tag{V, 100}$$

if the z axis is in the direction of the axis of symmetry. The first term on the right represents of course, the spherically symmetric field produced by the total charge if it were concentrated at the center; the second term represents the deviations from spherical symmetry. Q is the quadrupole moment which is defined by[23]

$$eQ = \int (3z'^2 - r'^2) \, de' \tag{V, 101}$$

where de' is an element of charge within the nucleus of coordinates x', y', z' and $r'^2 = x'^2 + y'^2 + z'^2$. Alternatively one may also write

$$Q = Z\langle 3z'^2 - r'^2 \rangle_{\text{average}} \tag{V, 102}$$

It is readily seen that for a spherical charge distribution $Q = 0$.

---

[23] It seems unfortunate (and contrary to the usage of earlier writers) that, in recent papers on the subject, Q rather than eQ is called the quadrupole moment, so that it is given the dimension cm². To be consistent one would then have to use the dimension cm, rather than charge $\times$ cm for the dipole moment, which is never done. However, in order to avoid confusion we follow here the usage of the recent literature of the subject.

It is clear that the electrostatic interaction of the nuclear quadrupole moment with the rest of the molecule will be different for different relative orientations of the axis of the quadrupole, that is, of I with respect to J. This interaction will therefore change the hyperfine structure splitting. Explicit formulae have been given only for the case in which one nucleus has zero spin and zero quadrupole moment. If I and Q are spin and quadrupole moment of the other nucleus, Casimir (841) found for the *interaction energy*[24] (assuming $\Lambda = 0$)

$$\Delta W = \frac{e^2 q Q[\frac{3}{8}G(G+1) - \frac{1}{2}I(I+1)J(J+1)]}{I(2I-1)(2J-1)(2J+3)} \tag{V, 103}$$

which for large J values ($\gg I$) goes over into [see Feld and Lamb (933)]

$$\Delta W = \frac{e^2 q Q[3(F-J)^2 - I(I+1)]}{4I(2I-1)} \tag{V, 104}$$

Here

$$G = F(F+1) - I(I+1) - J(J+1) \tag{V, 105}$$

and

$$eq = \left\langle \frac{\partial^2 V}{\partial z^2} \right\rangle_{\text{average}} = \int \frac{3z^2 - r^2}{r^5} \, de \tag{V, 106}$$

where V is the electrostatic potential at the nucleus and the integration is over all elements of charge de of the molecule other than the nucleus considered. The quantity eq is thus the average inhomogeneity of the electrostatic field at the position of the nucleus in the direction of the z axis (internuclear axis).

Formula (V, 104) shows that the shift produced by the quadrupole interaction for large J is independent of the sign of $(F - J)$. Thus, neglecting the magnetic interaction, instead of a splitting into $2I + 1$ components, a splitting into only $I + 1$ or $I + \frac{1}{2}$ components results depending on whether I is integral or half-integral. For smaller J values there are $2I + 1$ components (unless $J < I$) which, except at the lowest J values, occur in pairs. Fig. 149(b) shows the energy levels resulting from (V, 103) for $J = 5$, $I = \frac{5}{2}$. Unlike the splitting produced by the interaction of the nuclear magnetic moment with the magnetic moment of the molecule the magnitude of the quadrupole splitting is essentially independent of J. The same type of splitting arises also on account of the interaction of the two nuclear spins in a molecule with identical nuclei [see Rabi and collaborators (1127)].

Equations (V, 103) and (V, 104) show that in order to determine the nuclear quadrupole moment Q from observed hyperfine structures it is necessary to know q from data about the electronic wave functions of the molecule. This is similar to the situation that is met when one attempts to determine nuclear magnetic moments from observed hyperfine structures in atomic spectra [see for example Bethe and Bacher (88)].

From the above formulae for the energy levels we obtain the theoretical spectra by applying the *selection rule*

$$\Delta F = 0, \pm 1 \tag{V, 107}$$

in addition to the usual selection rules. No case of a complete resolution of such hyperfine structures in ordinary band spectra has as yet been reported. However, Hulthén and Heimer (343a) (286a) (288) found in a $^1\Sigma$—$^1\Pi$ transition of BiH indications of a splitting of the first few lines of the Q branch. Such a splitting would be expected on account of the large nuclear spin ($I = \frac{9}{2}$) and nuclear magnetic moment ($4.0\mu_{0n}$) of Bi^{209}. According to the preceding discussion it should decrease rapidly with increasing J, as is observed. Mrozowski (1251) has found in the $^2\Pi$—$^2\Sigma$ bands of HgH a splitting of the lines belonging to the odd isotopes of Hg into at least two components.

---

[24] This is the form given by Feld (932). For higher order correction terms see Bardeen and Townes (769).

The best possibility of studying the hyperfine structure of diatomic molecules presents itself in the rotation spectra of suitable molecules which lie in the microwave region ($\Delta J = +1$ in Fig. 149). While such hyperfine structures have been found for many polyatomic molecules thus far only one case[24a] of a diatomic molecule—ICl—has been studied [Townes, Merritt, and Wright (1469), Weidner (1521)]. Since in this case both nuclei have a large spin and quadrupole moment each rotational line is split into a considerable number of rather widely spaced hyperfine structure components. All of these can be accounted for by extending the above discussed theory to the case in which both nuclei have non-zero spin and quadrupole moment [compare the recent summary by Gordy (994a)].

Transitions between the hyperfine structure components of a given J have been observed for the sodium halides in radiofrequency magnetic resonance spectra by Nierenberg and Ramsey (1283).

Zeeman effect of hyperfine structure. In a strong magnetic field H a Paschen-Back effect of hyperfine structure (see A.A., p. 192) arises so that in a zero-order approximation T and J are independently space-quantized with respect to the field with components M_T and M_J. The energy to this approximation is given by

$$W^{(0)} = W_0 - \mu_{0n} g_I M_T H - \mu_{0n} g_r M_J H \qquad (V, 108)$$

where g_r is defined by (V, 84) and g_I by the similar relation $\mu_I = g_I \sqrt{I(I+1)}\, \mu_{0n}$. Instead of this magnitude μ_I of the nuclear magnetic moment usually the *maximum component* $\tilde{\mu}_I = g_I I \mu_{0n}$ is given.

The magnetic quantum numbers M_T and M_J can have the values

$$M_T = T, T-1, \cdots, -T \qquad (V, 109)$$

$$M_J = J, J-1, \cdots, -J \qquad (V, 110)$$

Thus there are $(2T+1) \times (2J+1)$ *component levels* in the field for a given J and T, that is, for each group of levels such as that in Fig. 149.

The energy formula (V, 108) holds only under the assumption that M_T is defined. This is always the case for homonuclear molecules since for them different T values correspond in general to different rotational levels. If the nuclear spins are uncoupled by the field their M_I values must be such as to give the proper M_T values so that (V, 108) still holds. For heteronuclear molecules for large fields the two nuclear spins are independently space-quantized and (V, 108) must be replaced by

$$W^{(0)} = W_0 - \mu_{0n} g_{I_1} M_{I_1} H - \mu_{0n} g_{I_2} M_{I_2} H - \mu_{0n} g_J M_J H \qquad (V, 111)$$

In order to obtain a better approximation for the energy in a magnetic field the following interactions must be taken into account: (a) the interaction between the nuclear magnetic moments and the magnetic moment of rotation of the whole molecule (*spin-orbit interaction*), (b) the mutual magnetic interaction of the nuclear spins (*spin-spin interaction*), (c) the interaction of the nuclear quadrupole moments with the rest of the molecule (*quadrupole interaction*). All these interactions are dependent on the relative orientation of the angular momenta and in a first approximation their effect on the energy of the system is obtained by averaging over all orientations occurring in a given quantum state. The result for a homonuclear molecule is the addition of the following expression to the energy [see Kellogg, Rabi, Ramsey, and Zacharias (1127)]

$$W^{(1)} = -\mu_{0n} g_I H' M_T M_J + a_{TIJ} \frac{\tilde{\mu}_I{}^2}{r^3} [3M_J{}^2 - J(J+1)][3M_T{}^2 - T(T+1)]$$

$$+ b_{TIJ} e^2 q Q_I [3M_J{}^2 - J(J+1)][3M_T{}^2 - T(T+1)] \qquad (V, 112)$$

Here H' is the magnetic field produced by the molecular rotation at the position of the nucleus; a_{TIJ} and b_{TIJ} are rational numbers determined by the values of the quantum numbers

---

[24a] More recently ClF was studied by Gilbert, Roberts and Griswold (985a).

T, I and J; $\tilde{\mu}_I{}^2/r^3$ may also be written $\tilde{\mu}_I H''$ where $H'' = \tilde{\mu}_I/r^3$ is the magnetic field produced by one nucleus at the position of the other (r = internuclear distance in the molecule); Q_I is the quadrupole moment of the nucleus of spin I. It is to be noted that the dependence on M_J and M_T is the same for the spin-spin and the quadrupole interaction.

For H_2 the third term in (V, 112) does not occur since $Q_I = 0$ (see above) and the constant a_{TIJ} for $J = 1$ ($I = \frac{1}{2}$, $T = 1$) has the value $\frac{2}{5}$.[25] For D_2 the two constants a_{TIJ}

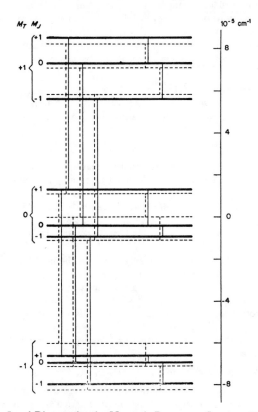

Fig. 150. Energy Level Diagram for the Magnetic Resonance Spectra of H_2 in the State $J = 1$, $T = 1$ at 500 gauss (to scale). The full lines represent the actual levels and transitions, the broken lines those obtained in zero approximation [equation (V, 108)].

and b_{TIJ} in the state $J = 1$ ($I = 1$, $T = 1$) have the values $\frac{1}{5}$ and $-\frac{1}{4}$, respectively. Fig. 150 gives an energy level diagram of the case $J = 1$, $T = 1$. The figure is drawn to scale for H_2 but applies qualitatively also to D_2 in the $J = 1$ level. The zero approximation levels are indicated by broken lines.

In the case of heteronuclear molecules formula (V, 112) for the first order energy correction has to be replaced by one in which each nucleus acts independently [see Kellogg, Rabi, Ramsey, and Zacharias (1127)]. Moreover both for homonuclear and heteronuclear molecules second and third order corrections become of importance, particularly for small fields (incomplete Paschen-Back effect).

---

[25] This is the value given in Kellogg, Rabi, Ramsey, and Zacharias's first paper (1126). It does not agree with the general formula in their second paper (1127).

Because of the weakness of the coupling of T (or I) and J there can be no transitions in which both M_T and M_J change. Thus the *selection rules* are [compare (I, 71)][26]

$$\Delta M_T = (0), \pm 1, \quad \Delta M_J = 0 \qquad (V, 113)$$

or

$$\Delta M_T = 0, \quad \Delta M_J = (0), \pm 1 \qquad (V, 114)$$

The transitions possible according to (V, 113) are indicated at the left, those possible according to (V, 114) at the right in Fig. 150 in which $J = 1$ and $T = 1$.

The six transitions possible in each case would coincide if the zero-order formula (V, 108) represented the levels exactly (broken-line levels in the figure). They would then occur at the wave numbers $\mu_{0n} g_I H / hc$ and $\mu_{0J} g_J H / hc$, respectively, which for H_2 in a field of 10000 oersted is $1.416_1 \times 10^{-3}$ and 0.2234×10^{-3} cm^{-1} respectively (42.45 and 6.698 megacycles/sec). Actually on account of the above-mentioned interactions the levels are not exactly equidistant and a splitting into six components occurs. This is indeed observed as shown by the molecular beam magnetic resonance spectra of H_2 reproduced in Fig. 35, p. 60. These spectra refer to the state $J = 1$. From the frequencies of the centers of these patterns, according to the above, accurate values for $\tilde{\mu}_I$ and $\tilde{\mu}_J$ can be obtained. The observed splittings can be used by comparison with (V, 112) to obtain the internal fields H' and H'' ($= \tilde{\mu}_I / r^3$). The third term in (V, 112) vanishes since the H nucleus, having $I = \frac{1}{2}$, cannot have a quadrupole moment. It is found that the two sets of values of H' and H'' obtained from the splittings in the two spectra ($\Delta M_I = \pm 1$ and $\Delta M_J = \pm 1$) agree exactly, particularly if the second and third order corrections (not considered here) are taken into account. Moreover, since $\tilde{\mu}_I$ is obtained from the position of the whole group with $\Delta M_J = 0$ (see above) and since the average internuclear distance r in the lowest state of H_2 is known from ultraviolet spectra, the value of $H'' = \tilde{\mu}_I / r^3$ can be predicted and excellent agreement with the value obtained from the splittings in Fig. 35 is found. The experimental values are $H' = 27.0$, $H'' = 34.1$ oersted.

If the nuclear quadrupole moment is not zero (as for D_2) the third term in (V, 112) must be included. Since the dependence of this term on M_J and M_T is the same as for the second term one obtains from the spacing in the fine structure only the combination $a_{TIJ} \tilde{\mu}_I^2 / r^3 + b_{TIJ} e^2 qQ$. However, since the value of $\tilde{\mu}_I / r^3$ is known (see above) the existence of a quadrupole moment can be readily demonstrated and the quantity qQ determined. For D_2 the quantity q has been determined from the electronic wave functions of the D_2 molecule and $Q = +2.73 \times 10^{-27}$ cm^2 has been obtained [Kellogg, Rabi, Ramsey, and Zacharias (1127)].

Excellent confirmations of the above results have been obtained from the study of the HD molecule [Ramsey (1127)(1332)].

Thus far in no case other than H_2, HD, and D_2 has a complete resolution of the magnetic resonance spectra been obtained. This is mainly due to the fact that at temperatures at which molecular beams are obtainable a large number of rotational levels are excited and their spectra are superimposed. In addition, the number of line components increases with J Feld and Lamb (933) [see also Foley (933) and Ramsey (1333)] have made detailed computations of the splittings for large J values and of the shape of unresolved splitting patterns. On account of the quadrupole moment subsidiary resonance minima are expected near the main resonance corresponding to the magnetic moment of the nucleus considered. Such subsidiary minima have been found, for example, for LiF by Kusch and Millman [quoted by Feld and Lamb (933)], and for Na^{23}Br, Na^{23}Cl and Na^{23}I by Nierenberg and Ramsey (1283).

It must be emphasized that the hyperfine structures discussed here are observable only for the undisturbed molecule as it occurs in molecular beam experiments or at extremely low pressure. This is why the investigation of a *nuclear radiofrequency resonance absorption* in gases of high pressure or liquids and solids by the nuclear induction method or related methods

---

[26] For heteronuclear molecules (V, 113) must be replaced by $\Delta M_{I_1} = 0, \pm 1, \Delta M_{I_2} = 0$, $\Delta M_J = 0$ or $\Delta M_{I_1} = 0, \Delta M_{I_2} = 0, \pm 1, \Delta M_J = 0$.

[Bloch (816), Purcell, Torrey and Pound (1325)(1324), Roberts (1350)] is not applicable to their study.   The frequent collisions of gas molecules at higher pressures completely disturb and average out the interactions considered in (V, 112), the frequency of collision, for example, at 1 atm being far greater than the frequencies corresponding to the hyperfine structures.

The Stark effect of the hyperfine structure of the energy levels of diatomic molecules has been discussed theoretically by Fano (926a) and Nierenberg, Rabiand Slotnick (1282a) [see also Trischka (1471)].

Note added in proof: Beringer and Castle (795a) have recently observed in a magnetic resonance absorption experiment at very low pressure transitions between the Zeeman components of the rotational levels of O_2 and NO in the microwave region.   These spectra are analogous to the H_2 magnetic resonance spectra discussed above (p. 300), except that since O_2 and NO have a magnetic moment of the order of a Bohr magneton rather than a nuclear magneton they are shifted from the radio-frequency into the microwave region.

CHAPTER VI

BUILDING-UP PRINCIPLES, ELECTRON CONFIGURATIONS, AND VALENCE

There are as yet only comparatively few diatomic molecules for which a large number of electronic states have been established on the basis of their observed band spectra. For most molecules only two or three electronic states have been found. Nevertheless it is clear that in all cases the number of states is of the same order as for atoms. It is the object of this chapter to discuss *what electronic states are to be expected on the basis of wave mechanics* and to compare this theoretical prediction with observation. In addition, in order to determine which electronic transitions may be expected to be observable it will be necessary to consider on the basis of wave mechanics the question of the *stability of the electronic states* as well as the question of the *absolute intensity* of the electronic transitions. The discussion of stability will of necessity include a discussion of *chemical valence*. The theoretical considerations of this chapter will prove to be of great value in the study of heats of dissociation and dissociation processes to be treated in Chapter VII.

The *manifold of electronic states*—that is, the totality of the electronic terms of a molecule—can be obtained, as for atoms, by the successive bringing together of the parts (*building-up principle*). However, whereas the building up of the atom can be accomplished only in one way, for a molecule there are three different possibilities:

(1) The molecule may be built up by *bringing together the whole atoms* of which it consists; that is, we can investigate the question of what kind of molecular states result from given states of the separated atoms. When we have carried this out for all possible combinations of atomic states, we obtain the complete manifold of the states of the molecule.

(2) Instead of beginning with infinitely large separation, as in (1), we may start with zero nuclear separation; that is, we may *split the so-called united atom*. This procedure is of course purely hypothetical, but, in spite of that, it is suitable for the determination of the term manifold.

(3) Finally, we may employ a procedure analogous to that used for atoms: We *add the individual electrons* one after another *to the nuclei*, which are regarded as fixed, and consider in which "orbits" or *orbitals* (see p. 22 and below) the electrons will arrange themselves. The different possible electronic arrangements (*electron configurations*) then give the possible states, just as for atoms.

1. Determination of the Term Manifold from the States of the Separated Atoms

Wigner and Witmer (712), on the basis of quantum mechanics, have derived rules for determining what types of molecular states result from given states of

315

the separate atoms.   These *correlation rules* have been derived under the assumption of an *adiabatic change of internuclear distance*, that is, they hold for the (adiabatic) potential curves of the electronic states.   As long as Russell-Saunders coupling is valid for the separated atoms as well as the molecule, the correlation rules are rigorously valid and complete—that is, give all molecular electronic states.   For other coupling cases different correlation rules, also given by Wigner and Witmer, hold.   Thus far no exceptions to these correlation rules have been found empirically, in agreement with the general validity of wave mechanics for all molecular problems.   Assuming this validity, the correlation rules supply a perfectly reliable guide in correlating the observed molecular states with those of the separated atoms.   While Wigner and Witmer used the method of group theory in the derivation of the correlation rules, we shall use here more elementary considerations which give most of the rules.   For those that cannot be derived in this way we shall refer the reader to Wigner and Witmer's original paper.

Unlike atoms.   When two atoms with L and S values equal to L_1, S_1, and L_2, S_2, respectively, are brought up to each other, an (inhomogeneous) electric field arises in the direction of the line joining the nuclei, producing a space quantization of L_1 and L_2 with reference to this direction such that the components are M_{L_1} and M_{L_2} (see p. 28).   The resultant orbital angular momentum about the line joining the nuclei is therefore $M_{L_1} + M_{L_2}$, and the quantum number Λ of the molecule thus formed is

$$\Lambda = \left| M_{L_1} + M_{L_2} \right|. \tag{VI, 1}$$

By combination of all the possible M_{L_i} values we obtain all the possible Λ values.

In general, the different orientations of the L_i and thus the different Λ values correspond to different energies in the electric field; in fact, the difference in energy will be greater the stronger the field—that is, the smaller the internuclear distance.   Thus, from a given combination of states of the separated atoms there results a number of different states of the whole system.   The quantum number Λ of the resultant orbital angular momentum retains its meaning even for a close approach of the two atoms, such as in the actual molecule, where the L_i and M_{L_i} of the individual atoms have completely lost their meaning as angular momenta.   Thus in this way, by starting out with large internuclear distance where the L_i and M_{L_i} are well defined, we can derive the number and type of the resulting molecular states.

It should be noted that states which differ only in the sign of both M_{L_1} and M_{L_2} (and thus also of $M_{L_1} + M_{L_2}$) have equal energies if $M_{L_1} + M_{L_2} \neq 0$. They correspond to the two components of a molecular state with $\Lambda \neq 0$ which are degenerate with each other (see p. 213).   For $\Lambda = M_{L_1} + M_{L_2} = 0$, however, each combination corresponds to a different molecular state (Σ state).

Let us consider as an example the case of an atom in an S state ($L_1 = 0$) approaching an atom in a D state ($L_2 = 2$).   The possible orientations of L_2

are indicated in Fig. 151 (*a*). We have $M_{L_1} = 0$ and the M_{L_2} values 2, 1, 0, -1, -2; therefore according to (VI, 1) Λ assumes the values 2, 1, 0, 1, 2. The two values $\Lambda = 2$ and the two values $\Lambda = 1$ belong to a doubly degenerate Δ and Π state, respectively, while $\Lambda = 0$ represents a Σ state. These three states have, of course, different energies. In a similar way, from an S state and P state we obtain the molecular states Π and Σ.

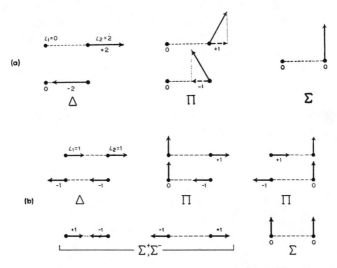

FIG. 151. **Vector Diagrams Showing the Determination of Molecular States (a) for the Combination $S + D$, (b) for the Combination $P + P$ of the Separated Atoms.** The arrows represent the orientations of the L_i of the separated atoms. The numbers $+2$, $+1$, 0, -1, -2 indicate the M_{L_1} and M_{L_2} values. For simplicity the approximate values, $L_i (h/2\pi)$, have been used for the magnitude of the L_i [not the rigorous values $\sqrt{L(L+1)}(h/2\pi)$; see p. 26].

Fig. 151 (*b*) gives similar vector diagrams for the combination $P + P$ of the separated atoms. Here, combining $M_{L_1} = +1$, 0, -1 with $M_{L_2} = +1$, 0, -1 gives the nine $M_{L_1} + M_{L_2}$ values $+2$, $+1$, 0, $+1$, 0, -1, 0, -1, -2 which yield one Δ, two Π, and three Σ states. Two of the configurations yielding Σ states differ only in the sign of both M_{L_i} values (see lower left part of the figure). In zero approximation they are degenerate with each other, just as the two configurations belonging to the Δ state as well as those belonging to the two Π states. However, while the Δ and Π states retain their twofold degeneracy even when the finer interaction of the two atoms is taken into account, the two coinciding Σ states split into two different Σ states of different energy. One of these is Σ^+, the other Σ^-. It must be noted that neither the Σ^+ nor the Σ^- state resulting in this way can be ascribed to a particular one of the two configurations in Fig. 151 (*b*) (compare the similar situation discussed on p. 217 f.).

The molecular states arising from other combinations of the separated atoms may be obtained by the same methods. The results for all cases of practical importance are collected in Table 26.   When several states of the same kind result their number is given in parentheses after the term symbol.   The number of Σ states that arise is always odd.   There is always one Σ state (with $M_{L_1} = 0$, $M_{L_2} = 0$) in addition to pairs of Σ^+ and Σ^- states similar to that discussed above for $P + P$.   The symmetry of this single Σ state cannot be obtained by elementary means.   According to Wigner and Witmer it is Σ^+ when $L_1 + L_2 + \Sigma l_{i_1} + \Sigma l_{i_2}$ is even and it is Σ^- when this sum is odd.   It will be remembered that Σl_i determines the *parity*, even or odd, of an atomic state.   Thus the symmetry of the Σ state in question depends on the L values as well as the parities of the atomic states from which it results.   It is for this reason that the parities are indicated by subscripts g and u in Table 26.

TABLE 26.   MOLECULAR ELECTRONIC STATES RESULTING FROM GIVEN STATES
OF THE SEPARATED (UNLIKE) ATOMS

[According to Wigner and Witmer (712); see also similar tables in Mulliken (514).]

States of the Separated Atoms	Molecular States
$S_g + S_g$ or $S_u + S_u$	Σ^+
$S_g + S_u$	Σ^-
$S_g + P_g$ or $S_u + P_u$	Σ^-, Π
$S_g + P_u$ or $S_u + P_g$	Σ^+, Π
$S_g + D_g$ or $S_u + D_u$	Σ^+, Π, Δ
$S_g + D_u$ or $S_u + D_g$	Σ^-, Π, Δ
$S_g + F_g$ or $S_u + F_u$	Σ^-, Π, Δ, Φ
$S_g + F_u$ or $S_u + F_g$	Σ^+, Π, Δ, Φ
$P_g + P_g$ or $P_u + P_u$	$\Sigma^+(2)$, Σ^-, $\Pi(2)$, Δ
$P_g + P_u$	Σ^+, $\Sigma^-(2)$, $\Pi(2)$, Δ
$P_g + D_g$ or $P_u + D_u$	Σ^+, $\Sigma^-(2)$, $\Pi(3)$, $\Delta(2)$, Φ
$P_g + D_u$ or $P_u + D_g$	$\Sigma^+(2)$, Σ^-, $\Pi(3)$, $\Delta(2)$, Φ
$P_g + F_g$ or $P_u + F_u$	$\Sigma^+(2)$, Σ^-, $\Pi(3)$, $\Delta(3)$, $\Phi(2)$, Γ
$P_g + F_u$ or $P_u + F_g$	Σ^+, $\Sigma^-(2)$, $\Pi(3)$, $\Delta(3)$, $\Phi(2)$, Γ
$D_g + D_g$ or $D_u + D_u$	$\Sigma^+(3)$, $\Sigma^-(2)$, $\Pi(4)$, $\Delta(3)$, $\Phi(2)$, Γ
$D_g + D_u$	$\Sigma^+(2)$, $\Sigma^-(3)$, $\Pi(4)$, $\Delta(3)$, $\Phi(2)$, Γ
$D_g + F_g$ or $D_u + F_u$	$\Sigma^+(2)$, $\Sigma^-(3)$, $\Pi(5)$, $\Delta(4)$, $\Phi(3)$, $\Gamma(2)$, H
$D_g + F_u$ or $D_u + F_g$	$\Sigma^+(3)$, $\Sigma^-(2)$, $\Pi(5)$, $\Delta(4)$, $\Phi(3)$, $\Gamma(2)$, H

We have yet to determine the *multiplicity* of the resulting molecular states. Let us assume that the coupling of the L_i to the field between the nuclei is strong compared to the coupling between L_i and S_i.   Then, since the spin is not influenced by an electric field, the two spin vectors S_1 and S_2 of the separated atoms add together forming a resultant S, the resultant spin vector of the molecule. For the corresponding quantum number S we have (see p. 25)

$$S = (S_1 + S_2), (S_1 + S_2 - 1), (S_1 + S_2 - 2), \cdots, |S_1 - S_2|. \quad \text{(VI, 2)}$$

For a given orientation of the L_i, each of the values of S in (VI, 2) is possible; that is, *each of the states given in Table 26 can occur with each of the multiplicities*

$(2S + 1)$ *given by* (VI, 2). Naturally the states with different multiplicities have different energies.

If, for example, one of the two atomic states is a singlet state $(S_1 = 0)$, only one S value of the molecule is possible, $S = S_2$, and therefore in this case the molecular states given in the particular row in Table 26 occur only with the multiplicity $(2S_2 + 1)$. (They are singlets if both atomic states are singlets.) On the other hand, if $S_1 = S_2 = \frac{1}{2}$, according to (VI, 2), $S = 1$ or 0; that is, the states resulting on the basis of the L_i according to Table 26 occur as singlets as well as triplets. There are then just twice as many states. These and other cases are collected in

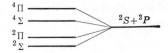

Fig. 152. **Molecular States Resulting from** $^2S + ^3P$ **(Schematic).** To the right, for large internuclear distances; to the left, in the molecule.

Table 27. As an illustration the molecular states resulting from an atom in a ^2S state and an atom in a ^3P state are represented schematically in an energy level diagram in Fig. 152.

TABLE 27. MULTIPLICITIES OF MOLECULAR ELECTRONIC STATES FOR GIVEN MULTIPLICITIES OF THE SEPARATED ATOMS

Separated Atoms	Molecule
Singlet + singlet	Singlet
Singlet + doublet	Doublet
Singlet + triplet	Triplet
Doublet + doublet	Singlet, triplet
Doublet + triplet	Doublet, quartet
Doublet + quartet	Triplet, quintet
Triplet + triplet	Singlet, triplet, quintet
Triplet + quartet	Doublet, quartet, sextet
Quartet + quartet	Singlet, triplet, quintet, septet

It can be seen from these considerations that the number of molecular states, though of the same order as, is appreciably larger than, that of the states of the separated atoms.

If the coupling between the L_i and S_i in the separated atoms is strong compared to the coupling of the L_i to the internuclear axis, a space quantization of the J_i's $(J_i = L_i + S_i)$ rather than of the L_i's takes place in the electric field produced when the atoms approach each other. This may be called (J_1, J_2) *coupling.* The electronic angular momentum along the internuclear axis of the molecule (inclusive of spin)—that is, Ω—is then obtained by adding the M_{J_i} values. Analogous to (VI, 1), we have

$$\Omega = \left| M_{J_1} + M_{J_2} \right| \qquad (VI, 3)$$

To every combination of two M_{J_i} values there corresponds a different molecular state, except that states differing only in the sign of both M_{J_1} and M_{J_2} form a degenerate pair as long as $\Omega \neq 0$. Usually in this case for the molecule Hund's case (c) applies (see p. 224) that is, only Ω and not Λ is defined.

When both J_i are integral, the resulting Ω values for a given combination of J_i values are the same as the Λ values for the same combination of L_i values (see Table 26). Also the conditions for 0^+ and 0^- states are the same as for Σ^+ and Σ^-. For example, for $J_1 = 1$ and $J_2 = 1$, if both atomic states are even we obtain the Ω values 2, 1, 1, 0^+, 0^+, 0^- for the resulting molecular states.

When both J_i are half-integral, the Ω values can be obtained in an analogous way from (VI, 3). In this case the number of states with $\Omega = 0$ is even, and half of them are 0^+ and the other half 0^-. For example, for $J_1 = \frac{3}{2}$, $J_2 = \frac{3}{2}$, we obtain the Ω values 3, 2, 2, 1, 1, 1, 0^+, 0^+, 0^-, 0^-; or, for $J_1 = \frac{3}{2}$, $J_2 = \frac{1}{2}$, we obtain the Ω values 2, 1, 1, 0^+, 0^-.

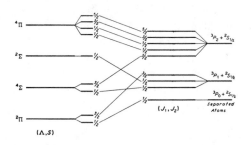

Fig. 153. **Transition from** (J_1, J_2) **to** (Λ, S) **Coupling for the Molecular States Arising from** $^3P + {}^2S$ **of the Separated Atoms (Schematic).** If both atomic states have the same parity the Σ states are Σ^-, otherwise Σ^+.

When one J_i is integral and the other half-integral, again it is easy to form all possible $M_{J_1} + M_{J_2}$ values and determine Ω from (VI, 3). Here $\Omega = 0$ does not occur. As an example we obtain for $J_1 = 2$, $J_2 = \frac{1}{2}$ the Ω values $\frac{5}{2}, \frac{3}{2}, \frac{3}{2}, \frac{1}{2}, \frac{1}{2}$.

Even when the molecular states belong to case (a) or (b), for sufficiently large internuclear distances, (J_1, J_2) coupling applies since then the multiplet splitting of the separated atoms is larger than the energy of interaction of the two atoms. Naturally, for given L_1, S_1 and L_2, S_2 of the separated atoms the possible Ω values resulting from (VI, 3) are the same as those obtained from the Λ and S values resulting from (VI, 1) and (VI, 2). As an example, in Fig. 153 there is shown for the case $^2S + {}^3P$ the correlation of the states obtained for large internuclear distance according to (J_1, J_2) coupling and the states resulting from (Λ, S) coupling at small internuclear distance. It is seen that some of the multiplet components of a given state at small internuclear distance must be correlated with components of different mutliplets of the separated atoms.

If according to strict (Λ, S) coupling an intersection of potential curves of two states arises (for example of a $^3\Pi$ and a $^1\Sigma^+$ state) it may happen that on account of a transition to (J_1, J_2) coupling the intersection for some of the component states may be avoided (in the example the intersection between $^3\Pi_{0^+}$ and $^1\Sigma^+$). Usually, however, in such cases there will be a near-intersection of the type discussed previously (p. 297).

Like atoms. If the nuclei of the two atoms that are brought together have the same charge, the symmetry property even or odd of the resulting states must also be determined. In the case that the two atoms are *in the same state* some of the resulting states, obtained in the same way as before (Table 26) are even, others odd. Which are even and which odd has been determined from group theory by Wigner and Witmer (712) [see also Mulliken (514)]. We give in Table 28 only the results for all cases of practical interest.

If the two atoms (for equal nuclear charge) are *not in the same state*, there is for large separation of the nuclei a degeneracy between those two states of the whole system for which either the one or the other atom is in the higher excited state.   According to Table 26 and relation (VI, 2) the same molecular states result from both combinations of atomic states.   For example, from $^1S_g + {}^1P_u$ we get $^1\Sigma^+$ and $^1\Pi$; but the same states arise from $^1P_u + {}^1S_g$—that is, when the

TABLE 28.   MOLECULAR ELECTRONIC STATES RESULTING FROM IDENTICAL
STATES OF THE SEPARATED LIKE ATOMS

[According to Wigner and Witmer (712); see also Mulliken (514).]

States of the Separated Atoms*	Molecular States
$^1S + {}^1S$	$^1\Sigma_g{}^+$
$^2S + {}^2S$	$^1\Sigma_g{}^+, \, {}^3\Sigma_u{}^+$
$^3S + {}^3S$	$^1\Sigma_g{}^+, \, {}^3\Sigma_u{}^+, \, {}^5\Sigma_g{}^+$
$^4S + {}^4S$	$^1\Sigma_g{}^+, \, {}^3\Sigma_u{}^+, \, {}^5\Sigma_g{}^+, \, {}^7\Sigma_u{}^+$
$^1P + {}^1P$	$^1\Sigma_g{}^+(2), \, {}^1\Sigma_u{}^-, \, {}^1\Pi_g, \, {}^1\Pi_u, \, {}^1\Delta_g$
$^2P + {}^2P$	$^1\Sigma_g{}^+(2), \, {}^1\Sigma_u{}^-, \, {}^1\Pi_g, \, {}^1\Pi_u, \, {}^1\Delta_g, \, {}^3\Sigma_u{}^+(2), \, {}^3\Sigma_g{}^-, \, {}^3\Pi_g, \, {}^3\Pi_u, \, {}^3\Delta_u$
$^3P + {}^3P$	Singlet and triplet terms as for $^2P + {}^2P$; in addition, $^5\Sigma_u{}^+(2), \, {}^5\Sigma_u{}^-, \, {}^5\Pi_g, \, {}^5\Pi_u, \, {}^5\Delta_g$
$^1D + {}^1D$	$^1\Sigma_g{}^+(3), \, {}^1\Sigma_u{}^-(2), \, {}^1\Pi_g(2), \, {}^1\Pi_u(2), \, {}^1\Delta_g(2), \, {}^1\Delta_u, \, {}^1\Phi_g, \, {}^1\Phi_u, \, {}^1\Gamma_g$
$^2D + {}^2D$	Singlets as for $^1D + {}^1D$; in addition, $^3\Sigma_u{}^+(3), \, {}^3\Sigma_g{}^-(2), \, {}^3\Pi_u(2), \, {}^3\Pi_g(2), \, {}^3\Delta_u(2), \, {}^3\Delta_g, \, {}^3\Phi_u, \, {}^3\Phi_g, \, {}^3\Gamma_u$
$^3D + {}^3D$	Singlets as for $^1D + {}^1D$, triplets as for $^2D + {}^2D$, and quintets like singlets

* Whether the atomic state is even or odd is of no importance here, since both atoms are in the same state.

excitation energies of the two like atoms are interchanged.   Thus we have two $^1\Sigma^+$ and two $^1\Pi$ states that have equal energies for large internuclear distance. On account of the interaction of the two atoms, as the internuclear distance becomes smaller we get four different states of different energies.   A detailed consideration [Wigner and Witmer (712)] shows that one of each of the pairs of like states is even and the other is odd.   Thus, in the example $^1S_g + {}^1P_u$, for like atoms, we obtain the states $^1\Sigma_g{}^+, \, {}^1\Sigma_u{}^+, \, {}^1\Pi_g$, and $^1\Pi_u$.   Similarly, in more complicated cases *each of the states occurring for unlike atoms* [see Table 26 and relation (VI, 2)] *occurs twice for like atoms, once as an odd and once as an even state.*   It is therefore not necessary to give a special table.   It must be noted that even if the two atomic states are of the same type but have different energy the last described method has to be applied and not Table 28.

For the case that the L_i and S_i are strongly coupled [case (c)], again analogous relations hold.   If the atoms are in the same state and if J is integral, we can use Table 28 if we substitute J for L and Ω for Λ and consider the singlet states only.   For half-integral J we supplement the table by the two following examples: If both atoms are in the same state with $J = \frac{1}{2}$, we obtain the molecular states $1_u, \, 0_u{}^+, \, 0_u{}^-$, where as usual in case (c) the numbers give the Ω values. For $J_1 = J_2 = \frac{3}{2}$ we obtain 3_u, 2_u, 2_g, 1_u, 1_g, 1_u, 0_g^+, 0_u^-, 0_g^+, 0_u^-. If the atoms are in unlike states, as before, each of the states occurring for unlike atoms occurs twice, once as an odd and once as an even state.

2. Determination of the Term Manifold from the States of the United Atom

Unlike atoms. If we imagine the nucleus of an atom to be split into two unlike nuclei an (inhomogeneous) electric field is produced in the direction of the internuclear axis. In general this field will be strong enough to uncouple the spin S from the orbital angular momentum L. The magnitude of the component M_L of L is Λ and we have for the resulting molecule (see Chapter V, section 1):

$$\Lambda = |M_L| = L, L - 1, L - 2, \cdots, 0. \tag{VI, 4}$$

The spin S of the molecule remains the same as in the atomic state from which the molecular state results. Thus, for example, from a ^3D_g state of the magnesium atom, we get the molecular states $^3\Delta$, $^3\Pi$, and $^3\Sigma$ if we imagine the nucleus to be divided into a Be and an O nucleus or into a B and an N nucleus, and so on. Whether the resulting Σ state is a Σ^+ or a Σ^- state depends, as a more detailed consideration shows, upon whether $L + \Sigma l_i$ for the united atom is even or odd. Since $L + \Sigma l_i$ is even for ^3D_g the $^3\Sigma$ state of the example is $^3\Sigma^+$.

If L and S are very strongly coupled in the united atom, there is no uncoupling of L and S when the nuclei are separated. A state of the united atom with given J will then split into molecular states with

$$\Omega = |M_J| = J, J - 1, J - 2, \cdots, \tfrac{1}{2} \text{ or } 0, \tag{VI, 5}$$

where the lowest Ω value is $\frac{1}{2}$ or 0, depending on whether J is half-integral or integral. In the latter case the state with $\Omega = 0$ is a 0^+ or 0^- state according as $J + \Sigma l_i$ is even or odd.

Like atoms. For a splitting of the united atom into atoms with nuclei of equal charge—for example, splitting of the Mg atom into C + C—we get exactly the same states as for unlike atoms, except that the symmetry property even or odd has to be added. Wigner and Witmer (712) have shown that the resulting molecular states are all even or all odd according as the state of the united atom is an even or an odd one. In the above example we get, therefore, for the C_2 molecule the states $^3\Delta_g$, $^3\Pi_g$, and $^3\Sigma_g^+$.

3. Determination of the Term Manifold from the Electron Configuration

(a) Quantum Numbers of the Individual Electrons

As mentioned previously (p. 34) for some diatomic molecules, especially He_2, Rydberg series of bands instead of, or in addition to the usual vibrational progressions have been observed. Obviously, these band series result from transitions from a *Rydberg series of electronic states* (see Chapter I, p. 8) to a lower electronic state.   Similar considerations apply, of course, to the Rydberg series of band *systems* observed in some cases (for example H_2, see below).

For atoms, Rydberg series of terms correspond to the excitation of a single electron to orbits with different principal quantum numbers.   It is natural to assume that a similar explanation is applicable to molecules.   In order to verify this assumption we must investigate in somewhat greater detail the motions of the individual electrons in the molecule and the corresponding quantum numbers.

Single electron in an axially symmetric electric field.   We shall first consider the motion of a single outer electron about a molecular core consisting of the two nuclei and the remaining electrons (for example, an electron moving about He_2^+).   We shall suppose the nuclei to be held fixed at a certain distance from each other.   Such a system is called a *two-center system*.   The field of force acting on the outer electron may be regarded, to a good approximation, as axially symmetric about the internuclear axis.   We shall assume here that it has exact axial symmetry.   For large distances of the electron from the core the potential energy V of this field approaches zero.

The investigation of the Schrödinger equation (I, 12) for the motion of an electron in such a field of force shows that, as for atoms, it is soluble for every positive energy value E, but only for certain discrete negative E values. $E = 0$ corresponds to an infinitely large separation of the electron from the core, without relative kinetic energy.

As for an atom, the continuous region of positive E values corresponds to the *removal or capture of an electron* (ionization or recombination) with various amounts of kinetic energy even at an infinite distance from the core.   On the other hand for the discrete negative E values the electron cannot leave the field of the core;   they correspond to the *stationary electronic states* of the system.

As will be shown in more detail below on the basis of wave mechanics, if we disregard electron spin, the stationary states can be characterized by three quantum numbers, of which, however, only one is precisely defined for all separations of the two nuclei.   The latter quantum number is the exact analogue of the quantum number Λ, previously introduced for the electronic state of the whole molecule, and is designated by λ.   It is the quantum number of the *component of the orbital angular momentum* (λ) *of the electron about the internuclear axis* whose magnitude is $\lambda(h/2\pi)$.   The possible values of λ are 0, 1, 2, 3, $\cdots$. According to which of these values an electron has, it is called a σ, π, δ, φ, $\cdots$ electron, respectively.   If $\lambda \neq 0$ the electron revolves about the internuclear axis, although according to wave mechanics there is no definite orbit (see below).

The other two quantum numbers can be defined only approximately, and, in fact, this can be done in two different ways, depending on whether we start out from the case of very small internuclear distance ($r \rightarrow 0$) or very large internuclear distance ($r \rightarrow \infty$).

In the united atom—that is, for $r = 0$—the possible states of an electron are defined by the quantum numbers n and l (see p. 22).   If the united atom is split in such a way that the distance between the two nuclei is still small, the possible electronic states are the same as for the united atom in an electric field (Stark effect).   In this case (*small* internuclear distance) the quantum numbers

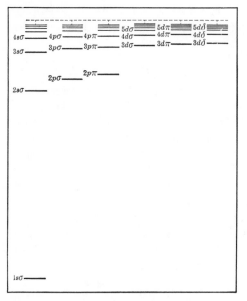

n and l are still approximately defined. The orbital angular momentum l of the electron is space quantized in the field and precesses with constant component $m_l(h/2\pi)$ about the field direction (see Fig. 95, which shows the corresponding behavior for L).[1]   States with different $|m_l|$ have somewhat different energies. $|m_l|$ is the quantum number λ, already introduced above, which retains its meaning as orbital angular momentum about the internuclear axis for any value of r. Thus λ, for a given value of l, can take the values

$$\lambda = l, l - 1, l - 2, \cdots, 0.   \text{(VI, 6)}$$

States with $\lambda \neq 0$ are doubly degenerate, since a positive and a negative m_l having the same magnitude correspond to one and the same λ.

Fɪɢ. 154. **Energy Levels of a Single Electron in the Field of Two Fixed Centers at a Small Distance from Each Other (Schematic).** Further term series with higher values of l and λ would join on to the right.   The broken line above represents the ionization limit.

The energy level diagram for an electron in the case of small internuclear distance is drawn schematically in Fig. 154.   In this diagram the values of the three quantum numbers are given by symbols such as $2s\sigma, 3p\sigma, 4d\pi$, and so on, in which the number gives the n value, the Roman letter gives the l value (as for atoms), and the Greek letter gives the λ value (see above).   For $n = 2$ there are according to the restrictions on l and λ, three states: $2s\sigma, 2p\sigma, 2p\pi$; for $n = 3$ there are six: $3s\sigma, 3p\sigma, 3p\pi, 3d\sigma, 3d\pi, 3d\delta$ which may be expected to lie in this order.   With increasing separation of the two fixed centers (nuclei) the splitting between the states with different λ for the same n and the same l (which is zero for $r = 0$) becomes greater and n and l lose more and more their meaning as quantum numbers.   However, according to the Ehrenfest adiabatic law (see A.A., p. 86), the number of states is not altered by this change.   Therefore, in the manner described, we can at least obtain the *manifold of the states of an electron* even for intermediate internuclear distances.

[1] On the old Bohr model this would mean that the plane of the elliptical orbit (which is perpendicular to l) can have only certain inclinations to the internuclear axis and rotates about this axis in such a way that the angle is kept constant.

For *very large separation* of the two nuclei, A and B, the one electron considered may be either with nucleus A or with nucleus B. Suppose it has the quantum numbers n and l in the particular atom. As the atoms approach each other, an electric field is produced, in which again a space quantization of l takes place such that its component in the internuclear axis is equal to $m_l(h/2\pi)$. $|m_l| = \lambda$ is again the orbital angular momentum of the electron about the internuclear axis. In designating an electron the n and l values that it has in the separated atom ($1s, 2s, 2p, \cdots$) are usually added after the symbol that indicates the λ value, thus: $\sigma 1s$, $\sigma 2p$, $\pi 2p$, and so on [some authors write $\sigma(1s)$, $\sigma(2p), \cdots$]. Frequently, the atom from which the electron under consideration is derived is also indicated—for example, $\sigma 1s_A$, $\sigma 1s_B$, and so on. It should be noted that the quantum numbers n and l that an electron has in one of the separated atoms are in general not the same as those it would have in the united atom.

The wave-mechanical treatment of the one electron system supplies not only the energy values and the angular momenta in the various possible states but also the *eigenfunctions* whose magnitude square gives the probability of finding the electron at the various positions in the molecule. In order to illustrate the form of the eigenfunctions of the lowest states, their *nodal surfaces*—that is, the surfaces on which they have the value zero—are given in two cross sections in Fig. 155 according to Weizel (703). Strictly speaking, the figure holds only for the case of an electron moving in the field of two bare nuclei; but qualitatively it holds also for the more general case of an electron moving in an axially symmetric electric field. The nodal surfaces are planes through the internuclear axis, or rotational ellipsoids or hyperboloids with the nuclei as foci. The hyperboloids degenerate in some cases into planes perpendicular to the internuclear axis. The sign of the eigenfunction is indicated in Fig. 155 by different hatching; it is opposite on the two opposite sides of a nodal surface.

If the two nuclei in the core have *unequal charges*, a given eigenfunction has different magnitudes at two points symmetrically placed with respect to the midpoint. It is obvious that the probability density in the neighborhood of the more strongly charged nucleus is the greater.

If, however, the two nuclei have *equal charges*, the eigenfunction has the same magnitude at two points placed symmetrically with respect to the midpoint (see also p. 218 f.) and can at most differ in sign. Corresponding to the earlier nomenclature, the states of the electron (for equal nuclear charge) are called *even* (g) or *odd* (u) according as the eigenfunction remains unaltered or changes sign by reflection at the origin. Symbolically, we write σ_g, σ_u, π_g, π_u, $\cdots$. We can easily verify from Fig. 155 that for an even l of the united atom ($s, d, \cdots$ electrons) the eigenfunctions have the property g and for odd l ($p, f, \cdots$ electrons) they have the property u. At infinite internuclear distance always two states, such as $\pi 2p_A$ and $\pi 2p_B$, have the same energy if $A = B$. As the atoms approach each other, this degeneracy is removed and we obtain in the example a π_g and π_u state. Other states of the electron behave correspondingly. Therefore, for large internuclear distances and equal nuclear charge we use designations like $\pi_u 2p$, $\pi_g 2p$, and so on. The property g–u for equal nuclear charge is exactly defined for any internuclear distance as long as the field in which the electron moves has exact axial symmetry.

The eigenfunctions considered here refer to the motion of the electron apart from spin. They are also called *orbital wave functions*, and the corresponding states are called *orbitals* [see Mulliken (517)].

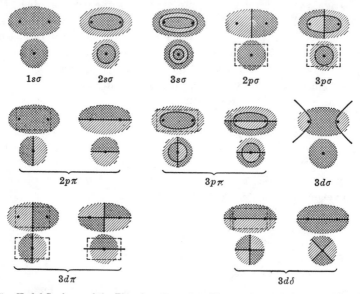

Fig. 155. **Nodal Surfaces of the Eigenfunctions of an Electron in the Field of Two Fixed Centers** [after Weizel (703)] (**Schematic**). In each case, two cross sections are drawn, the one containing the line joining the nuclei and the other at right angles to it. The sign of the eigenfunction (− or +) is indicated by single and cross hatching. When the plane of the drawing is a nodal plane, this is indicated by a broken-line rectangle. In this case, the hatching gives the sign immediately in front of the plane of the drawing. For π and δ electrons there are two eigenfunctions corresponding to the twofold degeneracy.

If the spin is to be included we must add to the three quantum numbers λ, n, and l (the last two corresponding either to the separated atoms or to the united atom) the *quantum number s of the electron spin*, which behaves in quite the same way as for atoms for any distance between the two nuclei.

For the subsequent discussion it is important to establish *in what way the orbitals alter in the transition from small to large internuclear distances* and, in particular, into which orbitals of the one limiting case those of the other limiting case go over. Thus far, an exact calculation for intermediate internuclear distances has been carried out only for the case of H_2^+, for which the core consists of only the two (equal) nuclei [see Morse and Stueckelberg (505), Teller (663), and Hellmig (296)]. However, for the more general cases we can obtain at least an approximate idea of the relative energies of the orbitals and the dependence of the energy on the internuclear distance when we interpolate between the limiting cases in the following way:

In Fig. 156, for the case of *unequal nuclear charges* the relative energies of the possible orbitals of an electron in the two limiting cases are represented by horizontal lines to the right and the left, respectively. We have now to consider that a σ orbital to the left can go over only into a σ orbital to the right, a π orbital can go over only into a π orbital, and so on. Also, theory shows that two different σ orbitals (and correspondingly for π, δ, and so on) cannot intersect if the internuclear distance is changed (see also p. 295) except when there are quantum numbers other than λ that are well defined for intermediate r values. While such an exception does arise for H_2^+ (see below) in all other cases the correlation between left and right in Fig. 156 is unambiguously given: the lowest σ orbital to the right goes into the lowest σ orbital to the left, the second lowest σ orbital to the right goes into the second lowest to the left, and so on. π, δ, $\cdots$ orbitals behave correspondingly. This correlation is indicated in Fig. 156 by the connecting lines.

In the case of *equal nuclear charges*, in addition, the symmetry property g–u must be conserved in going from left to right in the diagram. σ_u can go over only into σ_u, and so on. Therefore the order of the orbitals for intermediate internuclear distances is somewhat altered compared to the case of unequal nuclear charges. This altered correlation is shown in Fig. 157.

In this way it can be seen that in both cases the designations at small internuclear distances are unambiguously correlated with those at large internuclear distances. Therefore, in the one electron problem under consideration, both systems of nomenclature are equivalent.

It should be emphasized that the straight connecting lines in Figs. 156 and 157 do not indicate more than the qualitative behavior of the energies of the orbitals for various internuclear distances. No detailed calculations giving exact energy curves are available; but Mulliken (514), on the basis of empirical data has given semi-quantitative curves in place of the straight lines.

In the neighborhood of the two limiting cases, states which differ only in the principal quantum number (for example, $2p\sigma$, $3p\sigma$, $4p\sigma$, $\cdots$) form approximately a Rydberg series, since in the limiting cases they correspond to such a series. In particular, for large n—that is, when the distance of the electron from the atomic core is large compared to the internuclear distance—the system approximates the limiting case of the united atom, and therefore, in most cases, for large n, we should expect *Rydberg series*. Accordingly, the observed Rydberg series of electronic states for He_2, N_2, and others are indeed to be interpreted as terms of a single electron (emission electron) moving in the field of a molecular core, as suggested at the beginning of this section. From the limit of the Rydberg series we can obtain an accurate value for the ionization potential of the molecule under consideration (just as for atoms).

In solving the wave equation (I, 12) for an electron in an axially symmetric field, it is advantageous to introduce cylindrical co-ordinates z, ρ, and φ, where the z axis is the internuclear axis, ρ is the perpendicular distance from the internuclear axis, and φ is the azimuth

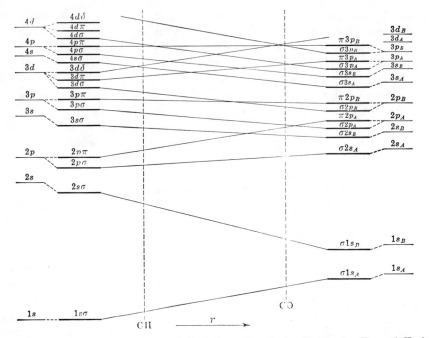

FIG. 156. **Correlation of Molecular Orbitals in a Two-Center System for Unequal Nuclear Charges.** To the extreme left and the extreme right are given the orbitals in the united and separated atoms, respectively, and, beside them, those in the molecule for very small and very large internuclear distances, respectively. The region in between corresponds to intermediate internuclear distances. The vertical broken lines give the approximate positions in the diagram that correspond to the molecules indicated. It should be noticed that the scale of r in this and the following figure is by no means linear but becomes rapidly smaller on the right-hand side.

measured from a fixed plane through the internuclear axis. The expression $\dfrac{\partial^2\psi}{\partial x^2} + \dfrac{\partial^2\psi}{\partial y^2}$ in

(I, 12) then goes over into $\dfrac{\partial^2\psi}{\partial\rho^2} + \dfrac{1}{\rho}\dfrac{\partial\psi}{\partial\rho} + \dfrac{1}{\rho^2}\dfrac{\partial^2\psi}{\partial\varphi^2}$ (see any text on partial differential equations) and the wave equation becomes

$$\frac{\partial^2\psi}{\partial z^2} + \frac{\partial^2\psi}{\partial\rho^2} + \frac{1}{\rho}\frac{\partial\psi}{\partial\rho} + \frac{1}{\rho^2}\frac{\partial^2\psi}{\partial\varphi^2} + \frac{8\pi^2 m}{h^2}(E-V)\psi = 0, \tag{VI, 7}$$

The potential energy V depends only on z and ρ. In consequence the solution can be written as a product of a function of z and ρ, and a function of φ alone:

$$\psi = \chi(z,\rho)\cdot f(\varphi). \tag{VI, 8}$$

Substituting in (VI, 7) and multiplying by $\rho^2/[\chi(z,\rho)\cdot f(\varphi)]$, we obtain

$$\frac{\rho^2}{\chi}\frac{\partial^2\chi}{\partial z^2} + \frac{\rho^2}{\chi}\frac{\partial^2\chi}{\partial\rho^2} + \frac{\rho}{\chi}\frac{\partial\chi}{\partial\rho} + \frac{8\pi^2 m\rho^2}{h^2}[E-V(z,\rho)] = -\frac{1}{f}\frac{\partial^2 f}{\partial\varphi^2}, \tag{VI, 9}$$

where the left-hand side depends on ρ and z only, whereas the right-hand side depends on φ only. Therefore both sides must be equal to a constant, which we call λ^2. Thus for $f(\varphi)$ we have the differential equation

$$\frac{d^2 f(\varphi)}{d\varphi^2} + \lambda^2 f(\varphi) = 0,$$

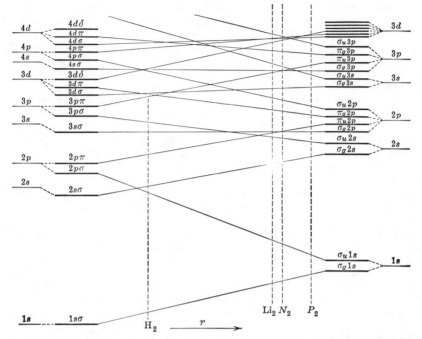

FIG. 157.   Correlation of Molecular Orbitals in a Two-Center System for Equal Nuclear Charges. See the remarks for Fig. 156.

whose solution is

$$f(\varphi) = e^{\pm i\lambda\varphi}. \tag{VI, 10}$$

Since ψ is to be single-valued everywhere, we must require that $f(\varphi + 2\pi) = f(\varphi)$. This, according to (VI, 10), is possible only when λ is an integer. Substituting $e^{\pm i\lambda\varphi}$ in (VI, 8), we obtain

$$\psi = \chi(z, \rho)e^{\pm i\lambda\varphi} \tag{VI, 11}$$

The possibility of taking out the factor $e^{\pm i\lambda\varphi}$ is due to the fact that the potential energy of the system is independent of φ.

The magnitude of the *orbital angular momentum about the internuclear axis* is given in wave mechanics by the eigenvalues of the equation (see p. 16)

$$P_z\psi = k\psi \tag{VI, 12}$$

Substituting P_z from (I, 38) and ψ from (VI, 11) it is immediately seen that $k = \pm\lambda(h/2\pi)$, that is, λ gives the magnitude of the orbital angular momentum in units $h/2\pi$; it is identical with the previously introduced quantum number. We see also that λ is exactly defined for any internuclear distance as long as the potential energy does not depend on the azimuthal angle φ. For $\lambda > 0$ there is, even wave-mechanically, a rotation of the electron about the internuclear axis, although there are no definite orbits.

The expression $-(1/f)\partial^2 f/\partial\varphi^2$ at the right of (VI, 9) is equal to λ^2 independent of the sign in (VI, 10). Therefore the same energy value is obtained from (VI, 9) for the two eigenfunctions (VI, 11) if $\lambda \neq 0$. Such states are *doubly degenerate*. Instead of the two functions (VI, 11), we may also use any linear combination of them (see p. 214), for example $\chi(z, \rho)\cos\lambda\varphi$ and $\chi(z, \rho)\sin\lambda\varphi$, both of which have λ nodal planes through the internuclear axis (at different angles for the former and the latter; see Fig. 155). For $\lambda = 0$ there is only one eigenfunction, which is axially symmetric about the internuclear axis. Such states are *non-degenerate*.

Quantum numbers corresponding to the function χ can be exactly defined only when χ can be resolved into a product of two functions, each depending on only one co-ordinate. This resolution is possible in the limiting cases (united atom or separated atoms), in which χ can be resolved into a product of two functions, one of r alone and one of ϑ alone, where r and ϑ are polar co-ordinates. Therefore the quantum numbers n and l (see above) are rigorously defined in these limiting cases and approximately also in the neighborhood of the limiting cases. However, in the region of intermediate internuclear distances, n and l cannot in general be exactly defined. Only when the core consists of two bare nuclei (for example, H_2^+) can χ be resolved into a product $f(\mu) \cdot g(\nu)$, where μ and ν are elliptical co-ordinates [see, for example, Weizel (22)]. Correspondingly, in this case, in addition to λ, we can exactly define, for any internuclear distance, two further quantum numbers, which are in a fixed relation to the n and l values of the united atom or of the separate atoms. In this case these additional quantum numbers must also be conserved in the correlation between the separated atoms and the united atom leading to somewhat different correlations from those in Figs. 156 and 157. However, we shall not discuss this in detail here.

In many cases the eigenfunctions of a single electron in a molecule (*molecular orbitals*) can be approximated by *linear combinations* of the orbital eigenfunctions of the separate atoms (*atomic orbitals*). In the case of *equal nuclear charges*, for example, the orbital $\sigma 2p$ for large internuclear distance has the same energy for each of the two atoms (see above). For smaller internuclear distances there is a splitting into $\sigma_g 2p$ and $\sigma_u 2p$. Let $\psi_A(\sigma 2p)$ and $\psi_B(\sigma 2p)$ be the orbital wave functions of the $\sigma 2p$ electron in the atoms A and B, respectively. For equal nuclear charges, these two functions are of exactly the same form and can be brought into coincidence by a shift AB. By the usual methods of perturbation theory (see p. 13) it is then easily seen that the molecular orbital wave functions of $\sigma_g 2p$ and $\sigma_u 2p$ are given to a first approximation by

$$\psi_{AB}(\sigma_g 2p) = N_g[\psi_A(\sigma 2p) + \psi_B(\sigma 2p)],$$
$$\psi_{AB}(\sigma_u 2p) = N_u[\psi_A(\sigma 2p) - \psi_B(\sigma 2p)],$$

(VI, 13)

where

$$N_g = \frac{1}{\sqrt{2 + 2T}}, \qquad N_u = \frac{1}{\sqrt{2 - 2T}}, \quad \text{with} \quad T = \int \psi_A \psi_B d\tau \qquad \text{(VI, 14)}$$

are normalization factors introduced so that all wave functions are normalized to unity (see p. 13). For other orbitals, similar relations hold, except that for π, φ, $\cdots$ orbitals, $+$ and $-$ in (VI, 13) are exchanged. In order to introduce this mode of representation of molecular orbital eigenfunctions directly into the symbol, Mulliken (519) writes, for example, for the above two orbitals, instead of $\sigma_g 2p$ and $\sigma_u 2p$, $(\sigma 2p + \sigma 2p, \sigma_g)$ and $(\sigma 2p - \sigma 2p, \sigma_u)$, or, in short, $(2p + 2p, \sigma_g)$ and $(2p - 2p, \sigma_u)$, and similarly in other cases.

For *unequal nuclear charges* at very large internuclear distances, the eigenfunctions $\psi_A(\sigma 2p)$ and $\psi_B(\sigma 2p)$ (and correspondingly for other orbitals) are no longer of the same form. For smaller internuclear distances the orbital wave function is, as in the previous case, to a first approximation a mixture of the two eigenfunctions (linear combination). However, in place of (VI, 13) we have

$$\psi_{AB}(\sigma 2p_A) = a\psi_A(\sigma 2p) + b\psi_B(\sigma 2p),$$
$$\psi_{AB}(\sigma 2p_B) = c\psi_A(\sigma 2p) - d\psi_B(\sigma 2p),$$

(VI, 15)

where $a > b$ and $d > c$. The greater the difference in nuclear charges the greater is the difference between a and b (and c and d). Following Mulliken the eigenfunctions (VI, 15) may also be indicated in the symbols for the orbital by writing $(\sigma 2p_A + \sigma 2p_B, \sigma)$ and $(\sigma 2p_A - \sigma 2p_B, \sigma)$, respectively. Other cases are similar.

As has been emphasized above the linear combinations (VI, 13) and (VI, 15) are only first approximations to the true molecular orbitals. In higher approximations finer interactions must be taken into account. A first step in this direction is the introduction of the mutual perturbation of states of the same species that lie fairly close together. For example in Fig. 157 for large internuclear distance the states $\sigma_g 2s$ and $\sigma_g 2p$ are fairly close together. The actual molecular orbitals are therefore mixtures (hybrids) of these two, of the type

$$a\psi_{AB}(\sigma_g 2s) + b\psi_{AB}(\sigma_g 2p) \tag{VI, 16}$$

where the ψ_{AB} are given by formulae similar to (VI, 13) or (VI, 15). Similar considerations apply to other pairs, such as $\sigma_u 2s$, $\sigma_u 2p$, and so on. For certain purposes it is convenient to use, in place of cumbersome symbols indicating all contributions to a given orbital, short-hand symbols such as $z\sigma_g$, $y\sigma_u$, $v\pi_g$, $w\pi_u$, especially when the symbols used in Figs. 156 and 157 are not sufficient for a description of the particular orbitals (see also below). Similarly for unequal nuclei orbitals of different type in the separated atoms may have similar energies and give molecular orbitals that are mixtures of them. For example, in diatomic hydride molecules MH the 1s orbit of hydrogen may have an energy similar to the 2p orbit in the atom M. Therefore the molecular orbitals $\sigma 1s_H$ and $\sigma 2p_M$ will mix in the same way as $\sigma 2p_A$ and $\sigma 2p_B$ in (VI, 15).

Several electrons. To a certain approximation the considerations of the foregoing section can be applied to all the electrons of a molecule, since to a first approximation the motion of each individual electron can be regarded as a motion in an axially symmetric field—namely, that obtained by averaging over all possible positions of the other electrons. However, actually, the field acting on a given electron naturally depends on the *momentary* positions of the other electrons and is therefore not strictly axially symmetric. This deviation is less and less important for larger and larger distances of the given electron from the remaining electrons.

Thus to a certain approximation we can characterize each individual electron in the molecule by the quantum numbers introduced above (see also Figs. 155–157). The electrons may be in any of the possible orbitals. A certain *electron configuration* of the molecule is defined by stating the quantum numbers of all the electrons in the molecule. States with different electron configurations have different, frequently widely different, energies. As we shall see, for a given electron configuration in general several electronic states of the molecule result. They would all have the same energy if the electrons were all independent of one another. But since they are not these states arising from a given electron configuration have somewhat different energies.

If two or more electrons in a molecule have the same quantum numbers (apart from spin) and the same symmetry g or u, they are called *equivalent electrons*.

(b) The Pauli Principle in the Molecule

In the molecule, as well as in the atom, the number of electrons that can be in the same orbital is limited by the Pauli principle. In the atom this principle requires that no two electrons can have the same set of the four quantum numbers n, l, m_l and m_s, or, since m_s can take only the values $+\frac{1}{2}$ and $-\frac{1}{2}$, that not more than two electrons can have the same set of the three quantum numbers n, l, and m_l.

The adaptation of this principle to the molecule is immediately clear from the foregoing considerations if we remember that the number of states of an atomic system is not altered by an alteration of the coupling conditions. Therefore we can start out from the behavior for very small internuclear distances, where the quantum numbers n and l are well defined. Since in this case $\lambda = |m_l|$, it follows that for given n and l there can be only two electrons with $\lambda = 0$ ($m_l = 0$, $m_s = \pm\frac{1}{2}$) that is, a σ shell (orbital) *is closed with two electrons*. On the other hand there can be four electrons for each $\lambda \neq 0$ ($m_l = +\lambda$, $m_s = \pm\frac{1}{2}$ and $m_l = -\lambda$, $m_s = \pm\frac{1}{2}$) that is, π, δ, $\cdots$ *shells are closed with four electrons each*. In Table 29 these relations are represented for the different n and l values in a manner analogous to that used for atoms in Table 3. The energy increases from left to right.

The same results are obtained if we start out from widely separated nuclei. Then, for unequally charged nuclei, the order of the shells is given by

$$\sigma 1s_A, \ \sigma 1s_B, \ \sigma 2s_A, \ \sigma 2s_B, \ \sigma 2p_A, \ \sigma 2p_B, \ \pi 2p_A, \ \pi 2p_B, \cdots, \qquad \text{(VI, 17)}$$

and as before, on the basis of the Pauli principle, each σ shell can contain only two electrons and each π, δ, $\cdots$ shell only four electrons. To be sure for equal

TABLE 29. PAULI PRINCIPLE IN THE MOLECULE (FOR SMALL INTERNUCLEAR DISTANCE)

n	1	2			3						4
l	0	0	1		0	1		2			$\cdots$
λ	0	0	0	1	0	0	1	0	1	2	$\cdots$
m_l	0	0	0	$+1$ -1	0	0	$+1$ -1	0	$+1$ -1	$+2$ -2	$\cdots$
m_s	↑↓	↑↓	↑↓	↑↓ ↑↓	↑↓	↑↓	↑↓ ↑↓	↑↓	↑↓ ↑↓	↑↓ ↑↓	$\cdots$
	$1s\sigma$	$2s\sigma$	$2p\sigma$	$2p\pi$	$3s\sigma$	$3p\sigma$	$3p\pi$	$3d\sigma$	$3d\pi$	$3d\delta$	$\cdots$

nuclear charges, we have at first, for example, four equivalent $1s$ electrons of the two separated atoms; however, as soon as the atoms approach each other, the distinction between g and u appears. Two of the $1s$ electrons become σ_g, and the other two become σ_u, which go over into different orbitals of the united atom (see Fig. 157), and again only two σ electrons are equivalent. The other electrons behave correspondingly.

Thus, when we have, for example, a system with six electrons, for small internuclear distance these cannot all go into the $1s\sigma$ orbital but give as the lowest state $1s\sigma^2 2s\sigma^2 2p\sigma^2$. For large internuclear distance, when the nuclear charges are equal, the lowest configuration is $(\sigma_g 1s)^2 (\sigma_u 1s)^2 (\sigma_g 2s)^2$ (see Fig. 157).

As we have seen in Chapter I, p. 24 the wave-mechanical formulation of the Pauli principle is: The eigenfunctions of any atomic system must be antisymmetric with respect to an exchange of any two electrons. As long as the motions of the various electrons can be considered as independent of one another the electronic eigenfunctions of a diatomic molecule are simply products of the eigenfunctions of the individual electrons, for example, in the case of three electrons

$$\psi_I = \varphi_{a_1}(x_1)\varphi_{a_2}(x_2)\varphi_{a_3}(x_3) \tag{VI, 18}$$

Here a_i stands for a set of quantum numbers (for example, n, l, m_l, and m_s in the united atom) and x_i stands for the set of coordinates of the ith electron (including the spin coordinate). On account of the identity of the electrons we may exchange any two electrons without changing the energy value, that is, functions like

$$\psi_{II} = \varphi_{a_1}(x_2)\varphi_{a_2}(x_1)\varphi_{a_3}(x_3) \tag{VI, 19}$$

and other permutations as well as any linear combinations of them belong to the same eigenvalue. However, only one of these functions is *antisymmetric*, that is, fulfills the Pauli principle. In the example this function is (omitting a normalization factor)

$$
\begin{aligned}
\psi_a = \ &\varphi_{a_1}(x_1)\varphi_{a_2}(x_2)\varphi_{a_3}(x_3) - \varphi_{a_1}(x_2)\varphi_{a_2}(x_1)\varphi_{a_3}(x_3) - \varphi_{a_1}(x_1)\varphi_{a_2}(x_3)\varphi_{a_3}(x_2) \\
&- \varphi_{a_1}(x_3)\varphi_{a_2}(x_2)\varphi_{a_3}(x_1) + \varphi_{a_1}(x_2)\varphi_{a_2}(x_3)\varphi_{a_3}(x_1) + \varphi_{a_1}(x_3)\varphi_{a_2}(x_1)\varphi_{a_3}(x_2)
\end{aligned} \tag{VI, 20}
$$

Here we are assuming the presence of a strong magnetic field so that all spatial degeneracies are removed. The function (VI, 20) shows that strictly speaking we cannot ascribe quantum numbers to the individual electrons but only state that there are three (or more) electrons with such and such quantum numbers (a_1, a_2, a_3). In the example one sixth of the time electrons 1, 2, 3 have quantum numbers a_1, a_2, a_3, respectively; but another sixth of the time they have quantum numbers a_2, a_1, a_3, respectively, and so on. Nevertheless for most practical purposes it is permissible to talk about electron configurations and $1s\sigma$, $2p\pi$, $\cdots$, $\sigma_g 1s$, $\cdots$ electrons in the way we have done above as long as it is realized that actually it is the orbitals not the electrons that are meant, and that the eigenfunctions are of the type (VI, 20).

(c) Derivation of the Term Type (Species) from the Electron Configuration in Russell-Saunders Coupling

In deriving the types of the molecular electronic states from the electron configurations it is necessary to make assumptions about the coupling of the electronic motions. Just as for atoms, in very many cases the assumption of Russell-Saunders coupling gives a satisfactory approximation. In what follows we shall therefore assume that the orbital angular momenta of the individual electrons are more strongly coupled with one another and the spins with one another than each individual orbital angular momentum with the corresponding spin.

Terms of nonequivalent electrons. The assumptions just mentioned lead immediately to the following rules: The *resultant orbital angular momentum about the internuclear axis*, Λ, *is equal to the sum of the individual orbital angular momenta* λ_i; that is,

$$\Lambda = \Sigma\lambda_i. \tag{VI, 21}$$

The *resultant spin is equal to the sum of the individual spins;* that is,

$$S = \Sigma s_i. \tag{VI, 22}$$

All the vectors in (VI, 21) lie along the internuclear axis, and therefore we have a simple *algebraic* addition. In this addition we have to take into account the fact that two opposite directions are possible for each individual vector. In (VI, 22) we have to add *vectorially* in the same way as for atoms (see p. 25 f.). S is integral or half-integral according as the number of electrons is even or odd. For nonequivalent electrons, the Pauli principle is satisfied *ipso facto* and need not be considered in the vector addition.

For example, for the case of a single π electron we have $\Lambda = \lambda = 1$ and $S = \frac{1}{2}$; that is, we have a $^2\Pi$ state. If there is a π and a σ electron, Λ is again 1 but S can now, according to (VI, 22), take the values 1 and 0 and we have a $^3\Pi$ and a $^1\Pi$ state. Similar considerations hold for a single δ electron and for the configuration $\sigma\delta$.

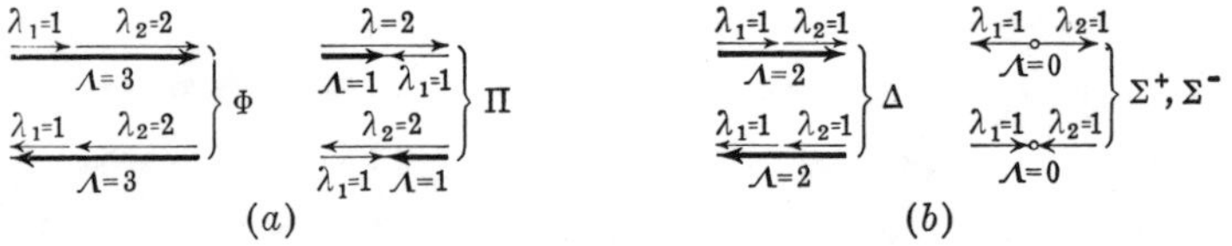

FIG. 158.   **Vector Addition of the λ_i to Give Λ.**   (a) $\lambda_1 = 1$, $\lambda_2 = 2$.   (b) $\lambda_1 = 1$, $\lambda_2 = 1$. The λ_i are represented by light arrows, and the Λ by heavy arrows.

In the case of the configuration $\pi\delta$ the two λ_i's can have either the same or opposite directions—that is, $\Lambda = 3$ or 1. The vector addition of the λ_i for this case is represented diagrammatically in Fig. 158(a). To zero approximation—that is, neglecting the finer interaction of the electrons (see p. 331)—the states corresponding to the four vector diagrams have equal energies. In higher approximation, when this interaction is taken into account, the energy of the two states with $\Lambda = 3$ (Φ state) is somewhat different from that of the states with $\Lambda = 1$ (Π state). However, the degeneracy of the two components of the Π state or of the Φ state, remains even when the interaction of the electrons is taken fully into account. Introducing the spin which can be 0 or 1 we obtain from $\pi\delta$ the four (doubly degenerate) states $^1\Pi$, $^3\Pi$, $^1\Phi$, and $^3\Phi$.

In Fig. 158(b) the addition of the λ_i vectors for two π electrons is represented. The result is that Λ can be 2 or 0. Each of these two values results in two ways. The degeneracy of the two states with $\Lambda = 0$, in contrast to that of the two states with $\Lambda = 2$, does not, however, persist when the finer interaction of the electrons is taken into account, but the states split into a Σ^+ and a Σ^- state (Σ states, as we have seen, are always non-degenerate). The resultant spin S as before is 1 or 0. We therefore obtain for the terms of two nonequivalent π electrons, $^1\Sigma^+$, $^3\Sigma^+$, $^1\Sigma^-$, $^3\Sigma^-$, $^1\Delta$, and $^3\Delta$.

These examples, together with some others, are collected in Table 30.

The energy difference between corresponding states of different multiplicity is due to a Heisenberg resonance (as for atoms)—that is, to the electrostatic interaction of the electrons. In spite of that, we can always proceed *as though* it were due to a coupling energy of the spins.

TABLE 30.   TERMS OF NONEQUIVALENT ELECTRONS
[After Hund (347)]

Electron Configuration	Molecular Electronic Terms
σ	$^2\Sigma^+$
π	$^2\Pi_r$
$\sigma\sigma$	$^1\Sigma^+, \ ^3\Sigma^+$
$\sigma\pi$	$^1\Pi, \ ^3\Pi_r$
$\sigma\delta$	$^1\Delta, \ ^3\Delta_r$
$\pi\pi$	$^1\Sigma^+, \ ^3\Sigma^+, \ ^1\Sigma^-, \ ^3\Sigma^-, \ ^1\Delta, \ ^3\Delta_r$
$\pi\delta$	$^1\Pi, \ ^3\Pi, \ ^1\Phi, \ ^3\Phi_r$
$\delta\delta$	$^1\Sigma^+, \ ^3\Sigma^+, \ ^1\Sigma^-, \ ^3\Sigma^-, \ ^1\Gamma, \ ^3\Gamma_r$
$\sigma\sigma\sigma$	$^2\Sigma^+, \ ^2\Sigma^+, \ ^4\Sigma^+$
$\sigma\sigma\pi$	$^2\Pi, \ ^2\Pi, \ ^4\Pi_r$
$\sigma\sigma\delta$	$^2\Delta, \ ^2\Delta, \ ^4\Delta_r$
$\sigma\pi\pi$	$^2\Sigma^+(2), \ ^4\Sigma^+, \ ^2\Sigma^-(2), \ ^4\Sigma^-, \ ^2\Delta(2), \ ^4\Delta_r$
$\sigma\pi\delta$	$^2\Pi(2), \ ^4\Pi, \ ^2\Phi(2), \ ^4\Phi_r$
$\pi\pi\pi$	$^2\Pi(6), \ ^4\Pi(3), \ ^2\Phi(2), \ ^4\Phi_r$
$\pi\pi\delta$	$^2\Sigma^+(2), \ ^4\Sigma^+, \ ^2\Sigma^-(2), \ ^4\Sigma^-, \ ^2\Delta(4), \ ^4\Delta(2), ^2\Gamma(2), \ ^4\Gamma_r$

The numbers in parentheses give the number of states of the type given if this number is not 1. The subscripts r indicate regular (normal) multiplets (see p. 216).

This procedure is usually applied for atoms (see A.A., p. 129) and has also been used implicitly in the above discussion for molecules.

Terms of equivalent electrons.   If the electrons whose terms we wish to determine are equivalent, we have to take account of the *Pauli principle* when adding the $\boldsymbol{\lambda}_i$ and the $\boldsymbol{s}_i$; that is, the electrons must differ either in m_l or in m_s.

For *two equivalent σ electrons*, since the m_l are equal, the spins must be antiparallel, so that only a single term results—namely, $^1\Sigma$ ($\Lambda = 0, S = 0$)—whereas for two nonequivalent σ electrons we obtain $^1\Sigma$ and $^3\Sigma$.

For two equivalent π electrons, if the two $\boldsymbol{\lambda}_i$ vectors are parallel [see Fig. 158(b)], on the basis of the Pauli principle the spins must be antiparallel. (If the spins were parallel, the two electrons would be alike in all four quantum numbers, n, l, m_l, and m_s.) Thus only a $^1\Delta$ and no $^3\Delta$ state can result. On the other hand, if the $\boldsymbol{\lambda}_i$ have opposite directions, the spins may be parallel or antiparallel and there results a $^1\Sigma$ as well as a $^3\Sigma$ state. However, as shown below, unlike the case of two nonequivalent π electrons, there are only two, not four Σ states; the Σ^+ state can occur only as singlet, the Σ^- state as triplet. We have thus the three states $^3\Sigma^-$, $^1\Delta$, and $^1\Sigma^+$. On the basis of the Hund rule, which holds for molecules as well as for atoms (see A.A., p. 135), the state with greatest multiplicity, $^3\Sigma^-$, lies lowest.

For the Σ states of π^2 there are the four spin configurations $\uparrow\uparrow, \uparrow\downarrow, \downarrow\uparrow,$ and $\downarrow\downarrow$ where the first arrow gives the spin direction of the electron with $m_l = +1$, the second that of the electron with $m_l = -1$. These four spin configurations correspond to just one triplet state ($M_S = +1$,

0, -1) and one singlet state ($M_S = 0$). The spin functions of the former are symmetric, that of the latter is antisymmetric (see A.A., p. 124). The positional eigenfunctions of Σ^- states are antisymmetric with respect to an exchange of any two electrons (compare the form of these functions discussed on p. 217). Therefore, since the total eigenfunction must be antisymmetric (Pauli principle), Σ^- occurs only as triplet, and similarly Σ^+ only as singlet. For the Σ states of two nonequivalent π electrons there are eight spin configurations, one set like the above with $m_{l_1} = +1$ and $m_{l_2} = -1$ and another set with $m_{l_1} = -1$ and $m_{l_2} = +1$. Therefore there are $^3\Sigma^-$, $^1\Sigma^+$ as well as $^1\Sigma^-$ and $^3\Sigma^+$.

For *three equivalent π electrons* (π^3), the three λ_i can have only the two relative orientations given in Fig. 159 with resultants -1 and $+1$ respectively, giving one Π term ($\Lambda = 1$). Since, owing to the Pauli principle, in both cases two of the electrons must have antiparallel spin directions, the total spin can be only $S = \frac{1}{2}$; that is, only a $^2\Pi$ state results.

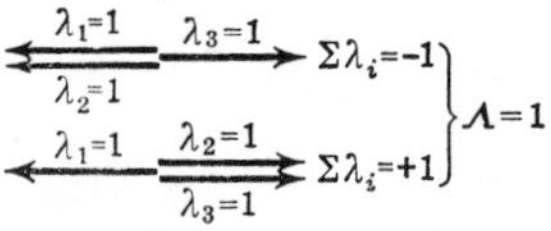

TABLE 31.   TERMS OF EQUIVALENT ELECTRONS

FIG. 159. **Vector Diagram** for the Configuration π^3. In order to avoid confusion, the resultant Λ has not been drawn. Since the electrons are equivalent, orientations of the λ_i, in which only the subscripts i are interchanged, are not to be counted as different from the two given.

[ʌ't r Hu d (34˙)]

Electron Configuration	Molecular Electronic Terms
σ^2	$^1\Sigma^+$
π^2	$^1\Sigma^+$, $^3\Sigma^-$, $^1\Delta$
π^3	$^2\Pi_i$
π^4	$^1\Sigma^+$
δ^2	$^1\Sigma^+$, $^3\Sigma^-$, $^1\Gamma$
δ^3	$^2\Delta_i$
δ^4	$^1\Sigma^+$

The subscripts i indicate inverted multiplets.

If *four equivalent π electrons* are present (π^4), the shell under consideration is closed (see p. 332) In consequence of the Pauli principle the λ_i must form antiparallel pairs; therefore $\Lambda = 0$. Similarly, the s_i must form antiparallel pairs, and, as a result, S is also equal to zero. We therefore obtain only one $^1\Sigma$ state.

These and other examples are collected in Table 31. The number of equivalent electrons is indicated by the exponent of the symbol in question. *Closed shells* always give a single $^1\Sigma$ state (similar to atoms, where they give a ^1S state).

Electron configurations with equivalent and nonequivalent electrons. If equivalent as well as nonequivalent electrons are present in an electron configuration, we first form separately the resultant Λ and S values for the equivalent electrons (Λ_e, S_e) and those of the nonequivalent electrons (Λ_n, S_n) and then add the Λ_e, S_e, Λ_n, and S_n in the same manner as discussed above for the addition of the λ_i and the s_i. In forming these resultants, we can always leave closed shells out of account, since they give $\Lambda = 0$ and $S = 0$. Table 32 gives the results for the most important cases.

TABLE 32. TERMS OF ELECTRON CONFIGURATIONS WITH EQUIVALENT AS WELL AS NONEQUIVALENT ELECTRONS

[After Hund (347) and Mulliken (514)]

Electron Configuration	Molecular Electronic Terms
$\pi^2\sigma$	$^2\Sigma^+,\ ^2\Sigma^-,\ ^2\Delta,\ ^4\Sigma^-$
$\pi^2\pi$	$^2\Pi_r,\ ^2\Pi_i(2),\ ^2\Phi_r,\ ^4\Pi_r$
$\pi^2\delta$	$^2\Sigma^+,\ ^2\Sigma^-,\ ^2\Delta_r,\ ^2\Delta_i,\ ^2\Gamma_r,\ ^4\Delta_r$
$\pi^2\sigma\sigma$	$^1\Sigma^+,\ ^1\Sigma^-,\ ^1\Delta,\ ^3\Sigma^+,\ ^3\Sigma^-(2),\ ^3\Delta,\ ^5\Sigma^-$
$\pi^2\sigma\pi$	$^1\Pi(3),\ ^1\Phi,\ ^3\Pi_r(2),\ ^3\Pi_i(2),\ ^3\Phi_r,\ ^5\Pi_r$
$\pi^2\sigma\delta$	$^1\Sigma^+,\ ^1\Sigma^-,\ ^1\Delta(2),\ ^1\Gamma,\ ^3\Sigma^+,\ ^3\Sigma^-,\ ^3\Delta(3),\ ^3\Gamma,\ ^5\Delta$
$\pi^2\pi\pi$	$^1\Sigma^+(3),\ ^1\Sigma^-(3),\ ^1\Delta(4),\ ^1\Gamma,\ ^3\Sigma^+(4),\ ^3\Sigma^-(4),\ ^3\Delta(5),\ ^3\Gamma,\ ^5\Sigma^+,\ ^5\Sigma^-,\ ^5\Delta$
$\pi^2\pi^2$	$^1\Sigma^+(3),\ ^1\Sigma^-,\ ^1\Delta(2),\ ^1\Gamma,\ ^3\Sigma^+(2),\ ^3\Sigma^-(2),\ ^3\Delta(2),\ ^5\Sigma^+$
$\pi^3\sigma$	$^1\Pi,\ ^3\Pi_i$
$\pi^3\pi$	$^1\Sigma^+,\ ^1\Sigma^-,\ ^1\Delta,\ ^3\Sigma^+,\ ^3\Sigma^-,\ ^3\Delta$
$\pi^3\delta$ (or $\pi^3\delta^3$)	$^1\Pi,\ ^1\Phi,\ ^3\Pi,\ ^3\Phi$
$\pi^3\sigma\sigma$	$^2\Pi,\ ^2\Pi,\ ^4\Pi$
$\pi^3\pi^2$	$^2\Pi_i,\ ^2\Pi_r,\ ^2\Pi,\ ^2\Phi_i,\ ^4\Pi_i$
$\pi^3\pi^3$	$^1\Sigma^+,\ ^1\Sigma^-,\ ^1\Delta,\ ^3\Sigma^+,\ ^3\Sigma^-,\ ^3\Delta_i$
$\pi^3\pi^2\sigma$	$^1\Pi(3),\ ^1\Phi,\ ^3\Pi_i(2),\ ^3\Pi_r(2),\ ^3\Phi_i,\ ^5\Pi_i$
$\pi^3\pi^3\sigma$	$^2\Sigma^+(2),\ ^2\Sigma^-(2),\ ^2\Delta,\ ^2\Delta_i,\ ^4\Sigma^+,\ ^4\Sigma^-,\ ^4\Delta_i$

The subscripts r and i indicate regular (normal) and inverted multiplets, respectively.

If the species of the equivalent electrons is Σ^+ and that of the nonequivalent electrons Σ^- (or vice versa) the resultant state is Σ^-; if both are Σ^- (or both Σ^+) the resultant state is Σ^+.

Like atoms. For molecules consisting of atoms of equal nuclear charge, everything is just as before (Tables 30–32) except that in addition we have to determine whether the resulting states are even or odd. According to what has previously been said, it is clear that the terms of a given electron configuration must either all be even or all be odd. They are *even if the number of "odd" electrons* ($\sigma_u,\ \pi_u,\ \cdots$) *is even*, whereas they are *odd if the number of "odd" electrons is odd*.

(d) Derivation of the Term Type (Species) for Other Types of Coupling

(ω, ω) **coupling.** If the coupling of spin and orbital motion is much stronger than the coupling of spins, for each electron, λ_i will be coupled with s_i to form a resultant ω_i. The ω_i of the various electrons are coupled more weakly and form the resultant Ω. This coupling is analogous to the $(j\ j)$ coupling in atoms (see A.A., p. 174) and is called (ω, ω) coupling [Mulliken (1263)]. In a similar way the previously considered Russell-Saunders coupling is called (Λ, S) coupling. In (ω, ω) coupling Λ and S are not defined and the resultant electronic states can only be distinguished by their Ω values corresponding to Hund's case (c) (see p. 224). The resultant Ω is obtained from the ω_i simply by algebraic addition

$$\Omega = \Sigma \omega_i \tag{VI, 23}$$

similar to (VI, 21) for Λ. Whether a 0 state that results is 0^+ or 0^- is determined by the same considerations as for Σ^+ and Σ^- above.

As an example consider the configuration $\sigma\pi$. The value of ω_1 is $\frac{1}{2}$ while ω_2 may be $\frac{1}{2}$ or $\frac{3}{2}$. The combination $(\frac{1}{2}, \frac{1}{2})$ yields, according to (VI, 23) the Ω values 1, 0^+, 0^-; while the combination $(\frac{1}{2}, \frac{3}{2})$ yields $\Omega = 2, 1$. In (ω, ω) coupling the two groups of states, $(\frac{1}{2}, \frac{1}{2})$ and $(\frac{1}{2}, \frac{3}{2})$, have fairly different energies while the splitting within each group is small. It should be noted that the same Ω values result from Russell-Saunders coupling which gives ${}^3\Pi_2$, ${}^3\Pi_1$, ${}^3\Pi_{0^+}$, ${}^3\Pi_{0^-}$, ${}^1\Pi_1$. But in the latter coupling the spacing of the levels is entirely different, the levels $2, 1, 0^+, 0^-$ lying close together and at a considerable distance below the other $\Omega = 1$ state (${}^1\Pi_1$).

As a second example consider the configuration $\pi\pi$. It gives rise to the groups of levels $(\frac{1}{2}, \frac{1}{2})$, $(\frac{1}{2}, \frac{3}{2})$, $(\frac{3}{2}, \frac{1}{2})$, and $(\frac{3}{2}, \frac{3}{2})$ whose Ω values are $1, 0^+, 0^-$; $2, 1$; $2, 1$; $3, 0^+, 0^+$, respectively. In the case of two equivalent π electrons the Pauli principle requires that the two electrons cannot have the same m_j value and therefore only the Ω values $2, 1, 0^+, 0^+$ occur. In comparing either case with Russell-Saunders coupling it must be remembered that a ${}^3\Sigma^+$ (${}^3\Sigma^-$) state in case (c) goes over into a 1 and a 0^- (0^+) state (see p. 237).

(Ω_c, ω) coupling. More frequent than the (ω, ω) coupling is a related coupling in which an outer electron is so little coupled with the rest of the molecule that it forms its own ω and this is then coupled less strongly with the resulting Ω_c of the rest (core) of the molecule forming the resultant Ω of the whole molecule. This coupling is called (Ω_c, ω) coupling. It is assumed that for the molecular core Russell-Saunders coupling holds.

Consider for example the case of a molecular ion in a ${}^2\Pi$ ground state. If a σ electron is added to this in a fairly highly excited state its spin will be quantized with respect to the internuclear axis with component $m_s = \pm\frac{1}{2}$ giving $\omega = \frac{1}{2}$. Adding this algebraically to the Ω_c values ($\frac{3}{2}$ and $\frac{1}{2}$) of the ${}^2\Pi$ state of the core, we obtain the Ω values 2, 1 and 1, 0^+, 0^-, respectively, just as for a $\sigma\pi$ configuration in (ω, ω) coupling. Similarly if a π electron is added to the core in a ${}^2\Pi$ state we obtain the Ω values $1, 0^+, 0^-, 2, 1, 2, 1, 3, 0^+, 0^-$ just as for a $\pi\pi$ configuration in (ω, ω) coupling. The spacing of the levels in both examples is similar to that for (ω, ω) coupling.

Other types of coupling. In addition to the couplings discussed above all sorts of intermediate couplings may occur. Furthermore it must be realized that at large separations of the atoms the electron configurations of the separated atoms become better approximations than molecular electron configurations. The resultant molecular states are obtained in such cases by first determining the resultant states of the atoms and then combining them according to the Wigner-Witmer rules [see section (1)].

(e) Term Manifold of the Molecule, Examples

General considerations. The representation of the electronic structure of a molecule by electron configurations is only an approximation—sometimes, in fact, only a poor approximation. But, at any rate, on the basis of the Ehrenfest adiabatic law we can derive from them the correct manifold of electronic states of any molecule.

In order to obtain this term manifold for a given molecule—that is, for a given number of electrons—we begin by placing the electrons in the lowest possible orbitals (shells) as far as is allowed by the Pauli principle, thus obtaining the ground state. Then we transfer one or more of the outer electrons into higher orbitals, and in each case determine the resulting term types from Tables 30–32.

In this way we can obtain all the excited electronic states of the molecule under consideration. Since we know at least approximately the order of the orbitals (see Figs. 156 and 157) we can predict in addition, even though quite roughly, the *relative positions* of the electronic states.

It will be realized that this procedure corresponds exactly to the procedure for determining the term manifold of atoms (Bohr's building-up principle). However, the difference is that for molecules the order of the orbitals *depends on the internuclear distance and on the nuclear charge*. If these are altered—that is, in different molecules—considerable alterations in the order of the orbitals result, as is illustrated by Figs. 156 and 157. The larger the internuclear distance and the larger the nuclear charges, the farther to the right lies in these figures the vertical line that gives the order of the orbitals. For unlike atoms (Fig. 156), in addition, the relative positions of the orbitals at large internuclear distances depend greatly on the degree of dissimilarity of the atoms. The order holding for some molecules is indicated by broken vertical lines in Figs. 156 and 157. Representation by such a vertical line is, of course, a very rough approximation. Frequently the higher orbitals of a given molecule will correspond to a different abscissa in Figs. 156 and 157 than the lower ones [see also Mulliken (514)]. The position of the "vertical line" for a given molecule cannot be obtained from theory with any certainty but is in general better taken from experiment. Naturally, these lines must be so drawn that for atoms of the second period of the periodic system, Li to F, the K electrons in the molecule have practically the same energy as in the separated atoms and correspondingly the L electrons for the atoms of the next period.[2]

H_2 and the hydrides. For the H_2 molecule and the diatomic hydrides the internuclear distances are small (for H_2, 0.74 Å), and so it can be assumed that they approach fairly closely the united atom. Thus for the order of the electronic shells we have to use the left sides of Figs. 156 and 157. On this assumption there is an excellent agreement between the observed electronic states and those predicted theoretically from the building-up principle. In addition, the magnitudes of the Λ doublings and of the multiplet splittings fit very well with this assumption [see Mulliken and Christy (523)].

The ground state of H_2 is the state in which both electrons are in the lowest orbital, $1s\sigma$ $(=\sigma_g 1s)$. This is a $^1\Sigma_g^+$ state (see Table 31). Most of the excited states result from one of the electrons going from the lowest orbital to one of the higher orbitals $2s\sigma$, $2p\sigma$, $2p\pi$, $\cdots$. In each case a singlet and triplet state result whose Λ value, according to Table 30, is equal to the λ of the one excited electron (emission electron). The analysis of the observed H_2 spectrum proved

---

[2] Strictly speaking, owing to the splitting $\sigma_g 1s - \sigma_u 1s$, we have in a homonuclear diatomic molecule two X-ray K levels instead of the one for the atom. However, the splitting is so small in comparison to the total energy of the X-ray levels that it has not hitherto been observed. The observed fine structures of X-ray absorption edges of molecular gases are due to other causes [see Kronig (1144), Petersen (1312), Shaw (1411), and Lindh and Nilsson (1169)].

at first very difficult, since no pronounced bands appear (many-line spectrum). However, using the theoretical results, Richardson, Dieke, Weizel and their co-workers, in the years 1928–1936, succeeded in analyzing a large number of bands and discovering in this way many electronic states [see Richardson (32)]. Fig. 160 is a diagram of the observed electronic states of the H_2 molecule (see also Table 39). It was found that most of the observed states can be explained without difficulty as being due to the excitation of one electron to various orbit-

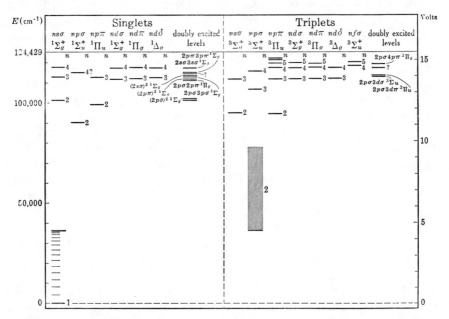

FIG. 160. **Diagram of the Observed Electronic States of the H_2 Molecule.** The data for most of the ordinary states are taken from Richardson (32), and those for the doubly excited states from Richardson (586), and Richardson and Rymer (587). For the ground state, the observed vibrational levels are also indicated (shorter horizontal lines). Above each column (with the exception of the doubly excited states) the orbital of the excited electron and the term type is given. The numbers beside the levels are the n values. The state $2p\sigma\ ^3\Sigma_u^+$ is the unstable state resulting from two normal atoms. The corresponding continuum is indicated by hatching. It should, of course, extend up to infinity.

als. A comparison of Fig. 160 with Fig. 154 shows how good the agreement between theory and experiment is. In addition to these normal terms a number of states have been found that correspond to the excitation of both electrons. They are also included in Fig. 160, together with the electron configurations which give rise to them. A few observed states are as yet unexplained because insufficient data are known for them. For some of the normal states longer Rydberg series have been found. Using them, an accurate determination of the ionization potential of H_2 has been made (see Table 37, p. 459).

For most of the diatomic hydride molecules only a few electronic states have been found. We consider here only those states that arise from the two lowest electron configurations. The electron configurations predicted on the

TABLE 33.   ELECTRON CONFIGURATIONS AND TERM TYPES OF THE LOWEST STATES
OF DIATOMIC HYDRIDES

Molecule	Lowest Electron Configuration	First Excited Electron Configuration
LiH, BeH$^+$	$K(2s\sigma)^2\ ^1\Sigma^+$	$2s\sigma2p\sigma\ ^1\Sigma^+\ [^3\Sigma^+]$
NaH, MgH$^+$	$KL(3s\sigma)^2\ ^1\Sigma^+$	$3s\sigma3p\sigma\ ^1\Sigma^+\ [^3\Sigma^+]$
KH, CaH$^+$	$KLM_{sp}(4s\sigma)^2\ ^1\Sigma^+$	$4s\sigma3d\sigma\ ^1\Sigma^+\ [^3\Sigma^+]$
CuH, ZnH$^+$	$KLM(4s\sigma)^2\ ^1\Sigma^+$	$4s\sigma4p\sigma\ ^1\Sigma^+\ [^3\Sigma^+]$
RbH	$KLMN_{sp}(5s\sigma)^2\ ^1\Sigma^+$	$5s\sigma4d\sigma\ ^1\Sigma^+\ [^3\Sigma^+]$
AgH, CdH$^+$	$KLMN_{spd}(5s\sigma)^2\ ^1\Sigma^+$	$5s\sigma5p\sigma\ ^1\Sigma^+\ [^3\Sigma^+]$
CsH	$KLMN_{spd}O_{sp}(6s\sigma)^2\ ^1\Sigma^+$	$6s\sigma5d\sigma\ ^1\Sigma^+\ [^3\Sigma^+]$
AuH, HgH$^+$	$KLMNO_{spd}(6s\sigma)^2\ ^1\Sigma^+$	$6s\sigma6p\sigma\ ^1\Sigma^+\ [^3\Sigma^+]$
BeH, BH$^+$	$K(2s\sigma)^2\ 2p\sigma\ ^2\Sigma^+$	$(2s\sigma)^2\ 2p\pi\ ^2\Pi_r$
MgH, AlH$^+$	$KL(3s\sigma)^2\ 3p\sigma\ ^2\Sigma^+$	$(3s\sigma)^2\ 3p\pi\ ^2\Pi_r$
CaH	$KLM_{sp}(4s\sigma)^2\ 3d\sigma\ ^2\Sigma^+$	$(4s\sigma)^2\ 3d\pi\ ^2\Pi_r$
ZnH	$KLM(4s\sigma)^2\ 4p\sigma\ ^2\Sigma^+$	$(4s\sigma)^2\ 4p\pi\ ^2\Pi_r$
SrH	$KLMN_{sp}(5s\sigma)^2\ 4d\sigma\ ^2\Sigma^+$	$(5s\sigma)^2\ 4d\pi\ ^2\Pi_r$
CdH	$KLMN_{spd}(5s\sigma)^2\ 5p\sigma\ ^2\Sigma^+$	$(5s\sigma)^2\ 5p\pi\ ^2\Pi_r$
BaH	$KLMN_{spd}O_{sp}(6s\sigma)^2\ 5d\sigma\ ^2\Sigma^+$	$(6s\sigma)^2\ 5d\pi\ ^2\Pi_r$
HgH	$KLMNO_{spd}(6s\sigma)^2\ 6p\sigma\ ^2\Sigma^+$	$(6s\sigma)^2\ 6p\pi\ ^2\Pi_r$
BH, CH$^+$	$K(2s\sigma)^2(2p\sigma)^2\ ^1\Sigma^+$	$(2s\sigma)^2\ 2p\sigma2p\pi\ ^1\Pi,\ ^3\Pi$
AlH	$KL(3s\sigma)^2(3p\sigma)^2\ ^1\Sigma^+$	$(3s\sigma)^2\ 3p\sigma3p\pi\ ^1\Pi,\ ^3\Pi$
InH	$KLM_{spd}(5s\sigma)^2(5p\sigma)^2\ ^1\Sigma^+$	$(5s\sigma)^2\ 5p\sigma5p\pi\ \ ^1\Pi\ [^3\Pi]$
TlH	$KLMNO_{spd}(6s\sigma)^2(6p\sigma)^2\ ^1\Sigma^+$	?
CH	$K(2s\sigma)^2(2p\sigma)^2\ 2p\pi\ ^2\Pi_r$	$(2s\sigma)^2\ 2p\sigma(2p\pi)^2\ [^4\Sigma^-],\ ^2\Delta,\ ^2\Sigma^+,\ ^2\Sigma^-$
SiH	$KL(3s\sigma)^2(3p\sigma)^2\ 3p\pi\ ^2\Pi_r$	$(3s\sigma)^2\ 3p\sigma(3p\pi)^2\ [^4\Sigma^-],\ ^2\Delta,\ [^2\Sigma^+],\ [^2\Sigma^-]$
SnH	$KLMN_{spd}(5s\sigma)^2(5p\sigma)^2\ 5p\pi\ ^2\Pi_r$	$(5s\sigma)^2\ 5p\sigma(5p\pi)^2\ [^4\Sigma^-],\ ^2\Delta,\ [^2\Sigma^+],\ [^2\Sigma^-]$
PbH	$KLMNO_{spd}(6s\sigma)^2(6p\sigma)^2\ 6p\pi\ ^2\Pi_r$	$(6s\sigma)^2\ (6p\sigma)(6p\pi)^2\ [^4\Sigma^-],\ ^2\Delta,\ ^2\Sigma^+,\ ^2\Sigma^-$
NH, OH$^+$	$K(2s\sigma)^2(2p\sigma)^2(2p\pi)^2\ ^3\Sigma^-,\ ^1\Delta,\ ^1\Sigma^+$	$(2s\sigma)^2\ 2p\sigma(2p\pi)^3\ ^3\Pi,\ ^1\Pi$
PH	$KL(3s\sigma)^2(3p\sigma)^2(3p\pi)^2\ ^3\Sigma^-,\ [^1\Delta],$ $[^1\Sigma^+]$	$(3s\sigma)^2\ 3p\sigma(3p\pi)^3\ ^3\Pi,\ [^1\Pi]$
BiH	$KLMNO_{spd}(6s\sigma)^2(6p\sigma)^2(6p\pi)^2\ [^3\Sigma^-],$ $[^1\Delta],\ ^1\Sigma$	?
OH	$K(2s\sigma)^2(2p\sigma)^2(2p\pi)^3\ ^2\Pi_i$	$(2s\sigma)^2\ 2p\sigma(2p\pi)^4\ ^2\Sigma^+$
HS, HCl$^+$	$KL(3s\sigma)^2(3p\sigma)^2(3p\pi)^3\ ^2\Pi_i$	$(3s\sigma)^2\ 3p\sigma(3p\pi)^4\ ^2\Sigma^+$
HBr$^+$	$KLM(4s\sigma)^2(4p\sigma)^2(4p\pi)^3\ ^2\Pi_i$	$(4s\sigma)^2\ 4p\sigma(4p\pi)^4\ ^2\Sigma^+$
HF	$K(2s\sigma)^2(2p\sigma)^2(2p\pi)^4\ ^1\Sigma^+$	$(2s\sigma)^2(2p\sigma)^2(2p\pi)^3\ 3s\sigma\ [^3\Pi],\ [^1\Pi]$
HCl	$KL(3s\sigma)^2(3p\sigma)^2(3p\pi)^4\ ^1\Sigma^+$	$(3s\sigma)^2(3p\sigma)^2(3p\pi)^3\ 4s\sigma\ ^3\Pi,\ ^1\Pi$
HBr	$KLM(4s\sigma)^2(4p\sigma)^2(4p\pi)^4\ ^1\Sigma^+$	$(4s\sigma)^2(4p\sigma)^2(4p\pi)^3\ 5s\sigma\ ^3\Pi,\ ^1\Pi$
HI	$KLMN_{spd}(5s\sigma)^2(5p\sigma)^2(5p\pi)^4\ ^1\Sigma^+$	$(5s\sigma)^2(5p\sigma)^2(5p\pi)^3\ 6s\sigma\ ^3\Pi,\ ^1\Pi$
MnH	$KLM_{sp}(3d\sigma)^2(3d\pi)^2(3d\delta)^2\ 4s\sigma4p\sigma\ ^7\Sigma$	$(3d\sigma)^2(3d\pi)^2(3d\delta)^2\ 4s\sigma4p\pi\ ^7\Pi$
CoH	$KLM_{sp}(3d\sigma)^2(3d\pi)^4(3d\delta)^2(4s\sigma)^2$ $[^3\Sigma^-],\ [^1\Sigma^+],\ ^1\Gamma$	$(3d\sigma)^2(3d\pi)^3(3d\delta)^3(4s\sigma)^2\ [^1\Pi,\ ^1\Phi,\ ^3\Pi],\ ^3\Phi$
NiH	$KLM_{sp}(3d\sigma)^2(3d\pi)^4(3d\delta)^3(4s\sigma)^2\ ^2\Delta$	$(3d\sigma)^2(3d\pi)^4(3d\delta)^3\ 4s\sigma4p\sigma\ ^2\Delta,\ ^2\Delta,\ ^4\Delta$

For the excited electron configurations the closed atomic shells have not been repeated. The states in brackets have not been observed. M_{sp} means that in the M shell only the subgroups $3s$ and $3p$ are closed, and similarly in other cases.

basis of Fig. 156 as well as the resulting states are given in Table 33 for all diatomic hydrides for which sufficient observations for comparison with the theoretical predictions are available (see also Table 39). LiH, for example, has four electrons for which, according to the Pauli principle, the energetically lowest configuration is $(1s\sigma)^2(2s\sigma)^2$. This configuration gives a $^1\Sigma^+$ state, which is indeed observed as the ground state of LiH. The two $1s$ electrons form the K shell of the Li atom, which is only slightly influenced by the molecule formation. Because of this, only the symbol K is written for it in Table 33. The excited states of LiH are obtained if one electron is brought from the $2s\sigma$ shell into higher shells. The lowest excited state is $K\ 2s\sigma\ 2p\sigma$, which gives a $^1\Sigma^+$ and a $^3\Sigma^+$ state, of which only the former has been observed thus far. The electronic states of the other alkali hydrides and of the ions of the alkaline earth hydrides are the same except that the principal quantum number is increased correspondingly (see Table 33). An extended discussion of the lower states of the alkali hydrides has been given by Mulliken (520).

For CH, with seven electrons, the lowest electronic configuration is $K(2s\sigma)^2(2p\sigma)^2(2p\pi)$, which gives a $^2\Pi$ state as ground state, in agreement with experiment. Since the energy difference between $2p\sigma$ and $2p\pi$ is small (they coincide in the united atom) the lowest excited states are obtained not by bringing the $2p\pi$ electron to higher orbitals but by transferring an electron from the $2p\sigma$ to the $2p\pi$ orbital. The resulting configuration $(2s\sigma)^2(2p\sigma)(2p\pi)^2$, according to Table 32, gives rise to the four states given in Table 33. Three of these have been observed. The electronic states of SiH are similar except that the principal quantum number is one higher. However, for SiH, apart from the $^2\Pi$ ground state, only the excited $^2\Delta$ state is known.

In a similar manner the normal and excited states of the other hydrides may be obtained. They are given in Table 33. Predicted states that have not been observed are put in square brackets. As far as the observations go the agreement with theoretical expectation is most satisfactory. Only in the case of MnH was it necessary to select among the various conceivable electron configurations for the ground state, one that would give the observed $^7\Sigma$ state. In none of the other cases is there any ambiguity in the theoretical electron configurations. Only in a few cases have higher excited states also been observed, although theoretically, as for H_2, a large number of term series with higher n values would be expected.

Additional confirmation of the closeness to the united atom which is assumed in Table 33 comes from a more detailed investigation of Λ-type and spin doublings. It was found by Mulliken and Christy (523) that for example for the hydrides BeH, $\cdots$, HgH the first excited and the ground states have splittings in agreement with the assumption of "pure precession" (see p. 229) as would be expected from the electron configurations given but not from separated atom configurations.

While throughout Table 33 Russell-Saunders coupling has been assumed there are a few cases where (Ω_c, ω) coupling actually represents a better approximation. This applies particularly to the excited states of the hydrogen halides [see Price (570) and Mulliken (1266)] and to the states of the heavier hydrides BiH, PbH, CoH, NiH. For example for the former, two

TABLE 34.   ELECTRON CONFIGURATIONS AND TERM TYPES OF THE GROUND STATES OF MOLECULES COMPOSED OF ATOMS WITH EQUAL OR NEARLY EQUAL NUCLEAR CHARGE

Z	Molecule	Lowest Electron Configuration	State	P_b	P_a	Diff.	D_0^0 (Volts)
1	H_2^+	$\sigma_g 1s\ (=1\sigma)$	$^2\Sigma_g^+$	$\frac{1}{2}$	0	$\frac{1}{2}$	2.648
2	H_2	$(\sigma_g 1s)^2$	$^1\Sigma_g^+$	1	0	1	4.476
3	He_2^+	$(\sigma_g 1s)^2(\sigma_u 1s)$	$^2\Sigma_u^+$	1	$\frac{1}{2}$	$\frac{1}{2}$	(2.6)
4	He_2	$(\sigma_g 1s)^2(\sigma_u 1s)^2$	$^1\Sigma_g^+$	1	1	0	0
6	Li_2	$KK(\sigma_g 2s)^2$	$^1\Sigma_g^+$	1	0	1	1.03
8	$[Be_2]$	$KK(\sigma_g 2s)^2(\sigma_u 2s)^2$	$[^1\Sigma_g^+]$	1	1	0	
10	B_2	$KK(\sigma_g 2s)^2(\sigma_u 2s)^2(\pi_u 2p)^2$	$^3\Sigma_g^-$	2	1	1	(3.6)
12	C_2(BN, BeO)	$\left\{\begin{array}{l} KK(\sigma_g 2s)^2(\sigma_u 2s)^2(\pi_u 2p)^3\sigma_g 2p \\ KK(\sigma_g 2s)^2(\sigma_u 2s)^2(\pi_u 2p)^4 \end{array}\right.$	$^3\Pi_u$ $^1\Sigma_g^+$	3	1	2	(3.6)
13	N_2^+(CO^+, CN, BO, BeF)	$KK(\sigma_g 2s)^2(\sigma_u 2s)^2(\pi_u 2p)^4\sigma_g 2p$	$^2\Sigma_g^+$	$3\frac{1}{2}$	1	$2\frac{1}{2}$	6.341
14	N_2(CO)	$KK(\sigma_g 2s)^2(\sigma_u 2s)^2(\pi_u 2p)^4(\sigma_g 2p)^2$	$^1\Sigma_g^+$	4	1	3	7.373
15	O_2^+(NO)	$KK(\sigma_g 2s)^2(\sigma_u 2s)^2(\sigma_g 2p)^2(\pi_u 2p)^4\pi_g 2p$	$^2\Pi_g$	4	$1\frac{1}{2}$	$2\frac{1}{2}$	6.48
16	O_2	$KK(\sigma_g 2s)^2(\sigma_u 2s)^2(\sigma_g 2p)^2(\pi_u 2p)^4(\pi_g 2p)^2$	$^3\Sigma_g^-$	4	2	2	5.080
18	F_2	$KK(\sigma_g 2s)^2(\sigma_u 2s)^2(\sigma_g 2p)^2(\pi_u 2p)^4(\pi_g 2p)^4$	$^1\Sigma_g^+$	4	3	1	<2.75
20	Ne_2	$KK(\sigma_g 2s)^2(\sigma_u 2s)^2(\sigma_g 2p)^2(\pi_u 2p)^4(\pi_g 2p)^4(\sigma_u 2p)^2$	$^1\Sigma_g^+$	4	4	0	0
22	Na_2	$KKLL(\sigma_g 3s)^2$	$^1\Sigma_g^+$	1	0	1	0.73
28	Si_2	$KKLL(\sigma_g 3s)^2(\sigma_u 3s)^2(\sigma_g 3p)^2(\pi_u 3p)^2$	$[^3\Sigma_g^-]$	3	1	2	
30	P_2 (SiS)	$KKLL(\sigma_g 3s)^2(\sigma_u 3s)^2(\sigma_g 3p)^2(\pi_u 3p)^4$	$^1\Sigma_g^+$	4	1	3	5.031
32	S_2	$KKLL(\sigma_g 3s)^2(\sigma_u 3s)^2(\sigma_g 3p)^2(\pi_u 3p)^4(\pi_g 3p)^2$	$^3\Sigma_g^-$	4	2	2	≤4.4
34	Cl_2	$KKLL(\sigma_g 3s)^2(\sigma_u 3s)^2(\sigma_g 3p)^2(\pi_u 3p)^4(\pi_g 3p)^4$	$^1\Sigma_g^+$	4	3	1	2.475

Z = number of electrons, P_b = number of bonding electron pairs, P_a = number of antibonding electron pairs, and D_0^0 = dissociation energy.   Molecules and states in square brackets have not been observed.   The D_0^0 values do not refer to molecules in parentheses.

excited states with $\Omega = 1$ are found which are clearly the $\Omega = 1$ states resulting from the $^2\Pi_{1/2}$ and $^2\Pi_{3/2}$ core by the addition of the $ns\sigma$ electron and corresponding to $^1\Pi_1$ and $^3\Pi_1$ in Russell-Saunders coupling.

Molecules with nuclei of equal charge. In order to predict the ground states (and low excited states) of molecules with nuclei of equal charge from the electron configurations it is necessary to use the orbitals of Fig. 157. If the nuclear charge is higher than two, the united atom is not a good approximation since the K shells and possibly higher shells of both atoms are not affected by the molecule formation [Lennard-Jones (441)]. Therefore in Table 34 which gives the electron configurations of the ground states of the most important molecules of this class the nomenclature adapted to the separated atoms is used. Atomic shells that are not influenced by the molecule formation are indicated by the symbols K, L, $\cdots$.

For He_2 in the lowest state the four electrons go into the orbitals $\sigma_g 1s$ and $\sigma_u 1s$, which they just fill. However, as we shall see later, this state is unstable. But a large number of excited states, in which an electron goes from the $\sigma_u 1s$ orbital into higher orbitals, are stable. Actually in this case, since the K shells do not remain unaffected and the internuclear distance is small, the designations adapted to the united atom are mostly used. Similar to the case of H_2, series of levels $(1s\sigma)^2(2p\sigma)(ns\sigma)\,^3\Sigma_u^+$, $^1\Sigma_u^+$; $(1s\sigma)^2(2p\sigma)(np\sigma)\,^3\Sigma_g^+$, $^1\Sigma_g^+$; $(1s\sigma)^2(2p\sigma)(np\pi)\,^3\Pi_g$, $^1\Pi_g$, and so on, are to be expected and have indeed been found by Weizel, Dieke, and their collaborators. The detailed agreement of the observed with the theoretical electronic structure is most satisfactory [see for example Weizel (22)]. The observed Rydberg series correspond to series of levels of the above-mentioned type. The energy level diagram is closely similar to that of H_2 given in Fig. 160.

The ground state of the Li_2 *molecule* has the electron configuration $KK(\sigma_g 2s)^2$. Excited states corresponding to the configurations

$$KK(\sigma_g 2s)(\sigma_u 2s), \; KK(\sigma_g 2s)(\pi_u 2p), \text{ and } KK(\sigma_g 2s)(\sigma_u 2p)$$

have been found (see Table 39).

The lowest configuration with 10 electrons (B_2 *molecule*) is

$$KK(\sigma_g 2s)^2(\sigma_u 2s)^2(\pi_u 2p)^2$$

which according to Table 31 gives rise to the states $^3\Sigma_g^-$, $^1\Delta_g$, and $^1\Sigma_g^+$ of which $^3\Sigma_g^-$ is lowest. It was only several years after this prediction was made by Mulliken (514) (and included in the first edition of this book) that a spectrum of B_2 was found [Douglas and Herzberg (900)] which is in all probability a $^3\Sigma_u^- - ^3\Sigma_g^-$ transition involving the predicted $^3\Sigma_g^-$ ground state. The excited state results from the configuration $KK(\sigma_g 2s)^2(\sigma_u 2s)(\pi_u 2p)^2(\sigma_g 2p)$.

For the C_2 *molecule*, with its 12 electrons, two different electron configurations are given for the ground state in Table 34. The first (giving $^1\Sigma_g^+$) corresponds to the one that would be expected if the electrons are brought into the lowest possible orbitals. Accordingly, the second (giving $^3\Pi_u$) should correspond to an excited state. Actually, however, it is found that the absorption spectrum of the C_2 molecule consists of the Swan bands [Klemenc (405)], whose lower state is $^3\Pi$. Thus $^3\Pi$ is the ground state of C_2. The explanation of this apparent contradiction is to be sought in the fact that, as shown by Fig. 157, the orbitals π_u2p and σ_g2p have not very different energies near the equilibrium internuclear distance of this group of molecules, so that when the interaction of the electrons is taken into account, it may very well happen that the $^3\Pi_u$ state resulting from $\pi_u^3\sigma_g$ lies lower than the $\pi_u^4\ ^1\Sigma_g^+$ state.[3] The same electron configuration that gives rise to the $^3\Pi_u$ ground state gives also a $^1\Pi_u$ state which has been observed as the lower state of the Deslandres-d'Azambuja bands. The upper states of both these bands and the Swan bands, $^1\Pi_g$ and $^3\Pi_g$ respectively, clearly belong to the configuration $KK(\sigma_g2s)^2(\sigma_u2s)(\pi_u2p)^3(\sigma_g2p)^2$. The $(\pi_u2p)^4\ ^1\Sigma_g^+$ state mentioned above has been observed as the lower state of the Mulliken bands at 2300 Å whose upper state is $KK(\sigma_g2s)^2(\sigma_u2s)(\pi_u2p)^4(\sigma_g2p)$ $^1\Sigma_u^+$. Recently a transition from the above-mentioned $^1\Pi_u$ state to the $(\pi_u2p)^4\ ^1\Sigma_g^+$ state has been found by Phillips (1314). For a more complete discussion of the electronic states of C_2 see Mulliken (1260). (Compare also the energy level diagram Fig. 199, p. 454.)

For the N_2 *molecule*, with 14 electrons, in the lowest state both the π_u2p and σ_g2p orbitals are completely filled (see Table 34), and therefore the ground state is a $^1\Sigma_g^+$ state, in agreement with observation. The first excited electron configurations (both with fairly high excitation energy, see Fig. 157) are $KK(\sigma_g2s)^2(\sigma_u2s)^2(\pi_u2p)^4(\sigma_g2p)(\pi_g2p)$ and $\cdots(\pi_u2p)^3(\sigma_g2p)^2(\pi_g2p)$. They account for the lowest observed excited states $^3\Pi_g$, $^1\Pi_g$, and $^3\Sigma_u^+$, respectively [see Herzberg (1041) and Fig. 196]. The second configuration, in addition, gives rise to the states $^1\Sigma_u^+$, $^1\Delta_u$, $^3\Delta_u$, $^1\Sigma_u^-$, $^3\Sigma_u^-$. Higher excited states are obtained by bringing the emission electron (taken from the σ_g2p or the π_u2p or the σ_u2s shells) to higher and higher orbitals. The number of states resulting even from a few of these configurations is very large and does not allow of such a unique correlation with the numerous observed states as is possible for H_2 and He_2. The observed Rydberg series of states correspond to high n values of an electron taken from the σ_g2p and the σ_u2s orbitals. Their series limits correspond to the states $^2\Sigma_g^+$ and $^2\Sigma_u^+$ of N_2^+ which form the lower and the upper states, respectively, of the negative nitrogen bands. The $^2\Sigma_g^+$ state is the ground state of the N_2^+ molecule. A third low-lying state ($^2\Pi_u$) of N_2^+ is obtained by removing a π_u2p electron from N_2. No Rydberg series of N_2 corresponding to this state of N_2^+ has as yet been found and its presence in N_2^+ has been established only indirectly by perturbations which it produces in the $^2\Sigma_u^+$ state. Apparently it is very close to the ground state.

In the lowest electron configuration of the O_2 *molecule* there are, in addition to the electrons present in N_2, two electrons in the π_g2p shell. According to Table 31 this gives rise to the three states $^3\Sigma_g^-$, $^1\Delta_g$, $^1\Sigma_g^+$. All three states have

---

[3] A similar situation arises in some cases in applying the building-up principle to atoms, although for larger numbers of electrons only.

been found (see Fig. 195), with $^3\Sigma_g^-$ as the ground state. The $^1\Delta_g$ state is 0.98 e.v., the $^1\Sigma_g^+$ state 1.63 e.v. above the ground state. These differences indicate incidentally the order of magnitude of the splittings of states of the same electron configuration. It should be noted in Table 34 that, from O_2^+ on, the order of the two orbitals $(\pi_u 2p)(\sigma_g 2p)$ is inverted. This may be expected from Fig. 157, since the internuclear distance is increasing, and is in agreement with the observed excited states of both O_2 and O_2^+. The first excited electron configuration of O_2 on this basis is $(\sigma_g 2p)^2(\pi_u 2p)^3(\pi_g 2p)^3$ which according to Table 32 gives rise to the states $^1\Sigma_u^+, \ ^1\Sigma_u^-, \ ^1\Delta_u, \ ^3\Sigma_u^+, \ ^3\Sigma_u^-, \ ^3\Delta_u$. Of these $^3\Sigma_u^+$ and $^3\Sigma_u^-$ have been observed (see Fig. 195). In addition two Rydberg series have been found [Price and Collins (571), Tanaka and Takamine (1458)] which converge to the O_2^+ states $(\sigma_g 2p)^2(\pi_u 2p)^3(\pi_g 2p)^2 \ ^4\Pi_u$ and $(\sigma_g 2p)(\pi_u 2p)^4(\pi_g 2p)^2 \ ^4\Sigma_g^-$. These are the lower and the upper state, respectively, of the red O_2^+ bands [see Mulliken (514) and Nevin (530)(1277)]. The ground state of O_2^+ is $(\sigma_g 2p)^2(\pi_u 2p)^4(\pi_g 2p) \ ^2\Pi_g$ which is the lower state of the ultraviolet O_2^+ bands. Their upper state $^2\Pi_u$ has the same configuration as the $^4\Pi_u$ state mentioned above.

The ground state of the F_2 *molecule* is $\cdots (\sigma_g 2p)^2(\pi_u 2p)^4(\pi_g 2p)^4 \ ^1\Sigma_g^+$. The first excited states are $\cdots (\pi_g 2p)^3(\sigma_u 2p) \ ^3\Pi_u, \ ^1\Pi_u$. The observed continuous absorption (see p. 390) corresponds very probably to the transition $^1\Pi_u \leftarrow \ ^1\Sigma_g^+$. However, for Cl_2 and the other halogens the triplet splitting of $^3\Pi_u$ becomes large and case (c) is approached. The observed absorption bands which have the structure of $^1\Sigma - ^1\Sigma$ bands must be interpreted as transitions to the 0_u^+ component of $^3\Pi_u$ as was first shown by Mulliken (516)(514)(519).

The ground state of the P_2 *molecule* is similar to that of N_2. However, because of the changed order of the orbitals the first excited electron configuration is different from that of N_2, namely $\cdots (\pi_u 3p)^3\pi_g 3p$. Of the various states of this configuration only one, $^1\Sigma_u^+$, can combine with the ground state. A $^1\Sigma_u^+ - ^1\Sigma_g^+$ absorption band system (rather than $^1\Pi_g - ^1\Sigma_g^+$ as for N_2) has indeed been found for P_2.

Other molecules. If the nuclei of a diatomic molecule have unequal charges, the distinction between g and u disappears. In consequence there is a slight alteration of the order of the orbitals (Fig. 156 instead of Fig. 157). However, the differences thereby introduced are small if the difference in nuclear charge amounts to only a few units. The electron configurations of the ground states and often also of the excited states are then the same as for the corresponding molecules having equal nuclei and the same number of electrons. Correspondingly, some molecules with unlike nuclei have been added in parentheses at appropriate places in Table 34. Their electron configurations are obtained from those given simply by omitting g and u.

The case of the molecules with 13 electrons, CO^+, CN, BO, and BeF, is particularly characteristic. For them, not only the ground states, $^2\Sigma$, but also the first two excited states, $^2\Pi$ and $^2\Sigma$, are the same as those of N_2^+. Likewise, a great similarity (not only spectroscopic) exists between N_2 and CO, both of which have 14 electrons, as well as between O_2^+ and NO, which have 15 electrons. To be sure the higher excited states of these molecules are different, since a slight difference in charge affects the relative order of the higher electron configurations more than the low ones. In the case of SiS and P_2 (unlike CO and N_2) such a difference arises already for the first excited singlet state which is $^1\Pi$ for the former and $^1\Sigma$ for the latter, as might have been predicted from a closer study of Figs. 156 and 157.

There is a significant difference even in the ground states for the molecules with 12 electrons C_2, BN, BeO, since here there are two low lying electron configurations of nearly the same energy (see above). While for C_2 the second configuration in Table 34 gives the ground state $(^3\Pi_u)$, for BeO it is the first one $(^1\Sigma_f^+)$, apparently because the two orbitals $\pi2p$ and $\sigma2p$ have a greater energy difference in BeO (compare Fig. 156). For BN the ground state has not yet been definitely established but the lowest observed state is $^3\Pi$ as for C_2.

Molecules with the same number of electrons are called *isoelectronic molecules*. The similarity of their empirical energy level diagrams and the similarity of their physical and chemical properties can easily be understood theoretically, as we have seen here, from a consideration of the electron configurations. The electron configurations and the resulting electronic energy levels of molecules with not too different nuclei are determined essentially by the *number of electrons* and are therefore very similar for isoelectronic molecules just as they are for isoelectronic sequences of atoms and ions (for example Li, Be^+, B^{++}, $\cdots$).

On the other hand, if we build up the molecule from the separate atoms, the similarity of isoelectronic molecules is by no means so easily understood. N_2 and CO, for example, result from the atomic states $^4S + {}^4S$ and $^3P + {}^3P$, respectively. From $^4S + {}^4S$ four different molecular states result, from $^3P + {}^3P$ eighteen (see Tables 28 and 26). That ultimately similar low electronic states should result for N_2 and CO is not obvious when considered in this way.

If the two atoms that form a molecule belong to different periods of the periodic system and thus have very different nuclear charges, the correlations in Fig. 156 will be radically changed. For example $3s_B$ at the right may be close to $2s_A$ and similarly for the other orbitals. Instead of plotting a new correlation diagram like Fig. 156 it is then more convenient to proceed in the following way: We take no account of those closed shells of the separate atoms that we should expect to be practically uninfluenced by molecule formation and consider only the electrons outside these closed shells [see Lennard-Jones (441)]. It is to be expected that these electrons will arrange themselves in orbitals similar to those for the corresponding molecules consisting of atoms from the second period of the periodic system (Fig. 156). However, these orbitals can no longer be designated in the same way. Following Mulliken (514), we introduce the nomenclature $z\sigma$, $y\sigma$, $w\pi$, $x\sigma$, and $v\pi$ for the orbitals previously called σ_g2s, σ_u2s, π_u2p, σ_g2p, and π_g2p. A molecule such as SiO has the same number of outer electrons as CO or N_2. We should therefore expect for the electron configuration and term type of the ground state

$$(KKL)(z\sigma)^2(y\sigma)^2(w\pi)^4(x\sigma)^2 \; {}^1\Sigma^+$$

and of the first excited state

$$(KKL)(z\sigma)^2(y\sigma)^2(w\pi)^4(x\sigma)(v\pi) \; {}^1\Pi, \, {}^3\Pi,$$

and indeed $^1\Sigma^+$ is observed as the ground state and $^1\Pi$ as the first excited state.   The same holds for the molecules CS and PN, which are isoelectronic with SiO.

For the molecules SiN and CP there is one outer electron less, and we obtain, similar to N_2^+ and CN, the three low-lying states $^2\Sigma$, $^2\Pi$, and $^2\Sigma$, according as the electron is removed from the $x\sigma$, $w\pi$, or $y\sigma$ orbital of the above configuration.   For CP all three states, and for SiN the two $^2\Sigma$ states, have been observed, the energy difference being very similar to that for CN.   The halides of the alkaline earths and the oxides of the earths also have the same number of outer electrons.   Thus for them the same low-lying states are to be expected.   For many of these molecules they have indeed been observed (see Table 39).

Other groups of molecules with the same number of outer electrons can be treated in a similar manner.   However, for the heavier molecules the theoretical predictions are more ambiguous, particularly for the excited states, since more and more orbitals in the separated atoms and therefore a fortiori in the molecule have approximately equal energy.

In order to arrive at a better theoretical understanding of the electronic structure of diatomic molecules, particularly of their excited states, much more experimental material is needed.   At present, in most cases only a few electronic states are known for a given molecule.   A great deal of further detailed work is necessary before the knowledge of the electronic structure of molecules reaches the stage already attained by our knowledge of the electronic structure of atoms.

4. Stability of Molecular Electronic States; Valence

According to any one of the three methods treated in the foregoing (sections 1, 2, and 3), a very large number of molecular electronic states results.   However, thus far we have not considered the question of *which of these numerous molecular states are stable and which are unstable*—that is to say, how the potential energy varies with changing internuclear distance in the individual cases— whether there is a minimum or not.   This question of stability is, of course, of particular interest to the spectroscopist and the chemist.   It is, however, very much more difficult to answer from theory than the question of which states arise.

Closely connected with the question of the stability of the individual molecular states is the question of the stability of the molecule itself.   We say that a molecule is *physically stable* if its ground state is stable—that is, if the ground state has an appreciable potential minimum such that the energy of the lowest vibrational level ($v = 0$) is lower than the energy of the separated atoms in their ground states.   This physical stability is to be distinguished from *chemical stability*, which is possessed by a given molecule only if, even on collision with like molecules at low temperatures, it is stable for an appreciable length of time.   Molecules such as CN, CH, OH, P_2, and others are physically but not chemically stable, since, although they appear in chemical reactions and in electric discharges, they do not form a stable gas at ordinary temperatures.   The He_2 molecule is an example of a molecule that is physically (and chemically) unstable.   While it has many stable excited states its lowest state has no potential minimum of the type described.

The fact that for many molecules only a few electronic states are observed appears at first to be in contradiction to the large manifold of terms demanded by theory. Although this has its basis partly in the inadequacy of the observational material obtained up to the present time, the main reason is that a large number of the states theoretically predicted are unstable and thereby escape observation. However, it should be noted that *the unstable states* (without a potential minimum) are *no less real than the stable states* and under suitable conditions may make themselves known by continuous or diffuse spectra (see Chapter VII).

The question of stability and instability of molecular states is obviously closely connected with the question of the *nature of chemical valence*. We shall therefore discuss this topic simultaneously, insofar as it refers to diatomic molecules.

Also closely related to the foregoing are these questions: Which of the theoretically predicted electronic states corresponds to a given observed molecular state? Into which atomic states does a given molecular state dissociate (see Chapter VII)?

In discussing the stability of molecules, three types of binding are generally distinguished: *homopolar or atomic binding, heteropolar or ionic binding*, and *polarization or van der Waals binding*. However, there are also many transition cases between these types.

In the chemical literature, the atomic binding is frequently referred to as *covalent binding*, and the ionic binding as *electrovalent binding*. Unfortunately, the same nomenclature is not used by all authors in this connection. In particular, sometimes under the name *homopolar molecules* are understood only those diatomic molecules that consist of like atoms, while all others are called *heteropolar*. However, it appears to be advantageous, and corresponds to the usage of many authors, to use *homopolar molecule* and *homopolar binding* as synonymous with *atomic molecule* and *atomic binding;* and *heteropolar molecule* and *heteropolar binding* as synonymous with *ionic molecule* and *ionic binding*, as is always done in the following. Sometimes, in addition, the names *polar* and *nonpolar molecules* are used—that is, molecules with or without a dipole moment. The latter are those with like atoms.

(a) Homopolar Binding (Atomic Binding)

While ionic binding can be explained on a classical basis [see subsection (b)], the fact that neutral atoms can attract one another strongly and form very stable (homopolar) molecules, such as, for example, H_2, N_2, and CO, has first been explained on the basis of quantum mechanics.

Except for the simplest cases (H_2^+ and H_2), the theoretical discussion of the stability of molecular states deals with the problem of a system consisting of a rather large number of particles (electrons and nuclei). There is no hope of solving this problem rigorously, but various factors must be neglected in order to arrive at a solution. According to the type of approximation used, we are led to different valence theories. Often the approximations are very questionable, and in such cases the results obtained can only be called daring extrapolations. However, when the different valence theories agree in their predictions in a given case, we can suppose that the rigorous solution would also lead to the same result [see, for this point, Van Vleck and Sherman (683)].

Two methods for the theoretical discussion of homopolar binding have proved to be the most important in the course of the development of the subject—the *method of Heitler and London*, which starts out from the separated atoms, and the *method of Hund and Mulliken*, which starts out from the orbitals of the individual electrons in the nuclear frame.   The former method has been extended and modified by Slater and Pauling, the latter by Herzberg and Lennard-Jones.

Treatment of the H_2 molecule according to Heitler and London.   Heitler and London (295) have solved the wave equation for the H_2 molecule by starting out from the state of the separated atoms as zero approximation and then introducing the interaction of the two atoms as a perturbation.   In this way

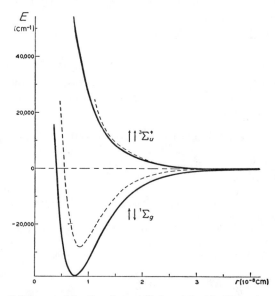

Fig. 161.   **Potential Curves of the Two Lowest States of the H_2 Molecule.**   The full line curves are the final curves resulting from the calculations of James and Coolidge (361)(859), which agree with the observed data; the broken line curves are first approximations as given by Heitler (1024a) (see p. 353 f.).

they obtained the theoretical potential curves for the two states $^1\Sigma_g{}^+$ and $^3\Sigma_u{}^+$, which result from two normal H atoms in the ^2S state according to the Wigner-Witmer correlation rules (see Table 28).   Their theoretical curves are represented as the broken-line curves in Fig. 161.   The $^1\Sigma_g{}^+$ state has a potential curve with a deep minimum and is thus a stable state.   On the other hand, the $^3\Sigma_u{}^+$ state is unstable.   We can also say that *two hydrogen atoms attract each other when their spins are antiparallel* $(^1\Sigma)$ *and repel each other when the spins are parallel* $(^3\Sigma)$.   The observed spectra show indeed that the ground state of the molecule is a $^1\Sigma$ state.[4]   A great deal of work has been done in carrying the

[4] Without an assumption about the statistics of the nuclei the spectrum does not allow us to decide what kind of a $^1\Sigma$ state the ground state of H_2 is.   But on the assumption, which

original calculations of Heitler and London to higher and higher approximations. These efforts have culminated in the work of James and Coolidge (361)(859) whose theoretical values for dissociation energy, equilibrium internuclear distance, vibrational frequency, and anharmonicity agree in a most satisfactory way with the empirical values[5] [see Richardson (585), and Beutler and Jünger (96)]. For example, the best experimental value (see Chapter VII) for the heat of dissociation of H_2 in the ground state is $D_0^0(H_2) = 36,116 \pm 6$ cm^{-1}, while the best theoretical value is $36,104 \pm 105$ cm^{-1}. The full line curves in Fig. 161 represent the final potential curves obtained by James and Coolidge. The unstable $^3\Sigma_u^+$ state has also been observed. It is the lower state of the extensive continuous spectrum of hydrogen (see Chapter VII).

As we shall see in more detail below, the essential reason for the strong attraction (or repulsion) of two H atoms according to the Heitler-London theory is the *"exchange degeneracy"*—that is, the fact that, for very large internuclear distance, by exchange of the two electrons of the two atoms a configuration results that is indistinguishable from the original configuration. As a result, when the two atoms approach each other an interaction between them arises which can be mathematically described in terms of "electron exchange" and which leads to a splitting into two states of different energy (Fig. 161). For this reason, the homopolar forces between two neutral atoms are sometimes also called *"exchange" forces*.[6] The exchange forces are attractive for antiparallel spin orientation and repulsive for parallel spin orientation of the electrons. However, it must be emphasized that the strong attraction (or repulsion) is not due to the mutual interaction of the spins, which is very weak. The spin comes in only through its effect on the Pauli principle. It acts as a sort of indicator for the sign of the interaction of the atoms. This is similar to the relation of the spin to the energy difference of singlet and triplet states of the same electron configuration (see p. 334).

Calculation shows that the strong attraction or repulsion brought about by the exchange forces decreases very rapidly with increasing internuclear distance; in fact, it decreases exponentially. Thus, with increasing r, the potential energy curves in Fig. 161 approach the asymptote very rapidly. This fact has proved to be typical for homopolar binding.

The *probability density distribution* of the electrons in the H_2 molecule, calculated for the $^1\Sigma$ and $^3\Sigma$ states (for the same internuclear distance), ac-

is confirmed by scattering experiments, that the protons follow Fermi statistics it follows from the spectrum that the ground state of hydrogen is either $^1\Sigma_g^+$ or $^1\Sigma_u^-$, of which the former is in agreement with the Heitler-London theory as well as with the electron configuration (see p. 339; compare also the discussion on p. 247). Conversely, if the theoretical result that the ground state is $^1\Sigma_g^+$ is accepted, it follows immediately that the protons follow Fermi statistics.

[5] To be sure their calculations make use to some extent of the molecular orbital approach.

[6] It should be noted, however, that these forces are not a new type of force. They arise from substituting the ordinary Coulomb forces between the individual particles into the wave equation and solving for the resultant attraction of the two atoms.

cording to London, is illustrated graphically in Fig. 162.   It can be seen that in the case of the $^1\Sigma$ state (attraction) the electron clouds of the two atoms blend into each other, whereas for the $^3\Sigma$ state (repulsion) they remain almost completely separated.   In the first case there is a considerable electron density

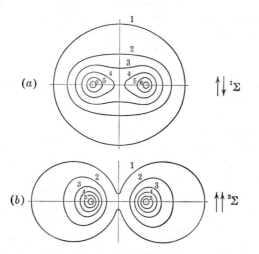

FIG. 162.   **Probability Density Distribution of the Electrons in the $^1\Sigma$ Ground State and the $^3\Sigma$ Repulsive State of the H₂ Molecule, Approximate [after London (452)].**   In both cases the probability distribution is drawn for the nuclear separation in the ground state of H₂ (0.74 Å).   The curves are curves of equal probability density in a section going through the internuclear axis. The numbers give relative values of the probability density.

between the nuclei leading to strong binding, while in the second case the electrons tend to avoid the region between the nuclei so that the repulsion of the nuclei predominates.

The potential energy of two electrons in the field of two protons A and B is given by

$$V = \frac{e^2}{R} - \frac{e^2}{r_{1A}} - \frac{e^2}{r_{2A}} - \frac{e^2}{r_{1B}} - \frac{e^2}{r_{2B}} + \frac{e^2}{r_{12}} \qquad (VI, 24)$$

where R, r_{1A}, $\cdots$ are defined in Fig. 163.   The Schrödinger equation of the H₂ molecule with the nuclei considered fixed at a distance R is according to (I, 13)

$$\nabla_1^2\psi + \nabla_2^2\psi + \frac{8\pi^2 m}{h^2}(E - V)\psi = 0 \qquad (VI, 25)$$

From the solution of this equation for various values of R the variation of the electronic energy with R (that is, the potential energy under which the nuclei move) may be found.   For very large R the potential energy V to be substituted in (VI, 25) reduces to $-(e^2/r_{1A}) - (e^2/r_{2B})$ that is, (VI, 25) is then the wave equation of two independent H atoms at a distance R.   The eigenfunctions are products $\varphi_A(1)\,\varphi_B(2)$, where $\varphi_1(1)$ is a hydrogen eigenfunction of electron 1 (with coordinates x_1, y_1, z_1) moving about nucleus A, and similarly for $\varphi_B(2)$.   Both functions are of course referred to the same coordinate system.   Let us consider only the case in which the two H atoms are in their ground states, so that $\varphi_A(1)$ is given by

$$\varphi_A(1) = \frac{1}{\sqrt{\pi a_H{}^3}} e^{-r_{1A}/a_H}, \qquad a_H = \frac{h^2}{4\pi^2 m e^2} \tag{VI, 26}$$

The energy of the system for large R is then simply $2E_H$ where E_H is the energy of the ground state of the H atom.

Since the electrons are indistinguishable they may be exchanged without affecting the energy of the system, that is, for large R we may use $-(e^2/r_{2A}) - (e^2/r_{1B})$ instead of the above, leading to the eigenfunction $\varphi_A(2)\,\varphi_B(1)$. Thus we have a double degeneracy (*exchange degeneracy*) and any linear combination of the two functions $\varphi_A(1)$ $\varphi_B(2)$ and $\varphi_A(2)\,\varphi_B(1)$ is a solution. If we introduce now the neglected terms in V (that is, bring the atoms closer together) a splitting of the degeneracy will arise. Since the potential energy remains unchanged for an exchange of the two electrons the eigenfunctions must either also remain unchanged or only change sign for such an exchange. In a zero-approximation the *eigenfunctions* are therefore the following linear combinations of those given above

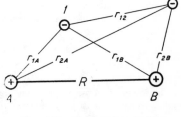

FIG. 163.   **Designations in the H₂ Molecule.**

$$\psi_s = N_s[\varphi_A(1)\varphi_B(2) + \varphi_A(2)\varphi_B(1)] \tag{VI, 27}$$

$$\psi_a = N_a[\varphi_A(1)\varphi_B(2) - \varphi_A(2)\varphi_B(1)] \tag{VI, 28}$$

The first of these is symmetric, remains unchanged for an exchange of the electrons; the second is antisymmetric, changes sign for such an exchange. The normalization factors N_s and N_a are given by

$$N_s = \frac{1}{\sqrt{2 + 2S}}, \qquad N_a = \frac{1}{\sqrt{2 - 2S}} \tag{VI, 29}$$

$$S = \int \varphi_A(1)\varphi_B(1)\varphi_A(2)\varphi_B(2)\,d\tau_1 d\tau_2 \tag{VI, 30}$$

If a first-order perturbation calculation (see Chapter I, p. 14) is carried out using as perturbation function W the deviation from the potential energy for large R

$$W = +\frac{e^2}{R} + \frac{e^2}{r_{12}} - \frac{e^2}{r_{1B}} - \frac{e^2}{r_{2A}} \tag{VI, 31}$$

one finds two *energy levels* E_s and E_a with eigenfunctions ψ_s and ψ_a given by (VI, 27) and (VI, 28). For the energies one finds

$$E_s = 2E_H + \frac{K + J}{1 + S} \tag{VI, 32}$$

$$E_a = 2E_H + \frac{K - J}{1 - S} \tag{VI, 33}$$

where

$$K = \int \varphi_A(1)\varphi_B(2)W\varphi_A(1)\varphi_B(2)\,d\tau_1 d\tau_2 \tag{VI, 34}$$

$$J = \int \varphi_A(1)\varphi_B(2)W\varphi_A(2)\varphi_B(1)\,d\tau_1 d\tau_2 \tag{VI, 35}$$

All three integrals, K, J, and S, occurring in the energy expressions depend on the value of the internuclear distance R. Actual evaluation of these integrals gives the lower broken-line curve in Fig. 161 for E_s and the upper for E_a (omitting the constant term $2E_H$), that is, for the state characterized by the symmetric eigenfunction ψ_s a potential minimum arises leading to molecule formation while for the state characterized by the antisymmetric eigenfunction ψ_a there is repulsion for all R values.

The integral K is called the *Coulomb integral* since it gives the interaction of the two charge clouds on a classical basis [$\varphi_A{}^2(1)$ and $\varphi_B{}^2(2)$ are the charge densities produced by electrons 1 and 2 around nuclei A and B respectively]. The integral J is called the *exchange integral* because the two functions $\varphi_A(1)\varphi_B(2)$ and $\varphi_A(2)\varphi_B(1)$ occurring in it are distinguished by an exchange of the two electrons. It arises because the electrons are indistinguishable and can be exchanged. This integral represents an entirely non-classical effect and, as it turns out, gives the major contribution. That is why the forces leading to the formation of homo-polar molecules are called exchange forces.

The exchange integral may also be written

$$ J = \int \varphi_A(1)\varphi_B(1)W\varphi_A(2)\varphi_B(2)d\tau_1 d\tau_2 \qquad (VI, 36) $$

Here $\varphi_A(1)\varphi_B(1)$ [and similarly $\varphi_A(2)\varphi_B(2)$] is a function depending on the overlapping of the two hydrogen eigenfunctions belonging to the two nuclei. For intermediate R this function [called exchange density by Heitler (16)] is largest near the two nuclei and consequently the terms $-(e^2/r_{1B}) - (e^2/r_{2A})$ of W give the largest contribution to J which is therefore nega-tive. Thus, since we always have $S < 1$, we see from (VI, 32 and 33) that for intermediate internuclear distances E_s is below the value $2E_H$ while E_a is above it, that is, we have attraction and repulsion, respectively (see Fig. 161). For small values of R the largest contribution to both K and J is that due to e^2/R which is positive, that is, we have repulsion in both states.

Thus far we have ignored both the *electron spin* and the *Pauli principle*. According to the Pauli principle only those states occur whose eigenfunctions are antisymmetric. There-fore if there were no spin, only the state E_a would occur and no stable H_2 molecule would be formed. However, if the spin is introduced the total eigenfunction is the product of the posi-tional eigenfunction discussed above and a spin function. For two electrons there is one antisymmetric spin function corresponding to antiparallel spins and three symmetric spin functions corresponding to parallel spins (see p. 335 and A.A., p. 124). Therefore the state E_s in which the H atoms attract each other can occur with antiparallel spins ($^1\Sigma$ state) while the state E_a in which the H atoms repel each other occurs with parallel spins ($^3\Sigma$ state).[7]

For the *excited states* of H_2 arising from a normal and an excited H atom the Pauli principle does not limit the spin orientation as it does for the ground state. In addition a complication arises on account of the *resonance degeneracy* (splitting into g and u states, see p. 321), which actually gives a greater contribution than the exchange degeneracy. The calculations show [see Heitler (293)] that some of the resulting triplet states have deeper potential minima than the singlet states. At the same time there are other states, both triplets and singlets, which have no minimum. Thus the previous rule for the attraction and repulsion of H atoms is not applicable when one of them is excited.

Generalization of the Heitler-London theory for more complicated cases.

The next simplest case is that of the interaction of a normal He atom with a

---

[7] A somewhat different explanation of homopolar attraction has been given by Hellmann (36b): In consequence of the Heisenberg uncertainty relation the electron in the H atom has a certain momentum distribution and therefore a certain zero-point energy given by the energy of the ground state of the H atom. If two H atoms are brought together, the space available for each of the electrons (if they satisfy the Pauli principle—that is, have antiparallel spins) is almost doubled. Consequently the minimum uncertainty of the momentum—that is, the zero-point energy—is reduced, or, in other words, the two atoms attract each other. Similar considerations may be applied to other atoms [see however Coulson and Bell (868)].

normal H atom. According to the Wigner-Witmer rules, only a single molecular state—namely, $^2\Sigma^+$—results from $He(^1S) + H(^2S)$. According to Heitler and London, this is a repulsive state, as we can see qualitatively from the fact that the electron of the H atom, as a result of the Pauli principle, is exchangeable only with that electron of He (in its ground state) whose spin orientation is parallel to its own:

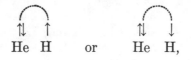

which gives a repulsive exchange force.[8] In a similar way, repulsion of an H_2 molecule and an H atom, as well as repulsion of two normal He atoms, follows from the theory. These results are in agreement with experiment and with the usual elementary chemical idea of valence.

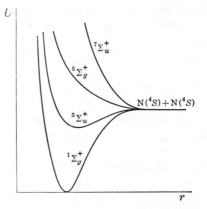

As we proceed to more complicated atoms, the calculations can only be carried out less and less rigorously. Nevertheless, in the case of the molecular states of different multiplicities resulting from two atoms in S states Heitler and London were able to show that *the states with the lowest total spin S always lie lowest* and the others lie in the order of their multiplicities. For example, two normal N atoms in ^4S states give, according to Wigner and Witmer (see Table 28), the molecular states $^1\Sigma_g^+$, $^3\Sigma_u^+$, $^5\Sigma_g^+$, and $^7\Sigma_u^+$. According to Heitler

Fig. 164. **Potential Curves of the Electronic States of the N_2 Molecule Arising from Normal Atoms, According to Heitler and London (Schematic).**

and London, they lie in this order. Fig. 164 shows qualitatively the corresponding potential curves. The states $^1\Sigma_g^+$ and $^3\Sigma_u^+$ are stable, and the states $^5\Sigma_g^+$ and $^7\Sigma_u^+$ are unstable. $^1\Sigma_g^+$ is the ground state of the N_2 molecule, in agreement with experiment. Here again it is the possibility of exchange between each newly formed pair of electrons with antiparallel spins that causes the attraction. Since for N_2 in the ground state there are three such pairs, the binding strength is very great. When two of the electrons have parallel spins, they give a repulsive exchange force, and therefore the binding strength of the $^3\Sigma_u^+$ state of N_2 is very much less than that of the ground state. Experimentally it is not yet certain whether or not this state is stable.

From this discussion we see that *the greater the number of electron pairs*

---

[8] In contrast to HeH, the ion HeH^+ is stable, as one would expect on the basis of its similarity to H_2 [see Glockler and Fuller (261), Beach (82a), Coulson and Duncanson (169a), and Toh (1463)].

formed from the unpaired electrons of the separate atoms the more stable the resulting molecular state will be. In the $^1\Sigma$ ground state of H_2, one new electron pair is formed and in the $^1\Sigma$ ground state of N_2, three are formed. According to elementary chemical ideas, a monovalent binding is assumed in H_2 and a trivalent binding in N_2. The dissociation energy of H_2 is 4.476, that of N_2 is 7.373 e.v. (or 9.756 e.v., see Table 39).

It seems natural to generalize these considerations for any atoms—that is, to assume the *valency of a bond* to be *equal to the number of newly formed electron pairs.* The larger the latter number is, the stronger is the binding and the more stable is the molecular state under consideration. Furthermore, we are led to assume that the *valency of an atom is equal to the number of unpaired electrons,* since only these can give pairs (with antiparallel spins) with the electrons of other approaching atoms and thereby an attractive exchange force. The valency is therefore $2S$, if S is the quantum number of the resultant spin of the atom. It is thus one less than the multiplicity. An N atom in the ^4S ground state can, for example, bind three H atoms and is trivalent, since each one of the three unpaired electrons of the N atom can form, with an electron belonging to an H atom, a pair with antiparallel spin orientations. A fourth H atom will not be attracted, because its electron cannot form an additional antiparallel pair with one of the electrons already present, and consequently there is no further gain in energy— that is, no bonding action. We see therefore that the *saturation of homopolar valencies* follows naturally from this representation as a *saturation of the spins*— a pairing off in antiparallel pairs.

The valencies obtained in the above-described way for the atoms in the different columns of the periodic system are given in the last row of Table 35. Both the valency—that is, the number of unpaired electrons—in the ground state (printed in heavy type) and that in the lowest excited states are given. On the Heitler-London theory the normal chemical valencies of the alkaline earths, the earths, and the elements of the carbon group must be ascribed to *excited* atomic states in which an electron is transferred from the 2s shell to the 2p shell. For the C atom, for example, the ground state $1s^2 2s^2 2p^2\ ^3P$ has only two unpaired electrons and is, therefore, according to this mode of representation, divalent. However, the excited state $1s^2 2s 2p^3\ ^5S$ has four unpaired electrons and can explain the *tetravalency of carbon.* It has an energy of 4.18 e.v. above the ground state [Shenstone (1415)].

It should be noted that, whereas oxygen and fluorine have only one valency —namely, 2 and 1, respectively—the other elements of the oxygen and fluorine groups exhibit in addition other valencies (4, 6, and 3, 5, 7, respectively). This is explained by the fact that in order to raise the multiplicities of O and F a very high excitation would be required, whereas this is not necessary for the other elements of these groups.

The valence theory described here is also called the *spin valence theory.* While it is attractively simple, we should not forget that implicitly in obtaining it a number of rather important points were disregarded. Above all, in the generalization we have completely neglected the circumstance that the calculation has been carried out only for atoms in S states. Furthermore, the theory holds only if the separations of neighboring atomic terms from those considered are large compared to the binding energies, a condition which is fulfilled for the H atoms and the inert-gas atoms but seldom for other atoms. Therefore, while the Heitler-London theory represents the essentials of chemical valence in broad outline, we must be prepared for deviations in the matter of detail.

TABLE 35. HOMOPOLAR VALENCY

(After Heitler and London)

Group in Periodic System	I Alkalis	II Alka-line Earths	III Earths	IV Car-bon Group	V Nitro-gen Group	VI Oxygen Group	VII Halo-gens	VIII Inert Gases
Multiplicity..	2	1 3	2 4	1 3 5	2 4 6	1 3 5 7	2 4 6 8	1
Valency.....	1	0 2	1 3	0 2 4	1 3 5	0 2 4 6	1 3 5 7	0

Examples of such deviations, which are obviously to be traced back to the nonfulfillment of the above two conditions, are the following: According to the elementary Heitler-London theory, one would expect $^1\Sigma_g^+$ as the ground state of the molecule C_2, whereas $^3\Pi_u$ is observed. Similarly, $^1\Sigma_g^+$ would be expected as the ground state of O_2 (all electrons paired off), whereas both the spectrum and the paramagnetism of O_2 gas (see p. 462) give very direct evidence that the resultant spin is not zero, the ground state being $^3\Sigma_g^-$. Furthermore according to the elementary Heitler-London theory, atoms such as Be, Mg, Ca, and so forth, in their ground states should not be able to bind another atom, since they contain only closed shells (^1S states), whereas, experimentally relatively stable diatomic hydrides and halides of these elements are known whose ground states arise from normal atoms.

These deviations between theory and experiment can, according to Heitler (293) and Nordheim-Föschl (533), be removed if account is taken of the interaction of states of the same species (p. 295 f.)—that is to say, when higher approximations are worked out.

As a simple example let us consider the BeH molecule. For the Be atom, the excited $1s^2 2s 2p$ ^3P state lies not very high above the ground state $1s^2 2s^2$ ^1S. From the latter state and the ^2S ground state of hydrogen only one molecular state, $^2\Sigma^+$, results; from the atomic states $^3P + {}^2S$ the molecular states $^2\Sigma^+$, $^2\Pi$, $^4\Sigma^+$, and $^4\Pi$ result (see Fig. 152). The potential curves for these states, as they would be given according to the elementary Heitler-London theory, are drawn quite roughly in Fig. 165 (broken-line curves). According to this theory, only the states $^2\Sigma^+$ and $^2\Pi$, resulting from the excited Be atom (^3P) and the normal H atom, would be stable.

However, if we take account of the interaction of states of the same species, the intersection of the two $^2\Sigma^+$ states (from $^1S + {}^2S$ and from $^3P + {}^2S$) "originally" present is avoided (see Chapter V, section 4). They repel each other quite considerably, and the full-line potential curves are obtained. In this higher approximation, the $^2\Sigma^+$ state resulting from normal atoms has a potential minimum—that is, it is a stable state, in agreement with experiment— while the $^2\Sigma^+$ state resulting from $^3P + {}^2S$ has a much shallower minimum than originally.

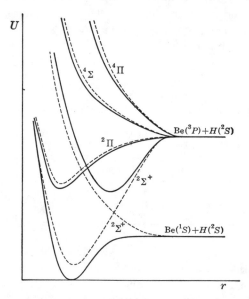

Fig. 165. **Potential Curves of the BeH Molecule; Broken-Line Curves, According to the Elementary Heitler-London Theory; Full-Line Curves, Allowing for the Mutual Interaction of the States.** The curves are drawn quite schematically, since accurate experimental data are not available.

The remaining states given are only slightly affected by taking into account the interaction of states of the same species, since no similar states lie in their neighborhood. The corresponding broken-line and full-line potential curves therefore almost coincide.

It should be noted that, near its minimum, the $^2\Sigma^+$ ground state has to a considerable degree the properties of the $^2\Sigma^+$ state resulting from $^3P + {}^2S$. In consequence of that, we should expect, for example, that the vibrational quanta in the ground state for small v would rather correspond to a dissociation into $^3P + {}^2S$, although in reality a dissociation into normal atoms takes place. This has actually been observed for some other hydrides of the elements of the second column of the periodic system, although not for BeH, where up to the present time sufficient data are not available.[9]

As a second example consider the O_2 molecule. Here we obtain from two normal O atoms (^3P) the 18 different states given in Table 28 (p. 321). Of these we expect, according to the elementary theory, the singlet states to be the lowest and, in particular, a $^1\Sigma_g^+$ state to be the ground state. However, not very far above $^3P + {}^3P$ there is the combination $^3P + {}^1D$ of the separated atoms, which gives rise to the states $^3\Sigma^+$, $^3\Sigma^-(2)$, $^3\Pi(3)$, $^3\Delta(2)$, $^3\Phi$ (see Table 26, p. 318), occurring both as u and as g states. A number of these states are of the same species

---

[9] For a more detailed discussion of the diatomic hydride molecules according to the Heitler-London theory see Stehn (650) and King (400a).

as some triplet states arising from $^3P + {}^3P$. Therefore in higher approximation these latter triplet states ($^3\Sigma_u^+$, $^3\Sigma_g^-$, $^3\Pi_g$, $^3\Pi_u$, and $^3\Delta_u$) will be shifted downward. In this way, one can see that at least it *may* happen that a triplet state becomes the ground state of the O_2 molecule. A more detailed calculation [Nordheim-Pöschl (533)] shows indeed that one of these triplet states, $^3\Sigma_g^-$, becomes the lowest-lying state, in agreement with experiment.

In a similar manner it is found that the ground state of C_2 is $^3\Pi_u$, in agreement with experiment, and not $^1\Sigma_g^+$, as would be expected on the basis of the elementary theory. On the other hand, for N_2 the conclusion of the elementary theory (see above) is not changed by introducing the interaction of the states resulting from normal atoms ($^4S + {}^4S$) with those resulting from excited atoms.

Thus, taking account of *higher approximations*, we obtain agreement between the Heitler-London theory and experiment even for the apparent exceptions. It must be mentioned, however, that the calculations often become very complicated and involved. Qualitative considerations such as the above then no longer suffice to make a prediction about the stability of the molecular states.

The refinement of the Heitler-London theory here briefly indicated is sometimes called the *theory of orbital valence*, or *l-valence*, since it takes account of the orbital angular momentum L of the two atoms as well as the spins [see Heitler (293), Bartlett (81), and Hellmann (36)].

Pauling (553) and Slater (638) have extended the Heitler-London theory in a somewhat different manner. Instead of using as zero approximation certain states of the separated atoms, they start out *from certain electronic configurations* of these atoms and thus at first neglect the finer interaction of the electrons that leads to a splitting of the terms of a given electronic configuration in the atom (for example, to the energy difference between p^2 ^3P, ^1D, and ^1S). For diatomic molecules this approach does not lead to essentially new results, but it does prove very fruitful for polyatomic molecules. A more detailed discussion of it will therefore be postponed until Volume III [see also Van Vleck and Sherman (683)].

Theory of bonding and antibonding electrons for equal nuclear charges.

The second method of treating the question of the stability of molecular states does not start out, as a zero approximation, from the states of the separate atoms but from the motion (orbitals) of the individual electrons in the field of the two nuclei and of the other electrons, for various distances between the nuclei (Mulliken, Hund, Herzberg, and Lennard-Jones). In other words, the molecular electron configurations are the basis of this method which is therefore also referred to as the *method of molecular orbitals*. By this treatment some experimental facts, explained by the Heitler-London theory only when higher approximations are taken into account, can be understood even in first approximation. Conversely, however, other facts are better understood (that is, in lower approximation) on the basis of the Heitler-London theory. For a recent exhaustive review of the molecular orbital method see Mulliken (1266a).

While the Heitler-London method was developed from a treatment of the H_2 molecule the molecular orbital method may be most readily developed from a study of the H_2^+ molecule. The systems H_2 and H_2^+ are the only ones which have up to the present time been worked out with adequate accuracy on the basis of wave mechanics. For H_2^+, the dependence of the pure electronic energy on internuclear distance has been calculated on the basis of the wave equation for a number of electronic states, including the ground state [Teller (663), Hylleraas (350), Jaffé (360), Sandeman (614), Hellmig (296), and Johnson

(1107)]. If the potential energy e^2/R of the two nuclei is added to the pure electronic energy, the potential curves of the molecule in the various electronic states are obtained (see p. 148). Fig. 166 shows the curves obtained for the two

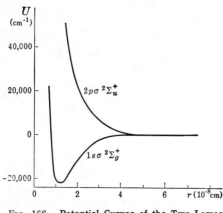

lowest states (both arising from normal $H + H^+$). The lowest state (ground state), $1s\sigma$, has a fairly deep potential minimum, whereas the second lowest state, $2p\sigma$, has a purely repulsive potential curve. The dissociation energy of the ground state obtained theoretically[10] is $D_0^0(H_2^+) = 21,345 \pm 20$ cm^{-1}, while experimentally an indirect method [from $D_0^0(H_2)$ and the ionization potentials of H_2 and H (see Chapter VII)] gives the value $21,366 \pm 15$ cm^{-1} [see Beutler and Jünger (96)]. The excellent agreement between theory and experiment, for H_2^+ as well as for H_2 (see p. 351), shows without any doubt that wave mechanics is quite adequate to deal theoretically with electronic motions in molecules. That deviations between theory and experiment appear for heavier molecules is solely due to the fact that inadequate approximations for the solutions of the wave equation are used.

FIG. 166. **Potential Curves of the Two Lowest States of H_2^+ [after Teller (663)] (Theoretical).**

The higher excited states of H_2^+ are in part stable and in part unstable, depending on the quantum numbers of the one electron present [see Teller (663) and Hellmig (296)]. The stable excited states have, however, only very shallow minima of their potential curves. That is the reason why a spectrum of H_2^+ has never been observed.

On the basis of a naïve application of the Heitler-London valence theory, a strong binding would not be expected in the ground state of H_2^+, since in order to form a strong homopolar bond this theory requires the formation of an electron *pair* from unpaired electrons of the separated atoms. An explanation of the fact that there is, in spite of that, a relatively firm bond produced by one electron only, can be given in two different ways, depending on which zero approximation is used to treat the H_2^+ molecule.

If we start from the situation at large internuclear distance as zero approximation, there is a *resonance degeneracy*, because the state "electron with the one nucleus" has the same energy as the state "electron with the other nucleus." At smaller internuclear distances this degeneracy is removed in such a way that of the two resulting components, σ_g and σ_u, the one is shifted downward and the

---

[10] This is the average of the values given by Sandeman and Johnson which differ by 36 cm^{-1}. If Johnson's value is accepted there is a very slight discrepancy with the experimental value. The error indicated for the experimental value is the maximum error.

other upward from the position for large internuclear distance. A stable state $(1s\sigma)$ and an unstable state $(2p\sigma)$ result (see Fig. 166).

However, in the case of H_2^+ we can also start out from small and intermediate internuclear distances and regard the relatively large binding strength of the ground state as due to the fact that the orbital $1s\sigma$ for small internuclear distance transforms into the ground state of the united atom (ion) He^+, whereas the orbital $2p\sigma$ transforms into an excited state of this ion. Fig. 167 shows schematically the variation of the purely electronic energy (disregarding nuclear repulsion) for the two lowest states of H_2^+. The ground state of He^+ lies 40 volts lower than the ground state of H, while the 2-quantum state of He^+ lies at nearly the same height as the ground state of the H atom. If now, in addition, we take account of the nuclear repulsion, it is understandable that (as the accurate calculation proves) the $2p\sigma$ state gives only repulsion and the $1s\sigma$ state gives a strong attraction.

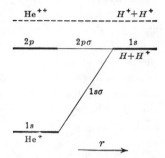

FIG. 167. **Correlation of the Two Lowest States of H_2^+ with Those of the United Atom.** To the left the nuclear separation is 0, and to the right it is ∞. The broken line gives the energy when the one electron is completely removed. The corresponding states $H^+ + H^+$ and He^{++} lie at the same height, since the nuclear repulsion is neglected in the figure.

The important point for the present considerations is that it depends on the quantum numbers of the electron in the molecule $(1s\sigma, 2p\sigma, \cdots)$ whether a stable or unstable state arises, or, in other words, that certain orbitals of H_2^+ are bonding, others antibonding, and some perhaps non-bonding.

It is natural to assume that, also if a number of electrons are present, *each individual electron*, when the nuclei are brought together, *gives either a positive or a negative contribution to the binding*—is either *bonding* or *antibonding* or possibly non-bonding. Whether an electron in a given orbital is bonding or antibonding is not necessarily determined by the same conditions as for H_2^+, since in the more general case the electron does not move in the field of two point charges but in the field of the nuclei plus that of the other electrons (see below). A condition that this method of bonding and antibonding electrons shall lead to usable results is, of course, that the representation of the electronic states by electron configurations is a good approximation and more particularly that, in the transition from the separated atoms to the molecule, each electron, in a way, retains its individuality; that is, for example (for equal nuclear charges), a σ_u electron remains a σ_u electron, and so on. Thus we assume that in the earlier scheme, Fig. 157 or 156, *a given state has the same electron configuration for large and medium internuclear distances.* This assumption is admittedly a rough approximation since it implies the neglect of the finer interaction of the electrons. Strictly speaking only the species of the resultant states is conserved in going from the separated atoms to the molecule and intersections of states of different electron configurations if they have the same species will be avoided (see p. 295).

However, if we do make this rough assumption and know which electrons are bonding and which are antibonding, we can find out which of the electronic states of the molecule are stable and which are unstable, since at least roughly we should expect that a *stable* molecular state would result *if the number of bonding electrons is greater than the number of antibonding electrons* [Herzberg (304)(308), and Hund (347)].

The question is now: When is an electron bonding and when is it antibonding? We should expect that *orbitals that move downward* relative to the others in Fig. 157 in going from the separated atoms to the united atom are *bonding*, whereas the *orbitals moving upward* are *antibonding*. In agreement with these expectations, calculation shows (see below) that, at least for large internuclear distances, the orbitals σ_g and π_u lie lower than the corresponding orbitals σ_u and π_g (that is, for example, $\sigma_g 1s$ lies lower than $\sigma_u 1s$, $\sigma_g 2p$ lies lower than $\sigma_u 2p$, and so on. See right-hand part of Fig. 157).

Using for the case of equal nuclear charges the approximation by the linear combinations (VI, 13) of atomic orbitals one finds from a first order perturbation calculation (closely similar to the Heitler-London treatment on p. 352 f.) for the energy changes of a σ_g and a σ_u orbital resulting from one and the same atomic orbital [$2p$ in (VI, 13)]

$$\Delta E(\sigma_g) = \frac{H_{AA} + H_{AB}}{1 + T}, \qquad \Delta E(\sigma_u) = \frac{H_{AA} - H_{AB}}{1 - T} \qquad \text{(VI, 37)}$$

Here

$$H_{AB} = \int \psi_A W \psi_B d\tau, \qquad H_{AA} = \int \psi_A W \psi_A d\tau = H_{BB}, \qquad \text{(VI, 38)}$$

T is the integral in (VI, 14) which represents the deviation from orthogonality and W is the deviation of the potential energy from that of the two separated atoms or ions. H_{AA} and H_{BB} are relatively small terms representing the Coulomb interaction of the second atom on the first and vice versa.

Since at least for large internuclear distance R the integral T is small compared to 1 it is clear that the dominant term in (VI, 37) is the *resonance integral* H_{AB} which, in the molecular orbital theory, plays a role similar to that of the exchange integral J in the Heitler-London theory. It should be noted however that H_{AB} unlike J involves the coordinates of one electron only. H_{AB}, just as H_{AA} and H_{BB}, is zero for $R \rightarrow \infty$ and becomes increasingly negative with decreasing R, until for small R it goes through zero again and assumes large positive values. Therefore the energy of the σ_g orbitals decreases to a minimum, that of the σ_u orbitals increases with decreasing R: σ_g orbitals are bonding, σ_u orbitals are antibonding (see Fig. 166 for H_2^+). The same applies to δ_g and δ_u orbitals. However, for π orbitals the relations are reversed since the signs in (VI, 13) are reversed (see p. 330), that is, π_u electrons are bonding; π_g electrons antibonding.

It may be noted that the bonding orbitals have no nodal surface between the nuclei while the antibonding orbitals have at least one (see Fig. 155). A nodal surface between the nuclei means a low electron density and therefore lack of attraction, while the absence of a nodal surface implies a high electron density and therefore a contribution to the attraction of the nuclei. It will be recalled that σ_g, σ_u, π_g, π_u, orbitals may be described more specifically by indicating the atomic orbitals from which they arise, for example $(\sigma 2p + \sigma 2p, \sigma_g)$,

$(\sigma 2p - \sigma 2p, \sigma_u)$, $(\pi 2p - \pi 2p, \pi_g)$, $(\pi 2p + \pi 2p, \pi_u)$ where the orbitals with a $+$ sign are bonding, those with a $-$ sign are antibonding.[11]

The manner in which the rules mentioned lead to an understanding of molecule formation is best made clear by some examples.

In the formation of the H_2 molecule from two H atoms, both electrons can go into the $\sigma_g 1s$ orbital, and, since for H_2^+ this orbital gives an appreciable attraction, we should expect, on the basis of the supposition that each electron contributes a certain amount to the binding, that with two electrons in this bonding orbital the attraction would be considerably strengthened, as is actually observed $[D_0^0(H_2) = 4.476$ volts, compared to $D_0^0(H_2^+) = 2.648$ volts].[12]   According to the Pauli principle, both electrons can be $\sigma_g 1s$ electrons only if their spins are antiparallel.   Thus a $^1\Sigma_g^+$ state results.   If the two electrons have parallel spins, they must be in different orbitals.   The only possibility for this, when both electrons in the separated atoms are $1s$ electrons, is $(\sigma_g 1s)(\sigma_u 1s)$ $^3\Sigma_u^+$ (see Fig. 157).   In this case we have one bonding and one antibonding electron, and we should therefore not expect a binding.   Thus for the molecular states resulting from two normal H atoms the molecular orbital theory leads qualitatively to the same conclusions as the Heitler-London theory and these agree with experiment.

The eigenfunction of the ground state of H_2 in the molecular orbital theory is in a first approximation

$$\Psi(^1\Sigma_g^+) = \psi_g(1)\psi_g(2) \qquad (VI, 39)$$

where ψ_g is a $\sigma_g 1s$ orbital wave function and where 1 and 2 refer to the coordinates of the two electrons.   The best approximation for ψ_g is obtained when the Schrödinger equation is solved for the motion of one electron in the field of the two centers modified by the average field of the other electron (Hartree field).   However, for our present purposes it is sufficient to use as a rough approximation for ψ_g a linear combination of atomic orbitals similar to the first equation (VI, 13).   Substituting in (VI, 39) we obtain

$$\Psi(^1\Sigma_g^+) = N_g^2[\psi_A(1)\psi_A(2) + \psi_B(1)\psi_B(2) + \psi_A(1)\psi_B(2) + \psi_A(2)\psi_B(1)] \quad (VI, 40)$$

The function $\psi_A(1)\psi_A(2)$ represents a situation in which both electrons are with nucleus A, that is, it represents an ionic state H^-H^+.   Similarly $\psi_B(1)\psi_B(2)$ represents an ionic state H^+H^-.   The function $\psi_A(1)\psi_B(2) + \psi_A(2)\psi_B(1)$ is entirely similar to the Heitler-London wave function (VI, 27) and represents an electron distribution belonging to both nuclei.   The presence of the ionic terms in the wave function Ψ with an amplitude equal to that of the homopolar terms does not of course correspond to reality.   It is due to the partial neglect of the repulsion of the electrons, only the average effect being considered.   For this reason the electrons can be much more frequently with one nucleus than they would if their repulsion

---

[11] Originally Mulliken (509a) assumed that all those electrons were antibonding that had higher quantum numbers in the united atom than in the separated atoms.   He called them *promoted electrons*.   This assumption was later modified by him as given in the text.

[12] If the two electrons were completely independent of each other, we should expect a doubling of the dissociation energy.   That the true value is smaller by about 1 volt is due to the interaction (repulsion) of the electrons.   This influence may also quantitatively be taken into account.

were properly taken into account. On the other hand in the Heitler-London theory the ionic terms are entirely neglected while in actual fact there is a small contribution of these terms to the ground state of H_2 (see below).

The *excessive emphasis on the ionic terms* in the molecular orbital theory becomes particularly glaring when the internuclear distance R approaches infinity. Then the states $(\sigma_g 1s)(\sigma_u 1s)\ ^1\Sigma_u^+$ and $(\sigma_u 1s)^2\ ^1\Sigma_g^+$ would have the same energy as the states $(\sigma_g 1s)^2\ ^1\Sigma_g^+$ and $(\sigma_g 1s)(\sigma_u 1s)\ ^3\Sigma_u^+$, all four having $1s$ electrons only in the separated atoms. The former two states must therefore be derived from $H^+ + H^-$ (in agreement with the Wigner-Witmer rules) which is the only state other than $H + H$ in which there are only $1s$ electrons. Indeed the eigenfunction of the $(\sigma_g 1s)(\sigma_u 1s)\ ^1\Sigma_u^+$ state is easily seen to be given by

$$\Psi(^1\Sigma_u^+) = N_g N_u[\psi_A(1)\psi_A(2) - \psi_B(1)\psi_B(2)] \qquad (VI, 41)$$

that is, it is purely ionic. The eigenfunction of the state $(\sigma_u 1s)^2\ ^1\Sigma_g^+$ is similar to (VI, 40) except that the last two terms have a negative sign, that is, in this approximation the two states $(\sigma_g 1s)^2\ ^1\Sigma_g^+$ and $(\sigma_u 1s)^2\ ^1\Sigma_g^+$ have the same proportion of ionic and non-ionic terms.

While according to the molecular orbital theory the states $H^+ + H^-$ and $H + H$ for $R \to \infty$ have the same energy (since $\sigma_g 1s$ and $\sigma_u 1s$ have the same energy) actually the state $H^+ + H^-$ is almost 13 e.v. above $H + H$. This large discrepancy exists however, only for very large R, for which in any case the Heitler-London method is the better approximation. For smaller R the states $(\sigma_g 1s)(\sigma_u 1s)\ ^1\Sigma_u^+$ and particularly $(\sigma_u 1s)^2\ ^1\Sigma_g^+$ move rapidly upward according to Fig. 157 so that the discrepancy in the energies disappears. Similar conclusions apply to other molecules so that for many purposes, particularly in discussing the ground states it is permissible to neglect the ionic states completely if in assigning electron configurations to the states resulting from the separated atoms one fills first the bonding orbitals and uses antibonding orbitals as little as possible even though for $R \to \infty$ they have the same energy.

For H_2 the ionic state $(\sigma_g 1s)(\sigma_u 1s)\ ^1\Sigma_u^+$ for small R goes over into $1s\sigma 2p\sigma\ ^1\Sigma_u^+$. This state has been observed as the lowest stable excited state of H_2 (see Fig. 160). It does indeed have many characteristics of an ionic state [see Mulliken (520) and section 5 below].

Two normal He atoms give, at large internuclear distances, the electron configuration $(\sigma_g 1s)^2(\sigma_u 1s)^2$—that is, a bonding and an antibonding electron pair. Therefore, according to the above rule, no stable molecular state results when the atoms approach each other, in agreement with experiment. On the other hand, when we bring a normal and an excited He atom together [for example, He $(1s^2) +$ He $(1s\ 2s)$], we obtain a stable molecular state if the excited electron is bonding—for example, in the configuration $(\sigma_g 1s)^2(\sigma_u 1s)(\sigma_g 2s)$— since then three bonding electrons and one antibonding electron are present. It has indeed been found experimentally that the observed He_2 bands are to be ascribed to a molecule resulting from a normal and an excited He atom. The limiting case, He $+$ He^+, also gives a stable molecule, as may easily be seen.

According to Fig. 157 the state $(\sigma_g 1s)^2(\sigma_u 1s)^2\ ^1\Sigma_g^+$, formed from two normal He atoms, goes over into the state $1s\sigma^2\ 2p\sigma^2\ ^1\Sigma_g^+$ for close nuclei if we correlate terms of the same electron configuration with one another in accordance with the aforementioned correlation rule (p. 361). However, in the transition to close nuclei, an intersection with the state $1s\sigma^2\ 2s\sigma^2\ ^1\Sigma_g^+$ occurs which for small internuclear distance lies lower and for large internuclear distance lies considerably higher than $1s\sigma^2\ 2p\sigma^2\ ^1\Sigma_g^+$. This is shown in Fig. 168 (solid-line correlation). However, this intersection can occur only in the approximation in which the finer interaction of the electrons is neglected. In the actual molecule, since the two states are of the same species, this intersection is avoided (broken-line correlation in Fig. 168); that is, the state

$(\sigma_g 1s)^2 (\sigma_u 1s)^2 \; {}^1\Sigma_g^+$ of the separated atoms in actuality goes over into $1s\sigma^2 \; 2s\sigma^2 \; {}^1\Sigma_g^+$ for very close nuclei in spite of the fact that $\sigma_u 1s$ if considered alone would go over into $2p\sigma$. However, it can be seen from Fig. 168 that the strong repulsion between normal He atoms holds in any approximation, since the terms of the united atom lie considerably higher than those of the separated atoms if, as in Fig. 168, the nuclear repulsion is neglected, and therefore *a fortiori* if it is taken into account.

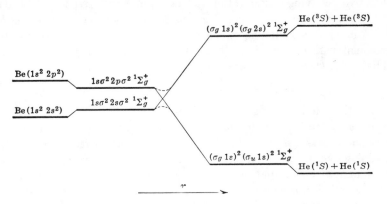

Fɪɢ. 168. **Correlation with the United Atom for He + He.**

From a He atom and a He$^+$ ion in their ground states ($^1S + {}^2S$) two molecular states arise, $^2\Sigma_g^+$ and $^2\Sigma_u^+$ whose electron configurations are $(\sigma 1s)(\sigma_u 1s)^2$ and $(\sigma_g 1s)^2(\sigma_u 1s)$, respectively. Since $(\sigma_g 1s)$ is bonding and $(\sigma_u 1s)$ antibonding it is immediately clear that the $^2\Sigma_u^+$ state is stable while $^2\Sigma_g^+$ is unstable, which is the reverse of the situation in H_2^+ (see Fig. 166). The observed electronic levels of He_2 are indeed found to approach a stable $^2\Sigma_u^+$ state of He_2^+. However, this state has not as yet been observed directly.

According to Fig. 157 two Li atoms have as their lowest electron configuration $(K)(K)(\sigma_g 2s)^2$ for large as well as medium internuclear distances. For the K electrons, no molecular symbol is given in Table 34, p. 343, since the K electrons remain practically uninfluenced in the molecule formation. They are *nonbonding* (inactive) electrons. Since the $\sigma_g 2s$ orbital is bonding, we should expect a stable $^1\Sigma_g^+$ ground state of Li_2, in agreement with experiment. If the two outer electrons of the separated Li atoms have parallel spins, they can form in the molecule only the configuration $\sigma_g 2s \; \sigma_u 2s \; {}^3\Sigma_u^+$, with one bonding and one antibonding electron. This state is repulsive, as for H_2.

Just as for H_2 there are two ionic states $^1\Sigma_u^+$ and $^1\Sigma_g^+$ resulting from $Li^+ + Li^-$ with configurations $(\sigma_g 2s)(\sigma_u 2s)$ and $(\sigma_u 2s)^2$, respectively. However, there is no experimental evidence for these states. They would be very high above the ground state for both large and small R.

For two normal Be atoms the behavior is quite similar to that for two normal He atoms. We have as many bonding as antibonding electrons and therefore do not expect a stable molecule. Excited stable molecular states of Be_2 are possible in the same way as for He_2 but none have been found as yet.

Two normal B atoms form a stable molecule (in agreement with experiment) since in the lowest electron configuration $KK(\sigma_g 2s)^2(\sigma_u 2s)^2(\pi_u 2p)^2$ there are four bonding electrons $(\sigma_g 2s)^2(\pi_u 2p)^2$ and only two antibonding electrons $(\sigma_u 2s)^2$. The binding energy is larger than that of Li$_2$ and the equilibrium internuclear distance is smaller [$r_e(\text{Li}_2) = 2.672$ Å, $r_e(\text{B}_2) = 1.589$ Å] since on account of the higher nuclear charge the K shells are smaller.

The two low-lying configurations of the C$_2$ molecule given in Table 34 have six bonding and two antibonding electrons, apart from the non-bonding K electrons. A much larger binding strength is therefore to be expected than for Li$_2$, in agreement with observation [$D_0{}^0(\text{C}_2) = 3.6$ and $D_0{}^0(\text{Li}_2) = 1.03$ e.v.].

The lowest state of the N$_2$ molecule,

$$KK(\sigma_g 2s)^2(\sigma_u 2s)^2(\pi_u 2p)^4(\sigma_g 2p)^2 \; {}^1\Sigma_g{}^+, \qquad (\text{VI, 42})$$

has two antibonding and eight bonding electrons, apart from the K electrons. We expect as a result a strong binding, in agreement with experiment. Unlike C$_2$, He$_2$, and others it is characteristic of N$_2$ and a further reason for its stability that in going over to very small internuclear distances the configuration (VI, 42) always remains the ground state (see Fig. 157). The next orbital, $(\pi_g 2p)$, above those filled in the ground state is an antibonding orbital. Its separation from the bonding orbital, $(\sigma_g 2p)$, is therefore large for medium internuclear distances and consequently the lowest excitation energy of N$_2$ is very large (inert-gas character). In the excited states with electron configurations $\cdots$ $(\pi_u 2p)^3(\sigma_g 2p)^2(\pi_g 2p)$ and $\cdots (\pi_u 2p)^4(\sigma_g 2p)(\pi_g 2p)$ the binding strength would be expected to be considerably smaller than in the ground state, since an antibonding electron has taken the place of a bonding electron. The dissociation energies and internuclear distances of the $A\ {}^3\Sigma_u{}^+$, $B\ {}^3\Pi_g$, $a\ {}^1\Pi_g$ states bear this out (see Table 39).

The two electrons that are added in going from the N$_2$ to the O$_2$ molecule must go into the antibonding $\pi_g 2p$ orbital (see Fig. 157). We have thus to expect that the binding in the ground state of the O$_2$ molecule will be weaker than that for the N$_2$ molecule, since we now have four antibonding electrons and eight bonding electrons. This conclusion is confirmed by experiment: $D_0{}^0(\text{N}_2) = 7.373$ e.v., and $D_0{}^0(\text{O}_2) = 5.080$ e.v. As we have seen, the lowest electron configuration of O$_2$ $\cdots (\sigma_g 2p)^2(\pi_g 2p)^2$ gives rise to three states ${}^3\Sigma_g{}^-$, ${}^1\Delta_g$, ${}^1\Sigma_g{}^+$ of which ${}^3\Sigma_g{}^-$ is the lowest, that is, is the ground state of the O$_2$ molecule in agreement with observation. This accounts also for the paramagnetism of oxygen gas (see Chapter VIII, section 2). These results which are derived from the molecular orbital theory in a direct and natural way can be obtained from the Heitler-London theory only when higher approximations and exact quantitative calculations are used (see p. 358 f.).

The spectroscopic study of the O$_2{}^+$ molecule confirms the antibonding nature of the $\pi_g 2p$ orbital. In agreement with the fact that in the ground state of O$_2{}^+$ there is only one electron in the $\pi_g 2p$ orbital the dissociation energy is found to be greater than that of O$_2$ [$D_0{}^0(\text{O}_2{}^+) = 6.48$ volts], and the internuclear distance in the equilibrium position is appreciably smaller (1.123 Å, compared to

1.207 Å). Furthermore, when in O_2^+ an electron is brought from the bonding $\pi_u 2p$ orbital into the antibonding $\pi_g 2p$ orbital, giving rise among others to the observed $^2\Pi_u$ state, the internuclear distance is appreciably raised and the dissociation energy is considerably lowered.

For the F_2 molecule two more electrons go into the antibonding $\pi_g 2p$ shell, and correspondingly, in agreement with experiment, the binding strength in the ground state is further decreased (see Table 34). When finally two Ne atoms are brought together, the two additional electrons must go·into the antibonding $\sigma_u 2p$ shell. The number of antibonding electrons is then equal to the number of bonding electrons. Therefore we do not obtain a stable molecule.

The considerations for the binding strengths of the molecules Na_2, Si_2, P_2, S_2, and Cl_2, according to the method of bonding and antibonding electrons, correspond closely to those for Li_2, C_2, N_2, O_2, and F_2. Only the principal quantum numbers are raised by 1, which causes slight changes in the energy relationships.

When we compare the above results with the elementary chemical conceptions of valence, we are led to the following rule [Herzberg (304)(308)]: *The valency of a bond is equal to the number of bonding electron pairs minus the number of antibonding electron pairs.* This difference can at the same time be regarded as a *measure of the strength of the bond.* The N_2 molecule, for example, has four bonding electron pairs and one antibonding electron pair—that is, a trivalent bond, which is very strong. In Table 34 the corresponding numbers for the other molecules treated above are given, together with the observed values for the dissociation energies. For molecules such as N_2^+ and O_2^+ the valency of the bond is half-integral, corresponding to the fact that in the method of bonding and antibonding electrons each electron (that is, half an electron pair) gives a positive or a negative contribution. The above generalization about binding strength can of course be regarded only as a very rough rule of thumb, since in reality the bonding and antibonding actions of the individual electrons are not always equal in magnitude.

Unlike nuclear charges. As we have seen before, the electron configurations of molecules with unlike nuclear charges are very similar to those with equal nuclear charges as long as the charge difference is small (see p. 346). In this case essentially the same conclusions hold in regard to binding strength, even though the characterization of bonding and antibonding electrons is no longer possible in such a definite manner, since the symmetry property g–u is now only approximately defined.

For the CO molecule, for example, there are eight bonding and two antibonding electrons in the ground state, similar to N_2. According to the above rule, we could with some justification call the bond in CO trivalent, analogous to the bond in N_2. Actually, the bond strength of CO is even greater than that of N_2 [$D_0^0(CO) = 9.141$ volts, compared to $D_0^0(N_2) = 7.373$ volts[12a]].

[12a] The D_0^0 values quoted here are the lower of several alternative values in the literature, see Table 39 in the appendix.

The electron configuration of the NO molecule in its ground state is the same as that of O_2^+ (see Table 34). Owing to the addition of an antibonding electron, $\pi_g 2p$, a smaller binding strength is to be expected than for CO, in agreement with experiment $[D_0^0(NO) = 5.30 \text{ volts}^{12a}]$.

Similar considerations apply to other molecules with not very different nuclei (compare Table 34).

If one uses linear combinations of atomic orbitals as approximations for molecular orbitals

$$\psi_{AB} = a\psi_A + b\psi_B \tag{VI, 43}$$

unlike the case of equal nuclear charges now a and b are no longer equal in magnitude [compare (VI, 13) and (VI, 14)]. Depending on whether a and b have the same or opposite sign we have respectively, bonding or antibonding $\sigma, \pi, \delta, \cdots$ orbitals (see p. 362). Formulae for the energies of these orbitals similar to, but more complicated than (VI, 37) have been given by Mulliken (1259).

If two electrons 1 and 2 are in the same (bonding) molecular orbital the resultant eigenfunction is [instead of (VI, 40)]

$$\Psi = \psi_{AB}(1)\psi_{AB}(2) = a^2\psi_A(1)\psi_A(2) + b^2\psi_B(1)\psi_B(2) + ab[\psi_A(1)\psi_B(2) + \psi_A(2)\psi_B(1)] \tag{VI, 44}$$

Again the first two terms on the right represent the contribution of the ionic states A^-B^+ and A^+B^-, respectively. Since these two terms occur now with different coefficients the molecule is polar. However, as in the case of equal nuclear charges the ionic terms are overemphasized and in reality, except when the nuclear charges are very different, the third term on the right of (VI, 44) predominates. This term represents a real sharing of the two electrons by the two nuclei, that is, it implies homopolar binding.

As an example of how the theory of bonding and antibonding electrons gives significant results even for excited states of molecules with unequal nuclei, consider the NO molecule. The ground state has the electron configuration $\cdots (\pi_u 2p)^4 (\pi_g 2p)$, and for the first excited electronic states one may expect the configurations $\cdots (\pi_u 2p)^4 (\sigma_g 3s)$ and $\cdots (\pi_u 2p)^3 (\pi_g 2p)^2$. The first of these, since the outermost electron has changed from an antibonding to a bonding orbital, should have greater binding strength than the ground state while the second, with one more antibonding electron, should have a much smaller binding strength. Two such states have actually been found (see Fig. 198): The first excited $^2\Sigma$ state has a smaller internuclear distance and a larger dissociation energy than the ground state while the reverse is the case for the first excited $^2\Pi$ state. Actually, an extrapolation of the vibrational levels of the $^2\Sigma$ state indicates that it dissociates into a normal atom and an atom in a three-quantum state, as might have been expected from the fact that the electron configuration includes a $\sigma_g 3s$ electron. However, we must remember that the distinction between σ_g and σ_u is only approximate when the nuclear charges are not equal. As a result, the intersection of $\sigma_g 3s$ and $\sigma_u 2p$ occurring for like nuclei (for example O_2^+, see Fig. 157) is avoided here and thus for adiabatic correlation the $^2\Sigma^+$ state of NO on dissociation must give atoms in two-quantum states only. Apparently, since the observed vibrational levels do indicate dissociation into a three-quantum atom, this is a case in which an intersection of the potential curves *almost* takes place [compare Fig. 143(d) and the accompanying discussion].

Other interesting examples are the molecules BeF, MgF, CaF, SrF, and BaF which have the same number of outer electrons as N_2^+ [see Fowler (941)].

If the nuclei of a diatomic molecule have *very different charges*—for example, in the hydrides—we cannot use the method of bonding and antibonding electrons in the above form. The procedure then to be used will be explained by the ex-

---

[2a] The D_0^0 values quoted here are the lower of several alternative values in the literature, see Table 39 in the appendix.

ample of the NH molecule [see Mulliken (514)]. The potential curves of the different electronic states of this molecule are given in Fig. 169. The correlation with the united atom (here, oxygen), to which all the hydrides approach fairly closely, is indicated at the left. The electronic states corresponding to the two lowest electron configurations of the molecule (Table 33) have been

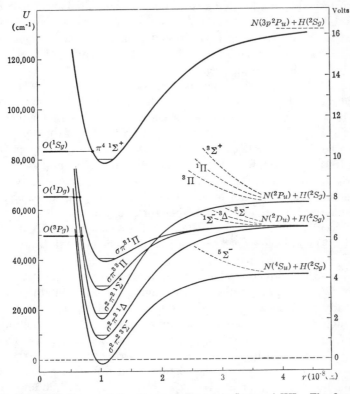

FIG. 169. **Potential Curves of the Observed Electronic States of NH.** The figure is similar to a figure of Mulliken's (514). The relative position of the singlet and triplet states is uncertain. The horizontal lines drawn somewhat above the potential minima are the levels $v = 0$. The energy scales to the left and to the right u e a z ro-poin J e le el $v = 0$ of the ground state $^3\Sigma^-$. A value of 4.2 e.v. has been assumed for the dissociation energy of the ground state. Since the figure was prepared King (400a) on the basis of theoretical considerations and by comparison with other hydrides has estimated $D_0^0(NH)$ to be 3.4 e.v. Neither value can be considered as reliable. It is therefore possible that the asymptotes of all the curves are to be shifted a certain amount upward or downward.

observed to be stable. We want to find a theoretical basis for this observation. Moreover, we shall determine the correlation with the terms of the separated atoms and obtain some information concerning the stability of other molecular states.

For the low-lying states of NH we can limit ourselves to the consideration of a normal H atom (^2S_g) and an N atom in one of the three low states ^4S_u, ^2D_u, and ^2P_u (all with the configuration $1s^2 2s^2 2p^3$); that is, we shall consider the three combinations $^2S_g + {}^4S_u$, $^2S_g + {}^2D_u$, and $^2S_g + {}^2P_u$, of which the first lies lowest. From $^2S_g + {}^4S_u$ the molecular states $^3\Sigma^-$ and $^5\Sigma^-$ result according to Table 26. Now, on the basis of the electron configuration, the ground state of the molecule is a $^3\Sigma^-$ state (see Table 33). Since we must avoid intersections,[13] we have to correlate this state with the $^3\Sigma^-$ state resulting from two normal atoms. We can expect that it will be stable, as demanded by experiment, since it corresponds to a low-lying state (^3P) of the united atom (see Fig. 169). Thus the stable ground state of NH dissociates into normal atoms. On the other hand, the $^5\Sigma^-$ state resulting from normal atoms must certainly be unstable, since the lowest state of the united atom that can give rise to such a quintet state lies very high above the ground state.

Since the $^3\Sigma^-$ state is stable, the two states $^1\Delta$ and $^1\Sigma^+$, which have the same electron configuration ($K\ 2s\sigma^2\ 2p\sigma^2\ 2p\pi^2$) in the molecule, must also be stable. However, they have to be correlated in a different manner with the separated atoms and the united atom. The lowest state of the separated atoms that can give rise to $^1\Delta$ is $^2D_u + {}^2S_g$, while $^1\Sigma^+$ can result only from $^2P_u + {}^2S_g$ (see Fig. 169). Both states, $^1\Delta$ and $^1\Sigma^+$, lead in the united atom to ^1D_g. From $^2D_u + {}^2S_g$, in addition to $^1\Delta$, the states $^3\Delta$, $^1\Pi$, $^3\Pi$, $^1\Sigma^-$, and $^3\Sigma^-$ result. Only two of these, $^1\Pi$ and $^3\Pi$, can be correlated with low-lying states of the united atom (namely, with ^1D_g and ^3P_g). Therefore, they alone are to be expected as stable molecular states. They belong to the configuration $K\ 2s\sigma^2\ 2p\sigma\ 2p\pi^3$ (see Table 33) and have indeed been observed. The other states resulting from $^2D_u + {}^2S_g$ can be correlated only with very highly excited states of the united atom and are therefore not stable (broken-line curves in Fig. 169).

From $^2P_u + {}^2S_g$ the states $^3\Sigma^+$, $^1\Pi$, and $^3\Pi$ result, in addition to the $^1\Sigma^+$ state already mentioned. All three states are unstable, since they correspond to very highly excited states of the united atom. Of the molecular states resulting from the low-lying states of the separated atoms or of the united atom, only the $^1\Sigma^+$ state, which results from the ^1S_g term of the united atom, remains to be discussed. It must have the configuration $K\ 2s\sigma^2\ 2p\pi^4$, since this is the only configuration left that has four $2p$ electrons, as have the low-lying states of the united atom O. We can see from Fig. 169 that this state cannot arise from any of the three low states of the separated atoms but must result from a three-quantum state. It is therefore a stable state and has indeed been observed [see Lunt, Pearse, and Smith (471a)].

Other hydrides or molecules with very different nuclear charges can be treated in a corresponding manner [see Mulliken (514)]. The above procedure can also be applied to molecules with equal nuclei, but the transition to the united atom leads to significant conclusions only for H_2, He_2, and their ions.

Another way of deriving the stability of the ground state, and a way that is more similar to the method applied to molecules with like nuclei, is as follows: For large internuclear

---

[13] For hydrides we are not likely to find that two states of the same species almost cross (see p. 364 f.) since the symmetry property g–u of the orbitals is no longer even approximately defined.

distance we have for NH the atomic orbital configuration

$$K_N(\sigma 2s_N)^2(\sigma 2p_N)(\pi 2p_N)^2(\sigma 1s_H) \tag{VI, 45}$$

for $^4S + {}^2S$. The two orbitals $\sigma 2p_N$ and $\sigma 1s_H$, being of the same type and having similar energies, will mix at smaller R values. Using linear combinations of atomic orbitals they form

$$\sigma 2p_N + \sigma 1s_H \quad \text{and} \quad \sigma 2p_N - \sigma 1s_H \tag{VI, 46}$$

with eigenfunctions similar to (VI, 15). The first of these is bonding, the second antibonding. The lowest state resulting from normal atoms is therefore

$$K(\sigma 2s_N)^2(\sigma 2p_N + \sigma 1s_H)^2(\pi 2p_N)^2 \,{}^3\Sigma^- \tag{VI, 47}$$

while the other state resulting from normal atoms is

$$K(\sigma 2s_N)^2(\sigma 2p_N + \sigma 1s_H)(\pi 2p_N)^2(\sigma 2p_N - \sigma 1s_H) \,{}^5\Sigma^- \tag{VI, 48}$$

The first state is stable since it has one bonding electron pair in addition to non-bonding electrons; it must clearly be identified with the state $\sigma^2\pi^2 \,{}^3\Sigma^-$ resulting for small R (Fig. 169 and Table 33).

On the whole, we can say that the molecular orbital method for estimating the stability of molecular electronic states gives just as good a·picture of homopolar chemical binding as the Heitler-London theory. While the results are not perhaps of such a general nature, the method is rather suitable for the discussion of special cases. A large number of such special cases have been discussed in detail by Mulliken (514)(519–521)(1259)(1260)(1266a).

Recent objections to the theory of bonding and antibonding electrons by Lessheim, Hunter, and particularly Samuel (442)(349)(613)(1368) have been refuted by Wheland (1529).

(b) Heteropolar Binding (Ionic Binding)

In heteropolar binding the molecules are held together simply by the classical electrostatic attraction between oppositely charged ions. For two pointlike ions of charge ε (= 4.8024×10^{-10} e.s.u.) at a distance r from each other, the potential energy, according to Coulomb's law, is

$$V = -\frac{\varepsilon^2}{r} = -\frac{11.615}{r} \times 10^4 \tag{VI, 49}$$

where the numerical factor gives V in cm^{-1} if r is substituted in Ångstrom units. As a result of the finite extent of the actual ions, a deviation from the simple Coulomb potential (VI, 49) takes place at small internuclear distances when the two electron clouds of the two ions start penetrating each other appreciably. This deviation is always in the sense of a repulsion. According to Born and Mayer (121), this repulsion can be represented by an exponential term, so that we obtain

$$V = -\frac{\varepsilon^2}{r} + Be^{-r/\rho}, \tag{VI, 50}$$

where B and ρ are constants. ρ is a measure of the sum of the radii of the ions under consideration.

In Fig. 170(a) and (b) the solid-line curves marked I give the potential energy of two singly charged ions, $H^+ + H^-$ and $Na^+ + Cl^-$, respectively. The broken-line branch of each of the curves corresponds to the pure Coulomb potential (VI, 49). In addition the potential curves of the lowest states arising from neutral atoms, $H + H$ and $Na + Cl$ respectively, are shown. It should be noted that the force of attraction extends to much greater distances for ions than for neutral atoms, for which it drops off exponentially (see p. 351). In Fig. 170 this makes itself felt by the fact that the ionic curves do not reach their asymptotes even at the extreme right of the figure (large r), whereas the two atomic curves reach them already at much smaller r values.

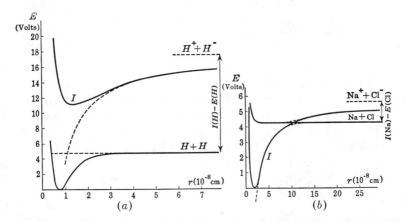

Fig. 170.   **Potential Curves of the States Resulting from Normal Atoms and Ions.**   (a) For H_2. (b) For NaCl.   Only the lowest of the states resulting from normal atoms is drawn.   The broken-line branch of the ionic curve corresponds to the pure Coulomb attraction.   For NaCl, the fact that the two curves actually do not cross is indicated by broken connecting lines.

Naturally, the Wigner-Witmer correlation rules hold also for the states that result from ions. If, as is often the case, the ions are in ^1S states [for example, $Na^+(^1S)$ and $Cl^-(^1S)$], only one $^1\Sigma$ molecular state, which has a potential curve of the type described, results from the ions. According to (VI, 50), the right-hand part of the curve is very nearly the same for all singly charged ions. For ions not in ^1S states [for example, $Be^+(^2S)$ and $O^-(^2P)$], we obtain a number of molecular states (see Table 26) whose potential curves, however, very nearly coincide at least for large r values and which, therefore, are all stable.

In order to obtain all the states of a diatomic molecule, we have to derive the states resulting from the singly and possibly multiply charged ions as well as from all possible combinations of the states of the neutral atoms. We distinguish between *atomic* and *ionic states* according as they result from neutral atoms or ions. The preceding considerations show that the *ionic states are stable without exception*.

Ionic molecules. Following Franck, we call a molecule an *ionic molecule* or an *atomic molecule* according as the *ground state* of the molecule is *derived from ions or neutral atoms*. This classification has proved to be very useful in practice, although there are some theoretical objections to it.

When this field was being developed, Franck and his co-workers gave some spectroscopic criteria for distinguishing between atomic and ionic molecules. Although at first these proved to be very fruitful, they were not subsequently confirmed by more detailed wave mechanical considerations [see, for example, Mulliken (520)(521)]. We can therefore omit their discussion here. However, we can in many cases decide in other ways whether the molecule under consideration is an atomic or an ionic molecule.

If A and B are the two atoms forming a molecule, the energy difference, for large internuclear distances, between the lowest ionic state $A^+ + B^-$ and the lowest atomic state $A + B$ is given by

$$I(A) - E(B),$$

where $I(A)$ is the ionization potential of A and $E(B)$ is the electron affinity[14] of B. Only when the difference $I(A) - E(B)$ is small can an ionic molecule occur; only then can the ground state dissociate into ions. In order to see this, let us consider two extreme examples, the molecules H_2 and NaCl, for which the potential curves of the states resulting from normal atoms and normal ions are given in Fig. 170.

For H_2 the state $H^+ + H^-$ lies above the state $H + H$ by an amount 13.60 − 0.75 = 12.85 e v. The point of intersection of the pure Coulomb curve with the asymptote of the $H + H$ potential curve lies at 1.1 Å—that is, in the region in which the attractive $H + H$ potential curve already lies considerably below the asymptote. At this distance, the ions can certainly no longer be regarded as pointlike, but we must take into account the repulsion between the ions. The ionic curve therefore has its minimum above the asymptote of the $H + H$ curve (actually very much above), and therefore the ionic state cannot be the ground state. Like all other elementary molecules, H_2 is a typical atomic molecule. However, there are two stable excited states of H_2 that are known to be ionic states [$1s\sigma 2p\sigma\ ^1\Sigma_u^+$ and $(2p\sigma)^2\ ^1\Sigma_g^+$; see p. 364 and Fig. 160].

The situation for NaCl is quite different. Here, at large internuclear distances, the state $Na^+ + Cl^-$ lies only 5.14 − 3.72 = 1.42 e.v. above the state $Na + Cl$, and the ionic curve intersects the asymptote of the atomic curve already at 10.5 Å. At this internuclear distance, the atomic curve is still perfectly horizontal, while the ionic curve does not yet deviate from the pure Coulomb curve (VI, 49). Therefore at smaller distances, which correspond to

---

[14] The electron affinity is the energy released when an atom combines with an electron to form a negative ion, or in other words, it is the ionization potential of the negative ion. For more details see A.A., p. 216 f. The value for the electron affinity of H given there should be corrected to 0.749 e.v. [Henrich (1026)].

the stable molecule, the ionic curve lies below the atomic curve (see Fig. 170). The molecule is therefore an ionic molecule. A similar situation exists for all other alkali halides.

It is clear that everything depends on the *difference* $I(A) - E(B)$. If it is small, we have an *ionic* molecule; if it is large, we have an *atomic* molecule. This purely qualitative rule can be expressed somewhat more quantitatively if we assume a knowledge of the ionic radii or if we know the internuclear distance of the molecule in the ground state: We can say that, if the intersection of the pure Coulomb curve (VI, 49) with the asymptote of the atomic curve (normal atoms) lies at an internuclear distance that is not greater than the sum of the ionic radii or is not greater than about 1.5 times the internuclear distance in the normal molecule, we can be certain that we have an atomic molecule (see the above example). On the other hand, if this point of intersection lies at an internuclear distance which is greater than roughly 1.5 times the sum of the ionic radii or twice the internuclear distance in the ground state, we can be certain that we have an ionic molecule. For, since the potential curve of the atomic state approaches its asymptote exponentially, at $2r_e$ it already lies fairly close to the asymptote. At the same time the ionic curve deviates very little from the pure Coulomb curve for such an internuclear distance. Therefore, when the intersection with the ionic curve takes place at an internuclear distance $> 2r_e$, the ionic curve is the lower one at smaller r and thus forms the ground state.

According to the above rough rule, if the point of intersection lies between $1.5r_e$ and $2r_e$, no definite decision is possible. This uncertainty has also a theoretical basis (see below). However, in most of the cases occurring in practice the point of intersection lies at such an r value that an unambiguous decision according to the rule is possible.

Let us consider the further example of the molecule HCl, which from a chemical standpoint might plausibly be assumed to be an ionic molecule, since HCl dissociates into ions in solution. For HCl the difference $I(A) - E(B)$ is 9.88 e.v. From this we might already conclude qualitatively that it is highly probable that we are dealing with an atomic molecule. A determination of the intersection of the Coulomb curve with the asymptote of the potential curve of the normal atoms according to (VI, 49) gives $r = 1.45$ Å. The ionic radius of Cl⁻ is 1.8 Å, and the internuclear distance in HCl is 1.27 Å. From this it follows that HCl in the gaseous state is certainly an atomic molecule. The same holds for the other halogen hydrides.

Transition cases. With one exception (CsF), for all diatomic molecules thus far known the electron affinity of the one atom is less than the ionization potential of the other. Therefore (except for CsF), *only if the ionic curve crosses the atomic curve* can the ionic state be the ground state of the molecule under consideration and thus can the molecule be an ionic molecule. However, as we have seen previously (see p. 295) for a strictly adiabatic correlation such intersections can occur only when the intersecting terms are of unlike species. Actually the ionic states are in general of the same species (usually $^1\Sigma$) as one of the molecular states resulting from normal atoms, or, if that is not the case, as one of the states resulting from an excited state of the atoms that still lies below the ionic state for large r. As a result, the intersection of the atomic with the ionic curve is always avoided [see the dotted connecting curves in Fig. 170(b)]. Therefore, strictly speaking, the ground state of a diatomic molecule always leads to neutral atoms, and thus according to the previous definition (if it is assumed to imply rigorous adiabatic correlation) there would only be atomic molecules (except CsF).

However, in the case of NaCl (see Fig. 170) and many other molecules, the point of intersection of the ionic and atomic curve lies at an internuclear distance at which the atomic curve scarcely deviates at all from the asymptote. Consequently, the interaction of the states is so small that a deviation from the "original" course occurs only in the immediate neighborhood of the point of intersection [see London (456)], since at these large separations the two atoms or ions can still be regarded approximately as separate systems. If a transition from the atomic to the ionic state is to take place, an electron must go from one atom to the other, and this exchange, for large internuclear distances, can occur only for extremely sharp resonance between the two states—that is, only in the immediate neighborhood of the point of intersection. Therefore, even at a very small distance from the point of intersection the potential curves run as if they actually had crossed, and consequently at smaller internuclear distances the properties of the lower state (eigenfunctions, behavior of vibrational quanta, and so on) are exactly what they would be for the ionic state. Thus in such cases we are doubtless justified in designating the molecule under consideration as an ionic molecule. Indeed we remain in agreement with the previous definition of ionic or atomic molecules if we interpret the dissociation into ions or atoms, respectively, not as an adiabatic dissociation (two-center system) but one in which the vibrational energy is increased in quantum-steps. In going through the point of intersection with a sufficient speed the molecule will remain in the ionic or atomic curve as the case may be and not switch over. For NaCl and similar molecules [see London (456)] thermal speeds are sufficient for this, as are the speeds acquired during a vibration; that is, for the higher vibrational levels of the ionic curve there is very little chance of switching over to the atomic curve during the vibrational motion.

The more the point of intersection shifts toward smaller r values, the stronger will be the interaction of the two states and the more will the two resulting curves draw away from each other. If the interaction is very strong there may be little sense in saying that "originally"—that is, in zero approximation—an intersection of the ionic and atomic curves was present and that therefore the molecule is an ionic molecule. Rather, it may be more appropriate to start out from a different zero approximation in which no intersection takes place and according to which the molecule will have to be regarded as an atomic molecule.

From the foregoing considerations it is seen that the *dividing line between atomic and ionic molecules is not at all rigid*. However, the border-line cases in which we can speak with equal justification of the molecule as an ionic or an atomic molecule are relatively rare.

Ionicity and polarity. Even if the ionic state is not the ground state (atomic molecule), in higher approximation it will influence the ground state. It forces the ground state downward and makes a contribution to its eigenfunction (see p. 290). Therefore, as was first recognized by Pauling (554)(37) there is some point in ascribing to a molecule in its ground state a certain *fractional ionic character*. Mulliken (520) has extended the discussion of this property to excited states and has called it *degree of ionicness* or *ionicity*.

If there were an unambiguous way of defining ionicity then the best way of distinguishing ionic and atomic (covalent) molecules would be on the basis of whether the ionicity of the ground states is larger or smaller than 50%. However, the numerical value of ionicity depends greatly on the particular zero approximation used in deriving it. For example if one were to use exact molecular orbitals the concept of ionicity would have no meaning [Mulliken (520)]. On the other hand, if one uses atomic orbitals or linear combinations of atomic orbitals, the wave functions have ionic contributions as we have seen previously (p. 363 f.); but the magnitude of these contributions depends on whether undisturbed or polarized atomic orbitals are used.

Pauling (37) assumes that the dissociation energy of a molecule AB, if it had no ionic character, is the algebraic mean of the dissociation energies of A_2 and B_2. He considers the deviation of the actual value from this theoretical (covalent) value as entirely due to interaction

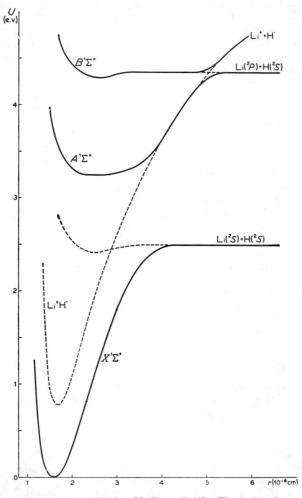

Fig. 171. **Potential Curves of LiH** [after Mulliken (520)]. The broken-line curves correspond to zero-order approximation. The state $B\,^1\Sigma^+$ has not yet been observed. The form of its potential curve has been predicted by Mulliken.

("*resonance*") with the lowest ionic state (A^+B^- or A^-B^+ as the case may be). *The magnitude of the deviation is a measure of ionicity.* The ionicity is 50% if the ionic state calculated in a certain way has the same energy as the atomic (covalent) state. This implies the neglect of the fact that even molecules with equal atoms (A_2 and B_2) have a certain ionicity in their ground states (see p. 363 f.).

In the majority of cases Pauling's criterion leads qualitatively to the same results as Franck's criterion. An example for which this is *not* the case is HF. Upon dissociation the ground state gives undoubtedly neutral atoms and thus according to Franck HF is an atomic molecule. On the other hand according to Pauling the zero-approximation ionic curve is lower than the atomic curve and therefore the ground state is predominantly ionic. In conformity with this the deviation of $D_0^0(\mathrm{HF})$ from $\frac{1}{2}[D_0^0(\mathrm{H}_2) + D_0^0(\mathrm{F}_2)]$ is very large. However, the position of the first excited state obtained in this way does not at all agree with experiment [see Mulliken (521)].

An interesting example showing the difficulties of obtaining a clear-cut distinction is provided by the LiH molecule. Fig. 171 shows the potential curves for the lowest states including the zero-approximation ionic and atomic curves (broken lines) as given by Mulliken (520). For small r values the zero-approximation ionic curve is the lowest and therefore it appears rather certain that the ground state is predominantly ionic. Yet on account of the strong interaction of the two states the ground state dissociates into normal atoms. The mean of the dissociation energies of Li_2 and H_2 is 2.80 e.v. while the observed $D_0^0(\mathrm{LiH})$ is 2.5 e.v. On this basis a negative ionicity would result. Even if the geometric rather than the algebraic mean is used [Pauling (37)] only a slight positive resonance energy results indicating only slight ionicity.

The first excited singlet state of LiH ($A^1\Sigma^+$ of Fig. 171 called V by Mulliken) has predominantly atomic character for small r values, but changes to ionic character for large r values where the potential curve coincides with the ionic curve $\mathrm{Li}^+\mathrm{H}^-$. It is on account of this rather sudden change from a weak atomic bond to a strong ionic bond that the $\Delta G_{v+\frac{1}{2}}$ and B_v values of this state first increase to a maximum before the regular decrease sets in. The $A^1\Sigma^+$ state would lead to $\mathrm{Li}^+ + \mathrm{H}^-$ if it were not for a further intersection with the $^1\Sigma^+$ state arising from $^2P + {}^2S$. However, since this intersection occurs at fairly large r values it is probably of a type similar to that discussed above for NaCl, that is, for most practical purposes, the $A^1\Sigma^+$ state leads to ions [see however Rosenbaum (1363)]. Very similar considerations and observations apply to other alkali hydrides [see Mulliken (520) and, for KH, Almy and Beiler (747)].

The attempt has also been made to consider the *dipole moment μ as a measure of the ionicity* [Pauling (37)]. It is assumed that for 100% ionicity the dipole moment would equal εr_e (where ε is the electronic charge). The ratio of μ to this maximum value gives then the ionicity. For HI, HBr, and HCl, this gives ionicities of 5, 11, and 17% respectively. However, even for a molecule like CsF which is certainly 100% ionic (see above) the recently determined dipole moment [Hughes (1076)] is only 59% of the value εr_e showing that polarity is not a very good measure of ionicity. It should also be noted that in the ionic states of molecules A_2 with equal nuclei (see the discussion of H_2 on p. 363) there is of course no polarity (no dipole moment) owing to the continual interchange between A^+A^- and A^-A^+.

(c) Van der Waals Binding

When no valence force either of homopolar or of heteropolar kind is acting between two atoms,—as, for example, between two inert-gas atoms,—a very small attraction between them still remains which is responsible for the deviations of the behavior of the gas from the ideal gas laws and for its liquefaction at sufficiently low temperatures. The deviations from the ideal gas laws are represented by the well-known van der Waals equation, and therefore the residual attraction is called a *van der Waals force*. Correspondingly, we speak of *van der Waals binding* (or sometimes of polarization binding) and of *van der Waals molecules* (or polarization molecules).

London (454)(455) has treated these forces on the basis of quantum mechanics and has shown that they are due (for example, for He + He) to the perturbation of the repulsive ground state by the higher electronic states of the system consisting of the two atoms.   As will be shown below this perturbation at large internuclear distances gives a *potential energy decreasing as* $-1/r^6$ *toward smaller r values.*   Naturally, at smaller separations r the strong repulsion of the zero-valent atoms sets in (see above), so that only a very shallow minimum at a relatively large internuclear distance results [compare the potential curves in Fig. 173(b) and (c)].   London has called these attractive forces *dispersion forces*, because they depend on the same quantities as the dispersion of light by the gas, namely the strengths of the transitions to all excited states.[15]

The London dispersion forces exist for every molecular state.   However, in general they are overshadowed by the valence forces.   They are noticeable only when the valence forces are very weak or zero—that is, in general, at large internuclear distances.   As a result of these dispersion forces, for example, two H atoms with parallel spins or two He atoms, which in consequence of valence forces repel one another strongly for small internuclear distances, attract one another very weakly at large internuclear distances.

If two atoms are at a distance r from each other which is so large that there is no appreciable overlapping of their charge distributions, the instantaneous (classical) potential energy of the system can be expressed as a series in inverse powers of r.   In the case of *two ions* this series would start with a term in $1/r$; in the case of *an ion and a neutral atom* it would start with a term in $1/r^2$ (since the neutral atom has an instantaneous dipole moment whose potential is proportional to $1/r^2$); in the case of *two neutral atoms*, the only case to be considered here in more detail, the series starts with a term in $1/r^3$ (since the potential of two dipoles is proportional to $1/r^3$), that is, we have

$$V = -\frac{a}{r^3} + \frac{b}{r^4} + \frac{c}{r^5} \qquad \text{(VI, 51)}$$

where [see for example Margenau (1196)]

$$a = \varepsilon^2 \sum_{ij} (2z_i^{(1)}z_j^{(2)} - x_i^{(1)}x_j^{(2)} - y_i^{(1)}y_j^{(2)}) \qquad \text{(VI, 52)}$$

Here $x_i^{(1)}$, $y_i^{(1)}$, $z_i^{(1)}$ are the coordinates of the electrons of the first atom referred to its nucleus and $x_j^{(2)}$, $y_j^{(2)}$, $z_j^{(2)}$ those of the second referred to its nucleus.   The constant b is cubic, the constant c quartic in the coordinates (see Margenau).   While the first term in (VI, 51) is due to *dipole-dipole interaction*, the second is due to *dipole-quadrupole interaction*, the third to *quadrupole-quadrupole interaction*.   Most of the present considerations will be concerned with the first term only.

What matters for the actual behavior of the atoms is not the instantaneous potential energy but its average over the electronic motions, that is, the energy of the system for various r values (the potential curve in the previous sense).   In order to obtain it the instantaneous potential energy V from (VI, 51) has to be introduced into the wave equation as a perturbation.

If the two atoms are in S states the first order perturbation energy vanishes.   The second order energy is given by (see p. 14)

$$\Delta E = \sum_{i(\neq n)} \frac{|V_{ni}|^2}{E_n - E_i} = \frac{1}{r^6} \sum_{i(\neq n)} \frac{|a_{ni}|^2}{E_n - E_i} \qquad \text{(VI, 53)}$$

---

[15] In the interaction of saturated *molecules* with one another the orientation effect and the induction effect are added to the dispersion effect (the first only for dipole molecules), both causing an additional van der Waals attraction.

where the V_{ni} are the matrix elements of V, and a_{ni} those of the quantity a given by (VI, 52); $E_n = E_{n_1} + E_{n_2}$ is the energy of the ground state of the two atoms, $E_i = E_{i_1} + E_{i_2}$ that of an excited state. Since $E_n < E_i$ the energy change ΔE is always negative and we have an attraction of the two atoms; their *potential energy is proportional to* $1/r^6$. If higher powers in (VI, 51) are taken into account terms in $1/r^8$ and higher ones appear in ΔE. It must be noted that the attraction is independent of the spin orientation; it would be the same, for example, for the $^1\Sigma$ and the $^3\Sigma$ state arising from two normal hydrogen atoms. If there is no valence attraction (as for example in the $^3\Sigma$ state of H + H) the potential energy of the two atoms with respect to each other may be written[16]

$$U(r) = Be^{-r/\rho} - \frac{C}{r^6} \qquad (VI, 54)$$

where the first term represents the repulsion at small r. The minimum of this potential function gives the dissociation energy and internuclear distance of the van der Waals molecule formed by the two atoms considered.

For the constant C in (VI, 54) which is given by the sum in (VI, 53) the following expression may be obtained[17] [see Margenau (1196)]

$$C = \frac{3\varepsilon^4 h^4}{32\pi^4 m^2} \sum \frac{f_{n_1 i_1} f_{n_2 i_2}}{(E_{n_1} - E_{i_1})(E_{n_2} - E_{i_2})(E_n - E_i)} \qquad (VI, 55)$$

where ε and m are the charge and mass of the electron and where $f_{n_1 i_1}$ and $f_{n_2 i_2}$ are the oscillator strengths (numbers of dispersion electrons; see section 5) of the transitions $E_{i_1} \to E_{n_1}$ and $E_{i_2} \to E_{n_2}$ in the two separate atoms, respectively. The optical dispersion of the gas is determined by the values of $f_{n_1 i_1}$ and $E_{n_1} - E_{i_1}$. That is why these attractive forces are frequently referred to as dispersion forces. We see from (VI, 55) that the van der Waals forces depend inversely on the separation of the ground state from the various upper states with which it can combine in ordinary dipole transitions.

For H + H the constant C in (VI, 54) has been calculated by Pauling and Beach (1306) to be

$$C_{\text{H+H}} = 31.3 \times 10^3$$

if r is in Å and $U(r)$ in cm^{-1}. In the case of He + He Margenau (1197) found it to be

$$C_{\text{He+He}} = 7.00 \times 10^3$$

The fact that the van der Waals attraction is much smaller in He than in H is due to the much higher excitation energies which enter (VI, 55). The authors mentioned have also calculated higher terms with $1/r^8 \cdots$. Slater (1425) has calculated the repulsive term for He, and, combining it with the van der Waals attraction, Margenau obtained for the resulting potential minimum a depth of 9.1 cm^{-1} and for the equilibrium internuclear distance the value 2.8 Å.

If the ground states of the atoms are *not S states* the preceding formulae for the van der Waals attraction do not hold. Knipp (1136) has shown that in this case a first order perturbation on account of the quadrupole-quadrupole interaction results, which is proportional to $1/r^5$.

In the case of *like atoms in unlike states* there is a degeneracy at large separations corresponding to the exchange of excitation energy and leading to the splitting into g and u states for smaller r values (see p. 321). Therefore introducing the interaction (VI, 51) into the wave equation leads to a first order perturbation which is proportional to $1/r^3$, that is, there is a van der Waals interaction at much greater separations than for atoms in their ground

---

[16] We use here U for the potential energy in order to distinguish it from V used above for the instantaneous potential energy.

[17] For a refinement taking account of retardation see Casimir and Polder (842).

states. The potential energy is given by

$$U(r) = E(r) + \frac{C_1}{r^3} \tag{VI, 56}$$

where $E(r)$ represents the attraction or repulsion due to exchange forces at smaller r values. For large r only the second term matters. If one atom is in an S, the other in a P state, neglecting the electron spin, four states result: Σ_g, Σ_u, Π_g, Π_u. According to King and Van

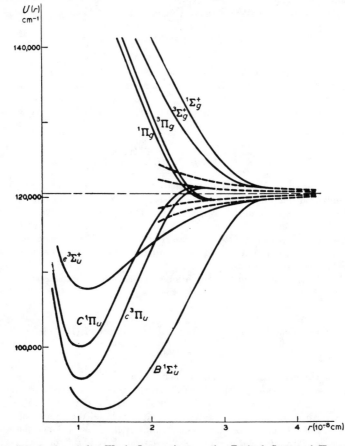

Fig. 172. **Effect of van der Waals Interaction on the Excited States of H_2 resulting from $1^2S + 2^2P$.** The theoretical van der Waals potential curves according to King and Van Vleck (1133) are indicated by broken lines. For smaller r values rough qualitative potential curves are given (full-line curves). In the case of the observed electronic states the variation of the potential energy near the minima is accurately drawn. However for intermediate r values the theoretical variation [King and Van Vleck (1133)] rather than Morse curves have been used. The latter would all lie below the lowest van der Waals curve in the range shown here. King and Van Vleck's $^3\Pi_u$, $^3\Pi_g$, $^3\Sigma_u$, $^3\Sigma_g$ are here assumed to be $^3\Pi_g$, $^3\Pi_u$, $^3\Sigma_g$, $^3\Sigma_u$ respectively. While the correlation of the Π states seems unambiguous it is not possible to ascertain whether the observed Σ states shown correspond to $1^2S + 2^2P$ or $1^2S + 2^2S$ since in hydrogen 2^2P and 2^2S have the same energy.

Vleck (1133) the constant C_1 in (VI, 56) for these four states has the values $+\frac{2}{3}(R^{SP})^2$, $-\frac{2}{3}(R^{SP})^2$, $-\frac{1}{3}(R^{SP})^2$ and $+\frac{1}{3}(R^{SP})^2$, respectively, where R^{SP} is the transition moment for the transition $S \to P$ which may be obtained theoretically from the eigenfunctions according to (I, 44) or empirically from the observed dipole strength f_{SP} of the transition (see section 5). Thus for Σ_u and Π_g there is attraction, for Σ_g and Π_u repulsion. The corresponding forces may be called *dipole-dipole resonance forces* since they are due to the dipole-dipole term of (VI, 51) combined with the effect of resonance of the two atoms.

As an example Fig. 172 shows the potential curves of the states arising from $1^2S + 2^2P$ of two H atoms. For smaller r values each of the above states splits into a singlet and a triplet state (on account of exchange forces). For the states $^3\Pi_u$ and $^1\Pi_u$ *potential maxima* result which are about 0.1 e.v. above the asymptote. Similar maxima will result in other cases (for example, Li_2, Cd_2) whenever the excited atomic state can combine with the ground state. In several instances the observed spectra confirm these predictions [see King and Van Vleck (1133)]. If the excited atomic states cannot combine with the ground state, for example for $H(1^2S) + H(2^2S)$, no such relatively strong van der Waals attraction or repulsion arises but only one that is proportional to $1/r^6$ and is of the same order as for two atoms in their ground states.

It was mentioned above that the instantaneous potential energy of *an ion and a neutral atom* can be expressed as a series starting with $1/r^2$ rather than $1/r^3$. If there is no first order perturbation the van der Waals potential $U(r)$ will then be a series starting with $1/r^4$. The coefficients for the case of a neutral H atom and a proton have been derived by Coulson (867). If the neutral atom is excited and the nuclei are alike a first order perturbation will arise if the excited state can combine with the ground state. In this case the van der Waals potential is a series starting with $1/r^2$, that is, forces of rather long range result. This case has been discussed in detail by Krogdahl (1143) and Coulson and Gillam (869). Krogdahl has also considered the case of unlike nuclei, in particular the interaction of a proton with an excited helium atom.

As was emphasized before, van der Waals forces of attraction or repulsion exist for every molecular state. If for a given molecular state the van der Waals force is the only cause of attraction, that is, if there is no valence force we speak of a *van der Waals binding* in the state considered. One might be tempted to regard the binding between a normal alkaline-earth atom and an H atom or halogen atom as van der Waals binding. We have seen previously that for this case the elementary Heitler-London theory gives no binding and that the binding actually observed may be explained as due to the influence of higher electronic states—that is, to the same influences that give rise to the dispersion effect. However, in the calculation of van der Waals forces between normal atoms it is implicitly assumed that all excited states are high above the ground state. In that case a large number of these states give contributions of a similar magnitude. On the other hand the relatively large binding strength of molecules such as BeH, MgF, and so on, is produced by the interaction of the ground state with one relatively low-lying excited state. The $1/r^6$ law does *not* hold in this case. It seems appropriate to *limit the designation of "van der Waals binding" to cases in which the attractive part of the potential energy is given completely by terms in $1/r^6$ or even lower powers of $1/r$* (except the first power which represents ionic binding). This excludes cases like BeH, but includes cases like He + He in the $^1\Sigma$ ground state, H + H and Na + Na in the $^3\Sigma$ repulsive state, A + Hg, and others.

5. Intensities of Electronic Transitions

In view of the large number of predicted molecular states it is of great importance, for a comparison with the observed electronic transitions, to obtain theoretical estimates of the intensities of the allowed transitions. In previous chapters the distribution of *relative* intensities within a band system or within a band was discussed. We must now consider the *absolute* intensities of band systems (electronic transitions). This field has been explored in a series of papers by Mulliken (1261)(1262)(1264)(1265).

General formulae. Since a band system corresponds to a line or multiplet of an atomic spectrum we must expect the sum of all band intensities to be of the order of the intensity of such an atomic line. We may also say that for fixed nuclei the intensity of an electronic transition of a molecule is of the same order as for a similar electronic transition in an atom. Introducing the vibration and rotation will not change this relation, that is, the same total intensity is simply distributed over a large number of lines and possibly a continuous spectrum.

The intensity of an electronic transition from an upper state n to a lower state m is determined by the *transition probability* A_{nm} in emission [equation (I, 46)] or B_{mn} in absorption [equations (I, 49 or 50)]. The transition probabilities A_{nm} and B_{mn} in turn are determined by the matrix element R^{nm} of the dipole moment according to (I, 47) and (I, 51), respectively. For electronic transitions

$$R_e{}^{nm} = \int \psi_m{}^* M_e \psi_n d\tau_e \tag{VI, 57}$$

where ψ_m and ψ_n are electronic eigenfunctions, M_e is the part of the dipole moment depending on the electrons (with components $\sum \varepsilon_i x_i$, $\sum \varepsilon_i y_i$, $\sum \varepsilon_i z_i$) and $d\tau_e$ is the volume element of the configuration space of the electrons. For nondegenerate electronic states the transition probabilities A_{nm} and B_{mn} are simply proportional to $|R_e{}^{nm}|^2$, for degenerate electronic states to $(1/d_n)\sum|R^{n_i m_k}|^2$ and $(1/d_m)\sum|R^{n_i m_k}|^2$, respectively, where d_n and d_m are the degrees of degeneracy of the upper and lower states and where the summation is over all combinations of the sublevels n_i and m_k [compare equations (I, 54) and (I, 55)].

We have seen previously (p. 200 f.) that the total matrix element for a given rotational and vibrational transition of a band system may be written

$$R = R_e{}^{nm} R_{\text{vib.}}{}^{v'v''} R_{\text{rot.}}{}^{J'J''} \tag{VI, 58}$$

The transition probability $B_{mn,v''v',J''J'}$ for absorption is given by [see equation (I, 55)]

$$B_{mn,v''v',J''J'} = \frac{8\pi^3}{3h^2c} \frac{\sum\limits_{M'M''}|R|^2}{2J''+1} = \frac{8\pi^3}{3h^2c}|R_e{}^{nm}|^2 |R_{\text{vib.}}{}^{v'v''}|^2 \frac{\sum\limits_{M'M''}|R_{\text{rot.}}{}^{J'J''}|^2}{2J''+1} \tag{VI, 59}$$

where the summation is over all values of the magnetic quantum numbers M' and M''.

If we sum the transition probability over all transitions that can occur from a given level v'', J'' we obtain on account of the sum rules (see pp. 208 and 203)

$$\sum_{J'}\sum_{M'M''}|R_{\text{rot.}}{}^{J'J''}|^2 = 2J''+1, \qquad \sum_{v'}|R_{\text{vib.}}{}^{v'v''}|^2 = 1 \tag{VI, 60}$$

the simple result

$$B_{mn} = \sum_{v'}\sum_{J'} B_{mn,v''v',J''J'} = \frac{8\pi^3}{3h^2c}|R_e{}^{nm}|^2 \tag{VI, 61}$$

that is, if we sum over all the transitions from a given level (within a given band system) we obtain the electronic transition probability. This result is of course based on the possibility of the resolution (VI, 58) which in turn is based on the possibility of resolving the eigenfunction into a product of electronic, vibrational and rotational functions. To this approximation the sum in (VI, 61) is independent of the values of v'' and J''. If therefore the molecules are distributed over a number of lower states the *total intensity of absorption, summed over all transitions of the band system that occur, is still proportional to B_{mn}*, that is, to the *electronic transition probability*, thus proving the previous statement. The same conclusions apply, as is readily seen, also for the transition probability A_{nm} for emission although here because of the occurrence of ν^3 in (I, 47) it must be assumed that the wave numbers of the transitions occurring in the summation do not differ too much. When transitions to a dissociation continuum are possible for the state considered appropriate integrals rather than sums must be included in (VI, 61). For degenerate electronic states the quantity $|R_e{}^{nm}|^2$ in (VI, 61) has to be replaced by $(1/d_m)\sum|R_e{}^{n_i m_k}|^2$ (see above).

Experimentally the absolute intensity of electronic transitions is usually determined from the absorption spectrum since for emission it is difficult to determine the number of molecules in the excited state. The primary experimental datum in absorption is the *absorption coefficient* k_ν which is defined by the relation

$$I_\nu = I_\nu^0 e^{-k_\nu \Delta x} \tag{VI, 62}$$

where I_ν and I_ν^0 are the intensities before and after transmission through a column of length Δx of the gas at $0°$ C and 1 atm pressure. The absorption coefficient k_ν varies across an absorption line (or band). It is readily seen that for small Δx the light absorbed by the transition $n \rightarrow m$ is given by

$$I_{\text{abs.}}{}^{nm} = \int (I_\nu^0 - I_\nu)d\nu = I_\nu^0 \Delta x \int k_\nu d\nu \tag{VI, 63}$$

where the incident intensity I_ν^0 is assumed to be constant over the width of the line or band. Substituting into equation (I, 50) and taking $I_0{}^{nm} \equiv I_\nu^0$ we obtain for the *integrated absorption coefficient*

$$\int k_\nu d\nu = N_m B_{mn} h\nu_{nm} = \frac{8\pi^3 \nu_{nm}}{3hc} N_m |R^{nm}|^2 \tag{VI, 64}$$

This equation allows one to compare the experimental quantity $\displaystyle\int k_\nu d\nu$ with the theoretical quantity R^{nm}.

Frequently the experimental data are expressed in terms of the *oscillator strength* f^{nm} (also called *number of dispersion electrons* or simply f-value). It represents the ratio of the quantum theoretical and classical contributions of the transition $n \rightarrow m$ to the refractivity $N - 1$ (N = index of refraction) where the classical value is calculated on the assumption of a single oscillating electron of frequency $c\nu_{nm}$. It can be shown that [see Condon and Shortley (10)]

$$f^{nm} = \frac{\mu h c^2 \nu_{nm}}{\pi \varepsilon^2} B_{mn} = \frac{8\pi^2 \mu c \nu_{nm}}{3h\varepsilon^2} |R^{nm}|^2 \tag{VI, 65}$$

where μ and ε are mass and charge of the electron. For a single emission electron there is a sum rule

$$\sum_n f^{nm} = 1 \tag{VI, 66}$$

For the strongest electronic transitions f may therefore be of the order of 1.

Rydberg transitions. Rydberg series of electronic transitions occur in molecules as in atoms. With few exceptions these transitions lie in the vacuum region where the available data are not very extensive. It is to be expected that the first member of such a series has an intensity similar to that in atomic Rydberg series (f value of the order 1, as, for example, for the sodium D lines) while the intensity falls off more and more for the higher members but still leaving an f value of the order 10^{-2} for the adjoining ionization continuum.

As an example the transitions from the ground state $1s\sigma^2$ $^1\Sigma_g^+$ of H_2 to the series of excited states $1s\sigma np\sigma$ $^1\Sigma_u^+$ or $1s\sigma np\pi$ $^1\Pi_u$ may be considered (compare Fig. 160). These are indeed exceedingly strong transitions. Even for $n = 3$ and 4 a path of 0.01 cm at atmospheric pressure is sufficient to produce strong absorption [Beutler (91)]. Similarly the transitions of the hydrogen halides to the second excited state (in the vacuum region) are observed to be exceedingly strong, 0.006 cm atm being sufficient for their production [Price (570)].

Transitions of this type are characterized by the fact that on dissociation the upper state yields one atom in a state in which the principal quantum number is greater than in the ground state, or, using Mulliken's expression, an atom in a *configurationally excited state*.   These Rydberg transitions must be distinguished from *sub-Rydberg transitions* for which the upper state does not lead to a configurationally excited atom.   Such sub-Rydberg transitions may be expected to have a low absolute intensity since in the limit of large internuclear distance they have vanishing intensity.   This expectation is borne out by more detailed considerations and observations except in the case of a special kind of transition of which there is no analogue in atomic spectra and to which we now turn our attention.

Charge-transfer spectra.   The transition of an electron in a diatomic molecule *from a bonding to the corresponding antibonding orbital* is always an allowed transition.   In the simplest case of the H_2^+ ion the transition (not yet observed) from the ground state $\sigma_g 1s \, ^2\Sigma_g^+$ to the state $\sigma_u 1s \, ^2\Sigma_u^+$ is of this type (see Fig. 166).   The intensity of this transition for very large internuclear distances clearly approaches zero, while in the molecule as we shall see presently and as was first shown by Mulliken (1261) its intensity is very high.

For a rough approximation we may use linear combinations of atomic orbitals for the molecular orbitals, that is, similar to (VI, 13),

$$\psi(\sigma_g 1s) = \frac{\varphi_A(1s) + \varphi_B(1s)}{\sqrt{2 + 2T}}$$

$$\psi(\sigma_u 1s) = \frac{\varphi_A(1s) - \varphi_B(1s)}{\sqrt{2 - 2T}}$$

(VI, 67)

where φ_A and φ_B are atomic $1s$ orbital functions (H atom eigenfunctions in the present case) and where $T = \int \varphi_A(1s)\varphi_B(1s)d\tau$.   Substituting (VI, 67) into R_x^{nm}, R_y^{nm}, R_z^{nm} we find that only the z component is different from zero (parallel transition, $\Delta\Lambda = 0$) and has the value

$$R_z^{nm} = \frac{\varepsilon}{2\sqrt{1 - T^2}}\left[\int \varphi_A(1s)z\varphi_A(1s)d\tau - \int \varphi_B(1s)z\varphi_B(1s)d\tau \right]$$

$$\approx \frac{\varepsilon(z_A - z_B)}{2\sqrt{1 - T^2}} = \frac{\varepsilon r}{2\sqrt{1 - T^2}}$$

(VI, 68)

where z_A and z_B are the z coordinates of the nuclei A and B and where r is the internuclear distance.   Thus, since T^2, for not too small r, is small compared to 1, we find that the transition moment is roughly equal to the dipole moment of an electron oscillating with an amplitude $r/2$.   Such an amplitude of the classical oscillator corresponding to the system under consideration results in a large absolute intensity.   Indeed if one substitutes $R^{nm} = \frac{1}{2}\varepsilon r$ into (VI, 65) one obtains with $r = 1.06 \times 10^{-8}$ (the equilibrium internuclear distance of H_2^+) and a corresponding value of $\nu \approx 100000$ cm^{-1} (see Fig. 166) an f value of 0.3.   This is an f value of the same order as for the first members of a Rydberg series (see above).

The result that the transition moment is roughly proportional to r is intimately connected with the fact that as shown by the eigenfunctions (VI, 67) the electron in the two states considered goes back and forth between the two nuclei.   Since this is characteristic for the type of transitions under consideration they have been called *charge-transfer spectra* by Mulliken.

A transition entirely similar to that discussed above for H_2^+ is the transition in H_2 from the ground state to the lowest excited singlet state: $(\sigma_g 1s)(\sigma_u 1s) \, ^1\Sigma_u^+ \leftarrow (\sigma_g 1s)^2 \, ^1\Sigma_g^+$ (see Fig. 160).   Since there are now two electrons that can carry out the transition from $\sigma_g 1s$ to $\sigma_u 1s$ the intensity is even greater than for H_2^+.   It is not exactly doubled because ν and r_e are changed and because of the interaction of the electrons [for more details see Mulliken (1261)].   This transition of H_2 has been observed in absorption as well as emission (so-called

Lyman bands) and although absolute intensity measurements have not been made qualitative estimates confirm that the intensity is high.

Another example of a charge-transfer spectrum is supplied by the Schumann-Runge bands of O_2. According to our previous discussion (p. 346) they represent a transition (in absorption) from the ground state of the O_2 molecule $\cdots (\pi_u 2p)^4 (\pi_g 2p)^2 \ ^3\Sigma_g^-$ to the excited state $\cdots (\pi_u 2p)^3 (\pi_g 2p)^3 \ ^3\Sigma_u^-$. In this transition an electron goes from the bonding $\pi_u 2p$ orbital to the corresponding antibonding $\pi_g 2p$ orbital and therefore, similar to the above, a considerable intensity will be expected, in spite of the fact that this is not a Rydberg transition. Indeed, quantitative intensity measurements by Ladenburg and Van Voorhis (1151) have yielded an f value of 0.193 for this transition. (Near the absorption maximum 0.0014 cm atm of O_2 are sufficient to absorb half of the incident intensity.) The same f value is also obtained from measurements of the dispersion of oxygen. Theoretical f values agree with these experimental data within the rather large uncertainty of the former [see Mulliken (1261)].

A second charge-transfer transition in O_2 is expected to be one in which an electron goes from the bonding $\sigma_g 2p$ to the antibonding $\sigma_u 2p$ orbital (compare Table 34). This transition would be at shorter wave lengths than the Schumann-Runge bands, but has not yet been definitely identified. The same transition should also occur for the halogens and has been observed for Br_2 and I_2 with correspondingly increased principal quantum numbers [see Mulliken (1261)].

Thus far we have used exclusively molecular orbitals for describing the states considered. According to the atomic orbital view point the energy difference of the two states $^2\Sigma_g^+$ and $^2\Sigma_u^+$ of H_2^+ is due to the resonance between the states H^+H and HH^+. The transition between $^2\Sigma_g^+$ and $^2\Sigma_u^+$ may therefore be called a *charge-resonance* (or electron-resonance) *spectrum*. The ultraviolet O_2^+ bands $^2\Pi_u$—$^2\Pi_g$ are another example of this type.

In the case of the $^1\Sigma_u^+$—$^1\Sigma_g^+$ transition of H_2 discussed above, on the atomic orbital view the ground state $^1\Sigma_g^+$ is the Heitler-London state described by the eigenfunction (VI, 27) while the upper state is one of the states resulting from H^+H^- and H^-H^+, that is, is an ionic state (see p. 364). For brief reference the ground state and the ionic state have been designated N and V, respectively, by Mulliken. The atomic orbital view makes it even more apparent why these spectra are called charge-transfer (or electron-transfer) spectra: for H_2 we have a transition from HH to H^+H^- involving the transfer of an electron from one H atom to the other. Calculations of R^{nm} on the atomic orbital view yield absolute intensities of the same order as the molecular orbital method. However, as might be expected, for large internuclear distances the atomic orbital approximation gives the better results [see Mulliken (1261)]. Similar to the V—N transition of H_2 the Schumann-Runge bands of O_2 may be considered as a V—N transition, that is, as a transition from the ground state to one of the states resulting from O^+O^- and O^-O^+. Furthermore the halogen spectra mentioned above are such V—N transitions.

Mulliken (1261) has discussed, in addition, charge-transfer spectra in which two or more electrons are simultaneously transferred.

Other sub-Rydberg transitions. The absolute intensities of sub-Rydberg transitions that are not of the charge-transfer type are generally much smaller. A large class of these is formed by those transitions in which an electron goes from a σ orbital to π orbital. These are in general transitions with $\Delta\Lambda = \pm 1$ for which the transition moment is at right angles to the internuclear axis (*"perpendicular" bands*). The long-wave-length absorption bands of the halogens are of this type. For example the ground state (N) and the first excited states (called Q by Mulliken) of F_2 have the electron configurations

$$N : \cdots (\sigma_g 2p)^2 (\pi_u 2p)^4 (\pi_g 2p)^4 \ ^1\Sigma_g^+$$

$$Q_a : \cdots (\sigma_g 2p)^2 (\pi_u 2p)^4 (\pi_g 2p)^3 (\sigma_u 2p) \ ^1\Pi_u, \ ^3\Pi_u$$

$$Q_b : \cdots (\sigma_g 2p)^2 (\pi_u 2p)^3 (\pi_g 2p)^4 (\sigma_u 2p) \ ^1\Pi_g, \ ^3\Pi_g$$

The transition $^1\Pi_u \leftarrow {}^1\Sigma_g^+ (Q_a \leftarrow N)$ is allowed according to the selection rules.   Unlike the charge-transfer spectra here an electron goes from one antibonding to another antibonding orbital.   Mulliken (1264) has calculated the matrix element using both molecular and atomic orbitals.   The agreement with the observed f value 2.6×10^{-4} is quite satisfactory for the atomic orbital calculations, but less so for the molecular orbital calculations.   This is probably due to the fact that linear combinations of atomic orbitals are not very good approximations to true molecular orbitals.

For the other halogens transitions to the $^3\Pi_u$ state become of increasing importance and for I_2 the transition to the $^3\Pi_{0^+}$ component predominates in the spectrum.   A full discussion of the intensities of these transitions may be found in Mulliken's papers (1263)(1264).   For the mixed halogens, transitions to the Q_b states also become of importance since the property g, u no longer exists.   However, these transitions also have small f values.

For the hydrogen halides somewhat similar transitions occur.   The two lowest electron configurations have been given in Table 33 using the united atom approximation.   From this approximation it might appear that the transition between them is a Rydberg transition.   However, on dissociation the $^3\Pi$ and $^1\Pi$ states go to normal (configurationally unexcited) atoms.   A better approximation to the electron configurations in the actual molecule would be

$$N: \; \sigma^2 \pi_X^4 \; {}^1\Sigma \qquad \text{and} \qquad Q: \; \sigma^2 \pi_X^3 \sigma^* \; {}^3\Pi, {}^1\Pi$$

where σ is a bonding, σ^* the corresponding antibonding orbital resulting from $\sigma 2p_X$ and $\sigma 1s_H$ of the separated atoms $(X + H)$ and where π_X is a non-bonding orbital which is essentially $\pi 2p_X$.   Mulliken (1265) has shown that with this assumption a small absolute intensity arises in agreement with quantitative measurements by Goodeve and Taylor (263)(264).   The observed f value for HBr is 0.035.   The halides of the alkalis and other univalent atoms have similar transitions of similar intensities [see Mulliken (1265)].

Further important examples of sub-Rydberg transitions in which the electron goes from a bonding σ to a non-bonding π orbital are the transitions from the ground state to the first excited state in OH, NH, CH, and BH (see Table 33).   Unlike the case of the hydrogen halides here even in the united atom as well as in the separated atoms, the upper state is configurationally unexcited, that is, the corresponding transitions in the united atom and in the separated atoms are forbidden.   Therefore, and for several other reasons [see Mulliken (1265)] the intensity will be expected to be appreciably less than for the hydrogen halides.   Precise intensity measurements in the case of OH by Oldenberg and Rieke (540a) as corrected by Dwyer and Oldenberg (922) have indeed yielded the low f value 0.0012 which agrees very satisfactorily with theoretical values of Mulliken (1265).   No precise measurements for NH and CH are available.   Theoretical f values have been given by Lyddane, Rogers, and Roach (1182) as 0.0030 and 0.0019, respectively.[18]

Finally the case of the violet CN bands $(^2\Sigma - {}^2\Sigma)$ may be mentioned.   They correspond to a transition in which an electron goes from a slightly antibonding σ orbital to a bonding σ orbital derived from different atomic orbitals.   Therefore an intensity intermediate between charge-transfer spectra and other sub-Rydberg spectra may be expected.   The experimental f value determined by White (1530) is 0.026.   Similar f values may be expected for the similar transitions in N_2^+, CO^+, $\cdots C_2$ and B_2.   For the red CN bands $(^2\Pi - {}^2\Sigma)$ see King and Swings (1132) and Herzberg and Phillips (1051).

The knowledge of absolute intensities of band spectra is of particular importance for many astrophysical problems (see Chapter VII).   Further theoretical and experimental work in this field would be very desirable.

---

[18] The same correction as for OH [Dwyer and Oldenberg (922)] has been applied since these values were relative theoretical values based on f(OH).

CONTINUOUS AND DIFFUSE MOLECULAR SPECTRA: DISSOCIATION AND PREDISSOCIATION

1. Continuous Spectra and Band Convergence Limits: Dissociation of Diatomic Molecules

As we have seen in Chapter II, continuous molecular spectra[1] frequently appear, both in absorption and in emission, in addition to discrete molecular spectra. Quite generally, such continuous spectra result from transitions between two states, one of which at least can assume a *continuous range of energy values*.

For any atomic system, continuous ranges of terms always correspond to *aperiodic motions* in which parts of the system approach one another or separate from one another with a relative kinetic energy which is not zero even at infinity (see Chapter I, p. 9). For atoms, a continuous range joins onto each series of electronic states and corresponds to the removal of an electron (*ionization*) with more or less relative kinetic energy or, conversely, to the capture of an electron by the ion (*recombination*). According to the old Bohr theory, the corresponding orbits are hyperbolic orbits; according to wave mechanics, the corresponding wave functions are outgoing or incoming spherical waves.

Such continuous ranges of energy levels, corresponding to ionization, are possible also for molecules. However, in addition there are continuous ranges that correspond to a *splitting of the molecule into atomic components* (normal or excited atoms, or positive and negative ions)—that is, correspond to a *dissociation*. As we have already seen in Chapter III, they join onto the series of vibrational levels of every electronic state and are the only feature present if the electronic state under consideration has no discrete vibrational levels at all (unstable state).

As was first pointed out by Franck (229) the study of continuous spectra is of great importance for an understanding of dissociation processes and for the determination of heats of dissociation of diatomic molecules.

(a) Absorption

Ionization continua. In a few cases Rydberg series of bands (or band systems) have been observed in absorption [for O_2 by Price and Collins (571); for N_2 by Hopfield (331), Worley and Jenkins (722)(1559), Takamine, Suga, and Tanaka (660*a*) (1455) (1457); for CO by Tanaka, Takamine, and Iwata

---

[1] A very detailed discussion of continuous spectra may be found in Finkelnburg's book (31).

(1456a) (1455a); for NO by Tanaka (1456b); for HI by Price (570)]. Only for one of the Rydberg series of O_2 and for one of N_2 has it been possible to observe the adjoining continuous spectrum that corresponds to ionization (*ionization continuum*). Absorption of light in these continuous regions leads to the formation O_2^+ and N_2^+ ions, respectively. The long-wave-length limit of the continuum—that is, the convergence limit of the Rydberg series—gives an exact value for one of the *ionization potentials* of the molecule. We say *one* of the ionization potentials, since the observed limit does not necessarily correspond to the transition to the ground state of the ion but may also correspond to an ionization leaving the ion in an excited state, as is indeed the case for several of the Rydberg series mentioned above. The ionization potential can of course be determined from a Rydberg series irrespective of whether or not the adjoining ionization continuum is observed.

The ionization potentials of most diatomic molecules that can be readily investigated in absorption are greater than 10 e.v. and therefore the Rydberg series and ionization continua lie in the far ultraviolet where investigations are difficult.

Dissociation continua. Continuous absorption spectra corresponding to a dissociation are much more frequently observed than the ionization continua, since, like the discrete band systems, they result from transitions between two electronic states and can therefore lie in the visible and near ultraviolet regions.

Three cases of dissociation continua are possible: those for which the upper state is continuous, those for which the lower state is continuous, and those for which the upper as well as the lower state is continuous.

Upper state continuous. The most important case of dissociation continua in absorption is that in which a transition takes place from a stable lower state to a continuous upper state. Absorption of a light quantum in the resulting continuous spectrum then leads to a dissociation of the molecule under consideration (*photodissociation*).

If the upper electronic state has discrete vibrational levels the continuous spectrum may join onto a series of converging bands (band convergence) as in Fig. 87(c), p. 194. On the other hand, the continuous spectrum may also appear without these accompanying bands. The latter is the case when the potential curve of the excited state has no minimum [Fig. 173(a)] or when this minimum lies at a much greater internuclear distance than that of the ground state (see below).

The I_2 absorption spectrum in the visible, reproduced in Fig. 15, p. 38, is an example of the first case (*band convergence with adjoining continuum*). It is the one that was first studied in this connection by Franck and his co-workers (229). Further spectra of this type have been found for Br_2 [Kuhn (426)], O_2 [Leifson (440)], and a few other molecules. The *convergence limit of the bands* (the beginning of the continuum), which can be located in these cases with great accuracy, gives the exact position of the asymptote of the potential curve of the upper state, that is, the position of the so-called *dissociation limit*. In the case of I_2 (and similarly of the other halogen molecules) it can be shown (see section 3) that at the convergence limit dissociation takes place into one normal atom in the $^2P_{3/2}$ state and one in the slightly excited metastable $^2P_{1/2}$ state; that is, the asymptote of the potential curve of the upper state does not coincide with that of the ground state [compare Fig. 88(c), p. 195].

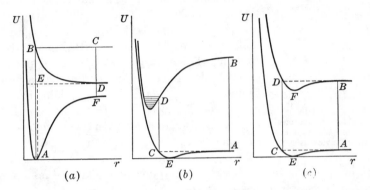

Fɪɢ. 173. **Potential Curves Explaining Continuous Absorption Spectra.** The lower curve in (b) and both curves in (c), which correspond to van der Waals' attraction, are actually much shallower than drawn.

That the absorption in the continuous region actually corresponds to a *dissociation into atoms* has been verified by a number of experiments which refer principally to the very readily accessible continuous absorption of the I_2 molecule. Dymond (196) has shown that no fluorescence appears on illumination with light in the continuous region, while, on the other hand, an intense fluorescence appears when the gas is illuminated with light in the region of discrete bands. Turner (669) established more directly the fact that atoms are formed in the following way: He found that, when I_2 is irradiated by light in the region of continuous molecular absorption, the gas absorbs the ultraviolet atomic lines of iodine, thus showing that iodine atoms are formed. Rabinowitch and Wood (575), by measuring the intensity of molecular absorption in the irradiated gas, detected the decrease in the number of molecules due to dissociation of some of the molecules. Senftleben and Germer (628) showed that the heat conductivity of the gas is changed in consequence of the dissociation of a certain fraction of the molecules. Finally the value of the dissociation energy of I_2 that follows from the position of the convergence limit agrees in a most satisfactory way with purely thermal determinations from the vapor density at high temperatures. In the most recent work of this type, by Perlman and Rollefson (1311), the agreement is within less than 0.1%.

The near ultraviolet absorption spectra of F_2 [Wartenberg, Sprenger and Taylor (689)] and the halogen hydrides [Goodeve and Taylor (263)(264)] are examples of the case where only the continuum, without the adjoining band series, is observed. In Fig. 174 the absorption coefficient of F_2 is plotted against wave length. It can be seen that, on coming from long wave lengths, the absorption at first increases, then reaches a maximum, and finally decreases. The extent of the continuum depends on the thickness of the absorbing layer.

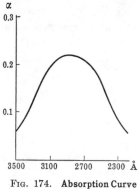

FIG. 174. Absorption Curve of F_2 [after von Wartenberg, Sprenger, and Taylor (689)].

The absorption spectrum of Cl_2 may be regarded as a *transition case* insofar as only continuous absorption is observed for thin and medium layers, whereas for thicker layers a series of converging bands appears at the long-wave-length limit of the continuum. This transition case makes it particularly clear that there can be no doubt that a pure continuous spectrum, without an adjoining band series, represents an electronic transition of the same type as a discrete absorption spectrum, except that the upper state is unstable (continuous). The existence of the continua shows at the same time that the unstable molecular states (without an appreciable potential minimum) are just as real as the stable molecular states.

For certain continua of the halides of gallium, indium, a: thallium electrical measurements have indicated a dissociation into positive and negative ions rather than neutral atoms [Terenin and Popov (666), Wehrli and Hälg (1520)]. The ionic states in these cases have undoubtedly deep potential minima which lie, however, at much larger internuclear distances than those of the ground states so that in absorption only the continuum is reached.

The *long-wave-length limit of the continuum* is obviously given theoretically by the height of the asymptote of the potential curve of the upper state above the vibrationless ground state [AE in Fig. 173(a)], corresponding to the band convergence limit in the case where discrete bands are also present. However, the observed long-wave-length limit of a continuum is often at considerably shorter wave lengths than the theoretical limit. The explanation for this and also for the variation of intensity with wave length is given by the Franck-Condon principle.

According to the Franck-Condon principle, the most probable transition in absorption is that going vertically upward from the minimum of the lower potential curve [AB in Fig. 173(a)]. This transition gives the maximum of the intensity in the absorption spectrum. In general, if the upper state is a repulsive state, this maximum lies at appreciably shorter wave lengths than the theoretical limit; that is, at the maximum of absorption the two atoms generally separate from each other with an appreciable kinetic energy [corresponding to CD in Fig. 173(a)]. The smaller the slope of the repulsive curve is, the more rapidly the intensity falls off to either side of this maximum. For a repulsive curve like that in Fig. 173(a), a transition in the neighborhood of the theoretical long-wave-length limit would correspond to a very large change in internuclear distance $(A-D)$. According to the Franck-Condon principle, such a transition practically does not occur. The continuum first begins at an appreciably shorter wave length but *without a sharp limit*. The theoretical long-wave-length limit can be observed only for cases such as that in Fig. 88(c), in which the band convergence is also observed, or for cases in which the repulsive curve runs quite flat up to relatively small internuclear distances. In the latter cases the continuum is rather narrow and possibly has the appearance of a diffuse band or even of a diffuse line [see also section 2(d)].

In the case of the photodissociation of NaI Hogness and Franck (324) have proven directly that there is a *large relative velocity of the two atoms* resulting from absorption in a continuous spectrum: One of the absorption continua of NaI leads to an excited Na atom in the ^2P state and a normal I atom as shown by the emission of the sodium D lines upon radiation of the vapor with light in the continuous region. It was found that the width of the sodium lines excited in this way is anomalously large on account of the *Doppler effect* of the motion of the atoms flying apart with large velocities after the light absorption. The width is found to be the larger the shorter the wave length of the light producing the dissociation, as is to be expected.

In order to determine quantitatively the *theoretical intensity distribution* we have to take into account the form of the eigenfunctions in the upper and in the lower state. According to the wave mechanical formulation of the Franck-Condon principle [see Chapter IV, section 4(a)], the intensity of a band of a given electronic transition in absorption is proportional to

$$ \nu \left[\int \psi_{v'} \psi_{v''} dr \right]^2 , \qquad \text{(VII, 1)} $$

where $\psi_{v'}$ and $\psi_{v''}$ are the vibrational eigenfunctions of the upper and lower states. It may be shown that the same expression holds for the present case of a continuous spectrum if it is applied to the intensity per unit wave-number interval. For an evaluation of the integral, in addition to a knowledge of the vibrational eigenfunction of the stable state (see Figs. 42 and 48, pp. 77 and 94), a knowledge of the "vibrational" eigenfunction of the unstable state is required.

As an example the *eigenfunctions of the nuclear motions* for three different energy values of the upper state of the main continuous absorption of the Cl_2 molecule are reproduced in Fig. 175. At some distance from the classical turning point of the motion we have a simple sine wave of constant amplitude whose wave length is the smaller the higher the energy lies above the asymptote of the potential curve, according to the de Broglie relation (I, 10). In the neighborhood of the classical turning point the amplitude and wave length of the wave are greater, and directly above the turning point there is a broad maximum which falls off exponentially to smaller r values. This broad and somewhat higher maximum corresponds to the fact that, according to classical theory, the atoms will stay for a longer time in the neighborhood of the turning point than at some distance from it.

The *overlap integral* $\int \psi_{v'} \psi_{v''} dr$ for a continuous absorption spectrum is the integral over the product of such a repulsive eigenfunction and a vibrational eigenfunction of the stable state. In the case of absorption at low temperatures the vibrational eigenfunction of the ground state $(v'' = 0)$ is a simple bell-shaped curve (given in the lower part of Fig. 175). The integral will have a maximum value when the broad maximum of the repulsive function lies approximately above the maximum of the bell curve. This is roughly the case for the repulsive eigenfunction drawn at 30,000 cm^{-1} in Fig. 175. However, owing to the factor ν in (VII, 1) the energy value for which the maximum of the overlap integral occurs does not coincide exactly with that for which the maximum of the intensity occurs: the intensity maximum is displaced a small amount toward shorter wave lengths. If the potential curve of the upper state is known, the repulsive eigenfunctions for various energy values and the cor-

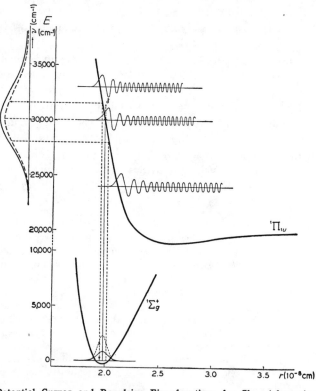

Fig. 175. **Potential Curves and Repulsive Eigenfunctions for Cl$_2$.** Adaptation of a similar figure by Gibson, Rice and Bayliss (259). The ordinate scale in cm^{-1} holds for the potential curves (heavy lines) as well as for the intensity distribution in the continuous spectrum drawn to the left. The eigenfunctions (thin full lines) have of course a different ordinate scale which is not given. The abscissa scale is the same for the potential curves and the eigenfunctions. The dotted curve denoted by δ is to be thought of as infinitely thin and infinitely high, so that the area under it is unity. The "reflection" of the curve $\psi_{v''}{}^2$ for the lower state $v'' = 0$ (lower broken-line curve) at the upper potential curve is indicated by the broken leading lines.

responding overlap integrals can be evaluated, that is, the intensity distribution in the continuous spectrum can be accurately predicted. Usually, however, the potential curve of the upper state is not known, but rather its slope and height near the equilibrium internuclear distance r_e'' of the lower state may be evaluated from the intensity distribution in the continuum. Such calculations have been carried out by Stueckelberg (1440) for O$_2$, by Gibson, Rice and Bayliss (259) for Cl$_2$, and by Bayliss (82) for Br$_2$. In each case very satisfactory agreement between the theoretical and the observed intensity distributions has been found. For both Cl$_2$ and Br$_2$ in addition to the main continuum which corresponds to a $^1\Pi_u \leftarrow {}^1\Sigma_g{}^+$ transition a much weaker continuum at longer wave length occurs. This is the continuum adjoining to the main system of discrete absorption bands which represent a $^3\Pi_{u0^+} \leftarrow {}^1\Sigma_g^+$ transition.

Frequently, an *approximation* first given by Winans and Stueckelberg (717) is used to derive the intensity distribution in continuous spectra: The repulsive eigenfunction is replaced by a so-called δ function, which is different from zero only at the classical turning point (dotted curve denoted by δ in Fig. 175). This would appear to be a very poor approximation for the repulsive eigenfunction. Actually, however, the results obtained with it deviate only very slightly from those obtained with the accurate repulsive eigenfunctions, as Coolidge, James, and Present (162) have shown in detail.

The construction of the theoretical intensity distribution according to this procedure is illustrated graphically in Fig. 175 for the case of the Cl_2 absorption spectrum. Owing to the assumed form of the repulsive eigenfunction, the overlap integral for a given energy of the

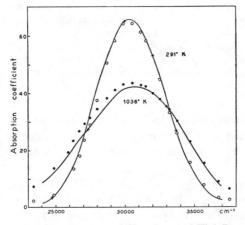

Fig. 176. **Continuous Absorption Spectrum of Cl_2 at Low and High Temperature** [after Gibson, Rice and Bayliss (259)]. The open circles represent observed absorption coefficients at 291° K, the full circles at 1038° K.

unstable state is simply equal to the value of $\psi_{v''}$ at the position vertically below the classical turning point for that energy.[1a] The intensity is therefore simply proportional to $\nu\psi_{v''}^2$, where $\psi_{v''}$ is to be taken for the r value that corresponds to the ν considered. We have therefore simply to "*reflect*" the function $\psi_{v''}^2$ (that is, the probability density distribution in the lower state) at the repulsive potential curve (obtaining the broken-line curve to the left in Fig. 175) and then to multiply by ν in order to obtain the theoretical intensity distribution (full-line curve to the left). Conversely, we can again use the observed intensity distribution to derive the course of the repulsive part of the potential curve. This simplified method has been applied by Goodeve and Taylor (263)(264) to HBr and HI.

From the above discussion it can be seen immediately that the continuous spectrum corresponding to transitions from the lower vibrational state $v'' = 1$ must show *two intensity maxima* (instead of the one for $v'' = 0$), since the probability density distribution curve has two maxima (see Fig. 42). Similarly, three maxima, of which the middle one has no analogue in the elementary classical description, are to be expected in the continuum corresponding to transitions from $v'' = 2$, and so on. In absorption at high temperatures the continua corresponding to $v'' = 0, 1, 2, \cdots$ are superimposed and in general, except when the potential curve of the upper state is very shallow (see p. 435) only a single maximum arises which is broader and less high than at low temperatures. In Fig. 176, as an example, the

---

[1a] Here it is assumed that the δ function is normalized so that $\int \delta \, d\tau = 1$.

observed absorption coefficient of Cl_2 as a function of wave number is plotted for two widely different temperatures. The widening of the absorption curve at the high temperature is clearly shown. The full-line curves represent the values calculated by Gibson, Rice, and Bayliss (259) in the way described above.

The theoretical long-wave-length limit of the absorption continuum corresponding to the state $v'' = 1$ lies $\Delta G_{\frac{1}{2}}''$ farther to long wave lengths than that corresponding to $v'' = 0$. Therefore it may happen that the absorption continuum for higher temperatures or for extremely thick layers extends beyond the theoretical limit corresponding to $v'' = 0$. This is an additional reason for the lack of sharpness of the long-wave-length limit of a continuum that does not join onto a band series.

It should be emphasized that the preceding discussion is based on the assumption, implicit in the Franck-Condon principle (see p. 200), that the *electronic transition probability is independent of r* in the range that matters for the continuous absorption considered. The excellent quantitative agreement between observation and theory, for example, in the case of Cl_2 (see Fig. 176) shows that this assumption is a good approximation. However, one must be prepared for deviations in cases in which a large range of r values is involved and when the electronic transition probability is widely different for the corresponding transition in the united atom and the separated atoms [see Coolidge, James and Present (162)]. Nevertheless even in these cases the qualitative features of the above discussion remain valid.

Lower state or both lower and upper state continuous. Cases of continuous absorption spectra for which the lower state or both the lower and the upper state belong to a continuous range of energy levels occur for inert gases and certain metal vapors. The potential curves for the two cases are drawn in Fig. 173(b) and (c).

In these cases, apart from the repulsive forces which come into play at smaller distances, only van der Waals forces act between the atoms in their ground states. These van der Waals forces give a very shallow minimum in the potential curve at rather large internuclear distances [van der Waals molecules; see Chapter VI, section 4(c)].

For not too low temperatures (if kT is larger than the heat of dissociation) and for low pressure, most of the molecules are dissociated, since thermal collisions are sufficient to throw the molecules out of the shallow minimum; that is, the gas is monatomic. The absorption spectrum then consists solely of atomic lines ($A \rightarrow B$ and transitions to higher levels) which correspond to the energy differences between the asymptotes of the potential curves.

However, when the pressure is sufficiently high, the atoms are relatively often in the *state of collision* (sometimes called the state of the *quasi molecule*). This means that the potential energy is different from that of the separated atoms. If absorption takes place during a collision, frequencies other than those for the widely separated atoms will be absorbed, as we can see immediately from Fig. 173(b) and (c) if we consider that, according to the Franck-Condon principle, mainly transitions vertically upward occur. Since the kinetic and potential energies of the colliding atoms can take any value within a certain range, a continuous absorption spectrum results.

We shall consider first of all the case where the *collision of the atoms* takes place *centrally or nearly centrally*. The turning point of the motion is then given by the point of intersection of the horizontal line corresponding to the total energy of the system with the potential curve [for example, C in Fig. 173(b) and (c)]. Both classically and quantum-theoretically the atoms spend most of the collision time near the turning point. Therefore according to the Franck-Condon principle we need only consider the transitions from this turning point vertically upward or at least nearly so.

If the *upper state* is *stable* [Fig. 173(b)], a transition into one of the discrete vibrational levels of this state takes place. Since, however, in different individual collisions there will be varying amounts of kinetic energy [that is, different heights of the point C in Fig. 173(b)], we obtain a continuous absorption spectrum for each given upper state. The different continua overlap one another and give an extended continuum—possibly with intensity fluctuations [see section 2(d)]. This continuum lies on the long-wave-length side of the atomic line under consideration. It may join directly onto the atomic line or lie at some distance from it, depending on the shape of the upper potential curve. The longest-wave-length region of the continuum will be absorbed by those atoms that collide with the greatest kinetic energy. The higher the temperature, the greater is this kinetic energy. We should therefore expect the extent and the intensity of the long-wave-length part of the continuum to be very sensitive to temperature.

If the *upper state* is a *repulsive state*, a continuum results in a corresponding manner [transitions $C–D$ in Fig. 173(c)], but it has in general a much smaller extent than when the upper state is stable, and it usually joins directly onto the atomic line. In this case the continuum extends to longer or shorter wave lengths from the atomic line, depending on the course of the upper potential curve relative to the lower. As we have seen in Chapter VI, section 4(c) the van der Waals forces in excited states are frequently larger than in the ground state and may be attractive or repulsive. The former case gives therefore usually a continuum extending to longer wave lengths; the latter gives one extending to shorter wave lengths. In this latter case there will frequently be a short-wave-length limit to the continuum since in general the potential curve of the ground state for $r < r_e$ will rise more steeply than that of the excited state so that there is a maximum vertical separation of the two curves at an intermediate r value.

Such continua having stable as well as unstable upper states, have been observed and investigated in detail for the vapors of Cd, Zn, and Hg [Rayleigh (581), Winans (716), Kuhn (433)(434), Mrozowski (1243) and others; see also the detailed discussion by Finkelnburg (31)]. In absorption at high pressures fairly extended continua join onto the resonance lines $^3P_1$—^1S and $^1P_1$—^1S of these atoms, to longer wave lengths. They correspond to transitions from the unstable ground state (two normal $^1S_0$ atoms in the state of collision) to a stable molecular state resulting from an excited $^3P_1$ or $^1P_1$ atom and a normal $^1S_0$ atom. As Kuhn and Freudenberg (434) have shown for Hg, the intensity of these continua increases quadratically with the pressure, as it must on the basis of the above explanation, since the number of atoms in the state of collision increases quadratically with the pressure. A temperature dependence of the long-wave-length part has also been observed by them. In addition narrow continuous extensions of atomic absorption lines have been found at high pressures, which are clearly due to transitions in the state of collision when both the upper and the lower molecular state formed are unstable. In some of these cases a short-wave-length limit, as expected from theory (see above), has also been observed.

Similar continua have been observed in mixtures of mercury vapor with inert gases of high pressure, in the neighborhood of the Hg lines [Oldenberg (537)]. They are due to absorption by Hg atoms in the state of collision with inert-gas atoms, or, in other words, by Hg–He, Hg–Ne, $\cdots$ quasi molecules.

Since in many cases several molecular states result from a given combination of the excited states of the separate atoms (see Chapter VI, section 1), a number of different continua can sometimes join onto or correspond to one and the same atomic line. For example the four states $^1\Pi_g$, $^1\Sigma_g{}^+$, $^1\Pi_u$, $^1\Sigma_u{}^+$, arise from $Hg(^1S_0) + Hg(^1P_1)$. Of these $^1\Pi_u$ and $^1\Sigma_u{}^+$ can combine with the ground state $^1\Sigma_g{}^+$ derived from $Hg(^1S_0) + Hg(^1S_0)$, and indeed two continua (one with fluctuations of intensity) have been observed to join onto the Hg line $^1P \leftarrow {}^1S$ at high pressure. According to the previous discussion of van der Waals molecules (p. 380) there is an attractive van der Waals force for $^1\Sigma_u{}^+$ but a repulsive one for $^1\Pi_u$. The short-wave continuum associated with the $^1P \leftarrow {}^1S$ transition does indeed seem to require such a repulsive van der Waals force, although there is probably a potential minimum (produced by exchange forces) at smaller internuclear distances [see Finkelnburg (31)]. A very similar situation exists for Cd_2 [see Cram (871)].

As mentioned before for continuous spectra with a stable lower state the condition of the applicability of the Franck-Condon principle, namely, the constancy of the electronic transition probability ($R_e{}^2$) in the range of r that matters, is in general quite well fulfilled. However, for an unstable lower state, that is, for collisions of atoms, such a large range of internuclear distances is covered that the condition just mentioned is often no longer fulfilled. In such cases the derivation of the intensity distribution cannot be carried out in the manner previously discussed. A particularly large variation of the electronic transition probability arises when a transition is forbidden for the separated atoms but allowed in the molecule. An example is the Hg_2 continuum at 1690 Å (and similar continua of Cd_2 and Zn_2) which must be interpreted as a transition from the $^1\Sigma_g{}^+$ ground state $[Hg(6^1S) + Hg(6^1S)]$ to the $^1\Sigma_u{}^+$ state which for large internuclear distance leads to one Hg atom in the 7^1S state. In the atom the transition $7^1S \leftarrow 6^1S$ is forbidden while in the molecule a $^1\Sigma_u{}^+ \leftarrow {}^1\Sigma_g{}^+$ transition is allowed. This continuum has therefore negligible intensity near the position of the (forbidden) atomic line and has a maximum of intensity at a considerable distance from it.

We shall now consider the effect of *non-central collisions* on the absorption spectrum. During non-central collisions the atoms do not approach each other so closely, and at no point in their path is their relative velocity zero. With an accuracy sufficient for many purposes we may simply consider the atoms as flying past each other. Thus in the application of the Franck-Condon principle the turning point is no longer favored. For each nuclear separation traversed, a transition to the upper state can take place with about the same probability. The *frequency absorbed corresponds to the separation of the two potential curves for the internuclear distance at which the quantum jump occurs*, since, according to the

Franck-Condon principle, the kinetic energy is practically unaltered by the electron jump. Therefore a continuous spectrum results, corresponding to transitions vertically upward from every point between A and E in Fig. 173(b) and (c), and joining directly onto the atomic line. It extends to longer or shorter wave lengths according to whether the potential curve has a deeper or a shallower minimum in the upper state than in the lower state. An extension toward longer wave lengths is more frequent, since excited states usually give a stronger van der Waals binding than does the ground state.

The intensity distribution in the continuum corresponding to the noncentral collisions can be easily derived, since the frequency of occurrence of the different values of the internuclear distance is proportional to r^2 and the wave number corresponding to each r can be taken from the potential curve diagram. According to the London theory of van der Waals binding, the potential energy for large r is given by $-C/r^6$. If this holds for the upper as well as the lower state the deviation $\Delta\nu$ from the corresponding atomic line for the part of the continuum absorbed at an internuclear distance r is

$$\Delta\nu = \frac{|C' - C''|}{r^6} \tag{VII, 2}$$

where C' and C'' are the van der Waals constants for the upper and lower state, respectively. Since the intensity in the interval $d\nu$ is proportional to the frequency of occurrence of an r value between r and $r + dr$, that is,

$$I(\nu)d\nu = A4\pi r^2 dr \tag{VII, 3}$$

where A is a constant, one finds by combination with (VII, 2) that

$$I(\nu) = \frac{2\pi A}{3} \sqrt{\frac{(C' - C'')}{(\Delta\nu)^3}} \tag{VII, 4}$$

that is, the *intensity of the continuum should vary inversely proportionally to* $(\Delta\nu)^{3/2}$ [Kuhn (430), see also Margenau and Watson (481)]. This continuum should form the long-wave-length wing of the atomic line if as usual $C' > C''$ (see p. 395).

The variation of the intensity with $(\Delta\nu)^{-3/2}$ demanded by theory has been quantitatively confirmed in experimental investigations of the intensity distribution in the wings of the Na resonance lines in absorption when they are broadened by the addition of argon of high pressure to the sodium [Minkowski (497)], and in the wings of the Hg line 2537 also broadened by argon [Kuhn (431), Rühmkorf (604b)]. Conversely the validity of the $(\Delta\nu)^{-3/2}$ law in these cases is a confirmation of London's r^{-6} law for van der Waals forces.

By an evaluation of the proportionality constant in (VII, 4) it is possible to obtain from the observed intensity distribution the difference $C' - C''$ of the van der Waals constants in the upper and lower states. In the case of Na + A the agreement between observed and theoretical values of these constants is very satisfactory. In the case of Hg + A agreement is obtained only if account is taken of the fact that two states arise from $Hg(^3P_1) + A(^1S_0)$ namely one with $\Omega = 1$ and the other with $\Omega = 0$, the former having $C' < C''$, the latter having $C' > C''$. Under this assumption only the $\Omega = 0$ state contributes to the long-wave-length wing and the intensity is one third of what it otherwise would be. The $(\Delta\nu)^{-3/2}$ variation in the short-wave-length wing is concealed by the continua produced by central collisions as well as those non-central collisions in which the quantum jump occurs for small r values where the r^{-6} law no longer holds.

In the case of collisions between like atoms with strong resonance lines (for example Na + Na), in the upper state the van der Waals potential is proportional to $-1/r^3$ rather than $-1/r^6$. Therefore one finds a *variation of the intensity proportional to* $(\Delta\nu)^{-2}$ in the wings of the line. Since the van der Waals attraction extends to larger r values much lower pressures are sufficient to produce these wings than in the case of interaction with a foreign gas. For experimental confirmations of these conclusions in the case of potassium and cesium vapor, see Hughes and Lloyd (341*a*) and Gregory (998), respectively.

The preceding considerations of the effect of non-central collisions account in a fairly satisfactory way for one aspect of the *broadening of spectral lines by pressure* [as was first recognized by Jabłoński (359) (1089) (1090)], namely the intensity distribution in the wings at separations $\Delta\nu$ from the center of the line of about 10 or 20 cm^{-1} to several hundred cm^{-1}. These wings appear only at rather high pressure when the duration of a collision is an appreciable fraction of the time between two collisions. The intensity distribution produced by the above discussed mechanism is usually referred to as the *statistical distribution* since it is determined solely by the positions of all the atoms with respect to the one emitting or absorbing the spectral line in accordance with statistical mechanics.

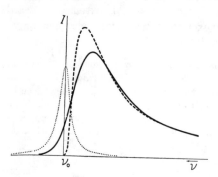

FIG. 177. **Intensity Distribution in a Spectral Line Broadened by Collisions, Schematic [after Margenau and Watson (481)].** The full-line curve arises from the broken-line curve, which corresponds to the statistical distribution, by "diffusing" it with the dotted curve which represents pure impact broadening.

The broken-line curve in Fig. 177 represents schematically a purely statistical distribution within a resonance line broadened by pressure of a foreign gas, assuming that the van der Waals attraction is stronger in the upper than in the lower state. This distribution is wholly on the long-wave-length side of the undisturbed line ν_0. It should be noted that for small $\Delta\nu$ the $(\Delta\nu)^{-3/2}$ law does not hold, but rather there is a rise from zero at ν_0 to a maximum whose separation from ν_0 is proportional to the square of the density of the foreign gas [see Margenau (480)].

Actually the statistical theory breaks down for small $\Delta\nu$ which are the only ones occurring at medium and low pressures. This is because the Franck-Condon principle in its elementary form does not hold rigorously: small changes of the kinetic energy in going from the lower to the upper state may indeed occur and therefore for each given r not a single frequency but a small range of frequencies is emitted or absorbed (see dotted curve in Fig. 177). This velocity distribution must be used for each ordinate of the statistical distribution in order to obtain a first rough approximation to the true distribution (full-line curve in Fig. 177). For low pressures only the *velocity broadening* remains which is symmetrical with respect to the undisturbed position and whose width is proportional to the density of the foreign gas. This velocity broadening is equivalent to the broadening resulting from a Fourier analysis of the finite wave trains emitted or absorbed between collisions and is frequently referred to as *Lorentz-Weisskopf* or *impact broadening* [see Weisskopf (702), Kuhn and London (435), Margenau and Watson (481)]. At high pressures the statistical effect predominates and produces the asymmetry of the line as well as the shift of the maximum which is also proportional to the pressure.

A more general treatment of line broadening combining the statistical and velocity effects in a more rigorous manner, has recently been given by Lindholm (1171) (1172) (1174), who has also carried out, together with Kleman (1135), measurements of the broadening of the sodium lines by argon over a wide range of density and found a very satisfactory agreement with the theory [see also Foley (938)].

Naturally, *both central and non-central collisions* take place. However, the former are much rarer, and it is therefore not surprising that the continuous extensions of the atomic lines (wings), particularly those toward longer wave lengths, can in most cases be explained quantitatively by the non-central collisions. The central collisions are responsible only for continua lying at a larger distance from the atomic line under consideration and in some cases also for the narrow short-wave-length extensions. These continua appear only at rather high pressures. While the intensity distribution in the wings, which is due to non-central collisions, is independent of the temperature since it depends only on the relative positions of the atoms, the intensity distribution in the continua due to central collisions does depend on the temperature.

At sufficiently low temperatures and high pressures, the *discrete vibrational levels* in the shallow minimum of the potential curve of the ground state are also populated [E in Fig. 173(b) and (c)]. The transition from these levels to the discrete levels of the upper state is a transition between two stable states and therefore, strictly speaking, does not belong here. However, owing to the shallowness of the potential curve of the lower state, the vibrational levels lie so close together that the individual bands of the resulting band system are frequently not resolved and only one or a few maxima of an apparently continuous spectrum are observed in the neighborhood of the corresponding atomic line. Thus for mercury vapor we find a narrow absorption maximum at 2540 Å, corresponding to the Hg line 2537 Å and representing the transition between the two corresponding van der Waals potential minima [$E–F$ in Fig. 173(c)]. The intensity of this band is a measure of the concentration of Hg_2 molecules. Correspondingly, its intensity decreases with increasing temperature. By accurate measurements of the intensity of this band at different temperatures, the heat of dissociation of the Hg_2 van der Waals molecule in the ground state has been estimated to be about 1.8 kcal. [see Kuhn (431)].

It will be remembered (p. 320 f.) that there are four states, 0_g^+, 0_u^+, 1_g, 1_u, arising from $Hg(^1S_0) + Hg(^3P_1)$ of which only 0_u^+ and 1_u combine with the ground state. The 0_u^+ state is assumed to be responsible for the band at 2540 Å and must therefore have a very shallow potential minimum while the 1_u state has a deep minimum and is responsible for the long wave extension of the line 2537 Å (see above).

Bands similar to 2540 Å of Hg_2 have been found close to the atomic absorption lines in alkali vapors [Kuhn (431), Ch'en (846), Gregory (998)]. They correspond to transitions from the repulsive state $^3\Sigma_u^+$ arising from normal atoms (with parallel spins of the electrons) to the van der Waals states arising from a normal and an excited atom. Furthermore such bands occur also near the atomic lines of mercury when an inert gas of high pressure is added, showing that Hg atoms and inert-gas atoms can form weakly bound van der Waals molecules [Oldenberg (537), Kuhn (431), Rühmkorf (604b)].

It is clear from the preceding discussion that the investigation of the continuous spectra occurring in the neighborhood of atomic lines is of great importance in the study of van der Waals forces. Much work remains to be done in this field, particularly with regard to a quantitative confirmation of the theory and an evaluation of van der Waals constants.

(b) Emission

The emission continua of molecules result from the converse process to that producing absorption continua.   Emission continua corresponding to the ionization continua which are observed in absorption, that is, resulting from the recombination of molecular ions and electrons, have not yet been observed for molecules.   But numerous emission continua corresponding to the other absorption continua have been observed.

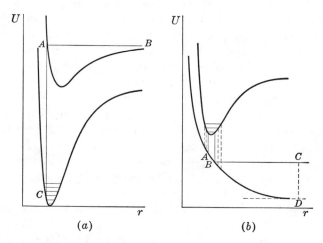

Fig. 178.   **Potential Curves Explaining Continuous Emission Spectra.**   In (*b*), *CD* gives the relative kinetic energy of the two atoms after the transition of the molecule from the upper state with $v = 0$ to the lower at *B*.

Upper state continuous (molecule formation in a two-body collision). The converse of the direct dissociation of a stable molecule by light absorption is the *recombination* of the two atoms *in a two-body collision* with emission of radiation.

Let us consider two atoms approaching each other in such a way that their potential energy in its dependence on the internuclear distance is given by the upper potential curve in Fig. 178(*a*).   The total energy of the two atoms is represented by the horizontal line *AB*.   At every point the relative kinetic energy is given by the vertical distance of this line from the potential curve. At the point *A*, the kinetic energy is completely converted into potential energy. Classically, it is the turning point for the motion.   After reaching it, the potential curve is traversed in the opposite direction, and the two atoms fly apart again (see p. 196).

A permanent recombination of the two atoms will take place only when energy is removed during the short time during which the potential energy is smaller than at infinite separation.   This time is of the order of a period of vibration, that is, of the order of 10^{-13} sec.[2] The removal of energy may occur in one of two ways: either by collision with a third particle during the collision time (*recombination by three-body collision*) or by radiation of the energy (*recombination by two-body collision*). In the latter case a transition takes place mainly at the classical turning point A [Fig. 178(a)] to one of the vibrational levels in the ground state, whose turning point lies nearly vertically below A. Since the kinetic energy of the colliding atoms may have values within a considerable range, a continuous spectrum results—the *recombination continuum*, corresponding to the dissociation continuum (see above). This recombination in a two-body collision is, however, a very rare process, since the time that elapses on the average before an excited molecule radiates (about 10^{-8} sec) is very large compared to the duration of the collision (10^{-13} sec), during which the electron jump must take place in order to lead to a recombination. As a result, recombination by a two-body collision takes place only for an extremely small fraction γ of the collisions (about 10^{-5}).

An additional requirement for the occurrence of such a *radiative recombination* is that the two atoms approach each other on a potential curve of an *excited* molecular state that combines with the ground state (or with another lower state). In many cases this is possible only if one of the atoms is excited, as, for example, for the case Fig. 178(a), and this leads to a further reduction of the frequency of occurrence of the recombination by two-body collisions. If the excited molecular state is formed from the atoms in their ground states, usually the electronic transition probability is low (see Chapter VI, section 5) so that the fraction of collisions which lead to molecule formation by radiation is much lower than the figure 10^{-5} given above. Moreover, it has to be considered that, of the various molecular states arising from the particular combination of the separated atoms, only one or a few are of such a nature that a transition to the discrete part of the ground state can occur. The probability that these particular molecular states are formed in the collision is proportional to their statistical weights and is, of course, smaller than unity.

If the two atoms approach on the potential curve of the ground state, molecule formation by a radiative two-body collision is practically impossible, since the transition probability for a vibrational transition, which would lead to a removal of energy, is very much smaller than that for electronic transitions. For two like nuclei, this possibility disappears altogether (see Chapter III, p. 80 f.).[3]

On the other hand, *recombination by three-body collision* without radiation of light is a much more frequent process, at least at high and medium pressures. If τ is the duration of a collision and Z the number of collisions suffered by a

---

[2] The collision time is somewhat less than the time obtained by dividing the distance $2 \times AB$ in Fig. 178(a) by the average speed of the free atoms, since in the region AB their kinetic energy is increased.

[3] Strictly speaking, for like nuclei a quadrupole rotation-vibration spectrum is possible (see p. 279), but since its intensity is 10^{-8} of that of ordinary infrared spectra, it need not be taken into account here.

given atom per second, it is immediately clear (since $1/Z$ is the average time between collisions) that τZ is the fraction of two-body collisions which are also three-body collisions. This fraction is proportional to the pressure, while the fraction γ of collisions leading to two-body recombinations is independent of the pressure. At atmospheric pressure for nitrogen for example, τZ is about 5×10^{-4}, that is, one in every 2000 collisions is a three-body collision, while even in the most favorable case only one in every 10^5 collisions would lead to a two-body recombination ($\gamma = 10^{-5}$). In this most favorable case, at a pressure of 0.02 atm the two types of recombinations would be about equal in number and at still lower pressures two-body recombinations would predominate.[4] However, in most cases the transition probability is less than in the most favorable case, and therefore the limiting pressure is lower, frequently much lower than 0.02 atm. Therefore, when molecule formation from free atoms plays a part in chemical elementary reactions or other collision processes it proceeds usually via the three-body mechanism.[5] But there are exceptions: for example, molecule formation in the upper atmosphere of the earth or in interstellar space proceeds exclusively via two-body recombinations.

In spite of their rare occurrence *two-body recombinations* can be observed spectroscopically by the continuous spectrum emitted. Thus far, this has been done only for the halogens and Te_2. Kondratjew and Leipunsky (410) found that, for example, I_2 vapor at high temperatures emitted a continuum as well as discrete bands. This continuum is to be explained as a recombination continuum (converse of the dissociation continuum). In consequence of the high temperature employed, atoms, some of which are in the excited state required, are continuously produced by dissociation of the molecules. In a small fraction of the cases, the atoms recombine by a two-body collision with radiation instead of by a three-body collision and this gives rise to the continuum. For Te_2 in an intense discharge, in which a strong dissociation of the vapor takes place, Rompe (596) found a continuum that is in all probability to be explained in a corresponding manner.

A case that may be of importance in certain astrophysical problems but has not yet been observed is the recombination continuum corresponding to the formation of H_2^+ from $H + H^+$. If H and H^+ on collision form the repulsive $2p\sigma\ ^2\Sigma_u^+$ state (see Fig. 166), a transition to the stable $1s\sigma\ ^2\Sigma_g^+$ ground state may occur with relatively high probability (see p. 384).

We shall see later that two-body recombinations accompanied by emission of discrete bands rather than of a continuous spectrum are possible as a result of inverse predissociation (see p. 414).

Lower state continuous. Emission continua for which the lower state belongs to a continuous range while the upper state is a discrete state are of

---

[4] It must of course be remembered that the absolute number of two-body collisions, and therefore of two-body recombinations decreases with the square of the pressure.

[5] For a more detailed discussion of three-body collisions, see Steiner (651).

much greater intensity and therefore more often observed than those considered in the preceding paragraphs. The potential curve diagram for this case is given in Fig. 178(b). According to the Franck-Condon principle, the transitions from the lowest vibrational level ($v' = 0$) of the excited electronic state take place to points on the lower potential curve that lie above the asymptote—that is, in the continuous range [roughly between A and B in Fig. 178(b)]. The resulting spectrum is therefore continuous. At the same time, when the system has once gone over to the lower state, a *dissociation* takes place (not recombination, as in the emission continua previously considered). The atoms fly apart with kinetic energies that depend on the magnitude of the emitted quantum—that is, on the height above the asymptote of the point reached after the quantum jump. When the lower potential curve is steep, the kinetic energy may be rather large. In such cases the extent of the continuous spectrum is considerable. If different vibrational states in the upper state are excited, the extent of the continuum is still further increased [see Fig. 178(b)].

The most important example of such a continuum is the well-known *continuous spectrum of the* H_2 *molecule*, which appears with great intensity in almost any electrical discharge in H_2 and is used widely as a continuous background for absorption spectra in the ultraviolet. As was first recognized by Winans and Stueckelberg (717), this spectrum corresponds to the transition from the lowest stable triplet state of the H_2 molecule ($1s\sigma2s\sigma\,^3\Sigma_g{}^+$) to the repulsive state ($1s\sigma2p\sigma\,^3\Sigma_u{}^+$) resulting from two normal atoms (see Figs. 160 and 161). The great extent of the continuum (1600 to 5000 Å) is to be explained in the way described above (a very steep potential curve in the lower state, and contributions from several vibrational levels in the upper state). The large intensity of the continuum is due to the fact that the upper $^3\Sigma_g{}^+$ state is a stable molecular state from which the only possible transition in emission is to the unstable $^3\Sigma_u{}^+$ state. Therefore the molecule will stay in the $^3\Sigma_g{}^+$ state until the transition to the repulsive state takes place. Thus the intensity is of the same order of magnitude as that of an intense band system.

The intensity distribution in the H_2 continuum has been studied experimentally by Hukumoto (1077), Chalonge (844), Smith (1429), and more recently by Coolidge (858). Coolidge, James and Present (162)(363) have carried out extensive calculations of the theoretical intensity distribution. Using the accurate electronic eigenfunctions and potential energy curves of the $^3\Sigma_g{}^+$ and $^3\Sigma_u{}^+$ states as calculated by Coolidge and James (859), they found that the electronic transition moment R_e is not constant, but decreases by a factor of about 2.5 in going from $r = 0.7$ to $r = 1.6$ Å. However, using an average R_e, as is usually done in applying the Franck-Condon principle, produces only a shift of 120 Å in the position of the intensity maximum corresponding to $v' = 0$, compared to the position calculated with the accurate R_e. The effect of the variability of R_e is much greater for higher v' values. The observed intensity distribution corresponds to the superposition of the continua corresponding to various v' values. But by using excitation by electrons of a velocity close to the minimum excitation potential of the $v' = 0$ level of the $^3\Sigma_g{}^+$ state Coolidge (858) obtained an intensity distribution approaching that predicted theoretically for transitions from this level only. Moreover, the effect of the higher vibrational levels can be taken into account by investigating the change in intensity distribution with increasing accelerating voltage of the electrons [compare also the earlier work of Finkelnburg and Weizel (936)]. The agreement between theory and experiment is very satisfactory.

Coolidge has also investigated the continuum of D_2 [see also Tournaire and Vassy (1465)] and compared it with the theoretical work of James and Coolidge (363) with equally satisfactory results. Because of the more rapid decrease of the vibrational eigenfunctions of the upper state outside the region of classical vibration (see p. 78), the long- and short-wave-length wings of the continuum of D_2 are relatively weaker than those of H_2.

A very intense continuous spectrum between 600 and 1000 Å has been discovered by Hopfield (330)(1061) in discharges in *helium*. This emission continuum is due to a transition of the He_2 molecule, similar to the one considered above for the H_2 molecule. As we have seen previously, the ground state $(\sigma_g 1s)^2 (\sigma_u 1s)^2\, {}^1\Sigma_g{}^+$ of He_2 which results from normal atoms, is unstable. However, stable excited states of He_2 arise from a normal and an excited He atom. The lowest stable singlet state is $(\sigma_g 1s)^2 (\sigma_u 1s)(\sigma_g 2s)\, {}^1\Sigma_u{}^+$. The transition from this state to the unstable ground state gives rise to the continuum [compare Fig. 178(b)]. Because of the high excitation energy of helium, the spectrum is in the far ultraviolet. [For a more recent investigation of this continuum see Tanaka (1456c)].

Similar continua have been observed for various metallic vapors—for example, for Hg, excited by an electrical discharge or by light. They represent exactly the converse of the absorption continua with stable upper states discussed previously (p. 395 f.). The excited atoms formed in a discharge or by light absorption, may combine during their lifetime with neutral atoms to form stable excited molecules. The latter in going over to the ground state emit the continuous spectrum and thereupon decompose again. The intensity distribution in these emission continua is in general different from that in the corresponding absorption continua. This is because in absorption only the right-hand turning points of the vibrational motion in the upper state lead to intensity maxima [see Fig. 173(b)], while in emission both right- and left-hand turning points do, the latter yielding intensity maxima at much longer wave lengths than the former. In addition, in emission the populations of the vibrational levels of the upper states matter. Thus the same stable excited state of Hg_2 [resulting from $Hg({}^3P_1) + Hg({}^1S_0)$] that is responsible for the absorption continuum joining onto the line 2537 Å accounts for the emission continuum at 3300 Å, which is observed in every mercury lamp.

Another emission continuum of Hg_2, at 4800 Å, is interesting, since no corresponding absorption continuum is observed. It has as upper state the stable molecular state resulting from $Hg({}^3P_0) + Hg({}^1S_0)$, which does not combine with the ground state at large internuclear distances (absorption) but does do so at small internuclear distances (emission), since then the molecular selection rules have to be used. This is a particularly characteristic example of a case in which the electronic transition probability varies with the internuclear distance (see p. 396). For a detailed discussion of the numerous other continua of Hg_2 and similar molecules, see Finkelnburg (31).

Recently a number of continuous emission spectra observed in I_2 and Br_2 have been interpreted by Venkateswarlu (1495)(1496) in a similar way as transitions from stable excited states to the various unstable states arising from normal atoms.

Both upper and lower state continuous. Continuous emission spectra corresponding to a transition between an unstable upper and an unstable lower state can occur when during the collision of excited atoms with normal atoms, a transition takes place to a lower unstable molecular state. This is the exact converse of the earlier case of absorption illustrated by Fig. 173 (*c*). We have again to distinguish between central and non-central collisions. The latter are the more important and result in a continuous spectrum that joins onto the long- or the short-wave-length side of the emission line. These continua were first investigated in detail by Oldenberg (537) in the fluorescence spectrum of Hg vapor to which inert gases at high pressure had been added [see also Kuhn and Oldenberg (436) and Preston (569*a*)]. It is indeed found that the emission spectrum is rather similar to the absorption spectrum under these conditions. Long- and short-wave wings are observed corresponding to the two molecular states arising from $Hg(^3P_1) + X(^1S_0)$ where X is an inert gas atom. In addition diffuse maxima close to the line 2537 Å are observed, some of which are due to central collisions while others, observed with the heavier rare gases, must be ascribed to transitions between the van der Waals minima; that is, they indicate the presence of weakly bound HgA, HgKr, $\cdots$ molecules.

Because of the differences in the van der Waals interaction in the upper and lower states the broadening of the emission lines thus produced, just as that of the absorption lines, is asymmetric and therefore in general a slight shift of the intensity maximum of the atomic line with pressure results. It is probable that this mechanism accounts for most cases of the pressure shift of emission lines [see Kuhn (432)].

2. Diffuse Molecular Spectra, Predissociation, and Related Topics

In this section we shall deal mainly with two characteristic phenomena in band spectra which have already been briefly described in Chapter II—namely, the diffuseness of certain band spectra and the breaking off in the rotational structure of some emission bands. It will appear that these two phenomena have in many cases the same origin: they are due to *predissociation*, or sometimes *pre-ionization*. Some cases of diffuse band spectra, however, are to be explained differently. The study of predissociation has led to many valuable conclusions concerning molecular structure, to an explanation of some photochemical processes, and to a determination of important heats of dissociation (see section 3).

(a) General Discussion of Spontaneous Radiationless Decomposition Processes

The Auger process. According to the previous discussion of perturbations (Chapters I and V), two energy levels of an atomic system which belong to different term series but happen to lie close together appear to influence each other when higher approximations are considered: there is a shift of the two levels in the sense of a repulsion and there is a mixing of the eigenfunctions of the two states, each of the actual levels being, so to speak, a hybrid of the two original nearly coinciding levels.

In general, only one or a few terms of a term series are influenced by such a perturbation (see Chapter V, section 4). If, however, one term in a discrete series has the same energy as a term of a *continuous* term spectrum, all the higher terms of the series have the same energy as correspondingly higher terms of the continuous range, and therefore all the higher terms of this series may be perturbed (see Fig. 179). In such a perturbation by a continuous term the shift of the originally discrete level can assume a continuous series of values. That is, the level becomes *diffuse;* the atom or molecule can assume all energy values in a more or less narrow region (depending on the strength of the perturbation). To the left in Fig. 179 the probability distribution of these energy values is indicated schematically. A spectral line corresponding to a transition to or from such a diffuse level is not sharp but more or less strongly broadened (diffuse).

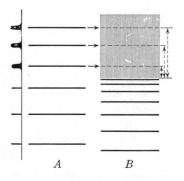

A   B

Fig. 179. **Energy Level Diagram for the Auger Process.** The three uppermost levels of the series *A* are overlapped by the continuum of the series *B*. To the extreme left the width of the levels is indicated schematically. The radiationless transitions from the discrete to the continuous state are indicated by horizontal arrows.

As in the case of ordinary perturbations a mixing of the eigenfunctions takes place here also, so that the true state is a hybrid. Part of the time the system is in the discrete "state," part of the time it is in the continuous "state." However, a continuous state means a splitting of the system and a flying apart of the parts with more or less kinetic energy (the eigenfunction is an outgoing spherical wave). Therefore, when, as a result of the mutual perturbation, the system has once gone from the discrete into the continuous "state," it cannot return to the discrete "state," since the parts are soon widely separated from each other. Thus, *if an atomic system is transferred to such a diffuse state*—for example, by light absorption—*it undergoes a radiationless decomposition after a certain lifetime.* This process was first observed by Auger in the X-ray region and is usually called the *Auger process.* It has to be carefully distinguished from the decomposition (ionization or dissociation) of the system by direct transition into a continuous state: in the case of the diffuse state, as an alternative to decomposition, a transition with emission of light into a lower-lying discrete state may take place; this is not possible in the continuous state.

In a somewhat less accurate but more descriptive way, we may consider the Auger process as a *radiationless transition* (quantum jump) from a discrete into a continuous state [see Kuhn and Martin (437)]. In the following we shall often use this mode of expression.

From the foregoing considerations there appear three criteria for an Auger process:

(1) *Radiationless decomposition* of the system (ionization or dissociation) after a mean life τ_l—that is, with a probability $\gamma = 1/\tau_l$.

(2) *Broadening* of the discrete levels under consideration and correspondingly of the spectral lines which have these levels as upper or lower states.

(3) *Weakening of the emission* from these levels, since only the molecules that do not decompose can radiate.

Let us consider an example: In an atom with a number of electrons the terms corresponding to the excitation of the outermost electron form several Rydberg series, onto which continuous term spectra join, as, for example, to the right in Fig. 179. If two outer electrons (or an inner electron) are excited, Rydberg series of terms lying considerably higher result, which, at least in part, lie above the limit of the first-mentioned series (see Fig. 179, left). Thus, in these states an Auger process—that is, a radiationless transition into the continuous state—and thereby an ionization of the atom can take place. Actually for spectral lines involving such levels, a broadening in emission and in absorption and an abnormally small intensity in emission have been observed for a number of atoms. In these cases the Auger process is also called *pre-ionization*, in analogy to the longer-known process for molecules, or also *auto-ionization* (see A.A., p. 171).

According to Heisenberg's uncertainty relation [equation (I, 35)] the half-width b (in ergs) of a state whose lifetime is τ is given by

$$b = \frac{h}{2\pi} \frac{1}{\tau} \qquad \qquad \text{(VII, 5)}$$

that is, *the greater the mean life of a state, the smaller is its width.* This relation holds quite generally: it determines the natural line width which is equal to the sum of the widths of the upper and lower states and for strong lines is of the order of 0.001 cm^{-1}; it also holds for the broadening produced by an Auger process—that is, the greater the probability $\gamma = 1/\tau_l$ of decomposition, the greater is the broadening of the discrete levels.

The magnitude γ of the *probability of a radiationless transition* depends on the strength of the mutual interaction of the two states, which can be calculated on the basis of their eigenfunctions (see below). This calculation is similar to that of the transition probability for transitions with radiation from the eigenfunctions of the two participating states.

A radiationless decomposition of the Auger type will of course be readily observable only if the transition probability γ for the radiationless quantum jump is not appreciably smaller than the transition probability β for the radiative transition into lower states. If γ is much smaller than β, the system will in general have returned to a stable state long before the radiationless decomposition would have taken place. Conversely, if γ is much larger than β practically no emission is to be expected.

According to Wentzel (1527) the radiationless transition probability is given by

$$\gamma = \frac{4\pi^2}{h} |W_{ni}|^2 \qquad \qquad \text{(VII, 6)}$$

where W_{ni} is the matrix element of the perturbation function and n and i refer to the discrete and the continuous state, respectively.   When the probabilities γ and β are of the same order, the width is given by [see equation (VII, 5)]

$$b = \frac{h}{2\pi}\,(\gamma + \beta) = \frac{h}{2\pi}\left(\frac{1}{\tau_l} + \frac{1}{\tau_r}\right) \tag{VII, 7}$$

where $\tau_r = 1/\beta$ is the mean life with respect to radiation.

The ratio of the number of light quanta emitted to the total number of atoms (molecules) in the diffuse state—that is, the yield of light emission—is obviously given by $\beta/(\beta + \gamma) = 1/[1 + (\gamma/\beta)]$, whereas the yield of the decomposition is given by $\gamma/(\beta + \gamma) = 1/[1 + (\beta/\gamma)]$. It is seen that the former is close to 1 if $\gamma \ll \beta$, whereas the latter is close to 1 if $\beta \ll \gamma$.

If the broadened level is excited by monochromatic light the kinetic energy of the decomposition products is determined by the energy of the incident light quanta which may or may not equal the energy corresponding to the intensity maximum of the broadened absorption line [see Kuhn and Martin (437)].

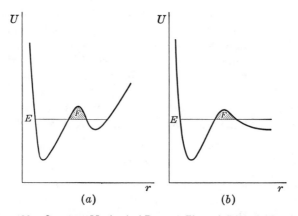

Fig. 180.   Quantum-Mechanical Passage Through Potential Barriers.

Passage through potential barriers.   Consider an oscillator with a potential energy curve of the shape indicated in Fig. 180(a), having a potential hill between two minima.   According to quantum mechanics if the energy of the oscillator (denoted by E in the figure) is smaller than the height of the potential hill but greater than the height of the two minima there is a finite probability that the mass point, when it is initially in the neighborhood of the left minimum, will be, after a time, in the neighborhood of the right minimum, and vice versa. This is due to the fact that the eigenfunction for this oscillator is different from zero in the region to the left as well as in the region to the right.   Thus, quantum-mechanically, a *passage through the potential barrier* takes place, whereas classically the mass point can go from the left into the right trough (and vice versa) only if the energy is greater than that of the maximum.   This passage through potential barriers is also referred to as *"tunnel-effect"* or *"tunneling."*

If the potential curve has a shape as in Fig. 180(b), approaching an asymptote below E at the right, the same phenomenon occurs, except that now a passage through the potential barrier from left to right means that the mass point will fly off to infinity; that is, a *radiationless decomposition* of the system results. This process is in many respects similar to the Auger effect: if we consider only the left-hand part of the potential curve we have a series of discrete energy levels, if we consider only the right-hand part we have a continuum; that is, as for the Auger process we have an overlapping of a discrete by a continuous range of energy levels. However, the passage through a potential barrier differs from an Auger process in that the former is a one-body process (the potential curve applies to a single mass point), while the latter is a many-body process (the interaction of several particles is necessary for it to occur). A broadening of the discrete levels arises on account of the passage through the potential barrier in the same way as on account of the Auger process, the width being inversely proportional to the mean life [equation (VII, 5)]. It is found that *the smaller the area of the slice F of the potential hill cut off by the line representing the energy level, and the greater the frequency of the vibration, the shorter is the mean life* and therefore the greater is the diffuseness of the levels.

The formula for the mean life is found to be [see for example Rice (1346)]

$$\tau = \tfrac{1}{2}\tau_0\, e^{(4\pi/h)\int \sqrt{2m(U-E)}\,dr} \tag{VII, 8}$$

where τ_0 is the period of oscillation and m the mass of the oscillator. The integration in the exponent is over the part of the potential hill cut off by the energy level E [Fig. 180 (b)]. The reciprocal of the exponential factor in (VII, 8) is the *transmission coefficient* which indicates the probability of passage through the barrier per vibration. It applies also for the inverse process in which the mass point hits the potential barrier from outside [from the right in Fig. 180 (b)], and has a certain probability of penetrating into the potential trough.

The phenomenon of going through a potential hill has proved to be of great importance in the explanation of many physical processes (for example, radioactivity, emission of electrons from metals, and so forth) and will also be important in the following discussion.

(b) Radiationless Decomposition Processes in the Molecule

Bonhoeffer and Farkas (116) and Kronig (421), on the basis of certain considerations put forward by Born and Franck (120) and by Polanyi and Wigner (566), were the first to realize that in the case of many diffuse molecular spectra the Auger effect is responsible for the diffuseness.

In a molecule the overlapping of discrete energy levels by a continuous range of levels, necessary for the occurrence of this process, is very often present. A *pre-ionization* is possible, exactly as for atoms (see p. 407), when discrete electronic states are overlapped by a continuous range of states that correspond to the separation of an electron (ionization). However, for molecules, we have in addition the much more frequent case of overlapping by one of the continuous ranges of levels that correspond to a dissociation into atoms (or ions) and that occur for every electronic state (compare Fig. 181) irrespective of whether it is a stable or an unstable state. The *possibility of going over without radiation from a discrete state into such a dissociated state* is, according to Bonhoeffer, Farkas, and

Kronig, the reason for the diffuseness of the bands. In Fig. 181, for example, the vibrational levels of the upper state B from $v = 4$ on are overlapped by the continuum of the lower state. As in Fig. 179, the system can go over, without radiating, from the discrete state into the continuous state lying at the same height, except that here the continuous state corresponds to a dissociation; that is, the molecule dissociates after the radiationless transition. This process is called *predissociation*.

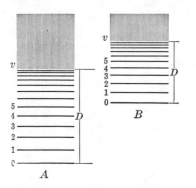

In the following paragraphs the correctness of this explanation of many cases of diffuse spectra will be proved with the help of the criteria previously given for the occurrence of an Auger process (p. 407).

Fig. 181. **Two Electronic States of a Molecule (with their Continuous Term Spectra) for which Predissociation is Possible.**

Diffuseness of the bands. The second criterion given on p. 407 —namely, the broadening of the lines—was the one by which the diffuse bands were first noticed [V. Henri (299)]. Kronig has, in fact, shown by a quantum-mechanical calculation that under certain conditions (see below) for a discrete state overlapped by a dissociation continuum the mean lifetime τ_l, until radiationless decomposition takes place, may be less than the period of rotation of the molecule ($\sim10^{-11}$ sec). The line widths of the corresponding spectral lines are then of the order of magnitude of the separation of successive lines in the fine structure. Thus in this case the band should have just the diffuse appearance observed experimentally (see Fig. 16, p. 38). However, in agreement with observations, the vibrational structure is in general not influenced by the possibility of a radiationless decomposition since the vibrational frequencies are ten to a hundred times greater than the rotational frequencies (see p. 82).

The diffuseness of the bands in absorption is itself a sufficient proof of the presence of predissociation if the bands are preceded by sharp bands in the same progression. Such a case was first observed for S_2 by Henri and Teves (301) (see Fig. 16). Another even more striking example was recently found by Froslie and Winans (946) in InCl.

Naturally, the radiationless transition probability need not always be so great that the bands are completely diffuse. Sometimes there is only a slight *broadening of the individual lines*. Both a lack of sharpness of the individual rotational lines and, for higher v' values, a complete washing out of the rotational structure have been observed in the S_2 absorption spectrum (see Fig. 16). The line shape of a line made diffuse by predissociation has been discussed by Rice (1346) and Stepanov (1437).

It may also happen that, although the transition probability for the radiationless decomposition is not zero, it is yet so small that the line broadening cannot be detected directly because of insufficient resolving power of the spectral apparatus used. In spite of this, the presence of a broadening and thereby of predissociation in such a case can be ascertained *indirectly* if the line width caused by the radiationless transition is greater than that caused by the Doppler effect of the thermal motion of the molecules. In a spectral apparatus of not too great resolving power, the sharper an absorption line is, the weaker it appears, since the parts of the continuous background to either side of the absorption line are less and less well separated. Therefore, if from a certain point on in a series of lines or bands there is a slight (not directly detectable) "broadening" it may be detected by the fact that the *broadened lines*

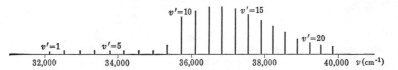

Fig. 182. **Anomalous Intensity Distribution Due to Predissociation in a Series of Absorption Bands, Assuming Small Resolving Power of the Spectrograph.** The figure refers to the case of the S_2 spectrum. The heights of the vertical lines give the estimated intensities of the bands. The jump in intensity occurs at $v' = 10$.

apparently have a greater intensity than the non-broadened lines. The intensity distribution in a band progression with $v'' = 0$ then has the anomalous appearance indicated schematically in Fig. 182, which should be compared with the normal distribution in Fig. 87, p. 194. This phenomenon has been used for the detection of predissociation in H_2 [Beutler, Deubner, and Jünger (92)] and in S_2 [Herzberg and Mundie (1050)]. In Fig. 16 it can be seen that, for S_2, the bands preceding the 10–0 band are all fairly weak whereas the 10–0 band itself and the following bands have considerable intensities (see also Fig. 182),[5a] although the lines of the 10–0 band (unlike those of the bands with $v' > 10$) do not appear to be broadened. However, from this intensity distribution we must conclude that there is already a small broadening present for it; that is, that the predissociation starts with $v' = 10$ not $v' = 11$. Subsequent investigations with very large dispersion by Olsson (547) have indeed shown that the lines of the 10–0 band are very slightly broadened, although they had always been considered as sharp before Olsson's investigation.

It may be noted that even with very high resolving power an anomalous intensity distribution will arise in a band progression with weak predissociation if there is complete absorption at the line centers. This is because the broadened lines will then have a broader region of complete absorption than the sharp lines. In other words, the former have a greater *"equivalent width"* than the latter where the equivalent width is the width (usually in Å) of a region of complete absorption which absorbs as much light as the line considered.

Photochemical decomposition. That a decomposition really does take place (first criterion on p. 407) when the molecule is brought by light absorption into the upper state of the diffuse bands was first shown by Bonhoeffer and Farkas (116). They found that a photochemical decomposition of NH_3 takes place on illumination with light of the wave lengths of the diffuse absorption bands. This decomposition (formation of H_2 and N_2) takes place even at very

[5a] This effect was much more pronounced in Fig. 16 of the first edition which was printed from a spectrogram of much lower dispersion (but which did not show the diffuseness of the predissociated bands as clearly as the new Fig. 16).

low pressures (0.001 mm), at which secondary processes cannot play any part. Moreover, the yield of this photochemical reaction (number of molecules decomposed per absorbed light quantum) was found to be independent of pressure. The initial step of the reaction must therefore be a *primary spontaneous decomposition* which is independent of collisions with other molecules (see also p. 478 f.). Another case in which the independence of pressure has been carefully established is that of NO [Flory and Johnston (225)(937)] which is found to decompose by radiation with light of the 1–0 band of the δ system (see Fig. 198). However, in this case a diffuseness has not been observed, possibly because of lack of resolving power of the spectrograph used (see, however, below).

Hipple, Fox, and Condon (1055) have detected by means of a mass spectrograph the spontaneous decomposition of molecular ions produced by electron impact. In order that the decomposition becomes noticeable, under these conditions, it is necessary that the lifetime is so long that the ions are accelerated in the mass spectrograph before decomposition; that is, only weak predissociations of metastable states can be detected in this way. No diatomic examples of this type have as yet been found.

Breaking-off of bands. If the explanation of the diffuse absorption bands by the process of predissociation (Auger process) is correct, the corresponding bands in emission must either be very weak or be completely missing (third criterion on p. 407). Indeed, for S_2 [van Iddekinge (352)], only those bands appear in emission that correspond to the *sharp* absorption bands, whereas the diffuse bands of the same band system (or other bands with the same upper levels) can in no way be obtained in emission either in electric discharges or in fluorescence. Thus *in emission the band system breaks off at a definite value of the vibrational quantum number v'*.

Conversely, we are led to conclude that also those cases of breaking-off for which an observation in absorption is not feasible are to be explained as predissociation. This explanation holds particularly for the characteristic cases in which the bands in emission *break off sharply at a certain value of the rotational quantum number* (see p. 51 f.). In one of these cases, that of AlH (see Fig. 28), it has actually been observed that the higher rotational lines that do not appear in emission do appear in absorption but are diffuse [Farkas (212)]. The correctness of this explanation of the breaking-off is further confirmed by the fact that for isotopic molecules—for example, CaH and CaD—the breaking-off occurs at different J values or even different v values but at very nearly equal energies above the minimum of the potential curve of the ground state, corresponding to the fact that the dissociation limits for isotopic molecules are the same.

Thus in general we can regard a predissociation as established either by the absorption bands becoming diffuse, or by the emission bands breaking off at a certain point in a series.

A diffuseness of the individual rotational lines is of course noticeable only when the line width is greater than the width produced by the Doppler effect due to the motion of the individual molecules—that is, greater than 0.01 to

0.1 Å.   On the other hand, the natural line width, if no decomposition is possible, is of the order of 0.001 to 0.01 Å, corresponding to a mean life of 10^{-8} sec (see p. 407).   Therefore, in order that the line width shall be greater than the Doppler width, the decomposition probability γ must be at least 10 to 100 times greater than the transition probability β for radiation.   Consequently if the band lines are observed to be diffuse *in absorption* the intensity of the corresponding transitions *in emission* is vanishingly small compared to the intensity of the transitions from the nonpredissociated levels.

On the other hand, a weakening of the intensity in a series of lines or bands is easily detectable when the intensity falls off suddenly to only, shall we say, one half.   In this case, even with very high resolution, no diffuseness would be detectable in absorption, since it would be hidden by the Doppler width.   The *breaking-off of emission bands* is thus a *much more sensitive criterion for the presence of a predissociation*.   A still more sensitive criterion is the occurrence of a photochemical decomposition, since under suitable conditions this can be established even when only a small fraction of the molecules brought into the excited state decomposes and when, therefore, there is no noticeable weakening in the intensity of the bands under consideration.   An example may be the case of NO mentioned above.   In no case can we conclude with certainty from the absence of a diffuseness in the absorption bands that there is no predissociation. This point is often overlooked.

It is hardly necessary to emphasize that the breaking-off is, of course, always observed at the same J' value for different bands with the same v'.   Moreover, if the predissociation occurs in the lower state of emission bands, no breaking-off occurs, but the bands become diffuse similar to absorption bands with predissociating upper states.

The above-mentioned indirect method of detecting the line broadening by the apparent strengthening of the absorption is not as sensitive a test for predissociation as the breaking-off of emission bands or as photochemical decomposition, although it is more sensitive than the directly visible broadening.

Different types of predissociation.   Corresponding to the three forms of energy of the molecule (electronic, vibrational, or rotational), three kinds of overlapping of molecular energy levels by a dissociation continuum—that is, three cases of predissociation—are possible:

I. Overlapping of a certain *electronic* state (that is, of its vibrational or rotational levels) by the dissociation continuum belonging to another electronic state—*radiationless transition into this other dissociated electronic state* (compare Fig. 181).

II. Overlapping of the higher *vibrational* levels of an electronic state of a polyatomic molecule by a dissociation continuum joining onto a lower dissociation limit of the same electronic state—radiationless splitting off of the particular atom or group of atoms (*predissociation by vibration*).

III.   Overlapping of the higher *rotational* levels of a given vibrational level of a diatomic molecule by the dissociation continuum belonging to the same electronic state—radiationless decomposition of the molecule with no change of electronic state (*predissociation by rotation*).

For diatomic molecules, case I is the most important. It always applies if the bands become diffuse or break off at a large distance from the point of convergence of the band system. This is the case for example for the diffuse S_2 bands as shown in Fig. 16, and for the CaH bands that break off in emission at $v' = 0$ (Fig. 27) but for which bands with high v' values have been found by Grundström (268). Case II applies to polyatomic molecules only.[6] We shall not discuss it further here. It will be considered in more detail in Volume III. Case III can occur for those vibrational levels of an electronic state that lie in the neighborhood of the dissociation limit, since the higher discrete rotational levels of such vibrational levels can lie above the dissociation limit. This case is most readily observed when the dissociation energy of the electronic state is small, as for example, for AlH (see Fig. 28) and HgH. Sometimes it is not possible to distinguish unambiguously between cases I and III.

From the foregoing discussion it is quite clear that the place in the energy level diagram of a molecule at which a predissociation begins (*predissociation limit*) gives at least an *upper limiting value for the corresponding dissociation limit*. However, we shall see that under certain circumstances the predissociation limit may lie appreciably higher than the dissociation limit belonging to it.

Pre-ionization. The same phenomena that are produced by predissociation—that is, diffuseness of the absorption bands and breaking-off of the emission bands—can also be brought about by *pre-ionization* of the molecule. As a condition for this, the upper state of the bands must, of course, lie higher than an ionization limit. In *absorption*, Henning (298) found diffuse bands of CO in the far ultraviolet at 785 to 750 Å whose diffuseness is very probably due to such a pre-ionization. Furthermore, Beutler and Jünger (94) have found diffuse bands in the far ultraviolet absorption spectrum of H_2 which they could ascribe with certainty to such a pre-ionization process and from whose long-wave-length limit they were able to determine a very reliable and accurate value for the ionization potential of the H_2 molecule. Breaking-off points in *emission* that are to be ascribed to pre-ionization have been found for the higher electronic states of H_2 by Beutler and Jünger (95).

Inverse predissociation. All cases of predissociation may also occur in reverse; that is, molecule formation by two-body collisions may result from inverse

---

[6] It may be mentioned that an important group of nuclear processes is of exactly the same type as case II of predissociation: When a nucleus is brought into an excited state it may instead of returning to the ground state with emission of γ radiation split off a neutron or other particle if the excitation energy is higher than the separation energy of the particle to be ejected. The probability of this process contributes to the width of the excited energy level in the same way as for predissociation.

predissociation.   The two atoms approach each other and during the collision a radiationless transition from the continuous range of energy levels to one of the discrete levels (see Fig. 179) may occur if the energy coincides with that of one of these levels within their width.   The molecule thus formed will predissociate again unless it loses its energy during the lifetime by emission of radiation.   The frequency of occurrence of the formation of a stable molecule in this way is of the same order as for the ordinary two-body recombinations discussed previously (p. 400) since the reduction of the number of favorable collisions by the energy condition is just about compensated by the increase in the effective collision time [remembering the relation (VII, 5) between level width and lifetime].

An example of inverse predissociation has been found by Stenvinkel (1435) for AlH.   He observed, under conditions that strongly favored the formation of AlH molecules in two-body collisions, the emission of just those band lines that are missing in ordinary excitation on account of predissociation.   A selective emission of the $v' = 6$ Swan bands of C_2 when formation of C_2 molecules is indicated, has been tentatively explained by Herzberg (1042) as due to inverse predissociation although in this case the direct predissociation has not yet been established.

Accidental predissociation. Ittmann (358) has pointed out that a certain type of perturbations in band spectra can be explained by what he calls *accidental predissociation* [see also Kovács and Budó (1139)].   If a stable electronic state is perturbed by a diffuse (predissociating) state [compare Fig. 135(a), in which we have to assume in the present case that the term series to the right is diffuse], one or more of the rotational levels of the perturbed state assume the properties of the corresponding predissociating levels; that is, they also predissociate.   Therefore, in emission, the corresponding lines of the bands under consideration should be *missing* or have *abnormally small intensities*.   At the same time their positions may not differ noticeably from the normal positions.   Perturbations of this type have been observed, for example, in the second positive nitrogen group—an abnormally small intensity of one or two lines with certain J values in all the bands with the same v' [Coster, Brons, and van der Ziel (167)].   In absorption an accidental predissociation of the upper state should give rise to a broadening of a few successive lines in the band considered or possibly, if the broadening is slight, to an abnormally high intensity of these lines (see p. 411).   Thus far no such case has been observed.   It will be remembered that the diffuseness in absorption is a much less sensitive test for predissociation than the weakening of the emission lines.

Experimentally accidental predissociation may be distinguished from normal predissociation by establishing whether only a few lines or a large number of lines (or all lines beyond a certain limit) have abnormally low intensity.[7]   Theoretically the difference between the two phenomena is that for normal predissociation a radiationless transition takes place from a discrete level directly into the dissociated state, whereas for accidental predissociation this happens only after a detour through a third state.   It is clear that an accidental predissociation can be observed only when, on the basis of the selection rules (see the following subsection), a direct radiationless transition into the dissociated state is not possible.

It is readily seen from these considerations that accidental predissociation is not a very

---

[7] If the perturbation leading to accidental predissociation is of the vibrational type one or two bands in a v'- progression will show an anomalously small intensity.   Such a case has been observed for Te_2 by Rosen (1357).

frequent process; it is much less likely to occur than ordinary predissociation and ordinary perturbations.

It must be emphasized that there is another cause for intensity anomalies in band spectra which has nothing to do with predissociation but is due to the effect of conditions of excitation and pressure on the intensities of perturbed lines (see p. 291 f.). If the bands considered can only be studied in emission it is very difficult to distinguish the two causes of abnormal weakness of a few successive lines in a band and therefore to establish the reality of a suspected accidental predissociation. In fact in no case has this as yet been done unambiguously. However, in absorption a distinction of the two causes is readily possible if it can be ascertained whether the lines in question are broadened. No weakening of the lines will occur in absorption in either case.

(c) Selection Rules for Predissociation

As far as the energy is concerned, the possibility of predissociation exists for all the discrete states of a molecule that lie above its lowest dissociation limit (corresponding to dissociation into normal atoms). In spite of that, predissociation is, at least for diatomic molecules, a comparatively rare phenomenon. This is due to the fact that the probability (γ) of the radiationless transition into the dissociating state is usually so small that long before the decomposition would have taken place the molecule has already gone over into a lower-lying stable state with emission of radiation ($\gamma < \beta$). In order for the transition probability to be so large that the bands can be observed to become diffuse or to break off, still further conditions—*selection rules*—must be fulfilled in addition to the condition of equal energy.

Kronig's selection rules. Since predissociation is nothing but a special case of perturbations exactly the same selection rules hold as for perturbations [Kronig (421); see Chapter V, section 4]. For the two participating states we must have for any coupling conditions

$$\Delta J = 0, \qquad + \leftrightarrow -, \qquad s \leftrightarrow a, \qquad (VII, 9)$$

and in Hund's cases (a) and (b), in addition,

$$\Delta S = 0 \quad \text{and} \quad \Delta \Lambda = 0, \pm 1 \qquad (VII, 10)$$

If both states belong to case (a) or both to case (b), we have respectively

$$\Delta \Sigma = 0 \quad \text{or} \quad \Delta K = 0 \qquad (VII, 11)$$

In Hund's case (c) (VII, 10) must be replaced by

$$\Delta \Omega = 0, \pm 1 \qquad (VII, 12)$$

The validity of the selection rules (VII, 9) for predissociation implies of course that the *angular momentum J and the symmetry properties are defined in a continuous range of energy levels.* This is indeed the case, since the directional dependence of the Schrödinger equation can always be separated off (see p. 69) and leads to the properties J, $+$ or $-$, s or a, whether the radial part of the eigenfunction corresponds to a stable molecular state or to one in which the atoms fly apart or approach each other with a certain kinetic energy. In other words, a quantized rotation exists even when the atoms fly apart and conversely non-central collisions of the two atoms can only occur with certain values of the angular momentum J and with certain symmetry properties. Thus, for Σ states in the continuous as well as in the discrete region, the rotational levels are alternately positive and negative and, for identical nuclei, alternately symmetric and antisymmetric in the nuclei. For Π, Δ, $\cdots$ states there is a positive and a negative level for each J, which for identical nuclei are symmetric and anti-symmetric, or vice versa [see Chapter V, section 2(c)].

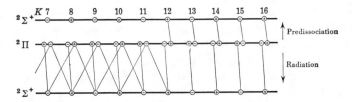

FIG. 183.   **Predissociation of MgH in the $B\,^2\Pi$—$X\,^2\Sigma$ band at 2430 Å**   The figure shows only the symmetry properties and K values.   The level $K = 12$ of the $^2\Pi$ state lies just above the predissociation limit.

For a large separation of the atoms, states with different J have practically the same energy (very large moment of inertia), or, conversely, all J values are possible for each energy value in the continuum. Therefore, if the energy condition is fulfilled, the selection rule $\Delta J = 0$ is always realizable. It is, however, not always compatible with the other selection rules. The fact that the possibilities for predissociation are very considerably restricted by the Kronig selection rules is best illustrated by a few examples.

If a stable Σ^+ state is overlapped by an unstable, continuous Σ^- state, predissociation cannot take place, owing to the rule $+ \leftrightarrow -$, even if all the other selection rules are fulfilled, since for rotational levels with equal J and K (necessary because of $\Delta J = 0$ and $\Delta K = 0$) the $+ -$ symmetries are always opposite (see Fig. 114, p. 238).

If a discrete Π state is overlapped by the continuum of a Σ state, only the molecules in the one Λ component of the Π state can decompose, since only for the one component are $\Delta J = 0$ (or $\Delta K = 0$) and $+ \leftrightarrow -$ simultaneously fulfilled (see Figs. 114 and 115). As a consequence, for example in a Π—Σ transition when the Π state is overlapped above a certain energy by the continuum of a Σ state, a breaking-off can appear either only in the P and R branches or only in the Q branch [Kronig (422)]. Such a case has been found for MgH by Pearse (556). In the 0–0 band of the $B\,^2\Pi$—$X\,^2\Sigma$ system at 2430 Å, the P and R branches break off at $K' = 11$ (that is, $J' = 10\frac{1}{2}$ and $11\frac{1}{2}$), while the Q branch does not break off. This case is represented schematically in Fig. 183. Since all three states belong to Hund's case (b), the quantum number K (rather than J) is relevant for the spectrum as well as for the predissociation. Bands with higher

v' have also been found [Turner and Harris (673a)].   In agreement with expectation they have only Q branches.   Another example of this type are the $b'\,^1\Pi_u$—$X\,^1\Sigma_g^+$ bands of N_2 which in emission, according to Watson and Koontz (697), consist of Q branches only.

Conversely, from the observation that in a Π state predissociation takes place for only one Λ-doublet component, we can conclude with certainty that the state causing the predissociation is a Σ state, whereas, if both Λ components break off, the state causing the predissociation must be a Π or Δ state.   Such conclusions are often of great importance in the interpretation of dissociation processes (see section 3).

Owing to the selection rule $s \leftrightarrow a$ the possibilities for predissociation are still further restricted for molecules with identical nuclei.   It can easily be seen (compare Figs. 114 and 115) that, according to this rule, a Σ_g^+ state, for example, can predissociate only into a Σ_g^+ or Π_g state but not into Σ_u^+, Σ_g^-, Σ_u^-, or Π_u; a Π_g state can predissociate only into Σ_g, Π_g, or Δ_g; a Π_u state only into Σ_u, Π_u, or Δ_u; and so on.   In addition, of course, the multiplicities of the two combining states must be the same.

From the preceding discussion it is readily seen that the same electronic transitions can occur without radiation (that is, as predissociation) as with radiation except that the rule for g and u is reversed ($g \leftrightarrow u$ for predissociation). Except for this reversal Table 24 (p. 243) giving allowed electronic transitions may also be applied to predissociation.   Taking account of this reversal the following rule [Herzberg (309)] is easily verified: *For molecules with identical nuclei, states that can combine with one another with emission or absorption of radiation cannot predissociate into one another, and vice versa.*   From this rule we can immediately see why, for example, the upper state of the visible I_2 bands does not predissociate into the continuum joining onto the ground state, although it has an energy that is greater than that of the separated normal atoms.

According to (VII, 6) the radiationless transition probability (γ) is determined by the same type of matrix element W_{ni} as the magnitude of perturbations (see p. 284 f.).   For this reason, as has been shown in more detail by Kronig (421), the same dependence on J arises: for $\Delta\Lambda = 0$ the transition probability is in a first approximation independent of J, whereas for $\Delta\Lambda = \pm 1$ it is approximately proportional to J and vanishes for $J = 0$.   In the first case we have therefore to expect in absorption a fairly uniform broadening of all the lines of a given band beyond the predissociation limit, and in the second case a broadening increasing linearly with increasing J.   This expectation is borne out by the observations, as far as they permit conclusions about the probability of the radiationless transitions.   A particularly clear case of a steady increase of broadening with increasing J, that is, a case with $\Delta\Lambda = \pm 1$ has been observed for TlH [Grundström and Valberg (273) and Herzberg and Mundie (1050)].   It should be noted that the radiationless transition probability depends in addition on the finer details of the electronic eigenfunctions of the states concerned, that is, on the electron configurations.   Thus two neighboring electronic states, which according to the selection rules as well as the Franck-Condon principle (see the next subsection), can predissociate into one and the same unstable electronic state, may still have very different lifetimes before decomposition.   This has not always been taken into account in the literature on the subject.

Forbidden predissociations. While the selection rules (VII, 9) hold rigorously, the validity of the rules (VII, 10–12) depends on the coupling conditions, and predissociations forbidden by them may occur with a small probability. Particularly the rule $\Delta S = 0$, which applies to both Hund's case (a) and (b), holds only to the same degree as it does for radiative transitions. Since a strong allowed predissociation has a probability $\gamma = 10^{11}$ sec^{-1} ($\tau_l = 10^{-11}$ sec, see p. 410) a predissociation with $\Delta S \neq 0$ (intercombination) may still have a probability of the order 10^8 sec^{-1} except for the lightest molecules. Such a predissociation probability may still compete with the probability of radiative transitions to a lower state, which is $\leq 10^8$ sec^{-1}, and may be sufficient to cause a breaking-off or at least a drop in intensity in emission bands even if it is not sufficient to cause a broadening in absorption bands.

The observed breaking-off in the ultraviolet P_2 bands [Herzberg (311)] can only be explained as such a forbidden predissociation (singlet-triplet intercombination). The same applies very probably to the breaking-off observed in the Ångstrom bands of CO [Coster and Brons (165)]. Another interesting example is the anomalous intensity distribution found by Schüler, Gollnow, and Haber (626a) in the $^1\Sigma^+$—$^1\Sigma^+$ CuH band at 4280 Å (see Fig. 19) when excited at very low pressure. Under these conditions the line $P(1)$, the only one with $J' = 0$, has predominating intensity, the intensity falling off rapidly for higher J' values. According to Herzberg and Mundie (1050) this anomalous intensity distribution is due to a forbidden predissociation of the upper $^1\Sigma^+$ state of the band into the $^3\Sigma^+$ state resulting from normal atoms. In such a predissociation the $J' = 0$ level cannot predissociate (since the level with $J = 0$ in $^3\Sigma^+$ has a parity opposite to that in $^1\Sigma^+$), but the levels with $J' > 0$ can predissociate and the predissociation probability increases with increasing J' (compare the behavior of perturbations with $\Delta\Lambda = \pm 1$, p. 288). The pressure dependence of this predissociation will be discussed in subsection 2(e).

Another type of forbidden predissociation is one produced by a magnetic field. It has been shown by Turner (670) that the observed quenching of iodine fluorescence by a magnetic field is due to such a mechanism. The theory of this phenomenon has been given by Van Vleck (680). He showed that the selection rule $\Delta J = 0$ no longer holds strictly in a magnetic field, and that, as a result, the upper state of the I_2 bands can go over into one of the states leading to normal atoms, into which it could not go in the absence of a field when the rule $\Delta J = 0$ holds strictly [see also Genard (247)(248), Scholz (625), Smoluchowski (639), and Rouppert (604a)].

Pre-ionization. Kronig's selection rules apply also to pre-ionization. For the quantum numbers and symmetry properties in the continuous range of energy levels one must use those referring to the system: *molecular ion plus electron*. Therefore the J and K values of the molecular ion (J_{ion}, K_{ion}) are not uniquely determined by the values in the neutral pre-ionizing molecule ($J_{mol.}$, $K_{mol.}$). Rather, all those values of the angular momentum of the ion are possible which, added vectorially to that of the electron, give the angular momentum of the pre-ionizing molecule. Neglecting electron spin and assuming that the electron which is ejected has an orbital angular momentum l we obtain from $\Delta K = 0$

$$K_{mol.} = K_{ion}, K_{ion} \pm 1, \cdots, K_{ion} \pm l \qquad (VII, 13)$$

In the case of identical nuclei the combined state of ion and electron must have the same (g, u) symmetry as the molecule; in addition, as is readily seen by considering the case of zero nuclear spin, the molecular ion and the neutral molecule from which it arises must have the same symmetry in the nuclei (both s, or both a). This leads to the elimination of the even or the odd $K_{mol.} - K_{ion}$ values of (VII, 13) depending on the type of the states involved.

The only case for which pre-ionization has been studied in detail and for which the selection rules have been tested is the H_2 molecule [Beutler and Jünger (94)(95)]. It will be realized that there are as many pre-ionization limits as there are rotational levels in the ion,

and above each of these limits according to (VII, 13) only certain states of the neutral molecule can pre-ionize. This makes possible a more accurate determination of the ionization
potential than would be possible from a single pre-ionization limit.

(d) The Franck-Condon Principle in Predissociation

Case I of predissociation. Although the Kronig selection rules restrict
the possibilities of predissociation very considerably, they are not sufficient to
exclude the theoretical possibility of its occurrence in all cases in which pre-

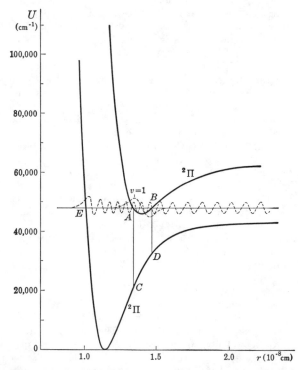

FIG. 184. **Potential Curves of the $^2\Pi$ Ground State and the First Excited $^2\Pi$ State of NO.**
AB is the vibrational level $v = 1$ of the upper $^2\Pi$ state. The broken-line curves are the eigenfunctions for this vibrational level and for the continuous level of the lower $^2\Pi$ state having the same
energy (only qualitatively drawn).

dissociation is not observed. Let us consider the NO molecule as an example.
Fig. 184 shows the potential curves of two $^2\Pi$ states of this molecule, the ground
state and the upper state of the so-called β bands which lies 45,500 cm^{-1} above
the ground state. Since the heat of dissociation for the ground state amounts
to only 42,730 cm^{-1}, a predissociation of the upper $^2\Pi$ state into the lower is
energetically possible for all vibrational and rotational levels of the upper state.
Moreover, such a predissociation is allowed by the Kronig selection rules (see
above). However, in this, as in many other similar cases, no predissociation is

observed.   The reason for this is that, apart from the Kronig selection rules, *the Franck-Condon principle has to be taken into account for radiationless as well as for radiative transitions* [Franck and Sponer (231) and Herzberg (307)].   From the semi-classical point of view which served as the original basis of the formulation of the Franck-Condon principle it is quite clear that, for a radiationless transition also, the position and velocity of the nuclei cannot alter appreciably at the instant of the transition.

If, in the example of NO (Fig. 184), a radiationless transition to the lower state were to take place during the vibrational motion AB in the upper state, the amount of energy AC or BD or some intermediate amount would have to be instantaneously converted into kinetic energy, or the internuclear distance would have to alter instantaneously by the amount AE or BE or some intermediate amount.   According to the Franck-Condon principle, this is impossible (see p. 194 f.).   Therefore, in this case, the radiationless transition cannot take place—in agreement with observation.

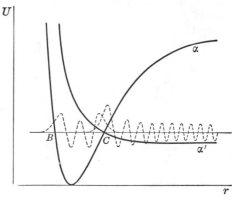

On the other hand, a radiationless transition is possible, even taking the Franck-Condon principle into account, *if the potential curves of the participating states intersect,* as in Fig. 185, or if at least they come very close to each other. When the molecule in the state α is in the neighborhood of the point of intersection C, obviously a transition to α' is possible without an ap-

FIG. 185. **Potential Curves Explaining the Action of the Franck-Condon Principle in Predissociation.** The broken-line curves represent the eigenfunctions.

preciable alteration of position and momentum, and thus a decomposition of the molecule may take place.   Naturally, the transition does not take place immediately the molecule is in the neighborhood of the point of intersection, but only with a certain probability, which depends on the types of the electronic states.   The molecule will in general carry out a number of vibrations in the stable state BC before it jumps over to the unstable state while traversing the point of intersection of the two potential curves.

On the basis of wave mechanics the proof that the Franck-Condon principle holds for predissociation proceeds in much the same way as for perturbations (see p. 286).   From the matrix element W_{ni}, which according to (VII, 6) determines the radiationless transition probability γ, one may split off the factor

$$W_{ni}{}^v = \int \psi_n{}^v W^v \psi_i{}^v dr \qquad\qquad \text{(VII, 14)}$$

which depends on the vibrational motion only.   Since W^v, the part of the perturbation function depending on the internuclear distance r, varies but slowly with r, the integral is essentially the *overlap integral* of the two vibrational eigenfunctions. In Fig. 184 are drawn schematically the vibrational eigenfunctions for a discrete level of the upper state ($v = 1$) and a continuous level of the lower state having the same energy. It is seen that, owing to the rapid oscillation of the repulsive eigenfunction $\psi_i{}^v$ in the region where $\psi_n{}^v$ is different from zero, the overlap integral will be vanishingly small, and therefore a predissociation is not to be expected. On the other hand, if the potential curves intersect (Fig. 185), the eigenfunctions for levels in the neighborhood of the point of intersection lie in such a way with respect to each other that the overlap integral has a considerable magnitude. Thus wave mechanics leads to essentially the same result as the elementary form of the Franck-Condon principle but the regions in which transitions are possible are somewhat broadened.

Three subdivisions of case I of predissociation may be distinguished according to whether the point of intersection of the two potential curves lies at about the height of the asymptote of the state giving rise to the predissociation or

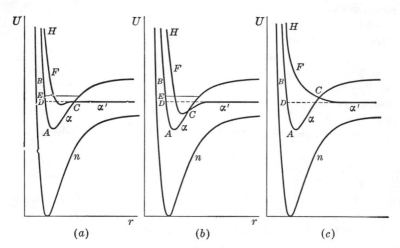

Fig. 186.   **Three Subcases of Case I of Predissociation.**   In each case, the broken line gives the position of the asymptote of the potential curve α'.

whether it lies below or above this asymptote. These subcases are represented by Fig. 186(a), (b), and (c). The potential curve n corresponds in each case to the normal state of the molecule, from which, for example, a transition to the excited state α is observed in absorption. α' is the potential curve of a third electronic state with a lower dissociation limit than α.

Let us consider first the absorption spectrum in the case of the potential curves, Fig. 186(a) and (b). On the basis of the Franck-Condon principle, as a result of light absorption, transitions take place from the vibrationless ground state n to the region A–B of the potential curve α and to the region F–H of the potential curve α'. Thus for the relative position of the potential curves drawn, in absorption the transition to α', if it is allowed at all, will be at much shorter wave lengths than that to α. The transition to the region F–H of α' gives a continuous spectrum, while the transition to α gives discrete bands which have quite a normal appearance as long as the upper vibrational level lies below the asymptote of α'—that is, below D. If a vibrational level of α somewhat above D is reached by light absorption, a radiationless transition into the state α' is energetically possible (while, of course, a direct transition to α' from the ground state, in the same wave-length region, is not possible). Moreover, owing to the intersection of the two potential curves, the predissociation is possible here without violating the Franck-Condon principle, so that, if the Kronig selection rules are also fulfilled, and the electronic part of the transition probability γ is not anomalously small, the bands will be observed to become diffuse in absorption and to break off in emission. The energy corresponding to the beginning of the diffuseness or in emission to the breaking-off point, is called the *predissociation limit*. In the subcases (a) and (b) considered here, the predissociation limit coincides with the *dissociation limit* of the state α' which causes the predissociation, that is, it has the energy of the asymptote of the potential curve α'.

The higher the level E lies above the limit D, the greater is the kinetic energy with which the atoms fly apart after the radiationless transition. At the same time, the velocity with which the point of intersection is traversed during the vibration in α becomes greater and greater, and therefore the probability of the radiationless transition becomes smaller and smaller. We should therefore expect that the absorption bands should *become sharp again* at shorter wave lengths (for larger v') or that, in emission, bands with higher v' should appear again after the breaking-off point. The former phenomenon has actually been observed for S_2 (see Fig. 16) and the latter for CaH [Grundström and Hulthén (272)(268)], and N_2 [van der Ziel (733)].

For S_2 the direct transition from the ground state n to α' has also been observed through the continuous spectrum at shorter wave lengths (see above). In a superficial observation, this continuum appears to join onto the series of bands and to belong to them. However, it begins long before the convergence point of the bands is reached, which shows that it is not the continuum belonging to these bands.

The third subcase, I(c), represented by Fig. 186(c) differs from I(a) and I(b) in that the point of intersection C lies *above* the asymptote of α'. Therefore predissociation actually sets in with appreciable probability only for those vibrational levels of α that lie above the point of intersection C, although the energy would be sufficient for predissociation immediately above the asymptote. Thus, here the *predissociation limit does not coincide with the asymptote of α'—that is, with the dissociation limit*—but lies higher. Below the point of intersection, but above the asymptote, the internuclear distance would have to alter quite appreciably in order for a radiationless transition to take place, and this is not possible according to the Franck-Condon principle.

The three subcases (a), (b), and (c) also include the cases in which the point of intersection lies on the left limb of the potential curve α—that is, in subcases (a) and (b), when α' has a deep minimum, and in subcase (c), when the asymptote of α' lies considerably lower than the minimum of α.

In subcase I(c) according to wave mechanics (see p. 408 f.) the possibility of a *passage through the potential hill* formed by the two intersecting potential curves must be considered. On account of this "tunnel effect" a transition can take place even somewhat below the point of intersection, since the eigenfunctions of the two states already overlap to some extent (see Fig. 185). In general, this correction is small, since for the motion of the heavy nuclei classical mechanics is a good approximation. However, if the potential hill is high and narrow, a passage through the hill may be quite frequent near its top and predissociation may take place appreciably below the point of intersection, although with a smaller probability than above it. In such a case the diffuseness in the absorption spectrum or the weakening in the emission spectrum *will not set in with maximum strength* but will be strongest at a small distance from the predissociation limit. An example is the first predissociation limit of S_2 [Herzberg and Mundie (1050)]. Here the diffuseness of the rotational lines of the 10–0 band is detectable with medium dispersion only by the strengthening of the absorption (see p. 411), whereas the lines of the next band, 11–0, do exhibit a broadening easily noticeable with the same dispersion (see Fig. 16). This seems to be the only well established case of this type. Usually the predissociation sets in with maximum probability even in subcase I(c), and, as in subcases I(a) and (b), the radiationless transition probability decreases with increasing energy of the vibrational levels above the limit.

For two states with $\Delta\Lambda = \pm1$ an intersection of the potential curves may occur even if all interactions are taken into account. However, according to the non-crossing rule (p. 295 f.) an intersection of the potential curves of two states which fulfill the Kronig selection rules and which have the same Λ can occur only in a certain approximation but not for the true adiabatic potential curves. In the discussion of case I of predissociation the intersection of potential curves is always meant (at least for $\Delta\Lambda = 0$) in the approximate sense—that is, *intersection in a certain low approximation.* Predissociation is brought about by the same originally neglected terms of the wave equation that produce the apparent repulsion of the states and lead to the avoidance of the intersection. The diffuseness of the levels just as the shift in the case of vibrational perturbations may be considered as due to the mutual interaction of the two states, which is large when the zero-approximation potential curves intersect. The levels in the region of "intersection" assume properties of both electronic states. This holds even if one would start out from the correct adiabatic (non-intersecting) potential curves. A predissociation would result just the same and might then be ascribed to the close approach rather than the intersection of the potential curves. However, if the interaction between the two "original" states is so strong that the resultant adiabatic potential curves are widely separated (that is, if the "original" curves represented a very poor approximation), the actual vibrational levels can be ascribed unambiguously to either the one or the other resultant electronic state with very little if any mixing; therefore the higher levels (above the lower of the two asymptotes) will not be appreciably broadened: there will be no appreciable predissociation. Thus only if the interaction between the original states is comparatively small or, in other words, if the resultant potential curves almost cross [(see Fig. 143(c) and (d)], will case I of predissociation occur. The simplified manner of expression used in the above is therefore justified.

An interesting example of an intermediate case [of type I(c); see Fig. 186(c)] was found by Brown and Gibson (137) for ICl. Here the interaction is strong enough to produce, in effect, two resulting states such as given in Fig. 143(d). But there is still a considerable mixing of the states, which produces diffuse levels (for details see the paper cited).

Though case I predissociation does not occur if the interaction between the two original states is strong, case III predissociation may still take place for each of the resultant electronic states. On the basis of the present discussion, it will be clear, however, that there is *not a sharp limit between case I and case III.* Which case we have in a specific example depends to a certain extent on the choice of the "original" states. Fortunately, in most cases thus far observed a decision seems to be possible [compare, for example, Büttenbender and Herzberg (147)].

Case III (predissociation by rotation); **effective potential curves.** For predissociation by rotation the Kronig selection rules can obviously always be fulfilled, since the molecule remains in the same electronic state. This means that always $\Delta\Lambda = 0$ and $\Delta S = 0$ and that the symmetry conditions are fulfilled for equal J. Accordingly, a predissociation could appear as soon as the dissociation limit is reached. In spite of that, in many cases sharp rotational

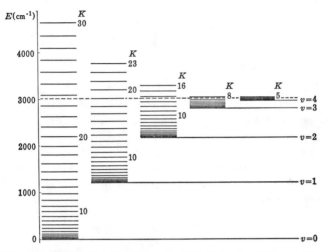

FIG. 187. **Observed (Stable) Rotational Levels of HgH for the Different Vibrational Levels of the $^2\Sigma$ Ground State [after Hulthén (342)].** The doublet splitting is not shown. The broken line gives the position of the dissociation limit.

levels have been observed considerably above the dissociation limit. Fig. 187 gives as an illustration the observed rotational levels of HgH. They do break off, but only somewhat above the dissociation limit, at a height differing for the different vibrational levels. The explanation for this is again to be sought in the Franck-Condon principle [Oldenberg (538)].

Let us consider, on the basis of classical mechanics, a rotating molecule in the vibrationless state. The internuclear distance r takes a value r_c such that the centrifugal force equals the restoring force; that is [compare equation (III, 106)]

$$\frac{P^2}{\mu r_c^3} = U_0'(r_c), \qquad (VII, 15)$$

where P is the angular momentum, μ is the reduced mass, and $U_0'(r_c)$ is the derivative of the potential energy $U_0(r)$ at r_c—that is, the restoring force in the non-rotating molecule for an internuclear distance r_c. From the relation (VII, 15) we can determine the equilibrium distance r_c of the rotating molecule, which of course is greater than r_e.[8]

---

[8] It will be recalled that the introduction of the rotational constant D for the non-rigid rotator is due to this difference between r_e and r_c (see p. 103 f.).

If the molecule carries out vibrations about the new equilibrium position, the restoring force in a displaced position is obtained by subtracting the centrifugal force from the restoring force for the non-rotating molecule; that is, the restoring force for the rotating molecule is (apart from sign)

$$U_0{'}(r) - \frac{P^2}{\mu r^3}.$$

This quantity is the derivative of

$$U(r) = U_0(r) + \frac{P^2}{2\mu r^2}, \tag{VII, 16}$$

which therefore, in its dependence on r, must have the same meaning for a rotating molecule as $U_0(r)$ has for the non-rotating molecule. It is the *effective potential energy* for the rotating system; its minimum gives the equilibrium position r_c, and its derivative gives the restoring force for the vibration.

The term $P^2/2\mu r^2$ of (VII, 16) is, according to (III, 1a), the kinetic energy of rotation, which is, according to wave mechanics, $(h/8\pi^2 c\mu r^2)J(J+1)$ (in cm^{-1}). Thus, when the angular momentum is J, we obtain for the effective potential energy (in cm^{-1})

$$U_J(r) = U_0(r) + \frac{h}{8\pi^2 c\mu r^2} J(J+1). \tag{VII, 17}$$

In Hund's case (b) J is of course replaced by K, since the rotational energy depends in first approximation only on K (see p. 222). Fig. 188 gives the effective potential curves for different K values for the ground state $(^2\Sigma)$ of HgH according to this formula. For large r, all the curves approach asymptotically the same value, $U_0(\infty)$, the dissociation limit of the state. However, unlike $U_0(r)$, the $U_J(r)$ [or $U_K(r)$] curves for J (or K) > 0, in coming from large r values, usually first go through a *maximum* and then through the minimum corresponding to the equilibrium position. With increasing J (or K), the minimum becomes shallower and finally coincides with the maximum at a point of inflection. Higher rotational states no longer have a maximum and a minimum. For these states the molecule is *mechanically unstable*, just as it is in unstable electronic states.

The actual rotational levels in Fig. 188 are obtained by adding the vibrational energy to the energy of the minimum of the particular $U_J(r)$ curve. It is seen that there are stable rotational levels considerably above the dissociation limit. · They are stable in the sense that in them the molecule is *separated by a high potential hill from the dissociated state*, a wave-mechanical passage through the hill being practically impossible; or, expressed in other words, if a transition were to take place into the dissociated state, the internuclear distance would suddenly have to be very much increased, and this is not possible according to the Franck-Condon principle.

Consideration of Fig. 188 shows furthermore that the more strongly the molecule vibrates the smaller is the J value at which the potential hill may be surmounted—that is, at which a decomposition is possible.   It follows that the *energy* (and the J value) *of the last stable rotational level decreases with increasing* v and for the last vibrational level lies only very slightly above the dissociation

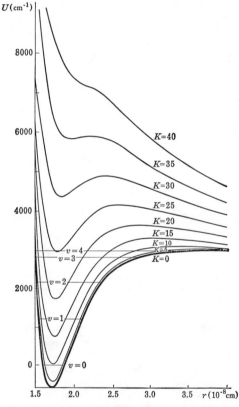

Fɪɢ. 188.   **Effective Potential Curves of HgH in the Ground State [after Villars and Condon (686)].** The vibrational levels indicated refer to the lowest curve with $K = 0$. For the remaining curves they are displaced upward by an amount equal to the energy difference of the minima.

limit.   This corresponds exactly to the observations made on HgH (see Fig. 187).   The observed breaking-off points agree indeed approximately with those that can be predicted on the basis of Fig. 188.

It must be emphasized that the maxima of the effective potential curves arise only on account of the fact that in general $U_0(r)$ approaches the asymptote more quickly than does the contribution $P^2/2\mu r^2$ of the centrifugal force.   If $U_0(r)$ is the potential energy of two ions [compare (VI, 49)] which varies as $1/r$ or if $U_0(r)$ behaves asymptotically as $1/r^2$ (as in the case discussed on p. 381), no potential maxima arise.

For HgH the decomposition takes place in the lower state (ground state) of the observed emission bands. That a breaking-off, and not, as one might expect, a diffuseness of the lines occurs, is due to the fact that the higher rotational states are completely unstable. A transition to one of them from a discrete rotational level of a higher electronic state therefore gives only a broad continuous band whose total intensity is equal to that of only one rotational line and which is therefore not observed.

In Fig. 189 the energies of the maxima of the effective potential curves are plotted as a function of $K(K+1)$. The curve obtained intersects the ordinate axis at the point $U_0(\infty)$, that is, at the dissociation limit. If all the maxima were at the same r value, the curve would be a straight line, as can be seen immediately from (VII, 17). Actually the maxima shift to increasing r values with decreasing K, thus producing the curve in Fig. 189 which has a horizontal tangent for $K(K+1) = 0$. As long as the quantum-mechanical passage through

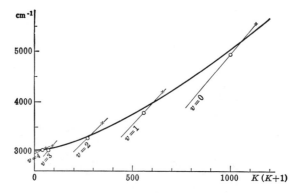

Fig. 189. **Limiting Curve of Dissociation for the Ground State $^2\Sigma$ of HgH.** The oblique straight lines give the rotational levels of the particular vibrational level (average of the two doublet components F_1 and F_2).

potential barriers is neglected (see below), the observed breaking-off points should evidently lie on this curve. The last observed rotational levels in each vibrational level of HgH are indicated by circles in Fig. 189, the first missing levels by crosses. For each v the true breaking-off point must lie between such a circle and cross. Conversely, from the observed breaking-off points plotted in this way, the energy of the asymptote $U_0(\infty)$ may be obtained by a short extrapolation to $K(K+1) = 0$ [see Büttenbender and Herzberg (147) and Schmid and Gerö (622)]. Curves such as that in Fig. 189 have been designated *limiting curves of dissociation* by Schmid and Gerö (622).

If the potential curve $U_0(r)$ of the rotationless state does not have the normal shape as in Fig. 188 but has a maximum (see p. 296), all the maxima of the corresponding effective potential curves at least for small K (or J), will be at the same r value, and the limiting curve of dissociation will be a straight line intersecting the ordinate axis at a non-zero angle.

Conversely if the course of the limiting curve of dissociation has been determined from the observed breaking-off points, it is possible to decide whether or not the electronic state considered has a potential curve $U_0(r)$ with maximum [Herzberg (313)]. For example, if the three breaking-off points (with $v = 0, 1, 2$) observed for AlD in the lowest $^1\Pi$ state are plotted against $J(J+1)$ they fall very nearly on a straight line proving that this state has a potential curve with maximum [Herzberg and Mundie (1050)]. The same has been found for the analogous state of BH. The slope of the straight line according to (VII, 17) is $h/8\pi^2 c\mu r^2$, that is, gives the r value of the maximum of $U_0(r)$; the intercept with the ordinate axis gives in this case the energy of the maximum, not of the asymptote of the potential curve $[U_0(\infty)]$. In the case of AlH only two breaking-off points have been found ($v = 0$ and 1); but the slope of the line connecting them is much too steep to be interpreted as part of a curve like that in Fig. 189 whereas on the assumption of a potential maximum it yields an r value of 2 64 Å for the maximum which agrees within less than 0.01 Å with the r value obtained for AlD. The value of r_e for the state in question is 1.647 Å.

Except in certain cases of van der Waals interaction [see Chapter VI, section 4(c)] a potential maximum may always be considered as due to the intersection of two zero-approximation potential curves [see Fig. 143(d) and accompanying discussion] and therefore the predissociation just considered may also be classified as case I(c) rather than III. However, this does not change the conclusions with respect to the height and internuclear distance of the potential maximum.

In predissociation according to case III, for sufficiently increased rotation, the molecule is to a zero approximation (that is, classically) unstable (exactly as in an unstable electronic state), even without the mutual interaction of the states necessary for predissociation according to case I. The decomposition thus results without any "transition" from one state to another.[9] However, we must now take account of the quantum-mechanical *passage through potential barriers*. As a result of this "tunnel effect," a "transition" into the dissociated state can take place even when the vibrational level is somewhat below the maximum of the effective potential curve under consideration (for example, $v = 1$ for $K = 20$ in Fig. 188). This transition makes itself manifest in a broadening of the level. Therefore, when this type of predissociation takes place in the upper state of a band system, in emission the breaking-off occurs at a somewhat smaller J value than would be expected on the basis of the effective potential curves, while in absorption all the possible rotational lines appear, but the last rotational lines which do not appear in emission are diffuse. The widths of the latter increase very rapidly with increasing J (or K), since the mean life is an exponential function of the area cut off from the potential hill [see the relation (VII, 8)].

The *influence of the tunnel effect* has been investigated in detail for AlH [Farkas and Levy (215)]. In emission (at medium pressure), the violet AlH bands with $v' = 1$ break off suddenly at $J' = 7$ (see Fig. 28, p. 52),[10] while they have been observed in absorption up to $J' = 16$, an increasing width being noticeable from $J' = 12$ up. It is clear that in this case for $J' \leq 16$ the levels lie below the maxima of the corresponding effective potential curves. The broadening of the absorption lines with $J' = 12, \cdots 16$ and the absence of the emission lines $J' = 7, \cdots 16$ is entirely due to the tunnel effect. The lines with $J' > 16$ would be absent even without tunnel effect. The difference between absorption and emission is due to the difference in sensitivity of detection of predissociation (see p. 413).

A striking confirmation of the influence of the tunnel effect is obtained when the intercepts on the ordinate axis in the diagrams of the breaking-off points are compared for AlH and AlD [see Olsson (548), Herzberg and Mundie (1050)]. Without the tunnel effect these intercepts should give the same energy above the minimum of the potential curve $U_0(r)$, that is, the energy of the potential maximum. Actually the intercept for AlD is higher than for AlH by 165 cm^{-1}: on account of the tunnel effect both are lower than corresponds to the maximum

---

[9] This process is therefore often called *"dissociation by rotation"* in the literature. The name *predissociation by rotation* is chosen here in order to indicate the close connection with predissociation and also because, as will be shown presently, a decomposition can take place somewhat below the classical instability limit as a result of a "transition" into the dissociated state. In addition, cases are possible and have indeed been observed in which a distinction between cases I and III is very difficult or impossible.

[10] More recently at much lower pressure ($\sim$0.1 mm) Schüler, Gollnow, and Fechner (1394) found the breaking-off to occur already at $J' = 5$. The effect of pressure will be discussed in more detail in subsection (e).

but that of AlH much more so than that of AlD, since H penetrates the barrier much more easily than D [compare the relation (VII, 8)].[11]

While the influence of the tunnel effect for AlH and other hydrides is appreciable, its exponential dependence on the mass causes it to be negligibly small for all case III predissociations of heavier molecules such as N_2, P_2, $\cdots$. For them it is quite sufficient to use simply the effective potential curves without considering the passage through the potential hill.

It should be added that the effective potential curves apply also to *non-central collisions* of two atoms (see p. 396 f.) for which of course the angular momentum (K or J) is different from zero. For a given relative kinetic energy of the atoms the curves in Fig. 188 (or similar ones for other cases) allow one to determine the maximum angular momentum (that is, the maximum impact parameter) for which the atoms will come into the region of mutual attraction. Considerations of this type are of importance for the discussion of two-body recombinations (see p. 400) since frequently the transition probability is large only for small r values. Again the tunnel effect must be taken into account: atoms with the proper kinetic energy may reach the potential minimum even if their energy is slightly below that of the maximum, and they will then stay together for several periods of vibration before flying apart again. During the oscillation a transition to a lower stable state may occur and thus a stable molecule be formed. This is precisely the mechanism of the inverse predissociation of AlH mentioned earlier (p. 415).

Influence of rotation in case I. In a number of examples undoubtedly belonging to case I of predissociation [as, for instance, the ultraviolet P_2 bands; see Herzberg (311)] a breaking-off of the rotational structure has been observed not only in one but in two or more successive vibrational levels, of course at a correspondingly higher J value for the lower vibrational levels.

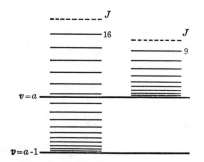

FIG. 190. **Breaking-Off in Two Successive Vibrational Levels in Case I of Predissociation.** The first level that does not occur in emission is indicated by a broken line.

In each case, when the energy values are determined from the spectrum, it is found that the breaking-off points in the lower vibrational levels lie at somewhat higher energy values, as indicated schematically in Fig. 190. The explanation for this behavior is readily found when we use effective potential curves, rather than ordinary potential curves, in case I also.

Fig. 191 shows the effective potential curves for a few J values for cases I(b) and I(c) [compare Fig. 186(b) and (c)]. When we take account of the selection rule $\Delta J = 0$, it is immediately evident that in both cases with increasing J the predissociation limit will be shifted to higher energy values. This means that *the smaller the vibrational energy is* (that is, the greater the necessary rotational energy to reach the limit), *the higher will be the predissociation limit*, as is indeed observed (compare Fig. 190). In case I(b) this is due to the fact that the effective potential curves belonging to α' have a maximum of increasing height

[11] Gerö and Schmid (978a) ascribe this effect to a difference in the energy of the potential minimum, that is to an electronic isotope effect. It appears to the writer that this explanation is not acceptable.

at large internuclear distances, while in case I(c) the intersection of corresponding α and α' curves lies higher and higher as J increases.

Let $v = a$ be the last observed vibrational level before a breaking-off point, J_a be the J value of the last observed rotational level, and E_a be its energy value; furthermore, let J_{a-1} be the J value and E_{a-1} be the energy value of the last rotational level in the vibrational state $v = a - 1$ (where $J_{a-1} > J_a$ and

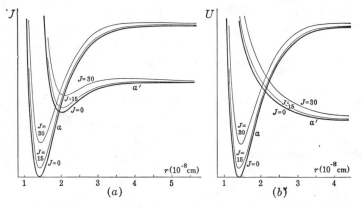

Fig. 191.   **Effective Potential Curves for Case I(b) and I(c) of Predissociation.**

$E_{a-1} > E_a$; see Fig. 190).   Then, according to (VII, 17), the energy difference between the two breaking-off points is given by

$$\Delta E = E_{a-1} - E_a = BJ_{a-1}(J_{a-1} + 1) - BJ_a(J_a + 1), \quad \text{(VII, 18)}$$

where it is assumed that the potential maxima for the two J values lie at about the same r value.   The constant B is $h/8\pi^2 c\mu r^2$ for this r value.   The values of J_a, J_{a-1}, and ΔE are obtained by observation.   Therefore, using (VII, 18) the constant B and from it the approximate r value of the maximum can be derived.[12]   When this has been done, we can immediately *distinguish between the two subcases* I(b) *and* I(c), since in subcase I(c) the r value from (VII, 18) has about the magnitude corresponding to the right limb of the potential curve of the discrete state, whereas in case I(b) the potential maximum lies at an appreciably greater r value (see Fig. 191).   In case I(a) the situation would be similar to that in case I(c).

Broadly speaking one may say that if ΔE, the energy difference of the two breaking-off points, is found to be small, subcase I(b) applies, whereas if ΔE is found to be large, subcase I(c) [or I(a)] applies.

In case I(c) the predissociation limit lies above the dissociation limit by an unknown amount, which may sometimes be rather large; in case I(b) on

---

[12] Since the "true" breaking-off point may be anywhere between the last observed and the first unobserved level, the best procedure is to substitute first the J values of the last observed levels in (VII, 18) and then those of the first missing levels.   The mean of the B (or r) values so obtained is then taken.

the other hand, it is only a very small amount, for which it is even possible to give an upper limit [Herzberg (311)]. According to (VII, 17), the maximum of the effective potential curve for J_a—that is, E_a—lies above the potential curve $U_0(r)$ by an amount $BJ_a(J_a + 1)$, where B is the rotational constant evaluated above, and where $U_0(r)$, for case I(b), is smaller than the energy of the dissociation limit $U_0(\infty)$. The energy $E_a - BJ_a(J_a + 1) = U_0(r)$ therefore represents a lower limiting value for the dissociation limit under consideration, whereas E_a is an upper limiting value. Since the r value of the maximum is usually large, $BJ_a(J_a + 1)$ is very small. The dissociation limit is therefore obtained within very narrow limits in this way. In a similar way, also in case III an upper and a lower limiting value may be obtained for the dissociation limit.

The two or more breaking-off points in case I(b) must lie on the limiting curve of dissociation for the state α' that produces the predissociation (see p. 428 and Fig. 189). The above procedure for determining a lower limit for the dissociation limit corresponds to approximating the limiting curve by a straight line. If more than two breaking-off points are observed, an even better approximation can be obtained by using a curve having a horizontal tangent at $J(J + 1) = 0$ instead of a straight line. In case I(c) the breaking-off points, plotted as a function of $J(J + 1)$, do not lie on such a limiting curve but on a straight line, just as in case III when the potential curve has a maximum (see above).

Most of the cases in which a breaking-off point has been observed in two or more successive vibrational levels have been shown to belong to case I(b) or III. This is understandable, since in case I(c), ΔE in (VII, 18) and therefore J_{a-1} are very large, usually so large that the rotational levels in question would not generally be observed even without predissociation.

The influence of the tunnel effect in case I(b) is very slight just as in case III when there is no maximum of the $U_0(r)$ curve. It may in general be neglected except for a very few levels in the immediate neighborhood of the breaking-off point. However, in case I(c) this influence may be considerable, particularly for hydrides, since here the potential hill may be fairly narrow depending on the form of the curve α' [Fig. 191(b)]. When only one breaking-off point is observed, nothing can be concluded about the extent to which it has been lowered on account of the tunnel effect. However, when a breaking-off is observed for two isotopes, an appreciable difference of the energies of the breaking-off points indicates a strong tunnel effect at least for the lighter isotope. An extreme case of this type has been found for the $D\,{}^1\Sigma$ state of AlH by Grabe and Hulthén (997). Emission bands with this state as upper state show a breaking-off at $v' = 0$, $J' = 10$, while for AlD no breaking-off is found in the $v' = 0$ level, but rather rotational lines are observed at 2500 cm^{-1} above the breaking-off point for AlH. In this case the predissociation is produced by the $C\,{}^1\Sigma$ state which intersects the $D\,{}^1\Sigma$ state on the left limb of the potential curve high above the breaking-off point, giving a very steep and narrow potential hill. For AlH a passage through this hill is possible at a very much lower energy than for AlD (see also the $A\,{}^1\Pi$ state of AlH discussed above, p. 429).

In absorption the presence of a tunnel effect can always be recognized by an increase of line width with increasing energy above the predissociation limit. The predissociation of S_2 discussed earlier (p. 424) is an example of this type. Here the potential hill must be rather narrow, since otherwise the mass factor in (VII, 8) would make the tunnel effect unobservable for a molecule as heavy as S_2.

(e) Pressure Effects in Predissociation

Induced predissociation. The selection rules for predissociation have been derived for the free undisturbed molecule. During collisions a number of these rules (for example $\Delta J = 0$)

need no longer hold.   Therefore, forbidden predissociations may occur *at high pressure:* we obtain *induced* predissociations.

The first case of this type was found by·Turner (672)(673), when he investigated the fluorescence of I_2 with the addition of argon of fairly high pressure ($\sim$30 mm).   He found a quenching of those I_2 bands whose upper states lie above the dissociation limit of the ground state and which have a normal intensity when no argon is added.   Furthermore Turner detected directly the presence of iodine atoms by the absorption of atomic lines under the same conditions under which the quenching occurred.   Finally, Loomis and Fuller (461) under similar conditions (but using oxygen instead of argon) detected a broadening of the corresponding absorption bands by the increase of their apparent intensity when the foreign gas was added but the number of absorbing molecules kept constant [as mentioned earlier (p. 411) broad lines under small or medium dispersion appear more intense than sharp ones].   Thus all three of the criteria for predissociation (p. 407) have been checked in this case of induced predissociation.   The weakening of the emission bands was found by Turner also in electric discharges through I_2 vapor when argon was added.   Kondratjew and Polak (413) have found three maxima in the strength of the induced predissociation of I_2 (at $v' = 22$, 29, and 39) which they interpret as due to three different unstable electronic states.

A number of other cases of induced predissociations have been found, for N_2 by Kaplan (395), for Br_2 by Kondratjew, Polak and Avramenko (412)(73), for NO by Wulf (723) and Kondratjewa and Kondratjew (414), for S_2 by Kondratjew and Olsson (411), Lochte-Holtgreven (448), and Rosen (597), for Te_2 by Kondratjew and Lauris (409) and Rosen (597), and for Se_2 by Rosen (597).   In all but the first of these cases the induced predissociation is detected by the anomalous intensity in absorption at high pressure of a foreign gas or of the absorbing gas itself.   Predissociations induced by the absorbing gas itself are recognized by the fact that the apparent absorption intensity increases faster than linearly with the pressure for a given absorbing path, that is, deviates from the Lambert-Beer law.   In the case of N_2 the induced predissociation has also been detected by the weakening of emission bands.

A detailed theoretical treatment of induced predissociation on the basis of wave mechanics has been given by Zener (1572).

It must be emphasized that there are of course other reasons of pressure broadening of band lines (some of them similar to those for atoms, see p. 397 f.) which have nothing to do with predissociation.   Usually this type of pressure broadening is small at pressures of less than 1 atm, but in special cases it may be considerable and is then difficult to distinguish from a broadening due to induced predissociation.   In one case, that of SO_2, a strong broadening of absorption bands (below 2800 Å) by pressure has been shown to be definitely not due to induced predissociation since it occurs below the lowest dissociation limit [Franck, Sponer, and Teller (232)].   In this case the broadening is apparently due to a transition into a neighboring discrete state made possible by the collisions.   No such case has yet been found for diatomic molecules but its possibility must be kept in mind.   For this reason the evidence for a dissociation limit supplied by an induced predissociation is never quite unambiguous unless the dissociated atoms are directly observed.

Suppression of breaking-off by pressure.   An effect that superficially appears to be the opposite of induced predissociation is the appearance under certain conditions, particularly at high pressures, of those band lines that are absent on account of predissociation in low pressure emission.   Mörikofer (503a) found that the AlH bands that break off under certain conditions of excitation (see Fig. 28) do not break off under other conditions of excitation.   Bengtsson and Rydberg (87), who first studied this effect in more detail, ascribed it to a pressure effect, because they found that the breaking-off occurs in an arc between Al electrodes at low pressure of H_2, whereas it does not at high pressure.   However, no breaking-off of the AlH bands occurs in the aureole of an arc even at fairly low pressure [Mörikofer (503a)], whereas they do break off in the core of the arc at the same pressure.

According to Farkas (212) the explanation of this suppression of the breaking-off is to be sought in the existence of approximate *thermal equilibrium* in the arc at high pressure or in the aureole even at comparatively low pressure. In thermal equilibrium, the population of the rotational levels of the upper state is entirely determined by the Boltzmann factor and thus there will be no breaking-off in thermal emission.[13] This has indeed been found to be the case down to quite low pressures ($\sim$1 mm) when the AlH bands and similarly the CaH bands (Fig. 27) were excited thermally in a high temperature furnace [Wurm (726), Olsson (546)]. In thermal equilibrium the number of predissociating molecules is exactly compensated by an equal number of new molecules formed by inverse predissociation (according to the principle of detailed balancing). On the other hand for excitation by electron collisions at low pressure such a compensation does not exist and the levels above the predissociation limit will be depopulated compared to the levels below in the ratio $\beta/(\beta + \gamma)$ (see p. 408), that is, a breaking-off will occur in emission. When the pressure is increased conditions in an electric arc approach thermal equilibrium and that is why the whole bands are observed in the core of the arc at high pressure. In the aureole the excitation is essentially thermal and therefore there is no breaking-off even at fairly low pressures.

As was first pointed out by Olsson (548) there is an additional cause suppressing the breaking-off for excitation by electron collisions at medium and high pressures which is particularly important in the glow discharge since here thermal equilibrium hardly ever obtains: the number of collisions removing the excitation energy before radiation has taken place (*deactivating collisions*) increases with the pressure; but they affect predissociating levels much less than the non-predissociating levels, since the former have a much shorter life. Therefore at sufficiently high pressure, even if there is no thermal equilibrium, the breaking-off will be suppressed because of the greater quenching of the non-predissociated lines. This effect occurs at a lower pressure the smaller the radiationless transition probability is. That is why in the $v' = 1$, $^1\Pi$ state of AlH for example, the breaking-off at a pressure of 45 mm occurs at $J' = 7$, whereas at 0.1 mm it occurs at $J' = 5$ (see p. 429). The levels $J' = 6$ and 7 have a longer life than the higher levels because of the greater height of the potential hill that has to be penetrated. Another striking example is provided by the visible CuH bands which show a normal intensity distribution down to about 2 mm pressure. But at a pressure of 0.02 mm Schüler, Gollnow and Haber (1395) found only the line $P(1)$ with $J' = 0$ with predominating intensity. According to Herzberg and Mundie (1050) this is due to a forbidden predissociation ($^1\Sigma$—$^3\Sigma$) which causes a breaking-off only at very low pressures because the radiationless transition probability is small.[14] Conversely these considerations show that in order to detect weak predissociations it is necessary to investigate discharges at very low pressures.

A suppression of the breaking-off has been found in a number of other cases: for CaH by Grundström and Hulthén (272) and Nilsson (1284a), for SrH by Humphreys and Frederickson (345a) and for TlH by Grundström and Valberg (273). A more detailed discussion of this effect has been given by Olsson (548).

It should perhaps be pointed out that at very low pressures ($<$1 mm) thermal equilibrium is not easily obtained. Lack of equilibrium is apparently the reason for the observation by Wurm (726) of a breaking-off in AlH in thermal emission at pressures less than 1 mm. At such a low pressure and the high temperature used, the equilibrium of the reaction AlH $\rightleftarrows$ Al+H is almost entirely on the right-hand side; but starting out with AlH, on the way to this equilibrium, emission from the non-predissociating levels will occur but not from the predissociating levels. Conversely, if at higher pressures, when the left-hand side of the reaction is favored, one starts out from Al + H, on account of inverse predissociation, emission from the predissociating levels will predominate as was found by Stenvinkel (1435) in the experiments referred to previously (p. 415).

---

[13] However, there is a broadening of the emission lines just as in absorption.

[14] The explanation put forward by Schüler and his collaborators for this effect seems to be untenable [see also Nilsson (1285)].

(f) Other Diffuse Molecular Spectra

It remains to discuss another type of genuinely diffuse bands whose diffuseness is not caused by predissociation. The Hg_2 spectrum reproduced in Fig. 17 affords an example of this type. In contrast to predissociation spectra, this type of diffuse spectrum consists in general of *symmetrical* fluctuations of intensity, which appear in emission as well as in absorption, and which do not form the continuation of a series of discrete bands but appear independently of them. Moreover, there is no breaking-off in emission.

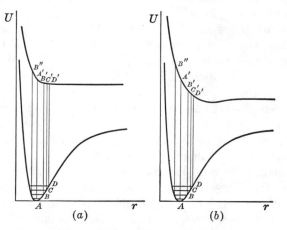

FIG. 192. Potential Curves Explaining Diffuse Bands.

It was shown by Winans (716), Sommermeyer (644), and Kuhn (428) that this type of diffuse band spectra is merely a *special case of continuous spectra* resulting from the combination of an unstable with a stable state.

Consider, for example, the transition in absorption from a stable ground state to an unstable state whose potential curve is nearly horizontal down to comparatively small r values, as shown in Fig. 192(a). In such a case the continuous spectrum corresponding to the transition from the vibrationless state will be very narrow, as a result of the Franck-Condon principle; that is, it will have the appearance of a diffuse band (see p. 391). At higher temperatures absorption from the higher vibrational levels must be considered. For a given vibrational level with $v'' > 0$, according to the Franck-Condon principle, the continuous absorption spectrum will consist of one narrow (and therefore fairly strong) peak, corresponding to the right-hand turning point, and one very broad (and therefore weak) maximum at shorter wave lengths which corresponds to the left-hand turning point (here we disregard weaker intermediate maxima, see p. 393). If the upper potential curve is sufficiently flat, the narrow continua (AA', BB', $\cdots$ in the figure) corresponding to the right-hand turning points of the successive vibrational levels will be separate from one another. We thus observe a *series of diffuse bands* whose separations decrease toward longer wave lengths and correspond to the vibrational quanta in the ground state of the molecule. At the same time, of course, the intensity in this series decreases rapidly because of the Boltzmann factor, unless the vibrational quanta are very small.

Since the potential curve of the upper state will in general be much steeper above the left- than above the right-hand turning points of the lower state (see Fig. 192), the continua corresponding to the former are very much broader. In most cases they will overlap one another to such an extent that they appear as one extended continuum, lying to the short-wave-length side of the diffuse bands. Such cases have been observed, for example, for the alkali halides by Sommermeyer (644) [see also Levi (443)]. The upper state in this case corresponds to dissociation into normal atoms [see p. 373 and Fig. 170(b)]. The wave number of the shortest-wave-length diffuse band, corresponding to the transition from the vibrationless ground state, gives approximately the heat of dissociation of the halide into normal atoms [see Fig. 192(a)]. The same bands have been observed in emission in alkali-halogen flames by Beutler, Josephy, and Levi (443)(93). They correspond to the converse of the photodissociation process, that is, formation of the alkali halide molecules by two-body collisions between alkali and halogen atoms.

It is comparatively rare that the upper potential curve runs nearly horizontally above the right-hand turning points of the lower curve, as in Fig. 192 (a).[15] But even if it has an appreciable slope as in Fig. 192(b), the preceding considerations remain practically unaltered. The only difference is that the individual diffuse bands are now broader and their separations decrease more rapidly than do the vibrational quanta in the ground state, as may be seen immediately from the diagram [Kuhn (428)].

Diffuse bands quite analogous to those described above are to be expected if the upper and lower states in Fig. 192 are interchanged. They have been observed both in absorption and emission for molecules that have only a van der Waals binding in the ground state. They are of course closely related to the previously discussed continuous spectra of this same group of molecules (see p. 395). An example are the diffuse bands of Hg_2 that occur near the long-wave-length end of the continuum joining onto the resonance line 2537 Å (see Fig. 17). Such diffuse bands correspond in absorption to transitions that take place during central or nearly central collisions of the normal atoms while in emission they correspond to transitions from the different vibrational levels of an excited stable molecular state to the unstable ground state. In emission such diffuse bands have also been found for molecules with stable ground states, particularly for the halogen molecules [compare the recent work of Venkateswarlu (1495)(1496)]. They correspond to transitions from a stable excited state to one (or more) of the unstable lower states that arise from normal atoms in addition to the ground state. Again only when the potential curve of

---

[15] If it were exactly horizontal the diffuse bands would degenerate into sharp lines.

the unstable state has a small slope to fairly small r values, will there be diffuse bands rather than a single continuum.

More details about this type of diffuse bands may be found in Finkelnburg's book (31).

3. Determination of Heats of Dissociation

On the basis of the considerations in sections 1 and 2, it is clear that the investigations of continuous and diffuse spectra, of the convergence of band progressions, and of the breaking-off in band systems can lead to important conclusions concerning the dissociation processes in diatomic molecules and in particular to the determination of their heats of dissociation.

In agreement with chemical usage we define the heat of dissociation (or dissociation energy) $D_0{}^0$ of a diatomic molecule AB as *the work required to dissociate the molecule from the lowest level* ($v = 0$, $J = \Omega$) *of the electronic ground state into normal atoms, $A + B$.* This corresponds in thermal measurements to $\Delta H_0{}^0$, the heat of the gas-phase reaction $A + B \rightleftarrows AB$ occurring under ideal-gas conditions at the temperature $T = 0°$ K.

In each of its stable electronic states the molecule has a certain dissociation energy D_0 (see p. 100) which corresponds to a dissociation from the lowest level of that state by increasing the vibrational energy (without altering the electronic state).[16] The products of dissociation may be normal or excited atoms. Every combination of states of the two atoms corresponds to a definite *dissociation limit*, to which, in general, several molecular states belong (see Chapter VI, section 1). In general the dissociation energy D_0 in the electronic ground state equals the heat of dissociation $D_0{}^0$ of the molecule. However, there are a few cases such as the BeO molecule in which the ground state does not dissociate into normal atoms and for which therefore $D_0{}^0$ is smaller than D_0 of the ground state.

According to the non-crossing rule (p. 295), the electronic ground state must always dissociate adiabatically into normal atoms except when the Wigner-Witmer rules (p. 315 f.) do not allow such a correlation. The case of BeO is such an exception, since the $^1\Sigma$ ground state cannot arise from $Be(^1S) + O(^3P)$. However, if one correlates on the basis of the vibrational levels, a crossing over of states of the same species is conceivable if the interaction of the two states is slight, that is, in principle it is possible to have excited dissociation products of the ground state even if that is not required by the Wigner-Witmer rules. In practice one would expect this to happen only in very exceptional circumstances and not, as Schmid and Gerö (1388)(1389) thought, in the majority of cases [see Hulthén (1080)].

The spectroscopic determination of the heat of dissociation of a molecule always involves two steps: (1) determination of the energy of a dissociation limit (or possibly several limits) above the ground state, and (2) determination of the products of dissociation at this limit. When the position E_d of a dissocia-

---

[16] Previously we have also considered the dissociation energy D_e which refers to the minimum of the potential curve and differs from D_0 by the zero-point energy (see p. 100). For our present considerations only D_0 matters.

tion limit and the dissociation products at this limit (excitation energy A) are known, the heat of dissociation of the molecule is immediately obtained. We have only to subtract the excitation energy of the dissociation products from the energy of the dissociation limit [see, for example, Fig. 88(c)]; that is,

$$D_0{}^0 = E_d - A. \tag{VII, 19}$$

In what follows we shall consider the two steps separately. For a much more detailed treatment the reader is referred to Gaydon's book (34).

(a) Determination of Dissociation Limits

Most of the methods for the determination of dissociation limits have already been dealt with. In the following they are summarized, and some further methods are added to them.

(1) **Band convergences.** A very accurate and reliable value for a dissociation limit is obtained when a convergence limit of bands, with its adjoining continuum, is observed in absorption. The position of the *convergence limit*—that is, the beginning of the continuum—corresponds to the *dissociation limit of the excited molecular state under consideration*. Such convergence limits have been observed, for example, for the halogen molecules (see Fig. 15, p. 38) and for O_2 and H_2.

In the case of H_2 a band convergence has also been found *in emission* corresponding to a transition from a fixed vibrational level of the $B\ {}^1\Sigma_u$ excited state to the various vibrational levels of the ground state [see Beutler (90)] and giving directly the dissociation energy of the ground state. This progression converges of course to longer wave lengths. A continuous spectrum extending to still longer wave lengths has not been observed. The D_0 value obtained agrees most satisfactorily with that obtained by methods (3) and (4) [see Beutler and Jünger (96)].

It must be kept in mind that the electronic state considered may have a *potential maximum* which may be present on account of an "original" intersection with another state (see p. 296) or on account of a repulsive van der Waals force for large r values (see p. 381). In such a case the band progression observed in absorption would end and the continuum would begin before the point of convergence is reached. Moreover, the rotational levels above the dissociation limit would show predissociation effects of a type somewhat different from the case without maximum (see p. 428). The $B \leftarrow X$ transition of ICl is of this type.

(2) **Extrapolation to convergence limits.** For the numerous cases in which no band convergence is observed, Birge and Sponer (113) have suggested an *extrapolation* to the position of the convergence limit from the observed bands. From cases in which a convergence is observed, we know approximately how the vibrational quanta ΔG (that is, the separations of successive bands in a progression) depend on the vibrational quantum number. Fig. 193 shows, for example, the behavior for the upper state of the ultraviolet O_2 bands (see also Fig. 51, p. 101, for the ground state of H_2). If, for example, we had observed only the first five vibrational quanta, we could have extrapolated the further vibrational

quanta—say, linearly—as indicated by the broken line.   It can be seen that all the extrapolated vibrational quanta are too high.

According to the relation (III, 93), the dissociation energy D_0 in a given electronic state is equal to the sum of all the vibrational quanta—that is, very nearly equal to the area under the ΔG curve.   From Fig. 193 it can be seen that the area under the linearly extrapolated ΔG curve is not very much greater than that under the true ΔG curve.   We can therefore expect that, even in those cases in which only the first few vibrational quanta are observed, extrapolation will

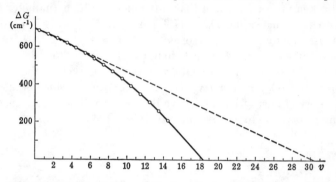

Fig. 193.   **Vibrational Quanta (ΔG) as a function of v in the Upper State ($^3\Sigma_u{}^-$) of the Ultraviolet O_2 Bands.**   The data are taken from Curry, Herzberg, and Herzberg (176) and Knauss and Ballard (406).

lead at least to an *approximate D_0 value* for the state under consideration. Analytically the linearly extrapolated dissociation energy is obtained from (III, 97):

$$D_0 = \frac{\omega_0{}^2}{4\omega_0 x_0} \tag{VII, 20}$$

If sufficient vibrational quanta are observed so that a curvature of the ΔG curve can be detected the extrapolation method can be considerably improved by using quadratic or higher terms in the formula for ΔG or corresponding curves in the graphical representation.[17]   It is clear that the reliability and accuracy of the D_0 values obtained in this way increases with the ratio of the number of observed ΔG values to their total number.   In the limiting case when almost all vibrational levels are observed and only a very short extrapolation is necessary we have substantially the first method, that is, an observed convergence limit, though without the continuum (compare the example of H_2 mentioned above and the corresponding ΔG curve in Fig. 51).

We obtain in this way the dissociation energy of the electronic state whose vibrational quanta are used.   In order to obtain the energy of the dissociation limit under consideration we have to add the excitation energy of the electronic state.

[17] Birge (104) and Rydberg (608) have suggested two further extrapolation methods, which, however, have thus far not found general application.

The advantage of the extrapolation method is that it can be applied to any electronic state for which several vibrational levels (at least three) are known.

There is unfortunately no great uniformity in the shape of the ΔG curves of different electronic states of the same molecule or of different molecules. While the curve for the $^3\Sigma_u^-$ state of O_2 (Fig. 193) is fairly typical there are cases like the ΔG curve for the $^1\Pi$ state of K_2 represented in Fig. 194 which deviates much more strongly from a straight line. If in this case we had observed, only, say, 10 vibrational quanta, we should have obtained by linear extrapolation (broken line) a D_0 value that was much too high (namely, $D_0 = 3860$, compared to the true value, $D_0 = 1790$ cm^{-1}). While this is perhaps an extreme case, for many "atomic" molecules such as O_2, N_2, HCl, Br_2, $\cdots$ the true heats of dissociation have been found to be up to 35% less than the D_0 values obtained by linear extrapolation of the ΔG curves of the ground states [for example, D_0^0 (O_2) = 5.08 e.v., $D_{\text{linear}} = 6.3$ e.v.]. Since in many of these cases the observed part of the ΔG curve is linear it follows that this curve must bend strongly for some higher v value, as in Fig. 194.

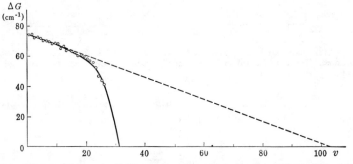

FIG. 194. Vibrational Quanta (ΔG) as a Function of v in the Upper State ($^1\Pi$) of the Red K_2 Bands. The data are taken from Loomis and Nusbaum (464).

Since the Morse potential function (III, 98) yields a linear ΔG curve (see p. 101) it is clear that when the actual ΔG curve deviates as in Figs. 193 and 194 from a straight line, the potential curve will have a steeper right-hand limb and reach the asymptote sooner than the Morse function. It is indeed found that more accurate potential functions of the ground states of molecules like N_2, HCl, I_2, $\cdots$ show precisely this behavior [see Hulburt and Hirschfelder (1078)].

The opposite type of deviation from a linear ΔG curve is to be expected for ionic molecules: for them there is an infinite number of vibrational levels just as there is an infinite number of electronic levels of an electron in a Coulomb field; therefore the ΔG curve reaches the v axis only asymptotically, that is, has a positive rather than a negative curvature. In no case have very extended ΔG curves as yet been observed for ionic molecules. However, in all cases in which the energy of dissociation into ions is known from other data the linear extrapolation does give much too small a value. The same is true for HF which, although not an ionic molecule has considerable ionicity (see p. 377). In this case the observed part of the ΔG curve shows a definite positive curvature.

As we have seen previously certain excited states of atomic molecules have ionic character, as, for example, the B $^1\Sigma$ state of H_2 (see p. 364). For these states ΔG curves with positive curvature are to be expected and indeed observed in a number of cases. Greatly anomalous

ΔG curves arise in the case of vibrational perturbations, that is, when two zero-approximation potential curves intersect (see p. 293 and Fig. 142). On account of the effect of both ionicity and perturbations it is clear that long linear extrapolations for excited electronic states are likely to lead to D_0 values that are far from the true values and may be either higher or lower than the latter.

In the case of the ground states vibrational perturbations are not likely to occur and, if the ionicity is small, one may expect the linear extrapolation to give an upper limit to the value of D_0. From the experience with molecules for which D_0 has been reliably determined by other methods one may conclude that a value of 20% less than the linearly extrapolated value is likely to be fairly close to the true value. On the other hand the energy of the last observed vibrational level of the ground state represents, of course, a lower limit to D_0 for this state. It should be emphasized that this last statement does not necessarily hold for excited electronic states since they may have a potential maximum.

A very detailed discussion of the extrapolation method of Birge and Sponer has recently been given by Gaydon (957). He has collected evidence showing that the percentage difference between the true and the linearly extrapolated heat of dissociation increases with atomic number in a group of similar molecules such as O_2, S_2, Se_2, $\cdots$. To be sure this conclusion is partly based on D_0^0 values that in the opinion of the writer are not yet definitely settled and are likely to be appreciably lower than assumed by Gaydon. Gaydon has also pointed out that in some cases near the point of convergence a slight bending up of the ΔG curve may be expected on account of van der Waals forces which have a greater range than the exponentially decreasing valence forces.

(3) **Long-wave-length limit of an absorption continuum.** According to the discussion of section 1(a), when a continuous absorption is observed, its long-wave-length limit is an *upper limiting value for the dissociation limit under consideration.* However, the true value may lie considerably below this upper limiting value if the potential curve of the upper unstable state is appreciably above its asymptote for r values in the neighborhood of the minimum of the ground state, as in Fig. 173(a); according to the Franck-Condon principle only transitions to this part of the upper potential curve can occur [see Franck and Kuhn (230)]. Usually in such a case the upper potential curve is fairly steep and the continuum is therefore very extensive and sets in gradually, as is observed, for example, for the halogen hydrides (see p. 389 f.). Conversely a large extent and gradual setting in of an absorption continuum may be taken as evidence that the long-wave-length limit is appreciably above the dissociation limit.[18] Thus for HBr Goodeve and Taylor (263) observed the continuum from higher wave numbers down to 35,000 cm^{-1} while the dissociation limit (from other data) is at 30,290 cm^{-1}.

On the other hand, when the continuum is of little extent, looking more like a diffuse band we may conclude that the upper potential curve has only a

---

[18] It should be noted that this holds for absorption from the lowest vibrational level $v'' = 0$ of the ground state. If molecules with $v'' = 1$ or higher ones are present the continuum will extend to appreciably smaller wave numbers, possibly even to below the dissociation limit for $v'' = 0$.

small slope and that therefore the dissociation limit is very close to the long-wave-length limit of the continuum, as, for example, for the alkali halides [Sommer-meyer (644)].

If the potential curve of the upper state has a minimum at large r values the point of intersection of the asymptote with the rising limb of the potential curve [point C in Fig. 88(c)] may or may not be within the Franck-Condon region reached from the ground state. In the former case we have a convergence of discrete bands with adjoining continuum [see method (1)]; in the latter case the situation is similar to that without minimum although now the long-wave-length limit may be close to the dissociation limit even if the continuum is fairly extended. An intermediate case has been found for H_2 and studied in detail by Beutler (91): There is an exceedingly sharp limit (slightly different for different rotational levels) of an absorption continuum but no preceding band convergence. Apparently here the intercept of the asymptote with the potential curve occurs just at the border of the Franck-Condon region, the last stable vibrational level being already beyond it. In such a case the sharp limit of the continuum obviously represents a very precise value for a dissociation limit if the possibility of a potential maximum is excluded. It is in this way that the most accurate value for $D_0^0(H_2)$ has been determined.

(4) **Predissociation limits.** The energy of a predissociation limit gives immediately an *upper limiting value for the dissociation limit of the state causing the predissociation.*[19] In case $I(c)$ of predissociation the dissociation limit may be appreciably below the predissociation limit; but in cases $I(a)$, $I(b)$, and III the two limits coincide, except for a slight rotational correction (see p. 425 f.). Which case applies can be decided *if a breaking-off in the rotational structure is observed not only in the last but also in the second last* (and possibly the third last, and so on) *vibrational level* (see p. 431 f.) since in cases $I(b)$ and III the two (or more) breaking-off points have a small energy difference, in cases $I(a)$ and $I(c)$ a comparatively large one. In the former cases the rotational correction can be estimated reliably and thus a very accurate value for the dissociation limit be determined. This method has been applied, for example, to P_2 by Herzberg (311), to N_2 by Büttenbender, Herzberg, and Sponer (147)(319), to SO by Martin (482), and to H_2 by Beutler (91).

When a breaking-off is observed in only one vibrational level, a distinction between cases $I(b)$ and III on the one hand and case $I(c)$ on the other cannot be made. The possibility that the dissociation limit lies considerably below the breaking-off point (predissociation limit) cannot then be excluded.

According to our previous discussion, we should expect a sharp breaking-off point in case $I(c)$ only when the potential hill is not too narrow. Conversely, it follows from this that, when a sharp breaking-off point is observed in a single vibrational level, the dissociation limit cannot in general lie very much below the breaking-off point. In the upper state of the first positive group of N_2, for example, a breaking-off point has been observed [van der Ziel (732)] which has been shown to lie 0.08 e.v. above the corresponding dissociation limit [Herzberg and Sponer (319)]. This is the order of magnitude of the differences to be expected.

When, however, no breaking-off point in the rotational structure but only a breaking-off

---

[19] This holds also for the case of accidental predissociation (see p. 415) and for induced predissociation (see p. 432).

of the vibrational structure for a given v' is observed, the corresponding dissociation limit may lie very considerably below this breaking-off point. The same conclusion must be drawn when, in absorption, the diffuseness sets in at a certain band but not at a certain line in the rotational structure, and in particular when the diffuseness does not start with maximum strength but reaches this maximum strength only some distance beyond the limit [see Turner (671) and p. 424]. An example for the latter case is the first predissociation of S_2 [see Fig. 16 and Herzberg and Mundie (1050)].

(5) **Excitation of atomic fluorescence.** When a continuous or diffuse absorption spectrum corresponds to dissociation into a normal and an excited atom, on illumination with light in the absorption region in question we may expect the corresponding atomic line (or lines) in fluorescence, provided the excited atomic state is not metastable. The *long-wave-length limit for the appearance of the atomic fluorescence* obviously gives an *upper limiting value for a dissociation limit* [Terenin (665)(1460)(1461)]. The advantage of this method is that it yields at the same time information about the kind of dissociation products produced at the limit and thus, according to (VII, 19), an upper limiting value for the heat of dissociation of the molecule. Terenin has applied this method particularly to the alkali halides. By irradiation of the alkali halides (in vapor form) with light of sufficiently short wave length, the resonance lines of the alkali atoms or even the second members of the principal series are observed in fluorescence.

(6) **Photodissociation.** Irradiation of a gas by light in a continuous or diffuse absorption region will lead to photodissociation which may be detected chemically or by other means. The longest wave length of incident light that will cause a photodissociation gives an upper limiting value for the dissociation limit. This method is useful in cases in which it is difficult to decide from the absorption spectrum whether or not there is predissociation or to detect a weak continuous absorption that may overlap a stronger discrete absorption. In this way Flory and Johnston (225)(937) were able to detect predissociation in NO and to establish an upper limit for its dissociation energy. By investigating the photoreaction over a wide range of pressures and finding it to be independent of pressure they were able to establish that the observed dissociation is not due to secondary reactions. Another example is the photodissociation of CO which Faltings, Groth, and Harteck (926) observed when irradiating CO with the strong xenon line 1295 Å but not when irradiating it with the xenon line 1470 Å. However, in this case the dependence on pressure has not been investigated and therefore the resulting dissociation limit is not uniquely established [see Schmid and Gerö (1390) and Gaydon (34)].

A photodissociation into positive and negative ions (see p. 390) which can be detected electrically may also be used to obtain a dissociation limit. In this case the excitation energy $[E_d$ in (VII, 19)] of the products of dissociation at the limit is at least equal to the difference of the ionization potential of the atom forming the positive ion and the electron affinity of that forming the negative ion. Thus far this method has only been applied to the halides of gallium, indium, and thallium [Terenin and Popov (666), Wehrli and Hälg (1520)].

(7) **Chemiluminescence.** The spectrum of the light emitted in the simplest type of chemical reaction, viz. recombination of atoms, may serve for the purpose of obtaining a dissociation limit. Usually the molecule formation proceeds via *three-body collisions*. The energy of dissociation set free in the formation of the molecule may be divided in different ways between the collision partners. It may be converted either into energy of excitation of the third collision partner or of the newly formed molecule or into excitation energy of both systems. Part of the energy set free may also be converted into translational energy of the products of the collision. However, according to the principle of resonance in collisions of the second kind, this part is always very small (see A.A., p. 231). If suitable excited states of the newly formed molecule or of the collision partner exist a chemiluminescence will accompany the

molecule formation.   The spectrum of the chemiluminescence will show an unusual intensity distribution in the line series or band progressions since the excitation energy is sharply limited. Conversely the *highest level* which is found to be excited in this spectrum *will correspond to the heat of dissociation of the molecule formed*, which can therefore be determined in this way.

As an example the luminescence accompanying the recombination of H atoms in the presence of metal and other vapors should be mentioned [Bonhoeffer (820)].   The highest level that is excited agrees satisfactorily with the now accurately known heat of dissociation of H_2.   For a more detailed understanding of the chemiluminescence spectrum the formation of H_2 molecules in excited vibrational levels must be taken into account [see Kaplan (1114)].

A similar process was first assumed by Sponer (646) for the recombination of atomic nitrogen in order to explain the afterglow of active nitrogen as well as the luminescence excited by it in various other substances.   But in this case an unambiguous determination of the heat of dissociation from these chemiluminescence spectra is not possible because of complications introduced by the formation of metastable atomic and molecular states.   However, these complications seem to be absent in the upper atmosphere (see p. 485).

If the recombination proceeds via *two-body collisions* it is always the newly formed molecule that emits the luminescence and the upper state of the luminescence must be very close to (but slightly above) the dissociation limit.   In the case of a recombination continuum, since it is in general impossible to ascertain what is the lower state, no conclusions about the dissociation limit can be drawn.   But in the case of an inverse predissociation this is possible. Thus from the observed preferential emission from the predissociated levels of AlH under conditions that favor formation from the separated atoms (see p. 415), an upper limit for the dissociation limit could have been derived if it had not already been obtained from the ordinary predissociation.

In the case of the selective excitation of the $v = 6$ $^3\Pi_g$ level of C_2 under conditions that favor formation of C_2 molecules (see p. 415), if the interpretation as inverse predissociation is correct, a dissociation limit at 3.6 e.v. above the ground state immediately follows.

(b) Determination of the Dissociation Products

The dissociation limits obtained in one of the ways described in subsection (a), correspond in general to dissociation into atoms which may be more or less highly excited.   In order to derive the heat of dissociation $D_0{}^0$ of the molecule in its ground state into normal atoms, we have to know the *state of excitation of the products of dissociation* [see equation (VII, 19)].   Unless this has been ascertained, all that can be concluded from an observed dissociation limit is that the heat of dissociation of the molecule is smaller than or equal to that limit. It is usually this step of determining the states of the resulting atoms that causes difficulties and is responsible for a number of ambiguities which still exist in the literature.   The following methods may be used in order to determine unambiguously the dissociation products in a given case.

(1) **Energy differences of dissociation limits.**   If two or more different dissociation limits of a molecule have been determined, their differences must be equal to *possible energy differences of the separate atoms*.   In some cases this condition alone is sufficient to eliminate all but one of the possibilities for the dissociation products; in other cases it will at least reduce considerably the number of possibilities.

In general, in deriving the possible energy differences of the separate atoms, we need take into account only their low-lying states, provided we are not dealing with highly excited molecular states.

Sometimes in order to eliminate various possibilities for the dissociation products, it is sufficient to know only one dissociation limit accurately and one or several others approximately.

(2) **Application of the Wigner-Witmer correlation rules.** In determining the dissociation products, the Wigner-Witmer correlation rules have to be taken into account (see p. 315 f.). We have to find out whether the molecular state under consideration, whose dissociation limit is considered, can actually result from the assumed atomic states. Such considerations have been very important, for example, in the determination of the heat of dissociation of O_2 (see below).

(3) **Application of the non-crossing rule.** In the correlation of molecular states with those of the separated atoms, account must be taken of the rule that the adiabatic potential curves of states of the same species cannot intersect (non-crossing rule). Thus of the various possible dissociation products at a given dissociation limit, those may be excluded that would lead to a dissociation of the ground state into excited atoms when according to the Wigner-Witmer rules it could arise from normal atoms (see p. 437). In the case of excited molecular states the non-crossing rule has little consequence for the present problem because of the possibility of appreciable maxima in the adiabatic potential curves and the possibility that below these maxima the vibrational quanta behave as if an intersection would occur (see p. 297).[20]

(4) **Observation of atomic fluorescence.** As has been mentioned above, when atomic lines are observed in fluorescence by irradiation of a molecular gas with light beyond a certain dissociation limit, the state of excitation of at least one of the dissociation products is established. In the case of the alkali halides this applies to the alkali atom; the halogen atom can only appear in one of the two components of the ^2P ground state (since the other excited states are too high), and thus, except for the doublet splitting, the products of dissociation are determined.

(5) **Use of thermochemical data.** In several cases approximate values for heats of dissociation have been obtained either by direct thermochemical measurements or by deriving them from a combination of the heats of appropriate chemical reactions. These values refer of course always to dissociation into normal atoms. If in such a case an accurate spectroscopic dissociation limit is known as well, the dissociation products at this limit are immediately obtained and consequently a $D_0{}^0$ value that is more accurate than the thermal value may be derived. It was in this way that in the case of the halogens the dissociation

[20] It is only by arbitrarily excluding these possibilities that Gaydon (34) is able to determine what he believes are final values of certain controversial heats of dissociation. The case of N_2, discussed below, will exemplify this.

products at the convergence limit of the strong absorption system in the visible region were found to be one normal $^2P_{3/2}$ and one excited $^2P_{1/2}$ atom (see p. 389). In the case of I_2 and Br_2 this conclusion was later confirmed by the observation, in a weak red band system, of a second lower convergence limit whose separation from the first is exactly equal to the excitation energy of the $^2P_{1/2}$ state. For I_2 the spectroscopic D_0^0 value agrees within 0.1% with more recent precise thermochemical measurements of Perlman and Rollefson (1311) but is still slightly more accurate.

In the case of the hydrogen halides for which as yet no precise spectroscopic dissociation limits have been found, the thermochemical heats of formation (reduced to 0° K), combined with the spectroscopic D_0^0 values of the hydrogen and the halogen molecule, may be used to find accurate heats of dissociation.

It may be mentioned that in a few cases (for example Hg_2 and Cd_2) a thermochemical determination has been made using the spectrum merely as an analytical tool for the measurement of molecular concentrations as a function of the temperature.

(c) Examples

In order to illustrate the methods described in subsections (a) and (b), we shall now discuss in some detail the way in which the heats of dissociation of oxygen and nitrogen have been determined. The former is a case about which there is general agreement while there is still some uncertainty about the latter.

O_2.   The potential curves of all electronic states of the oxygen molecule for which sufficient data are available are given in Fig. 195. The vibrational levels in each of the states are also indicated. The well-known ultraviolet O_2 absorp-

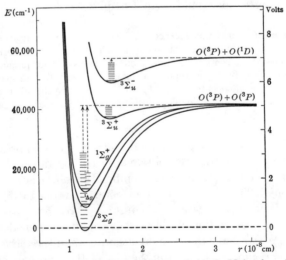

FIG. 195.   **Potential Curves of the Observed States of the O_2 Molecule.**   A number of states lying above 100,000 cm^{-1} [Price and Collins (571)] are not drawn, since sufficiently accurate data are not available for them.   The $^3\Sigma_u^+$ state has actually three more vibrational levels below the lowest one shown [see Herzberg (1044)].

tion bands (Schumann-Runge bands) correspond to the transition from the ground state $^3\Sigma_g^-$ to the highest state shown, $^3\Sigma_u^-$. These bands are found to converge to a very clear limit and are followed by a continuous spectrum. The convergence limit lies at 1759 Å, corresponding to 7.047 e.v. It represents a dissociation limit of the O_2 molecule (see Fig. 195).

The oxygen atom has three low-lying states: the ^3P ground state and the metastable states ^1D and ^1S, which lie 1.967 and 4.188 e.v., respectively, above the ground state. Higher excited states do not need to be considered in the discussion of dissociation products, since the lowest of these lies 9.1 e.v. above the ground state. If one of the atoms were in this state at the dissociation limit at 7.047 e.v., a negative value for $D_0{}^0(O_2)$ would be obtained.

The following combinations are conceivable as dissociation products at the convergence limit of 7.047 e.v.: $^3P + {}^3P$, $^3P + {}^1D$, $^3P + {}^1S$, $^1D + {}^1D$, $^1D + {}^1S$, and $^1S + {}^1S$. The last three possibilities can be eliminated immediately, since, according to Wigner and Witmer (see p. 318 f.), they cannot give a triplet state, and the upper state of the ultraviolet O_2 bands, as is shown by the rotational structure [Mulliken (509a) and Lochte-Holtgreven and Dieke (449)], is certainly a triplet state ($^3\Sigma_u^-$). It can be seen further from Table 28, p. 321, that a $^3\Sigma_u^-$ state cannot result from $^3P + {}^3P$ [Herzberg (305)]. Thus there remain only the two possibilities $^3P + {}^1D$ and $^3P + {}^1S$ for the dissociation products. From these, according to (VII, 19), we obtain for the heat of dissociation of O_2 either $7.047 - 1.967 = 5.080$ or $7.047 - 4.188 = 2.859$ e.v. The latter value is smaller than the energy of the highest observed vibrational level in the ground state, which lies 3.4 e.v. above the lowest level $v = 0$. Thus there remains only the value $D_0{}^0(O_2) = 5.080$ e.v. $= 117.2$ kcal/mol, corresponding to a dissociation at the convergence limit 7.047 e.v. into $^3P + {}^1D$. In this way the heat of dissociation of O_2 is unambiguously determined, and, in fact, from a single accurately known dissociation limit.

After the value of $D_0{}^0(O_2)$ had been established in the above described way a further convergence limit was found by Herzberg (310) in the forbidden bands $^3\Sigma_u^+ \leftarrow {}^3\Sigma_g^-$ (see Fig. 195). This limit lies at 5.116 volts and therefore obviously corresponds to a dissociation into two normal atoms $^3P + {}^3P$ (in agreement with the Wigner-Witmer rules). The small difference between the two values, 5.116 and 5.080 e.v., is probably due to the fact that at the dissociation limits different components of the ^3P term result whose total splitting amounts to 0.028 e.v.[21] The finding of the second convergence limit at the expected position demonstrates the reliability of the method. No doubt in the correct interpretation of the dissociation processes in O_2 is possible, and the accuracy of the spectroscopic $D_0{}^0(O_2)$ value (± 0.008 e.v.) is much greater than that of any $D_0{}^0(O_2)$ value obtained by thermal or chemical means.

---

[21] It should also be mentioned that the influence of rotation on the convergence limit has not yet been taken into account. If it were, the accuracy of the $D_0{}^0$ value could be further improved.

With the use of the second convergence limit of O_2 it is possible to determine the heat of dissociation without the implicit assumption made above that the ground state of O_2 dissociates into normal atoms: Since the energy difference of the two convergence limits equals the excitation energy of the ^1D state, it follows immediately that at the upper convergence limit one atom must be in the ^1D state. The other atom must be in the ^3P state, since otherwise the two atoms, according to the correlation rules, could not yield the $^3\Sigma_u^-$ molecular state. This is in agreement with the previous conclusion and confirms that the ground state dissociates into normal atoms.

N_2. While the heat of dissociation of the oxygen molecule is entirely based on observed convergence limits, that of the nitrogen molecule has been determined mainly from predissociation limits. The energy level diagram of N_2 is given in Fig. 196. In the $C\,^3\Pi_u$ state, a breaking-off was observed by Büttenbender and Herzberg (147) in three successive vibrational levels, $v' = 2$, 3, and 4. This has led to the exact determination of a dissociation limit [case I(b)] at 12.139 e.v. above the ground state.

TABLE 36.   DETERMINATION OF THE HEAT OF DISSOCIATION OF N_2

Possible Dissociation Products at 12.145 e.v.	$D_0{}^0$	
$^4S + {}^4S$	12.139	Not compatible with the type of predissociation at 12.139 e.v.
$^4S + {}^2D$	9.756	Not compatible with the type of the two predissociations at 9.8 e.v.
$^4S + {}^2P$	8.565	Not compatible with the energy of the predissociation at 9.839 e.v.
$^2D + {}^2D$	7.373	Actual value
$^2D + {}^2P$	6.182	Not compatible with the vibrational quanta of the ground state
$^2P + {}^2P$	4.991	Not compatible with the vibrational quanta of the ground state

The nitrogen atom has three low-lying states which alone need be considered as dissociation products: the ^4S ground state and the metastable states ^2D and ^2P, lying at 2.383 and 3.574 e.v. respectively, above ^4S. The various possible combinations of these states which may be considered as dissociation products at the limit 12.139 e.v. are given in the first column of Table 36. Depending on which pair we assume, we get one of the heats of dissociation $D_0{}^0$ given in the second column. The possibility $^4S + {}^4S$ is immediately eliminated by the consideration that it can give only Σ states, whereas it follows from the spectrum that the state causing the predissociation is a Π or a Δ state. (If it were a Σ state, only one of the Λ components of each level of the $C\,^3\Pi$ state could predissociate; see p. 417.) The possibilities $^2P + {}^2P$, and $^2P + {}^2D$ drop out, since the $D_0{}^0(N_2)$ values corresponding to them are smaller than the sum of the observed vibrational quanta in the ground state [6.6 e.v.; see Herman (1029)]. An elimination of the other possibilities, solely on the basis of the predissociation limit at 12.139 e.v., is not possible.

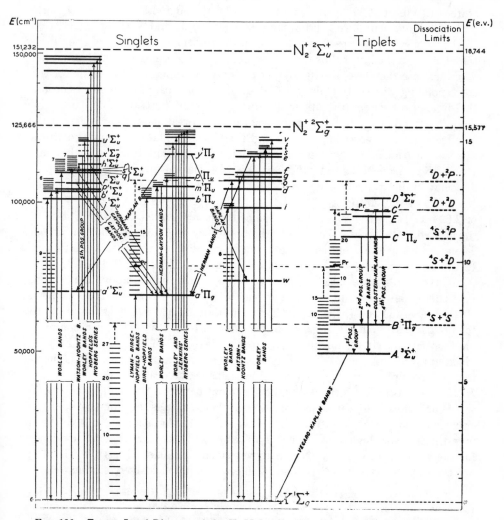

FIG. 196.   **Energy Level Diagram of the N₂ Molecule.**   The heavy (full) horizontal lines give the electronic states; the short thinner lines give the vibrational levels in each of them.   The four levels a', w, x, y whose heights above the ground state are not known are indicated by broken lines. The light broken horizontal lines ending in heavier lines at the right indicate dissociation limits.   The short heavy horizontal arrows designated Pr give the positions of observed predissociation limits. For the dissociation energy the value $D_0^0(N_2) = 7.373$ has been assumed (see text).   Two ionization limits are given (heavy broken horizontal lines across whole figure) corresponding to the two states $^2\Sigma_g^+$ and $^2\Sigma_u^+$ of N_2^+.   To avoid overcrowding of lines the higher members of the Rydberg series leading to the two ionization limits have been omitted as have the $v = 1$ vibrational levels observed for the Worley-Jenkins series [see Tanaka and Takamine (1457)].   An emission Rydberg series [Hopfield (331) and Takamine, Suga and Tanaka (660a)] has been observed very close to Hopfield's Rydberg series.   The upper states of this series (probably $^3\Sigma_u^+$) are not shown.

However, two other predissociations have been observed, one in the $B\,^3\Pi_g$ state with a limit at 9.839 e.v. [van der Ziel (732)], and another in the $a\,^1\Pi_g$ state with a limit at about 9.8 e.v. [Herman (1028)(1029), L. and G. Herzberg (1047)]. In the former a breaking-off is observed in the rotational structure of one vibrational level, in the latter thus far only a breaking-off of the vibrational structure. In either case therefore, the figure given is only an upper limiting value for the corresponding dissociation limit. But at least for the $B\,^3\Pi_g$ predissociation, the difference between the upper limiting value and the true value cannot be large (see p. 431 f.). In both predissociations the breaking-off is complete (that is, affects both Λ-type components) and therefore the states causing these predissociations must be Π or Δ states which cannot result from normal atoms ($^4S + {}^4S$). Depending on whether the dissociation products at the predissociation limit 9.839 e.v. are $^4S + {}^2D$, $^4S + {}^2P$, $^2D + {}^2D$, $^2D + {}^2P$, or $^2P + {}^2P$, we obtain for $D_0{}^0(N_2)$ the upper limiting values 7.46, 6.27, 5.07, 3.88, or 2.69 e.v., respectively. However, since the last observed vibrational state of the ground state lies at 6.6 e.v., only the first of these (7.46) need be considered. Therefore using the more accurate value resulting from the predissociation at 12.139 e.v. we find for the heat of dissociation of N_2 the value $D_0{}^0 = 7.373$ e.v.

Gaydon (34) has raised a number of objections to the above derivation. Particularly he has pointed out that the $B\,^3\Pi_g$ state is intermediate between Hund's cases (a) and (b), and that therefore the rule $\Delta K = 0$ does not apply to the predissociation in this state making it possible for a Σ state to cause complete predissociation. It is difficult to say to what extent the rule $\Delta K = 0$ may be violated for high K values when the $^3\Pi_g$ state approaches case (b). It is therefore not possible to meet Gaydon's objection unequivocally in the case of the $B\,^3\Pi_g$ predissociation. However, the argument based on the $a\,^1\Pi_g$ predissociation is not affected by this objection. Nevertheless it must be admitted that the situation is somewhat unsatisfactory, since this predissociation has not yet been investigated with high resolution. To the writer it appears that the evidence for the above choice of $D_0{}^0(N_2)$ is rather strong [see also Hagstrum (1016)]; but the possibility that $D_0{}^0(N_2) = 9.756$ e.v. is not conclusively eliminated.

The further objection by Gaydon that in the interpretation given here the non-crossing rule is not observed, must be rejected. It is entirely possible that the $^3\Sigma_u{}^+$ state resulting from normal atoms ($^4S + {}^4S$) goes over into the observed $A\,^3\Sigma_u{}^+$ state (see Fig. 196) via a considerable potential maximum (caused by the "original" intersection of two $^3\Sigma_u{}^+$ states, one with $^4S + {}^2D$ and the other with $^4S + {}^4S$ as a dissociation limit), thus accounting for the observation of stable vibrational levels of $A\,^3\Sigma_u{}^+$ considerably above the limit $^4S + {}^4S$ at 7.373 e.v.

The atomic excitation energies used in the above discussion have been taken from Moore (48c). A very useful list of the excitation energies of low excited states of all the more important atoms may also be found in Gaydon's book (34).

CHAPTER VIII

EXAMPLES, RESULTS, AND APPLICATIONS

1. Energy Level Diagrams; Molecular Constants

The energy level diagrams or potential curve diagrams of several important diatomic molecules have already been reproduced in previous chapters. They are those of H_2 (Fig. 160), Li_2 (Fig. 68), N_2 (Fig. 196), O_2 (Fig. 195), LiH (Fig. 171), and NH (Fig. 169). In addition, in order to have an example of each of the most important types of diatomic molecules, the energy level or potential curve diagrams of the molecules CO, NO, C_2, CN, Br_2, and CaH are reproduced in Figs. 197–202. In most cases all the observed vibrational levels of each of the different electronic states are indicated. Unfortunately, the heats of dissociation of CO, NO, C_2, and CN are still somewhat in doubt. In drawing the dissociation limits (broken horizontal lines) the $D_0{}^0$ values that seemed most probable to the author have been used, but an upward shift of some of these may be necessary [see Gaydon (34)].

The vibrational and rotational constants ω_e, $\omega_e x_e$, $\omega_e y_e$, and B_e, α_e, r_e of all diatomic molecules and all states for which adequate spectroscopic data are available are collected in the Appendix in Table 39. In addition the excitation energies of the excited states, the splittings of multiplet states and, for the ground states, the dissociation energies $D_0{}^0$ and higher rotational and vibrational constants are included wherever available. For more details of a given case the references cited in the second last column should be consulted. The force constants, which are not included, are readily obtained from the ω_e values according to the relation (III, 91).

From the data of Table 39, the *positions of the rotational and vibrational levels* in any of the electronic states may be derived with the aid of the formulae (III, 116) and (III, 121) or, for multiplet states, of the formulae given in Chapter V. The same data may also be used for a calculation of the *potential curves* either using the Morse function (III, 98) or more involved functions (see p. 102). Furthermore, from these data the positions of the bands and of the lines within the bands may be derived according to the formulae developed in Chapters III, IV, and V. For the wave lengths of the observed bands of the various molecules, the reader is referred to the extensive tables of Pearse and Gaydon (47).

The accuracy of the *internuclear distances* obtained from spectroscopic data is usually quite high (± 0.001 Å, or better). In a few cases these distances for the ground states have also been obtained by an entirely independent method, namely by electron or X-ray diffraction of the gas, liquid, or solid [see the review

451

articles by Brockway (128), Maxwell (1199), and Gingrich (986a)]. These cases are N_2 and O_2 [Gajewski (244)], S_2, Se_2, and Te_2 [Maxwell, Mosley and Hendricks (484a)], Cl_2 [Richter (589), Pauling and Brockway (555), Keesom and Taconis (1124)], Br_2 [Wierl (711), Pauling and Brockway (555), Vonnegut and Warren (687)], ICl [Pauling and Brockway (555)], and I_2 [Maxwell,

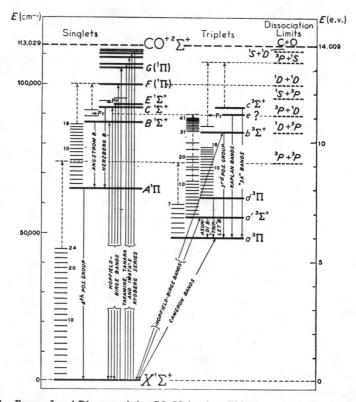

FIG. 197. **Energy Level Diagram of the CO Molecule.** This figure is similar to Fig. 196 for N_2. The dissociation limits are drawn under the assumption that $D_0^0(CO) = 9.141$ e.v. If instead the values 11.108 or 9.605 e.v. (see Table 39, p. 521) should prove to be correct the dissociation limits will have to be shifted upward by 1.967 or 0.464 e.v. respectively. The upper levels of two additional Rydberg series converging to the $A^2\Pi$ and $B^2\Sigma^+$ states of CO^+ [Tanaka (1456a)] and of several fragmentary band systems [Henning (298), Tanaka (1456a)] all lying above the first ionization potential have not been included.

Hendricks and Mosley (483)]. In all cases in which adequate band spectro-scopic data are available, the agreement is most satisfactory and within the accuracy of the diffraction method (which is appreciably lower than that of the spectroscopic method).[1]

---

[1] In the case of Te_2 a considerable disagreement was found; but the spectroscopic value in this case is unreliable.

Empirical relations.   In a class of similar molecules, such as the hydrides, it is clear that the molecular constants must vary in a periodic fashion as a function of the atomic number [Mecke (488)].   As examples, in Figs. 203 and 204

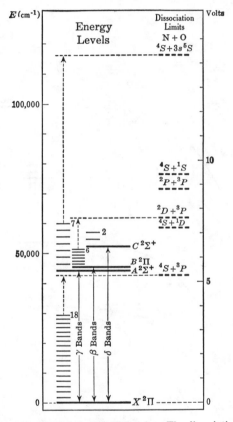

Fig. 198.   **Energy Level Diagram of the NO Molecule.**   The dissociation energy of NO has been assumed to be 5.296 e.v.   It is possible that the true value is 6.487 e.v.   In this case the dissociation limits at the right would have to be raised correspondingly.

are plotted the *internuclear distances* r_e (heavy curve) and *force constants* k_e (light curve) in the ground states of the hydrides and oxides respectively.   It is seen that for the hydrides the r_e values in each period of the periodic system decrease more or less regularly and then rise suddenly in going over to the next period.   This behavior corresponds exactly to that of the atomic radii.   In the case of the oxides, for which adequate data are available only in the second and third period, the curve of the r_e values has a minimum within each period.   This difference from the hydrides is clearly due to the fact that the oxygen atom has more outer electrons which can in favorable cases strengthen the binding (for example in CO, see p. 347).   The force constants show a behavior opposite to

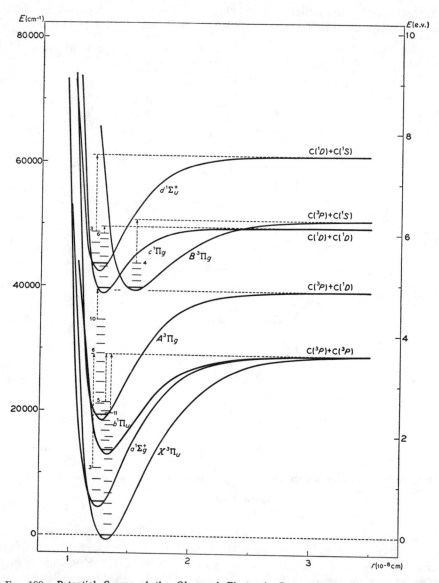

FIG. 199. Potential Curves of the Observed Electronic States of the C_2 Molecule. The relative positions of singlet and triplet levels is not definitely known. It has been assumed here that the excitation energy of the $a^1\Sigma_g^+$ state is 5300 cm^{-1} [see Herzberg and Sutton (1052)]. The dissociation energy, also, is not definitely established. The value $D_0{}^0 = 3.6$ e.v. has been assumed here (see Table 39, p. 513).

that of the internuclear distances (Figs. 203 and 204): for the hydrides they increase throughout each period and for the oxides they reach a maximum within each period.

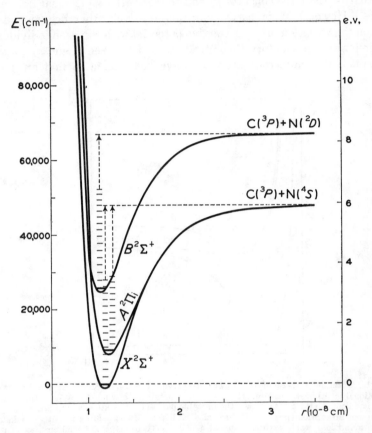

Fig. 200. **Potential Curves of the Observed Electronic States of the CN Molecule.** The magnitude of the heat of dissociation is not yet certain. If Gaydon's (34) high value for $D_0^0(\mathrm{CO})$ should turn out to be correct, a $D_0^0(\mathrm{CN})$ value of 7.6 e.v. would follow and the asymptotes in the figure would have to be altered correspondingly.

The internuclear distances and force constants of the elementary molecules (Li_2, B_2, C_2, $\cdots$) as well as of the nitrides and halides, so far as they are known, exhibit a similar behavior. The variation of k_e, r_e and other molecular constants within groups of homologous molecules has been studied by various investigators. For diatomic molecules formed from atoms of the carbon and oxygen groups see for example Vago and Barrow (1480).

The inverse behavior of internuclear distance and force constant is at least qualitatively in agreement with expectation from theory: with increasing strength of binding the potential minimum is shifted to smaller and smaller r values since the mutual repulsion of the nuclei balances the valence attraction at smaller and smaller r values; at the same time the minimum becomes narrower, that is, the force constant k_e and therefore the vibrational frequency ω_e increases. However it is clear that quantitatively this behavior depends greatly on the variation of the attractive potential with r in each individual case and therefore no satisfactory theoretical relation between r_e and k_e (or ω_e) has been found. But numerous attempts have been made to find an empirical

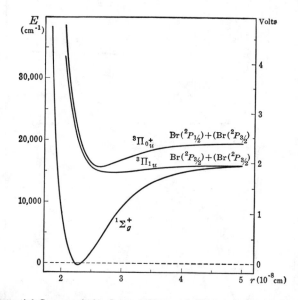

FIG. 201. **Potential Curves of the Lowest Observed Electronic States of the Br₂ Molecule.** The vibrational levels lie so close together that they cannot be drawn separately to the scale of the figure.   In addition to the states shown, the unstable part of a $^1\Pi_u$ state has been observed by means of a continuous absorption [Bayliss (82); see also Mulliken (519b)].   More recently Venkateswarlu (1496) has found four stable excited states at 47000, 55534, 61444, and 66500 cm^{-1} which are the upper states of a large number of diffuse emission bands, the lower states being the unstable states resulting from normal atoms in the $^2P_{3/2}$ or $^2P_{1/2}$ states.

relation between r_e and ω_e, mainly for the purpose of obtaining an approximate value of one of these quantities in cases in which only the other is known. We discuss here only a few of these attempts.

For the different electronic states of *one and the same molecule* the empirical relation

$$r_e{}^2\omega_e = \text{constant} \tag{VIII, 1}$$

holds fairly well [Birge (102) and Mecke (487)], and from it, if r_e is known for one state of a molecule, the r_e values for the other states can be derived approximately if only their ω_e values are known.

The r_e and ω_e values of the electronic states of *different molecules* are represented approximately by a formula first given by Morse (504) and later improved by Clark (157):

$$r_e^3 \omega_e \sqrt{n} = C. \qquad \text{(VIII, 2)}$$

Here n is the number of electrons outside the closed atomic shells, and C is a constant that is the same for molecules with the same closed atomic shells (for example, for molecules having only closed K shells, $C = 9550$; if the K shell is closed in one atom and the L shell in the other $C = 12,850$; in both cases it is assumed that ω_e is in cm^{-1} and r_e is in Ångstrom units). The relation (VIII, 2) does not represent the r_e and ω_e values of isotopic molecules. Such a representation can be accomplished if k_e is introduced in place of ω_e according to (III, 91). One obtains then [see Clark and Webb (851)]

$$r_e^6 k_e n = C' \qquad \text{(VIII, 3)}$$

where the constant C' is of course different from C in (VIII, 2). The relation (VIII, 3) without the n was suggested by Allen and Longair (61).

Another relation is *Badger's rule* (76) which is perhaps the one most frequently used particularly since it may readily be applied to polyatomic molecules:

$$k_e(r_e - d_{ij})^3 = C'' \qquad \text{(VIII, 4)}$$

Here d_{ij} is a constant that is different for each type of molecule (it is 0.68 Å if both atoms are in the first row of the periodic system, it is 1.25 if both are in the second row, and so on). C'' may be taken as the same for all molecules ($= 1.86$ if k_e is in 10^5 dynes/cm and r_e and d_{ij} in Å). However, a somewhat better agreement is obtained if C'' is given slightly different values for the different types of molecules [see also Glockler and Evans (988)].

Still other relations have been suggested by Huggins (1074), Linnett

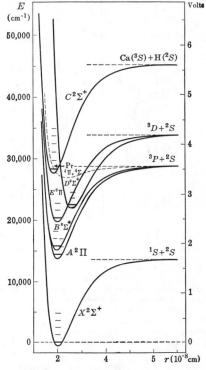

FIG. 202. **Potential Curves of the Observed Electronic States of the CaH Molecule after the Data of Grundström** (269). The correlation of the observed states with the dissociation products is not quite certain. It is conceivable that the predissociation in the $C\,^2\Sigma^+$ state leads to $^3D + {}^2S$ rather than $^3P + {}^2S$ as assumed here, in which case all dissociation limits would have to be shifted down by the difference between ^3D and ^3P of Ca. Grundström (1004) has assumed a different correlation [assuming Hund's case (c)], but with the same energies of the dissociation limits. Unlike Grundström we assume here that the predissociation is brought about by the $^4\Pi$ or $^4\Sigma$ state resulting from $^3P + {}^2S$ (broken-line potential curve).

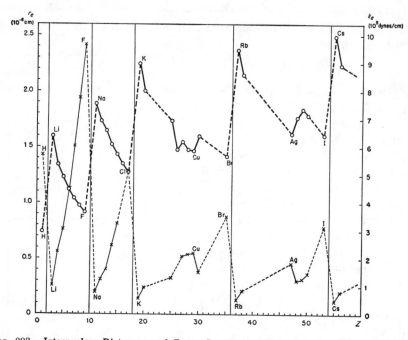

Fig. 203. **Internuclear Distances and Force Constants of the Hydrides as Functions of the Atomic Number.** The heavy curve gives the internuclear distances (left scale of ordinates), and the thin curve gives the force constants (right ordinate scale). The parts of the curves for which points are lacking are given by broken lines. The vertical lines indicate the periods in the periodic system. The few hydrides observed above $Z = 60$ have been omitted. The point indicated for FeH at $Z = 26$ should be deleted.

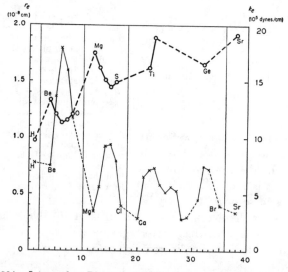

Fig. 204. **Internuclear Distances and Force Constants of the Oxides as Functions of the Atomic Number.** See caption of Fig. 203.

(1175)(1176), Guggenheimer (1006), Gordy (993), and Wu and Chao (1562). It should be emphasized that in all cases the agreement of the observed r_e and ω_e values with these relations is only within about $\pm 5\%$. The deviations in most cases are not due to experimental error (which is of the order of 0.1% or better) but due to shortcomings of the proposed formulae. Since the shape of a potential energy curve depends on the electron configuration as well as on the position of neighboring electronic states it is quite clear that any general relation between ω_e and r_e of different molecules cannot have exact validity. Nevertheless, as long as this is realized, the formulae are useful in cases in which experimental data are lacking.

Several authors have suggested empirical formulae relating r_e and k_e with the heat of dissociation for the ground states of various types of diatomic molecules [Huggins (1074), Sutherland (1443), Clark (850), Linnett (1175)(1176), Gordy (994), Bernstein (797)]. In view of the uncertainties with regard to the correct D_0^0 values of many important molecules, such attempts appear to be somewhat premature. Moreover, individual deviations from such formulae would be expected to be even more pronounced than for the relations between r_e and k_e, since D_0^0 depends on the individual force field for large as well as small internuclear distances.

Ionization potentials. Since comparatively little work has as yet been done in the vacuum region only a few ionization potentials of diatomic molecules have been determined spectroscopically. Those that have been derived in this way are given in Table 37. The method of determination is indicated in each

TABLE 37.   SPECTROSCOPICALLY DETERMINED IONIZATION POTENTIALS
OF DIATOMIC MOLECULES

Molecule	I (e.v.)	Determined from	References
H_2*	15.422	Pre-ionization (Rydberg series)	(94)
He_2*	4.251	Rydberg series	(33)
N_2	15.576	Rydberg series	(518)(722)(1559)
O_2	12.2	Cycle	(524)
CO	14.00_9**	Rydberg series	(1455a)(1456a)
CH	11.1	Cycle	(902)
HI	10.39	Rydberg series	(570)

* Refers to the lowest stable state $2\ ^3\Sigma_u{}^+$.

**A different value, 14.55 e.v. was given by Anand (750). It does not agree with the electron collision value of Hagstrum and Tate (1017).

case. For O_2 and CH the values given have been obtained from the heats of dissociation of ionized and neutral molecules and the lowest ionization potential of the atoms by means of the cycle

$$I(XY) = D_0^0(XY) + I(X) - D_0^0(XY^+) \qquad \text{(VIII, 5)}$$

For ionization potentials determined by electron collision experiments see Hagstrum and Tate (1017).

2. Applications to Other Fields of Physics

The following discussion of applications lays no claim to completeness. Only a few characteristic examples will be taken up.

Nuclear physics. As we have seen in Chapters III and V, the *nuclear spin* affects the observed band spectra in two ways: (1) it causes an *intensity alternation* with a ratio $(I + 1)/I$ in the ordinary rotational fine structure of bands of homonuclear molecules and (2) it causes a *hyperfine structure* of each otherwise single line in both hetero- and homonuclear molecules. The type of intensity alternation (whether the even or the odd lines are strong) is determined by another important nuclear property: the *statistics of the nuclei* which may be Einstein-Bose or Fermi-Dirac. The magnitude of the hyperfine structure splitting depends on the *nuclear magnetic moment* and the *nuclear quadrupole moment*. Conversely, all these nuclear properties can be determined from investigations of appropriate molecular spectra.

The intensity alternation has been studied for a considerable number of homonuclear molecules. The spin values and statistics determined in this way are collected in Table 38. Hyperfine structures in molecular spectra have been resolved and studied only in very few cases by the use of microwaves or of molecular beams (see Chapter V, section 6). The resultant spin values are included in Table 38 (some of them from the study of polyatomic molecules). For the numerous nuclei for which the spins have been obtained from hyperfine structures in atomic spectra, the reader is referred to Table 17 of A.A. It should be emphasized that the value $I = 0$ of a nuclear spin is very difficult to ascertain from hyperfine structure investigations (either in atomic or molecular spectra) since the apparent absence of a splitting may be due to the smallness of the nuclear magnetic or quadrupole moment. However, if a band of the corresponding homonuclear molecule shows alternate missing lines, the value $I = 0$ is immediately established without ambiguity. On the other hand, for high nuclear spin values the intensity alternation $(I + 1)/I$ is very slight and is difficult to measure with an accuracy sufficient for obtaining a unique value of I.

The nuclear statistics cannot be determined from hyperfine structure investigations. The only method other than the study of the intensity alternation in band spectra is the investigation of the collisions between identical nuclei (scattering). Thus far such measurements have been made only for H and He nuclei for which the results agree with those derived from band spectra.

Table 38 shows, as a result of the band-spectroscopic investigations of intensity alternation, that nuclei with even mass numbers (A), such as D^2, He^4, C^{12}, N^{14}, O^{16}, $\cdots$ follow Bose statistics, while, on the other hand, nuclei with odd mass numbers, such as H^1, Li^7, Na^{23}, $\cdots$ follow Fermi statistics. Moreover it is seen that the nuclear spin is integral for even mass number and half-integral for odd mass number. The fact that these two rules hold not only for nuclei with even $A - Z$ but also for those with odd $A - Z$ (for example D^2 and N^{14}) has been one of the essential reasons for rejecting the picture of the atomic nucleus consisting of electrons and protons and accepting the model of the nucleus consisting of protons and neutrons [see, for example, Rasetti (18a) and Bethe and Bacher (88)].

Nucleus	I*	Statistics	References
H^1	$\frac{1}{2}$	Fermi	(346)(397)(578)
D^2	1	Bose	(525)
T^3	$\frac{1}{2}$	Fermi	(897a)
He^3	$\frac{1}{2}$	Fermi	(902a)
He^4	0	Bose	(703)(33)
Li^7	$\frac{3}{2}$	Fermi	(282)
B^{10}	3(H)	—	(994b)
B^{11}	$\frac{3}{2}$(H)**	Fermi?	(900)(994b)
C^{12}	0	Bose	(511)
C^{13}	$\frac{1}{2}$	Fermi	(1098)(1470)(1468)
C^{14}	0	Bose	(1099)
N^{14}	1	Bose	(549)(294)(579)
N^{15}	$\frac{1}{2}$	Fermi	(424)(720a)
O^{16}	0	Bose	(510)(511)
O^{17}	$\frac{1}{2}$(H)?	—	(1179a)
O^{18}	0(H)	—	(1468)
F^{19}	$\frac{1}{2}$	—	(243)(51)
Na^{23}	$\frac{3}{2}$	Fermi	(388)(1104a)
P^{31}	$\frac{1}{2}$	Fermi	(368)(311)
S^{32}	0	Bose?	(545)
S^{33}	$\frac{3}{2}$(H)	—	(1466a)
S^{34}	0(H)	—	(1468)
S^{36}	0(H)	—	(1179a)
Cl^{35}	$\frac{3}{2}$(H)**	Fermi	(199)(511)(1467)
Cl^{37}	$\frac{3}{2}$(H)**	Fermi	(1417)(1467)
K^{39}	$>\frac{1}{2}$	Fermi	(468)(458)
Se^{77}	$\frac{1}{2}$(H)	—	(1437b)
$Se^{74,76,78,82}$	0(H)	—	(1437b)
Se^{80}	0	Bose?	(725)(542)
Br^{79}	$\frac{3}{2}$(H)	Fermi	(134)(1467)
Br^{81}	$\frac{3}{2}$(H)	Fermi	(134)(1467)
I^{127}	$\frac{5}{2}$	Fermi	(655)

* The values marked (H) are based on hyperfine structures in microwave absorption spectra. All others are based on intensity alternations.

** For B^{11}, Cl^{35}, and Cl^{37} measurements of the intensity alternation [Douglas and Herzberg (900), Elliott (199), Shrader (1417)] had given $I = \frac{5}{2}$. The deviation from the correct value illustrates the difficulty of measuring slight intensity alternations.

While a considerable number of nuclear magnetic moments have been determined by the use of molecular beams [see Kellogg and Millman (1125)], in all cases the magnetic fields used were so high that the spins were completely uncoupled from the intra-molecular fields and therefore the properties of the molecular energy levels involved were unimportant. However, for the determination of nuclear quadrupole moments these properties do matter: As we have seen in Chapter V, section 6, the magnitude of the splittings depends on the product of the quadrupole moment and the gradient of the electrostatic field within the molecule $(\partial^2 V/\partial z^2)$. Only in the case of D_2 has the latter been determined from theory with considerable precision and therefore has it been possible to obtain an accurate value for the quadrupole moment of the D nucleus from the observed magnetic resonance spectrum of the molecule.   This value is

$$Q(D^2) = +2.73 \times 10^{-27} \text{ cm}^2.$$

For the linear molecules ClCN and BrCN Townes (1466) has estimated $(\partial^2 V/\partial z^2)$ and from the observed hyperfine structures of the rotation spectra of these molecules obtains

$$Q(\text{Cl}^{35}) = -67 \times 10^{-27}, \qquad Q(\text{Cl}^{37}) = -51 \times 10^{-27},$$

$$Q(\text{Br}^{79}) = +210 \times 10^{-27}, \qquad Q(\text{Br}^{81}) = +160 \times 10^{-27} \text{ cm}^2$$

Somewhat smaller values were obtained by Gordy, Simmons and Smith (995).   The magnitude of the quadrupole moment of N^{14} revealed by the hyperfine structure in the inversion spectrum of NH_3 [Good (989)(855), Van Vleck and collaborators (879)], as well as in the rotation spectrum of ClCN [Townes, Holden, Bardeen, and Merritt (1467)] has not yet been determined with any accuracy.

Paramagnetism.   If the molecules of a gas have a permanent magnetic moment in the ground state, the gas will be paramagnetic and the *paramagnetic susceptibility* κ will depend on the temperature T according to Curie's law

$$\kappa = \frac{N\bar{\mu}^2}{3kT}. \tag{VIII, 6}$$

Here N is the number of molecules per cubic centimeter, and $\bar{\mu}$ is the average magnetic moment of the atom or molecule in question (see A.A., p. 205).   For diatomic molecules, as we have already seen (Chapter V, section 5), the average magnetic moment of the molecule is different for different J values, and therefore, strictly speaking, we should average over the different rotational levels.   However, it may be shown [see Van Vleck (35)] that, if the multiplet splitting is large compared to kT, we obtain a very good approximation if we simply substitute for $\bar{\mu}$ in (VIII, 6) the magnetic moment along the line joining the nuclei (see p. 300):

$$\bar{\mu} = (\Lambda + 2\Sigma)\mu_0. \tag{VIII, 7}$$

If, on the other hand, the multiplet splitting is small compared to kT, which is always the case in Hund's case (b) but will also apply in case (a) at sufficiently high temperatures, we have simply to sum the contributions to κ of the magnetic moment due to the electron spin and of that due to the orbital motion as if they were independent of each other [see also Nordheim (532)]; that is, we have to put

$$\bar{\mu}^2 = [\Lambda^2 + 4S(S + 1)] \mu_0^2 \tag{VIII, 8}$$

Most chemically stable diatomic molecules have a $^1\Sigma$ state as ground state (see Table 39) and therefore the magnetic moment $\bar{\mu}$ vanishes except for the exceedingly small contribution of rotational motion (see p. 299); such gases are diamagnetic.   But for the molecules O_2 and NO the ground state is not $^1\Sigma$ and these gases are therefore paramagnetic.

The ground state of O_2 is $^3\Sigma$; that is, $S = 1$, and therefore, according to (VIII, 8) $\bar{\mu}^2 = 8\mu_0^2$ has to be substituted in (VIII, 6).   This gives a susceptibility agreeing very well with the observed value [see Burris and Hause (835)].

The ground state of NO is a $^2\Pi$ state, whose splitting is 120.9 cm^{-1} and for which the $^2\Pi_{1/2}$ component lies lowest.   Since at room temperature the doublet splitting is of the order of magnitude of kT, neither of the above approximate formulae (VIII, 7 and 8) holds.   On the other hand, at low temperatures, when practically all the molecules are in the $^2\Pi_{1/2}$ state, the magnetic moment and therefore also the paramagnetic susceptibility is zero, since $\Lambda + 2\Sigma = 0$ (see p. 300).   At higher temperatures, when the multiplet splitting is small compared to kT, the $^2\Pi_{3/2}$ term, whose magnetic moment is not zero, is fully excited, and, according to (VIII, 8), $\bar{\mu}^2 = 4\mu_0^2$ has to be substituted in (VIII, 6)—that is, $\kappa = N \cdot 4\mu_0^2/3kT$.   The variation of the susceptibility between the two limiting values 0 and $N \cdot 4\mu_0^2/3kT$ for intermediate temperatures has been calculated in detail by Van Vleck (35). The agreement with experiment is very satisfactory [Scherrer and Stössel (616)(654), Burris and Hause (835)].

Collision processes. In the *excitation of atoms by collisions with other atoms or ions* a diatomic molecule (quasi molecule) is momentarily formed. It is therefore clear that the application of our knowledge of diatomic molecules can lead to a better understanding of these collision processes.

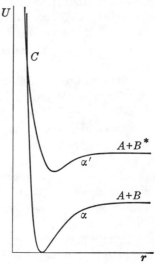

Let us consider the collision of two atoms in the ground state (Fig. 205). In general, the atoms remain on the same potential curve α while they approach each other and then separate again. However, if this potential curve crosses a potential curve α', leading to one normal and one excited atom, a transition (similar to predissociation; see Chapter VII, section 2) can take place such that after the collision one of the atoms is excited. But it can be seen that in the example shown in the figure the point of intersection can be reached and therefore a transition can take place only when the relative kinetic energy ($\frac{1}{2}\mu v_{rel.}^2$, where μ is the reduced mass) with which the two atoms (A and B) collide is considerably greater than the excitation energy of atom B.[2] As a result of the interaction of the two states (see p. 424 f.) and the overlapping of the eigenfunctions (tunnel effect), a transition

Fig. 205. Potential Curves Explaining the Excitation of Atoms by Collision with Other Atoms.

may take place somewhat below the point of intersection of the two potential curves, or even when they do not actually intersect but only come close together. The collision yield depends on the radiationless transition probability.

On the basis of such considerations, Weizel and Beeck (705) have discussed the experimental observations on the excitation and ionization of the inert gases by alkali ions [see also Rice (584), Jablónski (359), Stueckelberg (657), and Zener (1572)]. In a similar way one can understand the observation of Hanle (279a) that excitation of argon and helium by deuterium positive rays takes place at a higher kinetic energy than by hydrogen positive rays: although the potential curves and thus the energy of the point of intersection in Fig. 205 are essentially the same for H and D, the penetration of the narrow potential hill takes place at a lower energy for H than for D because of the smaller mass.

The same considerations may also be applied to the discussion of collisions

---

[2] It should be noted that even without this effect, on account of conservation of momentum not the total kinetic energy ($\frac{1}{2}m_A v_A^2 + \frac{1}{2}m_B v_B^2$) but only the relative kinetic energy ($\frac{1}{2}\mu v_{rel.}^2$) can be transformed into excitation energy.

of the second kind of the type

$$A^* + B \rightarrow A + B^*$$

Here the energies of initial and final states are in general more nearly alike and a crossing of the potential curves is therefore more likely. Application of the selection rules for predissociation (p. 416 f.) accounts immediately for the selection rules found to hold for these collisions [see A.A., p. 231 and Winans and McGowan (1546)(1204)].

A few other types of collision processes have been treated earlier in the discussion of continuous spectra (p. 394 f.) and of induced and inverse predissociations (pp. 432 and 414).

Nature of the liquid and the solid state. When molecules whose structure is known from spectroscopic investigations in the gaseous state are investigated in the liquid or solid state, changes in the spectra occur from which we can draw conclusions about the nature and properties of these states of aggregation. For example, McLennan and McLeod (485) have investigated the Raman spectra of liquid H_2, O_2, and N_2. In all three cases the vibrational bands observed for the corresponding gas appear in the liquid with only a slight shift, in agreement with the conclusion that in these liquids different molecules are bound to one another only very weakly, so that the binding within a molecule is not appreciably altered.

In H_2, even the rotational lines appear unaltered, showing that in the liquid the H_2 molecule can carry out *quantized rotations*. For all other diatomic molecules that have been investigated the rotational structure does not persist. However, in the Raman spectrum of liquid O_2 Saha (1367) has found an intensity distribution in the continuum adjoining the exciting line that is similar to that in the gas and which indicates some degree of free rotation in the liquid. In other liquids the intensity distribution in the continuous infrared or Raman bands is quite different from that in the corresponding discrete bands of the gas: Instead of two (or three) intensity maxima only one occurs. This would seem to indicate that the rotation of the molecules in the liquid is not even approximately free. In the case of HCl the transition from the spectrum of the gas at low pressure to that of the liquid has been investigated by West (1528) by observations at very high pressure (see Fig. 173 of Volume II). Above the critical pressure there is only one maximum in the infrared bands just as in the liquid.

The variations of the slight shifts of the vibration bands of HCl, HBr, and others in various solvents have been studied by a number of investigators both experimentally and theoretically [Plyler and Williams (563a), West, Arthur, and Edwards (707a), Leberknight and Ord (439b), Bauer and Magat (788)]. A theoretical relation between polarity of the solvent and the shift seems to be well confirmed by the experiments (see also Volume II, p. 534).

Electronic absorption spectra of diatomic molecules likewise are not changed very greatly when the gas is liquefied, solidified, or dissolved in another liquid. In the case of continuous spectra, for example of the halogens, there is in general a shift of the maximum to shorter wave lengths. This has been discussed in terms of the cage theory of liquids by Bayliss and Rees (791). The shift is fairly large in associated solvents such as H_2O. Frequently the absolute intensity is increased since the selection rules are less strict in the condensed phase (see p. 280). In the case of the halogens the vibrational structure in the excited state is completely washed out apparently because the internuclear distance is appreciably larger in the upper than in the lower state and since the time required to establish a larger "cage" for the absorbing molecule is much greater than the vibrational period [see Bayliss and Rees (791)]. On the other hand for the red and infrared absorption bands of O_2 for which the change of r_e is slight, the vibrational structure is well preserved, the individual bands occurring in the liquid and solid at very nearly the same wave lengths as in the gas. But no indication of rotational structure remains.

In addition to the gas bands of O_2 there occur much stronger bands due to O_4 in liquid and solid oxygen [Ellis and Kneser (202), Salow and Steiner (612), Prikhotko (572), Guillien (1007), and Smith and Barton (1427)]. Similarly for NO which is transparent at least to 2600 Å in the gaseous state new strong absorption continua above 5600 and below 4000 Å arise in the liquid state probably due to N_2O_2 [Vodar (1501), Bernstein and Herzberg (798)].

Determination of high temperatures. As already mentioned in Chapter IV, the intensity distribution in a band can be used for determining the temperature of the source of emission or absorption. The maximum of the intensity in a branch shifts toward higher J values with increasing temperature according to (III, 163). Thus from the position of the maximum the temperature of the source can be determined (compare Fig. 26). More accurate and reliable values are obtained when the whole intensity distribution (I) is measured: If in the case of $\Sigma-\Sigma$ bands the observed values of $\log I/(K' + K'' + 1)$ (or for other band types $\log I/S_J$, where S_J is the theoretical line strength) are plotted against $K(K + 1)$ [or $J(J + 1)$] a straight line is obtained (see Fig. 92) whose slope gives an accurate value of the temperature [see equation (IV, 80)]. Another method has been suggested and used by Knauss and McCay (407). They determine for which values of J (or K) the intensity in two overlapping branches of a band (for instance, a P and an R branch) is the same for equal or nearly equal wave number. This position shifts to larger J (or K) values with increasing temperature. This method has the advantage of not requiring precise intensity measurements but is limited to bands for which the head is not too far from the zero line. Even if the fine structure of the bands is not resolved the shape of the band profiles varies as a function of temperature and may be used for a determination of the temperature as has been shown in detail for the case of the violet CN bands by Smit-Miessen, Spier, and Smit (1430)(1432).

Instead of using the rotational intensity distribution one may also use the vibrational distribution using either the integrated intensities of (unresolved) bands or, with appropriate corrections [see Smit-Miessen, Spier, and Smit (1430)(1432)(1426)] the peak intensities of the band heads. As we have seen in Chapter IV, section 4(a), the logarithms of the sums of the strengths of all bands with a given initial state plotted against the vibrational energy of the initial state fall on a straight line whose slope is hc/kT; thus the temperature can be determined. This method was first applied to the CN bands in the carbon arc by Ornstein and Brinkman (1299)(1300) [see also Tawde and Trivedi (1459) and Voorhoeve (1504)]. If it is not feasible to measure the intensities of all the bands needed for this method one may yet determine the temperature if the intensities of at least two bands are measured and if the relative transition probabilities of these bands are known either from other experiments or from theory.

The band spectroscopic methods for the determination of temperature are naturally of particular importance when the usual methods cannot be used—for example, in the determination of the *temperature of the electric arc*. Ornstein and Brinkman (1299)(1300) were the first to apply this method to the arc using the CN and AlO bands. The temperature determined from the latter is much lower than that from the former since the AlO bands are emitted by the outer layers of the arc. A very careful investigation of the central part of the arc viewing it through a hole in the electrode and using the CN band at 3883 Å was carried out by Lochte-Holtgreven and Maecker (450) who obtained a temperature of 7600° K [compare also the more recent work of Duffendack and LaRue (912)].

It must be emphasized that the temperatures thus obtained are *effective* (rotational or vibrational) *temperatures*. They represent true temperatures only if either the excitation is strictly thermal or is of such a type that it does not affect the thermal distribution. A good indication that this condition is fulfilled in a given case is the agreement of the temperatures obtained independently from the rotational and vibrational intensity distributions. This is the case for the electric arc [see Ornstein and Brinkman (1299)(1300) and Tawde and Trivedi (1459)]. However, in flames frequently the rotational (and vibrational) distribution indicates a higher effective temperature than the true temperature since the excitation is due to chemical elementary reactions rather than to thermal collisions [Gaydon and Wolfhard (1549)(964)]. Similarly, in electric discharges frequently the vibrational distribution and sometimes also the rotational distribution deviates from the thermal distribution. The intensity distributions in absorption spectra of flames and discharges have been studied by Oldenberg (539) and White (1530).

The geophysical and astrophysical applications of the band spectroscopic temperature determinations will be discussed briefly in section 4.

Calculation of thermodynamic quantities. The heat content H^0 (*enthalpy*) of one mole of a perfect gas of pressure p, volume V, and temperature T is the sum of the external energy $pV = RT$ and the total internal energy E^0. Here R is the gas constant per mole which is equal to the product of the Avogadro number N and the Boltzmann constant k. The total internal energy E^0 is composed of the energies of translation, rotation, vibration and electronic motion of the molecules. The *translational part* can always be separated off and is given by the familiar expression

$$E_{\text{tr.}}^0 = \tfrac{3}{2}NkT \tag{VIII, 9}$$

which holds to an extremely good approximation even if wave mechanics is applied to the translational motion of the molecules.

For the contribution of the *internal degrees of freedom* (rotation and vibration) we have quite generally

$$E_{\text{int.}}^0 = E_0^0 + N_1\epsilon_1 + N_2\epsilon_2 + N_3\epsilon_3 + \cdots, \tag{VIII, 10}$$

where E_0^0 is the internal energy at $0°$ K (zero-point energy) and $N_1, N_2, \cdots$ are the numbers of molecules having internal energies $\epsilon_1, \epsilon_2, \cdots$ above the lowest energy. The summation is over all molecular states for which $N_n \neq 0$. Similar to (III, 165) the numbers N_n are given by

$$N_n = \frac{N}{Q_{\text{int.}}} g_n e^{-\epsilon_n/kT} \tag{VIII, 11}$$

where g_n is the statistical weight of the state n and $Q_{\text{int.}}$ is the *internal state sum* (or *partition function*):

$$Q_{\text{int.}} = g_0 + g_1 e^{-\epsilon_1/kT} + g_2 e^{-\epsilon_2/kT} + g_3 e^{-\epsilon_3/kT} + \cdots \tag{VIII, 12}$$

which is similar to (III, 164) except that the summation is now over all energy levels of the molecule while in (III, 164) it was only over the rotational levels of a given vibrational level.

Substituting (VIII, 11) into (VIII, 10), we obtain

$$E_{\text{int.}}^0 = E_0^0 + \frac{N\sum g_n\epsilon_n e^{-\epsilon_n/kT}}{Q_{\text{int.}}} = E_0^0 + NkT^2 \frac{d}{dT}(\log Q_{\text{int.}}) \tag{VIII, 13}$$

Thus we have for the *heat content of one mole*

$$H^0 = pV + E_{\text{tr.}}^0 + E_{\text{int.}}^0 = \tfrac{5}{2}RT + E_0^0 + RT^2 \frac{d(\log Q_{\text{int.}})}{dT} \tag{VIII, 14}$$

The *molar heat capacity* (for constant pressure) is the rate of change of heat content with temperature. From (VIII, 14) we obtain

$$C_p^0 = \frac{dH^0}{dT} = \tfrac{5}{2}R + R\frac{d}{dT}\left[T^2\frac{d(\log Q_{\text{int.}})}{dT}\right] \tag{VIII, 15}$$

According to (VIII, 14) and (VIII, 15), as was first suggested by Urey (1476) and Tolman and Badger (1464), the heat content, apart from the additive constant E_0^0, (that is, $H^0 - E_0^0$) and the heat capacity (C_p^0) of a molecular gas at any temperature can be predicted if the internal partition function $Q_{int.}$ is known. This function can be calculated according to (VIII, 12) if the energy levels of the molecule considered and their statistical weights have been derived from the spectrum.

The above formulae hold rigorously for an *ideal gas*. For the real gas they hold only at low pressure; at higher pressures small corrections must be added (see Volume II, p. 517). The evaluation of the ideal gas formulae can be greatly simplified if the interaction of rotation, vibration and electronic motion is neglected since then the internal partition function can be written as a product

$$Q_{int.} = Q_r Q_v Q_e \qquad \text{(VIII, 16)}$$

of functions depending on the rotational, vibrational and electronic energies and quantum numbers, separately. Consequently the contributions of the internal degrees of freedom to heat content and heat capacity can be resolved into a sum of terms depending on the rotation, vibration and electronic motion, respectively. One has a rotational and vibrational heat content and a rotational and vibrational heat capacity. The electronic contribution is negligible for molecules with a $^1\Sigma$ ground state if all excited electronic states are high so that only the first term in Q_e matters.

For sufficiently high temperatures the approximation (III, 166) can be used for Q_r and the *rotational heat content* and *heat capacity* become

$$H_r{}^0 = RT, \qquad C_{pr}{}^0 = R \qquad \text{(VIII, 17)}$$

which corresponds to classical equipartition. For the vibrational contribution frequently the harmonic oscillator with $\epsilon = hcG_0(v) = hc\omega v$ is a sufficiently good approximation yielding [compare (III, 159)]

$$Q_v = \frac{1}{1 - e^{-\omega hc/kT}} \qquad \text{(VIII, 18)}$$

and therefore we have for *vibrational heat content and heat capacity*

$$H_v{}^0 = R \frac{hc}{k} \frac{\omega e^{-\omega hc/kT}}{1 - e^{-\omega hc/kT}}, \qquad C_{pv}{}^0 = R \left(\frac{hc}{kT}\right)^2 \frac{\omega^2 e^{-\omega hc/kT}}{(1 - e^{-\omega hc/kT})^2} \qquad \text{(VIII, 19)}$$

Tables of these functions may be found in Aston (753). For better approximations to the rigorous formulae see Giauque (252a), Johnston and Davis (392a), Gordon and Barnes (264a), Kassel (398), and Mayer and Mayer (39). (Compare also Volume II, p. 505.)

The values for the heat capacity derived on the basis of the above formulae from band spectroscopic data agree extremely well with those observed by direct thermal measurements, as far as such measurements are available. As an illustration Fig. 206 shows the calculated variation of the molar heat capacity of chlorine, the thermal data being indicated by small circles. There is no question that if adequate spectroscopic data are available, the calculated heat capacities of diatomic gases are completely reliable and that in such cases a direct thermal determination which is very difficult at high temperatures is superfluous [see Eucken and Mücke (207)]. Moreover, the spectroscopic data allow one to predict the heat capacity even for chemically unstable molecules, such as OH, CH, and others, for which direct thermal measurements are impossible.

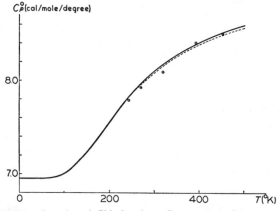

Fig. 206. **Molar Heat Capacity of Chlorine from Spectroscopic Data.** At low temperatures C_p° approaches the value $\frac{7}{2}R = 6.9552$. The deviation from this value at higher temperatures is due to the vibrational contribution C_{pv}°. The broken-line curve corresponds to the harmonic oscillator approximation, the full-line curve takes account of anharmonicity. The circles represent Eucken and Hoffmann's (924a) observed heat capacities (corrected for deviations from the ideal gas).

In calculating the heat content and heat capacity for high temperatures, account must be taken of the *low-lying electronic states* of the molecule considered. While different multiplet components of the lowest electronic state can be taken into account by calculating the electronic partition function Q_e in (VIII, 16) this is not possible for electronic states that have different rotational and vibrational constants from the ground state. Rather an additional term corresponding to all the rotational and vibrational levels of the excited electronic state has to be added to the partition function for the ground state. An example is the O_2 molecule with the two low-lying states $^1\Delta$ and $^1\Sigma$ (see Fig. 195). The effect of the $^1\Delta$ state on the heat capacity becomes noticeable at about 1500° K, that of the $^1\Sigma$ state at about 3000° K [see Johnston and Walker (1113), Lewis and von Elbe (444), Woolley (1558)]. Another example is the CH molecule for which the lack of any data about the low-lying $^4\Sigma$ state (see Table 33, p. 341) makes the calculated thermodynamic functions quite uncertain at high temperatures.

For heteronuclear molecules the *spins* of the two nuclei contribute a factor $(2I_1 + 1)(2I_2 + 1)$ to the statistical weight. This factor is the same for all levels and therefore appears as a constant factor in the state sum Q. It has no effect on heat content and heat capacity. However, for homonuclear molecules with non-zero spin there is a considerable effect for low temperatures when the classical equipartition value is not yet reached. In addition, the fact has to be taken into consideration, as was first pointed out by Dennison (181), that there are *two almost non-interconvertible modifications,* of which one has only the even and the other only the odd rotational levels [see Chapter III, section 2(f)]. Because of the extreme slowness of the interconversion (sym. ↔ antisym.) it is not sufficient simply to substitute the appropriate alternating statistical weights in (VIII, 12) [for H_2: $g_n = 1(2J + 1)$ for the even and $g_n = 3(2J + 1)$ for the odd rotational levels], but one has to calculate the heat capacity separately for each modification and then obtain the heat capacity for the mixture according to the abundance ratio of the two modifications.

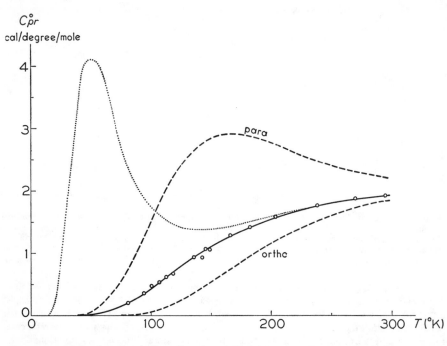

FIG. 207. **Molar Rotational Heat Capacity of Hydrogen as a Function of the Temperature.** The broken lines represent the heat capacities of ortho and para H_2, the full line that of the normal 3 : 1 mixture; the dotted curve represents the heat capacity of the equilibrium mixture (assuming no restriction of the ortho-para conversion). All curves are theoretical curves as obtained by Giauque (252a) [see also Davis and Johnston (883)] from the rotational constants of the H_2 molecule. The small circles represent heat capacity values observed by Cornish and Eastman (863a) and Eucken and Hiller (206) after subtraction of the translational contribution $\frac{5}{2}R$ (the vibrational contribution for H_2 is negligible in the temperature range considered).

Fig. 207 shows the variation with temperature of the heat capacity of the two separate modifications of H_2 (broken-line curves) as well as that of the ordinary 3 : 1 mixture (full-line curve) for which $C_p{}^0 = \frac{3}{4}C_p{}^0$ (ortho) $+ \frac{1}{4}C_p{}^0$ (para); that is, assuming that no interconversion takes place. The dotted curve represents the heat capacity curve that would be obtained if interconversion could take place freely, that is, if one would simply use the formula (VIII, 12) with appropriate weights corresponding to true equilibrium. As shown by Fig. 207 the observed points for ordinary hydrogen follow closely the full-line curve, confirming that no interconversion has taken place during the time of the measurements. On account of the large differences between the heat capacities of the two modifications and of the mixture, measurements of the heat capacity supply a convenient way of ascertaining whether or not a separation has been obtained. It is seen from Fig. 207 that at temperatures below $100°$ K if equilibrium is somehow established (see p. 139 f.), almost pure para-H_2 is obtained. For D_2, because of the smaller B value, the differences between the modifications occur at much lower temperatures [see Urey and Teal (677)]. For no other gas have they as yet been observed. Calculated curves for T_2 (tritium) have recently been given by Jones (1113a).

In addition to heat content and heat capacity, the *entropy* and *free energy* of a gas can be calculated once the partition function has been obtained from the spectroscopic data. According to statistical mechanics the entropy S^0 and the free energy F^0 of one mole of a perfect gas are given by

$$S^0 = S_{\mathrm{tr.}}{}^0 + R \log Q_{\mathrm{int.}} + RT \frac{d \log Q_{\mathrm{nt.}}}{dT} \qquad \text{(VIII, 20)}$$

$$F^0 = E_0{}^0 + \tfrac{5}{2}RT - S_{\mathrm{tr.}}{}^0 T - RT \log Q_{\mathrm{int.}} \qquad \text{(VIII, 21)}$$

Here $S_{\mathrm{tr.}}{}^0$ is the translational part of the entropy which can be readily calculated from the molecular weight, the temperature and the pressure (see Volume II, p. 520 f.). The parts of entropy and free energy that depend on the internal degrees of freedom can be resolved into a sum of terms due separately to rotation, vibration, and electronic motion if the resolution (VIII, 16) for $Q_{\mathrm{int.}}$ is a good approximation. Tables of the vibrational contribution for various values of ω/T may be found in Aston (753). Except at very low temperatures the rotational contribution may be calculated with the classical value (III, 166) for Q_r.

Experimentally values for the entropy are obtained from the observed low-temperature heat capacities of the solid, liquid, and gas and the heats of fusion and vaporization. As an illustration we consider the entropy of Cl_2 at $298.1°$ K. The calculated partial and total entropies are

$$S_{\mathrm{tr.}}{}^0 = 38.68 \qquad S_r{}^0 = 14.01 \qquad S_v{}^0 = 0.53$$

$$S^0 = 53.22 \text{ cal/degree/mole}$$

while experimentally [Giauque and Powell (984)] the value $S^0 = 53.32$ was obtained.

It must be emphasized that entropy and free energy according to (VIII, 20 and 21) do not remain unchanged as do heat content and heat capacity, when the partition function is multiplied by a constant factor. Thus the triplet character of the ground state of $O_2(^3\Sigma_g{}^-)$ has no effect on the heat capacity (except at extremely low temperatures when the triplet splitting is comparable to kT) but it produces a term $+R \log 3$ in the entropy and $-R \log 3$ in the free energy. Similarly the Λ doubling in the ground state of $NO(^2\Pi)$, even though the splitting is negligible, produces a term $R \log 2$. For all molecules with nuclei of non-zero spin the factor $(2I_1 + 1) \times (2I_2 + 1)$ in the partition function (see above) produces an additive term $\pm R \log (2I_1 + 1)(2I_2 + 1)$ in the entropy and free energy. However, this term is usually omitted since it does not produce a detectable effect except for H_2 and D_2. For homonuclear molecules the partition function at not too low temperatures is half that for a heteronuclear molecule with the same nuclear spin (as is most readily seen for $I = 0$) and therefore $-R \log 2$ has to be added to the entropy.

Statistical calculations using spectroscopic data have been carried out for the following diatomic molecules: H_2 [Giauque (982), Davis and Johnston (883), Woolley (1557)], D_2 and HD [Johnston and Long (1112), Urey and Teal (677)],[2a] C_2 [Gordon (991)], N_2 [Johnston and Davis (392a)], O_2 [Johnston and Walker (1113), Woolley (1558)], Cl_2 [Giauque and Powell (984)], Br_2 [Gordon and Barnes (992)], CN [Zeise (731)], CO [Johnston and Davis (392a)], NO [Johnston and Chapman (1109)], PN [McCallum and Leifer (1202)], OH [Johnston and Dawson (1110)], HCl [Giauque and Overstreet (983)], HBr [Gordon and Barnes (992)]. For some of these, extensive data (recalculated to modern fundamental constants) are given in Rossini's tables (50b) [see also Wagman, Kilpatrick, Taylor, Pitzer, and Rossini (1506)].

A more detailed presentation of the calculation of thermodynamic functions from spectroscopic data may be found in Aston (753), Wilson (1543), Zeise (731), and Jeunehomme (382a) (see also Volume II, Chapter V).

3. Applications to Chemistry

In addition to the problem of valence, or, quite generally, the problem of chemical attraction, already treated in detail in Chapter VI, the results of band spectroscopy are of importance in a number of other chemical problems.

Free radicals. Spectroscopic investigations have shown that many band spectra appearing in electric discharges, flames, and other sources are emitted by diatomic molecules that are not chemically stable under ordinary conditions—that is, by free radicals. While the existence of such free radicals in chemical reactions had been assumed long before the modern development of spectroscopy, their actual occurrence was first unambiguously demonstrated by the investigation of band spectra. Thus, for example, when water vapor is present in discharges or flames, bands often appear at 3060 Å whose analysis shows that they belong to the free OH radical. The occurrence of the radicals CH, C_2, CN, NH, and many others has been demonstrated in a similar manner. In addition to the proof of the existence and physical stability of these radicals, their internuclear distances, vibrational frequencies, heats of dissociation, and electronic structures have also been determined (see Table 39).

Emission spectra cannot in general serve as a *quantitative* measure of the concentration of free radicals, since, for the appearance of an emission spectrum, not only the occurrence of free radicals but also their excitation is necessary. However, the *absorption spectrum* can be used for such a quantitative test. Bonhoeffer and Reichardt (119) were the first to use the absorption of the OH bands for the detection of OH radicals in H_2O vapor heated to high temperatures (an emission of the OH bands does not occur under these conditions). The variation of the intensity with temperature indicates the variation of the OH concentration with temperature from which the heat of dissociation of H_2O into H + OH can be determined. This method has been considerably refined

---

[2a] See also the extensive tables recently published by Woolley, Scott and Brickwedde (1558a) and the calculations on T_2 and HT by Jones (1113a).

by Dwyer and Oldenberg (922) who obtained $D_0^0(H_2O \rightarrow H + OH) = 118.5$ kcal.[3] In addition, from this dissociation energy they obtained the absolute concentration of OH and therefore the absolute intensity (f-value) of the OH bands. Using this absolute intensity Oldenberg and his collaborators (540) (234)(540a) have determined by means of the absorption spectrum the concentration of OH radicals in discharges through H_2O and H_2O_2, and from the variation of this concentration, after shutting off the discharge, have drawn conclusions about the mechanism of the disappearance of these radicals.

Similar quantitative measurements were made on the CN radical by Kistiakowsky and Gershinowitz (1134) and White (1530). The free NH radical was detected by means of the absorption of the NH band at 3360 Å in the thermal decomposition of NH_3 by Funke (240)(948) and Franck and Reichardt (228). Absorption spectra of the radicals BCl, BBr, AlCl, AlBr, AlI, and SiCl were obtained after passage of a discharge through the vapors BCl_3, BBr_3, $AlCl_3$, $\cdots$ by Miescher (1231) and Maeder (1183).

The opposite of the detection of free radicals is the spectroscopic proof of the existence of stable diatomic molecules, for example, in the vapors of the alkali metals, previously regarded as monatomic.

Elementary chemical reactions in gases and chemiluminescence. The knowledge of the existence of free radicals in chemical reactions of gases is of great aid in the discussion of the individual steps of a given reaction and particularly of the chemiluminescence accompanying the reaction.

Let us consider as an example the *combustion of* H_2 described by the over-all formula

$$2H_2 + O_2 = 2H_2O \text{ (gas)} + 114.2 \text{ kcal} \qquad \text{(VIII, 22)}$$

It is clear that this reaction does not occur simply by the collision of two hydrogen molecules and an oxygen molecule. Indeed the observation of the OH bands in the oxy-hydrogen flame (as chemiluminescence) shows that OH radicals play a part in the reaction. Thus Bonhoeffer and Haber (117) were led to assume the following two elementary reactions in the formation of H_2O:

$$H + O_2 + H_2 = H_2O + OH + 98.9 \text{ kcal}, \qquad \text{(VIII, 23)}$$

$$OH + H_2 = H_2O + H + 15.3 \text{ kcal} \qquad \text{(VIII, 24)}$$

This means that, when once some H atoms have been formed—for example, by ignition—each of these produces an OH radical and an H_2O molecule according to (VIII, 23). Each OH radical in its turn, in addition to forming another H_2O molecule, reproduces an H atom according to (VIII, 24), and with the latter the process can begin again. We have a *chain reaction*. This mechanism has been supported further by Haber and his co-workers by experiments in which definite amounts of H atoms were added. While the presence of OH

---

[3] This value has been corrected slightly since Dwyer and Oldenberg employ an old conversion factor for their value of $D_0^0(H_2)$ used in obtaining $D_0^0(H_2O)$.

radicals has been established both by its emission and absorption spectrum in the oxy-hydrogen flame [see for example, Wolfhard (1549) and Kondratiev and Ziskin (1138)] no OH absorption was found in the slow thermal reaction at 550° C by Oldenberg and his collaborators (1292). However, this merely indicates that the OH radicals have a very short lifetime since immediately on being formed they react according to (VIII, 24). More details on the mechanism of the reaction between O_2 and H_2 and its modifications as well as references to more recent work may be found in Kassel (399), Oldenberg and Sommers (1293), and Hinshelwood (1054).

It is of particular importance that the *heats of the elementary reactions*, which obviously are not accessible to direct measurement, can be calculated accurately from the heats of dissociation of the molecules involved (see Table 39). For example, in reaction (VIII, 23) the energy required to dissociate all molecules on the left-hand side into atoms is $D_0^0(O_2) + D_0^0(H_2) = 220.4$ kcal, whereas on the right-hand side it is $D_0^0(H_2O) + D_0^0(OH) = 319.3$ kcal [for $D_0^0(H_2O)$ see below]. The difference of the two figures, 98.9 kcal, is evidently the heat of reaction. The heats of reaction for (VIII, 24) and for the reactions further below have been calculated in a similar way.

It is seen that both reaction (VIII, 23) and reaction (VIII, 24) are strongly exothermic. Actually, in (VIII, 23) the liberated energy is sufficient to excite OH to the upper state $^2\Sigma^+$ of the ultraviolet bands. It appears that the observed excitation of these bands in the oxy-hydrogen flame (as chemiluminescence) results directly from the elementary reaction (VIII, 23); that is, at least in a certain fraction of the successful collisions, excited and not normal OH radicals result [see L. Farkas (213)]. The excited OH can go over with emission of radiation to the ground state and then react further according to (VIII, 24), or else, while it is still excited, react with H_2. Since the energy that is carried by the excited OH is 92.8 kcal, the energies required for the reactions

$$OH^* + H_2 = OH + H + H - 10.4 \text{ kcal} \qquad (VIII, 25)$$

and

$$OH^* + H_2 = O + H + H_2 - 7.6 \text{ kcal} \qquad (VIII, 26)$$

are sufficiently small that they may occur at high temperatures. In addition to normal OH, reaction (VIII, 25) gives two H atoms, both of which can begin a new chain (*chain branching*), so that with rising temperature the reaction goes faster and faster (*explosion*). Similarly reaction (VIII, 26) may give rise to a chain branching since the O atoms can react further according to

$$O + H_2 = OH + H - 2.8 \text{ kcal} \qquad (VIII, 27)$$

In practice, however, this reaction seems to have a comparatively small probability.

The knowledge of the spectroscopically derived heats of reaction is of particular value in establishing *whether a postulated reaction actually can occur*.

For example, using the dissociation energies of H_2, O_2, and OH, the reactions

$$H + O_2 + H_2 = 2OH + H - 19.6 \text{ kcal} \qquad \text{(VIII, 28)}$$

and
$$H + O_2 = OH + O - 16.8 \text{ kcal} \qquad \text{(VIII, 29)}$$

are found to have the heats of reaction given. They are strongly endothermic and therefore do not occur at low temperatures. However, when once the temperature is raised by starting the reaction in the aforementioned manner, they may take place and then lead to chain branching.

Considerations similar to those for the combustion of hydrogen have been applied to many other combustion processes and chemical reactions [see the monographs by Kassel (40a), Lewis and von Elbe (42a), Gaydon (42b), and Schumacher (41)].

From the preceding discussion it is clear that a knowledge of the *excited states* of the radicals is of great importance not only for an understanding of the appearance or non-appearance of chemiluminescence but also of the occurrence of certain elementary processes. If a suitable excited state exists, as, for example, for OH, the radicals can carry with them the energy necessary for a reaction. Reactions (VIII, 25 and 26) would be quite impossible with un-excited OH, except at extremely high temperatures.

Atoms or molecules in *metastable states* are particularly effective as *energy carriers* in chemical reactions, since in general their energy cannot be given off as radiation. The action of metastable Hg atoms in the ^3P state (see A.A., p. 229) is, of course, well known. It is probable that the metastable $^1\Delta$ and $^1\Sigma$ states of the O_2 molecule (Fig. 195) play a similar part in some chemical reactions [see, for example, Hagstrum and Tate (1018), and Herman, Hopfield, Hornbeck, and Silverman (1035a)]. Moreover, the higher vibrational levels of a molecule must be regarded as metastable, and therefore they too may be of importance in chemical elementary processes. For example, the chemiluminescent emission of the sodium D lines observed in the reaction of chlorine and sodium vapor at low pressures was traced back by Beutler, Polanyi, and co-workers (98) to the strongly vibrating NaCl molecules resulting from the reaction

$$Na_2 + Cl = NaCl + Na + 80.5 \text{ kcal} \qquad \text{(VIII, 30)}$$

These NaCl molecules, on collision with Na atoms, may excite them to the upper state ^2P of the D lines [see also Gaydon (953)]. When no chemical reaction is possible the vibrating molecules may retain the vibrational energy over a large number of collisions as has been shown by the dispersion of sound in various gases [see the review by Oldenberg and Frost (1291)] and also by more direct absorption experiments [Dwyer (920)].

It must be emphasized that many exothermic elementary reactions do not occur for every collision but only for a certain fraction of these; for some reactions this fraction is so small that for practical purposes they may be disregarded. There are four main reasons that reduce the rate of elementary reactions:

(1) There may be an *energy of activation;* that is, in order that the reaction may take place, the energy of the system must be raised a certain amount above that of the initial state, or, in other words, a potential barrier must be overcome in going from the initial to the final state. This barrier is particularly high for reactions between saturated molecules (for example, $2H_2 + O_2 = 2H_2O$); however, it also occurs for reactions in which radicals or free atoms take part. Such activation energies were first treated theoretically on the basis of quantum mechanics by London (453). On the basis of a knowledge of the potential curves of the participating diatomic molecules it is possible to predict the approximate magnitude of the activation energy [see Eyring and Polanyi (209), Glasstone, Laidler, and Eyring (40b)].

(2) A *steric factor* may be present; that is, a reaction may take place only when the collision partners meet in space in a definite relative orientation; or the activation energy is much lower when they meet in one relative orientation than in another. For example, in the reaction (VIII, 27) the activation energy is least when the O atom approaches the H_2 molecule in the directon of the line joining the H nuclei while in reaction (VIII, 24) the activation energy is least when the H–H axis is approximately at right angles to the O–H axis and intersects it at the O nucleus (because in the H_2O molecule that is formed the valence angle is close to $90°$).

(3) For the formation of molecules from two atoms (or radicals) the *necessity of three-body collisions* reduces the rate very considerably. As we have seen in Chapter VII, section 1(b) two-body recombinations are in effect prohibited. For example, the reaction

$$H + H \rightarrow H_2 + 103.2 \, \text{kcal}, \qquad \text{(VIII, 31)}$$

which would lead to a breaking of the chain (VIII, 23 and 24), can take place only when a third atom or molecule is present at the moment of the collision. This reaction has therefore a very small rate; the chain (VIII, 23 and 24) is broken only after many links. To be sure the reaction (VIII, 23) is also a three-body collision. However, the reaction (VIII, 31) has a much smaller rate since it requires two H atoms [rather than one for (VIII, 23)] which are rare in the gas mixture. The association of atoms to form diatomic molecules has been thoroughly investigated both experimentally and theoretically by Steiner and Hilferding (651)(321), Rabinowitch and Wood (574)(576), Wigner (1540), and Rice (1347). These investigations have confirmed fully the validity of the prohibition of recombinations by two-body collisions and, in addition, have brought out interesting differences in the effectiveness of various atoms and molecules as third collision partners. As has been pointed out previously (p. 415) the possibility of an inverse predissociation does not increase the probability of two-body recombinations.

(4) If the reaction involves an electronic transition (see p. 463) the rate depends on the (radiationless) *transition probability.* On account of this the reaction will in general take place only in a small fraction of the otherwise favor-

able collisions. Examples are reactions like (VIII, 30) in which a transition from atomic to ionic binding occurs or reactions in which a transition from a singlet to a triplet state takes place. Such reactions are called *non-adiabatic*. Even for adiabatic reactions in which no change of electronic state takes place the crossing of a potential barrier implies that not every collision of sufficient energy leads to a completion of the reaction; the *transmission coefficient* is less than 1 [see Glasstone, Laidler, and Eyring (40*b*)].

Photochemical primary processes. A particularly important special case of chemical elementary reactions are the primary processes of photochemical reactions. The elucidation of these processes has been made possible in large measure by the investigations of molecular spectra [see, for example, Bonhoeffer and Harteck (38*a*), Rollefson and Burton (38*b*), and Noyes and Leighton (38*c*).

It is obvious that a photochemical reaction in a gas can be brought about only by light that is absorbed by the gas under consideration (law of Grotthus and Draper). The number of primarily influenced molecules or atoms is equal to the number of absorbed light quanta (Einstein's *law of photochemical equivalence*). The result of a photochemical reaction depends on the primary effect of the light (*primary process*) as well as on the subsequent reactions (*secondary processes*). The latter are of the same type as the chemical elementary processes just dealt with. We shall consider here only the primary process.

The kind of photochemical primary process occurring can frequently be deduced with certainty from the absorption spectrum of the gas in question on the basis of our earlier discussions. We have to distinguish three cases:

(1) *Absorption in a discrete band spectrum.* By light absorption the molecules are brought into a discrete excited electronic state, from which, in general, they return to the ground state (or another lower state) after a time of the order of 10^{-8} sec, with emission of radiation (fluorescence). The energy of excitation can be used for a chemical reaction only when a collision with another molecule takes place within the lifetime of the excited state. If this happens the excited molecule (a) may lose its excitation energy without producing any chemical reaction, (b) it may be dissociated by the collision with the other molecule, (c) it may react with the collision partner, or (d) it may simply transfer its energy to the collision partner, which then carries out the actual photochemical reaction (dissociates, for example). In all subcases (a)–(d) the yield of the photochemical reaction (referred to the number of absorbed quanta) depends strongly on the pressure and falls to zero with decreasing pressure.

The subcase (b) in which the absorbing molecule is itself dissociated by collision with like molecules or foreign gas molecules is essentially the same as the phenomenon of induced predissociation (p. 432 f.). For I_2 Rabinowitch and Wood (575) have studied this type of photoreaction in detail.

The subcase (c) in which a reaction with another molecule takes place has thus far been observed with certainty only when atoms are the absorbing systems. For example, Hg vapor mixed with O_2, on excitation with light of the Hg resonance line, gives the reaction

$$Hg^* + O_2 \rightarrow HgO + O \tag{VIII, 32}$$

or, mixed with H_2, gives

$$Hg^* + H_2 \rightarrow HgH + H \tag{VIII, 33}$$

An example in which a diatomic molecule is the absorbing system may be the photodissociation of CO by the Xe line 1295 Å [Faltings, Groth, and Harteck (926)] if Gaydon's high value of $D_0^0(CO)$ should prove to be correct. We would then have

$$CO^* + CO \rightarrow CO_2 + C \tag{VIII, 34}$$

In subcase (d) the absorbing molecule acts only as a sensitizer and suffers no lasting alteration in the reaction (*sensitized photochemical reaction*). This case also has been definitely identified only for atoms as absorbing systems. The best known example is the dissociation of H_2 by excited mercury or other excited atoms:

$$Hg^* + H_2 \rightarrow Hg + H + H \tag{VIII, 35}$$

(2) *Absorption in a continuous spectrum.* The absorbing molecule is raised by the incident light quantum into an unstable state and decomposes with more or less kinetic energy (see Chapter VII, section 1). This decomposition is independent of any secondary processes—that is, *independent of pressure*. The number of atoms formed is exactly twice as great as the number of absorbed light quanta. Naturally, the subsequent reactions of the atoms formed may depend on the pressure. In such cases also the total yield of the whole reaction will depend on the pressure.

The photodissociation of I_2 by light in the region of continuous absorption is an example that has already been discussed in detail (see p. 388 f.). Many photoreactions of iodine with other gases and vapors which have the same primary process have been investigated. The photoreaction that has perhaps been most extensively studied is the explosion of a hydrogen-chlorine mixture by irradiation. The primary process is here the dissociation of Cl_2 molecules by absorption of light quanta in the continuous region of the spectrum (see p. 390). The Cl atoms thus formed most probably react further according to the Nernst chain,

$$Cl + H_2 \rightarrow HCl + H \tag{VIII, 36}$$

$$H + Cl_2 \rightarrow HCl + Cl \tag{VIII, 37}$$

and thereby lead to explosion. Many similar photoreactions are known, in which the direct dissociation of the halogens, of oxygen, or of other molecules is the primary process [see the texts on photochemistry (38a)(38b)(38c)].

(3) *Absorption in a diffuse (predissociation) spectrum.* By light absorption the molecules are raised to a discrete state in which apart from the possibility of fluorescence there exists the possibility of a radiationless transition to an un-

stable dissociating state. Here also, as in (2), the decomposition is independent of collisions. However, not every molecule that has absorbed a light quantum in the diffuse absorption region dissociates, but only a certain fraction of them [given by $\gamma/(\beta + \gamma)$; see p. 408]. All the same, as we have seen previously, this fraction is very close to 1 if the absorption bands in question appear diffuse. Only if the absorption bands appear sharp even under large resolution and the predissociation is detected, say, by a breaking-off (or a weakening of the intensity) in emission, is a yield essentially different from 1 to be expected for the primary process. Indeed a photodecomposition with pressure independent yield is probably the most sensitive criterion for predissociation (see p. 413). The possibility of the presence of predissociation, that is, of a primary dissociation, even when there is no noticeable diffuseness in discrete absorption bands, has frequently been overlooked in the photochemical literature.

Only a few photochemical reactions brought about by predissociation have been investigated for diatomic molecules. The one best studied is the photodissociation of NO [Flory and Johnston (225)(937)] which has been discussed previously (p. 443). Unfortunately in this case there is no very strong spectroscopic evidence for predissociation. Other photoreactions of this type are: the photochemical formation of H_2S found when a mixture of H_2 and S_2 is irradiated by light in the region of the diffuse S_2 bands [see Henri (300)]; and possibly the photochemical ozone formation by irradiation of O_2 with light below 1900 Å [Flory (224)]. Many cases of this type are known for polyatomic molecules.

In several photodissociations of types (2) and (3) one (or both) of the atoms formed in the primary step is excited. If the excited atomic state is metastable (as for example the $^2P_{1/2}$ state of Cl or I) the excitation energy may be of importance for the subsequent reactions. If it is not metastable an atomic fluorescence will be observed accompanying the photodissociation. Similarly diatomic molecular fluorescences (for example CN, OH bands) have been found when appropriate polyatomic molecules are photodissociated [Neuimin and Terenin (1274), Jakovleva (1094)]. In addition a photodissociation into positive and negative ions has been found (see pp. 390 and 443).

Chemical equilibria. Not only is the study of elementary processes in chemical reactions advanced by band spectroscopy, but also the equilibria of chemical reactions in gases may be predicted on the basis of spectroscopic data. The equilibrium constant of a gas reaction

$$A + B + \cdots \rightleftarrows A' + B' + \cdots \qquad \text{(VIII, 38)}$$

is defined by

$$K_p = \frac{p_{A'}p_{B'}}{p_A p_B} \qquad \text{(VIII, 39)}$$

where $p_{A'}$, $p_{B'}$, $\cdots$ p_A, p_B $\cdots$ are the partial pressures of the produced molecules A', B', $\cdots$ and the reactant molecules A, B, $\cdots$ respectively.

According to a theorem of thermodynamics the equilibrium constant can be expressed in terms of the free energies F^0 of the gases involved as follows:

$$-RT \log K_p = \Delta E_0^0 + (F^0 - E_0^0)_{A'} + (F^0 - E_0^0)_{B'} + \cdots$$
$$- (F^0 - E_0^0)_A - (F^0 - E_0^0)_B - \cdots \qquad (VIII, 40)$$

where ΔE_0^0 is the total standard molar zero-point energy change which is positive when energy is absorbed in going from left to right in (VIII, 38).[4] The quantities $F^0 - E_0^0$ can be derived from spectroscopic data for every component of the system according to (VIII, 21). Tabulations of these quantities (divided by the temperature T) for various diatomic molecules may be found in the references quoted on p. 472.

If the heats of dissociation D_0^0 of the participating molecules are known from their spectra the zero-point energy change ΔE_0^0 (that is, the heat of reaction at $0°$ K) can be obtained from

$$\Delta E_0^0 = D_0^0(A) + D_0^0(B) + \cdots - D_0^0(A') - D_0^0(B') - \cdots \qquad (VIII, 41)$$

Thus in such cases the equilibrium and its dependence on the temperature can be completely calculated from spectroscopic data only. For polyatomic molecules the atomic heats of formation (see below) have to be substituted for D_0^0 in (VIII, 41).

In the case of a *dissociation equilibrium* of a diatomic gas $(A_2 \rightleftarrows A + A)$ the zero-point energy change ΔE_0^0 is simply equal to the heat of dissociation $[D_0^0(A_2)]$ of that gas. Using the spectroscopic D_0^0 of iodine Gibson and Heitler (257) were the first to calculate the dependence of the I_2 dissociation equilibrium on the temperature and found complete agreement with experiment [for more recent experimental determinations see Perlman and Rollefson (1311)]. Similarly, the dissociation equilibria of all other diatomic gases, so far as their heats of dissociation are known, can be predicted with certainty [see the tables of Lewis and von Elbe (445)].

For some equilibria, particularly those involving polyatomic molecules, only the $F^0 - E_0^0$ values of the participating gases can be calculated from spectroscopic data, while the heats of reaction must be taken from thermal measurements. However, it is just the former quantities that are in general not easily obtained from thermal data, whereas the heats of reaction can be relatively easily measured. The band-spectroscopic data are therefore of great value in calculating equilibria of gas reactions with considerable accuracy. More details about these calculations and some examples may be found in Volume II (p. 526 f.).

Atomic heats of formation and related topics. The atomic heat of formation of a polyatomic molecule is the energy released when the molecule in its lowest state is formed from free atoms. For diatomic molecules it is identical with the

---

[4] It may be noted that many authors use ΔH_0^0 in place of ΔE_0^0.

heat of dissociation. For polyatomic molecules the atomic heat of formation can be determined from the heat of combustion or the ordinary heat of formation (from the elements in their standard states) combined with the spectroscopic heats of dissociation of appropriate diatomic molecules. For example, the atomic heat of formation of the H_2O molecule is obtained by means of the following equations:

$$H_2O = H_2 + \tfrac{1}{2}O_2 - 57.10 \text{ kcal} \qquad \text{[see Rossini (50}b\text{)]}$$

$$\tfrac{1}{2}O_2 = O - 58.58 \text{ kcal} \qquad \text{(see Table 39)}$$

$$\underline{H_2 = H + H - 103.24 \text{ kcal} \qquad \text{(see Table 39)}}$$

$$H_2O = O + H + H - 218.92 \text{ kcal} \qquad\qquad\qquad \text{(VIII, 42)}$$

Thus the energy released in the formation of an H_2O molecule from free H and O atoms in their normal states (or conversely the energy required to split an H_2O molecule into free atoms) is 218.92 kcal/mole.

The atomic heats of formation of a large number of molecules, particularly of organic molecules, have been determined in a similar manner. However, in calculations involving organic molecules, one fact leads to difficulties—the *heat of sublimation L_1 of solid carbon* is thus far not known with certainty. This heat of sublimation could be obtained accurately if the heat of dissociation of CO were known with certainty, by using the following equations:

$$CO = C_{gas} + O - D_0^0(CO)$$

$$O = \tfrac{1}{2}O_2 + 58.58 \text{ kcal}$$

$$\underline{C_{solid} + \tfrac{1}{2}O_2 = CO + 27.20 \text{ kcal}}$$

$$C_{solid} \qquad = C_{gas} - D_0^0(CO) + 85.78 \text{ kcal} \qquad \text{(VIII, 43)}$$

In this way we see that $L_1 = D_0^0(CO) - 85.78$ kcal. At the present time three values of the heat of dissociation of CO seem possible: 9.141 e.v. = 210.82 kcal/mole [Herzberg (314)(1038), Long (1177)], 9.605 e.v. = 221.52 kcal/mole [Hagstrum (1015)], and 11.108 e.v. = 256.19 kcal/mole [Gaydon (34), Brewer, Gilles, and Jenkins (824)(825)]. The corresponding values for the heat of sublimation L_1 are 125.0, 135.7, and 170.4 kcal/mole respectively.

If L_1 were known uniquely we could calculate, for example, the *atomic heat of formation of methane* on the basis of the following equations:

$$CH_4 = C_{solid} + 2H_2 - 15.99 \text{ kcal}$$

$$2H_2 = 4H - 206.48 \text{ kcal}$$

$$\underline{C_{solid} = C_{gas} - L_1}$$

$$CH_4 = C_{gas} + 4H - (L_1 + 222.47 \text{ kcal}) \qquad \text{(VIII, 44)}$$

With the above L_1 values we obtain for the atomic heat of formation of methane

347.5, or 358.2, or 392.8 kcal/mole.   In a similar way one obtains for the atomic heat of formation of ethane 576.2, or 597.6, or 666.9 kcal/mole.

Since very few dissociation energies of polyatomic molecules are known it is convenient to introduce the concept of *bond energy* which is defined as the energy required to break a bond if all other similar bonds are simultaneously broken. These bond energies are immediately obtainable from the atomic heats of formation.   For example the bond energy of the OH bond in H_2O is one half of the atomic heat of formation of H_2O [see (VIII, 42)]—that is, 109.5 kcal/mole. Similarly the CH bond energy in CH_4 is one fourth of the atomic heat of formation of CH_4 [see (VIII, 44)]—that is, $\frac{1}{4}L_1 + 55.6$ kcal/mole.   Usually the assumption is made that these bond energies apply also to other molecules containing the same kind of bond.   If that is done many other bond energies may be calculated.   For example from the atomic heat of formation of C_2H_6 one obtains by subtracting six times the C–H bond energy a value $\frac{1}{2}L_1 - 7.4$ kcal/mole for the C–C bond energy.   Tables of bond energies determined in this way may be found in Pauling's book (37).

It must be emphasized that bond energies are only *average* dissociation energies and not individual dissociation energies.   For example in H_2O while the OH bond energy is 109.5 kcal/mole, the energy required to remove the first H atom is $D_0{}^0(H_2O \rightarrow OH + H) = 118.5$ kcal/mole (see p. 473) and the energy required to remove the second H atom—that is, $D_0{}^0(OH)$—is 100.4 kcal/mole. In some cases (for example NO_2) even larger differences between bond energies and dissociation energies occur.   This does of course not imply that a dissymmetry of the bonds exists in the polyatomic molecule.

4. Applications to Astrophysics

In recent years molecular spectra have become of increasing importance in the investigations of astrophysical problems.   The following résumé of this field is necessarily incomplete, but includes the few cases in which polyatomic molecular spectra are of importance in astrophysics.

Absorption spectrum of the earth's atmosphere.   In going through the earth's atmosphere the light from all extraterrestrial bodies suffers considerable absorption giving rise to *terrestrial Fraunhofer lines* or in certain spectral regions to the complete removal of all light.   This absorption is entirely molecular in origin.

In the *ultraviolet* there is a strong absorption below 3000 Å due to atmospheric ozone (O_3) which consists of a few diffuse bands followed by a very strong continuum which cuts off completely all stellar spectra below 2950 Å.   While at the earth's surface there is very little ozone, investigations of the solar spectrum at different altitudes of the sun and during balloon flights have shown that the ozone concentration rises to a maximum at a height of about 20 to 30 km and then falls off again.   The total amount of ozone corresponds to a layer of only 3 mm thickness at atmospheric pressure. The formation of ozone is brought about by the photodissociation of O_2 by the ultraviolet light of wave length below 2400 Å. More details about the formation of O_3 from O_2 in the upper atmosphere are given by Ladenburg (439), Wulf and Deming (724)(1563), and Chapman (845). The ozone absorption extends to about 2200 Å. But at 2400 Å the continuum joining onto the forbidden $^3\Sigma_u^+ \leftarrow {}^3\Sigma_g^-$ bands of O_2 sets in and prevents the reappearance of the solar spectrum below 2200 Å. Below 1950 Å a very much stronger transition of O_2, the Schumann-Runge bands $(^3\Sigma_u^- \leftarrow {}^3\Sigma_g^-)$ produces complete opaqueness of air even in very thin layers. This absorption extends to about 1300 Å. Below this wave length other band systems and continua of O_2 as well as of N_2 cause complete absorption. To be sure, in thin layers (of the order of 1 cm) air is transparent near 1200 Å [see Preston (1321)].

The absorption of light in the strong continuum of the Schumann-Runge bands causes a complete dissociation of oxygen in the top layers of the atmosphere. Nitrogen is dissociated to a much smaller extent by absorption in the forbidden $a\,^1\Pi_g \leftarrow X\,^1\Sigma_g^+$ bands below 1250 Å and subsequent predissociation [Herzberg and Herzberg (1047)]. Photo-ionization of N_2 by light of wave length below 800 Å and of N below 850 Å produces the ionized F layers of the atmosphere while photo-ionization of atomic oxygen produces the E layer. For a detailed discussion of the recombination processes in the ionospheric layers the reader is referred to Bates and Massey (786). In the region of the atmosphere where both N and O atoms are present there is of course the possibility of formation of NO molecules just as well as of O_2 and N_2 molecules. The fact that NO has not yet been detected is due to the occurrence of other much stronger absorptions at the wave lengths where NO would show absorption (N_2O has been detected, see below). It is significant that spectra taken from rockets above the ozone layer break off where the NO absorption due to the γ bands (see Fig. 198) would start [Baum, Johnson, Oberly, Rockwood, Strain, and Tousey (789), Hopfield and Clearman (1062)]. A detailed discussion of photochemical processes in an oxygen-nitrogen atmosphere has been given by Bamford (766).

In the *infrared and red* the most prominent absorptions are due to O_2 and H_2O. The strong red and infrared atmospheric oxygen bands due to $^1\Sigma_g^+ \leftarrow {}^3\Sigma_g^-$ and $^1\Delta_g \leftarrow {}^3\Sigma_g^-$ electronic transitions, respectively, have been discussed previously (p. 278 f. and Fig. 76). The strongest bands (0–0) occur at 7600 Å and 12680 Å, respectively. The absorption due to the rotation spectrum of H_2O blots out the solar spectrum completely above 24 μ and almost completely above 16 μ [see Adel (742)]. The vibration spectrum of H_2O gives complete absorption in the regions 7.5–5.0 μ and 3.0–2.5 μ. In addition there are very strong H_2O absorption bands at 3.2, 1.85, 1.45, 1.38, 1.13, 0.94 μ and less intense bands extending into the green part of the spectrum. A weak band of HDO at 7.12 μ has been found by Adel (739)(741). Narrow regions of complete absorption are also provided by the fundamentals of CO_2 at 15 ± 1 μ

and 4.25 ± 0.25 μ while weaker CO_2 bands occur at 12.6, 10.4, 9.4, 2.06, 2.01, 1.96, 1.64, 1.60, 1.57, 1.54, and 1.43 μ.   Two fairly strong bands due to O_3 occur at 9.6 and 4.75 μ [see Adel, Slipher, Barker, and Fouts (59)(60)].

The presence of a small amount of N_2O in the upper layers of the atmosphere was detected by Adel (738)(740)(741) by means of a weak atmospheric band at 7.78 μ and confirmed by Shaw, Sutherland, and Wormell (1412) who observed N_2O bands at 3.90, 4.06, and 4.50 μ in the solar spectrum.   More recently Migeotte (1234) discovered the presence of a small amount of CH_4 in the atmosphere by means of the P branch of the fundamental band at 3.3 μ whose Q and R branch is overlapped by the strong H_2O band at 3.2 μ.   Still more recently McMath, Mohler, and Goldberg (1217) have found all three branches of the first overtone of the CH_4 band at 1.67 μ in the solar spectrum at very high dispersion. At one time it was thought that a weak feature at 7.6 μ was due to N_2O_5 [Adel and Lampland (55b)], but it appears now that this is due to the fundamental ν_4 of CH_4 [Migeotte (1235)].   Mohler, Goldberg, and McMath (1240b) have identified lines of the 2.0 μ band of NH_3 in the solar spectrum; but Migeotte (1235a, c) was unable to find traces of the much stronger NH_3 band at 10.5 μ.[4a]

In addition to the strong infrared and ultraviolet bands ozone gives rise to a few very faint bands (the so-called Chappuis bands) in the visible part of the solar spectrum, at 6020, 5872, and 5750 Å.   For very low sun also the O_4 bands at 4770, 5770, and 6290 may be observed [see Dufay (907)].   All these bands are completely diffuse.

Emission spectrum of the earth's atmosphere.   Apart from the absorption bands occurring in the solar spectrum the emission band spectra exhibited by the aurora, the light of the night sky, and the twilight have been studied by a large number of investigators.   These spectra are produced in the upper layers of the atmosphere and their study gives important information about the physical conditions of these layers.

The spectrum of the *light of the night sky*, except for the green and red auroral lines of O I (see A.A., p. 158) and the sodium D lines, consists exclusively of bands.   The identification of these bands is a difficult problem because of the use of low dispersion necessitated by the faintness of the emission, and also because of the unusual conditions of excitation which favor forbidden transitions not readily observed in ordinary laboratory sources.   Recent lists of wave lengths in the night sky spectrum have been given by Dufay and Déjardin (910), Cabannes and Dufay (837)(838), and Barbier (767).   The strongest feature[4b]

[4a] Migeotte (1235b) has found indications of the lines of the CO band at 4.7 μ occurring with variable intensity.   This is possibly due to pollution of the atmosphere in cities where these observations were made.   No independent confirmation is as yet available.

[4b] Note added in page proof: Meinel (1222a) has recently obtained high dispersion night sky spectra in the region 7000–8800 Å which show strong well-resolved OH bands.   These bands are definitely to be identified as OH rotation-vibration bands with $\Delta v = 4$ and 5. The $\Delta v = 3$ sequence of this spectrum must be much stronger than the $\Delta v = 4$ sequence. It appears certain that the observed 10400 band is the strongest band, 4–1, of this sequence and that the interpretation given in the text must be abandoned.

(even stronger than the green auroral line) is the 0–0 band of the first positive group of N_2 ($B\,^3\Pi$—$A\,^3\Sigma$) at 10,440 Å [Herman, Herman, and Gauzit (1035), Stebbins, Whitford, and Swings (1434a)]. A few other much weaker bands of this system have also been found. The strongest bands in the blue and violet region and weaker bands in the whole accessible region have been definitely identified as belonging to the Vegard-Kaplan ($A\,^3\Sigma_u{}^+$—$X\,^1\Sigma_g{}^+$) system of N_2. The strongest ultraviolet bands, as was first suggested by Dufay (906)(909) are now definitely known to belong to the $A\,^3\Sigma_u{}^+$—$X\,^3\Sigma_g{}^-$ bands of O_2. The remaining bands, among them fairly strong ones, have not yet been identified with certainty [for various possibilities see Swings (1451)].

Chapman (153) was the first to suggest that the excitation of the light of the night sky is due to the three-body recombination of free oxygen atoms formed during the day by sunlight. If the similar recombination of nitrogen atoms is added [see Swings (1451)] the excitation of the definitely identified features of the night sky spectrum can be readily understood. The prominence of the forbidden transitions of N_2 and O_2 is due to the fact that the upper states have fairly low energy and that at the low pressures prevalent in the upper atmosphere deactivation by collisions is rare. On the basis of this theory one would expect NO to form just as well as N_2 and O_2 (see above). Possibly some of the unidentified features of the night sky spectrum are due to the forbidden $^4\Pi$—$^2\Pi$ bands of the NO molecule which have not yet been found in the laboratory. Difficulties of the recombination theory and various alternative theories have been summarized by Swings (1451).

Several attempts have been made to determine the height of the emitting layer from the variation of the intensity with zenith angle [see the reports by Kastler (1120) and Swings (1451), also Dufay and Mao-Lin (911)] but the results thus far obtained are rather divergent partly because different radiations are emitted at different heights and partly because the emission seems to be concentrated in certain patches which vary from night to night. It seems certain, however, that the height is not less than 80 km.

In the *twilight* there is a very considerable enhancement of the sodium D lines and an appreciable enhancement of the red O I lines. In addition, the violet $N_2{}^+$ bands are observed in emission. Both the Na and $N_2{}^+$ flash last only for a comparatively short time and are most probably due to direct excitation by sunlight from the ground states of Na and $N_2{}^+$, respectively (resonance fluorescence). Bricard and Kastler (826) have shown by studying the absorption of the Na emission by Na vapor that the line width corresponds to small atomic velocities and not to the large velocities that would arise, for example, by photodissociation of NaCl (see p. 391). The strengthening of the red O I lines may at least partly be due to the production of O ^1D atoms by photodissociation of O_2 on account of absorption of the Schumann-Runge continuum.

In the spectra of *aurorae* the forbidden OI lines are just as prominent as in the night sky, but otherwise the spectra are radically different. The forbidden lines of N I at 3466 Å (^2P—^4S) and at 5199 Å (^2D—^4S) have been observed in aurorae [Bernard (796)] but are absent in the night sky. The violet $N_2{}^+$ bands and the second positive group of N_2 in the ultraviolet are most prominent (next to the O I lines) in the aurorae but absent in the night sky. The red part of the first positive group of N_2 is much stronger, the Vegard-Kaplan bands of N_2 much weaker than in the night sky. Nothing is known about the infrared part of the first positive group in aurorae (0–0 band, see above). No bands of O_2 or O_2^+ have been definitely identified in the aurorae. Occasionally the D lines of sodium, the Balmer lines of hydrogen, and some He lines have been observed. A recent list of wave lengths has been given by Vegard and Kvifte (1491)(1489).[4c]

The main features of the auroral spectrum, except for the forbidden lines, are very similar, for example, to those of a hollow cathode discharge in nitrogen. This leaves little doubt that the excitation is similar, that is, by corpuscular rays [for more details also about variations of the spectrum and related topics, see Swings (1451) and Kastler (1120)]. There is no difficulty here about the height of the emission since it can be obtained by triangulation: the aurorae extend over a considerable range of heights whose lower limit is about 80 km.

The *temperature* of the atmospheric layers in which the auroral or night sky emission takes place can in principle be determined from the intensity distribution in the bands. In no case has it yet been possible to resolve the rotational fine structure.[4b] However, with a fairly high dispersion Vegard and Tonsberg (685)(1492) determined a temperature of about 230° K from the envelopes of the N_2^+ bands in aurorae. According to Vegard and Kvifte (1491) the temperature is essentially the same at the bottom and top of the aurorae and even for sunlit aurorae, that is, in daytime. Similar or slightly lower temperatures were obtained from less well resolved night sky spectra, using the $^3\Sigma_u^+-^1\Sigma_g^+$ N_2 bands and the $^3\Sigma_u^+-^3\Sigma_g^-$ O_2 bands, by Cabannes and Dufay (838), Barbier (768), and Swings (1450). These temperatures disagree strongly with those obtained from ionospheric data [see, for example, Bates and Massey (786)]. The reason for the discrepancy is not clear [see also Spitzer (1433a)].

Planetary atmospheres. In addition to the terrestrial absorption bands discussed above, the spectra of the planets Venus, Mars, Jupiter, Saturn, Uranus, and Neptune show further characteristic bands that obviously result from an absorption in the planetary atmosphere under consideration.

In the spectrum of *Venus*, Adams and Dunham (52) discovered three fairly strong bands degraded to the red with simple P and R branches and heads at 8689, 7883, and 7820 Å. Although these bands had not at that time been observed in the laboratory it was possible to conclude from their structure that they are rotation-vibration bands of the (linear) CO_2 molecule [see Adel and Dennison (55a)]. Subsequently the same bands were produced in the laboratory by Adel and Slipher (57) using a 45 m path in CO_2 at a pressure of 47 atm. At this pressure because of pressure broadening, the bands are quite diffuse and therefore not entirely comparable to the Venus bands. Using a path

---

[4c] Their identifications of certain features as due to O II and O III seem somewhat doubtful.

length of 2200 m Herzberg (1044) produced the bands in the laboratory with high dispersion at a pressure of 1 atm and even less. Under these conditions the appearance of the bands is closely similar to that in the Venus spectrum. Disregarding complications by multiple scattering it follows that there is about the equivalent of 1000 m atm of CO_2 in the Venus atmosphere above the reflecting layer. From the intensity maximum in the P branches of the Venus bands Adel (55) has estimated the average temperature of the Venus atmosphere (above the reflecting layer) to be about 50° C. More recently further much stronger CO_2 absorption bands have been found in the photographic infrared Venus spectrum (up to 1.25 μ) by Herzberg (1043) and in the infrared spectrum (up to 2.5 μ) by Kuiper (1147). For none of these bands has the fine structure been resolved. Thus far no gases other than CO_2 have been identified in the atmosphere of Venus. Particularly no rotation-vibration bands of CO were found as might have been expected on account of the photodecomposition of CO_2.

The strongest infrared bands of CO_2 (below 2.2 μ) which occur weakly in the solar spectrum on account of the CO_2 content of the earth's atmosphere have very recently been found by Kuiper (1148) to be appreciably stronger in the spectrum of *Mars* giving the first spectroscopic proof of a Martian atmosphere. The total amount of CO_2 is of the same order as that in the earth's atmosphere. On the other hand, the amounts of O_2 and H_2O on Mars are certainly much less, in the case of O_2 less than 1/1000 of the amount in the earth's atmosphere as was shown by Adams and Dunham (53)(917). In spite of the strong terrestrial O_2 and H_2O absorptions it is possible to ascertain these low limits in Mars since with the high dispersion possible in the region of these bands the planetary lines are shifted compared to the terrestrial lines on account of Doppler effect if the observation is carried out near quadrature (that is, when the angle planet-earth-sun is near 90°).

In the red and near infrared regions of the spectra of Jupiter, Saturn, Uranus, and Neptune [but not of Pluto, see Kuiper (1146)(1148)], several series of intense and broad absorption bands with complicated fine structures occur which for Uranus and Neptune extend with decreasing intensity to about 5000 Å. For Jupiter and Saturn the spectrum has been investigated by Kuiper (1147)(1148) up to 2.2 μ but above 1.6 μ there is almost complete absorption. By comparison with laboratory measurements (using CH_4 of high pressure and up to 45 m path) the observed absorption bands of the major planets have been unambiguously identified by Wildt (713a)(714) and Dunham (195) as rotation-vibration bands of *methane*. In the case of Jupiter and Saturn a few of the weaker bands have been shown to be due to *ammonia* [see also Adel and Slipher (56)]. The amounts of CH_4 which must be present in these atmospheres (above the reflecting layer) in order to account for the observed band intensities vary from about 0.15 km atm for Jupiter to 2.5 km atm for Neptune [Kuiper (1148)].[4d] The amount of NH_3 in Jupiter is only about 7 m atm, in Saturn still less. Unfortunately the fine structures of none of the bands have been analyzed so that

---

[4d] The earlier estimates of Adel and Slipher (58) were more than ten times as large.

it is not yet possible to determine the temperatures in these atmospheres.   Recently Kuiper (1148a) has found in the spectra of Uranus and Neptune a series of four (or five) nearly equidistant diffuse and very weak lines near 7500 Å. Their origin is obscure.   They are not due to CH_4, C_2H_4, C_2H_6, CO, CO_2, NH_3, H_2S, or N_2O.

The absence of hydrocarbons other than methane in the atmospheres of the major planets is easily understood.   Had they been present at one time they would either have frozen out on account of the low temperature or decomposed photochemically by absorption of sunlight [Wildt (715)].   The presence of large amounts of methane makes it very probable that there are also large amounts of molecular hydrogen present in these atmospheres [see, for example, Russell (607)].   A direct spectroscopic proof of the presence of H_2 is difficult, since the electronic absorption spectrum of H_2 lies below 1100 Å and since it has no normal rotation-vibration spectrum.   But for sufficiently large amounts of H_2, of the order of 10 km atm, it should be possible to observe its quadrupole rotation-vibration spectrum (see p. 279) [Herzberg (315)(1044)].   However, since the lines are very sharp and widely separated from one another, their detection is not easy and the fact that they have not been observed as yet need not mean that H_2 is absent.

A very important observation was made by Kuiper (1146) who found that the spectrum of Titan, the largest satellite of Saturn also shows the methane bands even though they are weaker than in Saturn.   None of the other large satellites in the solar system show evidence of an atmosphere in their spectra.

The subject of planetary atmospheres has been reviewed in much more detail than is possible here by Bobrovnikoff (819), Dunham (917), and Kuiper (1148).

Comets.   The spectra of comets consist of a large number of strong emission bands overlapped by a weak continuum with Fraunhofer lines which is due to sunlight reflected by the solid particles (meteorites) of the cometary nucleus. The only atomic lines that have been observed in emission in some comets when they come very close to the sun are the sodium D lines.   The band spectra occurring in the head (also called *coma*) and nucleus of the comet are very different from those occurring in the tail.   The most prominent bands of the tail are the $^2\Pi$—$^2\Sigma$ bands of CO^+ [known as comet-tail bands since they were discovered in comets before they were produced in the laboratory; see Fig. 11(a)]. In addition the violet N_2^+ bands are present.   In the heads of comets the strongest bands are the Swan bands ($^3\Pi_g$—$^3\Pi_u$) of C_2, the violet ($^2\Sigma$—$^2\Sigma$) and red ($^2\Pi$—$^2\Sigma$) CN bands and a group of bands near 4050 Å which have only recently been produced in the laboratory and which are very probably due to CH_2 [Herzberg (1039)(1040)].[5]   In addition, less strongly, appear the $^2\Sigma$—$^2\Pi$ bands of OH, the $^3\Pi$—$^3\Sigma^-$ bands of NH [Swings, Elvey, and Babcock (1452)], the

---

[5] It should be emphasized that while the identity of the 4050 Å group in comets with the laboratory spectrum is definite, the identification as CH_2 is not yet entirely conclusive. Note added in page proof: Recent experiments by Monfils and Rosen (1241) and Douglas (unpublished) show that this group is *not* due to CH_2.

$^2\Delta$—$^2\Pi$ and $^2\Sigma^-$—$^2\Pi$ bands of CH [Nicolet (1282), Dufay (905)], the $^1\Pi$—$^1\Sigma$ bands of CH$^+$ [Swings (1446)], and a group of bands near 6300 Å which are probably due to NH$_2$ [see Swings, McKellar, and Minkowski (1453)]. The last mentioned paper as well as one by McKellar (1209) contain extensive wavelength tables [see also Dufay and Bloch (909c)].

For all the band systems found in comets the lower state is the ground state of the particular molecule. This fact suggests strongly that the occurrence of these bands in comets is due to *resonance fluorescence* excited by sunlight. More detailed investigations have shown [Zanstra (730)] that this assumption can indeed explain the brightness of comets. It is also in agreement with the observed polarization of the cometary bands as shown by Ohman (1288) (see p. 305).

A further striking confirmation of the fluorescence mechanism has been obtained by a study of the intensity distribution in the rotational fine structure of the cometary bands [Swings (1447)(1448)(1449), McKellar (1207)(1208) (1210)]. Even though this structure is only incompletely resolved on the available cometary spectrograms certain intensity minima in the branches, for example, of the CN band at 3883 Å, are clearly visible. The positions of these minima correspond exactly to strong Fraunhofer (absorption) lines in the exciting solar radiation. Moreover, with a change of the velocity of the comet with respect to the sun these intensity minima change their positions in accordance with the shift of the Fraunhofer lines, as seen from the comet, on account of Doppler effect. These observations prove conclusively that all important cometary radiations are due to fluorescence.

If the effect of solar Fraunhofer lines is taken into account, the rotational intensity distributions allow the determination of *effective temperatures*. From the CN band 3883 Å McKellar (1207)(1210) found temperatures ranging from 200° K to 435° K for heliocentric distances ranging from 1.4 to 0.5 astronomical units [see also Dufay (904)]. This inverse dependence of temperature on distance is very marked and of course expected. However, it is very significant that from the CH bands appreciably lower temperatures are obtained (about 200° K at a distance of 0.5 a.u.) while from the C$_2$ bands very much higher temperatures are found (about 3000° K with only slight if any dependence on distance).

The explanation for these differences must be sought in the fact that the rotational distributions, on account of the extremely low pressure, are not produced by collisions with other molecules but are due to repeated excitations by solar radiation. For example, starting out with molecules with $J = 0$ or 1, after the first fluorescence process some molecules will be left in the states $J = 2$ and 3, after the second there will be some with $J = 4$ and 5, and so on. This accumulation of rotational energy will proceed almost indefinitely unless the molecule loses rotational energy by transitions in the far infrared rotation spectrum. The C$_2$ molecule has no rotation spectrum (except for a quadrupole spectrum which may be neglected) and therefore a very high rotational temperature results; but for CN and CH pure rotational transitions are possible and therefore a sort of equilibrium is established between the rotation-increasing effect of solar radiation and the rotation-decreasing effect of infrared emission leading to a fairly low rotational temperature whose value depends on the intensity of solar radiation (that is, on the distance from the sun), on the absolute intensity (f-value) of absorption of the particular band system,[6] and on the absolute intensity of the rotation spectrum (which is determined by the dipole moment of the molecule). On this basis the difference between the rotational temperatures of CH and CN is readily understood (for the f-values see p. 383).

Similar considerations apply to the excitation of the vibrational levels. In this way, combined with the Franck-Condon principle it is possible to account for the observations that the CN bands with $v' = 0$ have predominant intensity while the C_2 bands with $v' = 1, 2, 3, \cdots$ have similar intensities to those with $v' = 0$.

From the total intensity of the Swan bands in the head of a comet, assuming the fluorescence mechanism, it is possible to estimate the *partial density* of C_2 molecules. In this way Wurm (1566) obtained the density 10^5 molecules/cm^3. The other molecules mentioned above must have partial densities of the same order or less. Some gases that do not give observable fluorescence may conceivably have much higher partial densities. However, there is an upper limit of roughly 10^{10} molecules/cm^3 for the total density since otherwise the above discussed mechanism giving such a high rotational temperature of C_2 would not work because collisions between successive excitations would then reduce the rotational energy. The total intensity of the CO^+ emission of the tail gives in a similar way a partial density of 1 molecule/cm^3 for CO^+ [Wurm (1566)].

Let us now consider briefly the *formation of head and tail* of a comet. There can be no question that the molecules whose spectra we observe in comets are derived from gases released by the solid nuclei. Indeed when meteorites are heated in vacuum in the laboratory various gases such as H_2, H_2S, SO_2, CO, CO_2, CH_4 are given off. None of these gases have resonance band systems in the astronomically accessible spectral region and therefore they are not observed in cometary spectra. However, under the action of ultraviolet sunlight these molecules are dissociated or ionized giving rise eventually to the radicals and molecules whose spectra are observed. On account of the exceedingly low densities photochemical decomposition is the only way in which the various molecules can be formed. Thus the observation of CN and C_2 bands implies that the gases given out by the solid nucleus must contain molecules like HCN, C_2N_2, C_2H_6, $\cdots$ in which CN or C_2 occurs.

As was first pointed out by Wurm (728), in the act of photodissociation of the parent molecules considerable velocities are imparted to the products which accounts for the large extent of the head. On the other hand this extent is

---

[6] It should be noted that only absorption and fluorescence in the P and R branches but not in the Q branches leads to an increase of rotation. Therefore it is the absolute intensity of the P and R branches only that matters here.

limited by the fact that the fluorescing molecules (CN, C_2, CH_2, CH) in turn are photodissociated (or photo-ionized). For example, for C_2 a transition from the ground state to the $B\,^3\Pi_g$ upper state of the bands found by Fox and Herzberg (227) (see Fig. 199) will readily lead to a dissociation. On this basis it must be expected that the size of the head will decrease with increasing irradiation from the sun, that is, with decreasing heliocentric distance. This is indeed found to be the case [see Wurm (728)(1566)]. At the same time we understand why CN, C_2, CH_2, and CH bands are not observed in the comet tail [Wurm (728)].

Strictly speaking for each molecule there is a different radius of what might be called the half-value sphere, that is, the sphere which half of the molecules reach without decomposing. These radii are determined by the intensity of solar radiation in the photochemically active spectral region and by the f-value of the transition leading to photodissociation. For CN and C_2 these radii seem to be about the same, while for CH_2 the radius is distinctly smaller. If one assumes that for CO^+ and N_2^+ this radius is very much larger one can understand immediately that CO^+ and N_2^+ bands (and no others) occur far outside the head of the comet, namely in its tail. Laboratory spectra of CO^+ and N_2^+ do not contradict this assumption.

It remains to consider how the characteristic *shapes of head and tail* arise. For this purpose account must be taken of the recoil of the molecules following the act of absorption of light quanta (that is, the effect of radiation pressure must be introduced). The magnitude of this effect depends on the intensity of solar radiation and on the f-value of the transition. On account of this effect a motion away from the sun is superimposed on the motion away from the cometary nucleus giving rise to an elongated shape of the head as observed. In the case of the molecules CO and N_2 which do not arise by photodissociation of parent molecules the average velocities away from the cometary nucleus are small thermal velocities (corresponding to a very low temperature); moreover, upon photo-ionization to form CO^+ and N_2^+, the electron takes up most of the excess energy, that is, the radial velocities of these ions are much smaller than for CN, C_2, $\cdots$. Therefore under the action of radiation pressure narrower elongated formations, tails, arise which extend much farther from the nucleus because of the longer "life" of CO^+ and N_2^+ as compared to CN, C_2, $\cdots$.

Many more details about this interpretation of cometary forms may be found in Wurm's original papers (728)(1566). It must be emphasized that there are still considerable difficulties. The observed accelerations of some condensations in the tails are appreciably larger than seems possible on the basis of radiation pressure. The very sharp rays occurring in some tails are difficult to understand. It is not impossible that electric and magnetic fields play a part in the formation of comet tails.

Stellar atmospheres. Absorption bands of a number of diatomic molecules are found in the spectra of the sun and all stars of a temperature equal to or lower than that of the sun [see, for example, the reviews by Swings (660) and

Bobrovnikoff (115a)]. In the case of the sun, since very large dispersion can be used, weaker bands can be detected than in other stars of similar type and in addition different parts of the surface can be studied. According to the recent work of Babcock (763) [see also Stenvinkel, Svensson, and Olsson (1436)] bands of the following molecules have been definitely identified on the disk of the sun: CN, CH, C_2, NH, OH, SiH, MgH. In sunspots all these bands are strengthened and in addition bands of TiO, CaH, ZrO, YO, AlO, and BH are found of which the first two are particularly prominent. In disk spectra bands of these molecules are barely discernible. There is also some evidence for the molecules MgO, ScO, MgF, and SrF both in disk and spot spectra. The molecules H_2, N_2, O_2,[7] NO, CO, SiO and others which are certainly present with considerable abundance in the solar atmosphere, cannot be detected since their resonance systems lie beyond the spectral region that is accessible with ordinary means. Some of them will undoubtedly be detected as soon as solar spectra taken from rockets have been extended further. It is interesting to note that the CN bands have been observed in emission in the spectrum of the solar chromosphere (flash spectrum).

In stars of the same type (G) as the sun only the more prominent of the above-mentioned molecular bands have been observed, particularly CN and CH which have even been observed in slightly earlier types. As one goes to later types, that is, lower temperatures, the bands become more and more prominent, until for stars of types M, S, N, and R they form the most characteristic features of the spectrum. Even in K type stars the bands are sufficiently strong to be visible under low dispersion and are indeed used for the purpose of classification into the subdivisions of type K both with regard to spectral type ($K0$–$K5$) and luminosity (dwarfs, giants, supergiants) [see Morgan, Keenan, and Kellman (43)].

In M-type stars the most prominent bands are those of TiO in the visible, red, and near infrared spectral regions. Next in prominence are the hydrides MgH, SiH, and AlH followed by the oxides ZrO, ScO, YO, CrO, AlO and BO [see the recent high dispersion work on β Pegasi by Davis (884)]. A number of other molecules are probably present. Some stars of this type have fairly prominent VO bands [for example, R Leonis, see Merrill (44)]. In one M-type star (o Ceti) for a short time the AlO bands were observed in emission. Recently Nassau, van Albada and Keenan (1270a) have found new as yet unidentified bands in the region 7300—8000 Å of late M-type stars.

In S-type stars the ZrO bands are most prominent while TiO and YO bands are weaker but definitely present. Keenan (1122) has found very prominent LaO bands in some S-type stars.

In stars of types R and N, which may be combined in the one type C [Keenan and Morgan (1123)] the Swan bands of C_2 and the violet and red CN bands are of dominating intensity. These stars are therefore often referred to as

---

[7] Babcock (763) considers as established the presence of Schumann-Runge bands with very high v'' which lie in the violet region.

carbon stars. In addition the CH and NH bands are fairly strong [see Wildt (1541)]. Particularly the former vary greatly in intensity in different stars of this type and are unusually strong for a small group of peculiar carbon stars [CH stars, see Keenan and Morgan (1123)(1121)]. In another group of carbon stars a number of strong red shaded bands were found in the green part of the spectrum which up to now have not been identified [Sanford (1371), McKellar (1211)]. McKellar (1212) has recently observed in several late N stars the $\lambda 4050$ group first observed in comets (see above) and due to CH_2.[5] This is the only case thus far of the observation of a polyatomic molecule in a stellar atmosphere. Possibly the unidentified Sanford bands are also due to a polyatomic molecule [see McKellar and Swings (1216)].

From the observed band intensities in principle the relative (and absolute) *concentrations* of the respective molecules in stellar atmospheres can be derived if the absolute intensities (f-values) are known. Assuming all f-values to be of the same order Russell (606) has given observed relative abundances of the molecules in the solar reversing layer. With the few f-values that have more recently become available Lyddane, Rogers, and Roach (1182) have calculated the following relative logarithmic abundances (introducing the correction discussed on p. 386): 2.75 for OH, 1.59 for CN, 2.09 for NH, 2.76 for CH, and 1.27 for C_2.

Conversely, as was first done by Wildt (713) and in more detail by Russell (606) and Rosenfeld and Cambresier (600)(149), the molecular abundances can be predicted by calculating the *dissociation equilibrium in stellar atmospheres* using the relative abundances of the various elements known approximately from the line spectrum of the sun. For this purpose for a given molecule first the equilibrium constant K_p is calculated from equation (VIII, 40) and then the partial pressure of the molecular gas from (VIII, 39). For these calculations, of course, a knowledge of the dissociation energy as well as the vibrational and rotational constants of the molecule is required. The authors mentioned [see also Nicolet (1281)] have obtained in this way not only the relative concentrations for a given temperature and pressure but also the variations of these concentrations as a function of temperature (that is, spectral type) and pressure (that is, stellar mass).

Qualitatively there is fair agreement between the observed variations of band intensities in the sequence of G, K, M stars with those predicted. Thus in agreement with the observed band intensities the predicted concentration of molecules like CN, CH, and C_2 passes through a maximum with decreasing temperature while the concentration of TiO approaches a very large nearly constant value in late M stars. The calculated relative concentrations for a given temperature agree only in a very rough way with those observed. For example, Russell (606) obtained the following logarithmic relative abundances for the sun: 5.3 for OH, 2.3 for CN, 4.0 for NH, 3.1 for CH, and 0.9 for C_2 which should be compared with the observed values given above. The discrepancies are mainly due to uncertainties in the dissociation energies [thus according to Dwyer (921) with the new value of $D_0{}^0(OH)$ the logarithmic concentration of OH is 4.7 instead of 5.3] and partly to uncertainties of the f-values and the atomic abundances. The fact that, in spite of the great abundance of hydrogen, compounds other than hydrides have abundances comparable to those of the hydrides is due to the greater dissociation energies of these compounds.

In the case of the carbon stars the observed molecular abundances differ radically from those predicted on the basis of the solar abundances of the elements. However, according to Russell (606), rough agreement is obtained if the abundances of carbon and oxygen are exchanged, that is, if a large abundance of carbon and a fairly small abundance of oxygen is assumed. In this way one can immediately understand the fact that the TiO bands and other oxide bands. do not occur in these stars. On the other hand, the C_2, CN, and CH bands are intense, since the C atoms are no longer used up in forming the much more stable CO molecule.

In a similar way in other cases (for example, S stars, CH stars) from the observed band intensities, combined with equilibrium calculations, conclusions about deviations from the normal abundances of the elements can be drawn, or, in the case of elements such as B and F that do not show atomic absorption lines in the accessible region, even the "normal" abundances in the sun can be determined. Furthermore, the presence in great abundance of molecules such as H_2, CO, O_2, N_2, NO, SiO, which are not directly detectable spectroscopically, can be demonstrated with great certainty with the help of these equilibrium calculations. In all these considerations it is of course assumed that there is thermodynamic equilibrium in the stellar atmosphere considered. They would not be applicable, for example, to novae for some of which CN and C_2 bands have been observed.

The variation of band intensities with temperature demanded by the calculation of dissociation equilibria is apparent not only in stars of different spectral type but even in one and the same star if its temperature is variable. The spectra of *long-period variable stars* [see Merrill (44)] show clearly a much greater strength of the molecular absorptions near minimum light than near maximum light. Another phenomenon in these stars that may be related to molecular processes is the appearance of strong emission lines at certain phases in the cycle. Particularly prominent are the Balmer lines of hydrogen; in addition lines of Fe I, Fe II, Mg I, Si I and others appear. Photodissociation and chemiluminescence have been suggested by Wurm (729) as the origin of these emission lines. But other explanations are also possible [see Merrill (44)], and further investigations are necessary in order to ascertain the actual mechanism.

Another important astrophysical problem on which the study of band spectra can give information is the question of the *abundance ratios of isotopes*. Thus far only in one case have significant results been obtained: The Swan bands of C_2 and the violet CN bands are so strong in many carbon stars that the detection of bands due to the rare isotope C^{13} of carbon becomes possible.

Weaker heads lying at the longward side of the 1–0 band of the Swan system were long known in the spectra of some carbon stars and were first quantitatively explained as $C^{13}C^{12}$ and $C^{13}C^{13}$ 1–0 bands by Sanford (1372). Later on, similar isotopic bands were observed in the $\Delta v = -1$ and -2 sequences of the Swan system by Sanford (1373), Shajn (1404), and Herzberg (1043) and the 4–0 band of the red system of $C^{13}N^{14}$ was identified by McKellar (1213a). The isotopic bands show a relative intensity compared to the main bands that appears to be far greater than would correspond to the abundance ratio found on the earth (1 : 100). This is particularly clear in the $\Delta v = -2$ sequence where the 0–2 Swan band of $C^{13}C^{12}$ has about the same intensity as the neighboring 1–3 band of $C^{12}C^{12}$. McKellar (1213) has studied quantitatively the intensity ratios of the 1–0 bands and has obtained the remarkable result that in some carbon stars the $C^{13} : C^{12}$ abundance ratio is 1 : 4 while in other carbon stars it is of the same order as on the earth. No rare isotopes other than C^{13} have as yet been found in stellar spectra[8] nor have any abundance ratios of the more common isotopes been determined. Such investigations would clearly be of great importance for a better understanding of the nuclear processes leading to energy generation in the stars.

Finally we consider briefly the application of band spectroscopic methods to the determination of temperatures of stellar atmospheres. In principle the equilibrium calculations mentioned above could be used but in practice they are as yet too inaccurate. The determination of vibrational temperatures from the relative intensities of neighboring bands with different v'' (see p. 203) seems to be the most promising method since it does not require very high resolution. A beginning in this direction was made by Wurm (727) who used the relative intensities of several C_2 bands and of several CN bands in carbon stars. He obtained a temperature of about 1500° K for spectral type $N0$. In his calculations he used the harmonic oscillator approximation according to Hutchisson (see p. 202) and intensities estimated from reproductions of spectra. An obvious refinement would be the use of photometric intensities and of more accurate theoretical relative transition probabilities or of empirical transition probabilities from laboratory data. Such an attempt has been made by McKellar and Buscombe (1214) for three R-type stars, using CN and C_2 bands and obtaining temperatures ranging from 3900° to 6200° K. The uncertainties are still considerable. However by using bands of nearly equal intensities it should be possible to eliminate the effect of non-linear variation of absorption intensity with number of absorbing molecules (the so-called curve of growth) and to obtain fairly accurate temperatures. For the cooler stars a direct comparison with laboratory intensities in a furnace of known temperature would also be possible.

When high dispersion can be employed the intensity distribution in the rotational structure can be used for a determination of temperature (see p. 205).

---

[8] The observation of a high abundance of N^{15} by Daudin and Fehrenbach (881)(882) has been shown to be spurious by McKellar (1213a).

Thus far this method has only been applied to the sun, first by Birge (806). Various investigators have obtained rather discordant results for the temperature of the reversing layer but the three most recent determinations [Adams (735), Blitzer (815), Hunaerts (1081)] based on different molecules give within 200° the same result of 4500° K. This agrees fairly well with the excitation temperature obtained from atomic lines [see Wright (1561)].

As was first pointed out by Birge (806), in an atmosphere in which the temperature varies the higher rotational lines will indicate a higher temperature than the lower ones, corresponding to different layers of the atmosphere. In this way, using the CH band 4315 Å, Richardson (588) found a temperature of 6080° K for the lowest part of the reversing layer of the sun and 4430° K for the uppermost part. This corresponds to a temperature gradient of about 13°/km.

Interstellar space. In high dispersion spectra of distant stars a few exceedingly sharp lines occur which have been shown to be due to absorption in interstellar space [see the reviews by Beals (792) and Adams (737a)]. While some of these lines were readily identified as due to Ca II, Ca I, Na I, K I, Fe I, and Ti II (where for the last two only the lowest multiplet component of the ground state gives observable lines) several others remained unidentified for some time until it was realized, first by Swings and Rosenfeld (1454) and McKellar (1205) (1206) that they are due to *interstellar molecules* in their lowest rotational levels. Thus the lines 4300.32 Å and 3874.61 Å found by Dunham and Adams (918)(736) agree exactly with the lines $R_2(1)$ of the $^2\Delta$—$^2\Pi$ (0–0) band of CH and $R(0)$ of the $^2\Sigma$—$^2\Sigma$ (0–0) band of CN, respectively. Further lines of CH [$^PQ_{12}(1)$, $Q_2(1)$, $R_2(1)$ of the $^2\Sigma$—$^2\Pi$ (0–0) band] and of CN [$R(1)$ and $P(1)$ of the $^2\Sigma$—$^2\Sigma$ (0–0) band] predicted by McKellar (1205) were subsequently found by Adams (737).[9] Three of the remaining lines (4232.58, 3957.74, and 3745.33 Å) suggested a v'-progression of a diatomic molecule not known at the time [see McKellar (1206)]. It was later established by laboratory investigations of Douglas and Herzberg (902) to be the CH^+ molecule. At present only two sharp interstellar lines at 3934.29 and 3579.04 Å remain unidentified. The first of these may be due to NaH [see McKellar (1206)].

The observation that in interstellar space only the very lowest rotational levels of CH, CH^+, and CN are populated is readily explained by the depopulation of the higher levels by emission of the far infrared rotation spectrum (see p. 4) and by the lack of excitation to these levels by collisions or radiation. The intensity of the rotation spectrum of CN is much smaller than that of CH or CH^+ on account of the smaller dipole moment as well as the smaller frequency [due to the factor ν^4 in (I, 48)]. That is why lines from the second lowest level $(K = 1)$ have been observed for CN. From the intensity ratio of the lines with $K = 0$ and $K = 1$ a rotational temperature of 2.3° K follows, which has of course only a very restricted meaning.

It is interesting to observe that no lines due to $C^{13}H$ or $C^{13}N$ have been found. From the intensity of the observed $C^{12}H$ and $C^{12}N$ lines Wilson (1544a)

---

[9] This paper contains several beautiful spectrograms showing the interstellar lines.

obtained an upper limit of 1 : 5 for the abundance ratio of C^{13} : C^{12}. Thus the C^{13} abundance in interstellar space is probably not as anomalous as in some carbon stars (see above).

From the observed intensities of the interstellar lines and approximate f-values Dunham (916) has derived *concentrations*[10] of the order of 10^{-6} molecules/cm³ for CH. CH^+, and CN. However, this estimate assumes a uniform concentration along the line of sight when actually it is most probable that the molecules are concentrated in clouds [see Strömgren (1438)]. This would increase the concentration perhaps by a factor 10.

In order to understand the occurrence of molecules in such concentrations in interstellar space it is necessary to discuss the *formation and dissociation* of the interstellar molecules. This has been done by Swings (1446) and Kramers and ter Haar (1140). On the basis of atomic concentrations obtained from other data and assuming that a stationary state has been reached they obtain molecular concentrations that agree at least within one or two powers of 10 with the observed values. It is clear that only two-body recombinations are possible.[11] Spectroscopic information about the photodissociation of CH, CH^+, and CN is rather inadequate but it is likely that CH and CH^+ can be photodissociated with low probability by transitions to the continua of the $^2\Sigma$ and $^1\Pi$ states, respectively, requiring only light below 3300 Å while CN requires light of much shorter wave length. This difference accounts for the fact that CN and CH have similar concentrations in spite of the large abundance of H compared to C and N. According to Kramers and ter Haar (1013a) (1140) the formation of CH and CN in interstellar space is the first step in the formation of larger particles.

The relative concentration of CH and CH^+ is determined essentially by the probabilities of photoionization of CH and electron capture by CH^+ as well as by the electron concentration. With the help of the theoretical relations between these quantities it will be possible to obtain from the observed intensity ratio of the CH and CH^+ lines a value for the electron concentration in the interstellar clouds. However, more reliable f-values for the molecular absorptions involved will have to be obtained before a trustworthy figure for the electron concentration can be derived.

In addition to the sharp interstellar lines Merrill and Wilson (1223)(1224) have found several *diffuse interstellar lines* (4–10 Å in width) mainly in the yellow and orange part of the spectrum. These lines have thus far resisted all attempts at identification. It is likely that they are not due to absorption by free molecules but to absorption by solid particles.

The study of interstellar molecules is a very young field. Undoubtedly as stellar spectra of higher dispersion become available new interstellar lines will be found and, together with more laboratory data, considerable advances in our knowledge of the interstellar medium may be expected.

---

[10] Dunham (916) assumed $f(CH) = 0.06$, while in Chapter VI p. 386 the value 0.0019 is obtained; but Strömgren (1438) with $f = 0.002$ derives 10^{-6} molecules/cm³.

[11] According to Swings a three-body collision $C^+ + H + H$ would happen in one cm³ once in 10^{24} years.

APPENDIX

TABLE 39

VIBRATIONAL AND ROTATIONAL CONSTANTS FOR THE ELECTRONIC STATES
OF ALL KNOWN DIATOMIC MOLECULES

The molecules are given in alphabetical order. Data are presented for all reasonably well established electronic states. In general only the constants for the most abundant isotope are given except for the hydrides for which also the constants of the corresponding deuterides are included. The reduced masses μ_A (in atomic weight units, Aston's scale) have been calculated from precise atomic masses [Mattauch-Flügge (49), Cork (18b)] and are listed at the top of each part referring to a given molecule. The mass numbers assumed are added to the element symbols. If they are not given all data refer to the normal isotopic mixture.

All available data on the various molecules have been reviewed critically and those that appeared most trustworthy have been selected. In many instances data not given explicitly in the original papers had to be derived and corrections to those that were given had to be made. In addition to the constants given for each electronic state, for the ground state higher vibrational and rotational constants are given in the footnotes. The dissociation energy is given only for the ground state (immediately following the symbol for the molecule) and refers to the dissociation into normal products (D_0^0). These D_0^0 values are obtained are included in the column "remarks" or in the footnotes. Convergence limits and predissociation limits from which some of the D_0^0 values are obtained are included in the column "remarks" or in the footnotes. Convergence limits and predissociation limits from which some of the D_0^0 values are obtained are included considered and discrepancies occurring in the published data. These contain also other pertinent information such as coupling constants in multiplet states, unusual features of the states given by the use of equations (III, 91) and (III, 115) respectively. The force constants k_e and the moments of inertia I, which are not listed, can be readily obtained from the data given

All observed transitions are indicated and wherever possible the position of the 0–0 band (ν_{00}) is given. In general this has been taken from the formula representing all bands of the system so that there may be slight deviations from the observed ν_{00}. For short reference, in conformity with a fairly well established custom (see p. 245), the ground state is referred to as X, the excited states of the same multiplicity as $A, B, C, \ldots$, those of different multiplicity as $a, b, c, \ldots$. The only exception to this rule is N_2 where the custom of designating the excited singlet states $A, B, C, \ldots$ and the excited triplet states $a, b, \ldots$ appeared too well established to make a change. The order $a, b, c, \ldots A, B, C, \ldots$ is usually the order of increasing energy. But in some cases, particularly for H_2 and He_2, exceptions were made in order to be in agreement with earlier designations. Moreover for these two molecules in the column "remarks" the electron configurations are given. The arrows $\leftarrow, \rightarrow, \leftrightarrow$ for the transitions indicate whether they have been observed in absorption, emission or both.

At the dissociation energies D_0^0 (in e.v.) and internuclear distances r_e (in 10^{-8} cm) have been recalculated from the original data to conform with the new physical constants and conversion factors listed in Tables 1 and 2. As many significant figures are given as are compatible with the accuracy of the original spectroscopic data irrespective of the accuracy of the conversion factors. If the latter should be changed in the future, multiplication by the ratio of the new and the old factor would give the revised molecular constants.

Gaydon (34) has compiled a critical table of dissociation energies. Whenever the values given here differ from Gaydon's values or are taken from his list this is indicated by the name Gaydon (without reference number). Mostly such differences occur in cases for which long extrapolations of the vibrational quanta are required to obtain the D_0^0 values.

In these cases the extrapolated values are given without applying the plausible but uncertain correction introduced by Gaydon.

All numbers in the table except where otherwise indicated are in cm^{-1} units.

TABLE 39 (Continued)

State	T_e	ω_e	$\omega_e x_e$	$\omega_e y_e$	B_e	α_e	r_e (10⁻⁸ cm)	Desig-nation	ν_{00}	References	Remarks
Ag¹⁰⁹Br⁸¹											
	$D_0^0 = 2.6$ e.v., from thermochemical data (1462); $\mu_A = 46.436$										
	Fragments of two other band systems at 24000 and 49000 cm⁻¹										
		205.0 H	0.74							(1227) (781)	
C	43537.4	180.8 H	4.45	−0.060ᵃ				$C \leftarrow X$ R	43516.0	(779)	repr. (126)
B (³Π₀⁺)	31280.4	247.72 H	0.6795					$B \leftarrow X$ R	31246.0	(126) (521)	
X (¹Σ)	0										
Ag¹⁰⁷Cl³⁵											
	$D_0^0 = 3.1$ e.v., from thermochemical data (1462); $\mu_A = 26.358$										
	Continuous absorption above 47500 cm⁻¹										
D	48800	~290 H						$D \leftarrow X$ R		(376)	
C	43525.4	294.1 H	1.70					$C \leftarrow X$ R	43501.2	(376)	
B (³Π₀⁺)	31606.92	281.0 H	6.00	−0.095ᵃ				$B \leftarrow X$ R	31574.40	(125) (126)	repr. (125), Fig. 13
X (¹Σ)	0	343.6 H	1.163								
AgH¹											
	$D_0^0 = 2.5$ e.v., from thermochemical data (927); $\mu_A = 0.998800$										
A ¹Σ⁺	29959	1663.6 Z	87.0	ᵃ	6.265	0.203	1.641₅	$A \leftrightarrow X$ R	29897.9	(86) (979) (927)	
X ¹Σ⁺	0	1760.0 Z	34.05	*	6.453	0.348*ᵇ	1.617₄			(86) (979)	
AgH²											
	$\mu_A = 1.977798$										
A ¹Σ⁺	29959	ᶜ			3.1596	0.1161	1.6426	$A \rightarrow X$ R	29911.0Z	(415)	
X ¹Σ⁺	0				3.2595	0.0732*ᵈ	1.6172				
Ag¹⁰⁷I¹²⁷											
	$D_0^0 = 2.9_3$ e.v., from the convergence in the B state;ᵃ $\mu_A = 58.0439$										
	Continuous absorption above 24000 cm⁻¹										
E	47500	(130)						$E \leftarrow X$ R	(47460)	(1227)	
D	46000	(165)ᵇ H						$D \leftarrow X$ R	45471	(781)	
C	44720	155.7ᶜ H	1.9					$C' \leftarrow X$ R	44697	(1226) (779)	
B (³Π₀⁺)	31195.9	122.8 H	1.58	[−0.48]*				$B \leftrightarrow X$ R	31153.9	(1226) (779)	repr. (126)
A	23906	151.2 H						$A \leftarrow X$ R	23879	(126) (780) (1374)	
X (¹Σ)	0	206.18 H	0.4327							(1227)	
										(126)	
AgO¹⁶											
	$D_0^0 = (1.8)$ e.v.; $\mu_A = 13.9340$										
	Additional unclassified band system in the red										
B ²Σ⁻	28074.6	534.7 H	6.10					$B \rightarrow X$ V	28094.9	(466)	
A ²Π	{24310.9	[233.0] H						$A \rightarrow X$ R	{24181.8	(466)	
X ²Σ⁻	{24133.9	493.2 H	4.10						{24010.8	(466)	
	0										

Al²⁷Br⁷⁹ — $D_0^0 = (2.4)$ e.v.;[a] $\mu_A = 20.1129$

State	T_e	ω_e	$\omega_e x_e$	B_e	α_e	D_e	r_e	Transition	ν		Refs	Remarks
A ¹Π	35879.5	297.2 H[Q]	6.40	0.1555	−0.527	0.00216	2.322	A↔X	35837.8	R	(336)(1231) (1189)[b](1100)	{isotope eff. (1100)[c] {repr. (336)
X ¹Σ⁺	0	378.0 H[Q]	1.28	0.1591		0.000853[d]	2.295					

Al²⁷Cl³⁵ — $D_0^0 = (3.1)$ e.v.;[a] $\mu_A = 15.2350$

State	T_e	ω_e	$\omega_e x_e$	B_e	α_e	D_e	r_e	Transition	ν		Refs	Remarks
A ¹Π	38254.0	449.96 H[Q]	4.37	0.259	−0.216	0.006	2.06_7	A↔X	38237.7	Rv	(100)(327)(1059) (1231)(1188)	pred. in $v=10$ repr. (100)
X ¹Σ⁺	0	481.30 H[Q]	1.95	0.242		0.002	2.13_3					

Al²⁷F¹⁹ — $D_0^0 = (2.5)$ e.v.;[a] $\mu_A = 11.1521$

Another band system not related to the one given was found in emission by (1570) between 24000 and 31000 cm⁻¹

State	T_e	ω_e	$\omega_e x_e$	B_e	α_e	D_e	r_e	Transition	ν		Refs	Remarks
A ¹Π	43935	822.9 H	8.5		−0.187			A↔X	43939	Rv	(1352)	
X ¹Σ⁺	0	814.5 H	8.1									

Al²⁷H¹ — $D_0^0 < 3.06$ e.v. (from predissociation in A ¹Π state); $\mu_A = 0.971832$

State	T_e	ω_e	$\omega_e x_e$	B_e	α_e	D_e	r_e	Transition	ν		Refs	Remarks
E ¹Π	53396	(850)		[5.60]			$[1.76_0]$	E↔X	52987.5Z	R	(1284)	{pred. for $J=11$ & 12, ¹Π* of (325)
								E↔A	29516.4Z	R	(325)	
D ¹Σ⁺				[6.55]			$[1.62_1]$	D↔X	49288Z	V	(996)(84)(1284)	¹Σ** of (996)[g]
C ¹Σ⁺	44675.8	1575.3 Z	125.5	6.664		0.544	1.613_5	C↔X	44597.9Z	V	(1285)	¹Σ* of (996)
								C↔A	21127.4Z	V	(996)(85)	
B ¹Σ⁻	44132	(900)		[6.120]			$[1.683_7]$	B↔A	20277.16Z	V	(326)	¹Σ*** of (326)
b ³Σ⁽⁻⁾	a+26220	(1683)		[6.759]			$[1.602_1]$	b↔a	26217Z	V	(1057)	pred. above $K=25$
A ¹Π[a]	23763	[1083.1] Z		6.3935		0.7460	1.6466	A↔X	23470.9	R	(328)(212)(215) (548)	pred. above $v=1$[b]

AgBr; AgCl; [a]This state seems to have a potential maximum (521).

AgH: [a]There are strong vibrational perturbations in this state indicating a second excited ¹Σ state (see p. 293 f.).
[b]$D_v = (3.4 + \ldots) \times 10^{-4}$.
[c]Only 0-0 and 1-1 bands analyzed; therefore no vibrational constants.
[d]$D_0 = 0.87 \times 10^{-4}$.

AgI: [a]Thermochemical data (1227) give 2.93 ± 0.10 e.v. On the basis of an earlier thermochemical value (2.10 e.v.) Mulliken (521) assumed a potential curve with maximum in the B state. A low value for $D_0^0 (= 2.2$ e.v.) follows also from the atomic fluorescence observed by (1460). Gaydon gives 2.2 ± 0.3 e.v.
[b]Metropolis (1226) gives $\nu_e = 45487$, $\omega_e = 176$, $\omega_e x_e = 2.5$ but Barrow considers the vibrational analysis as doubtful (private communication).
[c]Average of (1226) and (779) as suggested by Barrow in private communication.

AlBr: [a]This is Gaydon's value, assuming predissociation into Al⁺ + Br⁻.
[b]Mahanti (1189) gives somewhat different constants.
[c]Predissociation for $v > 3$.
[d]$+0.0000014(v + \frac{1}{2})^2$; $D_v = 0.109 \times 10^{-6}$.

AlCl: [a]This is Gaydon's value, assuming predissociation into Al⁺ + Cl⁻.

AlF: [a]Gaydon gives 1.8 e.v.

AlH: [a]This state has a potential maximum (see p. 428 f.).
[b]See p. 429.

TABLE 39 (Continued)

State	T_e	ω_e	$\omega_e x_e$	$\omega_e y_e$	B_e	α_e	r_e (10^{-8} cm)	Designation	ν_{00}	References	Remarks
Al²⁷H¹ (Continued)											
a ³Πᵢ	a[c]	(1688)			[6.704]		$[1.608_7]$			(152)	
X ¹Σ⁺	0	1682.57 Z	29.145	+0.26	6.3962[d]	0.188[h]	1.6459_2[e]			(328)	
		$D_0^0 < 3.09$ e.v.; $\mu_A = 1.87478$									
Al²⁷H²											
D ¹Σ⁺	44685.7	[1004.2] Z			[3.45]	0.176	$[1.61_3]$	D→X V	44288	(99) (996)	no predissociation
C ¹Σ⁺		1014.89 Z	86.50		3.438	0.122₄	1.617	C→X V	44402.6	(996)	
A ¹Π	23653.2	1211.95 Z	15.138	+0.098	3.235[f]	0.0697*[i]	1.667	A→X R	23536.8	(328)	pred. above v=2
X ¹Σ⁺	0				3.3186[d]		1.6456[e]			(328) (1285)	
(Al²⁷H¹)⁺		$\mu_A = 0.971831$									
A ²Πᵣ	{(27689)[a]; (27596)}	(1753)[b]			6.851	0.248	1.591₃	A→X V	{27760.2Z; 27667.8Z}	(67) (1058)	(1058) finds breaking off in F and R at K=28
X ²Σ⁺	0	(1610)[b]			6.763	0.398[c]	1.601₆				
Al²⁷I¹²⁷		$D_0^0 = (2.9)$ e.v.; $\mu_A = 22.2578$									
		unstable; diffuse bands with various v''									
A ¹Π	30300	333.4 H	2.0					A←X V	{22098.2; 21899.6}	(1231)	
a ³Π₀,₁	{22089.5; 21889.3}	337.2 H	2.0					a↔X V	21899.6	(492) (1231)	
X ¹Σ⁺	0	316.1 H	1.0								
Al²⁷O¹⁶		$D_0^0 = (<3.75)$ e.v.;[a] $\mu_A = 10.0452$									
		Fragments of three other band systems are given by (852a)									
B	33151 H	845 H	4					B→X R	33085H	(852)	repr. (1398) (1365)
A ²Σ⁺	20699.2 H	870.2 H	3.80		0.60417	0.00453	1.6667	A↔X R	20635.3Z	(567) (1365)	
X ²Σ⁺	0	978.2 H	7.12		0.64148	0.00575[b]	1.6176				
As₂⁷⁵		$D_0^0 \le 3.96$ e.v.;[a] $\mu_A = 37.467$									
b (³Πᵤ)	40898							b→X R	42006	(401) (746)	
B ?		(302)[b] H	(5.8)					B↔X R	40833.0	(401) (985) (748)	only v=0 observed
A ¹Σᵤ⁺	40356	(280)[b] H	(1.0)					A↔X R	40260.9	(401) (985) (748)	pred. above v=9, breaking off in v=9
a (³Σᵤ⁺)	24644	337.0 H	0.83					a→X R	24598	(401) (746)	
X ¹Σ_g⁺	0	429.44 H	1.120	+0.005091[c]						(401)	

State	T_e	ω_e	$\omega_e x_e$	Transition	ν		Ref.	Remarks
(As$_2^{75}$)$^+$?		$D_0^0 = (2.4)$ e.v.; $\mu_A = 37.4668$						
A ($^2\Pi$)	{16359, 16197}	336.8 H	1.04	A→X	{16348, 16186}	V / V	(401)	
X ($^2\Sigma^+$)	0	314.8 H	1.25					
As^{75}N^{14}		$D_0^0 = (6.5)$ e.v.; $\mu_A = 11.8015$						
A $^1\Pi$	36005.0	871.3 H	8.24	A→X	35905.9	R	(645)	repr. (645)
X $^1\Sigma^+$	0	1068.0 H	5.36					
As^{75}O^{16}		$D_0^0 \leq 5.0$ e.v.;^{a} $\mu_A = 13.1848$						
B ($^2\Sigma$)	39864.4	1098.3 H^Q	6.1	B↔X	{39929.6, 38903.9}	V / V	(856a)	{pred. above $v = 0$ at low pressure
A ($^2\Sigma$)	31655.6	684.6^b H^Q	9.8	A→X	{31513.0, 30487.3}	R / R	(856a)(378) / (856a)(378)	repr. (856a)
X $^2\Pi$	{1025.7, 0}	967.4^b H^Q	5.3					
Au^{197}Cl^{35}		$D_0^0 = (3.5)$ e.v.;^{a} $\mu_A = 29.7055$						
B	19238.3	316.3 H	1.45	B→X	19205.0	R	(216)	
A	19113.8	312.0 H	0.70	A→X	19078.6	R	(216)	
X ($^1\Sigma^+$)	0	382.8 H	1.30					

AlH: ^{c}Coupling constant $A = +40.2$.

^{d}These are effective B values which differ noticeably from the true values [see (328)].

^{e}From the true B_e [see (328)] not the effective B_e given in the table.

^{f}Not comparable with AlH^1 since calculated from $v = 0, 1, 2$ while AlH value from $v = 0, 1$ only.

^{g}Predissociation above $J = 10$.

$^h D_0 = 3.72 \times 10^{-4}$.

$^i D_0 = 0.97 \times 10^{-4}$.

AlH$^+$: ^{a}Coupling constant $A = 108$ cm^{-1}.

^{b}From D_e.

AlO: ^{a}From a predissociation found by (852) above $K = 129$, $v = 0$ of $A\,^2\Sigma$. The low value of 0.93 e.v. obtained by (1356)(1357) from predissociation data does not seem acceptable to the writer in view of the occurrence of the AlO bands in stellar spectra. The data used by (1356)(1357) are not convincing.

^{b}Sen (1398) gives spin splitting constant $\gamma_0'' = 0.0211$; $D_v = [1.114 + 0.0087(v + \frac{1}{2})] \times 10^{-6}$.

$^c D_v = [4.79 - 0.09(v + \frac{1}{2})] \times 10^{-4}$.

As$_2$: ^{a}Assuming predissociation in the A state into $^4S + {}^2D$. A different formula applies for $v = 41$ to 70 [see (401)].

$^c \omega_e z_e = -0.00002563$; these vibrational constants hold from $v = 0$ to 40.

^{b}Very strong vibrational perturbations.

AsO: ^{a}Jenkins and Strait (378) and Gaydon give 4.93, but this refers to $v' = 0$ of B which is not predissociated.

^{b}Average of constants given by (856a) and (378).

AuCl: ^{a}Gaydon gives 2.8 ± 0.5 e.v.

TABLE 39 (Continued)

State	T_e	ω_e	$\omega_e x_e$	$\omega_e y_e$	B_e	α_e	r_e (10^{-8} cm)	Designation	ν_{00}		References	Remarks
Au¹⁹⁷H¹	$D_0^0 = 3.1$ e.v., from thermochemical data (927); $\mu_A = 1.002999$											
B ¹Σ⁺	27665.7				[5.754]		[1.709_2]	B→X	38232.3	R	(288)(289)	¹Σ** of (288)(289)
A ¹Σ⁺	0	1669.55 Z	55.06	−3.93*	6.0069	0.249*	1.6728_4	A→X	27344.5	R	(353)(927)	¹Σ* of (288)(289)^b
X ¹Σ⁺		2305.01 Z	43.12	−0.044	7.2401	0.2136^a	1.5237_3				(353)	^b
Au¹⁹⁷H²	$D_0^0 = 3.1$ e.v. (from the value for AuH¹); $\mu_A = 1.99433$											
B ¹Σ⁺	38538	1165.4 Z	21.85		2.961	0.079	1.690	B→X	38303	R	(288)(289)	repr. (353)^b
A ¹Σ⁺	27644.1	1195.24 Z	34.813	*	3.036	0.103*	1.669	A→X	27420.9	R	(353)(288)(289)	^b
X ¹Σ⁺	0	1634.98 Z	21.655	−0.0288	3.6415	0.07614*^c	1.5237				(353)	
B₂¹¹	$D_0^0 = (3.6)$ e.v.; $\mu_A = 5.50645$											
A ³Σ_u⁻	30573.4	937.4 Z	2.6		1.160	0.011	1.625	A→X	30518.1	R	(900)	repr. (900)
X ³Σ_g⁻	0	1051.3 Z	9.4		1.212	0.014	1.589					
BaBr⁷⁹	$D_0^0 = (2.8)$ e.v.; $\mu_A = 50.131$											
E ²Σ⁺	26865.9	219.9 H	0.35					E←X	26878.9	V	(1020)	
D ²Σ⁺	25670.9	209.1 H	0.53					D←X	25678.5	V	(1020)	
C ²Π	{19192.5	197.4 H	0.41					A↔X	{19194.3	R	(285)(1020)	4 heads
	18650.9}								18652.7}	R		
X ²Σ⁺	0	193.8 H	0.42									
Ba¹³⁸Cl³⁵	$D_0^0 = (2.7)$ e.v.; $\mu_A = 27.9022$											
G (²Σ)	32511.4	331.3	1.29					G←X	32537.3	V	(1020)	
F (²Σ)	29493.6	331.8 H	1.30					F←X	29519.7	V	(1020)	
E ²Σ	27064.8	311.5 H	0.93					E←→X	27080.9	V	(550)(1020)	repr. (550)
D ²Σ	25471.6	304.6 H	1.04					D←X	25484.2	V	(550)(1020)	repr. (550)
C ²Π	{19450.1	285.0 H^Q	0.79					C↔X	{19453.0	R	(550)(1020)	{violet shading for
	19062.9}	280.2 H^Q	0.79						19063.9}	R	(550)(1020)	high v, repr. (550)
A (²Π)	{11880.0	255.25 H	0.83					A↔X^a	{11868.0	R	(777)	4 heads
	10995.3}	256.35 H	0.73						10983.9}	R		
X ²Σ	0	279.3 H	0.89								(550)(1020)	
BaF¹⁹	$D_0^0 = (3.8)$ e.v.; $\mu_A = 16.6953$											
H ²Σ	31582	(509) H						H←X	31603	V^a	(941)	
G ²Σ	31451.9	510.4 H	0.83					G←X	31472.9	V^a	(941)	
F ²Σ	29411.3	529.9 H	2.00					F←X	29411.8	V^a	(941)	double heads

BaF

State	ω_e	$\omega_e x_e$	T_e	Transition	head	Sh.	References	Remarks
E′ ²Σ	538.4 H	1.90	28134.1	E←X	28168.8	V[a]	(941)	repr. (371)
D′ ²Σ	504.9 H	1.54	26222.3	D′←X	26240.4	V	(371)(941)	repr. (371)
D ²Σ	508.4 H	1.88	24152.3	D←X	24172.0	V	(371)(941)	repr. (371)
C′ ²Π	456.0 H	1.67	{20197.2 / 19998.2}	C↔X	{20190.8 / 19991.8}	R	(371)	repr. (371)
B ²Σ	424.4 H	1.88	{14070.6 / 14064.4}	B↔X	{14048.3 / 14040.1}	R	(1275)(371)	repr. (371)[b]
A ²Π	436.7 E[Q]	1.82	12281.1	A↔X	{12265.0 / 11630.7}	R	(1275)(371)	4 heads, repr. (371)
	435.5 P[Q]	1.68	11647.3					
X ²Σ	468.9 H[Q]	1.79	0					

BaH¹ — $D_0^0 \le 1.82$ e.v.; $\mu_A = 1.000788$

State	ω_e	$\omega_e x_e$	B_e	α_e	r_e	T_e	Transition	head	Sh.	References	Remarks
C ²Σ	1323 Z	22.5	3.55	0.045	2.18	23628	C↔X	23702.1	V	(241)	repr. (241)[a]
E ²Π	1231 Z	17.5	{3.559 / 3.486}	0.074	{2.176 / 2.198}	{15054 / 14600}	E←X	{15084 / 14629}	V	(239)	repr. (233b)
B ²Σ	1088 Z	15	3.267	0.070	2.271	11091	B→X	{11049.4Z / 9910Z}	R	(417)	
A ²Π_r			[3.280 / 3.12]		[2.266 / 2.32]		A→X	9285Z	R	(1514)(417)	
X ²Σ⁺	1172 Z	16	3.3823	0.0655[b]	2.2318	0				(270)(239)	

BaI¹²⁷ — $\mu_A = 65.979$

Other unclassified bands at 26000–27000 cm⁻¹ given by (1511)

State	T_e	Transition	head	Sh.	References	Remarks
A (²Π)	{(18569) / (17813)} H	A↔X	{18569 / 17813}	a	(1225)(1511)	only 0–0 sequence
X (²Σ)	0					

BaO¹⁶ — $D_0^0 = 4.7$ e.v. from thermochemical data (50a); $\mu_A = 14.3311$

State	ω_e	$\omega_e x_e$	B_e	α_e	r_e	T_e	Transition	head	Sh.	References	Remarks
A ¹Σ	498.8 Z	(1.5)	0.2582	0.0011	2.134	16807.7	A→X	16722.3Z	R	(778b)	repr. (476)
X ¹Σ	669.8 Z	2.05	0.3126	0.0014[b]	1.940	0					

BaS — $D_0^0 = (2.3)$ e.v. (Gaydon's estimate); $\mu_A = 26.004$

	References
Absorption continua with long wave length limits at 22510 and 28440 cm⁻¹	(1198)

[a] These bands have two heads because of the spin doubling in the ²Σ state.

AuH: [a] $D_v = [279 - 0.84(v + \tfrac{1}{2})] \times 10^{-6}$. [b] Heimer (288) (289) gives somewhat different constants. [c] $D_v = [70.9 - 0.34(v + \tfrac{1}{2})] \times 10^{-5}$.

BaCl: [a] In absorption according to private communication by Barrow. Although he has analyzed a few more bands his constants seem to be less accurately evaluated.

BaF: [a] Shading not clearly stated by (941) but reproduction indicates V.

BaH: [a] 0–0 band very weak in emission unlike absorption, perhaps due to predissociation. In addition breaking off above $K = 8$ $v = 1$ [see (241)]. [b] Spin splitting constant $\gamma = +0.186$ and $+0.182$ for $v = 0$ and $v = 1$ respectively; $D_v = [1.14 - 0.05(v + \tfrac{1}{2})] \times 10^{-4}$.

BaI: [a] No shading.

BaO: [a] Note that the ¹Σ ground state cannot dissociate into normal atoms ($^1S + {}^3P$); extrapolation of the vibrational quanta gives $D_0(X) = 6.7$ e.v. [b] $D_e = 0.25 \times 10^{-6}$.

TABLE 39 (Continued)

State	T_e	ω_e	$\omega_e x_e$	$\omega_e y_e$	B_e	α_e	r_e (10^{-8} cm)	Designation	ν_{00}	References	Remarks
B^{11}Br^{79}											
$D_0^0 = (4.1)$ e.v.; $\mu_A = 9.6644$											
A $^1\Pi$	33935.3	637.63 H^Q	17.58	+1.10*	0.501	0.0090*	1.86$_6$	A↔X	33908.6 V$_R$	(1229)(1364)	repr. (1364)(1229)^a
X $^1\Sigma^+$	0	684.31 H^Q	3.52		0.490	0.0035^b	1.88$_7$			(1231)	
B^{11}Cl^{35}											
$D_0^0 = (4.2)$ e.v.; $\mu_A = 8.37582$											
A $^1\Pi$	36750.9	849.04 H^Q	11.37	−0.100*	0.7054	0.00820*	1.689$_3$	A→X	36754.3 V$_R{}^a$	(1049)	repr. (1049)
X $^1\Sigma^+$	0	839.12 H^Q	5.11		0.6838	0.00646^b	1.715$_7$				
Be^{9}Cl^{35}											
$D_0^0 = (4.3)$ e.v.;^{a} $\mu_A = 7.16766$											
A $^2\Pi$	27970.5^b	824.19 H^Q	6.03					A→X	27959.1 R	(233a)	4 heads, repr. (233a)
X $^2\Sigma^+$	0	846.58 H^Q	5.11		(0.8)		(1.7)				
Be^{9}F^{19}											
$D_0^0 = (5.4)$ e.v.;^{a} $\mu_A = 6.11450$											
Continuous absorption at shorter wave lengths; see (941)											
A $^2\Pi_i$	33233.6 Z^b	1172.6 H^Q	8.78		1.41865	0.01610	1.3941$_6$	A↔X	33187.2Z R	(367)(1258)(941)	
X $^2\Sigma^+$	0	1265.6 H^Q	9.12		1.4877	0.01685^c	1.3614			(1103)(367)	
Be^{9}H^1											
$D_0^0 = (2.2)$ e.v.; $\mu_A = 0.906732$											
B $^2\Pi$	(50934)	[2133.4]			[10.65]		[1.321]	B→X	50980Z V	(696)	Q branches only, on account of pred.
A $^2\Pi$	20027.9^a	2087.7 Z	39.8$_5$	−0.5	10.470	0.329	1.3327	A→X	20041.3Z V^b	(541)	repr. (1515)
X $^2\Sigma^+$	0	2058.6 Z	35.5	−0.5	10.308	0.300^c	1.3431				
Be^{9}H^2											
$\mu_A = 1.646707$											
A $^2\Pi$					5.7583	0.1314	1.3334	A→X	20037.8 V^b	(416)	repr. (416)
X $^2\Sigma^+$					5.6807	0.1218^d	1.3425				
(Be^{9}H^1)$^+$											
$D_0^0 = (3.2)$ e.v. (Gaydon's value); $\mu_A = 0.906727$											
A $^1\Sigma^+$	39417.0	1476.1 Z	14.8	−0.038	7.1835	0.1249*	1.6088$_8$	A→X	39050.5 R	(696)	
X $^1\Sigma^+$	0	2221.7 Z	39.79	−0.021	10.7996	0.2935^a	1.3121$_6$				
(Be^{9}H^2)$^+$											
$D_0^0 = (3.2)$ e.v.; $\mu_A = 1.646689$											
A $^1\Sigma^+$	39416.2	1096.41 Z	8.49	(−0.16)	3.9712	0.0578	1.6057	A→X	39143.9 R	(416)	repr. (416)
X $^1\Sigma^+$	0	1647.64 Z	21.85	(−0.06)	5.9546	0.1233^b	1.3113				

Be⁹O¹⁶ $D_0^0 = (3.7$ or $3.0)$ e.v.;[a] $\mu_A = 5.76612$

[Fragments of additional singlet and triplet band systems in the region 29000–33000 cm⁻¹]

State	T_e	ω_e	$\omega_e x_e$	$\omega_e y_e$	B_e	α_e	r_e	Transition	ν_{00}		References	Remarks
$D\ ^1\Pi$	41365	1016 H	10			(1.308)		$D\rightarrow A$	31894	R	(849)(1156)	repr. (1156)
$C\ (^1\Sigma)$	39120.1	1081.5 (HQ)	9.1		(1.308)	(0.010)	1.495	$C\rightarrow A$	29683.0	R	(794)(1156)	doubtful according to (1156) & (849), repr. (1156)
$B\ ^1\Sigma^+$	21253.94	1370.817 Z	7.7455	−0.00027	1.5758	0.0154	1.3622	$\{B\rightarrow A$	11961.78	V	(1021)(1156)	repr. (1156)
								$\ B\rightarrow X$	21196.70	R	(1156)(1160)(602)	perturb. treated extensively in (1160)(1154)(1156)
$A\ ^1\Pi$	9405.61	1144.238 Z	8.4145	+0.03389	1.3661	0.01628*	1.4630	$A\rightarrow X$	9234.9₂	R	(320)(1159)(1155)(1156)(1160)	perturb. in (1159)
$X\ ^1\Sigma^+$	0	1487.323 Z	11.8297	+0.02235	1.6510	0.0190ᵇ	1.3308					

$^b D_v = 1.0 \times 10^{-6}.$
$^b D_v = [1.72 + 0.07(v + \tfrac{1}{2})] \times 10^{-6}.$

B¹¹F¹⁹ $D_0^0 = (4.3)$ e.v.; $\mu_A = 6.97245$

Other unclassified bands at 15500–22600 and at 37000–39000 cm⁻¹;

State	T_e	ω_e	$\omega_e x_e$	B_e	α_e	r_e	Transition	ν_{00}		References	Remarks
c	$a + 31870.0$ (only $v=0$ observed)						$c\rightarrow a$	37991.9	V	(1106)(913)(656) (656)	repr. (913)
$b\ ^3\Sigma$		1631.40 H	23.10	1.6338	0.0205ᵃ	1.2166	$b\rightarrow a$	32020.4	V	(656)(552a)	5 heads, repr. (656)(552a)
$a\ ^3\Pi$	a	1323.64 H	9.40	1.4120	0.0179	1.3086					
$D\ ^1\Pi$	72140	(1636) H	14				$D\rightarrow X$	72287	V	(1233)	
$C\ ^1\Sigma$	69010.1	1603.0 H	13.7				$C\rightarrow X$	69111.1	V	(1233)	
$B\ ^1\Sigma$	65334.8	1632.7 H	12.7	(1.20)	0.018	(1.66)	$B\rightarrow X$	65480.9	V	(1233)	
$A\ ^1\Pi$	65480.9	1214.4 H	14.9	1.304		1.423	$A\rightarrow X$	51085.0	R	(1233)(1233a)	
$X\ ^1\Sigma^+$	51150.1	1397.8 H	11.3	1.262	0.017	1.518					

BBr: ᵃIsotope effect. Predissociation above $v' = 4$ according to (1229).

BCl: ᵃBands with shading in both directions and two Q heads appear.

BeCl: ᵃGaydon gives 3.0 e.v.

BeF: ᵃGaydon gives 4 ± 1 e.v. ᵇRefers to Q_1 heads. Doublet splitting is $= 25$ cm⁻¹.
  ᵇCoupling constant $A = -16.46$; see (1258).
  $^b D_v = [8.225 - 0.0290(v + \tfrac{1}{2})] \times 10^{-6}$

BeH: ᵃCoupling constant $A = 2.14$ cm⁻¹.
  ᵇReversal of shading within some of the bands.
  $^c D_v = 9.8 \times 10^{-4}$

BeH⁺: $^a - 0.0049(v + \tfrac{1}{2})^2$; $D_v = [7.9 + 0.25(v + \tfrac{1}{2})] \times 10^{-4}.$
  $^d D_v = [3.01 + 0.13(v + \tfrac{1}{2})] \times 10^{-4}.$
  $^b D_v = 2.9 \times 10^{-4}.$

BeO: ᵃThe $X\ ^1\Sigma$ state cannot dissociate into normal atoms ($^1S + ^3P$). Its dissociation energy is 5.7 e.v. (by extrapolation) yielding $5.7 - 2.7 = 3.0$ e.v. or $5.7 - 2.0 = 3.7$ e.v. for D_0^0 depending on whether $X\ ^1\Sigma$ dissociates into $^3P + ^3P$ or $^1S + ^1D$. Gaydon gives 4.4 e.v.
  $^b D_v = [8.198 - 0.0096(v + \tfrac{1}{2})] \times 10^{-6}$

BF: ᵃPaul and Knauss (552a) give erroneously 0.025.

TABLE 39 (*Continued*)

State	T_e	ω_e	$\omega_e x_e$	$\omega_e y_e$	B_e	α_e	r_e (10^{-8} cm)	Observed Transitions			Remarks
								Designation	ν_{00}	References	
B^{11}H^1	$D_0^0 < 3.51$ e.v. from predissociation in A state; $\mu_A = 0.923585$										
C $^1\Sigma^+$	(55390)	(2230)			[12.23]		[1.222]	$C \to A$	32260.0Z V	(899)	{repr. (451), break-
B $^1\Sigma^+$	(52334)	(2400)	(65)		12.368	0.571	1.2149	$B \to A$	29272.8Z V	(899)	{ing off in $v=0, 1, 2$
b $^3\Sigma^{(-)}$					[12.126]		[1.2270]	$b \to a$	27060.8Z R	(63) (451)	
A $^1\Pi$	23105.0 Z	(2344)	(129)	a	12.152	0.521*	1.2257	$A \to X$	23074.0Z V$_R{}^b$	(63)	isotope eff. B^{10}, B^{11}
a $^3\Pi_r$	c				[12.667]		[1.2005]				
X $^1\Sigma^+$	0	(2366)	(49)		12.018	0.412^d	1.2325			(63) (1050)	
B^{11}H^2	$D_0^0 < 3.55$ e.v.; $\mu_A = 1.703147$										
A $^1\Pi$	(23143)	(1700)	(43)	a	6.653	0.280	1.219$_8$	$A \to X$	23098.75Z V$_R{}^b$	(667)	isotope eff. B^{10}, B^{11}
X $^1\Sigma^+$	0	(1780)	(27)		6.532	0.166^e	1.231$_1$				
(B^{11}H^1)$^+$	$\mu_A = 0.923581$										
A $^2\Pi$	a	(2235)			[11.565]	b	[1.2564]	$A \to X$	26376.2 R	(63)	
X $^2\Sigma^+$	0	(2435)			12.374		[1.2146]				
Bi$_2$:^{209}	$D_0^0 = 1.70$ e.v.;^{a} $\mu_A = 104.528$										
	In addition fragment of a system near 46000 cm^{-1} and three diffuse bands near 40200 cm^{-1}										
E	42252	129	9.7					$E \leftarrow B$	24485 R	(749) (746)	repr. (749)^b
D	36457	157	4.6					$D \leftarrow X$	36448 R	(749) (746)	repr. (749)^b
C	=32000	continuous absorption, repulsive state						$C \leftarrow X$		(749) (746)	
B	17742.3	132.21	0.3009	-0.000474				$B \leftrightarrow X$	17722.0 R	(749)	repr. (749)^b
X $^1\Sigma_g{}^+$	0	172.71	0.3227	-0.00232^c							
Bi:^{209}Br^{79}	$D_0^0 = 2.74$ e.v.;^{a} $\mu_A = 57.297$										
B	24710.9	265.34 H	1.956					$B \leftarrow X$	24738.5 V?	(501)	repr. (501)
A	20532.0	135.91 H	0.534					$A \leftarrow X$	20495.3 R	(501)	repr. (501)
X	0	209.34 H	0.468	-0.1030							
Bi:^{209}Cl^{35}	$D_0^0 = (3.0)$ e.v.; $\mu_A = 29.9651$										
B	25492.7	403.50 H	3.768	0.0016				$B \leftrightarrow X$	25539.8 V?	(501) (1342)	repr. (1342)
A	21801.8	220.3 H	2.50					$A \leftrightarrow X$	21757.6 R	(501) (1342)	repr. (501)
X	0	308.0 H	0.96								
Bi:^{209}F^{19}	$D_0^0 = (3.2)$ e.v.; $\mu_A = 17.4209$										
C ($^3\Pi$)	(36905)^a	[620] H						$C \to X$	36960^a V	(594)	[repr. (594)

A	22959.7	381.0 H	3.00	+0.10			A↔X	22894.6	R	(338)(501)	repr. (338)
X	0	510.7 H	2.05							(338)	
$Bi^{209}H^1$	$D_0^0 = (2.7)$ e.v.;[a] $\mu_A = 1.00329$										
$D\ ^1\Sigma$	?				[3.88]	[2.08]	D→C	20647		(286a)(288)	$^1\Sigma_x$ of (286a)(288); only $v=0$ obs.
$C\ ^1\Sigma$	?	[1313.6] Z			4.37	1.96					$^1\Sigma_y$ of (286a)(288)
$B\ 0^+$	21266.5	(1728) Z	(42.5)		5.307	1.780	B→X	21278.3	V	(286a)(288)	$^1\Sigma^*$ of (286a)(288) repr.
$A\ (1)$[b]	4923.2	1739.4 Z	35.4		5.259	1.788	B→A	16335.8	V	(286a)(288)	
$X\ (0^+)$[b]	0	1698.9 Z	31.6		5.137	1.809[c]					
$Bi^{209}H^2$	$D_0^0 = (2.7)$ e.v.; $\mu_A = 1.99549$										
$B\ (0^+)$	21263.7	1235.0 Z	25.0		2.669	1.779	B→X	21276.2	V	(287)(288)	no band lines given
$X\ (0^+)$	0	1205.5 Z	16.1		2.592	1.805[d]					
$Bi^{209}I^{127}$	$D_0^0 = (2.7)$ e.v.;[a] $\mu_A = 78.979$										
B	23388.9	198.6 H	1.44				B←X	23406.0	V?	(501)	repr. (501)
X	0	163.9 H	0.31								
$Bi^{209}O^{16}$	$D_0^0 = (2.9)$ e.v.; $\mu_A = 14.8625$										
A	15194.5	500.0 H	3.10				A→X	15094.0	R	(1401)	probably one comp. of $^2\Pi$—$^2\Pi$
X	0	702.1 H	5.20								

BH : [a]This state has a potential maximum see (1050). [b]Reversal of shading in 0–0 band. [c]Coupling constant $A = 5.95$ cm^{-1}.
[a]$D_0 = [12.4 - 0.4(v + \tfrac{1}{2})] \times 10^{-4}$.
[c]$D_v = [3.54 - 0.07(v + \tfrac{1}{2})] \times 10^{-4}$.

(BH)$^+$: [a]Coupling constant $A = 14.0$ cm^{-1}.
[b]$D_0 = 12.5 \times 10^{-4}$.

Bi$_2$: [a]Assuming that the B state dissociates into $^2D_{3/2} + {}^4S$. Gaydon gives 1.72 e.v. apparently overlooking the fact that Almy and Sparks (749) give D_e not D_0. From molecular beam experiments Ko (408) obtained 3.34. If the B state dissociates into normal atoms one obtains the D_0^0 value 3.62, if it dissociates into $^2D_{3/2} + {}^4S$ one obtains $D_0^0 = 2.20$ e.v. Extrapolation of the vibrational quanta of the ground state gives 1.85 e.v.
[b]Somewhat different constants are given by (526).
[c]$\omega_e x_e = +0.00001318$.

BiBr: [a]From band convergence in A state (at 22120 cm^{-1}) assuming dissociation into normal atoms.

BiCl: [a]Gaydon gives 2.9 ± 0.2.

BiF: [a]Central component of wide triplet the other two being separated from it by $+7330$ and -4780 cm^{-1}.

BiH: [a]Gaydon gives 2.5 e.v. [b]It is likely that the states A and X are the case (c) analogues of $^3\Sigma^-$ which is expected to be the ground state (see p. 236f. and 341).
[c]$D_v = [1.94 - 0.11(v + \tfrac{1}{2})] \times 10^{-4}$.
[d]$D_0 = 0.51 \times 10^{-4}$.

BiI: [a]Gaydon gives 2.5 ± 1.

TABLE 39 (*Continued*)

State	T_e	ω_e	$\omega_e x_e$	$\omega_e y_e$	B_e	α_e	r_e (10^{-8} cm)	Designation	ν_{00}	References	Remarks
B^{11}N^{14}											
		$D_0^0 = (5.0)$ e.v.;^{a} $\mu_A = 6.16550$									
		An incompletely analyzed singlet system occurs between 30900 and 34500 cm^{-1}									
A $^3\Pi$	27877.0	1317.5 H	14.9		1.555	0.010	1.326	$A \rightarrow X$	27775.8 R	(901) (901)	repr. (901)
(X) $^3\Pi^b$	0	1514.6 H	12.3		1.666	0.025	1.281				
B^{11}O^{16}											
		$D_0^0 = (9.1)$ e.v.;^{a} $\mu_A = 6.52305$									
		Diffuse emission bands in the visible region probably due to BO, see (935) and (1418)									
B $^2\Sigma^+$	43175.4	1280.3 Z	10.07		[1.503]		[1.311]	$B \rightarrow X$	42873.2 R	(923) (508)	β bands; repr.
A $^2\Pi_i$	$\begin{cases}23958.76^b\\23836.40^b\end{cases}$	1260.70 Z	11.157	$+0.049$	1.4132^c	0.0196	1.3524	$B \rightarrow A$	$\begin{cases}19226.6^d\\19348.9^d\end{cases}$ V V		
X $^2\Sigma^+$	0	1885.44 Z	11.769		1.7803	0.01648	1.2049	$A \rightarrow X$	$\begin{cases}23646.62\\23524.26\end{cases}$ R R	(373)	α bands; repr. (373)
Br^{79}Br^{81}											
		$D_0^0 = 1.971$ e.v.; $\mu_A = 39.958$									
		An additional system of discrete emission bands (15100–19600) not related to the others is given by (1475)									
F ($^2\Pi_{2,1g}$)	[66500]	(480)						numerous diffuse emission bands and continua in the visible and near ultraviolet with the repulsive states from $^2P_{3/2} + ^2P_{3/2}$, $^2P_{3/2} + ^2P_{1/2}$, $^2P_{1/2} + ^2P_{1/2}$ as lower states		(1496) (761)	$\begin{cases}\text{convergence limit}\\\text{at 19585 cm}^{-1}\end{cases}$
E ($^1\Pi_g$)	[61444]	(220)									repr. (179)
D ($^3\Pi_{1g}$)	[55534]	(330)									
C ($^3\Sigma_{1u}^+$)	[47000]										
		Absorption continua beyond 19585 cm^{-1} due to several electronic transitions see (863) (744) (1263) (1264) (1344)									
B $^3\Pi_{0u}^+$	15891.3^a	169.71 Z	1.913	*	0.0595	0.000625	2.66$_3$	$B \leftrightarrow X$	15814.3^a R	(134) (1474)	
A $^3\Pi_{1u}$	13814	170.7 H	3.69		0.08091	0.000275^d	2.283$_6$	$A \leftarrow X$	13737	(179)	
X $^1\Sigma_g^+$	0	323.2^c Z	1.07^c							(134)	
BrCl											
		$D_0^0 = 2.26$ e.v.;^{a} $\mu_A = 24.567$									
		Visible absorption bands found by Brown [see (163)] but not published									
D ($\Pi_{1,0}$)	(61563)	[504] H	9.5					$D \leftarrow X$	61600 V	(163) (519)	
C ($\Pi_{1,2}$)	(59318)	519 H	2.9					$C \leftarrow X$	59362 V	(163) (519)	
X $^1\Sigma^+$	0	[430]	3								
Br^{79}F^{19}											
		$D_0^0 = 2.19$ or 2.60 e.v.;^{a} $\mu_A = 15.3542$									
A $^3\Pi_0^+$	19335.2	363	9.5	^{b}	0.357165	0.005214	1.75555	$A \leftarrow X$	19179.6^c R	(829) (1426a)	repr.^{d}
X $^1\Sigma^+$	0	671	3					Rotation spectrum			

BrO[16] $D_0^0 = (2.2)$ e.v.;[a] $\mu_A = 13.3316$

State	T	ω_e	$\omega_e x_e$	Transition	ν	Type	Ref.	Bands
A	(24933)	(475) 713		$A \to X$	24813	R	(1483) (853)	
(X)	0	(9) 7					(853)	

Additional unclassified bands near 30000 cm⁻¹

C₂[12] $D_0^0 = (3.6)$ e.v.;[a] $\mu_A = 6.00194$

State	T	ω_e	$\omega_e x_e$	(*)	B_e	α_e	r_e	Transition	ν	Type	Ref.	Bands
d $^1\Sigma_u^+$	a+4240.23	1829.57 Z	13.97		1.8334	0.0204	1.2378	$d \to a$	43227.23	(V)[b]	(1164)	Mulliken bands
c $^1\Pi_g$	a+34261.9	1809.1 Z	15.81	−4.02*	1.7334	0.0180*	1.2730	$c \to b$	25969.19	V_R	(1052)	Deslandres–d'Azambuja bands
b $^1\Pi_u$	a+8391.66	1608.33 Z	12.14		1.6170	0.01720	1.3180	$b \to a$	8268.50	R	(1314)	Phillips bands
a $^1\Sigma_g^+$	a	1855.63 Z	14.08		1.8205	0.01832	1.2422				(1314)	
B $^3\Pi_g$	40080.41	1106.56 Z	39.26	+2.805*	1.1922	0.0242	1.5350	$B \to X$	39806.46	R	(227) (1316)	Fox-Herzberg bds.
A $^3\Pi_g$	19306.26[d]	1788.22 Z	16.44	−0.5067*	1.7527	0.01608*	1.2660	$A \to X$	19378.44	V_R	(1315)	Swan bands
X $^3\Pi_u$	0[e]	1641.35 Z	11.67		1.6326	0.01683[c]	1.3117				(1315)	

Ca₂ ? $\mu_A = 20.046$

Emission continuum from 19500 to 25000 cm⁻¹   (1018a)

CaBr[79] $D_0^0 = (2.9)$ e.v.;[a] $\mu_A = 26.557$

State	T	ω_e	$\omega_e x_e$	Transition	ν	Type	Ref.	Bands
D $^2\Sigma$	31190.8	326.6 H	1.02	$D \leftarrow X$	31211.4	V	(1020)	4 heads
C $^2\Pi$	25537.5 / 25514.0	265.2 H	0.97	$C \leftarrow X$	25527.4 / 25303.9	R / R	(1020)	double heads
B $^2\Sigma^+$	16380.0[a]	284.6[a] H	0.92	$B \leftarrow X$	16379.6	V	(1020)	4 heads
A $^2\Pi$	15985.8 / 15922.5	288.1 H	0.92	$A \leftrightarrow X$	15987.2 / 15923.9	V / V	(285) (1020)	
X $^2\Sigma^+$	0	285.3 H	0.86					

BN: [a]Gaydon gives 4.0 e.v. Thermochemical data (1378) give $D(BN) > 4.93$ e.v.

BO: [a]Gaydon gives 7 ± 1. [b]The doublet separation was made equal to A as given by (373).

BrBr: [a]Extrapolated from $v' = 8$. [b]Average B value for the two multiplet components. Only the value for $^2\Pi_{1/2}$ has been accurately determined (1.4277). [d]Calculated from $B{-}X$ and $A{-}X$. [c]Induced predissociation see (73).

BrCl: [a]From $D_0^0(Br_2)$, $D_0^0(Cl_2)$ and the thermochemical heat of formation of BrCl (50a). [c]Darbyshire (179) gives 15840 for the head of the 0-0 band; Rees (1344) gives 15749 for T_e but this seems to be an error. [d]$D_v = 2.0 \times 10^{-8}$. [c]Darbyshire (179) gives from band heads $\omega_e = 324.26$, $\omega_e x_e = 1.145$.

BrF: [a]Depending on whether the A state dissociates into $^2P_{1/2} + ^2P_{1/2}$ or $^2P_{3/2} + ^2P_{1/2}$.

BrO: [a]Coleman and Gaydon (853) give 1.75 e.v.

C₂: [a]Brewer, Gilles and Jenkins (825) give 4.95 ± 0.3 e.v. from the intensity variations of the Swan bands with temperature in a carbon tube furnace. [b]Headless bands. [c]Short extrapolation to convergence at 21370 cm⁻¹. [c]It is likely that the numbering in the upper state and therefore ν_{00} will have to be changed. [d]Depending on whether the A state dissociates into $^2P_{1/2} + ^2P_{1/2}$ or $^2P_{3/2} + ^2P_{1/2}$.

CaBr: [a]On account of large spin doubling in the upper $^2\Sigma$ state each band has two heads; the constants refer to the shortward one of these. [c]$D_v = [7.02 - 0.15(v + \tfrac{1}{2})] \times 10^{-6}$.

[b]It is not certain that this is the ground state.

[d]Dipole moment $\mu = 1.29$ debye.

[d]Coupling constant $A = -16.4$.

[e]Coupling constant $A = -16.0$.

TABLE 39 (*Continued*)

State	T_e	ω_e	$\omega_e x_e$	$\omega_e y_e$	B_e	α_e	r_e (10⁻⁸ cm)	Designation	ν_{00}		References	Remarks
CaCl³⁵											$D_0^0 \leq 2.76$ e.v.; $\mu_A = 18.6804$	
E (²Σ)	34266.4	413.3	1.68					E←X	34288.1	V	(1020)	
D (²Σ)	31107.8	423.4	1.61					D→X	31134.5	V	(551)(1020)	
C ²Π	{26574.7, 26498.9}	336.0 H	1.4		(0.24)		(1.9₄)	C↔X	{26558.8, 26483.0}	R, R	(551)(1511), (551)(1511)	4 heads
B ²Σ	16850.6	358.8ª H	1.2					B←Xᵇ	16849.1	V	(551)(1511)(297)	{double heads,pred. above $v=15$
A ²Π	{16162.8ᵈ, 16093.3}	364.9ᶜ H	1.16ᶜ	+0.0012ᶜ				A↔X	{16164.3ᵈ, 16094.8}	V, V	(755)(551)(1131), (1511)	4 heads
X ²Σ⁺	0	369.8ᵉ H	1.31ᵉ		(0.26)		(1.8₆)				(1020)	
Ca⁴⁰F¹⁹											$D_0^0 \leq 3.15$ e.v.; $\mu_A = 12.88080$	
F ²Π	37547.9	681.7	3.55					F←X	37595.0	V	(941)	repr. (941)
E ²Σ	34135.2	646.3	3.24					E←X	34164.7	V	(941)	repr. (941)
D ²Σ	30772.3	650.7	2.89					D←X	30804.1	V	(941)	repr. (941)
C ²Π	{30285.2, 30255.9}	481.7	2.02					C↔X	{30232.7, 30203.4}	R, R	(941)(391)	repr. (941)
B ²Σ	18844.4ª Z	566.7ᵇ H	2.82ᵇ	+0.0051				B↔X	18834.2Z	V_R	(391)(281)	pred. above $v=16$
A ²Π	{16557.2, 16482.1}	{593.4 HQ, 592.0 HQ}	3.113, 3.427	+0.0619	([0.326]), ([0.322])		([2.00]), ([2.02])	A↔X	{16560.3, 16484.4}	V_R, V_R	(391)(281), (391)(281)	{(297); repr. (281)
X ²Σ⁺	0	587.1ᶜ H	2.74ᶜ								(281)	
Ca⁴⁰H¹											$D_0^0 \leq 1.70$ e.v.;ª $\mu_A = 0.983332$	
C ²Σ	28281	1444 H	25		[4.79]		[1.89₂]	C→X	28352.5	V	(269)(509)	repr. (509)(268)ᵇ
D ²Σ	22603	1149 Z	32.5		2.50ᶜ	0.01	2.62	D→X	25525	R	(269)(1000)	
E ²Π	20417ᵈ	1248.6 Z	21.8		[4.2844]	0.058	[2.0005]	E→X	20391	V	(1517)	two Q heads
B ²Σ	15766	1285	20		4.492		1.954	B→X	15753.8	V	(691)(342a)	
A ²Π_r	14460	1333	20		4.34	0.10	1.99	A→X	{14472.2, 14392.3}	V, V	(342a)	
X ²Σ	14380	1299 Z	19.5		4.2778	0.0963ᵉ	2.0020				(691)(269)	
Ca⁴⁰H²											$\mu_A = 1.918055$	
C ²Σ					(2.35)ᶠ	0.045	(1.93)	C→X	28322	V	(1001)(691)	repr. (1001)(691)ᵍ
B ²Σ					2.282ᵈ	0.035ⁱ	1.963	B→X	15748.8	V	(691)	
X ²Σ	0				2.196		2.001					

$(Ca^{40}H^1)^+$ $\quad \mu_A = 0.983332$

State	B_e	r_e	Transition	ν	Refs	Remark
A $^1\Pi$	[6.08]	$[1.67_3]$	A→X V	41236.4 Z	(1005)	{ pred. in PR only, for $J' > 10$
(X) $^1\Sigma$ [a]	[5.71]	$[1.73_3]$				

CaCl $\quad$ $D_0^0 = (2.8)$ e.v.; $\mu_A = 30.468$

Other unclassified bands at 15000–16000 and 30000–32000 cm^{-1} given by (1511).

State	T_e	ω_e	$\omega_e x_e$	B_e	Transition	ν	Refs	Remark
A	23314.3	234.9 H	0.96	*	A↔X	23310.7	(1225)(1511)	{ narrow doublets, diffuse headless bands
X ($^2\Sigma$)	0	242.0 H	0.64					

CaF: $\quad D_0^0 = 5.9$ e.v. (Gaydon's thermochemical value);[a]

For fragments of other systems see (1167) and (129). The latter gives an entirely different analysis of the CaO spectrum.

$Ca^{40}O^{16}$ $\quad \mu_A = 11.4265$

State	T_e	ω_e	$\omega_e x_e$	B_e	α_e	r_e	Transition	shading	ν	Refs	System
F ($^1\Pi$)	44049	(550) H					F→B	R	28054	(1167)	ζ system
E ($^1\Pi$)	41174	(575) H		[1.91]			E→B	R	25191	(1167)	ϵ system
D ($^1\Sigma$)	29598	705.8 H	5.2	(0.405)			D→B	R	13682	(828)(1219)	β system
C ($^1\Sigma$)	(18260)						C→X	V_R	(18260)	(1165)(1131)	δ system
B ($^1\Sigma$)	15912	718.5 H	4.1	(0.464)[c]	(0.006)	1.78	B→X	V_R	15947	(1165)(1131) (1219)	γ system
A ($^1\Sigma$)	9474	684 H	4.7				A→X	V	9491	(1219)(1167)	α system
X ($^1\Sigma$)	0	650 H	6.6							(1167)(1219)	

Analysis of this green system is doubtful[b].

CaCl:

[a] This is from the $\Delta v = 0$ sequence assuming the vibrational constants of the lower state from the A–X system.[d] The vibrational constants for the lower state obtained from the $\Delta v = 0, +1, -1$ sequences of B–X do not agree well with those from the A–X system [see (551) and (755)] since the separation head-origin is very large and different for the different sequences.

[b] This system has been observed in absorption by (1511) and therefore contrary to Asundi's (755) assumption must include the ground state.

[c] The vibrational constants given here are those of (755); Parker (551) gives constants that are appreciably different.

[d] The vibrational constants for the lower state of the A–X system deviate considerably from those of the D–X and E–X systems since the separation head-origin is very large and variable. Asundi (755) gives $\omega_e'' = 361.8$, $\omega_e'' x_e'' = 1.01$. and $\omega_e'' y_e'' = -0.001$.

[e] These constants are from the D–X and E–X systems for which the separation head-origin is least and which are therefore likely to be the best; see [d].

CaF:

[a] There are two heads in each band on account of spin doubling and high K values of the heads. The T_e values for the two heads would be 18898.4 and 18904.9 cm^{-1}.

[b] These values have been derived by the author from the heads of the $\Delta v = 0$ sequence assuming the vibrational constants of the lower state from the other band systems. Because of the large separation between head and origin of these bands the differences between corresponding bands of the $\Delta v = 0$ and $\Delta v = +1$ sequences do not supply reliable ΔG values, particularly when the change of numbering in the $\Delta v = +1$ sequence suggested by (281) is taken into account. One would obtain in this way $\omega_e = 541.8$ and $\omega_e x_e = 2.11$ [see also Fowler's (941) ω_e value].

[c] Average from all systems except B–X.

CaH:

[a] Assuming predissociation in the $C\,^2\Sigma$ state into $^3P + ^2S$.

[b] At low pressure predissociation above $v = 0$, $K = 10$.

[c] Strong perturbations for higher K values.

[d] Coupling constant $A = 9.3$.

[e] Spin coupling constant $\gamma = 0.045$. $D_v = [1.85 - 0.02(v + \frac{1}{2})] \times 10^{-4}$.

[f] Very strong perturbations in this state.

[g] Breaking off above $K = 18$, $v = 0$.

[h] Large difference between true and effective B values (see p. 228): the effective value is given.

[i] Spin coupling constant $\gamma = 0.032$. $D_v = [0.42 + 0.13(v + \frac{1}{2})] \times 10^{-4}$. $^bD_0 = 2.0 \times 10^{-4}$.

CaH+:

[a] Not certain whether this is the ground state

CaO:

[a] Linear extrapolation in the ground state yields 2.0 e.v. Note that the ground state, if it is $^1\Sigma$, cannot dissociate into normal atoms ($^1S + ^3P$).

[b] It appears in absorption but the lower vibrational quanta given by (1165) do not agree with those of the B–X system by (828).

[c] Different constants are derived from the D–B system by (828).

TABLE 39 (*Continued*)

State	T_e	ω_e	$\omega_e x_e$	$\omega_e y_e$	B_e	α_e	r_e (10^{-8} cm)	Designation	ν_{00}	References	Remarks
CaS											
A		$D_0^0 \leq 5.2$ e.v.; $\mu_A = 17.819$									
X		Continuous absorption from 41860 cm⁻¹ to shorter wave lengths						A←X		(1198)	
C¹²Br		$\mu_A = 10.4367$									
		Diffuse bands between 27000 and 37000 cm⁻¹ tentatively assigned to this molecule								(918a) (853)	
C¹²Cl³⁵		$\mu_A = 8.93694$									
A ²Σ	35961	867.₅ H	1.5					A→X	{35972 V, 35834 V}	(1499a)	{repr. (1067)ᵃ, isotope eff.}
X ²Π	{138, 0}	846 H	1.0								
Cd₂		$D_0^0 = 0.087$ e.v.;ᵃ $\mu_A = 56.221$								(31)	
		Large number of continuous and diffuse spectra, the latter with stable upper states; see p. 395									
CdBr		$\mu_A = 46.722$									
		$D_0^0 = (3.3)$ e.v.;ᵃ									
C	31458.3	254.5 H	0.75					C→X	31470.5 R / V	(708) (1287), (708)	
(X)	0	230.0 H	0.50								
		Unclassified emission bands in the region 12300–28600 cm⁻¹ originally ascribed to CdBr₂									
CdCl³⁵		$\mu_A = 26.6793$									
		$D_0^0 = (2.8)$ e.v.;ᵃ $\mu_A = 26.6793$									
		Unclassified emission bands in the region 11500–30000 cm⁻¹ originally ascribed to CdCl₂									
E ²Σ	(45404)	[247.9] H						E→X	45363.4 R	(708) (1287)	repr. (164)
C ²Π	(32508)	(400) H						C↔X	{32543 R, 31428 V}	(164), (1510) (1068), (1287)	ᵇ
X ²Σ	31393	330.5 H	1.2								
CdCs¹³³ ?		$\mu_A = 60.919$									
		Diffuse V shaded absorption bands at 18810 and 19120 cm⁻¹								(769b)	
CdF¹⁹		$\mu_A = 16.2568$									
		Other bands found in emission in the same region (1308) (1068) have been shown by (1307a) to be peculiar Cd lines									
E (²Σ)	(34200)	(535) H						E←X	R	(942)	rather diffuse heads
X (²Σ)	0	(535) H									

CdH¹ $D_0^0 = 0.678$ e.v.; $\mu_A = 0.99917$

State	T_e	ω_e	$\omega_e x_e$	B_e	α_e	r_e	Transition	ν		Ref
B $^2\Sigma$	(24965)	[993.2] Z	*[d]	[2.95][b]	*	[2.39]	B→X	24749.0 Z	R	(658)
A $^2\Pi$	{ 23205.6[c] / 22104.9	1758.1 Z	38.7	6.099	0.196	1.663_3	A→X	{ 23271.2 / 22270.5	V / V	(658)(885)
X $^2\Sigma^+$	0	1430.7 Z	46.3	5.437	0.218[e]	1.761_7				

CdH² $D_0^0 = 0.704$ e.v., from D_0^0(CdH¹); $\mu_A = 1.97926$

State	T_e	ω_e	$\omega_e x_e$	B_e	α_e	r_e	Transition	ν		Ref
A $^2\Pi$		[1209.7][v] Z		3.071	0.066	1.665	A→X	{ 23233.4 / 22227.1	V / V	(658)(885)
X $^2\Sigma^+$	0			2.788	0.168 *[f]	1.748				

(CdH¹)⁺ $D_0^0 = (2.0)$ e.v.[a] $\mu_A = 0.99917$

State	T_e	ω_e	$\omega_e x_e$	B_e	α_e	r_e	Transition	ν		Ref
A $^1\Sigma^+$	42935.2	1252 Z	8.6	4.851	0.082	1.865	A→X	42680.6 Z	R	(659)
X $^1\Sigma^+$	0	1775.4 Z	37.3	6.071	0.189[b]	1.667_2	*			

(CdH²)⁺ $D_0^0 = (2.0)$ e.v.[a] $\mu_A = 1.97926$

State	T_e	ω_e	$\omega_e x_e$	B_e	α_e	r_e	Transition	ν		Ref
A $^1\Sigma^+$	42930.6	887.2 Z	3.44	2.451	0.027	1.864	A→X	42746.8 Z	R	(734)
X $^1\Sigma^+$	0	1262.5 Z	19.01	3.075	0.0682[c]	1.664				

CdI¹²⁷ $D_0^0 = (1.6)$ e.v.[a] $\mu_A = 59.624$

A band system in the region 15800–17600 has been found by (1442) but their analysis appears very doubtful; see also (1287)(1330).

State	T_e	ω_e	$\omega_e x_e$	Transition	ν		Ref
E ($^2\Sigma$)	41912.4	108.5 H	1.0	E→X	41978	R	(708)(1068)(1330) [b]
D ($^2\Pi_{3/2}$)[c]	29530.0	196.6 H	0.70	D↔X	29539.0	V	(708)(1538)
X $^2\Sigma$	0	178.5 H	0.625				

CdIn $\mu_A = 56.802$

Bands originally ascribed by (1511a) to CdIn have later been shown by (769c) to be due to Bi₂.

CCl: [a]Horie (1067) gives rather different constants.

Cd₂: [a]From temperature dependence of diffuse molecular absorption (433).

CdBr: [a]Gaydon gives 2.8 ± 1.0 e.v.

CdCl: [a]Gaydon gives 2.2 ± 1 e.v. [b]Not certain whether this system is due to CdCl.

CdH: [a]Large vibrational perturbations, therefore no formula given v; given by (658) even though levels up to $v = 13$ observed; $A_0 = 1012.8$. [b]Large rotational perturbations. [c]The doublet splitting increases slightly with v; $A_0 = 1012.8$. [d]Rapid convergence. [g]Refers to upper doublet component.
[e]$-0.095(v + \tfrac{1}{2})^2 + \ldots$; $D_v = [3.03 + 0.22(v + \tfrac{1}{2}) + \ldots] \times 10^{-4}$. [f]$D_0 = 0.76 \times 10^{-4}$, $D_1 = 1.78 \times 10^{-4}$.
[b]$D_v = [2.9 + 0.1(v + \tfrac{1}{2}) + \ldots] \times 10^{-4}$. [c]$D_v = [0.48 + 0.12(v + \tfrac{1}{2})] \times 10^{-4}$.

CdH⁺: [a]Gaydon gives 2.1 ± 0.4.

CdI: [a]Gaydon gives 1.4 ± 0.2 e.v. [b]Wieland (708) considered this system originally as due to CdI₂ but more recently to CdI (1534). [c]Only fragments of the second doublet component have been found at ~28230 cm⁻¹; see (1068) and (1330).

TABLE 39 (*Continued*)

State	T_e	ω_e	$\omega_e x_e$	B_e	α_e	r_e $(10^{-8}$ cm$)$	Observed Transitions		References	Remarks
							Designation	ν_{00}		
CdK ?		$\mu_A = 29.015$								
			Diffuse V shaded absorption bands at 23850 and 23960 cm^{-1}						(769b)	
CdNa^{23} ?		$\mu_A = 19.0918$								
			Diffuse V shaded absorption bands at 24990 and 25160 cm^{-1}						(769b)	
CdO^{16} ?		$\mu_A = 14.0069$								
			Bands originally ascribed by (1510) to CdO have later been shown by (769c) to be due to Bi$_2$							
CdRb ?		$\mu_A = 48.570$								
			Diffuse V shaded absorption bands at 22280 and 22600 cm^{-1} and other unclassified bands in the region 22500–24000 cm^{-1}						(769b)	
CdS		$D_0^0 \leq 3.9$ e.v.; $\mu_A = 24.956$								
			Two absorption continua with long wave length limits at 31700 and 42500 cm^{-1}						(1401a)	
CdSe		$D_0^0 \leq 3.2$ e.v.; $\mu_A = 46.393$								
			Two absorption continua with long wave length limits at 25500 and 43850 cm^{-1}						(1198a)	
CeO^{16}		$D_0^0 = (7.7)$ e.v.; $\mu_A = 14.3607$								
			Fragment of another band system near 13700 cm^{-1}							
E	$y + 20880.3$	807.9 H	2.04				$E \rightarrow (X')$ R	20864.3	(693)	[a]
D	$y + 20880.3$	791.7 H	1.72				$D \rightarrow (X')$ R	20556.3	(693)	[a]
(X')	y	840.2 H	2.58							[a]
B	13855.2	788.3 H	1.76				$B \rightarrow X$ R	13817.2	(693)	[b]
A	12803.9	785.3 H	2.13				$A \rightarrow X$ R	12764.3	(693)	[b]
(X)	0	865.0 H	2.99							[bc]
C^{12}H^1		$D_0^0 = 3.47$ e.v.; $\mu_A = 0.930024$								
C $^2\Sigma^{+\;a}$	(31821)	2824.1^b	105.8^b	14.629^g	0.744	1.1132	$C \rightarrow X$ V	31792.3	(974)	pred. in $v = 0$ and 1
B $^2\Sigma^{-\;a}$	(25949)	2542.5^b	373.8^b	12.887^g	0.485	1.1861	$B \rightarrow X^c$ R	25712.4	(974)	repr. (636)
A $^2\Delta$	(23150)	2921.0^b	90.4^b	14.912^g	0.670*	1.1026	$A \rightarrow X^c$ V	23173.5^d	(974) (1484)	repr. (47) (1484)
X $^2\Pi$	0^f	2861.6^b	64.3^b	14.457^g	0.534^h	1.1198			(974)	

C¹²H² $D_0^0 = 3.52$ e.v.; $\mu_A = 1.725173$

State	T_e	ω_e	$\omega_e x_e$	B_e	α_e	r_e	Transition	ν		Ref.
C ²Σ⁺	(31828)	2073.4[b]	57.0[b]	7.880	0.282*	1.113_7	C→X	31809.0	V	(973)
B ²Σ⁻	(25993)	1808.0[b]	201.5[b]	7.171[g]	0.528	1.167_4	B→X	25804.6	R	(973)
A ²Δ	(23183)	2144.5[b]	48.7[b]	8.032[g]	0.260	1.103_1	A→X	23201.2	V	(973)
X ²Π	0[f]	2101.0[b]	34.7[b]	7.808[g]	0.212[i]	1.118				

{ pred. in $v = 0$ and 1 repr. (636) }

(C¹²H¹)⁺ $D_0^0 = 3.6$ e.v.; $\mu_A = 0.930021$

Bands between 4²000 and 44000 cm⁻¹, originally assigned to CH⁺(1₂₃)(977), have recently been shown to be due to HgH⁺ (M.W. Feast, unpublished)

State	T_e	ω_e	$\omega_e x_e$	B_e	α_e	r_e	Transition	ν		Ref.
A ¹Π	(24145)	1850.02 Z	103.93	11.8977	0.9367*	1.23439	A→X	23596.94	R	(902)
X ¹Σ⁺	0	[2739.54] Z		14.1767	0.4898[a]	1.13083				

repr. (902)

Cl₂³⁵ $D_0^0 = 2.475$ e.v.; $\mu_A = 17.48942$

Additional diffuse bands are given by (839)

State	T_e	ω_e	$\omega_e x_e$	B_e	α_e	r_e	Transition	ν		Ref.
E (¹Π$_g$)	(75000)[b]									(1497)(761)
D (³Π$_{1u}$)	(67700)[b]									
C (³Σ$_{1u}^+$)	(58000)[b]									
A ³Π$_{0^+u}$	18310.5[c]	239.4[c] H	5.42	0.158[c]	0.003*	2.47	A↔X[d]	18147.4[c]	R	(1263)(60a)(863) (200)(22)(808) (990)
X ¹Σ$_g^+$	0	564.9 H	4.0[e]	0.2438	0.0017	1.988				

{ numerous continua in the visible and ultraviolet regions with the repulsive states from $^2P_{3/2} + {}^2P_{3/2}$, $^2P_{3/2} + {}^2P_{1/2}$, $^2P_{1/2} + {}^2P_{1/2}$ as lower states }

Absorption continua beyond 20850 cm⁻¹ due to several electronic transitions

{ convergence at 20893 cm⁻¹ for data on Cl₂³⁷ see (1417) }

CeO: [a]It has been assumed that Watson's systems D and E have the same lower state. [b]It has been assumed that Watson's systems A and B have the same lower state. [c]It is very doubtful whether this is the ground state.

CH: [a]Gerö interchanges Σ⁺ and Σ⁻, but since ²Σ⁻ arises from normal atoms and since the C state has vibrational levels above $^3P + {}^2S$ it must be B ²Σ⁻ and C ²Σ⁺. [b]These values have been derived from the observed $\Delta G_{1/2}$ values of CH and CD according to the isotope relations. [c]Observed in absorption in the solar spectrum and other stellar spectra (see Chapter VIII, section 4). [a]This is Fagerholm's (925) value since Gerö's (974) value is apparently erroneous. [e]Gerö and Schmid (978) have found a number of breaking-off points in this system which do not appear real. [f]Coupling constant $A = 27.95$ [from CD see (925)]; the separation of $^2\Pi_{1/2}(J = \frac{1}{2})$ and $^2\Pi_{3/2}(J = \frac{3}{2})$ in CH is 17.9 cm⁻¹, in CD it is 3.8 cm⁻¹ [see (973)]. [g]From Gerö's (973)(974) B_v values. Fagerholm (925) gives slightly different values but does not include the C ²Σ⁺ state. In order to have a consistent set Gerö's values were adopted.

[h]$D_v = [14.5 - 0.3(v + \frac{1}{2})] \times 10^{-4}$.
[i]$D_v = [4.4 - 0.3(v + \frac{1}{2})] \times 10^{-4}$.
[a]$D_v = [14.1 - 0.3(v + \frac{1}{2})] \times 10^{-4}$.

CH⁺: [a]This is Gaydon's value who has slightly changed Kuhn's (426) original value. [b]These are T_0 not T_e.

Cl₂: [a]This is Gerö's value who has slightly changed the change of numbering proposed by Birge (808) [not adopted by (23) and (24) but by (22)]. [b]These values take account of the change of numbering proposed by Birge (808) original value. [c]These are T_0 not T_e. [d]Observed in emission by (410) and (1474). [e]Levels observed only up to $v'' = 3$.

TABLE 39 (Continued)

State	T_e	ω_e	$\omega_e x_e$	$\omega_e y_e$	B_e	α_e	r_e (10^{-8} cm)	Designation	ν_{00}		References	Remarks
(Cl$_2^{35}$)$^+$		$D_0^0 = (4.4)$ e.v.; $\mu_A = 17.48928$										
		Fragments of another system in the same spectral region										
(A ^{2}H)	{20797.3	572.3 H	5.32	−0.013	0.186	0.0028	2.28	$A \rightarrow X$	{20760.2	R	(201)	repr. (201)
(X ^{2}H)$_i^a$	{20596.9	564.8 H	4.13	−0.038	0.183	0.0017	2.30		{20556.4	R		
	0	645.3 H	2.90		0.2697	0.0018	1.891					
Cl^{35}F^{19}		$D_0^0 = 2.616$ e.v.;^{a} $\mu_A = 12.31410$										
A ^{3}H$_0^+$	(18956)	313.48 Z	2.217	−0.400^b	0.372	0.014*	1.91$_8$	$A \leftarrow X$	18717.9^c	R	(1507) (1392)	repr. (1392)^d
X $^1\Sigma$	0	793.2^e Z	9.9^e		0.516508$_6$	0.004358$_5$	1.62813$_1$	Rotation spectrum			(985a)	^{f}
ClO^{16}		$D_0^0 = 1.9$ e.v., from predissociation of ClO$_2$ (Gaydon) ; $\mu_A = 11.026$										
B		(550) H						$B \rightarrow A$	(27600)	R	(1302)	
A		(780) H										
C^{12}N^{14}		$D_0^0 = ?;^b$ $\mu_A = 6.46427$										
B $^2\Sigma^+$	25751.8^c	2164.13^c Z	20.25^c	*	1.9701^c	0.02215^c	1.1506	$B \leftrightarrow X$	25797.85	V$_R^d$	(1141) (1530)	repr. (366).Fig.7,18
											(931a)	
A ^{2}H$_i$	9241.66^e	1814.43 Z	12.883	^{f}	1.7165	0.01746	1.2327	$A \leftrightarrow X$	9114.59	R	(377) (1051)	repr. Fig. 23
											(1303)	
X $^2\Sigma^+$	0	2068.705 Z	13.144		1.8996	0.01735^g	1.1718				(379)	^{a}
C^{12}O^{16}		$D_0^0 = ?;^a$ $\mu_A = 6.85841$										
		In addition Schmid and Gerö (1387) have derived, from perturbations in A ^{1}H, a $^3\Sigma^-$ state which they designate $e\,^3\Sigma^-$.^{b}										
c $^3\Sigma^+$	[93157.8]				[1.9563]		[1.1210]	$c \rightarrow a$	43602.1Z	V	(969)	"3A" bands;^{c}
e ?	[90973]							$e \rightarrow a$	41417H	V	(1115)	Kaplan bands
b $^3\Sigma^+$	(83804)	[2198] Z			2.075	0.033	1.088	$b \rightarrow a$	35358.5Z	V	(1385) (190) (793)	{3rd positive group
								$b \leftrightarrow X$	83831	V	(251)	and 5B bands^d
d ^{3}H$_i$	62299.4^e	1137.79 H	7.624	−0.1125	1.2615	0.0170	1.3960	$d \rightarrow a$	13310.7H	R	(756) (971) (1034)	Hopfield-Birge bds.
											(754) (793) (975)	"triplet" bands^f
a' $^3\Sigma^+$	(55901)	1218 H	9.5	^{o}	1.331	0.016	1.359	$a' \rightarrow a$	(6954)	R	(949b)	Asundi bands
								{$a' \leftarrow X$	(55380)	R	(332)	Hopfield-Birge bds.
a ^{3}H$_r$	48687.5$_b^h$ Z	1739.25^i H	14.47		1.6810	0.0193	1.2093	$a \leftrightarrow X$	48473.9;7Z	R	(139) (793) (252)	{Cameron bands;
											(298 (1456a)	repr. (252)}
		Fragments of other (singlet) band progressions in the region 113000–135000 cm^{-1}										^{jk}
		Tanaka's Rydberg series (β) of bands (absorption): $\nu = 158692 - R/(m + 0.32)^2$, $m = 3, 4, \ldots 10$										
T	(154781)	(1326)						$T \rightarrow X$	154362	R^l	(1456a)	Tanaka bands (P_5)
S	(145010)	(1620)						$S \leftarrow X$	144735	R^l	(1456a)	Tanaka bands (P_4)

Tanaka's Rydberg series (α) of bands (absorption): $\nu = 133380 - R/(m + 0.30)^2$, $m = 3, 4, \ldots 7$

State			Transition	ν		j
R	129154	1596 18.7	R←X	128866	R[l]	Tanaka bands (P_3) (1456a)
Q	129043	1558 10.6	Q←X	128738	R[l]	Tanaka bands (P_2) (1456a)
P	126708	1576 15.6	P←X	126410	R[l]	Tanaka bands (P_1) (1456a)

Takamine, Tanaka and Iwata's Rydberg series of bands (absorption): $\nu = 113029 - R/(m + 0.12)^2$, $m = 4, 5, \ldots 12$ (1455a) R[l]

State			Transition	ν		j
G ($^2\Pi$)	(105799)	(1097)	G←X	105266		Hopfield-Birge bands (332)
F ($^2\Pi$)	99805	2112 H 198	F←X	99730	R	Hopfield-Birge bands (332) (23)
E $^1\Sigma^+$	(92928)	[2134] H	E←X	92923[n]		Hopfield-Birge bands (332)

Cl$_2^+$: [a]Not certain whether this is the ground state. (However $^2\Pi$ ground state expected from electron configuration.)

ClF: [a]Assuming that at the convergence limit of $A\,^3\Pi_{0^+}$ the dissociation products are $^2P_{3/2} + {}^2P_{1/2}$. If they were $^2P_{1/2} + {}^2P_{3/2}$ one would obtain $D_0 = 2.557$ e.v. The products $^2P_{3/2} + {}^2P_{3/2}$ at this limit are excluded since there is a predissociation limit below it.
[b]Convergence limit at 21508 cm^{-1}.
[c]Extrapolated; Schmitz and Schumacher (1392) give $\nu_{00} = 19592.4$ but have not studied the isotope effect as Wahrhaftig (1507) seems to have done.
[d]Predissociation (diffuseness) for $v' = 11,12,13$; predissociation limit at 21200 ± 40 cm^{-1}.
[e]From measurements of heads Schmitz and Schumacher (1392) obtain $\omega_e = 780.4$, $\omega_e x_e = 4.0$. Since Wahrhaftig (1507) does not give bands with $v'' = 2$ it is difficult to see how his $\omega_e x_e$ was obtained. His $\Delta G_{1/2}$ is 773.4 cm^{-1}.
[f]Infrared rotation-vibration bands found by (1305a).

CN: [a]For data on C^{13}N^{14} see (379).
[b]This value can be obtained only indirectly. Depending on the choice of values for D_0^0(CO), D_0^0(N$_2$) and D(C$_2$N$_2 \rightarrow$2CN) a variety of D_0^0(CN) values is obtained ranging from 4.5 to 7.6 e.v. The upper limit is close to the convergence limit which Schmid, Gerö and Zemplén (624a) claim for the $^2\Pi$ state.
[c]These represent only $v = 0,1,2,3$ and not the tail bands. [d]See text p. 173.
[e]Coupling constant $A = -52.2$.
[f]Schmid, Gerö and Zemplén (624a) give a convergence of this state at 60500 cm^{-1} above the ground state derived from perturbations in the $B\,^2\Sigma$ state by means of a fairly long extrapolation.

CO: [a]The B_v'' values of the tail bands ($v'' \geq 8$) fall below the values extrapolated from the B_e and α_e values given. $D_v = [6.4 + 0.0095(v + \tfrac{1}{2})] \times 10^{-6}$ (calc.).
[b]Depending on the choice of dissociation products at the predissociation limit 89620 cm^{-1} or on the reality of other predissociations various values for D_0^0(CO) have been suggested. The three most discussed values are 11.108 e.v. [Gaydon (34)], 9.605 e.v. [Hagstrum (1015)] and 9.141 [Herzberg (314) (1038)]. Indirect evidence [Brewer, Gilles and Jenkins (825)] seems to favor 11.108 e.v.
[b]The lowest observed vibrational level of this state is at 64802 cm^{-1} with $B_v = 1.24$.
[c]Predissociation at about the same energy as for $C\,^1\Sigma$.
[d]Doubtful predissociation in $v = 0$ above $K = 55$ [see (966)].
[e]Coupling constant $A = -34.6$ for $v = 0$.
[f]Herman and Herman (1034) give predissociation above $v' = 10$ (breaking off) but this would require $D < 9.1$ e.v.
[g]The higher vibrational levels of this state cause many strong perturbations in the $b\,^3\Sigma^+$ state. These perturbations indicate a convergence limit at 89620 cm^{-1} [see (623)].
[h]Coupling constant $A = 41.5$.
[i]From zero lines $\Delta G_{1/2} = 1714.6_1$ [see (252) (1335)].
[j]Each member accompanied by two or more vibrational transitions (1-0, 2-0, ...).
[k]In addition a diffuse Rydberg series close to the β series was found.
[l]Not stated explicitly.
[m]The 1-0 bands of all members of this Rydberg series have also been observed with $\Delta G_{1/2} \approx 2175$ cm^{-1}.
[n]Read (582) gives 92841.9 which agrees only very poorly with the value of (332).

TABLE 39 (Continued)

State	T_e	ω_e	$\omega_e x_e$	$\omega_e y_e$	B_e	α_e	r_e (10⁻⁸ cm)	Designation	ν_{00}	References	Remarks
C¹²O¹⁵ (Continued)											
$C\ ^1\Sigma^+$	(91926)	[2133] H			[1.9422]		[1.1250]	$C \to A$	27174.0Z V	(619)	ᵒHerzberg bands
								$C \leftrightarrow X$	91920.5Z V	(582) (332)	Hopfield-Birge bands
$B\ ^1\Sigma^+$	(86948)	[2082.07] Z			1.961	0.027	1.120	$B \to A$	22171.3Z V	(619)	Ångström bandsᵖ
								$B \leftrightarrow X$	86917.8Z V	(582) (332)	Hopfield-Birge bands
$(I\ ^1\Sigma^-)$	from perturbations in $A\ ^1\Pi$									(1387)	q
$A\ ^1\Pi$	65074.8	1515.61 Z	17.2505		1.6116	0.02229*	1.2351	$A \leftrightarrow X$	64746.5ʳ R	(582) (621)	4th positive groupˢ
$X\ ^1\Sigma^+$	0	2170.21 Z	13.461	+0.0308	1.9313₉	0.01748₅	1.1281₉	rotation-vibration bands R		(1051a) (1335)	
(C¹²O¹⁶)⁺	$D_0^0 = (9.9)$ e.v.,ᵃ	$\mu_A = 6.85823$									
$B\ ^2\Sigma^+$	45876.7₀	1734.18 Z	27.927	+0.3283	1.7999₂	0.03025	1.1686₄	$B \to X$	45633.5₂ R	(1335a) (1555) (865) (1384)	first negative bands
$A\ ^2\Pi_i$	20733.1ᵇ	1562.06 Z	13.53₂	+0.0131	1.5894₀	0.01942	1.2436₃	$B \to A$	25225.9₇ V	(143)	Baldet-Johnson bands
$X\ ^2\Sigma^+$	0	2214.24 Z	15.164	−0.0007	1.9772₀	0.01896ᶜ	1.1150₆	$A \to X$	20407.5₅ R	(143) (1384) (865) (1555) (1335a)	comet-tail bands
CoBr	$\mu_A = 33.931$										
Fragments of two band systems in the region 17700–23100 cm⁻¹ with ω'' ≈ 310 cm⁻¹											
CoCl	$\mu_A = 22.145$									(490) (1225)	
Additional unassigned bands are given by (1225)											
F	$c + 22967.3$	420.0	1.66					$F \to C$	22966.3 R?	(498)	a
E	$b + 22404.3$	416.6	0.82					$E \to B$	22402.8 R?	(498)	a
D	$a + 22014.6$	420.0	1.14					$D \to A$	22013.9 R?	(498)	a
C	c	421.8	1.34								
B	b	419.4	0.28								
A	a	421.2	0.74								
CoH¹	$\mu_A = 0.99118$										
$A, \Omega = 4$	(22425)	[1528.1] Z			6.702	0.305	1.593₁	$A \to X$	22243.9 R	(288) (289)	repr. (228) (389)ᵃ
$X, \Omega = 4$	0	(1890)			[7.151]	b	[1.542₃]			(1358) (1192)	
CoO¹⁶	$\mu_A = 12.5847$										
Red-shaded bands between 10900 and 15900 cm⁻¹: Fragments of band systems with $\omega_e'' = 850$, $\omega_e'' x_e'' = 6$											

$C^{12}P^{31}$ $D_0^0 = (6.9)$ e.v.,[a] $\mu_A = 8.65196$

	T_e	ω_e	$\omega_e x_e$	B_e	α_e	r_e		ν_{00}			
B ²Σ⁺	29100.4	836.32 H	5.917	0.6829	0.00628	1.6892	B→X	28898.9	R	(77)	repr. (77)
A ²Π_i	{ 7053.2 / 6894.9	1061.99 H	6.035	0.698	0.0077	1.671	B→A	{ 21934.3 / 22092.6	R / R	(77)	repr. (77)
X ²Σ⁺	0	1239.67 H	6.86	0.7986	0.00597[b]	1.5621					

CrCl   Bands found by (1225) are due to AlCl according to (959)

A											repr. but no analysis
X											

CrH¹ $\mu_A = 0.98897$

A				27200			A→X	27200	V	(245a)	
X											

CrO¹⁶ $D_0^0 = (3.8)$ e.v.; $\mu_A = 12.23^m$

	T_e	ω_e	$\omega_e x_e$	B_e	α_e	r_e		ν_{00}			
A[a]	16594.7	750.7 H	8.9	0.7805	0.0085	1.5732	A→X	16520.0	R	(265)	repr. (265)
(X)[a]	0	898.8 H	6.5	0.8205	0.00624	1.5344					

$C^{12}S^{32}$ $D_0^0 = (7.8)$ e.v.;[a] $\mu_A = 8.728^{iv}$

A series of red shaded bands (39905.1, 38638.2, 37379.6, 36129.8) with P and R branches only has been found by (174)[b]

	T_e	ω_e	$\omega_e x_e$					ν_{00}			
A ¹Π	38912.15	1072.3 H^Q	10.3				A↔X	38804.8[c]	R	(174)(1137)(775)	repr. (174)[d]
X ¹Σ⁺	0	1285.1 H^Q	6.5								

CO: [o] Doubtful predissociation at $J = 28 - 29$ [see (1386)].

[p] Predissociation (sudden drop in intensity) between $J = 37$ and 38 for $v = 0$ and between $J = 17$ and 18 for $v = 1$, that is 2800 and 2700 cm⁻¹ respectively above $B\ ^1\Sigma\ (v = 0, J = 0)$ [see (619) (165)].

[q] The lowest observed vibrational level is at 66380 cm⁻¹ with $B_v \approx 1.48$ cm⁻¹. Schmid and Gerö (1387) assume a convergence at ~82000 cm⁻¹. The K state quoted in the literature (23) is probably spurious [see (1387)].

[r] Gerö (250) gives 64743.6 but no $G(v)$ formula.

[s] Double-headed bands (R and Q heads); for a reproduction see Fig. 14. Gerö (250) finds a predissociation for $v = 7$, 8 and 9 giving a dissociation limit at 77497 cm⁻¹. Attempts to reproduce this breaking off in the writer's laboratory have thus far been unsuccessful.

[t] $D_v = [6.43 + 0.04(v + \frac{1}{2})] \times 10^{-6}$.

CO⁺: [a] This is the extrapolated value which is very probably too high. From $D_0^0(CO) = 9.141$, $I(C) = 11.264$, $I(CO) = 14.009$ (see Table 37) one obtains $D_0^0(CO^+) = 6.40$ e.v.

[b] Coupling constant $A = 125$. Following (1335a) the v-numbering in the $A\ ^2\Pi$ state has been reduced by three units compared to (1037) and (143).

[c] $-0.000037(v + \frac{1}{2})$; $D_v = 6.40 \times 10^{-6}$.

CoCl:

CoH: [a] Heimer (288) (289) considers the two states as ⁴Φ₄ states, which however do not result from his assumed electron configurations. Probably this is case (c).

[b] $D_0 = 4.05 \times 10^{-4}$.

CP: [a] Gaydon gives 6 ± 1 e.v.   [b] Spin doubling constant $\gamma = -0.017$. $D_v = 1.33 \times 10^{-6}$.

CrO: [a] Probably triplet states; not certain whether lower state is ground state.

CS: [a] Gaydon gives 7.2 ± 1.0 e.v.   [b] The first two have been found in absorption by (1137). Their interpretation is uncertain. Howell (1071) suggests that this system is $^1\Pi - \,^1\Sigma$ and that the lower state is identical with the ground state. The latter conclusion also follows from the fact that they occur in absorption but the tentative fine structure analysis of (174) does not agree with it.   [d] Strong rotational and vibrational perturbations. Howell (1071) suggests that this is a $^3\Pi$ state. However not confirmed by (775).

[c] The observed ν_{00} is 38797.0 cm⁻¹.

TABLE 39 (Continued)

State	T_e	ω_e	$\omega_e x_e$	$\omega_e y_e$	B_e	α_e	r_e (10⁻⁸ cm)	Designation	ν_{00}	References	Remarks
Cs₂¹³³		$D_0^0 = 0.45$ e.v.; $\mu_A = 66.473$									
		Several other only partially analyzed band systems, as well as diffuse bands near atomic lines[a]									
D	16175.80	27.34 H	0.0733					D←X R	16168.47	(462) (1473)	repr. (462)
C	16066.03	29.38 H	0.0796					C←X V	16059.73	(1150)	repr. (462)
B (¹Πᵤ)	13043.87	34.230 H	0.07799	−0.0001881				B←X R	13039.99	(1150)	repr. (462)
A (¹Σᵤ⁺)		Extended system of absorption bands from 8800–11000 cm⁻¹, only partially analyzed						A←X R		(462) (223)	
X ¹Σg⁺	0	41.990 H	0.08005	−0.0001643							
Cs¹³³Br		$D_0^0 \geq 3.9_0$ e.v.; $\mu_A = 49.921$									
		Absorption maxima at 36350 and 39350 cm⁻¹ and a continuum above 31437 cm⁻¹								(1253)	
A		Unstable state giving diffuse bands below 31437 cm⁻¹					[3.14]ᶜ			(644) (1253)	
X ¹Σ⁺	0	(194)ᵇ	(2.0)							(644)	
Cs¹³⁵Cl		$\mu_A = 27.998$									
		Absorption continua with maxima at 40500 and 51500 cm⁻¹									
X ¹Σ⁺	0						[3.06]ᵃ			(1391)	
C¹²Se		$D_0^0 = (6.8)$ e.v.; $\mu_A = 10.4202$									
A ¹Π	35238.0	835.7 H	2.2					A→X R	35138.5	(771)	repr. (771)ᵇ
X ¹Σ⁺	0	1036.0 H	4.8								
CsF¹⁹		$D_0^0 = 5.6_5$ e.v.; $\mu_A = 16.6277$									
		Absorption continuum with maximum at 47700 cm⁻¹ and diffuse bands at smaller wave numbers								(842a)	ᵇ
								{radio frequency electric resonance spectrum}		(1075) (1076) (1471)	
X ¹Σ⁺	0	(270)			0.185	0.00185	2.34				
Cs¹³³H¹		$D_0^0 = (1.9)$ e.v.; $\mu_A = 1.000544$									
A ¹Σ⁺	18405.2	204.0 Z	−5.70	−0.350*ᵇ	1.126	−0.0185*	3.86₉	A←X R	18066.4	(65)	
X ¹Σ⁺	0	890.7 Z	12.6		2.709	0.057ᶜ	2.494				
Cs¹³³I¹²⁷		$D_0^0 = 3.3_7$ e.v.; $\mu_A = 64.935$									
		Several absorption continua with maxima at 30850, 38750, 41750, 47050, 50250 and 54050 cm⁻¹								(1391)	
A ¹Σ		Unstable state giving diffuse bands below and a continuum above 27166 cm⁻¹					[3.41]ᵇ	A←X	(27166)	(644)	
X ¹Σ	0	142	(1.2)								

CsRb $\mu_A = 52.0365$

State	T_e	ω_e	Transition	ν_{00}	R		Remarks
A	(13747.2)	38.5 H	A←X	(13741.7)[a]	R	(462)(438)	[b]
X	0	49.4 H					

Cu₂?[a] $\mu_A = 31.779$ $D_0^0 = (0.17)$ e.v.

State	T_e		Transition	ν_{00}			Remarks
B ($^1\Pi_u$)	19464	1.5	B→X	19435[b]		(1419)	
A ($^1\Pi_u$)	17713	0.5	A→X	17679[b]		(1419)	
X ($^2\Sigma_g^+$)	0	5					

Cu⁶³Br⁷⁹ $\mu_A = 35.022$ $D_0^0 = (2.5)$ e.v.

State	T_e	ω_e		Transition	ν_{00}	R		Remarks
C	23461.5	296.20 H	1.423	C↔X	23452.4	R	(591)	isotope effect
B	23044.4	284.22 H	1.323	B↔X	23029.3	R	(591)	isotope effect
A	20498.2	296.13 H	1.008	A↔X	20489.2	R	(591)	isotope effect
X $^1\Sigma^+$	0	314.10 H	0.865					

Cu⁶³Cl³⁵ $\mu_A = 22.4858$ $D_0^0 = (3.0)$ e.v.

State	T_e	ω_e		Transition	ν_{00}	R		Remarks
F	25282	383 H	2	F→X	25265[b]		(818)	
E ($^1\Sigma$)[c]	23076.8	405.9 H	1.82	E↔X	23071.2	R	(591) (1424b)	isotope effect
D ($^2\Pi$)[c]	22972.7	395.0 H	1.84	D↔X	22961.7	R	(591)	isotope effect
C	20634.9	399.1 H	1.65	C↔X	20626.0	R	(591)	isotope effect
B	20487.8	400.7 H	1.67	B↔X	20479.7	R	(591)	isotope effect
A	19001.3	409.8 H	1.90	A↔X	18897.7	R	(591)	isotope effect
X $^1\Sigma^+$	0	416.9[d] H	1.57[d]					

Cs₂: [a]See Chapter VII section 2(d). A system of sharp bands near 13700 cm⁻¹ found by (462) is due to CsRb [see (438) and (223)].

CsBr: [a]Refers to dissociation into normal atoms $^2S + {}^2P_{3/2}$ while the ground state on increasing the vibrational energy dissociates into ions. Thermochemical data (Gaydon) give 4.03 e.v. [b]Numbering not certain. [c]From electron diffraction (483).

CsCl: [a]From electron diffraction (483).

CSe: [a]Gaydon gives 5 ± 1 e.v. [b]Vibrational perturbation for $v' = 1$; Howell (1071) suggests that this state is $^3\Pi$.

CsF: [a]Refers to dissociation into normal atoms $^2S + {}^2P_{3/2}$. Thermochemical data (Gaydon) give $D_0^0 < 5.7$ e.v. [b]Dipole moment $\mu = 7.3 \pm 0.5$ debye.

CsH: [a]Gaydon gives 1.8 ± 0.3 e.v. [b]Slightly different constants are given by (65) for the higher vibrational levels. A shift in v' may be necessary. [c]$D_v = 1.0 \times 10^{-4}$.

CsI: [a]Refers to dissociation into normal atoms $^2S + {}^2P_{3/2}$ while the ground state on increasing the vibrational energy dissociates into ions. Thermochemical data (Gaydon) give 3.53 e.v. From excitation of atomic fluorescence Butkow and Terenin (836) obtain 3.21 e.v. which should be an upper limit. [b]From electron diffraction (483).

CsRb: [a]Not certain that this is 0–0 band. [b]This system was first ascribed to Cs₂ by (462) but later recognized by (438) to be due to CsRb [compare also its absence in the work of (223) on Cs₂].

Cu₂: [a]The evidence that Cu₂ is the carrier of these bands is rather indirect and not very strong. [b]Headless bands.

CuBr: [a]From thermochemical data on AgBr combined with an interpolation method (1462). Gaydon gives 2.7 ± 0.5 e.v.

CuCl: [a]From thermochemical data on AgCl combined with an interpolation method (1462). [b]Shading not stated but probably R. [c]From a partial rotational analysis by (1462). [d]Mean from B, C, D, E systems.

TABLE 39 (Continued)

State	T_e	ω_e	$\omega_e x_e$	$\omega_e y_e$	B_e	α_e	r_e (10^{-8} cm)	Designation	ν_{00}	References	Remarks
Cu^{63}F^{19} $\quad D_0^0 = (3.0)$ e.v.; $\mu_A = 14.5979$											
C $^1\Pi$	20260.6	640.9 H^Q	4.19		[0.3748]		[1.755]	C↔X	20269.62Z R	(1556)(591)	isotope effect
B $^1\Sigma$	19717.3	657.5[a] H	4.05[a]		0.3725	0.0050	1.761	B↔X	19734.66Z R	(1556)(591)	isotope effect
A $^1\Pi$	17543.7	648.0[a] H^Q	3.81[a]		[0.3675]		[1.773]	A↔X	17556.40Z R	(1556)(591)	isotope effect
X $^1\Sigma^+$	0	622.6[b] H^Q	3.95		0.3803	0.0046[c]	1.743				
Cu^{63}H^1 $\quad D_0^0 < 2.89$ e.v.; $\mu_A = 0.992242$											
D $^1\Pi$	(44710)	[1803.9] Z			7.74	0.30	1.48$_2$	D←X	44651.2 R	(1002)	
C $^1\Pi$	(26422)	[1562.3] Z			[6.388]		$[1.630_9]$	C→X	27107.7 R	(291)(292)	
B $^1\Sigma^+$					6.581	0.287	1.606$_9$	B→X	26281.7 R	(290)(231)(292)	$^1\Sigma$** of (291)(292)
A $^1\Sigma^+$	23433.8	1698.4 Z	44.0	*	6.875	0.263	1.572$_1$	A↔X	23311.1 R	(290)(291)(292)(1002)	$^1\Sigma$* of (291)(292)[e]
X $^1\Sigma^+$	0	1940.4 Z	37.0		7.938	0.249[d]	1.463$_1$				
Cu^{63}H^2 $\quad D_0^0 < 2.93$ e.v.;[e] $\mu_A = 1.95225$											
A $^1\Sigma^+$	23412.0	1213.16 Z	20.65	−0.41	3.5199	0.0898*	1.5664	A→X	23326.0 R	(382)	isotope effect
X $^1\Sigma^+$	0	1384.38 Z	19.14	+0.037	4.0375$_4$	0.09140[f]	1.46253				
(Cu^{63}H^1)$^+$? $\quad \mu_A = 0.992242$											
A $^2\Pi_i$		[1874]			[3.30]	[b]	[2.27]	A→X		(1186)	
X $^2\Sigma$											
Cu^{63}I^{127} $\quad D_0^0 = (3.0)$ e.v.;[a] $\mu_A = 42.084$											
E	24001.4	229.23 H	0.95					E↔X	23983.6 R	(591)(1255)	
D	22957.7	212.78 H	0.92					D↔X	22931.6 R	(591)(1255)	{ isotope effect,
C	21869.7	229.7 H	0.53					C↔X	21852.2 R	(591)(1255)	repr. (1255) }
B	21759.1	243.7 H	1.88					B→X	21748.3 R	(1255)	
A	19734.3	213.27 H	2.22					A↔X	19708.2 R	(591)(1255)	
X $^1\Sigma^+$	0	264.8 H	0.71								
CuO^{16} $\quad D_0^{0,a}$; $\mu_A = 12.7822$											
C	21382	(510) H						C→X	21570 R	(1358)(1166)	
B ($^2\Sigma^+$)		(274)[b] H						B→X	21324 R	(1358)(1166)	
A ($^2\Sigma^+$)	16398	628[b] H	3[b]					A→X	16222 R	(1358)(1166)	
X ($^2\Sigma^+$)											
DyO^{16} $\quad \mu_A = 14.566$											
Unclassified band systems near 16000–17600 and 18200–19300 cm^{-1}									R	(951)	

F_2 [19]

$D_0^0 < 2.75$ e.v.;[a] $\mu_A = 9.50227$

State		ω_e	$\omega_e x_e$		B_e	α_e				
$C\ ^1\Sigma$	$B+20901.0$	977.4 H	140.6		[0.815]		$C \to B$	20754.4[b]	R	(243)
$B\ ^3\Pi$	B	1139.8 H	9.7		1.080	0.014				(689)
$A\ ^3\Pi_u$	0	Continuous absorption only, with maximum at 34500 cm⁻¹			1.282		$A \leftarrow X$	Raman spectrum		(752b)
$X\ ^1\Sigma_u^+$	0	[8?1]			[1.435][d]					

$FeBr_2$?

$\mu_A = 32.884$

Unclassified emission bands in the regions 15400–17200 and 26900–27400 cm⁻¹, possibly due to $FeBr_2$ .... (1225)

$FeCl$ [16]

$\mu_A = 21.5105$

Additional bands between 16000 and 25000 cm⁻¹ (498)(1254)(1225)

State		ω_e	$\omega_e x_e$				
$B\ ^4\Pi$	X_2+ {28007.9, 27950.5, 27914.4, 27886.6} {29254.9, 29184.0, 29119.2, 29064.5, 29018, 28982.7}	431.0 H	2.9	$B \leftrightarrow X_2$	{28024.6, 27967.2, 27931.1, 27903.3} {29269.2, 29197.3, 29133.5, 29078.8, 29032, 28997.0}	V	(1254)
$A\ ^6\Pi$		435.7 H	2.3	$A \leftrightarrow X_1$		V	(494)(1225)
$X_2\ ^4\Sigma$ [a]		397.0 H	1.6	X_2	0		
$X_1\ ^6\Sigma$ [a]	X_2 28982.7	406.6 H	1.2	X_1	28982.7		

CuF: [a]Recalculated using ω_e'' and $\omega_e'' x_e''$ from C–X system. [Ritschl (591) had used B–X system.]
[b]From the C–X system. The values from the B–X system differ from this considerably since the separation head-origin is large and there are no Q heads. The A–X system shows $\Delta v = 0$ only. [c]$D_0 = 0.56 \times 10^{-6}$.

CuH: [a]From predissociation in $A\,^1\Sigma$ state. Gaydon estimates 2.7 ± 0.5 e.v.
[b]Heimer (290)(291)(292) gives two further ²Π states in this region at $\nu_0 = 27963.4$ and 28555.7 which are probably higher vibrational transitions of $C \to X$ although according to Heimer the isotope effect does not fit this interpretation. [e]Predissociation above $v = 0$, $J = 0(1050)$.
[d]$D_v = [5.22 - 0.07(v + \frac{1}{2})] \times 10^{-4}$. [a]Doubtful whether this is carrier of observed band system. Only one paragraph published about this system. [e]From predissociation in CuH¹ (which is not observed for CuH²). [f]$D_v = [1.374 - 0.0144(v + \frac{1}{2})] \times 10^{-4}$.

CuH⁺: [a]Gaydon gives 1.9 ± 0.2 e.v. from thermochemical data (1462). But this is based on an old value for $D(AgI)$. [b]$D_0 = 0.416 \times 10^{-4}$.

CuI: [a]Gaydon gives 4.5 ± 1.5, but vibrational analysis is too uncertain.

CuO: [a]Gaydon gives $\omega_e' = 285.6$, $\omega_e' x_e' = 9.2$, $\omega_e'' = 318.6$, $\omega_e'' x_e'' = 4.4$, $\nu_e = 16273.9$, $B_e' = 0.591$, $B_e'' = 0.632$, but according to a private communication by Hulthén this analysis is probably incorrect.
[b]Guntsch (1012)(1013) gives for the A–X system $\omega_e' = 9.2$, $\omega_e' x_e' = 4.4$, $\nu_e = 4.4$, but vibrational analysis is too uncertain.

F_2: [a]From $D_0^0(ClF)$, $D_0^0(Cl_2)$ and the fact that the heat of formation of ClF is positive (1507). If the heat of formation of ClF given recently by (1393) is accepted (15 ± 0.5 kcal), $D_0^0(F_2) = 1.45$ e.v. follows, that is, considerably less than $D_0^0(Cl_2)$. Caunt and Barrow (842a) derive $D_0^0(F_2) = 2.17 \pm 0.2$ e.v. from RbF and CsF spectra and other evidence.
[b]The observed wave number of the 0–0 band is 20738.0 cm⁻¹ on account of a vibrational perturbation. The value given in the table is from the band formula.
[c]Strong perturbations, also in the intensity alternation.
[d]From electron diffraction (1354).

FeCl: [a]It is not certain whether X_1 or X_2 is the ground state since both band systems have been observed in absorption.

TABLE 39 (Continued)

State	T_e	ω_e	$\omega_e x_e$	$\omega_e y_e$	B_e	α_e	r_e $(10^{-8}$ cm$)$	Designation	ν_{00}	References	Remarks
FeH¹ ?ᵃ		$\mu_A = 0.990261$									
FeO¹⁶											
		$D_0^0 = (4.8)$ e.v.; $\mu_A = 12.4378$									
		Fragments of two additional band systems									
E	(22495)	(540) H						$E \to X$ R	22326	(1193) (1358)	
D	(18014)	(667) H						$D \to X$ R	17908	(1193) (1358)	repr. (886)
C	(17917)	(660) H						$C \to X$ R	17808	(886) (1358)	repr. (886)
B	(17293)	(825) H						$B \to X$ R	17267	(886) (1358)	repr. (886)
A	(4970)	(955) H						$D \to A$ R	12900	(1193) (1358)	repr. (1193)
X	0	880 H	5								
Ga⁶⁹Br⁸¹	36000		$\mu_A = 37.232$								
		$D_0^0 = (2.7)$ e.v.;ᵃ									
		diffuse absorption bands indicating shallow upper state potential curve									
C $^1\Pi_1$	36000							$C \leftarrow X$		(495)	isotope effect
B $^3\Pi_1$	28532.0	271.6 H	2.50					$B \leftrightarrow X$ V	28535.9	(495)	isotope effect
A $^3\Pi_0{}^+$	28161.8	272.2 H	2.53					$A \leftrightarrow X$ V	28166.0	(495)	
X $^1\Sigma^+$	0	263.0 H	0.81								
Ga⁶⁹Cl³⁵			$\mu_A = 23.2069$								
		$D_0^0 = (3.7)$ e.v.;ᵃ									
		Narrow continuous absorption near 41330 cm⁻¹									
C $^1\Pi_1$	40261	[120] H						$C \leftarrow X$ R	40139	(495)	repr. (495)ᵇ
B $^3\Pi_1$	29955.7	395.1 H	2.5					$B \leftrightarrow X$ V	29870.4	(495)	repr. (495)
A $^3\Pi_0{}^+$	29524.1	395.8 H	2.5					$A \leftrightarrow X$ V	29539.2	(495)	repr. (495)
X $^1\Sigma^+$	0	365.0 H	1.1								
Ga⁶⁹I¹²⁷			$\mu_A = 44.682$								
		$D_0^0 \leq 2.88$ e.v.;ᵃ									
		Continuous absorption with maximum at 32600 cm⁻¹									
C $^1\Pi_1$								$C \leftarrow X$		(495)	
B $^3\Pi_1$	25900.6	185.0 H	2.7					$B \leftrightarrow X$ Rvᵇ	25584.4	(495)	repr. (495)
A $^3\Pi_0{}^+$	25571.0	193.2 H	2.4					$A \leftrightarrow X$ Rvᵇ	25558.9	(495)	repr. (495)
X $^1\Sigma^+$	0	216.4 H	0.5								
GaO¹⁶			$\mu_A = 13.0142$								
		$D_0^0 = (2.9)$ e.v.;ᵃ									
		A second unclassified band system in the region 20000–23000 cm⁻¹									
B $^2\Sigma$	$\begin{cases} 25709.0^b \\ 25706.4 \end{cases}$	763.63ᶜ H	3.89					$B \to X$ R	$\begin{cases} 25707.6 \\ 25705.0 \end{cases}$	(278) (278)(1069)	repr. (278)
X $^2\Sigma$ᵈ	0	767.69ᶜ H	6.34					R		(1397)	

GdO¹⁶ $D_0^0 = (5.9)$ e.v.; $\mu_A = 14.520$

C	A+21700.3	Fragments of further band systems at longer wave lengths	749.2 H	3.30	$C \rightarrow A$	21659.6	R (564)	repr. (951)
B	20470.3		771.3 H	5.45	$B \rightarrow (X)$	20435.0	R (564)	repr. (951)
A	A		830.0 H	2.25				
(X)ᵃ	0		841.0 H	3.70				

GeBr $D_0^0 = (3.0)$ e.v.;ᵃ $\mu_A = 38.052$

A ²Σ	33413.4	383.7 H	0.7	$A \rightarrow X$	33457.0	V (386)	repr. (386)
X ²Π	{ 1150.0 0	296.6 H	0.9				

Ge⁷⁴Cl³⁵ $D_0^0 = (4.0)$ e.v.;ᵃ $\mu_A = 23.7466$

B ²Σ	33992.2	526.6 H	0.3	$B \rightarrow X$	{ 33078.0 34051.9	V V	(386)
A (²Δ)	{ 29561.7 29499.3	335.5 H 342.1 H	4.5₁ 4.6₅	$A \rightarrow X$	{ 28550.9 29466.2	R R	(778a)
X ²Π	975.0	405.5 H 407.6 H	1.2₃ 1.3₆				

GeF¹⁹ $D_0^0 = (4.9)$ e.v.; $\mu_A = 15.0627$

B ²Σ	35007.4	800.4 H	4.1₅	$B \rightarrow X$	{ 34139.7 35074.7	V V	(752)
A (²Σ)	:3316.8	413.5 H	1.0₅	$A \rightarrow X$	{ 22256.4 23191.4	R R	(752)
X ²Π	{ 935 0	665.2 H	2.7₉				

ᵇ$D_0 = 5.0 \times 10^{-4}$.

FeH: ᵃThe band originally identified as due to FeH by (386) has later been found to be due to CuH (A. Heimer, Nature i.., 6:0, 1936).

GaBr: ᵃFrom diffuse character of $C \leftarrow X$ transition follows $D_0^0 < 4.4$ e.v.

GaCl: ᵃPredissociation gives D_0^0(GaCl) < 5.0.   ᵇPredissociation for $v = 1$ (diffuse bands).

GaI: ᵃFrom atomic fluorescence of Ga (561). Gaydon gives 2.85 from the same data.
ᵇBoth directions of shading occur, even in one and the same band.

GaO: ᵃGaydon gives 2.5 ± 0.5.
ᵇTwo heads on account of spin doubling of ²Σ according to (278). Sen (1397) assumes isotope splitting due to Ga⁶⁹ and Ga⁷¹. The splitting is much too constant for either interpretation. Probably it is a result of both effects. Since the separation head-origin is large and probably changes rapidly from band to band the constants can only be considered as very approximate.
ᶜSen (1397) gives slightly different constants.
ᵈNot quite certain that this is ground state.

GdO: ᵃVery uncertain whether this is ground state.
GeBr: ᵃGaydon gives 2.5 ± 0.4 e.v.
GeCl: ᵃGaydon gives 2.7 ± 0.4 e.v.

Table 39 (Continued)

State	T_e	ω_e	$\omega_e x_e$	$\omega_e y_e$	B_e	α_e	r_e (10^{-8} cm)	Designation	ν_{00}		References	Remarks
Ge^{74}O^{16}												
$D_0^0 = (6.9)$ e.v.;[a] $\mu_A = 13.1540$												
A $^1\Sigma^+$	37762.4	651.3 H	4.24		0.4018	0.0037	1.786_1	A→X	37595.2	R	(387)(631a)	repr. (387), isotope
X $^1\Sigma^+$	0	985.7 H	4.30		0.4704	0.002_3)[b]	1.650_7					effect (387)(631a)
Ge^{74}S^{32}												
$D_0^0 = (5.6)$ e.v.;[a] $\mu_A = 22.3266$												
B	38890.0	310.4 H	1.35					B←X	38757.4	R	(632)(773)	isotope effect
A $^1\Sigma^+$	32889.5	374.99 H	1.514	*				A↔X	32789.2	R	(632)(773)	
X $^1\Sigma^+$	0	575.8 H	1.80									
Ge^{74}Se^{80}												
$D_0^0 = (4.1)$ e.v.;[a] $\mu_A = 38.415$												
Additional bands between 32000 and 37700 cm^{-1} not analyzed												
A $^1\Sigma^+$	30431.9	272.4 H	1.05					A→X	30364.7	R	(778)(778)	repr. (778)
X $^1\Sigma^+$	0	406.8 H	1.2									
Ge^{74}Te^{130}												
$D_0^0 = (3.2)$ e.v.;[a] $\mu_A = 47.129$												
A $^1\Sigma^+$	27969.5	213.4 H	1.3					A→X	27917.4	R	(778)	repr. (778)
X $^1\Sigma^+$	0	323.4 H	1.0									
H$_2^1$												
$D_0^0 = 4.476_3$ e.v.;[a] $\mu_A = 0.504066$												
Fragments of five other band systems $^3Y \to a$, $^3D \to a$, $^3B \to$, $^3C \to c$, $7p\pi \to a$.												
v ($^3\Pi_g$)	(118293)	(2339)			[29.1]		$[1.07_2]$	v→c	22487	R	(587)(32)	$1s\sigma 4p\pi$; ^3A of (587,
u $^3\Pi_u$	[123485]				[29.3]		$[1.06_9]$	u→a	26232.5	R	(587)	$1s\sigma 6p\pi$; (δ bands)
t $^3\Sigma_u^+$	121295	(2661.4)			[31.5]		$[1.03_1]$	t→a	25343[b]	R	(587)	$1s\sigma 5f\sigma$; ^3F of (587)
q ($^3\Sigma_g^+$)	121039	[2172.6] Z	(121.9)		[30]		$[1.0_6]$	q→c	(25224)	R	(587)	$1s\sigma 5d\sigma$?; ^3G of (587)
n $^3\Pi_u$	120954	2322 Z	62.9		29.93	1.3	1.057_1	n→a	24847.5	R	(32)	$1s\sigma 5p\pi$; (γ bands)
m $^3\Sigma_u^+$	119319	[2457.1] Z			[36]		$[0.96]$	m→a	23295.1		(587)	$1s\sigma 4f\sigma$; ^3E of (587)
s $^3\Delta_g$	118502	[2170]	*					s→c	22626		(32)	$1s\sigma 4d\delta$
r $^3\Pi_g$	118575	[2170]	*		irregular structures, perturbations			r→c	22699[c]		(32)	$1s\sigma 4d\pi$
p $^3\Sigma_g^+$	118475	[2147.7]						p→c	22588		(32)	$1s\sigma 4d\sigma$
k $^3\Pi_u$	118371	2336 Z	(60)		29.40[d]	1.58	1.066_6	k→a	22271.5	R	(32)	$1s\sigma 4p\pi$; (β bands)
f $^3\Sigma_u^+$	116708	[2140.1][e] Z	*		29.61	2.18	1.062_9	f→a	20526.0	R	(32)	$1s\sigma 4p\sigma$
j $^3\Delta_g$	113198	2265[c]	58[c]					j→c	17355[c]		(32)	$1s\sigma 3d\delta$[f]
i $^3\Pi_g$	113008	2268	(75)		not given			i→c	17162		(32)	$1s\sigma 3d\pi$
g $^3\Sigma_g^+$	112777	2395.5 Z	89		(10.0?)		(1.83?)	g→c	16926		(32)(587)	$1s\sigma 3d\sigma$
h $^3\Sigma_g^+$	112770	2395.2 Z	64.2		30.6	1.2_6	1.04_5	h→c	16990	R	(587)	$1s\sigma 3s\sigma$[a]
d $^3\Pi_u$	112702[h]	2371.58[i] Z	66.27	+0.88	30.364[i]	1.545	1.0496	d→a	16619.0	R	(895)	{$1s\sigma 3p\pi$; Fulcher (α) bands[j]

State	T_e	ω_e	$\omega_e x_e$	$\omega_e y_e$	B_e	α_e	r_e	Transition	ν_{00}		Ref.	Electron configuration
$e\ ^3\Sigma_u^+$	107777	2195.8^e Z	65.80	−0.433	27.30	1.515	1.106$_9$	$e\rightarrow a$	11605.7^e	R	(32)	1sσ3pσ; 4^{1}O of (32); 4^{1}K of (32) {doubly excited states}
$a\ ^3\Sigma_g^+$	95938^h	2664.83 Z	71.65	+0.92	34.216	1.671	0.9887$_2$	$a\rightarrow b$	see text			1sσ2sσ
$c\ ^3\Pi_u$	95744^{kl}	2465.0 Z	61.4$_0$		31.07	1.42$_5$	1.037$_6$					1sσ2pπ
$b\ ^3\Sigma_u^+$	Unstable, lower state of continuous spectrum of H_2. Fragments of two other singlet band systems: $x^1\Sigma_u^+ \rightarrow E^1\Sigma_g^+$ near 13700 cm^{-1} and the λ4142.8 progression (586)											1sσ2pσ
$O\ ^1\Sigma_g^+$	[119851]	only $v = 0$			(32)		(1.0$_6$)	$O\rightarrow B$	27487.4		(32)	1sσ4sσ; 4^{1}χ of (32)
$T\ ^1\Sigma_g^+$	[119494]	only $v = 0$			(~19)		(~1.3$_3$)	$T\rightarrow B$	27130.4^m		(32)	
$N\ ^1\Sigma_g^+$	[116268]	[1983.3] Z			(~17.5)		(~1.38)	$N\rightarrow B$	24896.4^m		(32)	
$M\ ^1\Sigma_g^+$	[114654]	[2176] Z			(~13)		(~1.6)	$M\rightarrow B$	23190.0^m		(32)	
$L\ ^1\Sigma_g^+$	[114500]	([1835])			(~10)		(~1.8)	$L\rightarrow B$	23054.8^m		(32)	
$S\ ^1\Delta_g$	[119820]	only $v = 0$			[28.8]		[1.07$_7$]	$S\rightarrow B$	27460		(32)	
$R\ ^1\Pi_g$	[118690]	[2142]^i Z			(30)		(1.0$_6$)	$R\rightarrow C$	(18400)		(32)	1sσ4dσ; 4^{1}B of (32)
								$R\rightarrow B$	(27400)			
$P\ ^1\Sigma_g^+$	[119512]	only $v = 0$			(30)		(1.0$_6$)	$P\rightarrow C$	18260		(32)	1sσ4dσ; 4^{1}C of (32)
								$P\rightarrow B$	27148			
$H\ ^1\Sigma_g^+$	113889	2538 Z	124		(29.5)		(1.065)	$H\rightarrow C$	13866.6	R	(32)	1sσ3sσ; 3^{1}O of (32)
								$H\rightarrow b$	22754.1	V	(586)	
$D\ ^1\Pi_u$	113888	2525.1^c Z	[52 25]*		31.30^{vc}	1.468^e	1.034	$D\rightarrow E$	13713.3	R	(586)	{1sσ3pπ; ^{n}repr. (92)} (1456d)
								$D\rightarrow X$	112869.0	R	(1101)(92)(1456d)	
$J\ ^1\Delta_g$	113404	[2220] Z			[28.8]		[1.07$_7$]	$J\rightarrow C$	13264	R	(32)	1sσ3dδ
								$J\rightarrow B$	22150	V	(32)	

GeO: ^{a}Gaydon gives 5.5 ± 1 e.v.

$^b D_0 = 0.43 \times 10^{-6}$.

GeS: ^{a}Gaydon gives 5.0 ± 0.5 e.v.

GeSe: ^{a}Gaydon gives 3.8 ± 0.5 e.v.

GeTe: ^{a}Gaydon gives 2.9 ± 0.4 e.v.

H_2^1: ^{a}Gaydon gives 4.4776 but at no place gives the value in cm^{-1}. Here $D_0^0 = 36116$ cm^{-1}.

^{b}There is an inversion of the rotational levels, $K = 0$ being at about the level of $K = 5$.

^{c}For one Λ component only (Π$^-$).

^{d}Mean of two Λ components.

^{e}Corrected for the difference between the use of $B(J + \frac{1}{2})^2$ in the original paper and the use of $BJ(J + 1)$ in this book.

^{f}Preionization above $v' = 3, K' = 4$ (95).

^{g}Predissociation above $v' = 3, K = 1$ (95).

^{h}Beutler and Jünger (95) give for $a^3\Sigma_g^+(v'' = 0, K'' = 0)$ an energy 29344 ± 2 cm^{-1} below the ionization limit [for $d^3\Pi_b(v = 0, K = 0)$ 12724.7 cm^{-1}] i.e. 95085 cm^{-1} above $X^1\Sigma_g^+ v = 0, J = 0$.

^{i}Given only for the Λ component of type II$^-$ since the other is strongly perturbed.

^{j}Preionization above $v' > 3$ (591) (PR levels only).

Preionization for $K > 1$ of $v' = 7, 8, 9$ (95).

$^k v = 0, K = 0$ is 29635 cm^{-1} below ionization limit (32)(95) i.e. 94794 cm^{-1} above $X^1\Sigma_g^+ v = 0, J = 0$.

^{l}One component is separated from the other two unresolved ones by 0.20 cm^{-1} for $K = 1$, the splitting decreases for higher K values.

^{m}These apply to the actual $J = 0$ levels whether or not they are perturbed.

^{n}While $\omega_e, \omega_e x_e, B_e, \alpha_e$ are from Jeppesen (1101) since Tanaka (1456d) gives a discrepancy of ~10 cm^{-1} with the $D-E$ system. ν_{00} is from Tanaka (1456d).

^{o}Emission bands observed only with $v' < 3$ on account of preionization; limit $I(H_2) = 124429$ cm^{-1}.

TABLE 39 (*Continued*)

State	T_e	ω_e	$\omega_e x_e$	$\omega_e y_e$	B_e	α_e	r_e (10^{-8} cm)	Observed Transitions: Designation	Observed Transitions: ν_{00}	References	Remarks
H_2^1 (*Continued*)											
$I\ ^1\Pi_g$	113065	2265.2[i] Z	78.47		29.79[i]	1.515	1.059_6	{ $I \rightarrow C$ { $I \rightarrow B$	12925[i] R 21813[p] V	(32) (188)	$1s\sigma 3d\pi$; 3^1B of (32)
$G\ ^1\Sigma_g^+$	112793	{ 2404.3 { or 2341.1[q]	88.8 55.6		([28.4])		($[1.08_5]$)	{ $G \rightarrow C$ { $G \rightarrow B$	12725 R 21609.2[r] V	(32)	$1s\sigma 3d\sigma$; 3^1C of (32)
K ?	112657	2293 Z	30		([~11])		([~1.7])	{ $K \rightarrow C$ { $K \rightarrow B$	12541.2 R 21425.4 R	(32)	3^1K of (32); PR only
Q ($^1\Pi_g$)	(113144)	[742] Z			([16.3])		([1.43])	$Q \rightarrow B$	$21151._1$ R	(587)	?
$F\ ^1\Sigma_g^+$	(103480)	(1000)[s] Z			(6.24)		(2.32)	$F \rightarrow B$	[ν_{20} = 13615.7][s] R	(894a)	$(2p\sigma)^2$
$E\ ^1\Sigma_g^+$	1000 2.8	2588.9[t] Z	[130.5]		32.68[t]	[1.818]	1.011_7	$E \rightarrow B$	8961.2 V	(892)	$1s\sigma 2s\sigma$; ^1X of (32)
$C\ ^1\Pi_u$	100043.0	2442.72[i] Z	[67.03]		31.340[i]	1.626*	1.0331	$C \leftrightarrow X$	99080.3[eu] R	(380) (893) repr.	{ $1s\sigma 2p\pi$; Werner { bands
$B\ ^1\Sigma_u^+$	$91689._9$	1356.90[evw] Z	19.932[e]	+0.4029[e]	20.0159[ev]	1.1933*[e]	1.29270	$B \leftrightarrow X$	90196.1[e] R	(380) repr.	{ $1s\sigma 2p\sigma$; Lyman { bands
$X\ ^1\Sigma_g^+$	0	$4395.?_4$[x] Z	$11?.??_5$	[$+0.29_3$]	60.80	2.993[y]	0.7416_6	{ Quadrupole rotation-vibra- { tion bands		(1046)	$1s\sigma^2$ [z]
H^1H^2	$D_0^0 = 4.511_2$ e.v. [from $D_0^0(H_2)$]; $\mu_A = 0.671917$										
$d\ ^3\Pi_u$	112705[a]	2054.59 Z	49.74	+0.58	22.810	1.020	1.0488	$d \rightarrow a$	16640.6[b] R	(895)	{ $1s\sigma 3p\pi$; Fulcher { (α) bands
$e\ ^3\Sigma_u^+$	107764[a]	1905.17 Z	51.70	+0.522*	20.766[c]	1.010[c]	1.0993	$e \rightarrow a$	11624.6 R	(891)	$1s\sigma 3p\sigma$ [d]
$a\ ^3\Sigma_g^+$	(95938)[a]	2308.44 Z	53.77	+0.60	25.685	1.099	0.9884_0				$1s\sigma 2s\sigma$
	Several other incompletely classified singlet bands									(188)	
$I\ ^1\Pi_g$	113080	1962.1[e] Z	58.21		22.36[e]	1.21	1.059	$I \rightarrow B$	21749.7[f] V	(188)	$1s\sigma 3d\pi$
$G\ ^1\Sigma_g^+$	(112839)	[1879.9] Z	*(vibrational perturbations)			perturbed		$G \rightarrow B$	21492[g]	(188)	$1s\sigma 3d\sigma$
$E\ ^1\Sigma_g^+$	100115	2204.4 Z	[81.6]	*	24.57	[1.288]*	1.011	$E \rightarrow B$	8901.7	(892)	$1s\sigma 2s\sigma$
$C\ ^1\Pi_u$	100058[h]	2139.1[i] Z	66.55	+3.980*	23.64[i]	0.938*	1.030	$C \rightarrow X$	99226[j]	(1228) (238)	{ $1s\sigma 2p\pi$; Werner { bands
$B\ ^1\Sigma_u^+$	91711[h]	1179.22[k] Z	16.091	+0.528*	15.055	0.781*	1.2910	$B \rightarrow X$	90410	(1228) (238)	{ $1s\sigma 2p\sigma$; Lyman { bands
$X\ ^1\Sigma_g^+$	0	3817.09 Z	94.958	+1.4569[l]	45.655[m]	1.9928[n]	0.7413_6	Raman lines		(662) (802)	$(1s\sigma)^2$
H_2^2	$D_0^0 = 4.553_6$ e.v. [from $D_0^0(H_2)$]; $\mu_A = 1.007363$										
$d\ ^3\Pi_u$	112707[a]	1678.22 Z	32.94	+0.24	15.200	0.5520	1.0493	$d \rightarrow a$	16666.0[b] R	(895)	{ $1s\sigma 3p\pi$; Fulcher { (α) bands
$e\ ^3\Sigma_u^+$	107751[a]	1556.64 Z	34.51	+0.287*	13.856[c]	0.541	1.0991	$e \rightarrow a$	11649.2 R	(891)	$1s\sigma 3p\sigma$ [o]
$a\ ^3\Sigma_g^+$	(95938)[a]	1885.84 Z	35.96	+0.34	17.109	0.606	0.9891				$1s\sigma 2s\sigma$ [p]

H₂¹: Several other incompletely classified singlet bands

State	ν	ω_e	ω_ex_e (perturbed)		14.739^e	B_v	B_e	Transition	Type	Ref / Configuration
$I\,^1\Pi_g$	113054.7	1600.1 Z	39.42		14.739^e	1.0656	0.526	$I\rightarrow B$	V	(188) $1s\sigma 3d\pi$
$G\,^1\Sigma_g^+$	(112891)	[1440.8]				1.011	0.682	$G\rightarrow B$	R	(188) $1s\sigma 3d\sigma$
$D\,^1\Pi_u$	(114029)	1638.5 H	30.40$_5$	+0.0591	16.37	1.0336	0.5739*	$D\leftarrow X$	V	(188) $1s\sigma 3p\pi$; repr. (92)
$E\,^1\Sigma_g^+$	100125.2	1784.5 Z	[48.1]	*				$E\rightarrow B$	R	(892) $1s\sigma 3p\sigma$
$C\,^1\Pi_u$	100092.1	1735.8$_5$^{r} Z	39.05$_9$	+1.338$_1$^{r}^{*o}	15.665^e			$C\rightarrow X$	R	(381) $\{1s\sigma 2p\pi$; Werner bands$\}$
$B\,^1\Sigma_u^+$	91698.4	964.40^k Z	11.269	+0.410*	10.0716	1.2891$_0$	0.429*	$B\rightarrow X$	R	(381) $\{1s\sigma 2p\sigma$; Lyman bands$\}$
$X\,^1\Sigma_g^+$	0	3118.4$_6$^{r} Z	64.09$_7$	+1.254$_2$*	30.429	0.7416$_4$	1.0492^t	Raman lines		(662)(801) $(1s\sigma)^2$

H¹H³ $D_0^0 = 4.524_1$ e.v. [from $D_0^0(\mathrm{H_2^1})$]; $\mu_A = 0.755637$

State	ν	ω_e	ω_ex_e			B_v	B_e	Transition		Ref
$d\,^3\Pi_u$	a + 16767.05	1936.93	43.439	+0.459*	20.219	1.0505	0.823*	$d\rightarrow a$		(897a)
$a\,^3\Sigma_g^+$	a	2177.01	47.84	+0.502*	22.819	0.9888	0.9182*			
$X\,^3\Sigma_u^+$	0	3608.3	87.58$_5$		40.5747		1.67048*^{u}			(1113$_1$) calculated^w from H₂¹

H₂³ $D_0^0 = 4.588_1$ e.v. [from $D_0^0(\mathrm{H_2^1})$]; $\mu_A = 1.508517$

State	ν	ω_e	ω_ex_e			B_v	B_e	Transition		Ref
$d\,^3\Pi_u$	a + 16770.61	1372.11	22.135	+0.159	10.150	1.0494	0.3050*	$d\rightarrow a$		(897a)
$a\,^3\Sigma_g^+$	a	1541.57	24.47	+0.312	11.4374	0.9885	0.3258*			
$X\,^3\Sigma_u^+$	0	2553.8	43.872		20.3243		0.59222*^{v}			(1113$_3$) calculated^w from H₂¹

^{p}Obtained from $Q(1)$ which has an anomalous position. ^{q}Refers to $J = 0$ which has an anomalous position.

^{s}There are strong vibrational and rotational perturbations in this state which make it impossible to give a reliable value of ω_e. Two states $\Sigma(a)$ and $\Sigma(b)$, found by Richardson (586) and interpreted as $(2p\sigma)^2$, are probably spurious according to Dieke (894a) except for one level which is identical with $v = 4$ of $F\,^1\Sigma_g^+$.

^{r}From the first two vibrational quanta: strong perturbations for the higher ones. ^{u}Includes $-B\Lambda^2$.

^{t}Slightly different constants have been given by (188). They are probably more reliable. But the main difference is due to their use of a correction $-2B_v$ to the vibrational levels. Still another set of constants for this state has been given recently by (1456d).

^{v}Note that Jeppesen (380) assumes erroneously that $\Delta G_v = \omega_e - 2\omega_e x_e(v + \tfrac{1}{2}) + 3\omega_e y_e(v + \tfrac{1}{2})^2 + \ldots$. This explains the difference between his figures and those given here.

^{z}From the quadrupole spectrum (1046) combined with Raman data (662). The ultraviolet spectrum [see (677)(662)(1456d)] gives constants which differ appreciably from these, but are considered to be less reliable.

^{w}Frequently referred to as state A.

H₂¹: $v: +0.025(v + \tfrac{1}{2})^2$; $D_v = 0.0164_8 - 0.00134(v + \tfrac{1}{2})$. $H_v = (0.0000518)$.

^{a}This is based on T_e for $a\,^3\Sigma_g^+$ of H₂¹, that is, it is assumed that there is no electronic isotope effect. ^{b}Unlike in (895) this does not include the term $-B\Lambda^2$ (see p. 169). ^{c}Corrected for interaction with $^3\Pi$ (891) i.e. true, not effective B value. ^{d}Refers to II^- (upper state of Q branch) since II^+ is strongly perturbed. ^{e}From $\nu_0 - 2B'$. ^{f}This refers [unlike (188)] to the level $J = 0$, obtained from $Q(1)$. ^{g}Predissociation for $v > 3$ (PR levels only).

^{a}From $P(1) = \nu_0 - 2B$. ^{h}From Mie's (1228) and Fujioka and Wada's (238) observed values with correction for different formula used by them. ^{i}Calculated from Jeppesen's (380) data on H₂ by (1228) and (238) with corrections for different formula used [see ^{w} and ^{e} of H₂¹]. ^{k}These refer not to the levels $J = 0$ but to levels obtained by subtracting $2B$, from them [see (677)]. ^{m}Calculated from Jeppesen's (380) data on H₂ by (677). ^{o}Predissociation for $v > 4$ (891) (PR levels only). ^{r}Only the bands with $v' \geq 4$ have been observed in absorption. ^{s}Corrected according to ^{w} of H₂¹. ^{t}Slightly different constants, evaluated with the help of Dunham's formulae (see p. 109) have been derived by (1369). ^{w}Using older constants.

^{u}Includes $-B\Lambda^2$.

$^t+0.03043(0 + \tfrac{1}{2})^2 - 0.0024(v + \tfrac{1}{2})^3$; $D_v = 0.02068 - 0.00369(v + \tfrac{1}{2})$; $H_e = 1.5378 \times 10^{-5}$.

$^u+0.03850(v + \tfrac{1}{2})^2 + \ldots$; $D_v = 0.02602 - 0.000658(v + \tfrac{1}{2})$. $\omega_e z_e = -0.07665$. $\omega_e z_e = 0.02602 - 0.000655(v + \tfrac{1}{2})$. $H_e = 0.19328 \times 10^{-5}$.

H₂¹, H₂³: $\omega_e z_e = -0.10612$. $\omega_e z_e = -0.07665$.

H¹H³, H₂³: $v: +0.00579(v + \tfrac{1}{2})^2 + \ldots$; $D_v = 0.01159 - 0.000118(v + \tfrac{1}{2})$.

H¹H³, H₂³: $^v+0.007636(v + \tfrac{1}{2})^2 - 0.00042(v + \tfrac{1}{2})^3$; $D_v = 0.005190 - 0.0000655$.

TABLE 39 (*Continued*)

State	T_e	ω_e	$\omega_e x_e$	$\omega_e y_e$	B_e	α_e	r_e (10^{-8} cm)	Designation	ν_{00}	References	Remarks
(H₂¹)⁺											
		$D_0^0 = 2.648_1$ e.v.; $\mu_A = 0.503928$									
X ²Σ_g⁺	124429^a	2297^b	$6_?^b$		29.8^b	1.4^b	1.06_0			(96) (585) (1348)	1sσ
H¹Br											
		$D_0^0 = 3.75_4$ e.v.; $\mu_A = 0.99558$									
		Emission continuum with fluctuations from 27000 cm⁻¹ to higher wave numbers									
		Additional absorption bands at higher wave numbers not explicitly given									
C (²Π)	70617	2506	58					C←X R	70542	(1525) (570)	
B (²Π)								B←X ?	67084	(570)	
A (²Π)								A←X		(570)	
X ¹Σ⁺	0	2649.67 Z	45.21		8.473	0.226^b	1.413_3	rotation–vibration bands R		(263) (787) (563) (1273a)	
		Continuous absorption starting at ~35000 and with maximum at ~55000 cm⁻¹									
(H¹Br)⁺											
		$D_0^0 = 3.5_0$ e.v.; $\mu_A = 0.99558$									
A ²Σ⁺	$\begin{cases}2653\\0^+\end{cases}$				6.10	0.25	1.66_6	A→X R	$\begin{cases}29227\\26574\end{cases}$	(534)	
X ²Π_i					[7.95₅]		[1.459]				
H¹C³⁵											
		$D_0^0 = 4.430$ e.v.; $\mu_A = 0.979889$									
		Additional absorption bands at shorter wave lengths not explicitly given									
C (²Π)	77612	2710 H	20					C←X^b R	77480	(570)	
B (²Π)								B←X^b ?	75134	(570)	
A (²Π)								A←X		(440) (1502)	
X ¹Σ⁺	0	2989.74 Z	52.05	+0.056	10.5909	0.3019^c	1.27460	rotation–vibration bands R		(1170) (318)	repr. (318) see tables 8 and 13
		Continuous absorption starting at 44000 cm⁻¹									
H²C³⁵											
		$D_0^0 = 4.481$ e.v.; $\mu_A = 1.905000$									
X ¹Σ⁺	0	$[2090.78]^e$			5.445^e	0.1118^e	1.274_9	rotation–vibration band R		(280)	
(H¹Cl³⁵)⁺											
		$D_0^0 = 4.48$ e.v.; $\mu_A = 0.979889$									
A ²Σ⁺	28637.2	1605.79 Z	39.58	+0.342	7.5077	0.3417*	1.5138_7	A→X R	28119.3^b	(535)	
X ²Π_i	0^c	2675.4 Z	53.5		9.9463	0.3183^d	1.3152_6				
(H²Cl³⁵)⁺											
		$D_0^0 = 4.5_3$ e.v.; $\mu_A = 1.904998$									
A ²Σ⁺	28630.9	[1111.66] Z			3.8566	0.1184	1.5149	A→X R	28254.7^b	(535)	
X ²Π_i	0^c	[1863.96] Z			5.1158	0.1170^f	1.3153_0				

He₂⁴ $[D_0(a\,{}^3\Sigma_u^+) = (2.6)\ \text{e.v.}]$; $\mu_A = 2.00193$

State	Term	$7.x$	$(0.xx)$	r	Transition	R/V	ν	Ref.	Config
$u\ {}^3\Pi_g$	[a+33189.0]	[7.06]		[1.09₂]	$u\to a$		33189.0[a]	(877)(1084)	$10p\pi$ q; repr. (1084)
$t\ {}^3\Pi_g$	[a+32926.4]	7.24	(0.16)	1.08₃	$t\to a$		32926.4[a]	(877)(1084)	$9p\pi$; repr. (1084)
$s\ {}^3\Pi_g$	[a+32556.8]	7.23	(0.27)	1.07₉	$s\to a$		32556.8[a]	(877)(1084)	$8p\pi$; repr. (1084)
$r\ {}^3\Pi_g$	[a+32016.9]	[7.092][b]	(0.23)	1.07₉	$r\to a$		32016.9[a]	(877)(1084)	$7p\pi$; repr. (1084)
$q\ {}^3\Delta_u$	[a+(31248)]			[1.089₇]	$q\to c$		(20305)	(896)	$6d\delta$
$q\ {}^3\Pi_u$	[a+(31193)]				$q\to b$		(26425)	(896)	$6d\pi$
$p\ {}^3\Sigma_u^+$	[a+31180.0]	7.22	(0.23)	1.08₀	$p\to a$	R	31180.0[a]	(877)(1084)	$6d\sigma$
$o\ {}^3\Sigma_g$	[a+31058]	[7.109]		[1.088₄]	$o\to c$	V	20169	(896)	$6p\pi$
					$o\to b$	R	26290.3	(875)	$6s\sigma$
$n\ {}^3\Sigma_u^+$	[a+(29896)]	[7.09]		[1.09₀]	$n\to a$	R	(25128)	(896)	$6p\sigma$
$m\ {}^3\Delta_u$	[a+(29868)]	[7.07]		[1.09₁]	$m\to c$	R	(25100)	(896)	$5d\delta$
$m\ {}^3\Pi_u$	[a+(29804)]				$m\to b$	R	(25036)	(896)	$5d\pi$
$l\ {}^3\Pi_g$	[a+29785.5]	7.22₃[c]	0.22	1.080	$l\to a$	V	29785.5[a]	(877)	$5d\sigma$
$k\ {}^3\Sigma_u^+$	[a+29573]	[7.117]		[1.087₃]	$k\to c$	R	18684	(896)	$5p\pi$
					$k\to b$	R	24805.0	(896)	$5s\sigma$
$j\ {}^3\Delta_u$	[a+(27466)]	[7.088][b]		[1.090₀]	$j\to c$	R	(22698)	(896)	$4d\delta$
$j\ {}^3\Pi_u$	[a+(27297)]	[7.067][d]		[1.091₇]	$j\to b$	R	(22529)	(896)	$4d\pi$
$j\ {}^3\Sigma_u^+$	[a+(27208)]					R	(22440)	(896)	$4d\sigma$
$i\ {}^3\Pi_g$	[a+27193.1]	7.245[c]	0.23	1.078₁	$i\to a$	R	27193.1[a]	(877)(897)(707)	$4p\pi$

Fragment of this transition.

He₂⁺: [a]Above $X\,{}^1\Sigma_g^+$ of H₂.

[b]These are indirectly observed values. Theoretical values that are probably somewhat more accurate have been listed by (1348). $D_0 = 0.0198$.

HBr: [a]From $D_0^0(\text{H}_2)$, $D_0^0(\text{Br}_2)$ and the heat of formation of HBr. In the first edition it was overlooked that the heat of formation refers to the standard state which for Br₂ is liquid. Therefore the value given here is slightly higher. The same error occurs in Gaydon.

[b]Coupling constant $A = 2683.7$ cm⁻¹; large Λ-doubling in ${}^2\Pi_{1\frac{1}{2}}$.

HBr⁺: [a]$D_p = [3.72 + 0.02(v + \tfrac{1}{2})] \times 10^{-4}$.

[b]$D_p = [3.72 + 0.02(v + \tfrac{1}{2})] \times 10^{-4}$.

HCl: [a]From $D_0^0(\text{HBr})$, $I(\text{HBr}) = 12.1_0$ e.v. [Price (570), corrected] and $I(\text{Br}) = 11.84$ e.v.

[a]From $D_0^0(\text{H}_2)$, $D_0^0(\text{Cl}_2)$ and the heat of formation of HCl taken from (50b).

[b]There are weaker subsidiary bands 200–300 cm⁻¹ to the high frequency side of each of these bands.

[c]$D_v = [5.32 - 0.04(v + \tfrac{1}{2})] \times 10^{-4}$.

[d]From $D_0^0(\text{H}^1\text{Cl})$.

[e]These constants are from a theoretical equation derived from the constants of H¹Cl and shown by (280) to agree very well with the observed H²Cl lines.

H¹Cl⁺: [a]From $D_0^0(\text{HCl})$, $I(\text{HCl}) = 12.9_1$ e.v. [Price (570), corrected] and $I(\text{Cl}) = 12.96$ e.v.

[b]From ν_e given by (536).

[c]Coupling constant $A_0 = -643.4$ (=doublet splitting) (536).

[d]$D_v = 5.38 \times 10^{-4}$.

[e]From $D_0^0(\text{H}^1\text{Cl}^+)$.

[f]$D_v = [1.30 + 0.08(v + \tfrac{1}{2})] \times 10^{-4}$.

He₂: [a]Refers to head of Q branch $(J = 0)$. Imanishi (1084) gives slightly larger values since he includes the term $-B\Lambda^2$. In Weizel (890), but for the average the usual formulae hold (890). In Weizel (890) the values for $\Pi^-\Delta^-$ and $\Sigma^+\Pi^+\Delta^+$ are exchanged, but this does not agree with Dieke (890) from whom their values are taken as are those given here.

[b]Average of Π^- and Δ^-; the individual rotational series are complicated on account of l-uncoupling (see p. 229), but for the average the usual formulae hold (890). Imanishi (1084) gives slightly larger values since he includes the term $-B\Lambda^2$. (22) and Sponer (24) the values for $\Pi^-\Delta^-$ and $\Sigma^+\Pi^+\Delta^+$ are exchanged.

[c]Refers to Π^- since Π^+ is more strongly influenced by uncoupling.

[d]Average of Σ^+, Π^+, Δ^+, see [b].

[e]Dieke, Takamine and Suga (897) give 1634.4.

Table 39 (Continued)

State	T_e	ω_e	$\omega_e x_e$	$\omega_e y_e$	B_e	α_e	r_e (10^{-8} cm)	Designation	ν_{00}	References	Remarks
He₄ (*Continued*)											
h $^3\Delta_u$	[a+26760.7]	[1636.2]			7.263	0.228	1.076_1	{ h→c, h→b }[h]	15871 (V); 21992.6 (R)	(896); (875)	4 σ[f]
$^3\Pi_u$	[a+22205]	[1636]			[7.21] [b]	0.27	1.08_1	(f→c)[h]	17437 (R)	(875)	3dδ[g]
f $^3\Pi_u$	[a+21752]	[1569]	36.0		[7.12] [d]	0.18	1.08_3	f→b	16984 (R)	(875)	3dπ[g]
$^3\Sigma_u^+$	[a+21551]	[1549]							16783 (R)	(877)(897)	3dσ[g]
e $^3\Pi_g$	[a+21507.3]	1724.6			7.315_0	0.284	1.073_0	e→a	21507.3[a] (R)	(896)(897)	3pπ
d $^3\Sigma_u^+$	[a+20392.1]	[1654.0]			7.342	0.230	1.071	d→b[i]	15624.0 (R)	(1220)	3σ
c $^3\Sigma_g^+$	[a+10889.6]	[1481.0]			7.00_2	0.150	1.097	c→a	10889.6 (R)	(897)	3pσ
b $^3\Sigma_g^+$	[a+4768.1][j]	[1697.7]			7.45_4[c]	0.24_2	1.063				2pπ
a $^3\Sigma_u^+$	[a][k]	1811.2	39.2		7.66_4	0.131*	1.048_3				2σ
								Fragment of 4dδ $^1\Delta_u \to$ B $^1\Pi_g$ at 21630 cm⁻¹ (876)			
S $^1\Pi_g$	[A+30228.8]				[7.11]		$[1.08_9]$	S→A	30228.8 (R)	(1083)	8pπ
R $^1\Pi_g$	[A+29696.8]				[7.11]		$[1.08_9]$	R→A	29696.8 (R)	(1083)	7pπ
P $^1\Pi_g$	[A+28874.3]				[7.11]		$[1.08_9]$	P→A	28874.3 (R)	(877)(707)	6pπ
$^1\Delta_u$	[A+(27605)]				[7.092] [b]		[1.089][l]		(24070)[l] ((R))	(896)(890)	5dδ
M $^1\Pi_u$	[A+(27535)]				[7.082] [d]		$[1.090_5]$	M→B	(24000)[l] ((R))	(896)(890)	5dπ
$^1\Sigma_u^+$	[A+(27485)]								(23950)[l] ((R))	(896)(890)	5dσ
L $^1\Pi_g$	[A+27507.8]				[7.12]		[1.08]	L→A	27507.8 (R)	(877)(707)	5pπ
I $^1\Pi_g$	[A+24979.6]				[7.131] [c]		$[1.086_7]$	I→A	24979.6 (R)	(877)	4pπ
H $^1\Sigma_u^+$	[A+24698.3]				[7.11]		$[1.08_9]$	H→B	21163.6 (R)	(1522)	4sσ
$^1\Delta_u$	[A+19895]				[7.079] [b]		$[1.090_7]$		16360 (R)	(896)	3dδ
F $^1\Pi_u$	[A+19544]				[7.039] [d]		$[1.093_3]$	F→B	16009 (R)	(875)	3dπ
$^1\Sigma_u^+$	[A+19378]								(15843) (R)	(875)	3dσ
E $^1\Pi_g$	[A+19476.5]	[1650.9][n]			7.28_6[c,n]	[0.224]*	1.076	E→A	19476.5[a] (R)	(1083)	3pπ
D $^1\Sigma_u^+$	[A+18695.3]	n			[7.274]		$[1.076_0]$	D→B	15160.6 (R)	(897)	3sσ
B $^1\Pi_g$	[A+3534.7][m]				[7.296] [c,n]		$[1.074_4]$				2pπ
A $^1\Sigma_u^+$	[A][o]	[1790.8]			7.77_4	0.20_6*	1.041	A→X	continuum[p]; 89000–200000	(1061)(22)	2sσ
X $^1\Sigma_g^+$	0	unstable									2pσ
(He₂⁴)⁺											
		$D_0^0 = (3.1)$ e.v.;[a] $\mu_A = 2.00179$									
X $^2\Sigma_u^+$	0	[1627.2]			7.22	(0.23)	1.08_0			(707)	[b]
HF¹⁹											
		$D_0^0 \leq 6.40$ e.v.;[a] $\mu_A = 0.957347$									
X $^1\Sigma^+$	0	4138.52 Z	90.069	+0.980[b]	20.939	0.770_5[c]	0.9171	rotation-vibration bands (R)		(403)(1455b)	

State			ω_e	$\omega_e x_e$		B_e	α_e	r_e		Observed transitions	References
H²F¹⁹ $\mu_A = 1.82161$											
X $^1\Sigma^+$	0		2998.25	45.71		11.007	$0.293_5{}^d$	0.9170		rotation-vibration bands R	(1455b)
HfO $\mu_A = 14.685$											
	Several systems of red degraded bands between 15700 and 30000 cm⁻¹, no constants given										(1130)(1413)
Hg₂ $D_0^0 = 0.060$ e.v.; $\mu_A = 100.33$											
	Numerous continuous and diffuse spectra in absorption and emission (see p. 395 f.)^a										(31)
X $^1\Sigma_g^+$	0	(36)							3.3^b		(431)(1547)
Hg₂⁺ $\mu_A = 100.332$											
	A group of bands near 40300 cm⁻¹ tentatively ascribed to Hg₂⁺ but not fully analyzed										(1545)
Hg²⁰²Br⁸¹ $D_0^0 = 0.7$ e.v.; $\mu_A = 57.785$											
E ($^2\Pi_{3/2}$)	40720	40710	166	1.1					V	E→X	(1336)(1536)
D ($^2\Pi_{1/2}$)	38574.4	38595.5	228.5 H	0.950						D→X	(708)(1536)
C ($^2\Sigma$)	Unclassified bands in the region 34000–37000^a										(1068)(708)
B	23486	23460	$135._0$ H	0.30					R	B→X	(1533)(1536a)
X ($^2\Sigma$)	0		$186._{2_5}$ H	0.97_5	−0.009						diffuse bands

$^b\omega_e z_e = -0.025.$

$^c +0.005(v + \tfrac{1}{2})^2;\ D_v = 22 \times 10^{-4}.$

$^dD_v = 6.5 \times 10^{-4}.$

He₂: ^{f} Fragment of a band involving $4p\sigma$ is mentioned by Sponer (24).
^{g} These states are called $3d\delta$, $3d_\pi$, $3x$, $3y$, $3z$ in the older literature.
^{h} Only $^3\Pi_u$ and $^5\Sigma_u^+$ are found to combine with $c\,^3\Sigma_g^+$ since the uncoupling is not complete.
^{i} A fragment of the $d \to c$ transition has been found (1220).
^{j} The triplet splitting has been partially resolved by (1267); one component ($J = 1$) for $K = 1$ is separated from the other two by 0.34 cm⁻¹. The splitting decreases somewhat irregularly for higher K-values.
$^k a\,^3\Sigma_u^+$ is 34301.8 cm⁻¹ below the ionization limit (897).
^{l} Very strong L-uncoupling makes accurate determination of band origins impossible.
^{m} This value is from extrapolation of the $ns\sigma$ and $np\pi$ Rydberg series.
^{n} The values given here for $E\,^1\Pi_g$ are quoted by Weizel (22) and Sponer (24) for $B\,^1\Pi_g$. This seems erroneous [see (1083)].
^{o} This state is 31958.1 cm⁻¹ below the ionized state [(875), adopted by Sponer; but Weizel (22) gives 31967]. That is, this state is 2343.7 cm⁻¹ above $a\,^3\Sigma_u^+$.
^{p} Higher odd singlet states may also contribute to this continuum. Compare also the continuous bands at 166660, 154450, and 150800 cm⁻¹ discussed by (1280).
^{q} These designations refer to the outermost electron. The configuration of the core is $1s\sigma^2 2p\sigma$ for all observed states.

He₂⁺: ^{a} From D_0(He₂, $a\,^3\Sigma_u^+$), I(He, ³S) and I(He, $a\,^3\Sigma_u^+$).
HF: ^{a} From D_0^0(H₂), D_0^0(F₂) and the heat of formation of HF taken from (50b). ^{b} Obtained from extrapolation of Rydberg series of He₂.

Hg₂: ^{a} See the potential curve diagram in (31) for the various excited states.
^{b} From data concerning liquid Hg (198).

HgBr: ^{a} These form probably the $^2\Pi_{1/2}$—$^2\Sigma$ component of the $^2\Pi$—$^2\Sigma$ transition of which the D–X system forms the $^2\Pi_{3/2}$—$^2\Sigma$ component. The doublet splitting is approximately 3950 cm⁻¹. The analysis of the C–X system by (1375) appears very doubtful.

Table 39 (*Continued*)

State	T_e	ω_e	$\omega_e x_e$	$\omega_e y_e$	B_e	α_e	r_e $(10^{-8}\,\mathrm{cm})$	Designation	ν_{00}		References	Remarks
HgCl³⁵												
	$D_0^0 = 1.0$ e.v.; $\mu_A = 29.7866$											
$D\ ^2\Pi_{3/2}$	39703.5	341.8						$D\rightarrow X$	39728.0	V	(164)	
$C\ ^2\Pi_{1/2}$	(35804)	a	1.87					$C\leftrightarrow X$	(35828)	V	(1068) (1534) (1337) (1535)	
$B\ ^2\Sigma^+$	23421.0	192.0^b H	0.50^b				[2.23]^e	$B\rightarrow X$	23371.0	R^c	(1534) (1536a)	
$X\ ^2\Sigma^+$	0	292.61 H	1.6025	0.01493^d							(1534)	
HgCs ?												
	$\mu_A = 79.967$										(769b)	
	|Diffuse V shaded bands and other unclassified bands in the region 18500–21000 cm^{-1}											
HgF¹⁹												
	$D_0^0 = (1.8)$ e.v.;^{a} $\mu_A = 17.3604$											
B^b	42999.6	469.4^c H^Q	10.05^c					$B\rightarrow X$	42987.4	V	(1068)	{double headed bands
A^b	(39060)	(495)						$A\rightarrow X$	39053	V	(1068)	
$(X\ ^2\Sigma)$	0	490.8 H^Q	4.05									
HgH¹												
	$D_0^0 = 0.376$ e.v.; $\mu_A = 1.00309$											
$D\ ^2\Sigma^+$					(4.7)		[1.8_9]	$D\rightarrow X$	(37040)	R	(1366)	{(1366) observed Q branches
$C\ ^2\Sigma^+$					[4.50]		[1.93]	$C\rightarrow X$	35599.0	R	(1366)	
$B\ (^2\Sigma^+)$					[4.09]^b		[2.03]	$B\rightarrow X$	33876.5	R	(1079)	
$A\ ^2\Pi_{3/2}$	28256.2	2066.9^c Z	41.85	[−0.43]*	6.730	0.214*	1.580_3	$A\rightarrow X$	28616.8	V	(1366)	
$A\ ^2\Pi_{1/2}$	24578.1	2065.8 Z	[63.55]	*^{d}	6.683	[0.242]*	1.585_9	$A\rightarrow X$	24932.8	V	(1366)	
$X\ ^2\Sigma^+$	0	1387.09 Z	83.01^e	−2.950^f	5.549_0	0.312^g	1.7404				(237)	
HgH²												
	$D_0^0 = 0.398$ e.v.;^{h} $\mu_A = 1.99470$											
$A\ ^2\Pi$	No constants evaluated									V	(237)	
$X\ ^2\Sigma$	0	995.15 Z	49.93	+1.113^i	2.7989	0.1133^j	1.7378				(237)	
(HgH¹)⁺												
	$D_0^0 = (2.3)$ e.v.;^{a} $\mu_A = 1.00309$											
$A\ ^1\Sigma^+$	44319.4	1621.0 Z	43.06	−7.653^b	5.870	0.208*	1.692_2	$A\rightarrow X$	44112.6	R	(506)	repr. (1063)
$X\ ^1\Sigma^+$	0	2033.87 Z	46.16	+1.681^{bc}	6.613	0.206^d	1.594_3					
(HgH²)⁺												
	$D_0^0 = (2.4)$ e.v.;^{a} $\mu_A = 1.99470$											
$A\ ^1\Sigma^+$	44307.9	1150.3^e Z	21.70	−2.733	2.953	0.074*	1.692	$A\rightarrow X$	44161.9	R	(506) (1245) (335)	
$X\ ^1\Sigma^+$	0	1442.15^e Z	23.24	+0.600^f	3.328	0.0736^g	1.593_7					

HgI[127] $D_0^0 = 0.36$ e.v.;[a] $\mu_A = 77.751$

Unclassified bands in the region 38000–40000 cm⁻¹ (E system)

State	(cm⁻¹)			Transition	(cm⁻¹)		References
							(709)
H	47110	97.1 H	1.65	H→X	47096	R	(1331) { pred. above $v' = 2$ · repr. (1331)
G	45542	88.4 H	0.2	G→X	45524	R	(1331) repr. (1331)
F₃	44530.6	85.5 H	0.8	F₃→X	44511	R	(1328)
F₁	40152.3	90.8 H	0.93	F₁→X	40135	R	(1328) [b]
				Bands in the region 22200–24200 cm⁻¹ have been partially analyzed by (1441) and (1537)(1536a)			
D (²Π₃/₂)	36269	178.0 H	1.14	D→X	36295	V	(709) (1338)
C (²Π₁/₂)	32730	235.6 H	2.20	C↔X	32785	V	(709) (1338)
							(1536) [c]
X (²Σ)	0	125.6 H	1.09				

HgK ? $\mu_A = 32.729$

Diffuse V shaded absorption bands in the region 16000–25000 cm⁻¹ | (769b)

HgNa[23] ? $\mu_A = 20.6320$

Diffuse V shaded absorption bands in the region 15000–23000 cm⁻¹ | (769b)

HgO[16] ? $\mu_A = 14.8185$

Absorption bands in the region 33900–36500 usually ascribed to Hg₂ are ascribed to HgO by (1510)

HgRb ? $\mu_A = 59.956$

Diffuse V shaded absorption bands and other unclassified bands in the regions 15500–20500 and 22500–24000 cm⁻¹ | (769b)

HgCl: [a]According to (1068) and (1534) the analysis of the C–X system by (164) is wrong and according to (1068) its upper state forms the second component of ²Π while the lower state is X ²Σ. [b]For $v' > 30$: $\omega_e = 186.2$, $\omega_e x_e = 0.40$. [c]A number of bands with high v'' are shaded to the violet, but their assignment is not quite certain. [d]$\omega_e z_e = -0.000033$. [e]From electron diffraction (1200).

HgF: [a]Gaydon gives 1.4 ± 0.5 e.v. [b]From limiting curve of dissociation (3030 ± 30 cm⁻¹). [c]Above $v' = 4$: $\omega_e' = 410$ and $\omega_e' x_e' = 1.5$.

HgH: [a]From Rydberg's (1366) data; he did not evaluate any vibrational constants. [b]Rydberg (1366) gives without explanation 4.02. [c]Howell (1068) considers these as components of a ²Π state. [d]Strong perturbations. [e]Fujioka and Tanaka (237) give 82.75 apparently due to a numerical error. [f]$\omega_e z_e = -1.588$. [g]$-0.070(v + \tfrac{1}{2})^2$; spin coupling constant $\gamma_0 = 2.15$ cm⁻¹; $D_v = [3.44 + 0.66(v + \tfrac{1}{2}) + \ldots] \times 10^{-4}$; [h]From $D_0^0(\text{HgH}^1)$. [i]$-0.0149(v + \tfrac{1}{2})^2$; spin coupling constant $\gamma_0 = 1.08$ cm⁻¹; $D_v = [0.79 + 0.23(v + \tfrac{1}{2}) + \ldots] \times 10^{-4}$. [i]$\omega_e z_e = -0.58$.

HgH+: [a]Gaydon gives 3 ± 1 e.v. [b]These and $\omega_e'' z_e''$ have been derived by (506) from the data on (HgH²)+. [c]$\omega_e z_e = -0.18300$. [d]$D_v = 2.85 \times 10^{-4}$. [e]Rather different constants are given by (335). [f]$\omega_e z_e = -0.04583$. [g]$D_v = 0.72 \times 10^{-4}$.

HgI: [a]From (1538a); Gaydon gives 0.30 ± 0.05 e.v. [b]It seems very doubtful whether this system has been correctly analyzed since no bands have been assigned to $v'' < 4$. Ramasastry (1328) assumes that F_1 and F_3 form a ²Π state. [c]Howell (1068) considers states A and B as components of one ²Π state. But it is difficult to account for the large difference in ω_e.

TABLE 39 (*Continued*)

State	T_e	ω_e	$\omega_e x_e$	B_e	α_e	r_e $(10^{-8}$ cm$)$	Designation	ν_{00}	References	Remarks
							Observed Transitions			
HgS	$D_0^0 \le 2.8$ e.v.; $\mu_A = 27.655$									
	Three absorption continua with long-wave-length limits at 22500, 32200, and 44400 cm^{-1}								(1401a)	
HgSe	$D_0^0 \le 2.7$ e.v.;[a] $\mu_A = 56.674$									
	Two absorption continua with long-wave-length limits at 22200 and 38630 cm^{-1}								(1198a)	
HgTl	$D_0^0 = (0.031)$ e.v.; $\mu_A = 101.27$									
E	D+22633.9	81.2	0.08				E→D	22621.5	(1548)	
D	D	106.0	0.27							
C	B+{ 15349.0 / 15298.5	82.2	0.15				C→B	{ 15339.9 / 15289.4	(1548)	
B	B+	101.4	2.19							
A	26258.7	15.6	0.06				A↔X	26253.2	(1548)	
X	0	26.9	0.69							
H^{1}I^{127}	$D_0^0 = 3.056_4$ e.v.;[a] $\mu_A = 1.000187$									
	Fragments of other Rydberg series, including some going to $^2\Pi_{3/2}$ of ion								(570)	bands with *P*, *Q*, and *R* branches
	Rydberg series of absorption bands: $\nu = 89130 - R/(m + 0.70)$, $m = 2, 3, \ldots 13$; corresponding to $^2\Pi_{1/2}$ of the HI$^+$ ion [1970]								(570)	
C	(62480)						C←X	62320	(570)	
B							B←X	56750	(570)	
	Continuous absorption starting at 27500 with maximum at ~48000 cm^{-1}						A←X		(264)	
X $^1\Sigma^+$	0	2309.5_3 Z	39.73	6.551	0.183[b]	1.604_1	rotation-vibration bands R		(531) (1273a)	
(H^{1}I^{127})$^+$	$D_0^0 = 3.11$ e.v.;[a] $\mu_A = 1.000187$ No spectrum observed									
HoO	$\mu_A = 14.585$									
	Two complicated groups of bands in the regions 16900–17970 and 18660–19730 cm^{-1}, not analyzed								(950)	V_R repr. (950)
H^{1}S^{32}	$D_0^0 < 3.8$ e.v.;[a] $\mu_A = 0.977325$									
A $^2\Sigma$	*b*			[8.30]		[1.44_2]	A↔X	30659.1	(445a) (963)	R
X $^2\Pi_i$				[9.47]		[1.35_0]				
I$_2^{127}$	$D_0^0 = 1.5417$ e.v.;[a] $\mu_A = 63.466_5$									
	Fragments of three absorption band systems at 55900–64800 cm^{-1}									
							{ numerous diffuse emission bands and continua with the repulsive states from $^2P_{3/2} + ^2P_{3/2}$, $^2P_{3/2} + ^2P_{1/2}$, $^2P_{1/2} + ^2P_{1/2}$ as lower states			
N ($^1\Pi_g$)	(58620)	(120)							(860)	
M ($^1\Pi_{2,1g}$)	(55930)	(360)							(1495)	
L ($^3\Pi_{1g}$)	(51528)	(215)							(1495)(874)(1499)	
K ($^1\Pi_{1,0u^+}$)	(62780)	(260)					K←X	62802	(1495) / (860)(1499)	V

I_2 $D_0^0 = 1.817_0$ e.v.;[a] $\mu_A = 48.6670$

State	T_e	ω_e	$\omega_e x_e$	$\omega_e y_e$	B_e	α_e	r_e (Å)	Transition	ν	V or R	References
$J\ (^2\Pi_{1,0u})$	(59218)	(212)						$J \leftarrow X$	59217	V	(860)(1499)
$I\ (^2\Pi_{1,2u})$	(57770)	(266)						$I \leftarrow X$	57794	(R)	(860)(1499)
$H\ (^2\Pi_{1,2u})$	(56637)	(206)						$H \leftarrow X$	56933	(R)	(860)(1499) (Cordes Bands)

Several diffuse emission bands with C and G as upper states and unstable states resulting from ^2P atoms as lower states (1494)(1495)

State	T_e	ω_e	$\omega_e x_e$	$\omega_e y_e$	B_e	α_e	r_e (Å)	Transition	ν	V or R	References
$G\ (^2\Sigma_u^+)$	51708	165.1 H	0.595	−0.0035				$G \leftarrow X$	51683	(R)	(1493)(860)(1434)
$F\ (^3\Pi_u)$	46488.7[b]	96.2 H	0.49					$F \leftarrow X$	46440.7[b]	R	(1512)(924)(1499)
$C\ (^3\Sigma_{1u}^+)$	(44960)	(90) H	(0.17)					$C \leftarrow X$	(44900)	(R)	(1498)(1323)(1128)(919)
$E\ (^1\Sigma_g^+)$	41407	102.2 H	0.34					$E \rightarrow B$	25757.2	R	(1512)(924)(1499) repr. (924)
$D\ (^1\Sigma_u^+)$	33744[c]	104	0.2					$D \rightarrow X$	33689[c]	R	(924)(1499) repr. (924)
$B\ ^3\Pi_{0u}^+$	15641.6	128.0 H	0.834	*	0.02920	0.00017*		$B \leftrightarrow X$	15598.3	R	(1218)(457)(807)(1056) {band convergence Fig. 15}
$A\ ^3\Pi_{1u}$	11888	44.0 H	1.0_1	+0.008			3.016	$A \leftarrow X$	11803	(R)[d]	(832)(457)(1334)
$X\ ^1\Sigma_g^+$	0	214.57[e]	0.6127	−0.000895[f]	0.03735_9	0.000117	2.666_6				

$I^{27}Br^{79}$

Diffuse emission bands between 19000 and 29000 cm^{-1} and 36000 and 43000 cm^{-1}

State	T_e	ω_e	$\omega_e x_e$	Transition	ν	V or R	References
$D\ (^2\Pi_{1,0})$	(56349)	[310] H		$D \leftarrow X$	56370	V	(760)(519)
$C\ (^2\Pi_{1,2})$	(51677)	[314] H		$C \leftrightarrow X$	51700	V	(163)(519)
$B'\ 0^+$	16880[b]	(60) H	(4.3)	$B' \leftarrow X$	(16776)	(R)[c]	(163)(863)(1178)(519) {partially diffuse bands[d]}
$B\ ^3\Pi_0^+$	(16155)	(150)[e] H	(1.9)	$B \leftrightarrow X$	16095	(R)[c]	(135)
$A\ ^3\Pi_1$	(12230)[b]	(140) H		$A \leftarrow X$	(12166)	(R)[c]	(135) {pred. at $v' = 6$}
$X\ ^1\Sigma^+$	0	268.4 H	0.78				(135)

HgSe: [a]Gaydon gives erroneously 2.0 e.v.

HI: [a]From $D_0^0(H_2)$, $D_0^0(I_2)$ and the heat of formation of HI taken from (50b). In the first edition it was overlooked that the heat of formation usually listed refers to the standard state which for I_2 is solid. Therefore the value given here is appreciably higher. The same error occurs in Gaydon.
[b]$D_v = 0.000213 + 0.000003(v + \tfrac{1}{2})$.

HI$^+$: [a]From D_0^0(HI), I(HI) = 10.39 e.v. [Price (570), corrected] and I(I) = 10.44 e.v.

HS: [a]Since the A–X transition is not usually observed in emission [except in flames at atmospheric pressure (963)] it is very probable that a predissociation limit is below ν_{00}.
[b]Coupling constant $A = -378.6$ cm^{-1}.

I_2: [a]From the convergence in the $B \leftarrow X$ bands at 20037 cm^{-1} assuming dissociation into $^2P_{1/2} + ^2P_{3/2}$.
[b]In view of the long extrapolations involved Venkateswarlu (1499) suggests that the F state is identical with the C state and has type $^1\Sigma_u^+$.
[c]According to (1499) this state is probably $^3\Sigma_u^-$ and is located at approximately 40000 cm^{-1}.
[d]Shading not stated explicitly but almost certainly R.
[e]These constants are for $J = 33$ (not $J = 0$) since they have been obtained from a resonance series [see (721) and (1334)].
[f]$\omega_e z_e = -0.0000187$.

IBr: [a]From the convergence points in the $A \leftarrow X$ and $B' \leftarrow X$ systems both of which give 14660 cm^{-1} (135).
[b]Not certain whether these numbers given by (135) are ν_{00} rather than T_e.
[c]Shading not clearly stated but probably R.
[d]This state is formed by intersection of $B\ ^3\Pi_0^+$ with 0^+ from $^2P_{3/2} + ^2P_{3/2}$.
[e]Brown (135) gives 140 apparently due to a mistake.

TABLE 39 (*Continued*)

State	T_e	ω_e	$\omega_e x_e$	$\omega_e y_e$	B_e	α_e	r_e (10^{-8} cm)	Designation	ν_{00}		References	Remarks
I^{127}Cl^{35}												
	$D_0^0 = 2.152$ e.v., from observed convergence limit in $A\ ^3\Pi_1$ state; $\mu_A = 27.4221$											
	Diffuse emission bands between 18700 and 27100 cm^{-1}											
	Continuous absorption from 38000 to 45000 with maximum at 41600 cm^{-1}										(760)	
D ($^3\Pi_{1,0}$)	58167	[431] H						$D \leftarrow X$	58191	V	(863) (805)	repr. (863)
C ($^3\Pi_{1,2}$)	53457	[426] H						$C \leftarrow X$	53478	V	(163) (519)	repr. (863)
$B'\ 0^+$	(18157)	([32])						$B' \leftarrow X$	17981	R	(163) (519)	{ partially diffuse bands^a
$B\ ^3\Pi_{0^+}$	(17332)	(242) H	13		0.090	0.0029	2.6$_1$	$B \leftarrow X$	(17258)	R	(137)	pred. at $v=4$
$A\ ^3\Pi_1$	13745.6^b	209.7$_5^b$ Z	1.947$_4$	-0.03366^c	0.08389$_0$	0.003828*	2.7072	$A \leftarrow X$	13658.3	R	(137) (177)	repr. (177)
$X\ ^1\Sigma^+$	0	384.18 Z	1.465		0.1141619^d	0.000536	2.32069$_6$	Rotation spectrum			(1469)	
I^{127}N^{14}												
	$\mu_A = 12.6154$											
	Unclassified bands in the region 21500–22500 probably due to this molecule									V	(924) (1512)	
In$_2$												
	$\mu_A = 57.396$											
	Several continua, diffuse and sharp bands in absorption, but no detailed analysis										(1508)	
In^{115}Br^{81}												
	$D_0^0 \leq 3.3$ e.v. [from (627)]; $\mu_A = 47.492$											
$C\ ^1\Pi_1$	(34260)	218.0 H	1.60					$C \leftarrow X$	27380.5	V_R	(701)	repr. (701)
$B\ ^3\Pi_1$	27382.2	227.4 H	1.58		diffuse absorption bands indicating shallow upper potential curve			$B \leftrightarrow X$		V	(701)	repr. (701)
$A\ ^3\Pi_{0^+}$	26596.0	221.0 H	0.65				[2.57]^a	$A \leftrightarrow X$	26599.0	V	(701)	repr. (701)
$X\ ^1\Sigma^+$	0											
In^{115}Cl^{35}												
	$D_0^0 \leq 4.54$ e.v.;^{a} $\mu_A = 26.8179$											
D	37483.6	177.3 H^Q	12.6		[0.1091]		2.40$_1$	$D \leftarrow X$	37410.6	R	(701)	{ pred. for $v' \geq 1^b$
$C\ ^1\Pi_1$		339.4 H	2.1					$C \leftrightarrow X$			(946)	repr. (946)
$B\ ^3\Pi_1$	28560.2	340.3 H	2.0					$B \leftrightarrow X$	28570.9	V	(701)	repr. (701)
$A\ ^3\Pi_{0^+}$	27764.7	317.4 H^Q	1.01		0.1170	0.0009^c	2.318^d	$A \leftrightarrow X$	27775.4	V	(946)	repr. (701)
$X\ ^1\Sigma^+$	0											
	Narrow continuum with maximum at 38260 cm^{-1}											
In^{115}H^1												
	$D_0^0 \leq 2.48$ e.v.;^{a} $\mu_A = 0.999366$											
$B\ ^1\Pi$	16916.5	1491 Z	103		5.4403	[0.3446]*	1.7610	$B \rightarrow X$	16905.00	$R_V{}^b$	(1003)	{ breaking off in $v=0, 1, 2$; different for Q, and P, R
$A\ ^1\Sigma^+$	16277.7	1458.4 Z	60.9	-6.85	5.3639	[0.3239]*	1.7735	$A \rightarrow X$	16259.64	$R_V{}^b$	(1003)	
$X\ ^1\Sigma^+$	0	1474.7 Z	24.7	$+0.16$	4.9959	0.14500^c	1.8376					

In¹¹⁵I¹²⁷ $D_0^0 \le 2.7$ e.v. [from (627)]; $\mu_A = 60.320_7$

State					Transition				
C $^1\Pi_1$	25050.5	146.7	2.3		$C \leftarrow X$	25034.8	$V_R{}^a$	(701)	repr. (701)
B $^3\Pi_1$	Continuum with maximum at 31400 cm⁻¹				$B \leftrightarrow X$		$V_R{}^a$	(701) / (701)	repr. (701); In¹¹³ { isotope effect (1519)
A $^3\Pi_{0^+}$	24401.6	158.5	1.7		$A \leftrightarrow X$	24392.0	$V_R{}^a$	(701) (700)	In¹¹³
X $^1\Sigma^+$	0	177.1	0.4	[2.86]ᵇ					

InO¹⁶ $D_0^0 = (1.3)$ e.v.; $\mu_A = 14.0427$

State					Transition				
$(A$ $^2\Pi)^a$	23595.1 / 23033.1	626.66 H^Q / 703.09 H^Q	3.40 / 3.71		$A \rightarrow X$	23557.0 / 22995.0	R_V	(699)	repr. (699)
$(X$ $^2\Sigma)^a$	0			−0.285					

I¹²⁷O¹⁶ ᵃ $D_0^0 = (1.9)$ e.v.; $\mu_A = 14.2090$

State				Transition				
B	21565	512ᵇ H	5	$B \rightarrow X^c$	21478	R	(1482) (814) (854)	{ bands are diffuse for $v \ge 4$; repr. (854)
${X}$	0	687 H	5					

K₂³⁹ $D_0^0 = 0.514$ e.v., from convergence limit of B $^1\Pi_u$ state; $\mu_A = 19.488$

Diffuse bands of van der Waals K₂ molecules close to lines of principal series of K and fragments of additional band systems (429) (843) (1289).

State				Transition			
G	28091	64.9₀ H	0.55	$G \leftarrow X$	28077	R	(1568)
F	27571	62.2₉ H	0.24	$F \leftarrow X$	27556	R	(1568)
E	26495	63.09 H	0.40	$E \leftrightarrow X$	26480	R	(1568)

ICl: ᵃThis state is formed by intersection of B $^3\Pi_{0^+}$ with 0^+ coming from $^2P_{3/2} + ^2P_{3/2}$.
ᵇThese values are based on bands with $v' > 7$. Darbyshire (880) gives from band heads of low v' values $T_e = 13742$, $\omega_e = 212.3$.
ᶜConvergence limit at 17365 cm⁻¹ (137) [Darbyshire (880) gives 17345] corresponding to dissociation into normal atoms.
ᵃThis value, obtained from microwave spectra (1469), agrees very closely with the older value of (177) obtained from the electronic band spectrum.

InBr: ᵃFrom electron diffraction (827).

InCl: ᵃFrom atomic fluorescence (1526); predissociation gives $D < 4.68$ e.v.; but excited products may arise in predissociation.
ᵇParticularly clear case of predissociation in absorption; the diffuseness is stronger for $v' = 2$ than for $v' = 1$.
ᶜ$D_v = 2 \times 10^{-8}$.
ᵈFrom electron diffraction $r_0 = 2.42$ Å (827).

InH: ᵃFrom predissociation, plotting limiting curve.   Shape of this curve indicates potential maximum.
ᵇReversal of shading within individual bands.
ᶜ$+0.00187(v + \frac{1}{2})^2$; $D_v = [2.248 - 0.033(v + \frac{1}{2}) + ...] \times 10^{-4}$.

InI: ᵃReversal of shading.
ᵇFrom electron diffraction (827).

InO: ᵃAccording to (1069) the upper state is $^2\Sigma$ the lower $^2\Pi$ or $^2\Pi_{1/2}$ and $^2\Sigma$.   But his arguments are not decisive.

IO: ᵃIt is not certain that the carrier of these bands is IO.
ᵇBlake and Iredale's (814) constants are rather different, but their analysis seems to be superseded by that of (854).
ᶜMethyl iodide flame bands.

TABLE 39 (*Continued*)

State	T_e	ω_e	$\omega_e x_e$	$\omega_e y_e$	B_e	α_e	r_e $(10^{-8}$ cm$)$	Designation	ν_{00}	References	Remarks
K₂³⁹ (*Continued*)											
D $(^1\Pi_u)$	24627.7	61.60 H	0.90	$+0.0010^{a}$*				$D\leftarrow X$	24611.0 R	(1422)	
C $(^1\Sigma_u^+)$	22970.0	60.60 H	0.20					$C\leftarrow X$	22954.0 R	(1422)	
B $^1\Pi_u$	15378.0	75.00	0.3376	$+0.004366^{b}$*	0.04824	0.000235	4.235	$B\leftarrow X$	15369.2 R	(464) (458)	
A $^1\Sigma_u^+$	11682.6^{c}	69.09 H	0.153					$A\leftarrow X$	11670.9 R	(872)	
X $^1\Sigma_g^+$	0	92.64	0.354		0.05622	0.000219^{d}	3.923			(464)	

$D_0^0 = 3.96$ e.v.;^{a} $\mu_A = 26.260$

State	T_e	ω_e	$\omega_e x_e$	$\omega_e y_e$	B_e	α_e	r_e $(10^{-8}$ cm$)$	Designation	ν_{00}	References	Remarks
KBr											
A $^1\Sigma^+$	31770	231	0.7	$+0.0011$			$[2.94]^{b}$	$A\leftrightarrow X$	(31654)	(443) (1253)	
X $^1\Sigma^+$	0	280	0.9	$+0.0011$			$[2.79]^{c}$			(443) (644) (93) (443)	

Absorption continua with maxima at 36100 and 39400 cm⁻¹, the first of which probably belongs to the $A\leftarrow X$ transition while the second corresponds to dissociation into $^2S + {}^2P_{1/2}$

Diffuse bands and continuum

$D_0^0 = 4.42$ e.v.;^{a} $\mu_A = 18.599$

State	T_e	ω_e	$\omega_e x_e$	$\omega_e y_e$	B_e	α_e	r_e $(10^{-8}$ cm$)$	Designation	ν_{00}	References	Remarks
KCl											
A $^1\Sigma^+$	36205^{b}							$A\leftrightarrow X$	36065	(443) (1253)	
X $^1\Sigma^+$	0									(443) (93)	

Absorption continua with maxima at 40100 and 41100 cm⁻¹

Diffuse bands and continuum

KCs¹³³ $\mu_A = 30.218$

Diffuse absorption band at 18558 cm⁻¹   — References (1509)

KF¹⁹

$D_0^0 \leq 5.9$ e.v.,^{a} from thermochemical data (50a) and $D_0^0(F_2)$; $\mu_A = 12.789_4$

State	T_e	ω_e	B_e	r_e $(10^{-8}$ cm$)$	Designation	References	Remarks
X $^1\Sigma^+$	0	(390)	[0.2022]	$[2.55_3]$	radio-frequency electric resonance spectrum	(1253) (997)	^{b}

Continuous absorption from 39000 cm⁻¹ to higher wave numbers

K⁽³⁹⁾H¹

$D_0^0 = 1.8_6$ e.v.;^{a} $\mu_A = 0.98271$

State	T_e	ω_e	$\omega_e x_e$	$\omega_e y_e$	B_e	α_e	r_e $(10^{-8}$ cm$)$	Designation	ν_{00}	References	Remarks
A $^1\Sigma^+$	19530.2	251.0 Z	$[-4.5]^{b}$	*	1.32	$[-0.04]^{c}$*	3.6_1	$A\leftrightarrow X$	19168.0^{d} R	(1065) (747) (334) (62)	partly ionic state
X $^1\Sigma^+$	0	985.0 Z	14.65		3.407	0.0673^{e}	2.244				

Fluctuating continuum in emission

K⁽³⁹⁾H²

$D_0^0 = 1.8_9$ e.v.;^{f} $\mu_A = 1.91570$

State	T_e	ω_e	$\omega_e x_e$	$\omega_e y_e$	B_e	α_e	r_e $(10^{-8}$ cm$)$	Designation	ν_{00}	References	Remarks
A $^1\Sigma^+$	19583	182.4	-2.067	-0.05008*	0.6685	-0.00858*	3.62_8	$A\rightarrow X$	19324 R	(1084a)	^{g}
X $^1\Sigma^+$	0	705.9_5	7.6		1.666_4	0.0280^{h}	2.29_8				^{g}

KI^{127} $D_0^0 = 3.33$ e.v.;[a] $\mu_A = 29.896$

Absorption continua with maxima at 30700, 38300 and 42700 cm^{-1} corresponding to dissociation into various pairs of excited atoms

		Diffuse bands and continuum	$A \rightarrow X$		
A	26850^b		(26744)		$(1253)\ (443)$
		212	0.7	+0.001	$(443)\ (644)\ (93)$
$X\ ^1\Sigma^+$	0		$[3.23]^c$		(443)

KRb ? $\mu_A = 26.834$

Diffuse absorption band at 20160 cm^{-1}

$X\ ^1\Sigma^+$		(1509)

La^{139}O^{16} $D_0^0 = (9)$ e.v.;[a] $\mu_A = 14.3479$

State				Transition			
D	(28045)			$D \rightarrow X$	28033	V	(1104)
$C\ ^2\Pi$	22876.8	$[785.7]$ H	2.52	$C \rightarrow X$	$\begin{cases}22865\\22625\end{cases}$	R	$\begin{cases}(1104)(1221)\\(1222)\end{cases}$
	22639.5	788.2 H	2.39				
$B\ ^2\Sigma$	17890.0	782.7 H	1.84	$B \rightarrow X$	17851	R	$\begin{cases}(1104)(1221)\\(1222)\end{cases}$
		732.9 H					
$A\ ^2\Pi$	13532.4	754.2 H	2.07	$A \rightarrow X$	$\begin{cases}13504\\12641\end{cases}$	R	$\begin{cases}(1104)(1221)\\(1222)\end{cases}$
	12668.1	756.5 H	2.18				
$X\ ^2\Sigma$	0	811.6 H	2.23				

K$_2$:   [a]Convergence at 25590 cm^{-1}.   [b]Convergence at 17160 cm^{-1} referred to $v'' = 0$.
[c]Crane and Christy (872) give 11683.6 which does not agree with their o–c values.
[d]$D_v = [8.3 + 0.08(v + \tfrac{1}{2})] \times 10^{-8}$ (calc.).

KBr: [a]Refers to dissociation into normal atoms $^2S + {}^2P_{3/2}$ while the ground state on increasing the vibrational energy dissociates into ions.   Thermochemical data (97) give 3.92 e.v.
[b]From electron diffraction (483).

KCl: [a]Refers to dissociation into normal atoms $^2S + {}^2P_{3/2}$ while the vibrational energy dissociates into ions.   Thermochemical data (97) give 4.39 e.v.
[b]Probably there are two unstable states near this energy.   [c]From electron diffraction (483).

KF: [a]From the longward limit of the continuum one obtains $D_0^0 = 4.8$ e.v.; however this limit was obtained at fairly high temperatures when vibrational excitation in the ground state can shift the limit beyond the point corresponding to D_0^0.
[b]Dipole moment $\mu = 7.33$ debye.

KH: [a]From extrapolation of A state.
[b]Maximum of the ΔG curve at $v = 9$.   A formula representing all ΔG values requires many terms.
[c]Maximum of the B_v curve at $v = 5$.
[d]Hori (334) gives 19170.2.
[e]$-0.00096(v + \tfrac{1}{2})^2$; $D_v = 1.57 \times 10^{-4}$.
[f]From D_0^0(KH$^+$).
[g]All constants are calculated from lines with $J > 25$ since for lower J values there appear strong perturbations for all vibrational levels of the upper and lower state.   This throws serious doubt on the correctness of the analysis which is enhanced by the poor agreement of r_e(KH^2) with r_e(KH^1).

KI: [a]Refers to dissociation into normal atoms $^2S + {}^2P_{3/2}$ while the ground state on increasing the vibrational energy dissociates into ions.   Thermochemical data (97) give 3.30 e.v.
[b]It is probable that there are two unstable states near this energy.
[c]From electron diffraction (483).
[d]$D_v = 0.24 \times 10^{-4}$.

LaO: [a]Gaydon gives 7 ± 2 e.v.

TABLE 39 (*Continued*)

State	T_e	ω_e	$\omega_e x_e$	$\omega_e y_e$	B_e	α_e	r_e (10^{-8} cm)	Designation	ν_{00}	References	Remarks
Li$_2^7$											
					$D_0^0 = 1.03$ e.v.;^{a} $\mu_A = 3.50908$						
			Fragments of other absorption band systems in the ultraviolet								
$C\ ^1\Sigma_u^+$	30558.5^b	231.5 H	1.5					$C\leftarrow X$ R	30498.8	(1486)	
$B\ ^1\Pi_u$	20439.4	269.69	2.744	-0.0637^*	$0.5572_3{}^d$	0.0080_4	2.936_4	$B\leftarrow X$ R	20398.5^e	(1423) (463) (282)	repr. (282)
$A\ ^1\Sigma_u^+$	14068.36	$255.45_6{}^f$ Z	1.574	$+0.0018$	$0.4974_9{}^f$	0.00541^*	3.107_7	$A\leftarrow X$ R	14020.62	(1215) (64)	{ repr. (1215); isotope effect
$X\ ^1\Sigma_g^+$	0	$351.43_5{}^f$ Z	2.592	-0.0058	$0.6727_2{}^f$	0.00704^g	2.672_5			(463)	
LiBr					$D_0^0 = 4.5_3$ e.v., from thermochemical data (50a) and $D_0^0(Br_2)$; $\mu_A = 6.3872$; no spectrum observed						
LiCl					$D_0^0 = 5.1$ e.v., from thermochemical data (50a) and $D_0^0(Cl_2)$; $\mu_A = 5.8056$						
					Continuous absorption with maximum at 41110 cm^{-1}					(1253)	
Li^{7}Cs^{133}					$\mu_A = 6.6663$						
B	(16477)	(77) H						$B\leftarrow X$	16432^a	(1509) (706)	fragment
$X\ ^1\Sigma^+$	0	(167) H									
LiF^{19}					$D_0^0 \le 6.6$ e.v., from thermochemical data (50a) and $D_0^0(F_2)$; $\mu_A = 5.0846$; no spectrum observed						
Li^{7}H^1					$D_0^0 = (2.5)$ e.v.; $\mu_A = 0.881506$						
$A\ ^1\Sigma^+$	26516.2_0	234.413^a Z	-28.947^b	-4.1849^*	2.8186^a	-0.07831^{b*}	2.5963^c	$A\leftrightarrow X$ R	25943.11	(173)	
$X\ ^1\Sigma^+$	0	1405.649^a Z	23.200	$+0.1633$	7.5131^a	0.2132^d	$1.5953_5{}^c$				
Li^{7}H^2					$D_0^0 = (2.5)$ e.v.; $\mu_A = 1.565354$						
$A\ ^1\Sigma^+$	$26513.7_0{}^e$	180.711^a Z	-13.987^f	-1.1784^*	1.6060^a	-0.01450^{f*}	$2.585_9{}^c$	$A\leftarrow X$ R	26083.14	(172)	
$X\ ^1\Sigma^+$	0	1055.12^a Z	13.228	$+0.1300^g$	$4.2338_4{}^a$	0.09198^h	1.5949^c				
LiI^{127}					$D_0^0 = 3.5_8$ e.v.;^{a} $\mu_A = 6.552$						
A		Diffuse bands and continuum from 18000 cm^{-1} to higher frequencies						$A\leftarrow X$		(443)	
$X\ ^1\Sigma^+$		450	1.5	$+0.0017$							
LiK					$\mu_A = 5.895$						
B	(17578)	(130) H						$B\leftarrow X$	17539	$(R)^a$ (1509) (706)	repr. (1509); fragm't
$X\ ^1\Sigma^+$	0	(207) H									

LiRb $\mu_A = 6.421$

State	T_e	ω_e		Transition	ν	References	fragment[b]
$B\,^1\Sigma$	(17552)	(81) H		$B \leftarrow X$	17500[a]	(1509)(706)	
$X\,^1\Sigma$	0	(185) H					

LuO¹⁶ $D_0^0 = (5.3)$ e.v.; $\mu_A = 14.6600$

Unclassified band system with 0–0 band at $\sim$19375 cm^{-1} [b]

State	T_e	ω_e	$\omega_e x_e$	Transition	ν	Shading	References	fragment[b]
A	21471.8	791.60	3.12	$A \to X$	21447.0	R	(951)	repr. (951)
X^a	0	841.66	4.07		[b]	R	(698)	repr. (951)

Mg₂ ? $\mu_A = 12.163_5$

Emission continua with fluctuations probably due to Mg₂ van der Waals molecules   (1018a)

Mg²⁴Br⁷⁹ $D_0^0 \le 3.35$ e.v., from predissociation; $\mu_A = 18.3998$

State	T_e	ω_e	$\omega_e x_e$	Transition	ν	Shading	References	fragment[b]
$C\,(^2\Pi)$	39309 / 39200	(222.5)	(0.8)	$C \leftarrow X$	39233 / 39126	R, R	(1020)	
$A\,(^2\Pi)$	25877 / 25766.9	393.9 H	2.04	$A \to X$	25887 / 25776.8	V, V	(503)(1294)(1020)	{4 heads; pred. above $v = 3$}
$X\,(^2\Sigma)$	0	373.8 H	1.34				(1020)	

Li₂: [a] From molecular beam (that is, thermochemical) data (1168). From the near convergence of the $B\,^1\Pi$ state a value of 1.12 e.v. is obtained (Gaydon). According to (1133) the difference indicates a van der Waals potential hump in the $^1\Pi$ state (see p. 381). [b] Sinha (1423) gives 30655.5 which is apparently a misprint. [c] This is calculated from band head formula. [d] From R and P lines only. [e] Very slightly different constants including higher powers apply. Harvey and Jenkins (282) observe $\nu_{00} = 20395.96$.
[a] $-0.00008(v + \tfrac12)^2$; $D_v = [9.8_6 + 0.028(v + \tfrac12)] \times 10^{-6}$.

LiCs: [a] Shading not stated.

LiH: [a] These are effective values (see p. 109); when the Dunham correction is introduced Crawford and Jorgensen (173) find for Li⁷H¹: $B_e' = 2.8375$, $\omega_e' = 236.225$, $B_e'' = 7.5149$, $\omega_e'' = 1405.401$ and for Li⁷H²: $B_e' = 1.6107$, $\omega_e' = 181.472$, $B_e'' = 4.2344$, $\omega_e'' = 1055.015$. [b] Maximum of ΔG and B_v curves at $v = 9$ and 5 respectively. [c] The r_e values given are derived from the true B_e values given in footnote [a] not from the effective B_e values given in the table. [d] $+0.00075(v + \tfrac12)^2$; $D_v = [8.617 - 0.159(v + \tfrac12) + \ldots] \times 10^{-4}$. [e] Crawford and Jorgensen (173) give 26512.05 based on earlier vibrational constants of the upper state. The difference from the value for LiH¹ is mainly due to the uncertainty of the extrapolation of the zero point energy of the upper state. [f] Maximum of ΔG and B_v curve at $v = 13$ and 4 respectively. [g] $\omega_e z_e = -0.00667$. [h] $+0.000671(v + \tfrac12)^2$; $D_v = [2.756 - 0.0663(v + \tfrac12) + \ldots] \times 10^{-4}$.

LiL: [a] Refers to dissociation into normal atoms $^2S + {}^2P_{3/2}$ while the ground state on increasing the vibrational energy dissociates into ions. Thermochemical data give the two rather divergent values 3.26 e.v. (97) and 3.78 e.v. (1201) due to uncertainty in the heat of sublimation of solid LiL.

LiK: [a] Shading not stated but reproduction suggests R.

LiRb: [a] Shading not stated.

LuO: [a] Not certain that this is ground state. [b] Comparatively large term in $v'v''$.

TABLE 39 (*Continued*)

State	T_e	ω_e	$\omega_e x_e$	B_e	α_e	r_e (10^{-8} cm)	Designation	ν_{00}		References	Remarks
Mg²⁴Cl³⁵											
	$D_0^0 = (3.2)$ e.v.;^{a} $\mu_A = 14.23132$										
	[Fragments of four other band systems are given at 12700 [by (1326)] at 25900 [by (551)] at 37060 [by (1020), absorption] and at 40850 cm⁻¹ [by (1020), absorption]										
A ²Π	$\begin{cases}26520.4\\26465.4\end{cases}$	$\begin{cases}491.5\text{ HQ}\\465.4\text{ HQ}\end{cases}$	$\begin{array}{c}2.53\\2.05\end{array}$				$A \leftrightarrow X$	$\begin{cases}26533.3\\26478.3\end{cases}$	V V	$(503)(551)(1020)$	$\begin{cases}4\text{ heads;}^b\\ \text{isotope effect}\end{cases}$
X ²Σ⁺	0										
MgCs¹³³ ?											
	$\mu_A = 20.564$										
	[One diffuse and one V shaded band at 17520 and 20659 cm⁻¹ respectively									$(769a)$	
Mg²⁷F¹⁹											
	$D_0^0 = (4.2)$ e.v.;^{a} $\mu_A = 10.60470$										
C ²Σ	42528.0 H	821.9 H	4.82	[0.537]		[1.72₀]	$C \leftarrow X$	42579.9	V	(941)	repr. (941)
B ²Σ	37151.7 H	757.8 H	6.24				$B \leftrightarrow X$	37187.4Z	V	$(1103)(941)(370)$	repr. (370) (941)
A ²Πᵢ	$\begin{cases}27849.5\text{ Z}\\27815.2\text{ Z}\end{cases}$	746.0 H	3.97	[0.529]		[1.73₃]	$A \leftrightarrow X$	$\begin{cases}27863.7Z\\27829.4Z\end{cases}$	V V	$(370)(941)$	repr. (370) (941)
X ²Σ	0	717.6 H	3.84	[0.518]		[1.75₂]				(370)	
Mg⁽²⁴⁾H¹											
	$D_0^0 \leq 2.49$ e.v.;^{a} $\mu_A = 0.967480$										
D ²Σ⁻	(42070)	(1620)^b		[6.296]		[1.664]	$D \to A$	$\begin{cases}22858\\22865\end{cases}$	V V	$(1010)(277)$	²Σ** of (277)
C ²Π	41120^d	1740	56	6.161	0.144	1.682	$\begin{cases}C \leftrightarrow A\\C \leftrightarrow X\end{cases}$	$\begin{cases}21955\\21958\\41235.9\end{cases}$	V c V	$(1010)(556)$ $(1010)(269)$ $(673a)(276)$	²Π* of (1010) repr. (556)^e
B ²Σ⁺	$\begin{cases}(38730)\\19224.7\end{cases}$	(990)^b	40.7₃	+1.48	[5.44₈]	0.1883*	[1.789]	$B \leftrightarrow X$	38485	Rv^f	$(1009)(269)$	²Σ* of (1009)^g
A ²Πᵣ	19221.2^h	1611.3 Z	40.7₃	+1.48	6.177₉	0.1883*	1.6795	$A \to X$	19280.0	V	$(277)(1010)$	
X ²Σ⁺	0	1495.7 Z	31.5	−0.15	5.818₁	0.1668^i	1.7306		19276.5			
Mg⁽²⁴⁾H²											
	$D_0^0 \leq 2.49$ e.v.;^{j} $\mu_A = 1.858651$										
D ²Σ⁻	(42070)	(1170)^b		[3.28]		[1.66]	$D \to A$	22858	V	(1010)	
⌐C ²Π	(41156)	[1178]		3.290	0.140	1.661	$\begin{cases}C \to A\\C \leftrightarrow X\end{cases}$	$\begin{cases}21935\\21945\\41210.5\end{cases}$	V V V	(1010) $(1010)(673a)$ (276)	^{k}
B ²Σ	(38735)	(715)^b		([2.93])		([1.76])	$B \to X$	38558	R	(1010)	

State	T_e	ω_e	$\omega_e x_e$	$\omega_e y_e$	B_e	α_e	r_e (Å)	Transition	ν_0	Obs.	References	Remarks
$A\ ^2\Pi_r$	(19243)[l] / 19223	1155.47[m] Z	17.47		3.2175	0.0672*	1.6791	$A \to X$	19282 / 19262[n]	V	(235)	repr. (235)
$X\ ^2\Sigma^+$	0	1077.76 Z	16.09		3.0307	0.0654[o]	1.7301					
(Mg(24)H)+	$D_0^0 = (2.1)$ e.v.; $\mu_A = 0.967479$											
$B\ ^1\Pi$					[3.39]		[2.27]	$B \to X$	49900	R	(1010)(1008)	
$A\ ^1\Sigma^+$	35505[a]	1132.7 Z	6.80	−0.36	4.330	0.076*	2.006	$A \to X$	35630	R	(275)(1010)	
$X\ ^1\Sigma^+$	0	1695.3 Z	30.2	−0.51	6.411	0.206[b]	1.649				(1307)(1318b)	
(Mg(24)H²)+	$D_0^0 = (2.1)$ e.v.; $\mu_A = 1.858648$											
$A\ ^1\Sigma^+$	35902.1	817.0[c] Z	3.47	−0.117	2.251	0.023*	2.007	$A \to X$	35700.5	R	(394)(1011)	
$X\ ^1\Sigma^+$	0	1226.6 Z	16.30	−0.167	3.321	0.064[d]	1.653					
MgI¹²⁷	$\mu_A = 20.415$											
B	25294	[304] H						$B \leftarrow X$	25290	(R)	(503)	
A	24319	[318] H						$A \leftarrow X$	24322	V		
$X\ (^2\Sigma)$		[312] H										
MgK ?	$\mu_A = 14.998$										(769a)	diffuse bands

One R shaded and two V shaded absorption bands at 15264, 19411 and 21678 cm⁻¹

MgCl: [a] Gaydon gives 2.7 e.v.

MgF: [a] Gaydon gives 3.2 ± 0.7 e.v.

MgH: [a] Assuming predissociation of $C^2\Pi$ into $^3P + {}^2S$ (which is lowest possible pair). Gaydon gives 2.0 ± 0.5 e.v.
[b] From the observed difference of zero point energy of MgH and MgD.
[c] No shading for 0–0 band since $B' = B''$.
[d] Splitting not determined; very close to Hund's case (b).
[e] Breaking off of P and R branches above $K' = 10(v = 0)$ i.e. above 41907 cm⁻¹. C–X bands with $v = 1$ and 2 have only Q branches (see p. 418). Predissociation probably caused by $B^2\Sigma^+$.
[f] Reversal of shading within the band.
[g] Occurs in emission only at high pressure indicating predissociation.
[h] Coupling constant $A = 35$ cm⁻¹ (1010).
[i] $-0.0073(v + \tfrac12)$; $D_v = [3.2_5 + 0.1(v + \tfrac12)] \times 10^{-4}$; spin splitting constant 0.02_5 cm⁻¹.
[j] From the predissociation in $C^2\Pi$ of MgH². From MgH¹ one obtains for MgH² $D_0^0 \leq 2.52$ e.v. It is not clear why the predissociation limit in MgH² is slightly lower than in MgH¹.
[k] Breaking off of P and R branches above $K' = 14(v = 0)$ i.e. above 41887 cm⁻¹. C–X bands with $v = 1$ have only Q branches.
[l] Coupling constant $A = 35.28$ for $v = 0$.
[m] Refers to $^2\Pi_{1/2}(J = \tfrac32)$.
[n] Refers to first line of Q_1 branch ($K = 1$).
[o] $D_v = [0.906 + 0.026(v + \tfrac12)] \times 10^{-4}$.

The emission measurements of (551) agree only poorly with the absorption measurements of (503). The latter were used here.

MgH+: [a] Even though the band lines are measured to 0.01 cm⁻¹ Guntsch (275) gives his band origins only to the nearest cm⁻¹.
[b] $+0.0065(v + \tfrac12)^2 - 0.00090(v + \tfrac12)^3$; $D_v = 3.7 \times 10^{-4}$.
[c] Guntsch (1011) does not evaluate all the necessary constants. Therefore we use the constants of (394) throughout except for D_v.
[d] $+0.0013(v + \tfrac12)^2$; $D_v = 0.96 \times 10^{-4}$.

TABLE 39 (Continued)

State	T_e	ω_e	$\omega_e x_e$	$\omega_e y_e$	B_e	α_e	r_e (10⁻⁸ cm)	Designation	ν_{00}	References	Remarks
MgNa²³ ?				$\mu_A = 11.822$							
		V shaded absorption band at 18895 cm⁻¹								(769a)	repr. (769a)
Mg²⁴O¹⁶				$D_0^0 = (3.7)$ e.v.; $\mu_A = 9.59888$							
		Additional strong sequence of headless bands at 26871 cm⁻¹									
H	$G + 26562$	906 H	2.1					$H\to G$	26542 V	(1500)	
G	G	945 H	1.3							(1500)	
F	$E + 26334$	899 H	4.5					$F\to E$	26314 V	(1500)	a
E	E	938 H	3.5								a
B ¹Σ	19984.0	824.1[b] Z	4.7_6		0.5822	0.0045	1.737	$B\to X$	20003.5_7Z V	(1158) (1153) (776)	repr. (1153) (776)
A ¹Π	3563.3	664.4 Z	3.9_1		0.5056	0.0046	1.864	$B\to A$	16500.2_9Z V	(1158) (1190) (1153)	repr. (475)
(X) ¹Σ[c]	0	785.1[b] Z	5.1_8		0.5743	0.0050[d]	1.749				
MgRb ?				$\mu_A = 18.939$							
		V shaded absorption band at 21144 cm⁻¹								(769a)	
MgS				$D_0^0 = (2.9)$ e.v.; $\mu_A = 13.834$							
B	23055.8	495.3 H	2.8					$B\to A$	23040.9 R	(1542)	
A[a]	0	525.2 H	2.93								
Mn⁵⁵Br				$D_0^0 = (2.9)$ e.v.; $\mu_A = 32.570$							
		Unclassified bands in the region 19600–20500 cm⁻¹ [ascribed to MnBr₂ by (1225)] are assigned to a ⁵Π—X⁷Σ transition of MnBr by (765)									
A ⁷Π	26307.7[a]	306.7	0.7					$A\leftrightarrow X$	26316.2[a] V$_R$	(1254) (1225)	repr. (765)
X ⁷Σ	0	289.7	0.9								
Mn⁵⁵Cl³⁵				$D_0^0 = (3.3)$ e.v.; $\mu_A = 21.3757$							
		Unclassified bands in the region 19600–20800 cm⁻¹ [ascribed to MnCl₂ by (1225)] are assigned to a ⁵Π—X⁷Σ transition of MnCl by (765)									
B (⁷Σ)	40808	[320] H						$B\leftrightarrow X$	40776 V	(765)	repr. (1353)
A (⁷Π)	27013.2[a]	410.9 H	1.5					$A\leftrightarrow X$	27026.2[a] V	(1254) (1225) (765)	repr. (765)
X (⁷Σ)	0	384.9 H	1.4								
Mn⁵⁵F¹⁹				$D_0^0 = (3.9)$ e.v.; $\mu_A = 14.1218$							
B (⁷Σ)	41231.5	637.2 H	4.46					$B\leftarrow X$	41240.3 V	(1353)	repr. (1353)
A (⁷Π)	28464.6[a]	677 H	6					$A\leftrightarrow X$	28493.0[a] V$_R$	(765) (1353)	repr. (765)
X (⁷Σ)	0	618.8 H	3.01								

Mn⁵⁵H¹ $D_0^0 < (2.4)$ e.v. (Gaydon); $\mu_A = 0.989974$

State					Transition	ν	V/R	Bands	Remarks
A ⁷Π[a]	[1490.58]	[6.332]	0.16079[c]	[1.640]	A↔X	17698[b]	V	(1278)(1023)	repr. (1278)
X ⁷Σ		5.68548		1.73075					

Mn⁵⁵P²⁷ $\mu_A = 38.3560$

State					Transition	ν	V/R	Bands	Remarks
A (⁷Π)	Incompletely analyzed band system					(25000)	V	(765)	repr. (765)
X (Σ)	0	(240)	(1.5)						

Mn⁵⁵O¹⁶ $D_0^0 = (4.4)$ e.v.; $\mu_A = 12.3926$

State					Transition	ν	V/R	Bands	Remarks
A	17922.5	792.0	18.30	+0.81	A→X	17894.9	R	(630)	repr. (630)
(X)	0	840.7	4.89						

MoO¹⁶ ? $\mu_A = 13.7138$

Emission bands in the region 15400–16400 cm⁻¹ observed in an arc with MoO₃ but not measured (1317a)

N₂¹⁴ $D_0^0 = 7.373$ or 9.756 e.v. (see p. 448 f.); $\mu_A = 7.00377$.

Six diffuse absorption bands between 57300 and 60600 cm⁻¹ which may be part of A←X
Fragments of several band systems in emission

State					Transition	ν	V/R	Bands	Remarks
D ³Σ_u⁺	[101873]		[1.961]	[1.108]	D→B	42559.3Z	V	(1060) (955)(954)(1029) (1033)(1116)	4th positive group
C'	[97584]				C'→B[a]	38270H	R	(976)	{ Goldstein-Kaplan bands; repr. (955)
E	(95770)	[2184.5] H			E→A	46014	V	(955)(1117) (1029)(1033) (1118)	γ bands of (1029)

MgO: [a]These analyses are somewhat uncertain [see (776)].
[b]From band head measurements Mahanti (475) and Barrow and Crawford (776) have given vibrational constants that do not agree well with those given here and obtained by Lagerqvist and Uhler (1158) from origins.
[c]Not certain whether this is the ground state.
[d]$D_e = [1.22 + 0.02(v + \frac{1}{2})] \times 10^{-6}$.

MgS: [a]Not certain that this is the ground state.

MnBr: [a]Refers to the central multiplet component (Q_4 head). The separation of successive multiplet components varies from 55 to 65 cm⁻¹. Coupling constant $A = 60$ cm⁻¹.

MnCl: [a]Refers to the central multiplet component (Q_4 head). The separation of successive multiplet components varies from 43 to 45 cm⁻¹. Coupling constant $A = 44$ cm⁻¹.

MnF: [a]Refers to central multiplet component (P_4 head) (Q heads not clearly observed). Coupling constant $A = 25$. Separation of successive multiplet components varies from 19 to 31 cm⁻¹.

MnH: [a]This is a normal case (b) state.
[b]This figure refers to the head of the Q_4 branch.

MnO: [c]$D_e = [3.0315 + 0.0146(v + \frac{1}{2})] \times 10^{-4}$.
Gaydon gives 4.0 ± 0.5 e.v.

N₂: [a]It is not certain that these bands have B ³Π_g as lower state since only $v'' = 9, \ldots 13$ have been observed.

TABLE 39 (Continued)

State	T_e	ω_e	$\omega_e x_e$	$\omega_e y_e$	B_e	α_e	r_e (10⁻⁸ cm)	Designation	ν_{00}		References	Remarks
N_2^{14} (Continued)												
C ³Π$_u$	89147.3	2035.1 Z[b]	17.08	−2.15	1.8259[c]	0.0197*	1.1482	C→B	29670.6[d]	V	(147)(167)	2nd positive group[e]
B ³Π$_g$	59626.3[f]	1734.11 Z	14.47	[g]	1.6380[c]	0.0184*	1.2123	B→A	9557.0[d]	V	(528)(23) (1096)(1030)	1st positive group[f] Vegard-Kaplan bands; repr. (1096)(1030)(1564)(1033a)
A ³Σ$_u^+$ (Z)	50206.0 (Z)	1460.37 H	13.891	−0.025*	1.440	0.013	1.293	A→X	49756.5(Z)	R	(1564)(1033a)	repr. (660a)

Hopfield's Rydberg series of bands (absorption) $\nu = 151240 - R/(m - 0.092)^2$, $m = 3, 4, 5 \ldots 10$. Upper (331)(518)(660a)
states probably $^1\Sigma_u^+$ [i]
Worley-Jenkins' Rydberg series of bands (absorption)
$$\nu = 125665.8 - R \Big/ \left(m + 0.3450 - \frac{0.1000}{m} - \frac{0.1000}{m^2} - \frac{0.100}{m^2} \right)^2 \quad m = 2, 3, \ldots, 26\,[j]$$

State	T_e	ω_e	$\omega_e x_e$	$\omega_e y_e$	B_e	α_e	r_e (10⁻⁸ cm)	Designation	ν_{00}		References	Remarks
y (¹Π$_g$)	a′ + (46327)	[1708]			[1.80]		[1.16]	y→a′	46420	V	(954)(1032)	Kaplan's first system; repr. (954)
x (¹Σ$_g^-$)	a′ + 45274	1910 H	20.5		1.74	0.01₅	1.18	x→a′	45463	V	(954)(1573)	Van der Ziel bands, 5th pos. group; repr. (954); Kaplan's second system
w	a′ + 3943	1560.1 H	11.9		1.48	0.01₅	1.28	y→w	42461	V	(1029)(1032)(1033)(954)	
a′ (¹Σ$_u^-$)	a′ ᵏ	1527 H	11.5									
v ¹Σ$_u^+$	(121961)	[925] H			[1.07]		[1.50]	v←X	121247	R	(1559)(1560)	repr. (1559)(1560)
u ¹Σ$_u^+$	121490	547 H	7		[1.16]		[1.44]	u←X	120585	R	(1559)(1560)	repr. (1559)(1560)
t	119422	[482] H			[1.11]		[1.47]	t←X	118487	R	(1559)(1560)	
s	117598	[522]						s←X	116683H	R	(1029)(1032)(1033)	η bands of (1029)[m]
e ¹ (eˡ)	(115650)	[2180]			[1.63]		[1.22]	{ e←X	46611	V	(954)(1029)	
								{ h→a	115564	(R)	(697)(1559)	
h ¹Σ$_u^+$ ⁿ								h→a	43817.6Z	V	(1560)	
								h←X	112774.2Z	R	(1559)(1560)	
r ¹Σ$_u^+$	(111800)	([640])			(1.07)		(1.50)	r←X	110944H	R	(954)	repr. (954)
s′ ¹Σ$_u^+$	[110662]				[1.58₅]		[1.23₅]	s′→a	41705.4Z	R	(697)	
f								f→X	110190H	R	(1559)(1560)	repr. (1559)(1560)
q ¹Σ$_u^+$	110654	715	9		1.13	(0.02)	1.46	q←X	109833H	R	(697)	repr. (954)
g ¹Π$_u$								g←X	108950H	R	(1559)(1560)	repr. (1559)(1560)
p ¹Π$_u$	(109174)	[749]			[1.21₅]		[1.40]	p→X	108372H°	R	(1559)(1560)	repr. (1559)(1560)
o ¹Σ$_u^+$	107862							o←X	107657Hᵖ	R	(1559)(1560)	repr. (1559)(1560)
r′ ¹Σ$_u^+$	[106373.5]	1951 H	16		1.67		1.20	r′→a	37416.9	V	(954)(965)	repr. (954)

State	T_e	ω_e	$\omega_e x_e$	$\omega_e y_e$	B_e	α_e	r_e	Transition	Origin / Head	Shading	References	Remarks
m $^1\Pi_u{}^t$	(106145)	[764]			[1.36₅]		[1.32₈]	$m\rightarrow a$	36394.5Z	R	(1029)(954)(965) / (1096)	repr. (954)
								$m\leftarrow X$	105351.1Z	R	(1559)(1560) / (1029)(1096)	repr. (1559)(1560); repr. (1029)
d'	[104718]							$d\rightarrow a$	35761H	R	(954)(1096)	repr. (954)
p' $^1\Sigma_u{}^+$	104400.4	(2217)	(19)		[1.93]		[1.12]	$p'\rightarrow a$	35371.3Z	V	(1559)(1560) / (1402)	
								$p'\leftarrow X$	104327.9Z	R	(1402)	
b' $^1\Sigma_u{}^+$	104480.5	751.7 Z	4.82		1.145	0.002₁	1.450	$b'\leftrightarrow X$	103678.9	R	(1402)(1472) / (1559)(1560)	s
b $^1\Pi_u$	102283	[698]			[1.41]		(1.31)	$b\leftrightarrow X$	101456.0	R	(111)(1559) / (1560)	repr. (1559)(1560)
j $^1\Sigma_u{}^+$ / i	(99327)	[670]			[1.46]		[1.28]	$j\leftarrow X$	100821.4H	R	(1559)(1560)	repr. (1559)(1560)
								$i\leftarrow X$	98486H[o]		(1559)(1560)	
a $^1\Pi_g$	69290.0 Z	1692.01 H	12.791	−0.3489*	1.637	0.0224	1.213	$a\leftrightarrow X$	68956.6Z'	R	(111)(1433)(697) / (1041)(648)	shading not stated [Lyman-Birge-Hopfield bands; repr. (111)(697)(648)]; Fig. 37
X $^1\Sigma_g{}^+$	0	2359.61 H	14.456	+0.00751[u]	2.010[v]	0.0187[w]	1.094	Raman spectrum			(578)(579)(1240)	

N₂: [b]The vibrational constants represent $v = 0, 1, 2, 3$ only; the last level $v = 4$ is perturbed [see (147)].

[c]Using the proper average of the three triplet components (see p. 271); the B_v values are from (139), but B_e and α_e were recalculated by the author.

[d]Refers to $^3\Pi_1$ component.

[e]For a reproduction see Figs. 8a and 22. Breaking off on account of predissociation in $v' = 0, 1, 2, 3, 4$ at $K = 65, 55, 43$ and 28 respectively and giving a dissociation limit at 97944 cm⁻¹ [(147); see also (866)]. A second predissociation in $v' = 2$ and 3 at $K = 80$ and 67 was found by (1066). Heads of satellite branches have been found by (1096). Intensity perturbations have been discussed by (167) (967) and (864).

[f]Coupling constant $A_e = 42.3$ cm⁻¹.

[g]Higher powers and slightly different ω_e and $\omega_e x_e$ are given by (1029) for the band *heads* of the first positive group.

[h]For a reproduction see (528) and Fig. 8a. Predissociation (drop in intensity) affecting all levels above $v = 12$. $K = 33$ [see (732)].

[i]Worley (1559) (1560) has tentatively identified a member $m = 2$ of this series with an absorption band at 120693 cm⁻¹. A Rydberg series of emission bands close to the absorption series and going to the same limit was observed by (331) and (660a). According to (1559) (1560) their upper states may be the $^3\Sigma_u{}^+$ states corresponding to the $^1\Sigma_u{}^+$ states.

[j]The first two members are the states p' and e. Transitions to $v' = 1$ of the Rydberg states have been observed for $m = 2, 3, 4, 6, 9, 10, 11, 12, 14, 15, 16, 17$ by (1457) (1559) (1560). For all these states $\Delta G = 2174$ which is very close to that in the ground state of $N_2{}^+$ (2174.8 cm⁻¹). The upper states have been provisionally identified by (1559) (1560) as $^1\Pi_u$ states. But the p' state ($m = 2$) is $^1\Sigma_u{}^+$.

[k]Gaydon (954) estimates a' to be 60000 cm⁻¹ above the ground state.

[l]Not to be confused with the e state of (697) which has been identified by (1402) as the $v = 3$ level of the b' state.

[m]The identification of the upper state of the η bands of (1029) with that of the $e\leftarrow X$ bands of (1559) (1560) has not previously been made in the literature.

[n]Identical with the l' state of (954) and (965) [see (1041)]. [o]It is not certain that this is the 0-0 band.

[p]Instead of this a weak band at 105693 may be the 0-0 band. [q]Identical with the q' state of (954) and (965) [see (1041)].

[r]Not to be confused with the d state of (697) which has been identified by (1402) with $v = 2$ of b' $^1\Sigma_u{}^+$.

[s]Birge and Hopfield's (111) c state (similar to the p' state, see (1402)). The fact that they appear as double headed bands under low dispersion is due to perturbations with the p' state (similar to the anomalous intensities observed in CN) (see p. 292).

[t]From band heads Birge and Hopfield (111) found 68962.7 cm⁻¹.

[u]Miller (1240) gives $B_0 = 1.980$ from an analysis of the Raman spectrum which agrees poorly with Rasetti's (578) Raman value and with the value from the a–X bands.

[v]$\omega_e z_e = -0.000509.$

[w]$D_e = [5.8 - 0.001(v + \tfrac{1}{2})] \times 10^{-6}$ (calc.).

TABLE 39 (Continued)

State	T_e	ω_e	$\omega_e x_e$		B_e	α_e	r_e $(10^{-8}$ cm$)$	Designation	ν_{00}	References	Remarks
$(N_2^{14})^+$		$D_0^0 = 6.341$ or 8.724 e.v.;[a] $\mu_A = 7.00363$									
$C\ ^2\Sigma^+$	(64622)	(2050)[b]	c		1.65	[0.05][d]	1.21	$C \to X$	64547 R	(697) (1402) (1455)	repr. (1455) (697)
$B\ ^2\Sigma_u^+$	25461.5	2419.84 Z	23.19_0	-0.5375*	2.083	0.0195*	1.075	$B \to X$	25566.0 V$_R$	(169) (1304) (873)	repr. (847) (303)[ef]
$X\ ^2\Sigma_g^+$	0	2207.19 Z	16.136	-0.0400	1.932_2	0.020_2[g]	1.116_2				
Na_2^{23}		$D_0^0 = 0.73$ e.v.;[a] $\mu_A = 11.49822$									
		Diffuse bands of van der Waals Na$_2$ molecules close to lines of principal series of Na. Several fragments of other ultraviolet emission and absorption band systems								(429) (1565) (1421) (1129) (1568)	[b]
E	33205	109.5 H	0.5					$E \leftrightarrow X$	33190	(1421)	
D	31954	112 H	0.7					$D \leftarrow X$	31930	(1421)	
$C\ ^1\Sigma_u^+$	29342.0	119.33 H	0.53		$(B_2 = 0.103)$		(3.7_7)	$C \leftrightarrow X$	29322.1[c] R	(1309) (1421) (1424a) (1424c)	
$B\ ^1\Pi_u$	20320.2	123.79[g]	0.6303	-0.00936[d]	0.12588	0.00094	3.413_0	$B \leftrightarrow X$	20302.5 R	(465) (467)	
$A\ ^1\Sigma_u^+$	14680.4	117.6[g]	0.38	-0.0027	0.1107[e]	0.00054	3.63_9	$A \leftrightarrow X$	14659.7 R	(233) (944)	
$X\ ^1\Sigma_g^+$	0	159.23[g]	0.726		0.15471	0.00079[f]	3.078_6				
$Na^{23}Br$		$D_0^0 = 3.85$ e.v.;[a] $\mu_A = 17.8558$									
		Absorption continua with maxima at 36000 and 40070 cm^{-1} the first of which probably belongs to the A–X transition while the second corresponds to dissociation into $^2S + {}^2P_{1/2}$ Diffuse bands and continuum									
A[b]	(30720)	315	1.15	$+0.0008$			$[2.64]$[c]	$A \leftrightarrow X$	(30562)	(443) (1253) (443) (93) (443)	
$X\ ^1\Sigma^+$	0										
$Na^{23}Cl$		$D_0^0 = 3.58$ e.v.;[a] $\mu_A = 13.9508$									
A		{Diffuse bands and continuum (in absorption from 27000–44000 cm^{-1} with maximum at 42700 cm^{-1}; in emission from 33000 to 18000 cm^{-1}) 380	(1.0)				$[2.51]$[b]	$A \leftrightarrow X$		(443) (1253)	
$X\ ^1\Sigma^+$	0									(443)	
$Na^{23}Cs^{133}$		$\mu_A = 19.6052$									
		Fragment of another system overlapping $C \leftarrow X$									
C	(24270)	(62) H						$C \leftarrow X$	24250 [a]	(1509) (706)	repr. (1509)
B	(18250)	(65) H						$B \leftarrow X$	18233 R	(1509) (706)	repr. (1509)
$X\ ^1\Sigma^+$	0	(98) H									
$Na^{23}F^{19}$		$D_0^0 \le 5.3$ e.v.,[a] from thermochemical data (50a) and $D_0^0(F_2)$; $\mu_A = 10.4054$									
		Continuous absorption from 41000 cm^{-1} to higher wave numbers								(1253)	

Na^{23}H^1 $D_0^0 = (2.2)$ e.v.;^{a} $\mu_A = 0.965792$

State	T_e	ω_e	$\omega_e x_e$	$\omega_e y_e$	B_e	α_e	r_e	ν_{00}		Transition	References
$A\,^1\Sigma^+$	22719.1	310.6 Z	-5.41	-0.197^{d*}	1.696^c	-0.0175^{d*}	3.20$_8$	22294.$_5$	R	A↔X	(1064)(333)(543) (1301) (543) repr. (333)(1064)
$X\,^1\Sigma^+$	0	1172.2 Z	19.72	+0.160^e	4.9012^f	0.1353^g	1.8873				(543)

Na^{23}H^2 $D_0^0 = (2.2)$ e.v.; $\mu_A = 1.852433$

State	T_e	ω_e	$\omega_e x_e$	$\omega_e y_e$	B_e	α_e	r_e	ν_{00}		Transition	References
$A\,^1\Sigma^+$	h	[255.0$_5$]^i Z			$B_7=1.010^j$		[3.00$_2$]	[$\nu_{00}=24106.2$]R		A←X	(543)
$X\,^1\Sigma^+$	0	[826.1$_0$] Z			2.5575	0.0520^k	1.8865				(543)

Na^{23}I^{127} $D_0^0 = 3.16$ e.v.;^{a} $\mu_A = 19.4692$

Absorption continua with maxima at 30800, 38700 and 47150 cm^{-1} corresponding to dissociation into various pairs of excited atoms

State	T_e	ω_e	$\omega_e x_e$	$\omega_e y_e$	B_e	α_e	r_e	ν_{00}		Transition	References
$A\,^1\Sigma^+$	25490^b						2.90^d	(25347)		A↔X	(1391) (443)(644)(93) (443)
Diffuse bands and continuum		286	0.75	+0.001^c							
$X\,^1\Sigma^+$	0										

N$_2^+$: ^{a}From $D_0^0(N_2)$, $I(N)$. and $I(N_2)$. The two values correspond to the two possible values for $D_0^0(N_2)$.
^{b}Estimated from Setlow's (1402) data who does not give any vibrational constants but did change the vibrational numbering compared to (697).
^{c}Strongly curved ΔG curve.
^{d}The B_v curve much flatter for higher v's than for first three for which the given α holds.
^{e}Numerous perturbations in this state are due to the $A\,^2\Pi_u$ which has not yet been observed directly. Estimates of the energies of some of the vibrational levels of the $^2\Pi_u$ state and their B_v values have been derived from the perturbations by (130) (131) (831) (847).
^{f}The spectrum of $(N^{14}N^{15})^+$ has been studied by (424) (720a) and $(N_2^{15})^+$ has been observed by (130) (131) (1552).
^{g}Spin doubling constant $\gamma = +0.002$ cm^{-1}; $D_v = [5.75 + 0.29(v + \frac{1}{2})] \times 10^{-6}$.

Na$_2$: ^{a}From molecular beam (that is, thermochemical) measurements (see also footnote ^{d}).
^{b}The data of (1568) diverge considerably from those of (1129) done earlier in the same laboratory.
^{c}This value obtained from the band formula of (1309) does not agree with ν_{00} given by (1421).
^{d}Not valid above $v' = 23$; apparent convergence at 23120 cm^{-1}. However thermochemical D_0^0 leads to the dissociation limit at 22952 cm^{-1} (1133) indicating a potential hill due to van der Waals forces (see p. 381).
^{e}From (233) but corrected for change of numbering by (944). ^{f}-0.00003$(v + \frac{1}{2})^2$; $D_v = [0.584 + 0.005(v + \frac{1}{2})] \times 10^{-6}$. ^{g}From magnetic rotation spectrum.

NaBr: ^{a}Refers to dissociation into normal atoms $^2S + ^2P_{3/2}$ while the ground state on increasing the vibrational energy dissociates into ions. Thermochemical data (97) give 3.80 e.v.
^{b}Possibly there are two unstable states near this energy. ^{c}From electron diffraction (483).

NaCl: ^{a}From thermochemical data; refers to dissociation into normal atoms $^2S + ^2P_{3/2}$ while the ground state on increasing the vibrational energy dissociates into ions.
^{b}From electron diffraction (483).

NaCs: ^{a}Shading not stated.

NaF: ^{a}Agrees with longward limit of continuum.

NaH: ^{a}Gaydon gives 2.05 ± 0.2 e.v. ^{b}Maximum of ΔG curve at $v = 9$. ^{c}These are Pankhurst's (1301) values. ^{d}B_v curve has maximum at $v = 6$.
$\omega_e z_e = -0.005$. ^{e}These are Olsson's (543) values. Pankhurst (1301) gives $B_e = 4.886$, $\alpha = 0.129$. He gives B_v'' values only from $v'' = 3$ up.
$^f T_e$ cannot be determined since ν_{00} and ω_e are not known. ^{g}This is $\Delta G_{1/2}$; no lower ΔG's are observed; for the higher ones there is a maximum at $v = 12$ and therefore no simple expression holds and ω_e cannot be extrapolated.
$^h B_e$ cannot be extrapolated since the B_v curve has a maximum at $v = 9$. $^j D_v = [0.915 + 0.009(v + \frac{1}{2})] \times 10^{-4}$. $^k D_v = [3.32 - 0.03(v + \frac{1}{2})] \times 10^{-4}$.

NaI: ^{a}Refers to dissociation into normal atoms $^2S + ^2P_{3/2}$ while the ground state on increasing the vibrational energy dissociates into ions. Thermochemical data (97) give 3.07 e.v.; Gaydon gives 3.11 ± 0.04 e.v. Excitation of atomic fluorescence (836) gives 3.0 e.v.
^{b}Possibly there are two unstable states near this energy.
$^c\omega_e z_e = -0.0002$. ^{d}From electron diffraction (483).

TABLE 39 (Continued)

State	T_e	ω_e	$\omega_e x_e$	$\omega_e y_e$	B_e	α_e	r_e (10^{-8} cm)	Designation	ν_{00}		References	Remarks
Na²³K												
$D_0^0 = 0.62_1$ e.v.;[a] $\mu_A = 14.481$												
E ¹Σ⁺	25228[b]	95.85 H	0.94					$E{\leftarrow}X$	25214	R	(1424)	repr. (1424)
D ¹Π	20090.6	82.17 H	0.350	-0.00814				$D{\leftarrow}X$	20070.0	R	(459)	repr. (1424)
C ¹Π	16991.4	72.60 H	1.475	$+0.02436^{c}$*				$C{\leftarrow}X$	16965.8	R	(459)	
A ¹Σ⁺	12139.7	79.85 H	0.0872	-0.00389				$A{\leftarrow}X$	12118.1	(R)[d]	(459)	
X ¹Σ⁺	0	123.29 H	0.400									
Na²³Rb												
$D_0^0 = (0.57)$ e.v.;[a] $\mu_A = 18.122$												
A further band system at about 18600 cm⁻¹ has not been analyzed												
A ¹Π	16421.8	61.49	0.945					$A{\leftarrow}X$	16399.1	[b]	(1509)	
X ¹Σ⁺	0	106.64	0.455								(438)	
N¹⁴Br												
$D_0^0 = (3.0)$ e.v.;[a] $\mu_A = 11.919$												
A	14791	785	4.45					$A{\rightarrow}(X)$	14837	V	(200a)	{isotope eff.; repr. (200a)}
(X)[b]	0	693	5.0									
NdO¹⁶												
$\mu_A = 14.403$												
Unclassified bands in the region 14800–24000 cm⁻¹											(1318)	
N¹⁴H¹												
$D_0^0 = (3.8)$ e.v.;[a] $\mu_A = 0.940447$												
d ¹Σ⁺	$a+$(70539)	(2640)			[14.085]		[1.1282]	$d{\rightarrow}c$	39510.4	V	(471a)	repr. (471a)
c ¹Π	$a+$(31289)	[2119][b]			[14.159]	(1.39)	[1.1252]	$c{\rightarrow}a$	30755.6[c]	R	(558) (187) (1270)	{repr. (558) (187)}
A ³Π_i	(29772.5)[d]	(3300)			16.67	0.74	1.037	$A{\leftrightarrow}X$	2³772.5		(240) (240a) (948)	no shading
b ¹Σ⁺	$a+$(8502)	[3480]			[16.401]	(0.17)	[1.045₅]	$c{\rightarrow}b$	22106.5	R	(1181)	repr. (1181)
a ¹Δ		[3186]			[16.45₃]	0.64[e]	[1.043₉]					
X ³Σ⁻	0	(3300)			16.65		1.038					
N¹⁴H²												
$D_0^0 = (3.9)$ e.v.; $\mu_A = 1.761383$												
c ¹Π	$a+$(31240)	(1550)[f]			[7.640]		[1.119₃]	$c{\rightarrow}a$	30849.0	R	(187)	repr. (187)
a ¹Δ	a	(2330)[f]			[8.836]		[1.040₈]					
NiBr												
$\mu_A = 33.848$												
Fragments of a third system near 25000 cm⁻¹											(235)	
B	(24000)							$B{\rightarrow}X$	23952	R	(1225) (1225)	

State	T_e	ω_e	$\omega_e x_e$	B_e	α_e	r_e		ν_{00}	R	Refs	Remarks
A	21796	302 H	7				A→X	21779	R	(1225)	
(X)[a]	0	334 H	5								

$D_0^0 = (7.3)$ e.v.;[b] $\mu_A = 22.110$

NiCl

Additional unassigned bands

State	T_e	ω_e	$\omega_e x_e$	B_e	α_e	r_e		ν_{00}	R	Refs	Remarks
B ($^2\Sigma$)	24623.2	400.7 H^Q	1.58				B→X	24613.8	R	(498) (1225)	
A ($^2\Sigma$)	23232.7	397.3 H^Q	1.18				A→X	23221.7	R	(498)	
(X $^2\Pi$)[a]	{ 485 / 0 }	419.2 H^Q	1.04						R	(498)	

$D_0^0 \le 3.1$ e.v.;[a] $\mu_A = 0.99111$

NiH¹

State	T_e	ω_e	$\omega_e x_e$	B_e	α_e	r_e		ν_{00}	R	Refs	Remarks
C $^2\Delta_{5/2}$				[6.140]		[1.665]	C→X	23760.8	R	(288)	$^2\Delta$*** of (288)[bc]
B $^2\Delta_{5/2}$		1587.4	47.5 *	(4.97)	(−0.26)	(1.85)	B→X	15977.3	R	(288) (1022) (245)	$^2\Delta$* of (288)[b]; repr. (245)
A $^2\Delta_{5/2}$				[6.283]		[1.645]	A→X	15520.1	R	(288) (1022) (245)	$^2\Delta$* of (288)[b]; repr. (245)
X $^2\Delta_{5/2}$		[1926.6]		7.823	0.248[d]	1.474_6					

$\mu_A = 12.573$

NiO(16)

Fragments of three additional systems in the same spectral region

State	T_e	ω_e	$\omega_e x_e$	B_e	α_e	r_e		ν_{00}	R	Refs	Remarks
b	a + (21262)	(570)					b→a	21135		(1358) (1192)	
a	a	(825)									
B	(16447)	(560)					B→X	16420	R	(1358) (1192)	
A	(12725)	(475)					A→X	12655		(1358) (1192)	
(X)	0	(615)									

NaK: [a]From convergence of C state, assuming no potential hill. [b]Sinha (1424) gives 25201 which does not fit with his observed band wave numbers. [c]Convergence at 18025 cm⁻¹ above $X(v=0)$. [d]Not stated explicitly.

NaRb: [a]From Gaydon.

NBr: [a]Gaydon gives 2.5 ± 0.5. [b]Shading not given, but probably R.

NH: [a]See legend of Fig. 169. Gaydon gives 3.4 ± 1 e.v. [b]Nakamura and Shidei's (1270) value 2612 does not agree with their own data. [c]Pearse (558) gives 30704.13 which must be erroneous since Q(2) is at 30741.94. [d]Coupling constant $A = -36.2$ [from (240a)]; Budó (139) gives -35.0]. [e]Spin coupling constants $\lambda = 0.45$, $\gamma = -0.04$. $D_v = [16.8 - 0.2(v + \frac{1}{2})] \times 10^{-4}$. [f]From the values for NH¹ by means of the isotope relations.

NiBr, NiCl: [a]Not certain whether this is the ground state.

NiH: [a]From predissociation; Gaydon gives 2.6 ± 0.3 e.v. [b]There are strong perturbations in all excited states. In some of them the Λ type doubling is anomalously large [b]Gaydon gives 5 ± 2 e.v. [c]Predissociation (diffuseness of emission lines in thermal emission) above 24900 cm⁻¹. [d]$D_v = [4.71 + 0.23(v + \frac{1}{2})] \times 10^{-4}$.

TABLE 39 (Continued)

State	T_e	ω_e	$\omega_e x_e$	$\omega_e y_e$	B_e	α_e	r_e (10^{-8} cm)	Designation	ν_{00}	References	Remarks
$N^{14}O^{16}$											
	$D_0^0 = 5.29_6$ or 6.48_7 e.v.; [a] $\mu_A = 7.46881$										
	Tanaka's Rydberg series of bands (absorption): $\{\nu = 147759 - R/(m + 0.22)^2,\ m = 2, 3, \ldots 9\ (v = 0);\ \nu = 147417 - R/(m + 0.16)^2,\ m = 3, \ldots 8\ (v = 0)\}$									(1456b)	γ series of Tanaka
N	133432	1155 H	(4)					$N\rightarrow X$	133160	(1456b)	β series of Tanaka
	Tanaka's Rydberg series of bands (absorption): $\nu = 133596 - R/(m + 0.29)^2,\ m = 3, 4, \ldots 10\ (v = 0, 1)$ [b]									(1456b)	P_4 of Tanaka
M	125242	1300 H	(12)					$M\rightarrow X$	124940	(1456b)	P_3 of Tanaka
L	118015	1170 H	(5)					$L\rightarrow X$	117650	(1456b)	P_2 of Tanaka
	Tanaka's Rydberg series of bands (absorption): $\nu = 114742 - R/(m + 0.91)^2,\ m = 4, 5, \ldots 10\ (v = 0)$									(1456b)	α series of Tanaka [c]; repr. P_1 of (1456b);
	Fragment of an absorption band system near 113000										
$E\ {}^2\Sigma^+$	60628.5	2373.6	15.8_5		1.9863	0.0182	1.0661	$E\rightarrow A$	16663.6	(931b)	headless bands [l]; repr. (192)
$D\ {}^2\Sigma^+$	53083 [d]	2327	23		[1.9917]		[1.0646]	$D\leftrightarrow X$	53171.7 V / 53292.6 V	(981) (956)	ε bands; repr. (440)(956)
$C\ {}^2\Sigma^+$	[52148]	(2347)			[1.955]		[1.075]	$C\leftrightarrow X$	52252.5 H^q V / 52373.4 V	(1382) (618) (440)	δ bands [e]; repr. (1382)(440)(956)
$B\ {}^2\Pi_r$	45946.7 / 45918.0	1038.41 Z / 1036.96 Z	7.603	+0.0967 [f]	1.177 / 1.076	0.0189 / 0.0116	1.385 / 1.448	$B\leftrightarrow X$	45394.6 R / 45486.1 R	(369) (618) (440)	β bands; repr. (440)(956) [k]
$A\ {}^2\Sigma^+$	43965.7	2371.3 Z	14.48	−0.28	1.9952	0.0164*	1.0637	$A\leftrightarrow X$	44078.3 V / 44199.2 V	(618) (980) (440)	γ bands; repr. (440), Fig.8b [a]
$X\ \{{}^2\Pi_{3/2}\ ;\ {}^2\Pi_{1/2}\}$	{121.1 [hi] / 0}	1903.68 Z / 1904.03 Z	13.97	−0.00120*	1.7046	0.0178 [j]	1.1508	rotation-vibration bands	{120.9 / 0} R	(986)	
$(N^{14}O^{16})^+$											
	$D_0^0 = 9.4$ or 10.6 e.v.; [a] $\mu_A = 7.46869$										
	Bands at 16670 cm^{-1} ascribed to NO$^+$ by (192) have recently been shown by (931b) to be due to NO										
$N^{14}S^{32}$											
	$D_0^0 = (5.9)$ e.v.; [a] $\mu_A = 9.74115$										
$B\ {}^2\Sigma$	[43389]							$B\rightarrow X$	43166 H V / 43389 H V	(226)	γ bands; repr. (226)
$A\ {}^2\Pi_r$	{40055 / 40007}	944.0 H / 967.0 H	5.5 / 10.5					$A\rightarrow X$	39694 R / 39880 R	(226)	β bands; repr. (226)
$X\ {}^2\Pi_r$	{223 / 0}	1220.0 H	7.75								
O_2^{16}											
	$D_0^0 = 5.080$ e.v.; [a] $\mu_A = 8.000000$										
	Fragments of several absorption band systems in the region 111000–142000 cm^{-1} and of a Rydberg series with vibrational structure and a limit (for $v' = 0$) at 163802 cm^{-1}									(571) (1458)	repr. (571)(1458)

Tanaka and Takamine's Rydberg series (absorption): $\nu = 146548 - R/(m + 0.32)^2$, $m = 3, 4, \ldots 15$ and corresponding series with $v' = 1, 2$ and limits 147705 and 148831 cm^{-1} converging to the $^4\Sigma_g^-$ state of O_2^+

State	T	ω_e	const					Transition	Origin		Refs	Reproduction
Q^b	136750	1205	(20)					$Q \leftarrow X$	136559		(571i) (1458)	repr. (571) (1458)
P^b	123430	1385	(22)					$P \leftarrow X$	123330		(571) (1458)	repr. (571) (1458)
N^c	121390	935	(17)					$N \leftarrow X$	121064		(571) (1458)	repr. (571) (1458)
l^{de}	118765, 117465	1050	15					$I \leftarrow X$	118500, 117200	^{f}	(571) (1458)	repr. (571) (1458)
M^{cd}	107405, 104725	1080	(28)					$M \leftarrow X$	107150, 104470	^{f}	(571) (1458)	repr. (571) (1458)
H^{de}	101675, 99865	1115	(19)					$H \leftarrow X$	101440, 99630	^{f} V	(331) (571)	repr. (571) (1061)
											(571)	repr. (571)
$B\,^3\Sigma_u^-$	49802.1	700.36	8.002$_3$	-0.3753_5^h	0.819^i	0.011*	1.60$_4$	$B \leftrightarrow X$	49363.1^j	R	(406)(175)(931)	Schuman-Runge bands^k; repr. (175) (406)

Fragment of an emission band system with 0-0 band at 49224 and $\Delta G'_{1/2} = 49224$ and $\Delta G''_{1/2} = 1085$ cm^{-1}

Unclassified absorption bands from 80500-98000 cm^{-1}

^{a}Depending on the choice for $D_0^0(N_2)$, from thermochemical heat of formation of NO taken from (50b). $^b\Delta G_{1/2} = 1341$ cm^{-1}

^{c}This series converges to an excited state of NO$^+$ since the ionization potential of NO from electron collision experiments (1017) is 9.5 e.v. (=76800 cm^{-1})

^{d}Herzberg and Mundie (1050) have suggested that the ϵ bands belong to the γ bands since the $v = 0$ level of the D state agrees closely (even though not exactly) with the extrapolated position of the $v = 4$ level of the A state and since no other γ bands with $v \geq 4$ have been observed in absorption or emission. They also suggested that the high intensity of the ϵ bands in absorption found by (440) was only apparent and due to predissociation. The observation of the ϵ bands in emission by (956) and (981) seems to contradict this interpretation. However recent high dispersion absorption spectra of NO in the vacuum region by Tanaka (1456e) do show a slight broadening of the fine structure lines and no anomalously large intensity of the first (0-0) ϵ band compared to the 3-0 γ band. It is possible that this is a case of induced predissociation. There are probably strong vibrational perturbations between the D and C states which may account for deviations of the D levels from the extrapolated levels of the A state.

^{e}Predissociation for $v = 1$, see (225) (937).

^{f}Recent measurements of the β bands in absorption (1456e) strongly suggest a negative $\omega_e y_e$ giving a convergence in agreement with the lower D_0^0 value.

^{g}Isotope effect, see (527). Induced pred dissociation, see (723).

^{h}Coupling constant $A = 124.2$. see (986).

^{i}Corrected for difference in zero-point vibration (refers to $J = 0$). $^i D_0 = 5 \times 10^{-6}$.

^{j}Geró and Schmid (980) report a predissociation (breaking off) above $v' = 4$ corresponding to 50500 cm^{-1} which is less than the higher of the two $D_0^0(NO)$ values.

^{l}These bands were previously ascribed by (192) to NO$^+$ and by (1396) to NO$_2$. The new assignment (931b) is based on an accurate agreement of combination differences with the γ bands.

NO$^+$: ^{a}From $D_0^0(NO)$, $I(O)$ and $I(NO) = 9.5$ e.v. [see (1017)].

NS: ^{a}Gaydon gives 5.0 ± 1.

O$_2$: ^{a}Gaydon gives 5.084 since he assumes a slightly different convergence limit in the B state.

^{b}The states P and Q are the first members of the Rydberg series corresponding to the $^4\Sigma_g^-$ state of O_2^+.

^{c}First members of a Rydberg series possibly corresponding to the $^2\Pi_u$ state of O_2^+.

^{d}The apparent doublet character of these states is due to (Ω, ω) coupling (see p. 338). The doublets correspond probably to $^5\Pi$ and $^3\Pi$ arising from $^4\Pi$ of the O_2^+ ion.

^{e}The states H and I are the first members of a Rydberg series with a limit at 130800 cm^{-1} and corresponding to the $^4\Pi_u$ state of O_2^+.

^{f}Headless bands with two maxima and zero gap.

^{g}Possibly the lower state is H. Another emission band system was found by (1031) in the region 20000 - 18500 cm^{-1}. ^{h}Convergence at 56859 cm^{-1}.

^{i}Somewhat different constants that represent the higher levels better but not the first levels are given by (406).

^{j}From Knauss and Ballard's (406) formula. The observed value 49357.5 deviates much more than all other vibrational levels from the formula.

^{k}Bands with $v' = 8$ have been observed in fluorescence by (1340).

TABLE 39 (*Continued*)

State	T_e	ω_e	$\omega_e x_e$	$\omega_e y_e$	B_e	α_e	r_e (10^{-8} cm)	Designation	ν_{00}	References	Remarks
O_2^{16} (*Continued*)											
$A\ ^3\Sigma_u^+$	36096	819.0	22.5_0	*[l]	[1.05]		[1.42]	$A \leftrightarrow X$	35713[m] R	(310) (1027) (909) (1044)	"Herzberg bands
$b\ ^1\Sigma_g^+$	13195.222_1	1432.687_4	13.9500_8	-0.01075	1.400041_6	0.01817_0*	1.22675_0	$b \leftrightarrow X$	13120.9080 R	(764)	atmospheric oxygen bands; repr. (764) Fig. 76[o]
$a\ ^1\Delta_g$	7918.1	1509.3[p]	12.9		1.4264	0.0171	1.2155	$a \leftarrow X$	7882.39 R	(1053)	infrared atmospheric oxygen bands; repr. (1053)
$X\ ^3\Sigma_g^-$	0	1580.361_3	12.0730	$+0.0546$[q]	1.44566_6	0.01579_1[r]	1.20739_8			(175) (764)	
	$D_0^0 = 6.48$ e.v.; $\mu_A = 7.99986$										
O_2^+											
$b\ ^4\Sigma_g^-$	$a + 16587.88$	1196.77 Z	17.09		1.28729	0.02206*	1.27953	$b \to a$	16666.74 V	(823) (1420) (1277) (1279) (530)	1st negative bands[a]
$A\ ^2\Pi_u$	38795[b]	900 H	13.4		1.0617	0.01906*	1.4089	$A \to X$	38108H R; 38308H R	(123) (204) (1161) (1277)	2nd neg. bands;[a] repr. (204)
$a\ ^4\Pi_u$	a[c]	1035.69 Z	10.39		1.10466	0.01575	1.38126				
$X\ ^2\Pi_g$	0[d]	1876.4 H	16.53		1.6722	0.01984[e]	1.1227				
	$D_0^0 = 4.35$ e.v.;[a] $\mu_A = 0.948376$										
$O^{16}H^1$											
C	$(x + 23792)$[g]	[660] H						$C \to B$	23053 R	(1395a) (1393a)	[b]
B	x	(2210) H	[145]	[-0.647]							
$A\ ^2\Sigma^+$	32682.5	3180.5_6 Z	94.93		17.355	0.807*	1.0121	$A \leftrightarrow X$	32402.1 R	(895a) (661) (1290) (1111)	repr. (661)[c]
$X\ ^2\Pi_i$	0[d]	3735.21 Z	82.81		18.871	0.714[e]	0.9706				
	$D_0^0 = 4.39$ e.v.;[f] $\mu_A = 1.789403$										
$O^{16}H^2$											
C	$(x + 23435)$[g]	(735)	[85]					$C \to B$	23014 R	(1395a) (1393a)	[b]
B	x	(1570)	[72]								
$A\ ^2\Sigma^+$	35875.0	2319.9 Z	52.0	*	9.19_4	0.32_9	1.012_3	$A \to X$	35472.5[h] R	(355) (1086) (1376) (1377) (1108) (1300a)	repr. (1085)
$X\ ^2\Pi_i$	0[i]	2720.9 Z	44.2		10.01_6	0.29_5[j]	0.9699				

State	T_e	ω_e	$\omega_e x_e$	B_e	α_e	D_e	r_e	Transition	ν_{00}	deg.	References / Remarks
(O^{16}H^1)$^+$ $D_0^0 \geq 4.4$ e.v.;^{a} $\mu_A = 0.948374$											
$A\,{}^3\Pi_i$	$(28937)^b$	[1986] Z		13.756	0.845		1.1368	$A\rightarrow X$	27952	R	(460)(152)
$X\,{}^3\Sigma^-$	0	[2955] Z		16.793	0.732^c		1.0289				
P$_2^{31}$ $D_0^0 = 5.031$ e.v.; $\mu_A = 15.49221$											
Fragment of a band system (emission) in the region 23600–25200 cm⁻¹											
$A\,{}^1\Sigma_u^+$	46942.7	475.24 H	2.633	0.2416_6	+0.0217	0.00165*	2.121_1	$A\leftrightarrow X$	46790.1	R	(311)(1195)(311) { repr.(1195)(311), $(69)^a$
$X\,{}^1\Sigma_g^+$	0	780.43 H	2.804	0.3032_7	−0.00533	0.00142^b	1.894_3				(1093)
Pb$_2$ $D_0^0 = (0.7)$ e.v.;^{a} $\mu_A = 103.63_3$											
Some unclassified bands in the Schumann region											
A	19570.8	159.22 H	0.882		+0.00518			$A\leftrightarrow X$	19522.7	R	(1142)(6334)(1271) repr.$(6334)^b$
X	0	256.5 H	2.96								
PbBr^{79} $D_0^0 = 2.9_7$ e.v.;^{a} $\mu_A = 57.161$											
$B\,({}^2\Sigma)$	34523	25.2 H	0.6		−0.028			$B\leftarrow X$	34548	V	(1279a) pre-liss.?
$A\,({}^2\Sigma)$	20884.3	152.5 H	0.40					$A\leftarrow X$	20856.8	R	(502) isotope effect
$X\,({}^2\Pi_{1/2})$	0	207.5 H	0.50								

O₂: ᶠObserved convergence at 41280 cm⁻¹ [slightly higher than $D(O_2)$ from convergence in B state, probably corresponding to energy difference of components of O ^3P]. ᵐIt is not certain that this is the 0–0 band [see (909) and (1044)]. ᵐ'Occur in emission in the night sky (see p. 485). ⁿIsotope effect (764). Observed in emission by (1035a). ᵖ$\Delta G_{1/2} = 1483.5_0$. ᵠ$\omega_e z_e = -0.00143$. ᵒSpin coupling constants $\lambda = 1.984$, $\gamma = -0.00837$; $D_v = [4.95; -0.08(v + \tfrac{1}{2})] \times 10^{-6}$.

O₂⁺: ᵃSome authors reverse the usage of the names 1st and 2nd negative bands. ᵇCoupling constant $A = +8.2$ [see (653)]. ᶜCoupling constant $A = +47.96$. From the Rydberg series of (1458) and the ionization potential of O_2 (see Table 38) one obtains $a = 31900$ cm⁻¹. ᵈCoupling constant $A = +195$ [see (653)]. ᵉ$D_v = 6.85 \times 10^{-6}$.

OH: ᵃThis is the value of (922) but corrected for the new values of the fundamental constants. Gaydon gives 4.40 ± 0.05 e.v. ᶜAll constants of the A–X system are from (895a). ᵇThese bands were originally ascribed to H_2O but more recently (13'43a) to OH or possibly OH⁺ or OH⁻. ᶜCoupling constants $A_0 = -139.7_5$, $A_1 = -140.2_5$. ᵈ$+0.0035(v + \tfrac{1}{2})^2$; $D_v = [18.8 - 0.3(v + \tfrac{1}{2})] \times 10^{-4}$. ᶠFrom $D_0^0(OH^1)$. ᵍThe fact that this agrees so poorly with the value for OH¹ throws some doubt on the correctness of the preliminary analysis. ʰThis is Sastry's (1376) ν_{00} for $^2\Pi_{3/2}$ less $\tfrac{1}{2}A$. ⁱCoupling constant $A = -139.6$ [(1376); see also (1108)].

OH⁺: ᵃ$D_v = [4.5 - 0.65(v + \tfrac{1}{2})] \times 10^{-4}$. ᵇFrom $D_0^0(OH)$, $I(H)$ and $I(OH)$, the latter from (1194). ᶜCoupling constant $A = -80.4$ (for $v = 0$), see (152). $D_v = [19.57 - 0.53(v + \tfrac{1}{2})] \times 10^{-4}$.

P₂: ᵃStrong vibrational and rotational perturbations (311) (1195). Breaking off (predissociation) at $J' = 58$, $v' = 10$ and $J' = 38$, $v' = 11$ (52068 and 51983 cm⁻¹) corresponding to dissociation into $^4S + {}^2D$. ᵇ$-0.0000052(v + \tfrac{1}{2})^2$; $D_v = [0.21 - 0.001(v + \tfrac{1}{2})] \times 10^{-6}$ (calc.).

Pb₂: ᵃGaydon gives 0.6 ± 0.2 e.v. ᵇThe assignments of (1271) seem rather doubtful since they include bands of very high v' without intermediate ones (e.g. 13–0, 42–0, 48–0 but no others with $v'' = 0$).

PbBr: ᵃFrom (1279a); Gaydon gives 2.2 ± 0.4 e.v.

TABLE 39 (Continued)

State	T_e	ω_e	$\omega_e x_e$	B_e	α_e	r_e (10^-8 cm)	Designation	ν_{00}		References	Remarks
PbCl[35]		$D_0^0 = 3.1_2$ e.v.;[a] $\mu_A = 29.9281$									
B	35199	382.1 H	1.05				$B \leftarrow X$	35238	V	(1279a)	prediss.?
A ($^2\Sigma$)	21865.0	228.7 H	0.78				$A \leftrightarrow X$	21827.9	R	(592)(502)	{ repr. (592); isotope effect
X ($^2\Pi_{1/2}$)	0	303.8 H	0.88								
PbF[19]		$D_0^0 = 3.4_5$ e.v.;[a] $\mu_A = 17.4083$									
F $^2\Pi$	47870	[628] H					$F \leftarrow X_1$	47930	b	(592)	
E	45400	(565) H					$E \leftarrow X_1$	45430	V	(592)	repr. (592)
D $^2\Pi$	43820	[597] H					$D \leftarrow X_1$	43865	b	(592)	
C $^2\Delta$	38046	594.0 H	2.50				$C \leftarrow X_1$	38089	V	(592)	diffuse bands,pred.?[c]
B $^2\Sigma$	35642.8	612.8 H	3.42				$B \leftarrow X_1$	35695.3	V	(592)	repr. (592)
A $^2\Delta$	22566.6	397.8 H	1.77				$A \leftrightarrow X_1$	22511.9	R	(592)	repr. (592)
X_2 $^2\Pi$	8266	531.1 H	1.50				$B \rightarrow X_2$	27417	V	(592)	
X_1	0	507.2 H	2.30								
PbH[1]		$D_0^0 \leq 1.5^7$ e.v.;[a] $\mu_A = 1.003251$									
		Additional band (not analyzed) at 26205 cm^{-1}									
B $^2\Sigma$	(18031)	(535)	(15)	2.48[b]	0.08	2.60	$B \rightarrow X$	17520	R	(1516)(1070)	
A $^2\Sigma$	(17590)	(500)	(10)	(3.02$_5$)	(0.05)	(2.35$_7$)		(17060)		(694)(1516)(972)[c]	
X ($^2\Pi_{1/2}$)[c]	0	1564.1	29.75	4.971	0.144[f]	1.839					[d]
PbI[27]		$D_0^0 = 2.8_4$ e.v.; $\mu_A = 78.722$									
B	33488	198.7	0.35				$B \leftarrow X$	33508	V	(1279a)	prediss.?
A	20528.5	142.0	1.5				$A \rightarrow X$	20520	R	(1279a)	
X ($^2\Pi_{1/2}$)	0	160.5	0.25								
PbO[16]		$D_0^0 = (4.3)$ e.v.;[a] $\mu_A = 14.8534$									
		Unclassified bands 22800-33100 cm^{-1} [b]									
E	34900	430 H	perturbations				$E \leftarrow X$	34755	R	(337)	
D	30198.7	530.5 H	2.92	[0.2707][e]		[2.048]	$D \leftrightarrow X$	30103.2	R	(337) (1479) (337)	repr. (1479)
C	24864.0	518.0 H	3.90				$C \leftarrow X$	24762.0	R	(156)(817) (337) (1414)	
B	22284.9	498.0 H	2.20				$B \leftrightarrow X$	22173.4	R	(337)(1414) (337)(817)	

	T_e	ω_e	$\omega_e x_e$	B_e	α_e	r_e	$A \leftrightarrow X$	ν_{00}	R	References	
A	19863.3	451.7 H	3.33	$[0.2579]^c$	0.0019^d	$[2.098]$	$A \leftrightarrow X$	19728.3	R	(817)(156)(337)	repr. (1479)
$X\ ^1\Sigma^+$	0	721.8 H	3.70	0.3073^c		1.922					

$\mathrm{Pb^{208}S^{32}}$ $D_0^0 = (4.7)$ e.v.;^{a} $\mu_A = 27.7213$

Fragment of an infrared system between 13000 and 15100 cm^{-1}
Additional ultraviolet absorption bands (unclassified) between 32200 and 36400 cm^{-1}
strong perturbations

	T_e	ω_e	$\omega_e x_e$	B_e	α_e	r_e	Transition	ν_{00}	R	References	
F	47770	370 H	(7.8)				$F \leftarrow X^b$	47739	R	(595)	
E	29650.5	299.34 H	1.574				$E \leftarrow X$	29586.0	R	(1479)	
D	25024.4	283.95 H	1.171				$D \leftarrow X$	24952.3	R	(595)	
C	23212.9	303.93 H	1.436				$C \leftarrow X$	23150.7	R	(595)	
B	21847.7	282.17 H	0.856				$B \leftarrow X$	21774.8	R	(595)	
$A\ ^1\Sigma^+$	18851.3	261.09 H	0.365	0.08560	0.000296	2.666	$A \leftarrow X$	18768.0	R	(83)(595)	repr. (595)
$X\ ^1\Sigma^+$	0	428.14 H	1.201	0.10605	0.000873	2.394$_8$				(83)	{repr. (595); (isotope eff. (83)}

PbSe $D_0^0 = (4.7)$ e.v.;^{a} $\mu_A = 57.189$

	T_e	ω_e	$\omega_e x_e$	B_e	α_e	r_e	Transition	ν_{00}	R	References	
F	45220.9	224.8 H	0.50				$F \leftarrow X$	45194.5	R	(1479)(1410)	repr. (1479)
D	28418.0	190.4 H	0.53				$D \rightarrow X$	28374.4	R	(1479)(784)	repr. (784)
C	23315.7	183.0 H	0.25	-0.004			$C \leftarrow X$	23268.5	R	(688)	
B	21005.8	184.8 H	0.43				$B \leftarrow X$	20959.4	R	(688)	
A	18716.8	166.9 H	0.14				$A \leftarrow X$	18661.5	R	(688)	
$X\ ^1\Sigma^+$	0	277.6 H	0.51						R	(784)(688)	

PbCl: ^{a}From (1279a); Gaydon gives 2.6 ± 0.4 e.v.

PbF: ^{a}From (1279a); Gaydon gives 3.2 ± 0.5 e.v.; from predissociation $D \leq 4.5$ e.v.
^{b}Shading not given but probably V.
^{c}Breaking off in emission above $v = 1$.

PbH: ^{a}From predissociation in B state assuming dissociation at that limit into $^3P_1 + {}^2S$. The single state that arises from $^3P_0 + {}^2S$ is very probably the ground state. Gaydon gives 1.8 ± 0.2 e.v.
^{b}This state is very strongly perturbed; the constants are those of (972) who has analyzed these perturbations. Watson and Simon (1516) give rather different constants with negative α.
^{c}There is a predissociation (breaking off) at $K' = 30$, 24 and 20 for $v' = 3$, 4, 5 respectively (694). The energy of $K = 20$, $v' = 5$ is 20550 cm^{-1}, of $K = 21$, $v' = 5$ is 20740 cm^{-1}.
^{d}The data for this state have been obtained entirely from the strong perturbations occurring in the B state [see (972)].
^{e}This state is given as $^2\Sigma$ by (1516); but Howell (1070) has given various arguments that it is $^2\Pi_{3/2}$, the $^2\Pi_{1/2}$ component being about 8200 cm^{-1} above it and the $B\ ^2\Sigma \rightarrow {}^2\Pi_{3/2}$ transition being in the not yet investigated infrared.

PbO: ^{a}Gaydon gives 3.3 ± 0.4 e.v.
^{b}Shawhan and Morgan (1414) report a band system with $\nu_e = 30899.0$ which according to (337) is not real.
^{c}Rotational constants refer to Pb^{206}O^{16}.
^{d}$D_0 = 0.22 \times 10^{-6}$.
^{f}$D_0 = 2.01 \times 10^{-4}$.

PbS: *Gaydon gives 3.0 ± 0.5 e.v.
^{b}Called $D \leftarrow X$ by (1479).

PbSe: ^{a}Gaydon gives 3.5 ± 1 e.v.

TABLE 39 (Continued)

State	T_e	ω_e	$\omega_e x_e$	$\omega_e y_e$	B_e	α_e	r_e (10^{-8} cm)	Designation	ν_{00}		References	Remarks
PbTe				$D_0^0 = (3.5)$ e.v.;[a]	$\mu_A = 78.996$							
G	46541.7	159.6 H	1.4					$G \leftarrow X$	46515.3	R	(1479)	repr. (1479)
F	41658.8	176.4 H	1.0					$F \leftarrow X$	41640.9	R	(1479)	repr. (1479)
D	27176.5	142.6 H	1.58					$D \leftarrow X$	27141.5	R	(1479)	repr. (1479)
B	19736.4	145.51 H	0.464					$B \leftarrow X$	19703.2 H	R	(688)	
A	16362.3	141.37 H	0.224					$A \leftarrow X$	16327.1	R	(688)	
X $^1\Sigma$		211.8 H	0.12									
P^{31}H^1		$\mu_A = 0.976363$										
A $^3\Pi_i$	(29560)[a]	(1910)			[8.025]		[1.466$_9$]	$A \to X$	29321.9[b]	R	(1087)	repr. (1087)
X $^3\Sigma^-$		(2380)			[8.412]	c	[1.432$_8$]					
P^{31}H^2		$\mu_A = 1.891718$										
A $^3\Pi_i$	a				[4.175]		[1.461]	$A \to X$	d	R	(1087)	repr. (1087)
X $^3\Sigma^-$					[4.363]	e	[1.429]					
P^{31}N^{14}		$D_0^0 = (6.3)$ e.v.;	$\mu_A = 9.64651$									
A $^1\Pi$	39805.7	1103.09 Z	7.222		0.7307$_1$	0.00663	1.5466	$A \leftrightarrow X$	39688.5	R	(176) (1242)	repr. Fig. 9, (176)
X $^1\Sigma^+$	0	1337.24 Z	6.983		0.7862$_1$	0.00557[a]	1.4910					
P^{31}O^{16}		$D_0^0 = (6.2)$ e.v.;	$\mu_A = 10.55138$									
		Two fragmentary band systems in the region 15500–21000									(605)	
A $^2\Sigma$	40408.6	1391.0$_0$ H	7.65		0.8121	0.0056	1.402$_7$	$A \to X$	40264.7H / 40488.5H	V / V	(267) (631)	repr. (267) (1313)
B $^2\Pi$	30846[a]	1157.$_2$	6.6					$B \to X$	30584Z / 30811Z	V_R / V_R	(176)	repr. (1313)[b]
X $^2\Pi_r$	223.8[c] H / 0	1230.6$_4$ H	6.52		0.7645 / 0.7613	0.0055[d]	1.445$_7$ / 1.448$_8$					
Pr^{141}O^{16}		$\mu_A = 14.369$										
A	18704.5	769.0	1.92					$A \to (X)$	18679.4	R	(693)	
(X)[a]		818.9	1.20									
Rb$_2$^{85}		$D_0^0 = 0.49$ e.v.;	$\mu_A = 42.469$									
		Diffuse absorption bands corresponding to van der Waals molecules										
C	22777.5	40.42 H	0.0745	−0.00144				$C \to X$	22769.3	R	(1473) (668)	

State		ω_e	B_e	α_e	System	ν_{00}		References	
B	20835.1				B←X	20824.9	R	(668)	$\Big\{$ magnetic rot. spectrum
A $^1\Pi_u$	14662.6	**36.46** H	0.124		A←X	14658.2	b	(438)	
X $^1\Sigma_g^+$	0	48.05	0.191	-0.00083				(438)(668)	
		57.28 H	0.96						

RbBr $\quad D_0^0 = 3.9_3$ e.v. from thermochemical data (50a) and $D_0^0(\mathrm{Br_2})$; $\mu_A = 41.313$

Continuous absorption with maxima at 35770 and 38820 cm^{-1}

RbCl $\quad D_0^0 > 3.96$ e.v.;^{a} $\mu_A = 25.068$

Unstable state giving diffuse bands at wave numbers smaller than and a continuum at wave numbers larger than 31939 cm^{-1} (maximum at 40230 cm^{-1})

State					System	ν_{00}		References	
A	0	(253)^b			A←X	>31939		(644)(1253)	
X $^1\Sigma$	0		$\{12.89\}^c$					(644)	

RbCs^{133} $\quad \mu_A = 52.036$

RbF^{19} $\quad D_0^0 = 5.4_5$ e.v.;^{a} $\mu_A = 15.5486$

State		ω_e			System	ν_{00}		References	
A	13747.2	38.46 H			A←X	13741.7	R	(438)	
X $^1\Sigma^+$	0	49.41 H						(842a)	

Continuous absorption with maximum at 46500 cm^{-1} and diffuse bands at smaller wave numbers

RbH^1 $\quad D_0^0 = (1.9)$ e.v.;^{a} $\mu_A = 0.99638$

State		ω_e				System	ν_{00}			Shading	References
A $^1\Sigma^+$	18906.4	244.6 Z	-4.1	-0.169^*	1.231	A→X	18564.9	-0.023^*	3.70_8	R	(960)
X $^1\Sigma^+$	0	936.77 Z	14.15	$+0.075$	3.020			$+0.072^b$	2.367		repr. (960)

PbTe: ^{a}From Gaydon; linear extrapolation gives 11.5 e.v.

PH: ^{a}Coupling constant $A = -115.5$ cm^{-1}.
^{b}This is $Q_2(1)$ of $^3\Pi_1 \to {}^3\Sigma$ given by (557).
^{b}Spin coupling constants $\lambda = +0.733$, $\gamma = -0.078$ from (356); $D_0 = 4.3 \times 10^{-4}$.
^{c}Spin coupling constants $\lambda = +0.743$, $\gamma = -0.042$ from (356); $D_0 = 1.20 \times 10^{-4}$.

PN: $^a D_v = 1.09 \times 10^{-6}$.

PO: ^{a}The magnitude of the doublet splitting cannot be obtained from the available data, but it is likely to be small.
^{d}No band origin given nor sufficient data to calculate an approximate value.
^{b}This band system has red and violet shaded bands as well as headless bands. The ν_{00} values refer to the observed zero gaps of the 0-0 band. The vibrational constants are provisional and likely to be somewhat inaccurate because of the change of shading. To the writer it seems quite certain that both violet and red shaded bands belong to one and the same system (compare also the corresponding band system of NS). Ramanadham, Rao and Ramasastry (1327) have attempted to assign the bands to two different systems one red and the other violet shaded.
$^d D_v = 1.16 \times 10^{-6}$.
^{c}In another part of their paper Ghosh and Ball (267) give 221.0.

PrO: ^{a}Doubtful whether this is ground state.

Rb$_2$: ^{a}Gaydon gives 0.48 ± 0.05 e.v. $\quad$ ^{b}Shading not given but probably R.

RbCl: ^{a}This refers to dissociation into normal atoms $^2S + {}^2P_{3/2}$ while the ground state on increasing the vibrational energy dissociates into ions. Thermochemical data (see Gaydon) give 4.38 e.v.
^{b}This is probably not the first vibrational quantum. $\quad$ ^{c}From electron diffraction (483).

RbF: ^{a}Refers to dissociation into normal atoms $^2S_0 + {}^2P_{3/2}$ while the ground state on increasing the vibrational energy dissociates into ions. Thermochemical data (50a) and $D_0^0(\mathrm{F_2})$ give $D_0^0 \leq 5.7$ e.v.

RbH: ^{a}Gaydon gives 1.7 ± 0.2 e.v. $\quad$ $^b +0.0003(v + \tfrac{1}{2})^2$; $D_v = [1.22 - 0.01(v + \tfrac{1}{2})] \times 10^{-4}$.

TABLE 39 (Continued)

State	T_e	ω_e	$\omega_e x_e$	$\omega_e y_e$	B_e	α_e	r_e (10^-8 cm)	Observed Transitions		References	Remarks
								Designation	ν_{00}		
RbI^127											
$X\ ^1\Sigma$	\multicolumn{8}{l}{$D_0^0 = 3.29$ e.v.;[a] $\mu_A = 51.089$}			(1391)	
	\multicolumn{7}{l}{Continuous absorption with maxima at 30850, 38750, and 47150 cm^-1}	[3.26][b]				
S_2^32											
$X\ ^1\Sigma$	\multicolumn{8}{l}{$D_0^0 \leq 4.4$ e.v.;[a] $\mu_A = 15.99126$}				
	\multicolumn{9}{l}{Fragments of band systems in the regions 12300–14400, 35000–37000, 38000–40100 cm^-1}		
$D\ ^3\Pi_u$	58976.8 / 58707.2 / 58515.1	793.9 H	4.00					$D \leftarrow X$ V	59010.6 / 58741.0 / 58548.9	(1359)(1360)	[b] repr. (1184)
$C\ (^3\Pi_u)$	55598.2	830.2 H	3.75					$C \leftarrow X$ V	55650.2	(1184)(1185)	bands with 4 heads, S^34 isotope effect
$B\ ^3\Sigma_u^-$	31835	434.0 H	2.75		0.2219	0.0018	2.180	$B \leftrightarrow X$ R	31689[c]	(1272)(1273)	repr. Fig. 16[d]
$X\ ^3\Sigma_g^-$	0	725.68 Z	2.852		0.2956	0.0016_0[e]	1.889			(548)(545)(1297)	
Sb_2											
	\multicolumn{7}{l}{$D_0^0 = (3.7)$ e.v.;[a] $\mu_A = 60.89_7$}					
F	44780	226.0	1.17					$F \leftarrow X$	44758	(526)	[b]
D	(32084)	(215)[c]						$D \leftarrow X$	32057	(528a)	[b]
B	19069	216.8	0.40					$B \leftrightarrow X$	19043	(746)	[b] weak in absorption
A	14991	217.2	0.44					$A \leftrightarrow X$	14965	(746)	[b] weak in absorption
$X\ ^1\Sigma_g^+$	0	269.85	0.59								
SbBi^209											
	\multicolumn{7}{l}{$D_0^0 = (3.0)$ e.v.; $\mu_A = 76.95_8$}					
A	40647[a]	190.2 H	0.73					$A \rightarrow X$ R	40632	(526)	
$X\ ^1\Sigma^+$	0	220.0 H	0.50								
SbCl^35											
	\multicolumn{7}{l}{$D_0^0 = (4.6)$ e.v.; $\mu_A = 27.174_4$}					
B	25855	240.2[a] H	2.19					$B \rightarrow X$ R	25791	(934)	isotope effect,
A	22395	244.4[a] H	2.28					$A \rightarrow X$ R	22333	(934)	repr. (934)
$X\ ^b$	0	369.0 H	0.92								
SbF^19											
	\multicolumn{7}{l}{$D_0^0 = (4.2)$ e.v.; $\mu_A = 16.439_4$}					
C_3	(44757)	(700) H						$C_3 \rightarrow X$ V	44801	(340a)	double heads for strongest bands
C_2	43513.7	698.8 H	1.93					$C_2 \rightarrow X$ V	43356.2	(340a)	
C_1	37937.6	696.9 H	1.09					$C_1 \rightarrow X$ V	37979.4	(340a)	
A_3	27912	412.0 H	2.35					$A_3 \rightarrow X$ R	27811	(594)(340a)	
A_2	23992.5	420.0 H	1.75					$A_2 \rightarrow X$ R	23895.7	(594)(340a)	

	T_e	ω_e (H)	$\omega_e x_e$	System	ν_{00}	Shading	References	
A_1[a]	21887.5	411.3 H	1.71	$A_1 \to X$	21786.3	R	(594)(340a)	
X[a]	0	614.2 H	2.77					

SbN¹⁴ $D_0^0 = (4.8)$ e.v.;[a] $\mu_A = 12.562_7$

	T_e	ω_e (H)	$\omega_e x_e$	System	ν_{00}	Shading	References	
$A\ ^1\Pi$	34465	830.7 H	6.0	$A \to X$	34409	R	(870)	
$X\ ^1\Sigma$	0	942.0 H	5.6					

SbO¹⁶ $D_0^0 = (3.8)$ e.v.;[a] $\mu_A = 14.142_1$

	T_e	ω_e (H)	$\omega_e x_e$	System	ν_{00}	Shading	References	
$A\ ^2\Sigma$	39767.7	879.5 H^Q	5.85	$A \to X$	39798.7 / 37526.7	V / V	(1400)	repr. (1399)
$C\ ^2\Sigma$	26594.0	582.0 H^Q	6.50	$C \to X$	26476.1	R	(1399)	
$B\ ^2\Pi$	20800.5 / 20667.5	569.0 H^Q	5.00		24204.1	R		
$X\ ^2\Pi$	2272.0 / 0	817.2 H^Q	5.30	$B \to X$	18404.5 / 20543.5	R / R	(1399)	

Sc⁴⁵O¹⁶ $D_0^0 = (7)$ e.v.;[a] $\mu_A = 11.8012$

	T_e	ω_e (H)	$\omega_e x_e$	System	ν_{00}	Shading	References	
$B\ ^2\Sigma$	20652.6	825.0 H	4.5	$B \to X$	20579.2	R	(389)(1221)(23)	repr. (951)(1317)
$A\ ^2\Pi$	16610.9 / 16492.9	874.8 H	4.99	$A \to X$	16562.3 / 16444.3	R / R	(389)(1221)(23)	repr. (951)(1317) / (1221)
$X\ ^2\Sigma$	0	971.55 H	3.95					

RbI: [a]From thermochemical data (Gaydon); refers to dissociation into normal atoms $^2S + {}^2P_{3/2}$.   [b]From electron diffraction (483).

S₂: [a]This corresponds to the first predissociation limit. It is probable but not certain that $D_0^0(S_2) \leq 3.6$ e.v. [see (1050)].

[b]An absorption band system found in gases drawn from a discharge through S and SO₂ is ascribed by (861) to metastable S₂ molecules in the $^1\Sigma_g^+$ state, but this interpretation has been refuted by (1379). It seems now certain that these bands are due to S₂O₂ [see (1095)(1136a)(1136b)].

[c]The "observed" position of the 0–0 band is ~31659 indicating a strong vibrational perturbation.

[d]Strong perturbations and two predissociations one above $v' = 9$ the other above $v' = 17$ [see (548)]. Rosen, Désirant and Duchesne (1361) have observed a sudden termination of the branches in the bands with $v' = 7$ and 8 but since no complete rotational analysis has been possible their conclusion that case I(b) applies (see p. 431) does not seem established. Kondratjew and Olsson (411) found induced predissociation above $v' = 4$ (see p. 432 f.).

[e]$D_v = 0.20 \times 10^{-6}$.

Sb₂: [a]Gaydon gives 3.0 ± 0.5 e.v.   [b]Shading not stated but probably R.   [c]Strong vibrational perturbations.

[d]Nakamura and Shidei (526) give appreciably different constants for the $D–X$ system both in upper and lower state. Their assignments of individual bands are entirely different from (528a).

SbBi: [a]Nakamura and Shidei (526) give erroneously 40617.

SbCl: [a]The observed $\Delta G'$ values are very poorly represented by these $\omega_e, \omega_e x_e$ values. No bands with $v'' < 3$ are observed in $A–X$ and no bands with $v'' < 6$ in $B–X$. This throws doubt on the whole analysis.

[b]It is possible that the ground state is a multiplet state [Sb(^4S) + Cl(^2P) cannot give a singlet state].

SbF: [a]It is probable that this state is a triplet state, that is, that part of the splittings of the $A–X$ and $C–X$ systems is accounted for by the splitting of X. This would change the T_e values for the excited states.

SbN: [a]Gaydon gives 3.1 e.v.

SbO: [a]Gaydon gives 3.2 e.v.

ScO: [a]Gaydon gives 6 ± 1 e.v.

Table 39 (*Continued*)

State	T_e	ω_e	$\omega_e x_e$	B_s	α_e	r_e (10^{-8} cm)	Designation	ν_{00}		References	Remarks
$\mathrm{Se_2^{80}}$		$D_0^0 \leq 3.55$ e.v.; $\mu_A = 39.971$									
D	54272	396 H					$D \leftarrow X$	54274	V	(1416)	[b]
C	52434	204[c] H	0.6				$C \leftarrow X$	52340	V	(1416)	
		Several other band systems found by various authors in the region 19000–36000 cm^{-1} have been shown by (598) to belong to the main B–X system with the possible exception of the fluctuations discussed by (1362) [see however (71)]. Some other bands were found to be due to impurities [see (598)]									
B (Σ_u^+)	26035	281.11 H	2.65	0.0703	0.0002	2.45	$B \leftrightarrow X$	25979	R	(598)(548)(1276)	isotope eff. (1276)
									R	(1298)	(598)[d]
A	6296	326 H	1				$B \rightarrow A$	16706	R	(1362)	[f]
(X) (Σ_g^+)	0	391.77 H	1.06 +0.002	0.0907	0.00027[e]	2.15_7					
$\mathrm{SeO^{16}}$		$D_0^0 = (5.4)$ e.v.; $\mu_A = 13.304_7$									
A[d]	(33302)[b]	[518][c] H					$A \rightarrow X$	33109	R	(1416)(757)	repr. (1416)(757)
(X)[d]	0	907.1 H	4.61								
$\mathrm{Si_2}$?		$\mu_A = 14.03_4$									
A	19000	(1050)					$A \rightarrow X$	(19150)	V	(903)	[a]
(X)	0	(750)									
SiBr		$D_0^0 = (3.7)$ e.v.; $\mu_A = 20.774$									
		Fragments of two other band systems not involving A or X and possibly due to SiBr$^+$									
A $^2\Sigma$	33571.0	573.6[b] H	3.1				$A \rightarrow X$	33226.7	V	(758)	repr. (1230)
								33644.7	V	(385)	
X $^2\Pi$	{418.0[b]; 0}	425.4[b] H	1.5								
$\mathrm{Si^{(28)}Cl^{35}}$		$D_0^0 = (4.0)$ e.v.; $\mu_A = 15.5474$									
D	44941.5	[663] H					$D \rightarrow X$	45005.9	V	(384)	
C ($^2\Delta$)	{41178.4; 41165.4}	674.2 H	2.20				$C \rightarrow X$	41039.9	V	(384)	repr. (384)
								41234.8	V	(384)	repr. (384)
B ($^2\Sigma$)	34102.7	701.5[b] H	1.40[b]				$B \rightarrow X$	34186.0	V	(384)	
X $^2\Pi_r$	{207.9; 0}	535.4 H	2.20								
$\mathrm{Si^{(28)}F^{19}}$		$D_0^0 = (4.8)$ e.v.; $\mu_A = 11.3187$									
		Several unclassified bands and fragments of band systems									
C $^2\Sigma$	39513.6	891.7 H	6.2				$C \rightarrow X$	39369.9	V	(1105)(1569)	[e]
								39530.7	V	(72)	γ system

State	T_e	ω_e	$\omega_e x_e$	$\omega_e y_e$	B_e	α_e	D_e	r_e (Å)	Trans.	ν (head)	References
$B\ {}^1\Sigma$	34640.2	1011.2 H	4.8		[0.622]			[1.55]	$B\rightarrow X$	34556.6 V; 34717.4 V	$\{$ β system; repr. (1105) $\}$ (72)
$A\ {}^2\Pi_r$	22942.4[b]	710.9 H	8.6		[0.5743]			[1.611]	$A\rightarrow X$	22707.7 R; 22868.5 R	$\{$ α system; repr. (1105) $\}$ (210)
$X\ {}^2\Pi_r$	{160.8[c]; 0}	856.7 H	4.7		[0.5795]		[d]	[1.603]			(72)(210)

Si(28)H¹ $\mu_A = 0.973080$

State	T_e	ω_e	$\omega_e x_e$	$\omega_e y_e$	B_e	α_e	D_e	r_e (Å)	Trans.	ν (head)	References
$A\ {}^1\Delta$	(24269)[a]	(1870)			7.503	0.427		1.519_6	$A\rightarrow X$	24164 R	repr. (593)(1092)
$X\ {}^2\Pi_r$	0[b]	(2080)			7.49_6	0.213[c]		1.520			(593)(1092)

Si(28)H² $\mu_A = 1.879430$

State	T_e	ω_e	$\omega_e x_e$	$\omega_e y_e$	B_e	α_e	D_e	r_e (Å)	Trans.	ν (head)	References
$A\ {}^2\Delta$	[d]				[3.801]	[e]		[1.536]	$A\rightarrow X$	(24186) R	repr. (593)
$X\ {}^2\Pi_r$	0[d]				[3.842]			[1.528]		(24186)	(593)

Si(28)N¹⁴ $D_0^0 = (4.5)$ e.v.;[a] $\mu_A = 9.33526$

State	T_e	ω_e	$\omega_e x_e$	$\omega_e y_e$	B_e	α_e	D_e	r_e (Å)	Trans.	ν (head)	References
$B\ {}^2\Sigma^+$	24299.4_0	1031.00_7 Z	16.742_8	$+0.1172_2$*	0.7235	0.01037		1.580_0	$B\rightarrow X$	24236.5_3 R	$\{$ isotope eff.; repr. (372)(1256) $\}$ (372)
$X\ {}^2\Sigma^+$	0	1151.68_0 Z	6.560_0		0.7310	0.00567[b]		1.571_8			

Se₂:
[a] From the predissociation in the B state. If the products of predissociation are $^3P + {}^1D$ the upper limit would be reduced to 2.37 e.v. Gaydon from an extrapolation of the B state vibrational levels, assuming dissociation into $^3P + {}^1D$ obtains 2.8 ± 0.1 e.v.
[b] This system is ascribed by (1416) to a polyatomic selenium molecule.
[c] A different analysis with $\omega' = 600$ is proposed by (1236).
[d] Strong vibrational perturbations. Predissociation (breaking off) for $v' = 11, 12, 13, 14$ at lower K values. In addition Rosen (598) gives a second much stronger predissociation for $v' > 22$. Predissociation (breaking off) for $v' = 10$, $K' = 49$ corresponding to 28670 cm⁻¹ [see (548)]. Rosen (598) assumes predissociation for $v' > 22$.
[e] $D_v = 1.8 \times 10^{-8}$.
[f] While Olsson (548) did not find any indication of triplet structure in the B–X bands Bhatnagar, Lessheim and Khanna (804) have found selenium vapor to be paramagnetic indicating the ground state (or a low-lying state) to be $^3\Sigma$ similar to the ground states of O_2 and S_2.

SeO:
[a] Gaydon gives 3.5 ± 1.
[b] Shin-Piaw (1416) gives a second much weaker system with $\nu_e = 33375.9$ and similar vibrational constants. It seems probable that these heads are formed by satellite branches of the main system.
[c] A second vibrational quantum 525 has been observed; one of the two is probably perturbed.
[d] It is not entirely certain that this is the ground state.

Si₂:
[a] The identification of these bands as due to Si₂ is tentative. It is not certain that the lower state is the ground state. There is a possibility of a somewhat different assignment which would lead to slightly altered constants [see (385)].

SiBr:
[a] Gaydon gives 3 ± 0.5 e.v.

SiCl:
[a] Gaydon gives 3.3 ± 0.5 e.v.
[b] The identification of the 2–0 and 3–0 bands is doubtful and therefore $\omega_e x_e$ may have to be changed [see (384)].

SiF:
[a] Gaydon gives 3.8 ± 0.4 e.v.
[b] Very close to case (b).
[c] Coupling constant $A = 162.0$.

SiH:
[a] Yuasa's (1569) system II is the $B\rightarrow X$ system of PO.
[b] Coupling constant $A = 142$ cm⁻¹.
[c] Coupling constant $A = 3.4$ cm⁻¹.
[d] Coupling constants not determined, but presumably very close to those of SiH¹.
[c] $D_v = [3.89 + 0.03(v + \tfrac{1}{2})] \times 10^{-4}$.
[d] $D_0 = 1.02 \times 10^{-4}$.
[d] $D_0 = 1.12 \times 10^{-6}$.

SiN:
[a] From Gaydon.
[b] $D_v = [1.18_2 + 0.001(v + \tfrac{1}{2})] \times 10^{-6}$.

TABLE 39 (Continued)

State	T_e	ω_e	$\omega_e x_e$	$\omega_e y_e$	B_e	α_e	r_e $(10^{-8}$ cm)	Designation	ν_{00}	References	Remarks
									Observed Transitions		
Si²⁸O¹⁶	\multicolumn{7}{l}{$D_0^0 = (7.4)$ e.v.;^{a} $\mu_A = 10.18013$}					
A ¹Π	42835.3	851.51 Z	6.143	+0.0437	0.6303	0.00657	1.621	$A \leftrightarrow X$	42640.0 R	(615)(1405)	
X ¹Σ⁺	0	1242.03 Z	6.047	+0.00329	0.7263	0.00494^b	1.510₁			(1102)	
(Si²⁸O¹⁶)⁺	\multicolumn{7}{l}{$\mu_A = 10.18006$}					
A ²Σ	(26035)	(811)			0.7250	0.0140	1.511₄	$A \rightarrow X$	26015.05Z R	(1554)	repr. (1554)
X ²Σ	0	(851)			0.7320	0.0133^a	1.504₂				
Si²⁸S³²	\multicolumn{7}{l}{$D_0^0 = (6.6)$ e.v.;^{a} $\mu_A = 14.92589$}					
	\multicolumn{7}{l}{Incompletely analyzed bands between 16000 and 29000 cm⁻¹}					
E	41923.5^b	403.5₄ H^b	1.40	−0.0329*	0.2663₆	0.0020₃*	2.059₃	$E \leftarrow X$	41750.8 R	(80)(782)	repr. (1477)
D ¹Π	35028.8	512.0 H	2.38	−0.045	0.3036₃	0.00149^d	1.928₈	$D \rightarrow X$	34910.1 R	(1477)(782)	repr. (80)
X ¹Σ⁺	0	749.5 H^c	2.56^c							(80)(774)(782)	
Si²⁸Se	\multicolumn{7}{l}{$D_0^0 = (5.8)$ e.v.;^{a} $\mu_A = 20.664₆$}					
E	38505.9	308.8 H	1.95	−0.032				$E \leftarrow X$	38370.3 ^{b}	(1477)	repr. (79)
D ¹Π	32450.3	399.8 H	1.93					$D \rightarrow X$	32360.2 R	(1477)(79)	
X ¹Σ	0	580.0 H	1.78								
Si²⁸Te	\multicolumn{7}{l}{$D_0^0 = (5.5)$ e.v.;^{a} $\mu_A = 22.954₃$}					
E	33991	242 H	(3.63)	(+0.13)				$E \leftarrow X$	33871 ^{b}	(1477)	repr. (79)
D ¹Π	28661.8	338.6 H	1.70					$D \rightarrow X$	28590.4 R	(1477)(79)	
X ¹Σ⁺	0	481.2 H	1.30								
SmO¹⁶	\multicolumn{7}{l}{$\mu_A = 14.4622$}					
	\multicolumn{7}{l}{Unclassified bands in the region 14800–23300 cm⁻¹}				(1318a)	repr. (1318a)
SnBr	\multicolumn{7}{l}{$D_0^0 = (3.0)$ e.v.;^{a} $\mu_A = 47.774$}					
B ²Σ	[33218.2]	[163.6]						$B \rightarrow X$	{30627.5 33094.5} V V	(385)	repr. (385)
A ²Δ	{(27067) 26695.1}	169.1	6.8					$A \rightarrow X$	{24557.9 26654.3} R R	(385)	repr. (385)
X ²Π$_r$	{2467.0 0}	247.7	0.62								

SnCl³⁵ $D_0^0 = (3.6)$ e.v.;[a] $\mu_A = 27.0190$

Strong absorption continua at 29000–33000 and 40000–53000 cm⁻¹

State			Transition		V/R		notes
C ($^2\Pi_r$)	$\begin{cases}43724.0\\43656.1\end{cases}$	392.8 H — 1.40	C←X	$\begin{cases}41383.8\\43676.2\end{cases}$	V	(942) (942)	repr. (383); isotope eff.
B $^2\Sigma$	33582.6	432.5 H — 1.2	B←→X	$\begin{cases}31262.5\\33622.6\end{cases}$	V	(383) (942)	
A $^2\Delta$	$\begin{cases}28965.7\\28691.8\end{cases}$	301.0 H^Q — 4.15	A←→X	$\begin{cases}26579.1\\28665.3\end{cases}$	R	(217) (942)	isotope eff.
X $^2\Pi_r$	$\begin{cases}2360.1\\0\end{cases}$	352.5 H — 1.06					

SnF¹⁹ $D_0^0 = (3.9)$ e.v.;[a] $\mu_A = 16.3823$

Strong continua with maxima at 41000 and 53000 cm⁻¹

State			Transition		V/R		notes
D $^2\Delta_r$	$\begin{cases}46425^b\\46343^b\end{cases}$	$\begin{cases}[607]\ \text{H}\\ [598]\ \text{H}\end{cases}$	D←X	$\begin{cases}44118\\46351\end{cases}$	V	(375)	repr. (375); Sn isotope eff.
C $^2\Pi_i$	$\begin{cases}45498.9\\44161.6\end{cases}$	688.2 H — 4.65	C←X	$\begin{cases}45551.0\\41894.1\end{cases}$	V	(375)	
B $^2\Delta$	40834	[594] H	B←X	$\begin{cases}38521\\(40840)^c\end{cases}$	V	(375)	
A $^2\Sigma$	34108.4	676.7 H — 2.65	A←X	$\begin{cases}31835.7\\34155.3\end{cases}$	V	(375)	
E $^2\Sigma$	20132	418 H — 1	E→X	$\begin{cases}17730\\20050\end{cases}$	R	(1571)	repr. (1571)
X $\begin{cases}^2\Pi_{3/2}\\^2\Pi_{1/2}\end{cases}$	$\begin{cases}2317.3\\0\end{cases}$	$\begin{cases}587.6\ \text{H} — 2.65\\582.9\ \text{H} — 2.69\end{cases}$					

SiO: [a]Gaydon gives 8 ± 1 e.v.

  [b]$D_v = [0.993 + 0.00155(v + \tfrac{1}{2})] \times 10^{-6}$ (calc.).

SiO⁺: [a]Spin doubling constant $\gamma_0 = 0.002$ or 0.022; $D_v = 2.11 \times 10^{-6}$.

SiS: [a]Gaydon gives 6.4 ± 0.5 e.v.

  [b]These are the data of (782) who have shifted the numbering compared to (1477).

  [c]From band origins one obtains 749.65 and 2.575.

  [d]$D_v = [0.201 - 0.001(v + \tfrac{1}{2})] \times 10^{-6}$.

SiSe: [a]Gaydon gives 5.3 ± 0.5 e.v.

  [b]Shading not given but probably R.

SiTe: [a]Gaydon gives 4.5 ± 0.3 e.v.

  [b]Shading not given but probably R.

SnBr: [a]Gaydon gives 2.0 ± 1 e.v.

SnCl: [a]Gaydon gives 3.2 ± 0.5 e.v.

SnF: [a]Gaydon gives 3.3 ± 0.5 e.v.

  [b]Jenkins and Rochester (375) give 46427 and 46326 which fits poorly with the observed 0-0 bands.

  [c]The transition B←X ($^2\Pi_{3/2}$) is not observed because of overlapping by continuum.

TABLE 39 (*Continued*)

State	T_e	ω_e	$\omega_e x_e$	$\omega_e y_e$	B_e	α_e	r_e (10⁻⁸ cm)	Desig-nation	ν_{00}		References	Remarks
SnH¹				$D_0^0 < 3.2$ e.v.; $\mu_A = 0.999643$								at
$B\ ^2\Delta_i$	a						[1.853]	$B\to X$	$\begin{cases}22480\text{H}\\24660\text{H}\end{cases}$	R	(699a)	{breaking off at
$A\ ^2\Sigma$	(16557)	([1270])			[4.911]			$A\to X$	14220H	R	(699a) (1516)	{$K'=17, v=0$
$X\ ^2\Pi_r$	$\begin{cases}2182^b\\0\end{cases}$	(1580)			[5.331]\\[5.293]	^{e}	[1.779]\\[1.785]		16402H		(1070)	Fragment
SnO¹⁶				$D_0^0 = (5.6)$ e.v.;^{a} $\mu_A = 14.0999$								
	Fragment of another system [Mahanti's (1187) C system] near 24400 cm⁻¹ whose upper state may be D											
$E\ ^1\Sigma^+$	36295	508.0 H	2.9	*	0.3103₈	0.00475	1.962₈	$E\to X$	36138.0	R	(1179) (1406) (922a)	
D	29624.9	582.6 H	3.08	−0.135				$D\to X$	29505.1	R	(1187) (856) (478)	A system of (1187)^b
											(384a)	
$B\ ^1\Sigma^+$	25418.6	637.0 H	8.0		0.3540₀	0.00450^d	1.837₉	$B\to X\ ^c$	25322.8	R	(1187)	
											(384a)	
SnS				$D_0^0 \leq 3.0$ e.v.;^{a} $\mu_A = 25.253$							(1351)	
	Further unassigned bands											
E	33035.0	295.05 H	1.09					$E\leftarrow X$	32938.8	R	(1351) (1407)	repr. (1351)
D	28337.9	331.9 H	1.25	−0.012				$D\leftrightarrow X$	28260.0	R	(1351) (1351) (633)	repr. (1351)
$A\ (^1\Sigma^+)$	(23591.8)	[367.5] H			(0.140)		(2.18)	$A\leftarrow X$	23532.0	R	(633) (1351)	^{b} repr. (1351)
$X\ ^1\Sigma^+$	0	487.68 H	1.34		(0.157)		(2.06)					
SnSe				$D_0^0 = (4.6)$ e.v.; $\mu_A = 47.430$								
F	(47850)	(290) H						$F\leftarrow X$	(47830)	R	(1478)	repr. (1478)
E	30738.9^a	196.6^a H	0.77	−0.0016*				$E\leftrightarrow X$	30671.6	R	(1478) (1408)	repr. (1478)
											(783)	
D	27549.6	225.1 H	0.69					$D\leftrightarrow X$	27496.6	R	(688) (783) (1478)	(783)
											(1408)	
C	22750	(220)						$C\leftarrow X$	22695	R	(1481)	^{b}
B	21850	(220)						$B\leftarrow X$	21795	R	(1481)	^{b}
A	19353.3	223.8 H	0.88					$A\leftarrow X$	19299.3	R	(688) (783)	
$X\ ^1\Sigma^+$	0	331.2 H	0.736								(1478)	
SnTe				$D_0^0 = (4.2)$ e.v.;^{a} $\mu_A = 61.514$								
J	(47260)	(230) H	incompletely analyzed					$J\leftarrow X$	(47245)	R	(1478)	repr. (1478)
I	44033.5	229.7 H	1.25	−0.003				$I\leftarrow X$	44018.4	R	(1478)	repr. (1478)

State	T	ω_e	$\omega_e x_e$		Transition	ν		Ref.	Repr.
H	30818.3	201.0	0.6		$H \leftarrow X$	30789.0	b	(1409)	repr. (1478)
G	29071.8	200.8	0.3		$G \leftarrow X$	29042.5	b	(1409)	
F	28545.9	98.0	1.0		$F \leftarrow X$	28465.0	b	(1409)	
E	27642.8	135.0 H	2.5		$E \leftarrow X$	27580.2	R	(772)(1478)(1409)	
D	25444.3	179.1 H	0.40		$D \leftrightarrow X$	25404.1	R	(772)(785)	repr. (772)(785)
C	21418.6	218.1 H	0.98		$C \leftrightarrow X$	21397.8	R	(785)	repr. (785)
B	20394.9	230.3 H	1.53		$B \leftarrow X$	20380.0	R	(785)	repr. (785)
A	16844.0	178.5 H	0.44	-0.013	$A \leftarrow X$	16803.5	R	(785)	repr. (785)
$X\ ^1\Sigma$	0	259.5 H	0.50						

$S^{32}O^{16}$

$D_0^0 = 4.001$ or 5.146 e.v.;[a] $\mu_A = 10.66472$

State	T	ω_e	$\omega_e x_e$	B_e	α_e		Transition	ν		Ref.	Repr.
$B\ ^3\Sigma^-$	39356.3	628.7 H	[5.65]	0.5020	0.0062	1.775	$B \rightarrow X$	39108.9	R	(582)(1025)	repr. (582)(1025)[b]
$(X)\ ^3\Sigma^-$ c	0	1123.7_3 Z	6.116	0.7089_4	0.005562[d]	1.4933					

$D_0^0 = 1.94 \times 10^{-4}$.

$SrBr^{79}$

$D_0^0 = (2.8)$ e.v.; $\mu_A = 41.532$

State	T	ω_e	$\omega_e x_e$		Transition	ν		Ref.	Repr.
$D\ ^2\Sigma$	28958.2	247.8	0.55		$D \leftarrow X$	28973.8	V	(1020)	(1020)
$C\ ^2\Pi$	{ 24665.8 / 24343.7 }	205.2 H	0.49		$C \leftrightarrow X$	{ 24660.2 / 24338.1 }	R / R	(1295)(1020)	bands with 4 heads
$B\ ^2\Sigma$	15352.0	222.0 H	0.55		$B \leftarrow X$	15354.7	V	(285)(1020)	(285)(1020)
$A\ ^2\Pi$	{ 15000.7 / 14699.4 }	222.1 H	0.53		$A \leftrightarrow X$	{ 15003.5 / 14702.2 }	V / V	(285)(1020)	bands with 4 heads
$X\ ^2\Sigma^+$	0	216.5 H	0.51						

SnH: [a]Coupling constant $A = -1.75$.
[b]Coupling constant $A = 2182.7$.

SnO: [a]Gaydon gives 3.2 ± 1 e.v.
[b]Reproduction in (1187) and (856); isotope effect see (1191).
[c]The fact that this system has not been observed in absorption is not necessarily an argument [as assumed by (856)] against its lower state being the ground state.
[d]$D_e = [0.26 - 0.04(v + \frac{1}{2})] \times 10^{-6}$.

SnS: [a]Assuming that predissociation is genuine. Extrapolation gives 5.5 e.v.
[b]No bands with $v' > 1$, predissociation (?).

SnSe: [a]The constants given by (1408) are rather different due to appreciable differences in his assignments compared to (1478).

SnTe: [b]Walker, Straley and Smith (688) gave only one system in this region with $\nu_e = 22579.5$, $\omega_e' = 218.8$, $\omega_e' x_e' = 0.50$ [see (783)].
[a]Gaydon's value 1.7 e.v. is based on Barrow's (772) old data.
[b]Shading not given but probably R.

SO: [a]Depending on whether predissociation in B state occurs into $^1D + {}^3P$ or $^3P + {}^3P$. The writer favors the lower value, Gaydon the higher one.
[b]Strong perturbations for $v' = 1, 2, 3$. Predissociation (breaking off) for $v' = 0, 1, 2, 3$ above $K' = 66, 53, 39$ and 6 respectively. No bands with $v' > 3$ are observed.
Energy of $v' = 3$, $K = 6$ is 41520 cm⁻¹.
[c]It is not quite certain whether this is the ground state. Absorption experiments by (862) under conditions which suggest the existence of free SO have led to the discovery of a band system in the region 29000 − 40200 cm⁻¹ which is not identical with the $B-X$ system. Cordes (861) has interpreted this system as due to metastable S₂; but Schenk (1379) (1380) has refuted this interpretation which indeed seems most doubtful. It is probable that the absorption is due to S₂O₂ [see (1095) (1136a) (1113b)].
[d]The triplet splitting has been measured but not evaluated according to the formulae of Schlapp (617). $D_v = [1.13 + 0.00032(v + \frac{1}{2})] \times 10^{-6}$ (calc.).

TABLE 39 (Continued)

State	T_e	ω_e	$\omega_e x_e$	$\omega_e y_e$	B_e	α_e	r_e (10⁻⁸ cm)	Designation	ν_{00}	References	Remarks
SrCl³⁵		$D_0^0 = (3.0)$ e.v.;[a]					$\mu_A = 25.001_8$				
F ²Σ	34256.7	364.6 H	1.08					F←X	34287.8 V	(1020)	
E ²Σ	32201.8	346.3 H	1.10					E←X	32223.8 V	(1020)	
D ²Σ	28822.9	344.8 H	1.04					D←X	28844.1 V	(1020)	bands with 4 heads
C ²Π	25401.7 / 25245.5	279.9 H	0.68					C↔X	25390.6 R / 25234.4 R	(551)(499)	repr. (951)
B ²Σ	15719.5	306.4 H	0.98					B↔X	15721.5 V	(1020)(551)	repr. (951); bands with 4 heads
A ²Π	15112.6 / 14818.4	309.4 H / 302.3 H	0.98 / 0.95					A↔X	15116.1 V / 14821.9 V	(1020)(551)(499) / (1020)	
SrF¹⁹		$D_0^0 = (3.5)$ e.v.;[a]					$\mu_A = 15.6183$				
G (²Π)	34759.2	573.9 H	1.28					G←X	34796.3 V	(941)	
F ²Σ	32820.3	598.5 H	3.42					F←X	32869.2 V	(941)	
E ²Π	31614.8 / 31528.7	564.4 H	3.20					E←X	31646.7 V / 31560.6 V	(941)	
D ²Σ	28296.6	552.1 H	2.15					D←X	28322.6 V	(941)	
C ²Π	27041.3 / 26977.8	452.1 H	(0.27)					C↔X	27017.9 R / 26954.4 R	(391)(941)	
B ²Σ	17303.4[b]	488.9 H	1.86	−0.007				B↔X	17297.9[b] R	(391)(941)	repr. (281); isotope eff. (743)
A ²Π	15349.0 / 15069.7	505.9 H / 506.1 H	2.26 / 2.21	+0.002 / +0.004*				A↔X	15351.9 V / 15072.7 V	(391)(941)	repr. (281)
X ²Σ+	0	500.1 H	2.21								
SrH¹		$D_0^0 \leq 1.68$ e.v.;[a]					$\mu_A = 0.996668$				
C ²Σ+	26230	1347 H	23.5		4.008	0.132	2.054	C↔X	26298.5Z V	(499)(345a)(943)	perturbations[b]
D ²Σ+	20847.6	1014.1 Z	15.4		1.925	0.024	2.964	D↔X	20752.0 R	(499)	perturbations
E ²Π,	c				[3.8687] / [3.6388]		[2.0911] / [2.1561]	E←X	18960Z V / 18860 V	(943)	
B ²Σ+	d				3.8788	0.0930	2.0884	B→X	14352 V	(695)	repr. (694a)
A ²Π					[3.6787] / [3.6683]	0.0814[e]	[2.1444] / [2.1474]	A→X	13650 V / 13360 V	(694a)	large Λ doubling; repr. (694a)
X ²Σ+	0	1206.2 Z	17.0		3.6751		2.1455				
SrH²		$D_0^0 \leq 1.70$ e.v.;[f]					$\mu_A = 1.969457$				
C ²Σ+					[1.95]		[2.10]	C→X	26287 V	(695)	very strong (perturb.[g])

State	T_e	ω_e	$\omega_e x_e$	B_e	α_e	r_e	Transition	ν_{00}	V/R	Refs	perturbations
B $^2\Sigma^+$				1.9427	0.0349	2.0992	B→X	14344	V	(695)	
X $^2\Sigma^+$				1.8609	0.0292[A]	2.1448					
SrI^{127} $D_0^0 = (2.2)$ e.v.; $\mu_A = 51.849$ — Other unclassified bands at 14000–16500 and 29000–30000 cm⁻¹ given by (1511)											
B	23226.0	168.6	0.50				B→X	23223.4		(1225)(1511)	} headless bands
A	22666.0	170.5	0.42				A↔X	22664.3		(1225)(1511)	with two maxima
X ($^2\Sigma$)	0	173.9	0.42								
SrO^{16} $D_0^0 = (4.5)$ e.v.; $\mu_A = 13.5302$ — Unclassified band system near 16400–16800											
D	C + 28622.2	497.8 H	5.97				D→C	28532.3	(V)	(951)	repr. (479)
C	24702.8	679.1 H	9.13						R	(475)	
B		519.1 H	3.50	0.3047	0.0011	2.022	B→X	24635.7	R	(475)	
A $^1\Sigma$	10885.8	624 H	(2)	0.3378	0.0020[b]	1.921	A→X	10871.5_3	R	(745a)	
(X)$^1\Sigma$	0	653.5 H	4.0								
SrS $D_0^0 \leq 2.7$ e.v.; $\mu_A = 23.482$											
B — Continuous absorption from 28030 cm⁻¹ to shorter wave lengths							B←X			(1198)	
A — Continuous absorption from 22200 cm⁻¹ to shorter wave lengths							A←X			(1198)	
X	0										
TaO^{16} $\mu_A = 14.7000$ — Unclassified emission bands (shaded to the red) in the regions 9900–10900 and 17800–28200 cm⁻¹ interpreted as recombination spectrum (see p. 402)										(1127a)	
Te$_2$ $D_0^0 \leq 3.18$ e.v.; $\mu_A = 63.823$ — Emission continuum with maximum at 19200 cm⁻¹ interpreted as recombination spectrum (see p. 402). Fragment of an absorption band system in the region 49600–51400 and unclassified absorption bands in the region 41200–45500 cm⁻¹										(596)	
F	50714[b]	347 H	3				F←X	50761	V	(1238)(1416)	
										(1238)(1416)	

ᵃThere are two heads in each band on account of spin doubling and high K values of the heads. The T_e and ν_{00} values refer to the heads closest to the origin. But the distance to the origin is large [about 37 cm⁻¹ for the 0–0 band, see (281)].

ᵃBreaking off (predissociation) in C state assuming $^3P_0 + {}^2S$ as dissociation products. At high pressure (345a) and in absorption (499) higher levels are observed. ᵇSpin doubling constant $\gamma_0 = 0.122$; $D_v = [1.376 - 0.055(v + \frac{1}{2})] \times 10^{-4}$. ᶜCoupling constant $A = 299$.

ᵃFrom predissociation in C state assuming $^3P_0 + {}^2S$ as dissociation products. ᵇ$D_v = [0.34 + 0.024(v + \frac{1}{2})] \times 10^{-6}$.

SrCl: ᵃGaydon gives 2.5 ± 1 e.v.

SrF: ᵃGaydon gives 2.7 ± 1 e.v.

SrH: ᵃBreaking off (predissociation) in emission at $K' = 19$, $v' = 0$ at low pressure (943). ᵇCoupling constant $A = 117$ [see (943)]. ꟲFrom predissociation in C state assuming $^3P_0 + {}^2S$ as dissociation products [agrees with the value obtained from D_0^0(SrH)]. ᵈBreaking off in emission at $K' = 29$.

SrO: ᵃFrom Gaydon, thermochemical. Note that the $^1\Sigma$ ground state cannot dissociate into normal atoms $^1S + {}^3P$. ᵇSpin coupling constant $\gamma_0 = +0.0613$; $D_v = [0.339 + 0.017(v + \frac{1}{2})] \times 10^{-4}$.

SrS: ᵃFrom longward limit of continuum; Gaydon estimates 2.3 e.v.

Te₂: ᵃGaydon gives 2.3 ± 0.2 from dissociation limit of B state and assumption of ^3P state and assumption of $^3P + {}^1D$ as dissociation products. Therefore considerable changes in T_e are possible. ᵇThe numbering in this system does not appear to be certain.

TABLE 39 (Continued)

State	T_e	ω_e	$\omega_e x_e$	$\omega_e y_e$	B_e	α_e	r_e $(10^{-8}$ cm)	Observed Transitions		References	Remarks
								Designation	ν_{00}		
Te₂ (Continued)											
E	46278	381 H	1.0					$E' \leftarrow X$	46343 V	(1238) (1416)	
B	22714	164ᶜ H						$B \leftrightarrow X$	22670 R	(1355) (544) (548)	ᵈ
										(1237)	
										(888)	
		Fluctuation bands with estimated ν_{00} = 16370 and 18900									
X	0	251ᶜ H	0.55				[2.59]ᵉ				
TeO¹⁶											
	D_0^0 = 2.728 or 3.453 e.v.;ᵃ μ_A = 14.2169										
	Fragment of another absorption band system in the region 32700–35700 cm⁻¹ *ᵇ										
A	29504.3	370.1 H	5.10					$A \leftrightarrow X$	29291.₀ R	(1416)	ᶜ
X	0	796.0 H	3.50							(1416)	
Ti⁴⁸Cl³⁵											
	D_0^0 = (1.0) e.v.;ᵃ μ_A = 20.2278										
	Additional unclassified bands between 24600 and 27000 cm⁻¹										
B	23820.0	503.4 H	2.5					$B \rightarrow A$	23844.5 V	(500)	
Aᵇ	0	456.4 H	6.3								
Ti⁴⁸O¹⁶											
	D_0^0 = (6.9) e.v.;ᵃ μ_A = 11.9979										
c ¹Φ	⎰19565.2ᶜ				[0.5212]		[1.642]	$c \leftrightarrow a$	17840.60Z R	(1316a) (1180)	repr. (1316a)ᵇ
b ¹Π	⎱19499.4				[0.5120]		[1.657]	$b \leftrightarrow a$	11272.7₇Z R	(1316a)	repr. (1316a)ᵇ
a ¹Δ	19434.6				[0.5362]		[1.619]				
C ³Πᵣ		837.86 H	4.54₆		0.4889	0.0029	1.695	$C \rightarrow X$	⎰19338.6 R ⎱19347.4 R / 19349.3 R	(155)	repr. (155)
B ³Σ	14242.6	866.3 H	3.83		[(0.505)]		[(1.67)]	$B \leftrightarrow X$	⎰14030.4 R / 14105.0 R ⎱14171.7 R	(848) (1180)	repr. (848)
(X) ³Πᵣ,ᵈ	⎰141.3ᵉ H / 66.7 H ⎱0	1008.4 H	4.61		0.5355	0.0031ᶠ	1.620				
Tl₂ ?		μ_A = 102.223									
	Several emission continua and diffuse emission bands in the region 15000–36500 cm⁻¹									(1018a)	
TlBr⁸¹											
	$D_0^0 \leq$ 3.19 e.v.; μ_A = 57.979										
	Several absorption continua [72] H										
D	54550							$D \rightarrow A$	25340.2 R	(1073)	repr. (1073)

TlBr (continued)

State	T_e	ω_e	$\omega_e x_e$	$\omega_e y_e$	ν_{00}	System	r_e (Å)	Refs	Remarks
$A\ ^3\Pi_{0^+}$[b]	29191.5	108.32 H	5.15	−0.22	29148.4 R	A↔X	[2.68][d]	(146)(1073)	repr. (146)(1073)[c]
$X\ ^1\Sigma^+$	0	192.11 H	0.39						

$D_0^0 = 3.75$ e.v. from thermochemical data (50a)(34); $\mu_A = 29.869$

TlCl³⁵

Unclassified emission bands 23300–25700 cm⁻¹
Continuum near 39700 cm⁻¹ with preceding diffuse bands
Narrow continuum at 32145 cm⁻¹ with preceding diffuse bands

State	T_e	ω_e	$\omega_e x_e$		ν_{00}	System	r_e (Å)	Refs	Remarks
$C\ ^3\Pi$						C←X		(1232)	repr. (1072)
$B\ ^3\Pi_1$	31054.2	216.91[c]	6.80			B←X		(1072)	repr. (1072)
$A\ ^3\Pi_{0^+}$[b]					31017.5 R	A↔X	[2.55][d]	(1072)(1232)	repr. (1072)(1232)
$X\ ^1\Sigma^+$	0	287.47[c]	1.24						

$D_0^0 < 4.72$ e.v.; $\mu_A = 17.3882$

TlF¹⁹

Several continua

State	T_e	ω_e	$\omega_e x_e$		ν_{00}	System	Refs	Remarks
$C\ ^1\Pi_1$	(45545)	[346] H			45481 h	C←X	(339)	only $v = 0$ & 1 appear
$B\ ^3\Pi_1$[b]	36869.5	360.65 H^Q	12.25		36809.7 R	B←X	(339)	repr. (339)
$A\ ^3\Pi_0$[b]	35180.7	439.87[c] Z	8.60		35161.5 V$_R$	A↔X	(339)	repr. (339)
$X\ ^1\Sigma^+$	0	475.00 H^Q	1.89					

Te₂: [c]These are substantially Rosen's (1355) original constants. The modification suggested by (544)(548) has been shown to be doubtful by (1237) [see also (1239)].
[d]Predissociation above $v' = 20$ [in Migeotte's (1237) numbering] i.e. at 25650 cm⁻¹, found by (1056). For induced predissociation below this limit see (548)(1296) and (409)
An accidental predissociation of the vibrational type was observed by (1239) for $v' = 14$ and 16 [see also (1357)].

TeO: [a]From electron diffraction experiments by (484).

TeCl: [a]Depending on whether dissociation products of A state are $^1D + ^3P_2$ or $^3P_0 + ^3P_2$.

TiCl: [a]Convergence at 32570 cm⁻¹.
[b]Predissociation (breaking off in emission, diffuseness in absorption) above $v' = 6$ i.e. limit between 31300 and 31600 cm⁻¹.
[c]Gaydon gives ?0.8 ± 0.5 e.v.

TiO: [a]Gaydon gives 5.5 ± 1 e.v.
[b]Doubtful whether this is the ground state.
[a]This system was considered to be a triplet system involving the $X\ ^3\Pi$ state by (898). But each band has only one head.
[b]Coupling constant $A = 88$ [from (139)].
[c]It is not certain whether $(X)\ ^3\Pi$ or $a\ ^1\Delta$ is the ground state. Transitions from both states appear in absorption in low temperature stars.
[d]Coupling constant $A = 100$ [from (139)].
[e]$/D_e = [0.60 + 0.02(v + \tfrac{1}{2})] \times 10^{-6}$.

TlBr: [a]From excitation of atomic fluorescence (836); thermochemical data give values between 3.17 and 3.34 (Gaydon).
[b]Howell and Coulson (1073) assume $^3\Pi_1$ (see TlCl).
[c]Isotope effect. Bands are diffuse above $v' = 3$.
[d]From electron diffraction (999).

TlCl: [a]From long wave limit of continuum one obtains $D \leq 3.97$ e.v.
[b]Howell and Coulson (1072) assume $^3\Pi_1$ but according to (1232) Q branches are definitely absent.
[c]From observed zero gaps; Miescher (1232) gives somewhat different values, but does not give ν_e.
[d]From electron diffraction (999).

TlF: [a]From longward edge of continuum at 38050 cm⁻¹. Linear extrapolation of ground state gives 3.7 e.v.
[b]Howell (339) exchanges $^3\Pi_1$ and $^3\Pi_0$ but the B←X system has strong Q branches and must therefore have $\Delta\Omega = \pm 1$.
[c]From observed zero gaps.

TABLE 39 (Continued)

State	T_e	ω_e	$\omega_e x_e$	$\omega_e y_e$	B_e	α_e	r_e (10⁻⁸ cm)	Designation	ν_{00}	References	Remarks
TlH[1]				$D_0^0 \leq 2.18$ e.v. from predissociation; $\mu_A = 1.003184$							
E 0+ a		[1244.6] Z			3.126	0.289	2.319	E→X	19146.3 R	(273)	repr. (273); diffuse lines due to pre-dissociation; observed in thermal emission b
D 0+ a		[1210.8] Z			3.412	0.421	2.219	D→X	18753.9 R	(273)	
C 0+ a		[1269.5] Z			4.446	1.060	1.944	C→X	18279.0 R	(273)	
B 0+ a					[4.617]		[1.908]	B→X	17519.9 Z	(273)	
(X) 1Σ+	0	1390.7	22.7		4.806	0.154 c	1.870				
TlI[127]				$D_0^0 \leq 2.64$ e.v.; $\mu_A = 78.312$							
A ³Π₀+		(30)	Continuous absorption with maximum near 33050 cm⁻¹				[2.87]b	A↔X	(26250)	(146)	
X ¹Σ+		150 c	Broad fluctuations in absorption							(836) (146)	
V(51)O16				$D_0^0 = (6.4)$ e.v.;a $\mu_A = 12.1768$							
A (²Δ)	17501.3	863.5 H	5.4		0.3349 b	0.0027	2.033	A→X	17426.6 R	(477)	repr. (477)
(X ²Δ)c	0	1012.7 H	4.9		0.3876 b	0.0024 d	1.890				
WO16 ?				$\mu_A = 14.7198$							
			Unclassified R shaded emission bands in the region 18500–24400 cm⁻¹							(47)	
YbCl				$D_0^0 = (1.2)$ e.v.; $\mu_A = 29.435$							
			Fragment of a band system between 19600 and 20300 cm⁻¹ with similar constants								
B (²Π)	{19369.3a, 17881.7b}	314.5₀	1.25	(+0.0072)				B→A	{19379.7 V, 17892.1 V}	(952) (952)	repr. (952)
A (²Σ)	0	293.6₁	1.23								
YbO				$\mu_A = 14.646$							
			Unclassified red shaded bands between 18200 and 21200 cm⁻¹							(952)	
Y89O16				$D_0^0 = (9)$ e.v.;a $\mu_A = 13.5606$							
B ²Σ	20799.6	765.03 H	7.75					B→X	20754.5 R	(389) (23) (1221)	repr. (1317) (951)
A ²Π	{16760.8, 16324.2}	{808.9 H, 812.7 H}	{2.96, 2.80}					A→X	{16738.9 R, 16304.2 R}	(389) (23) (1221)	repr. (1317) (951) (1221)
X ²Σ	0	852.5 H	2.45								
Zn₂				$D_0^0 = (0.25)$ e.v.; $\mu_A = 32.699$							
			Large number of continua and diffuse bands; see Finkelnburg (31)							(716)	

ZnBr $\mu_A = 35.970$

Unclassified emission bands in the region 11800–30000 cm⁻¹ originally ascribed to ZnBr₂

State	ν_{00}	ω_e	Transition	ν	R	Ref.	Remarks
(C ²Π)	{32538 / 32130}	(250)	C←X	{32553 / 32145}	R / R	(708) ; (1510)(1068)	{ not certain that these bands are due to ZnBr
(X ²Σ)	0	(220)					

ZnCl³⁵ $\mu_A = 22.790$ $D_0^0 = (3.0)$ e.v.;[a]

State	ν_{00}	ω_e	$\omega_e x_e$	Transition	ν	R	Ref.	Remarks
E ²Σ	33978.1	382.9 H	1.05	E→X	48163.0	R	(164)	{ only $\Delta v = 0$ observed
C ²Π	33593.2	390.5 H	1.55	C↔X	{33974.4 / 33589.5}	R / R	(164)(1510)	
X ²Σ	0							

ZnCs¹³³ ? $\mu_A = 43.835$

Diffuse V shaded absorption bands at 19363 and 19503 cm⁻¹   (769b)

ZnF¹⁹ $\mu_A = 14.725$

State	ν_{00}	ω_e		Transition	ν	R	Ref.	Remarks
D ²Σ	{37359}			D←X	{38633}	R	(1353)	{ diffuse bands,
C ²Π	{36989}	([599]l)		C←X	{37344 / 36974}	R / R	(1353)(1353)	pred.
X ²Σ	0	(630) H	(3.5)					

ZnH¹ $\mu_A = 0.992826$ $D_0^0 = 0.851$ e.v.;[a]

State	ν_{00}	ω_e	$\omega_e x_e$	B_e	α_e	r_e	Transition	ν	R	Ref.	Remarks
B ²Σ⁺	27587.7	1020.7	16.5[b]	(3.38)[c]	(0.05)*	(2.24)	B→X	27303.9	R	(652)	{ Zn isotope eff. incl. "nuclear" isotope shift; see (1250)
A ²Π_r	23269.6[d]	1910.2	40.69	7.4397[e]	0.2422	1.5108_3	A→X	23424.5	V	(652)	
X ²Σ	0	1607.6	55.14	6.6794	+0.398[f]	1.5945_0				(652)	

TlH: [a]It seems quite possible that all these levels arise from two mutually perturbing electronic states with Ω = 0. [b]Diffuseness increases with J indicating ΔΩ = ±1 for the predissociation (see p. 418).   [c]+0.0044(v + ½)²; $D_0 = 3.25 \times 10^{-4}$ D_v non-linear.

TlI: [a]From excitation of atomic fluorescence (836). Thermochemical values of 2.52 and 2.91 e.v. have been suggested by Butkow (146) and Gaydon. [b]From electron diffraction (999). [a]Recently Rao and Rao (1339a) have derived $\omega_e'' = 121.2$ and $\omega_e'' x_e'' = 0.09$ from two emission band systems of TlI. However their analysis appears somewhat doubtful. [b]The agreement of combination differences from which the B values have been derived is poor.

VO: [a]Gaydon gives 5.5 e.v.

YbCl: [a]It is doubtful whether this is the ground state. The classification of the observed transition as ²Δ – ²Δ is also doubtful.   [d]$D_v = 0.23 \times 10^{-6}$ [c]It is possible that the lower rather than the upper state has the doublet splitting or that both are split.

YO: [a]Gaydon gives 2.5 ± 1 e.v. [b]Gatterer, Piccardi and Vincenzi (952) give 17800.86 which does not fit their data.

ZnCl: [a]Gaydon gives 0.845 ± 0.02.

ZnH: [a]Coupling constant A = 341.20 for v = 0, very slightly lower for higher v. Fujioka and Tanaka (236) give $A_0 = 342.35$. [b]Many vibrational perturbations. [c]Large irregularities; $B_0 = 3.2877$. [d]Gaydon gives 0.845 ± 0.02. [e]Stenvinkel (652) gives $B_e = 7.4361$ and $\alpha = 0.236$ yielding negative o–c for all his B_v. [f]$\omega_e z_e = -0.4339$, very rapid convergence. $\gamma_0 = 0.254$. $D_v = [4.50 + 0.34(v + \tfrac{1}{2}) + \cdots] \times 10^{-4}$.

$\nu = [-0.03765(v + \tfrac{1}{2})^2 + 0.00897(v + \tfrac{1}{2})^3 - 0.001479(v + \tfrac{1}{2})^4]$. Spin doubling constant for v = 0: $\gamma_0 = 0.254$.

TABLE 39 (*Continued*)

State	T_e	ω_e	$\omega_e x_e$	$\omega_e y_e$	B_e	α_e	r_e (10^{-8} cm)	Observed Transitions		References	Remarks
								Designation	ν_{00}		
ZnH²		$\mu_A = 1.954512$									
A $^2\Pi_r$	h				[3.7342]		[1.5199]	$A \rightarrow X$	$\begin{cases}(23549) & V \\ (23225) & V\end{cases}$	(236)	
X $^2\Sigma$	0				[3.3497]^i		[1.6048]				
(ZnH¹)⁺		$D_0^0 = (2.5)$ e.v.;^{a} $\mu_A = 0.992826$									
A $^1\Sigma^+$	46700	1365^b Z	15		5.768	0.105	1.716	$A \rightarrow X$	46431 R	(84)	
X $^1\Sigma^+$	0	1916^b Z	39	−0.2	7.403	0.236^e	1.514$_6$				
(ZnH²)⁺		$D_0^0 = (2.5)$ e.v.; $\mu_A = 1.954511$									
A $^1\Sigma^+$	46693.9	974.4 Z	7.60		2.931^c	0.043	1.716	$A \rightarrow X$	46501.7 R	(242)	
X $^1\Sigma^+$	0	1364.8 Z	19.8		3.767^c	0.107^d	1.513				
Zn⁶⁴I¹²⁷		$D_0^0 = (2.0)$ e.v.;^{a} $\mu_A = 42.528$									
		Other bands between 16700 and 19200 cm^{-1} not yet analyzed									
E	44114	142.0	3.0					$E \rightarrow X$	44073 R	(708) (1534)	isotope effect
F	(39911)	80.0	1.25					$F \rightarrow X$	(39839)	(1329)	^{c}
										(1329a)	
D $^2\Pi_{3/2}$	30506.3	211.7	2.5					$D \rightarrow X$	30500	(1339)	diffuse bands
C $^2\Pi_{1/2}$	30117.6	248.2	0.70					$C \rightarrow X$	30130.0 V	(708)	
$(X$ $^2\Sigma)^b$	0	223.4	0.75								
ZnK ?		$\mu_A = 24.473$									
		V shaded diffuse absorption band at 24107 cm^{-1}								(769b)	
ZnRb ?		$\mu_A = 37.056$									
		Unclassified absorption bands in the region 22500–24000 cm^{-1}								(769b)	
ZnS		$D_0^0 \leq 4.4$ e.v.; $\mu_A = 21.520$									
		Two absorption continua with long wave length limits at 35700 and 46500 cm^{-1}								(1401a)	
ZnTe		$D_0^0 \leq 2.2$ e.v.; $\mu_A = 43.243$									
		Two absorption continua with long wave length limits at 17830 and 31480 cm^{-1}								(1198a)	

Zr⁹⁰O¹⁶ $D_0^0 = (7.8)$ e.v.;[a] $\mu_A = 13.5836$

State	T	ω_e	$\omega_e x_e$	B_e	α_e	r_e	Transition	ν_0	R	Ref.	[b]
$b\ (^1\Sigma)$	$a + 27217.9$	843.7_5	3.15	+0.072			$b \leftrightarrow a$	27151.2	R	(742a)(470)	(742a)
$a\ (^1\Sigma)$	a	978.0_7	5.04								
Fragments of two other triplet band systems near 28700 and 33800 cm⁻¹											
$C\ (^3\Pi)$	22206.5 / 21911.5 / 21698.1	820.6 H	3.31	0.533_3[c]	0.012_8	(1.526)	$C \rightarrow X$	21543.4 / 21556.3 / 21640.1	R	(470)(471)	α bands; (470)(471) repr.
B	18134.3 / 18103.2 / 18053.0	845.4 H	3.63				$B \rightarrow X$	17483.5 / 17760.3 / 18007.3	R	(470)	β bands; (470) repr.
$A\ (^3\Sigma)$	16088.9	853.9 H	3.14	0.565_8[c]	0.0077	(1.481)	$A \rightarrow X$	15442.5 / 15750.4 / 16047.6	R	(470)	γ bands; (470) repr.
$(X)\ ^3\Pi$	605.1 / 297.2 / 0	936.6 H	3.45	0.618_7[c]	0.0070[d]	(1.416)					

[a] The zero lines have not been determined very accurately.
[b] Gabel and Zumstein (242) give $B_v'' = 3.818 - 0.13(v + \tfrac{1}{2})$ and $B_v' = 2.944 - 0.46(v + \tfrac{1}{2})$ which do not fit their observed B_v.
[c] $D_e = 1.0 \times 10^{-4}$.
[d] $D_e = 4.8 \times 10^{-4}$.

ZnH: [h] $A_0 = 341.04$.
[i] Spin doubling constant $\gamma_0 = 0.131$ cm⁻¹; $D_0 = 1.24 \times 10^{-4}$.
(ZnH)⁺: [a] Gaydon gives 2.8 e.v.

ZnI: [a] Gaydon gives 1.8 ± 0.6 e.v.
[b] It is doubtful whether this is the ground state.
[c] Only bands with $v'' \geq 15$ and $v' \leq 2$; analysis appears doubtful.

ZrO: [a] Gaydon gives 6.5 ± 1.5 e.v.
[b] The fact that these bands have been observed in absorption in low temperature stars (1026a) indicates that possibly their lower state rather than (X) ³Π is the ground state [even though Lowater (471) and Tanaka and Horie (1456) agree for the ground state] since the r_e values derived from them are much smaller than expected for as heavy a molecule as ZrO. They are even smaller than those of TiO.
[c] These rotational constants appear somewhat doubtful.
The individual α for $^3\Pi_{0,1,2}$ vary from 0.0045 to 0.0107. Tanaka and Horie (1456) give 0.0038.

BIBLIOGRAPHY

I. HANDBOOKS, MONOGRAPHS, TEXTBOOKS, TABLES

1. H. Kayser and H. Konen, Handbuch der Spektroskopie (Hirzel, Leipzig, 1905–33).
2. E. C. C. Baly, Spectroscopy (Longmans, Green and Co., London, 1924–27).
3. W. R. Brode, Chemical Spectroscopy, 2nd edition (Wiley, New York, 1943).
4a. K. W. Meissner, Spektroskopie, Sammlung Göschen Nr. 1091 (de Gruyter, Berlin, 1935).
4b. R. A. Sawyer, Experimental Spectroscopy (Prentice-Hall, New York, 1944).
4c. G. R. Harrison, R. C. Lord, J. R. Loofbourow, Practical Spectroscopy (Prentice-Hall, New York, 1948).
5. A. Eucken and K. L. Wolf, Hand- und Jahrbuch der chemischen Physik (Akad. Verlagsges., Leipzig, 1932–), vol. 9: Die Spektren.
   (a) Part I: H. Kuhn, Atomspektren (1934).
   (b) Part II: W. Finkelnburg, R. Mecke, O. Reinkober, and E. Teller, Molekül- und Kristallgitterspektren (1934).
   (c) Part III: W. Hanle, Anregung von Spektren (1936).
   (d) Part IV: G. Scheibe and W. Frömel, Molekülspektren von Lösungen und Flüssigkeiten (1936).
   (e) Part V: K. Phillipp, Kernspektren (1937).
   (f) Part VI: K. W. F. Kohlrausch, Ramanspektren (1943).
6a. R. W. Wood, Physical Optics (Macmillan, New York, 1934).
6b. F. A. Jenkins and H. E. White, Physical Optics (McGraw-Hill, New York, 1937).
7. M. Born, Optik (Springer, Berlin, 1933).
8. A. E. Ruark and H. C. Urey, Atoms, Molecules and Quanta (McGraw-Hill, New York, 1933).
9a. L. Pauling and S. Goudsmit, The Structure of Line Spectra (McGraw-Hill, New York, 1930).
9b. H. E. White, Introduction to Atomic Spectra (McGraw-Hill, New York, 1934).
9c. G. Herzberg, Atomic Spectra and Atomic Structure, 2nd edition (Dover Publications, New York, 1944).
9d. A. C. Candler, Atomic Spectra and the Vector Model (Cambridge University Press, 1937).
10. E. U. Condon and G. H. Shortley, The Theory of Atomic Spectra (Cambridge University Press, 1935).
11a. E. U. Condon and P. M. Morse, Quantum Mechanics (McGraw-Hill, New York, 1929).
11b. A. Sommerfeld, Wave Mechanics (Methuen, London, 1930).
12a. W. Heisenberg, Physical Principles of the Quantum Theory (Chicago University Press, 1930).
12b. P. A. M. Dirac, Principles of Quantum Mechanics, 2nd edition (Oxford University Press, 1947).
13. L. Pauling and E. B. Wilson, Introduction to Quantum Mechanics (McGraw-Hill, New York, 1935).
14. E. C. Kemble, Fundamental Principles of Quantum Mechanics (McGraw-Hill, New York, 1937).
15a. S. Dushman, Elements of Quantum Mechanics (Wiley, New York, 1938).
15b. V. Rojansky, Introductory Quantum Mechanics (Prentice-Hall, New York, 1938).

16. W. Heitler, Elementary Wave Mechanics (Oxford University Press, 1945).
17a. E. P. Wigner, Gruppentheorie und ihre Anwendung auf die Quanten- mechanik der Atomspektren (Vieweg, Braunschweig, 1931).
17b. B. L. van der Waerden, Die gruppentheoretische Methode in der Quantummechanik (Springer, Berlin, 1932).
18a. F. Rasetti, Nuclear Physics (Prentice-Hall, New York, 1936).
18b. J. M. Cork, Radioactivity and Nuclear Physics (Van Nostrand, New York, 1947).
19. E. C. Kemble, R. T. Birge, W. F. Colby, F. W. Loomis, and L. Page, Molecular Spectra in Gases, National Research Council Bulletin 57 (Washington, 1930).
20a. R. Mecke, Bandenspektra und ihre Bedeutung für die Chemie (Born- traeger, Berlin, 1929).
20b. R. Ruedy, Bandenspektren auf experimenteller Grundlage (Vieweg, Braunschweig, 1930).
21a. R. de L. Kronig, Band Spectra and Molecular Structure (Cambridge University Press, 1930).
21b. R. de L. Kronig, The Optical Basis for the Theory of Valency (Cambridge University Press, 1935).
22. W. Weizel, Bandenspektren (Akad. Verlagsges., Leipzig, 1931).
23. W. Jevons, Band Spectra of Diatomic Molecules (Physical Society, London, 1932).
24. H. Sponer, Molekülspektren und ihre Anwendung auf chemische Prob- leme, I. Tabellen, II. Text (Springer, Berlin, 1935–36).
25. P. Pringsheim, Fluoreszenz und Phosphoreszenz (Springer, Berlin, 1928).
26a. J. Lecomte, Le Rayonnement Infrarouge (Gauthier-Villars, Paris, 1948).
26b. F. J. G. Rawlins and A. M. Taylor, Infrared Analysis of Molecular Structure (Cambridge University Press, 1929).
26c. C. Schaefer and F. Matossi, Das ultrarote Spektrum (Springer, Berlin, 1930).
27a. G. B. B. M. Sutherland, Infrared and Raman Spectra (Methuen, London, 1935).
27b. S. Bhagavantam, Scattering of Light and the Raman Effect (Andhra University, 1940).
28. G. Herzberg, Molecular Spectra and Molecular Structure II. Infrared and Raman Spectra of Polyatomic Molecules (Van Nostrand, New York, 1945).
29. J. P. Mathieu, Spectres de Vibration et Symétrie des Molecules et des Cristaux (Hermann, Paris, 1945).
30. K. W. F. Kohlrausch, Der Smekal-Raman-Effekt (Springer, Berlin, 1931). Supplement (Springer, Berlin, 1938).
31. W. Finkelnburg, Kontinuierliche Spektren (Springer, Berlin, 1938).
32. O. W. Richardson, Molecular Hydrogen and Its Spectrum (Yale Uni- versity Press, 1934).
33. A. Farkas, Orthohydrogen, Parahydrogen and Heavy Hydrogen (Cam- bridge University Press, 1935).
34. A. G. Gaydon, Dissociation Energies and Spectra of Diatomic Molecules (Chapman and Hall, London, 1947).
35. J. H. Van Vleck, The Theory of Electric and Magnetic Susceptibilities (Oxford University Press, 1932).
36a. A. E. van Arkel and J. H. de Boer, Chemische Bindung als elektro- statische Erscheinung (Akad. Verlagsges., Leipzig, 1931).
36b. H. Hellmann, Einführung in die Quantenchemie (Deuticke, Leipzig, 1937).
37. L. Pauling, The Nature of the Chemical Bond, 2nd edition (Cornell University Press, 1941).
38a. K. F. Bonhoeffer and P. Harteck, Grundlagen der Photochemie (Stein- kopff, Dresden, 1933).
38b. G. K. Rollefson and M. Burton, Photochemistry (Prentice-Hall, New York, 1939).

38c. W. A. Noyes, Jr. and P. A. Leighton, Photochemistry of Gases (Reinhold, New York, 1941).

39. J. E. Mayer and M. G. Mayer, Statistical Mechanics (Wiley, New York, 1940).

40a. L. S. Kassel, The Kinetics of Homogeneous Gas Reactions (Chemical Catalog Co., New York, 1932).

40b. S. Glasstone, K. J. Laidler and H. Eyring, The Theory of Rate Processes (McGraw-Hill, New York, 1941).

41. H. J. Schumacher, Chemische Gasreaktionen (Steinkopff, Dresden, 1938).

42a. B. Lewis and G. von Elbe, Combustions, Flames and Explosions of Gases (Cambridge University Press, 1938).

42b. A. G. Gaydon, Spectroscopy and Combustion Theory (Chapman and Hall, London, 1942).

43. W. W. Morgan, P. C. Keenan and E. Kellman, An Atlas of Stellar Spectra (University of Chicago Press, 1943).

44. P. W. Merrill, Spectra of Long-Period Variables (University of Chicago Press, 1940).

45. H. Kayser, Tabelle der Schwingungszahlen (Hirzel, Leipzig, 1925).

46. G. R. Harrison, M. I. T. Wavelength Tables (Wiley, New York, 1939).

47. R. W. B. Pearse and A. G. Gaydon, The Identification of Molecular Spectra (Wiley, New York, 1941).

48a. R. F. Bacher and S. Goudsmit, Atomic Energy States (McGraw-Hill, New York, 1932).

48b. C. E. Moore, A Multiplet Table of Astrophysical Interest (Princeton University Observatory, 1945).

48c. C. E. Moore, Atomic Energy Levels (Washington, 1949).

49. J. Mattauch and S. Fluegge, Nuclear Physics Tables (Interscience Publishers, New York, 1946).

50a. F. R. Bichowsky and F. D. Rossini, The Thermochemistry of the Chemical Substances (Reinhold, New York, 1936).

50b. F. D. Rossini, National Bureau of Standards Tables of Selected Values of Chemical Thermodynamic Properties (Washington, 1947).

II. References to Individual Papers

51. J. Aars, Z. Physik **79**, 122 (1932).
52. W. S. Adams and Th. Dunham, Publ. Astron. Soc. Pac. **44**, 243 (1932).
53. ———— ————, Astrophys. J. **79**, 308 (1934).
54. A. Adel, Astrophys. J. **85**, 345 (1937).
55. ————, Astrophys. J. **86**, 337 (1937).
55a. ———— and D. M. Dennison, Physic. Rev. **43**, 716, **44**, 99 (1933).
55b. ———— and C. O. Lampland, Astrophys. J. **87**, 198 (1938).
56. ———— and V. M. Slipher, Physic. Rev. **46**, 902 (1934); **47**, 651 (1935).
57. ———— ————, Physic. Rev. **46**, 240 (1934).
58. ———— ————, Physic. Rev. **47**, 787 (1935).
59. ———— ———— and E. F. Barker, Physic. Rev. **47**, 580 (1935).
60. ———— ———— and O. Fouts, Physic. Rev. **49**, 288 (1936).
60a. R. G. Aickin and N. S. Bayliss, Trans. Faraday Soc. **33**, 1333 (1937).
61. H. S. Allen and A. K. Longair, Phil. Mag. **19**, 1032 (1935).
62. G. M. Almy and C. D. Hause, Physic. Rev. **42**, 242 (1932).
63. ———— and R. B. Horsfall, Physic. Rev. **51**, 491 (1937).
64. ———— and G. R. Irwin, Physic. Rev. **49**, 72 (1936).
65. ———— and M. P. Rassweiler, Physic. Rev. **53**, 890 (1937).
66. ———— and H. A. Schultz, Physic. Rev. **51**, 62 (1937).
67. ———— and M. C. Watson, Physic. Rev. **45**, 871 (1934).
68. E. Amaldi, Z. Physik **79**, 492 (1932).
68a. F. L. Arnot and M. B. M'Ewen, Proc. Roy. Soc. London **171**, 120 (1939).
69. M. F. Ashley, Physic. Rev. **44**, 919 (1933).
70. R. K. Asundi and S. M. Karim, Proc. Ind. Acad. **6A**, 328 (1937).
71. ———— and Y. P. Parti, Proc. Ind. Acad. **6A**, 207 (1937).
72. ———— and R. Samuel, Proc. Ind. Acad. **3A**, 346 (1936).
73. L. Avramenko and V. Kondratjew, Physik. Z. d. Sowjetunion **10**, 741 (1936).
74. H. D. Babcock, Proc. Nat. Acad. Amer. **15**, 471 (1929).
75. ———— and W. P. Hoge, Physic. Rev. **37**, 227 (1931).
76. R. M. Badger, J. Chem. Phys. **2**, 128 (1934); **3**, 710 (1935).
77. H. Baerwald, G. Herzberg, and L. Herzberg, Ann. Physik **20**, 569 (1934).
78. F. Baldet, Ann. Observ. Astronomie physique de Meudon **7**, 53 (1926).
79. R. F. Barrow, Proc. Phys. Soc. London **51**, 267 (1939).
80. ———— and W. Jevons, Proc. Roy. Soc. London **169**, 45 (1938).
80a. E. Bartholomé, Z. physik. Chem. B **23**, 131 (1933).
81. J. H. Bartlett, Physic. Rev. **37**, 507 (1931).
82. N. S. Bayliss, Proc. Roy. Soc. London **158**, 551 (1937).
82a. J. Y. Beach, J. Chem. Phys. **4**, 353 (1936).
83. H. Bell and A. Harvey, Proc. Phys. Soc. London **50**, 427 (1938).
84. E. Bengtsson-Knave, Nova Acta Reg. Soc. Scient. Ups. **8**, No. 4 (1932).
85. ———— and E. Hulthén, Z. Physik **52**, 275 (1928).
86. ———— and E. Olsson, Z. Physik **72**, 163 (1931).
87. ———— and R. Rydberg, Z. Physik **59**, 540 (1930).
87a. R. Bernard, Z. Physik **110**, 291 (1938).
88. H. A. Bethe and R. F. Bacher, Rev. Mod. Phys. **8**, 82 (1936).
89. H. Beutler, Z. Physik **50**, 581 (1928).
90. ————, Z. physik. Chem. B **27**, 287 (1934).
91. ————, Z. physik. Chem. B **29**, 315 (1935).
92. ————, A. Deubner, and H. O. Jünger, Z. Physik **98**, 181 (1935).
93. ———— and B. Josephy, Z. Physik **53**, 747 (1929).
94. ———— and H. O. Jünger, Z. Physik **100**, 80 (1936).
95. ———— ————, Z. Physik **101**, 285 (1936).
96. ———— ————, Z. Physik **101**, 304 (1936).
97. ———— and H. Levi, Z. physik. Chem. B **24**, 263 (1934).
98. ———— and M. Polanyi, Z. physik. Chem. B **1**, 3 (1928).
99. O. Bewersdorff, Z. Physik **103**, 598 (1936).
100. B. N. Bhaduri and A. Fowler, Proc. Roy. Soc. London **145**, 321 (1934).

101. S. Bhagavantam, Physic. Rev. **42**, 437 (1932).
102. R. T. Birge, Physic. Rev. **25**, 240 (1925).
103. ———, Physic. Rev. **28**, 1157 (1926).
104. ———, Trans. Faraday Soc. **25**, 707 (1929).
105. ———, Trans. Faraday Soc. **25**, 718 (1929).
106. ———, Physic. Rev. **37**, 841 (1931).
107. ———, Rev. Mod. Phys. **1**, 1 (1929).
108. ———, Physic. Rev. **40**, 319 (1932).
109. ———, Nature (London) **134**, 771 (1934).
110. ———, Physic. Rev. **52**, 241 (1937).
111. ——— and J. J. Hopfield, Astrophys. J. **68**, 257 (1928).
112. ——— and J. D. Shea, Univ. Cal. Publ. Math. **2**, 67 (1927).
113. ——— and H. Sponer, Physic. Rev. **28**, 259 (1926).
114. H. Biskamp, Z. Physik **86**, 33 (1933).
115. L. and E. Bloch and Ch. Shin-Piaw, C. R. Paris **201**, 824 (1935).
115a. N. T. Bobrovnikoff, Astrophys. J. **89**, 301 (1939).
116. K. F. Bonhoeffer and L. Farkas, Z. physik. Chem. A **134**, 337 (1927).
117. ——— and F. Haber, Z. physik. Chem. A **137**, 263 (1928).
118. ——— and P. Harteck, Z. physik. Chem. B **4**, 113 (1929).
119. ——— and H. Reichardt, Z. physik. Chem. A **139**, 75 (1928).
120. M. Born and J. Franck, Z. Physik **31**, 411 (1925).
121. ——— and J. E. Mayer, Z. Physik **75**, 1 (1932).
122. ——— and R. Oppenheimer, Ann. Physik **84**, 457 (1927).
123. L. v. Bozóky, Z. Physik **104**, 275 (1937).
124. W. H. Brandt, Physic. Rev. **50**, 778 (1936).
125. B. A. Brice, Physic. Rev. **35**, 960 (1930).
126. ———, Physic. Rev. **38**, 658 (1931).
127. L. O. Brockway, J. Amer. Chem. Soc. **60**, 1348 (1938).
128. ———, Rev. Mod. Phys. **8**, 231 (1936).
129. P. H. Brodersen, Z. Physik **104**, 135 (1936).
130. H. H. Brons, Dissertation (Groningen, 1934).
131. ———, Proc. Amst. **38**, 271 (1935).
132. W. G. Brown, Physic. Rev. **38**, 709 (1931).
133. ———, Physic. Rev. **38**, 1179 (1931).
134. ———, Physic. Rev. **39**, 777 (1932).
135. ———, Physic. Rev. **42**, 355 (1932).
136. ———, Z. Physik **82**, 768 (1933).
137. ——— and G. E. Gibson, Physic. Rev. **40**, 529 (1932).
138. A. Budó, Zeeman Verh. (1935), 166.
139. ———, Z. Physik **96**, 219 (1935); **98**, 437 (1936).
140. ———, Z. Physik **105**, 73 (1937).
141. ———, Z. Physik **105**, 579 (1937).
142. ——— and I. Kovács, Z. Physik **109**, 393 (1938); **111**, 633 (1939).
143. H. Bulthuis, Dissertation (Groningen, 1935).
144. ———, Physica **1**, 873 (1934).
145. ——— and D. Coster, Zeeman Verh. (1935), 135.
146. K. Butkow, Z. Physik **58**, 232 (1929).
147. G. Büttenbender and G. Herzberg, Ann. Physik **21**, 577 (1935).
148. W. Burmeister, Ber. deutsch. phys. Ges. **15**, 595 (1913).
148a. J. Cabannes and J. Dufay, C. R. Paris **203**, 903 (1936).
148b. ——— ——— and J. Gauzit, Nature **142**, 718, 755 (1938).
149. Y. Cambresier and L. Rosenfeld, Monthly Notices Roy. Astron. Soc. **93**, 710 (1933).
150. T. Carroll, Physic. Rev. **52**, 822 (1937).
151. H. Casimir, Physica **1**, 1073 (1934).
152. C. N. Challacombe and G. M. Almy, Physic. Rev. **51**, 63, 930 (1937).
153. S. Chapman, Phil. Mag. **23**, 657 (1937).
154. W. H. J. Childs and R. Mecke, Z. Physik **68**, 344 (1931).
155. A. Christy, Physic. Rev. **33**, 701 (1929).
156. ——— and S. Bloomenthal, Physic. Rev. **35**, 46 (1930).

157. C. H. D. Clark, Physic. Rev. **47**, 238 (1935).
158. A. P. Cleaves and C. W. Edwards, Physic. Rev. **48**, 850 (1935).
159. K. Clusius and E. Bartholomé, Z. Elektrochem. **40**, 524 (1934).
160. E. U. Condon, Physic. Rev. **32**, 858 (1928).
161. ———, Physic. Rev. **41**, 759 (1932).
162. A. S. Coolidge, H. M. James, and R. D. Present, J. Chem. Phys. **4**, 193 (1936).
162a. ——— ——— and E. L. Vernon, Physic. Rev. **54**, 726 (1938).
163. H. Cordes and H. Sponer, Z. Physik **79**, 170 (1932).
164. S. D. Cornell, Physic. Rev. **54**, 341 (1938).
165. D. Coster and F. Brons, Physica **1**, 155 (1934).
166. ——— ———, Physica **1**, 634 (1934).
167. ——— ——— and A. van der Ziel, Z. Physik **84**, 304 (1933).
168. ——— and H. H. Brons, Z. Physik **70**, 492 (1931).
169. ———, Z. Physik **73**, 747 (1932).
169a. C. A. Coulson and W. E. Duncanson, Proc. Roy. Soc. London **165, 90** (1938).
170. N. H. Coy and H. Sponer, Physic. Rev. **53**, 495 (1938).
171. F. H. Crawford, Rev. Mod. Phys. **6**, 90 (1934).
172. ——— and T. Jorgensen, Physic. Rev. **47**, 358 (1935).
173. ——— ———, Physic. Rev. **47**, 932 (1935); **49**, 745 (1936).
174. ——— and W. A. Shurcliff, Physic. Rev. **45**, 860 (1934).
175. J. Curry and G. Herzberg, Ann. Physik **19**, 800 (1934).
176. ———, L. Herzberg, and G. Herzberg, Z. Physik **86**, 348 (1933).
177. W. E. Curtis and J. Patkowski, Phil. Trans. Roy. Soc. London **232, 395** (1934).
178. M. Czerny, Z. Physik **34**, 227 (1925).
179. O. Darbyshire, Proc. Roy. Soc. London **159**, 93 (1937).
180. G. Déjardin, Rev. Mod. Phys. **8**, 1 (1936).
181. D. M. Dennison, Proc. Roy. Soc. London **115**, 483 (1927).
182. H. Deslandres, C. R. Paris **103**, 375 (1886).
183. G. H. Dieke, Physic. Rev. **38**, 646 (1931).
184. ———, Physic. Rev. **47**, 661 (1935).
185. ———, Physic. Rev. **47**, 870 (1935).
186. ——— and H. D. Babcock, Proc. Nat. Acad. Amer. **13**, 670 (1927).
187. ——— and R. W. Blue, Physic. Rev. **45**, 395 (1934).
188. ——— and M. N. Lewis, Physic. Rev. **52**, 100 (1937).
189. ——— and W. Lochte-Holtgreven, Z. Physik **62**, 767 (1930).
190. ——— and J. W. Mauchley, Physic. Rev. **43**, 12 (1933).
191. O. S. Duffendack, R. W. Revans, and A. S. Roy, Physic. Rev. **45, 807** (1934).
192. M. Duffieux and L. Grillet, C. R. Paris **202**, 937 (1936).
193. J. DuMond and V. Bollmann, Physic. Rev. **51**, 400 (1937).
194. J. L. Dunham, Physic. Rev. **41**, 721 (1932).
194a. ———, Physic. Rev. **34**, 438 (1929).
195. Th. Dunham, Publ. Astron. Soc. Pac. **45**, 42 (1933).
196. E. G. Dymond, Z. Physik **34**, 553 (1925).
197. L. T. Earls, Physic. Rev. **48**, 423 (1935).
198. H. Ekstein and M. Magat, C. R. Paris **199**, 264 (1934).
199. A. Elliott, Proc. Roy. Soc. London **123**, 629 (1929).
200. ———, Proc. Roy. Soc. London **127**, 638 (1930).
200a. ———, Proc. Roy. Soc. London **169**, 469 (1939).
201. ——— and W. H. B. Cameron, Proc. Roy. Soc. London **164**, 531 (1938).
202. J. W. Ellis and H. O. Kneser, Z. Physik **86**, 583 (1933).
203. ——— ———, Publ. Astron. Soc. Pac. **46**, 106 (1934).
204. V. M. Ellsworth and J. J. Hopfield, Physic. Rev. **29**, 79 (1927).
205. W. Elsasser, Z. Physik **81**, 332 (1933).
206. A. Eucken and K. Hiller, Z. physik. Chem. B **4**, 142 (1929).
207. ——— and O. Mücke, Z. physik. Chem. B **18**, 167 (1932).
208. R. S. Estey, Physic. Rev. **35**, 309 (1930).

209. H. Eyring and M. Polanyi, Z. physik. Chem. B **11**, 97 (1930).
210. E. H. Eyster, Physic. Rev. **51**, 1078 (1937).
211. A. and L. Farkas and P. Harteck, Proc. Roy. Soc. London **144**, 481 (1934).
212. L. Farkas, Z. Physik **70**, 733 (1931).
213. ——, Von den Kohlen- und Mineralölen, Vol. IV (1931), 35.
214. ——, Erg. d. exakt. Naturwiss. **12**, 163 (1933).
215. —— and S. Levy, Z. Physik **84**, 195 (1933).
216. W. F. C. Ferguson, Physic. Rev. **31**, 969 (1928).
217. ——, Physic. Rev. **32**, 607 (1928).
218. W. Finkelnburg, Physik. Z. **31**, 1 (1930); **34**, 529 (1933).
219. ——, Z. Physik **90**, 1 (1934).
220. ——, Z. Physik **96**, 699 (1935).
221. ——, Acta Phys. Pol. **5**, 1 (1936).
222. ——, Z. Physik **99**, 798 (1936).
223. —— and O. T. Hahn, Physik. Z. **39**, 98 (1938).
224. P. J. Flory, J. Chem. Phys. **4**, 23 (1936).
225. —— and H. L. Johnston, J. Amer. Chem. Soc. **57**, 2641 (1935).
226. A. Fowler and C. J. Bakker, Proc. Roy. Soc. London **136**, 28 (1932).
227. J. G. Fox and G. Herzberg, Physic. Rev. **52**, 638 (1937).
228. H. H. Franck and H. Reichardt, Naturwiss. **24**, 171 (1936).
229. J. Franck, Trans. Faraday Soc. **21**, 536 (1925).
230. —— and H. Kuhn, Naturwiss. **20**, 923 (1932).
231. —— and H. Sponer, Gött. Nachr. 1928, p. 241.
232. —— —— and E. Teller, Z. physik. Chem. B **18**, 88 (1932).
233. W. R. Fredrickson, Physic. Rev. **34**, 207 (1929).
233a. —— and M. E. Hogan, Physic. Rev. **46**, 454 (1934).
233b. —— and W. W. Watson, Physic. Rev. **39**, 753 (1932).
234. A. A. Frost and O. Oldenberg, J. Chem. Phys. **4**, 642, 781 (1936).
235. Y. Fujioka and Y. Tanaka, Sci. Pap. Inst. Physic. Chem Res. (Tokyo) **30**, 121 (1936).
236. —— ——, Sci. Pap. Inst. Physic. Chem. Res. (Tokyo) **32**, 143 (1937).
237. —— ——, Sci. Pap. Inst. Physic. Chem. Res. (Tokyo) **34**, 713 (1938).
238. —— and T. Wada, Sci. Pap. Inst. Physic. Chem. Res. (Tokyo) **27**, 210 (1935).
239. G. W. Funke, Z. Physik **84**, 610 (1933).
240. ——, Z. Physik **96**, 787 (1935).
240a. ——, Dissertation (Stockholm, 1936).
241. —— and B. Grundström, Z. Physik **100**, 293 (1936).
242. J. W. Gabel and R. V. Zumstein, Physic. Rev. **52**, 726 (1937).
243. H. G. Gale and G. S. Monk, Astrophys. J. **69**, 77 (1929).
244. H. Gajewski, Physik. Z. **33**, 122 (1932).
245. A. G. Gaydon and R. W. B. Pearse, Proc. Roy. Soc. London **148**, 312 (1935).
245a. —— ——, Nature (London) **140**, 110 (1937).
246. —— ——, Nature (London) **142**, 291 (1938); Rep. Progress in Physics **5**, 252 (1938).
247. J. Genard, C. R. Paris **197**, 1402 (1933).
248. ——, Physica **1**, 849 (1934).
249. L. Gerö, Z. Physik **99**, 52 (1936).
250. ——, Z. Physik **100**, 374 (1936).
251. ——, Z. Physik **101**, 311 (1936).
252. ——, G. Herzberg, and R. Schmid, Physic. Rev. **52**, 467 (1937).
252a. W. F. Giauque, J. Amer. Chem. Soc. **52**, 4808, 4816 (1930).
253. —— and H. L. Johnston, Nature (London) **123**, 318 (1929); J. Amer. Chem. Soc. **51**, 1436 (1929).
254. —— ——, Nature (London) **123**, 831 (1929); J. Amer. Chem. Soc. **51**, 3528 (1929).
255. G. E. Gibson, Z. Physik **50**, 692 (1928).
256. —— and N. S. Bayliss, Physic. Rev. **44**, 188 (1933).
257. —— and W. Heitler, Z. Physik **49**, 465 (1928).

258. G. E. Gibson, and R. C. Ramsperger, Physic. Rev. **30**, 598 (1927).
259. ———, O. K. Rice, and N. S. Bayliss, Physic. Rev. **44**, 193 (1933).
260. C. Gilbert, Physic. Rev. **49**, 619 (1936).
261. G. Glockler and D. L. Fuller, J. Chem. Phys. **1**, 886 (1933).
262. P. Goldfinger, W. Jeunehomme, and B. Rosen, Nature (London) **138**, 205 (1936).
263. C. F. Goodeve and A. W. C. Taylor, Proc. Roy. Soc. London **152**, 221 (1935).
264. ——— ———, Proc. Roy. Soc. London **154**, 181 (1936).
264a. A. R. Gordon and C. Barnes, J. Chem. Phys. **1**, 297 (1933).
265. C. Ghosh, Z. Physik **78**, 521 (1932).
266. ———, Z. Physik **86**, 241 (1933).
267. P. N. Ghosh and G. N. Ball, Z. Physik **71**, 362 (1931).
268. B. Grundström, Z. Physik **69**, 235 (1931).
269. ———, Dissertation (Stockholm, 1936).
270. ———, Z. Physik **99**, 595 (1936).
271. ———, Nature (London) **141**, 555 (1938).
272. ——— and E. Hulthén, Nature (London) **125**, 634 (1930).
273. ——— and P. Valberg, Z. Physik **108**, 326 (1938).
274. A. Guntsch, Z. Physik **86**, 262 (1933).
275. ———, Z. Physik **87**, 312 (1934).
276. ———, Z. Physik **93**, 534 (1934).
277. ———, Z. Physik **104**, 584 (1937).
278. M. L. Gurnsey, Physic. Rev. **46**, 114 (1934).
279. O. Hahn, Ber. deutsch. chem. Ges. **71A**, 1 (1938).
279a. W. Hanle, Physik. Z. **38**, 995 (1937).
280. J. D. Hardy, E. F. Barker, and D. M. Dennison, Physic. Rev. **42**, 279 (1932).
281. A. Harvey, Proc. Roy. Soc. London **133**, 336 (1931).
282. ——— and F. A. Jenkins, Physic. Rev. **35**, 789 (1930).
283. L. B. Headrick and G. W. Fox, Physic. Rev. **35**, 1033 (1930).
284. M. H. Hebb, Physic. Rev. **49**, 610 (1936).
285. K. Hedfeld, Z. Physik **68**, 610 (1931).
286. A. Heimer, Naturwiss. **24**, 491 (1936).
286a. ———, Z. Physik **95**, 328 (1935).
287. ———, Z. Physik **103**, 621 (1936).
288. ———, Dissertation (Stockholm, 1937).
289. ———, Z. Physik **104**, 448 (1937).
290. ——— and T. Heimer, Z. Physik **84**, 222 (1933).
291. T. Heimer, Z. Physik **95**, 321 (1935).
292. ———, Dissertation (Stockholm, 1937).
293. W. Heitler, Marx's Handb. d. Radiol. VI, **2**, 485 (1934).
294. ——— and G. Herzberg, Naturwiss. **17**, 673 (1929).
295. ——— and F. London, Z. Physik **44**, 455 (1927).
296. E. Hellmig, Z. Physik **104**, 694 (1937).
297. K. H. Hellwege, Z. Physik **100**, 644 (1936).
298. H. J. Henning, Ann. Physik **13**, 599 (1932).
299. V. Henri, C. R. Paris **177**, 1037 (1923).
300. ———, Leipz. Vortr. **1931**, 131.
301. ——— and M. C. Teves, Nature (London) **114**, 894 (1924).
302. G. Herzberg, Ann. Physik **84**, 565 (1927).
303. ———, Ann. Physik **86**, 189 (1928).
304. ———, Z. Physik **57**, 601 (1929).
305. ———, Z. physik. Chem. B **4**, 223 (1929).
306. ———, Z. physik. Chem. B **9**, 43 (1930).
307. ———, Z. Physik **61**, 604 (1930).
308. ———, Leipz. Vortr. **1931**, 167.
309. ———, Erg. d. exakt. Naturwiss. **10**, 207 (1931).
310. ———, Naturwiss. **20**, 577 (1932).
311. ———, Ann. Physik **15**, 677 (1932).

312. G. Herzberg, Nature (London) **133**, 759 (1934).
313. ———, J. Phys. Chem. **41**, 299 (1937).
314. ———, Chem. Rev. **20**, 145 (1937).
315. ———, Astrophys. J. **87**, 428 (1938).
316. ———, Astrophys. J. **89**, 290 (1939).
317. ——— and L. Mundie (unpublished).
318. ——— and J. W. T. Spinks, Z. Physik **89**, 474 (1934).
319. ——— and H. Sponer, Z. physik. Chem. B **26**, 1 (1934).
320. L. Herzberg, Z. Physik **84**, 571 (1933).
320a. G. Hettner, Z. Physik **89**, 234 (1934).
321. K. Hilferding and W. Steiner, Z. physik. Chem. B **30**, 399 (**1935**).
322. E. L. Hill, Physic. Rev. **34**, 1507 (1929).
323. ——— and J. H. Van Vleck, Physic. Rev. **32**, 250 (1923).
324. T. R. Hogness and J. Franck, Z. Physik **44**, 26 (1927).
325. W. Holst, Z. Physik **90**, 728 (1934).
326. ———, Z. Physik **90**, 735 (1934).
327. ———, Z. Physik **93**, 55 (1934).
328. ——— and E. Hulthén, Z. Physik **90**, 712 (1934).
329. H. Hönl and F. London, Z. Physik **33**, 803 (1925).
330. J. J. Hopfield, Physic. Rev. **35**, 1130; **36**, 784 (**1930**).
331. ———, Physic. Rev. **36**, 789 (1930).
332. ——— and R. T. Birge, Physic. Rev. **29**, 922 (**1927**).
333. T. Hori, Z. Physik **71**, 478 (1931).
334. ———, Mem. Ryojun Coll. Eng. **6**, 1 (1933).
335. ——— and J. Huriuti, Z. Physik **101**, 279 (1936).
336. H. G. Howell, Proc. Roy. Soc. London **148**, 696 (**1935**).
337. ———, Proc. Roy. Soc. London **153**, 683 (1936).
338. ———, Proc. Roy. Soc. London **155**, 141 (1936).
339. ———, Proc. Roy. Soc. London **160**, 242 (1937).
340. ——— and N. Coulsan, Proc. Roy. Soc. London **166**, 925 (1938).
340a. ——— and G. D. Rochester, Proc. Phys. Soc. London **51**, 329 (**1939**).
341. I. Hudes, Physic. Rev. **52**, 1256 (1937).
341a. D. S. Hughes and P. E. Lloyd, Physic. Rev. **52**, 1215 (1937).
342. E. Hulthén, Z. Physik **32**, 32 (1925).
342a. ———, Physic. Rev. **29**, 97 (1927).
343. ———, Nature (London) **126**, 56 (1930).
343a. ——— and A. Heimer, Nature (London) **129**, 399 (1932).
344. ——— and G. Johannson, Z. Physik **26**, 308 (1924).
345. ——— and R. Rydberg, Nature (London) **131**, 470 (1933).
345a. R. F. Humphreys and W. R. Fredrickson, Physic. Rev. **50**, 542 (**1936**).
346. F. Hund, Z. Physik **42**, 93 (1927).
347. ———, Z. Physik **63**, 719 (1930).
348. ———, Handb. d. Phys. **24**, I, 561 (1933).
349. R. F. Hunter and R. Samuel, Nature (London) **138**, 411 (1936).
349a. E. Hutchisson, Physic. Rev. **37**, 45 (1931).
350. E. A. Hylleraas, Z. Physik **71**, 739 (1931).
351. ———, Z. Physik **96**, 643 (1935); Physik. Z. **36**, 599 (1936).
352. H. H. van Iddekinge, Nature (London) **125**, 858 (1930).
353. S. Imanishi, Sci. Pap. Inst. Physic. Chem. Res. (Tokyo) **31**, 247 (1937).
353a. ———, Nature (London **143**, 165 (1939).
354. E. S. Imes, Astrophys. J. **50**, 251 (1919).
355. M. Ishaq, Proc. Roy. Soc. London **159**, 110 (1937).
356. ——— and R. W. B. Pearse, Proc. Roy. Soc. London **156**, 221 (1936).
357. G. P. Ittmann, Z. Physik **71**, 616 (1931).
358. ———, Naturwiss. **22**, 118 (1934).
359. A. Jabłoński, Z. Physik **70**, 723 (1931).
360. G. Jaffé, Z. Physik **87**, 535 (1934).
361. H. M. James and A. S. Coolidge, J. Chem. Phys. **1**, 825 (1933); **3**, 129 (1935).
362. ——— ———, Astrophys. J. **87**, 438 (1938).

363. H. M. James and A. S. Coolidge, Physic. Rev. **55**, 184 (1939).
364. —— —— and R. D. Present, J. Chem. Phys. **4**, 187 (1936).
365. C. Jausseran, L. Grillet, and M. Duffieux, C. R. Paris **205**, 39 (1937).
366. F. A. Jenkins, Physic. Rev. **31**, 539 (1928).
367. ——, Physic. Rev. **35**, 315 (1930).
368. ——, Physic. Rev. **47**, 783 (1935).
369. ——, H. A. Barton, and R. S. Mulliken, Physic. Rev. **30**, 150 (1927).
370. —— and R. Grinfeld, Physic. Rev. **45**, 229 (1934).
371. —— and A. Harvey, Physic. Rev. **39**, 922 (1932).
372. —— and H. de Laszlo, Proc. Roy. Soc. London **122**, 103 (1929).
373. —— and A. McKellar, Physic. Rev. **42**, 464 (1932).
374. —— and L. S. Ornstein, Proc. Amst. **35**, 1212 (1932).
375. —— and G. D. Rochester, Physic. Rev. **52**, 1135 (1937).
376. —— ——, Physic. Rev. **52**, 1141 (1937).
377. ——, Y. K. Roots, and R. S. Mulliken, Physic. Rev. **39**, 16 (1932).
378. —— and L. A. Strait, Physic. Rev. **47**, 136 (1935).
379. —— and D. E. Wooldridge, Physic. Rev. **53**, 137 (1938).
380. C. R. Jeppesen, Physic. Rev. **44**, 165 (1933).
381. ——, Physic. Rev. **49**, 797 (1936).
382. M. A. Jeppesen, Physic. Rev. **50**, 445 (1936).
382a. W. Jeunehomme, Actualités scientifiques et industrialles No. 569 (1937);
383. W. Jevons, Proc. Roy. Soc. London **110**, 365 (1926).
384. ——, Proc. Phys. Soc. London **48**, 563 (1936).
384a.——, Proc. Phys. Soc. London **50**, 910 (1938).
385. —— and L. A. Bashford, Proc. Phys. Soc. London **49**, 554 (1937).
386. —— —— and H. V. A. Briscoe, Proc. Phys. Soc. London **49**, 532 (1937).
387. —— —— ——, Proc. Phys. Soc. London **49**, 543 (1937).
388. J. Joffe and H. C. Urey, Physic. Rev. **43**, 761 (1933).
389. L. W. Johnson and R. C. Johnson, Proc. Roy. Soc. London **133**, 207 (1931).
390. R. C. Johnson, Phil. Trans. Roy. Soc. London **226**, 157 (1927).
391. ——, Proc. Roy. Soc. London **122**, 161 (1929).
392. ——, Proc. Roy. Soc. London **122**, 189 (1929).
392a. H. L. Johnston and C. O. Davis, J. Amer. Chem. Soc. **56**, 271 (1934).
393. G. Joos and W. Finkelnburg, Die Physik **4**, 35 (1936).
394. H. Juraszyńska and M. Szulc, Acta Phys. Pol. **7**, 49 (1938).
395. J. Kaplan, Physic. Rev. **38**, 373, 1079 (1931).
396. ——, Physic. Rev. **44**, 947 (1933); **45**, 675 (1934).
397. W. Kapuszinski and J. G. Eymers, Proc. Roy. Soc. London **122**, 58 (1929);
398. L. S. Kassel, Chem. Rev. **18**, 277 (1936).
399. ——, Chem. Rev. **21**, 331 (1937).
400. A. S. King and R. T. Birge, Astrophys. J. **72**, 251 (1930).
400a. G. W. King, J. Chem. Phys. **6**, 378 (1938).
401. G. D. Kinzer and G. M. Almy, Physic. Rev. **52**, 814 (1937).
402. M. Kiuti, Jap. J. Phys. **1**, 29 (1922); **4**, 13 (1925).
402a.—— and H. Hasunuma, Proc. Phys. Math. Soc. Japan **19**, 821 (1937).
403. D. E. Kirkpatrick and E. O. Salant, Physic. Rev. **48**, 945 (1935).
404. O. Klein, Z. Physik **76**, 226 (1932).
405. A. Klemenc, R. Wechsberg, and G. Wagner, Z. Elektrochem. **40**, 488 (1934).
406. H. P. Knauss and H. S. Ballard, Physic. Rev. **48**, 796 (1935).
407. —— and M. S. McCay, Physic. Rev. **52**, 1143 (1937).
408. C. C. Ko, Journ. Frankl. Inst. **217**, 173 (1934).
409. V. Kondratjew and A. Lauris, Z. Physik **92**, 741 (1934).
410. —— and A. Leipunsky, Trans. Faraday Soc. **25**, 736 (1929).
411. —— and E. Olsson, Z. Physik **99**, 671 (1936).
412. —— and L. Polak, Z. Physik **76**, 386 (1932).
413. —— ——, Physik. Z. d. Sowjetunion **4**, 764 (1933).
414. E. Kondratjewa and V. Kondratjew, Acta Phys. Chim. **3**, 1 (1935).

415. P. G. Koontz, Physic. Rev. **48**, 138 (1935).
416. ———, Physic. Rev. **48**, 707 (1935).
417. ——— and W. W. Watson, Physic. Rev. **48**, 937 (1935).
418. H. Kopfermann and H. Schweitzer, Z. Physik **61**, 87 (1930).
419. I. Kovács, Z. Physik **106**, 431 (1937).
420. ———, Z. Physik **109**, 387 (1938); **111**, 640 (1939).
420a. H. A. Kramers, Z. Physik **53**, 422 (1929).
420b. A. Kratzer, Z. Physik **3**, 460 (1920); **4**, 476 (1921).
421. R. de L. Kronig, Z. Physik **50**, 347 (1928).
422. ———, Z. Physik **62**, 300 (1930).
423. ———, Physica **1**, 617 (1934).
424. H. Krüger, Naturwiss. **26**, 445 (1938); Z. Physik **111**, 467 (1939).
425. H. Kühl, VDI.-Forschungsheft 1935, 373.
426. H. Kuhn, Z. Physik **39**, 77 (1926).
427. ———, Naturwiss. **16**, 552 (1928).
428. ———, Z. Physik **63**, 458 (1930).
429. ———, Z. Physik **76**, 782 (1932).
430. ———, Phil. Mag. **18**, 987 (1934).
431. ———, Proc. Roy. Soc. London **158**, 212, 230 (1937).
432. ———, Physic. Rev. **52**, 133 (1937).
433. ——— and S. Arrhenius, Z. Physik **82**, 716 (1933).
434. ——— and K. Freudenberg, Z. Physik **76**, 38 (1932).
435. ——— and F. London, Phil. Mag. **18**, 983 (1934).
436. ——— and O. Oldenberg, Physic. Rev. **41**, 72 (1932).
437. W. Kuhn and H. Martin, Z. Physik **81**, 482 (1933).
438. P. Kusch, Physic. Rev. **49**, 218 (1936).
439. R. W. Ladenburg, J. Opt. Soc. **25**, 259 (1935).
439a. G. O. Langstroth, Proc. Roy. Soc. London **146**, 166 (1934); **150, 371** (1935).
439b. C. E. Leberknight and J. A. Ord, Physic. Rev. **51**, 430 (1937).
440. S. W. Leifson, Astrophys. J. **63**, 73 (1926).
441. J. E. Lennard-Jones, Trans. Faraday Soc. **25**, 668 (1929).
442. H. Lessheim and R. Samuel, Proc. Ind. Acad. **1**, 623 (1935).
443. H. Levi, Dissertation (Berlin, 1934).
444. B. Lewis and G. v. Elbe, J. Amer. Chem. Soc. **55**, 511 (1933); **57, 1399** (1935).
445. ——— ———, J. Amer. Chem. Soc. **57**, 612 (1935).
445a. M. N. Lewis and J. V. White, Physic. Rev. **55**, 894 (1939).
446. P. Lindau, Z. Physik **25**, 247; **26**, 343; **30**, 187 (1924).
447. W. Lochte-Holtgreven, Z. Physik **67**, 590 (1931).
448. ———, Z. Physik **103**, 395 (1936).
449. ——— and G. H. Dieke, Ann. Physik **3**, 937 (1929).
450. ——— and H. Maecker, Z. Physik **105**, 1 (1937).
451. ——— and E. S. van der Vleugel, Z. Physik **70**, 188 (1931).
452. F. London, Z. Physik **46**, 455 (1928).
453. ———, Sommerfeld-Festschrift, p. 104 (Leipzig, 1928); Z. Elektrochem. **35**, 552 (1929).
454. ———, Z. Physik **63**, 245 (1930).
455. ———, Z. physik. Chem. B **11**, 222 (1930).
456. ———, Z. Physik **74**, 143 (1932).
456a. F. W. Loomis, Astrophys. J. **52**, 248 (1920).
457. ———, Physic. Rev. **29**, 112 (1927).
458. ———, Physic. Rev. **38**, 2153 (1931).
459. ——— and M. J. Arvin, Physic. Rev. **46**, 286 (1934).
460. ——— and W. H. Brandt, Physic. Rev. **49**, 55 (1936).
461. ——— and H. Q. Fuller, Physic. Rev. **39**, 180 (1932).
462. ——— and P. Kusch, Physic. Rev. **46**, 292 (1934).
463. ——— and R. E. Nusbaum, Physic. Rev. **38**, 1447 (1931).
464. ——— ———, Physic. Rev. **39**, 89 (1932).
465. ——— ———, Physic. Rev. **40**, 380 (1932).

466. F. W. Loomis and T. F. Watson, Physic. Rev. **48**, 280 (1935).
467. —— and R. W. Wood, Physic. Rev. **32**, 223 (1928).
468. —— ——, Physic. Rev. **38**, 854 (1931).
469. W. Lotmar, Z. Physik **93**, 528 (1935).
470. F. Lowater, Proc. Phys. Soc. London **44**, 51 (1932).
471. ——, Phil. Trans. Roy. Soc. London A **234**, 355 (1935).
471a. R. W. Lunt, R. W. B. Pearse, and E. C. W. Smith, Proc. Roy. Soc. London **155**, 173 (1936).
471b. E. R. Lyman, Physic. Rev. **53**, 379 (1938).
472. J. K. L. MacDonald, Proc. Roy. Soc. London **123**, 103 (1929); **131**, 146 (1931).
473. ——, Proc. Roy. Soc. London **138**, 183 (1932).
474. P. C. Mahanti, Nature (London) **125**, 819 (1930).
475. ——; Physic. Rev. **42**, 609 (1932).
476. ——, Proc. Phys. Soc. London **46**, 51 (1934).
477. ——, Proc. Phys. Soc. London **47**, 433 (1935).
478. —— and A. K. Sen Gupta, Z. Physik **109**, 39 (1938).
479. K. Mahla, Z. Physik **81**, 625 (1933).
480. H. Margenau, Physic. Rev. **48**, 755 (1935).
481. —— and W. W. Watson, Rev. Mod. Phys. **8**, 22 (1936).
482. E. V. Martin, Physic. Rev. **41**, 167 (1932).
482a. L. D. Matheson, Physic. Rev. **40**, 813 (1932).
483. L. R. Maxwell, S. B. Hendricks, and V. M. Mosley, Physic. Rev. **52**, 968 (1937).
484. —— and V. M. Mosley, Physic. Rev. **51**, 684 (1937).
484a. —— —— and S. B. Hendricks, Physic. Rev. **50**, 41 (1936).
485. J. C. McLennan and J. H. McLeod, Nature (London) **123**, 160 (1929); Trans. Roy. Soc. Can. III, **22**, 413 (1928); **23**, 19 (1929).
486. R. Mecke, Z. Physik **31**, 709 (1925).
487. ——, Z. Physik **32**, 823 (1925).
488. ——, Z. Physik **42**, 390 (1927).
489. ——, Trans. Faraday Soc. **30**, 200 (1934).
490. P. Mesnage, C. R. Paris **204**, 1929 (1937).
490a. ——, C. R. Paris **206**, 1634 (1938).
491. C. F. Meyer and A. A. Levin, Physic. Rev. **34**, 44 (1929).
492. E. Miescher, Helv. Phys. Acta **8**, 279 (1935).
493. ——, Helv. Phys. Acta **8**, 486 (1935).
494. ——, Helv. Phys. Acta **11**, 463 (1938).
495. —— and M. Wehrli, Helv. Phys. Acta **7**, 331 (1934).
496. R. A. Millikan, Ann. Physik **32**, 34 (1938).
497. R. Minkowski, Z. Physik **93**, 731 (1935).
498. K. R. More, Physic. Rev. **54**, 122 (1938).
499. —— and S. D. Cornell, Physic. Rev. **53**, 806 (1938).
500. —— and A. H. Parker, Physic. Rev. **52**, 1150 (1937).
501. F. Morgan, Physic. Rev. **49**, 41 (1936).
502. ——, Physic. Rev. **49**, 47 (1936).
503. ——, Physic. Rev. **50**, 603 (1936).
503a. W. Mörikofer, Dissertation (Basle, 1925).
504. P. M. Morse, Physic. Rev. **34**, 57 (1929).
505. —— and E. C. G. Stueckelberg, Physic. Rev. **33**, 932 (1929).
505a. S. Mrozowski, Nature (London) **129**, 399 (1932).
506. —— and M. Szulc, Acta Phys. Pol. **6**, 44 (1937).
507. B. C. Mukherji, Z. Physik **70**, 552 (1931).
508. R. S. Mulliken, Physic. Rev. **25**, 259 (1925).
509. ——, Physic. Rev. **25**, 509 (1925).
509a. ——, Physic. Rev. **32**, 186 (1928).
510. ——, Physic. Rev. **32**, 880 (1928).
511. ——, Trans. Faraday Soc. **25**, 634 (1929).
512. ——, Rev. Mod. Phys. **2**, 60 (1930).
513. ——, Rev. Mod. Phys. **3**, 89 (1931).

514. R. S. Mulliken, Rev. Mod. Phys. **4**, 1 (1932).
515. ———, Physic. Rev. **36**, 611 (1930).
516. ———, Physic. Rev. **36**, 699 (1930).
517. ———, Physic. Rev. **41**, 49 (1932).
518. ———, Physic. Rev. **46**, 144 (1934).
519. ———, Physic. Rev. **46**, 549 (1934).
519a.———, J. Chem. Phys. **2**, 400, 712 (1934).
519b.———, J. Chem. Phys. **4**, 620 (1936).
520. ———, Physic. Rev. **50**, 1017, 1028 (1936).
521. ———, Physic. Rev. **51**, 310 (1937).
522. ———, J. Phys. Chem. **41**, 5 (1937).
523. ——— and A. Christy, Physic. Rev. **38**, 87 (1931).
524. ——— and D. S. Stevens, Physic. Rev. **44**, 720 (1933).
525. G. M. Murphy and H. L. Johnston, Physic. Rev. **46**, 95 (1934).
526. G. Nakamura and T. Shidei, Jap. J. Phys. **10**, 11 (1935).
527. S. M. Naudé, Physic. Rev. **36**, 333 (1930).
528. ———, Proc. Roy. Soc. London **136**, 114 (1932).
528a.———, South African J. Sci. **32**, 103 (1935).
529. J. v. Neumann and E. Wigner, Physik. Z. **30**, 467 (1929).
530. T. E. Nevin, Nature (London) **140**, 1101 (1937); Phil. Trans. Roy. Soc. London **237**, 471 (1938).
531. A. H. Nielsen and H. H. Nielsen, Physic. Rev. **47**, 585 (1935).
532. L. Nordheim, Müller-Pouillet's Lehrb. d. Phys. IV, **4**, 798ff. (1934).
533. G. Nordheim-Pöschl, Ann. Physik **26**, 258 (1936).
534. F. Norling, Z. Physik **95**, 179 (1935).
535. ———, Z. Physik **104**, 638 (1935).
536. ———, Z. Physik **106**, 177 (1937).
537. O. Oldenberg, Z. Physik **47**, 184 (1928); **55**, 1 (1929).
538. ———, Z. Physik **56**, 563 (1929).
539. ———, Physic. Rev. **46**, 210 (1934).
540. ———, J. Phys. Chem. **41**, 293 (1937).
540a.——— and F. F. Rieke, J. Chem. Phys. **6**, 439, 779 (1938).
541. E. Olsson, Z. Physik **73**, 732 (1932).
542. ———, Z. Physik **90**, 138 (1934).
543. ———, Z. Physik **93**, 206 (1935).
544. ———, Z. Physik **95**, 215 (1935).
545. ———, Z. Physik **100**, 656 (1936).
546. ———, Z. Physik **104**, 402 (1937).
547. ———, Z. Physik **108**, 40 (1937).
548. ———, Dissertation (Stockholm, 1938).
549. L. S. Ornstein and W. R. van Wijk, Z. Physik **49**, 315 (1928).
549a. C. H. Page, Physic. Rev. **53**, 426 (1938).
550. A. E. Parker, Physic. Rev. **46**, 301 (1934).
551. ———, Physic. Rev. **47**, 349 (1935).
552. J. Patkowski and W. E. Curtis, Trans. Faraday Soc. **25**, 725 (1929).
552a. F. W. Paul and H. P. Knauss, Physic. Rev. **54**, 1072 (1938).
553. L. Pauling, J. Amer. Chem. Soc. **53**, 1367, 3225 (1931).
554. ———, J. Amer. Chem. Soc. **54**, 988 (1932).
555. ——— and L. O. Brockway, J. Chem. Phys. **2**, 867 (1934).
556. R. W. B. Pearse, Proc. Roy. Soc. London **122**, 442 (1929).
557. ———, Proc. Roy. Soc. London **129**, 328 (1930).
558. ———, Proc. Roy. Soc. London **143**, 112 (1934).
559. ——— and A. G. Gaydon, Proc. Phys. Soc. London **50**, 201 (1938).
560. W. G. Penney, Phil. Mag. **11**, 602 (1931).
561. A. Petrowa, Acta Physicochim. **4**, 559 (1936).
562. E. Placzek, Marx's Handb. d. Radiol. VI, **2**, 205 (1934).
563. E. K. Plyler and E. F. Barker, Physic. Rev. **44**, 984 (1933).
563a.——— and D. Williams, Physic. Rev. **49**, 215 (1936).
564. G. Piccardi, Gazz. chim. Ital. **63**, 887 (1933).
565. ———, Atti Acad. Lincei **21**, 589 (1935).

566. M. Polanyi and E. Wigner, Z. Physik **33**, 429 (1925).
567. W. C. Pomeroy, Physic. Rev. **29**, 59 (1927).
568. G. Pöschl and E. Teller, Z. Physik **83**, 143 (1933).
569. R. D. Present, Physic. Rev. **48**, 140 (1935).
569a.W. M. Preston, Physic. Rev. **51**, 298 (1937).
570. W. C. Price, Proc. Roy. Soc. London **167**, 216 (1938).
571. ———— and G. Collins, Physic. Rev. **48**, 714 (1935).
572. A. Prikhotko, Physik. Z. d. Sowjetunion **11**, 465 (1937).
573. P. Pringsheim, Handb. d. Phys. **23**, I, 185 (1933).
574. E. Rabinowitch, Trans. Faraday Soc. **33**, 283 (1937).
575. ———— and W. C. Wood, J. Chem. Phys. **4**, 358 (1936).
576. ———— ————, J. Chem. Phys. **4**, 497 (1936).
577. H. M. Randall, Rev. Mod. Phys. **10**, 72 (1938).
578. F. Rasetti, Physic. Rev. **34**, 367 (1929).
579. ————, Z. Physik **61**, 598 (1930).
580. W. Rave, Z. Physik **94**, 72 (1935).
581. Lord Rayleigh, Proc. Roy. Soc. London **116**, 702 (1927).
582. D. N. Read, Physic. Rev. **46**, 571 (1934).
583. O. Reinkober, Hand- u Jahrb. d. chem. Phys. **9**, II, 1.
584. O. K. Rice, Physic. Rev. **37**, 1187, 1551 (1930).
585. O. W. Richardson, Proc. Roy. Soc. London **152**, 503 (1935).
586. ————, Proc. Roy. Soc. London **160**, 487 (1937); **164**, 316 (1938).
587. ———— and T. B. Rymer, Proc. Roy. Soc. London **147**, 24, 251, 272 (1934).
588. R. S. Richardson, Publ. Astron. Soc. Pac. **44**, 250 (1932).
589. H. Richter, Physik. Z. **33**, 587 (1932); **36**, 85 (1935).
590. O. Riechemeier, H. Senftleben, and H. Pastorff, Ann. Physik **19**, 202 (1934).
591. R. Ritschl, Z. Physik **42**, 172 (1927).
592. G. D. Rochester, Proc. Roy. Soc. London **153**, 407 (1936); **167**, 567 (1938).
593. ————, Z. Physik **101**, 769 (1936).
594. ————, Physic. Rev. **51**, 486 (1937).
595. ———— and H. G. Howell, Proc. Roy. Soc. London **148**, 157 (1935).
596. R. Rompe, Z. Physik **101**, 214 (1936); Physik. Z. **37**, 807 (1936).
597. B. Rosen, Acta Phys. Pol. **5**, 193 (1936).
598. ————, Physica **6**, 205 (1939).
599. ———— and M. Désirant, C. R. Paris **200**, 1659 (1935).
600. L. Rosenfeld, Monthly Notices Roy. Astron. Soc. **93**, 724 (1933).
601. J. E. Rosenthal, Proc. Nat. Acad. Amer. **21**, 281 (1935).
602. ———— and F. A. Jenkins, Physic. Rev. **33**, 163 (1929).
603. ———— ————, Proc. Nat. Acad. Amer. **15**, 381 (1929).
604. F. D. Rossini, Bur. Stand. J. Res. **6**, 1 (1931).
604a.S. Rouppert, Acta Phys. Pol. **6**, 228 (1937).
604b.H. A. Rühmkorf, Ann. Physik. **33**, 21 (1938).
605. K. Rumpf, Z. physik. Chem. B **38**, 469 (1938).
606. H. N. Russell, Astrophys. J. **79**, 317 (1934).
607. ————, Nature (London) **135**, 219 (1935).
608. R. Rydberg, Z. Physik **73**, 376 (1932).
609. ————, Z. Physik **80**, 514 (1933).
610. ————, Dissertation (Stockholm, 1934).
611. E. O. Salant and A. Sandow, Physic. Rev. **37**, 373 (1931).
612. H. Salow and W. Steiner, Z. Physik **99**, 137 (1936).
613. R. Samuel, Current Science **4**, 762 (1936).
614. I. Sandemann, Proc. Roy. Soc. Edinburgh **55**, 72 (1935).
615. P. G. Saper, Physic. Rev. **42**, 498 (1932).
616. P. Scherrer and R. Stössei, Helv. Phys. Acta **3**, 435 (1931).
617. R. Schlapp, Physic. Rev. **51**, 342 (1937).
618. R. Schmid, Z. Physik **49**, 428 (1928); **64**, 84 (1930).
619. ———— and L. Gerö, Z. Physik **93**, 656 (1935).
620. ———— ————, Z. Physik **94**, 386 (1935).
621. ———— ————, Z. Physik **101**, 343 (1936).

622. R. Schmid and L. Gerö, Z. Physik **104**, 724 (1937).
623. ——— ———, Z. Physik **105**, 36 (1937).
624. ——— ———, Z. physik. Chem. B **36**, 105 (1937).
624a. ——— ——— and J. Zemplén, Proc. Phys. Soc. London **50**, 283 (1938).
625. E. Scholz, Z. Physik **106**, 230 (1937).
625a. K. Scholz, Z. Physik **78**, 751 (1932).
626. H. Schüler and H. Gollnow, Z. Physik **108**, 714; **109**, 432 (1938).
626a. ——— ——— and H. Haber, Z. Physik **111**, 508 (1939).
627. J. Sedov and A. Filippov, C. R. Leningrad **4**, 376 (1934).
628. H. Senftleben and E. Germer, Ann. Physik **2**, 847 (1929).
629. A. K. Sen Gupta, Bull. Acad. Alahabad **2**, 245 (1933).
630. ———, Z. Physik **91**, 471 (1934).
631. ———, Proc. Phys. Soc. London **47**, 247 (1935).
631a. ———, Proc. Phys. Soc. London **51**, 62 (1939).
632. C. V. Shapiro, R. C. Gibbs, and A. W. Laubengayer, Physic. Rev. **40**, 354 (1932).
633. E. N. Shawhan, Physic. Rev. **48**, 521 (1935); **49**, 810 (1936).
634. ———, Physic. Rev. **48**, 343 (1935).
635. J. D. Shea, Physic. Rev. **30**, 825 (1927).
635a. P. E. Shearin, Physic. Rev. **48**, 299 (1935).
636. T. Shidei, Jap. J. Phys. **11**, 23 (1936).
637. C. Shin-Piaw, C. R. Paris **201**, 1181 (1935).
638. J. C. Slater, Physic. Rev. **37**, 481; **38**, 1109 (1931).
639. R. Smoluchowski, Z. Physik **85**, 191 (1933).
640. W. R. Smythe, Physic. Rev. **45**, 299 (1934).
641. H. Snell, Phil. Trans. Roy. Soc. London A **234**, 115 (1934).
642. C. P. Snow, J. F. G. Rawlins, and E. K. Rideal, Proc. Roy. Soc. London **124**, 453 (1929).
643. ——— and E. K. Rideal, Proc. Roy. Soc. London **125**, 462 (1929).
644. K. Sommermeyer, Z. Physik **56**, 548 (1929).
645. J. W. T. Spinks, Z. Physik **88**, 511 (1934).
646. H. Sponer, Z. Physik **34**, 622 (1925).
647. ———, Erg. exakt. Naturwiss. **6**, 75 (1927).
648. ———, Z. Physik **41**, 611 (1927).
649. ———, Leipz. Vortr. **1931**, 107.
650. J. R. Stehn, J. Chem. Phys. **5**, 186 (1937).
651. W. Steiner, Z. physik. Chem. B **15**, 249 (1932).
652. G. Stenvinkel, Dissertation (Stockholm, 1936).
653. D. S. Stevens, Physic. Rev. **38**, 1292 (1931).
654. R. Stössel, Ann. Physik **10**, 393 (1931).
655. L. A. Strait and F. A. Jenkins, Physic. Rev. **49**, 635 (1936).
656. H. M. Strong and H. P. Knauss, Physic. Rev. **49**, 740 (1936).
657. E. C. G. Stueckelberg, Helv. Phys. Acta **5**, 369 (1933).
658. E. Svensson, Dissertation (Stockholm, 1935).
659. ——— and F. Tyrén, Z. Physik **85**, 257 (1933).
660. P. Swings, Actualités scientifiques et industrielles Nos. 50 and 162 (1932–34).
660a. T. Takamine, T. Suga, and Y. Tanaka, Sci. Pap. Inst. Physic. Chem. Res. (Tokyo) **34**, 854 (1938).
661. T. Tanaka and Z. Koana, Proc. Phys. Math. Soc. Jap. **16**, 365 (1934).
662. G. K. Teal and G. E. MacWood, J. Chem. Phys. **3**, 760 (1935).
663. E. Teller, Z. Physik **61**, 458 (1930).
664. ———, Hand- u. Jahrb. d. chem. Phys. **9**, II, 43 (1934).
665. A. Terenin, Z. Physik **37**, 98 (1926).
666. ——— and B. Popov, Z. Physik **75**, 338 (1932); Physik. Z. d. Sowjetunion **1**, 307, **2**, 299 (1932).
667. S. F. Thunberg, Z. Physik **100**, 471 (1936).
668. N. Tsi-Ze and T. San-Tsiang, Physic. Rev. **52**, 91 (1937).
669. L. A. Turner, Physic. Rev. **27**, 397 (1926); **31**, 983 (1928); **37**, 1023 (1931).

670. L. A. Turner, Z. Physik **65**, 464 (1930).
671. ———, Z. Physik **68**, 178 (1931).
672. ———, Physic. Rev. **38**, 574 (1931).
673. ———, Physic. Rev. **41**, 627 (1933).
673a.——— and W. T. Harris, Physic. Rev. **52**, 626 (1937).
674. ——— and E. W. Samson, Physic. Rev. **37**, 1023 (1931).
675. H. S. Uhler and R. A. Patterson, Astrophys. J. **42**, 434 (1915).
676. H. C. Urey, F. C. Brickwedde, and G. M. Murphy, Physic. Rev. **39**, 164 (1932).
677. ——— and G. K. Teal, Rev. Mod. Phys. **7**, 34 (1935).
678. J. H. Van Vleck, Physic. Rev. **33**, 467 (1929).
680. ———, Physic. Rev. **40**, 544 (1932).
681. ———, Astrophys. J. **80**, 161 (1934).
682. ———, J. Chem. Physics **4**, 327 (1936).
683. ——— and A. Sherman, Rev. Mod. Phys. **7**, 167 (1935).
684. L. Vegard, Z. Physik **75**, 30 (1932).
685. ——— and E. Tønsberg, Geophys. Publ. **11**, No. 2 (1935).
686. D. S. Villars and E. U. Condon, Physic. Rev. **35**, 1028 (1930).
687. B. Vonnegut and B. E. Warren, J. Amer. Chem. Soc. **58**, 2459 (1936).
688. J. W. Walker, J. W. Straley, and A. W. Smith, Physic. Rev. **53**, 140 (1938).
689. H. v. Wartenberg, G. Sprenger, and J. Taylor, Z. physik. Chem. Bodenstein-Festb. **1931**, 61.
690. W. W. Watson, Physic. Rev. **32**, 600 (1928).
691. ———, Physic. Rev. **47**, 27 (1935).
692. ———, Physic. Rev. **49**, 70 (1936).
693. ———, Physic. Rev. **53**, 639 (1938).
694. ———, Physic. Rev. **54**, 1068 (1938).
694a.——— and W. R. Fredrickson, Physic. Rev. **39**, 765 (1932).
695. ——— ——— and M. E. Hogan, Physic. Rev. **49**, 150 (1936).
696. ——— and R. F. Humphreys, Physic. Rev. **52**, 318 (1937).
697. ——— and P. G. Koontz, Physic. Rev. **46**, 32 (1934).
698. ——— and W. F. Meggers, Bur. Stand. J. Res. **20**, 125 (1938).
699. ——— and A. Shambon, Physic. Rev. **50**, 607 (1936).
699a.——— and R. Simon, Physic. Rev. **55**, 358 (1939).
700. M. Wehrli, Helv. Phys. Acta **7**, 617, 673 (1934); **9**, 587 (1936).
701. ——— and E. Miescher, Helv. Phys. Acta **7**, 298 (1934).
702. V. Weisskopff, Physik. Z. **34**, 1 (1933).
703. W. Weizel, Z. Physik **54**, 321 (1929).
704. ———, Z. Physik **59**, 320 (1930).
705. ——— and O. Beeck, Z. Physik **76**, 250 (1932).
706. ——— and M. Kulp, Ann. Physik **4**, 971 (1930).
707. ——— and E. Pestel, Z. Physik **56**, 197 (1929).
707a. W. West, P. Arthur, and R. T. Edwards, J. Chem. Phys. **5**, 10, 14 (1937).
707b. J. W. White, J. Chem. Phys. **6**, 294 (1938).
708. K. Wieland, Helv. Phys. Acta **2**, 46 (1929).
709. ———, Z. Physik **76**, 801 (1932).
710. ———, Z. physik. Chem. B **42**, 422 (1939).
711. R. Wierl, Ann. Physik **8**, 521 (1931).
712. E. Wigner and E. E. Witmer, Z. Physik **51**, 859 (1928).
713. R. Wildt, Z. Physik **54**, 856 (1929).
713a.———, Gött. Nachr. **1932**, 87; Naturwiss. **20**, 851 (1932).
714. ———, Veröff. Univ. Sternw. Gött. **1932**, No. 22.
715. ———, Astrophys. J. **86**, 321 (1937).
716. J. G. Winans, Phil. Mag. **7**, 555 (1929).
717. ——— and E. C. G. Stueckelberg, Proc. Nat. Acad. Amer. **14**, 867 (1928).
718. E. E. Witmer, Physic. Rev. **28**, 1223 (1926).
719. R. W. Wood, Phil. Mag. **12**, 329, 499 (1906).
720. ——— and G. H. Dieke, Physic Rev. **35**, 1355 (1930).
720a.——— ———, J. Chem. Phys. **6**, 734, 908 (1938).

721. R. W. Wood and F. W. Loomis, Physic. Rev. **31,** 705 (1928); Phil. Mag. **6,** 231 (1928).
722. R. E. Worley and F. A. Jenkins, Physic. Rev. **54,** 305 (1938).
723. O. R. Wulf, Physic. Rev. **46,** 316 (1934).
724. ———, J. Opt. Soc. **25,** 231 (1935).
725. K. Wurm, Naturwiss. **20,** 85 (1932).
726. ———, Z. Physik **76,** 309 (1932).
727. ———, Z. Astrophysik **5,** 260 (1932).
728. ———, Z. Astrophysik **8,** 281; **9,** 62 (1934).
729. ———, Z. Astrophysik **10,** 133 (1935).
729a.———, Astrophys. J. **89,** 312 (1939).
729b. ———, unpublished.
730. H. Zanstra, Monthly Notices Roy. Astron. Soc. **89,** 178 (1928).
731. H. Zeise, Z. Elektrochem. **39,** 758, 895 (1933); **40,** 662, 885 (1934).
732. A. van der Ziel, Physica **1,** 353 (1934).
733. ———, Physica **4,** 373 (1937).
734. R. V. Zumstein, J. W. Gabel, and R. E. McKay, Physic. Rev. **51,** 238 (1937).

III. REFERENCES TO INDIVIDUAL PAPERS ADDED
IN THE SECOND EDITION

735. M. G. Adams, Monthly Notices Roy. Astron. Soc. **98,** 544 (1938).
736. W. S. Adams, Ann. Rep. Dir. Mt. Wilson Obs. 1938–1939.
737. ———, Astrophys. J. **93,** 11 (1941).
737a. ———, Publ. Astron. Soc. Pac. **60,** 174 (1948).
738. A. Adel, Astrophys. J. **90,** 627 (1939).
739. ———, Astrophys. J. **93,** 506 (1941).
740. ———, Astrophys. J. **93,** 509 (1941).
741. ———, Astrophys. J. **94,** 451 (1941).
742. ———, Astrophys. J. **96,** 239 (1942).
742a. M. Afaf, Nature **164,** 752 (1949).
743. L. H. Ahrens, Physic. Rev. **74,** 74 (1948).
744. R. G. Aickin and N. S. Bayliss, Trans. Faraday Soc. **34,** 1371 (1938).
745. ——— ——— and A. L. G. Rees, Proc. Roy. Soc. London **169,** 234 (1938).
745a. G. Almkvist and A. Lagerqvist, Arkiv Fysik **1,** 477 (1949).
746. G. M. Almy, J. Phys. Chem. **41,** 47 (1937).
747. ——— and A. C. Beiler, Physic. Rev. **61,** 476 (1942).
748. ——— and G. D. Kinzer, Physic. Rev. **47,** 721 (1935).
749. ——— and F. M. Sparks, Physic. Rev. **44,** 365 (1933).
750. B. M. Anand, Science and Culture **8,** 278 (1942).
751. D. H. Andrews, W. F. Brucksch, W. T. Ziegler and E. R. Blanchard, Rev. Sci. Instruments **13,** 281 (1942).
752. E. B. Andrews and R. F. Barrow, private comm.
752a. D. Andrychuk, to be published.
752b. ———, J. Chem. Phys. **18,** 233 (1950).
753. J. G. Aston, in Taylor-Glasstone's Treatise on Physical Chemistry **1,** 511 (Van Nostrand, 1942).
754. R. K. Asundi, Proc. Roy. Soc. London, **124,** 277 (1929).
755. ———, Proc. Ind. Acad. **1,** 830 (1935).
756. ———, Proc. Ind. Acad. **12A,** 491 (1940).
757. ———, M. Jan-Khan and R. Samuel, Proc. Roy. Soc. London **157,** 28 (1936).
758. ——— and S. M. Karim, Proc. Ind. Acad. **6,** 281 (1937).
760. ——— and P. Venkateswarlu, Ind. J. Phys. **21,** 76 (1947).
761. ——— ———, Ind. J. Phys. **21,** 101 (1947).
762. H. D. Babcock, Physic. Rev. **46,** 382 (1934).
763. ———, Astrophys. J. **102,** 154 (1945).
764. ——— and L. Herzberg, Astrophys. J. **108,** 167 (1948).
765. J. Bacher, Helv. Phys. Acta **21,** 379 (1948).
766. C. H. Bamford, Rep. Progress in Physics **9,** 75 (1943).
767. D. Barbier, Ann. d'Astrophys. **10,** 47 (1947).
768. ———, C. R. Paris, **224,** 635 (1947); Ann. d'Astrophys. **10,** 141 (1947).
769. J. Bardeen and C. H. Townes, Physic. Rev. **73,** 97, 627, 1204 (1948).
769a. S. Barratt, Proc. Roy. Soc. London **109,** 194 (1925).
769b. ———, Trans. Faraday Soc. **25,** 758 (1929).
769c. ——— and A. R. Bonar, Phil. Mag. **9,** 519 (1930).
770. H. Barrell and J. E. Sears, Phil. Trans. Roy. Soc. London **238,** 1 (1939).
771. R. F. Barrow, Proc. Phys. Soc. London **51,** 989 (1939).
772. ———, Proc. Phys. Soc. London **52,** 380 (1940).
773. ———, Proc. Phys. Soc. London **53,** 116 (1941).
774. ———, Proc. Phys. Soc. London **58,** 606 (1946).
775. ——— and T. J. Bowen, private comm.
776. ——— and D. V. Crawford, Proc. Phys. Soc. London **57,** 12 (1945).
777. ——— ———, Nature **157,** 339 (1946).
778. ——— and W. Jevons, Proc. Phys. Soc. London **52,** 534 (1940).
778a. ——— and A. Lagerqvist, Arkiv Fysik **1,** 221 (1949).

778*b*. R. F. Barrow, A. Lagerqvist and E. Lind, Nature **164**, 923 (1949).
779. ——— and M. F. R. Mulcahy, Nature **162**, 336 (1948).
780. ——— ———, Proc. Phys. Soc. London **61**, 99 (1948).
781. ——— ———, private comm.
782. ——— and D. M. Thomas, private comm.
783. ——— and E. E. Vago, Proc. Phys. Soc. London **55**, 326 (1943).
784. ——— ———, Proc. Phys. Soc. London **56**, 76 (1944).
785. ——— ———, Proc. Phys. Soc. London **56**, 78 (1944).
786. D. R. Bates and H. S. W. Massey, Proc. Roy. Soc. London **187**, 261 (1946); **192**, 1 (1947).
787. J. R. Bates, J. O. Halford and L. C. Anderson, J. Chem. Phys. **3**, 531 (1935).
788. E. Bauer and M. Magat, Physica **5**, 718 (1938).
789. W. A. Baum, F. S. Johnson, J. J. Oberly, C. C. Rockwood, C. V. Strain and R. Tousey, Physic. Rev. **70**, 781 (1946).
790. N. S. Bayliss and A. L. G. Rees, Trans. Faraday Soc. **35**, 792 (1939).
791. ——— ———, J. Chem. Phys. **8**, 377, 429 (1940).
792. C. S. Beals, Pop. Astron. **52**, 47 (1944).
793. B. S. Beer, Z. Physik **107**, 73 (1937).
794. E. Bengtsson, Ark. Mat. Astr. Fys. **20A**, No. 28 (1928).
795. R. Beringer, Physic. Rev. **70**, 53 (1946).
795*a*. ——— and J. G. Castle, Jr., Physic. Rev. **75**, 1963, **76**, 868 (1949).
796. R. Bernard, Ann. de Géophys. **3**, 63 (1947).
797. H. J. Bernstein, J. Chem. Phys. **15**, 284, 339 (1947).
798. ——— and G. Herzberg, J. Chem. Phys. **15**, 77 (1947).
799. H. Beutler and M. Fred, Physic. Rev. **61**, 107 (1942).
800. S. Bhagavantam, Ind. J. Phys. **5**, 35 (1930).
801. ———, Proc. Ind. Acad. **2A**, 303, 477 (1935).
802. ———, Proc. Ind. Acad. **2A**, 310 (1935).
803. ———, Proc. Ind. Acad. **10A**, 224 (1939).
804. S. S. Bhatnagar, H. Lessheim and M. L. Khanna, Nature **140**, 152 (1937).
805. J. L. Binder, Physic. Rev. **54**, 114 (1938).
806. R. T. Birge, Astrophys. J. **55**, 273 (1922).
807. ———, Int. Crit. Tables **5**, 409 (1929).
808. ———, Physic. Rev. **35**, 133 (1930).
809. ———, Rep. Progress in Physics **8**, 90 (1941).
810. ———, Rev. Mod. Phys. **13**, 223 (1941).
811. ———, Physic. Rev. **60**, 766 (1941).
812. ———, Rev. Mod. Phys. **19**, 298 (1947).
813. P. M. S. Blackett and F. C. Champion, Proc. Roy. Soc. London **130**, 380 (1931).
814. R. C. Blake and T. Iredale, Nature **157**, 229 (1946).
815. L. Blitzer, Astrophys. J. **91**, 421 (1940).
816. F. Bloch, Physic. Rev. **70**, 460 (1946).
817. S. Bloomenthal, Physic. Rev. **35**, 34 (1930).
818. ———, Physic. Rev. **54**, 498 (1938).
819. N. T. Bobrovnikoff, Rev. Mod. Phys. **16**, 271 (1944).
820. K. F. Bonhoeffer, Ergeb. exakt. Naturw. **6**, 201 (1927).
821. ———, A. Farkas and K. Rummel, Z. physik. Chem. B **21**, 225 (1933).
822. ——— and T. G. Pearson, Z. physik. Chem. B **14**, 1 (1931).
823. L. Bozóky and R. Schmid, Physic. Rev. **48**, 465 (1935).
824. L. Brewer, J. Chem. Phys. **16**, 1165 (1948).
825. ———, P. W. Gilles and F. A. Jenkins, J. Chem. Phys. **16**, 797 (1948).
826. J. Bricard and A. Kastler, Ann. de Géophys. **1**, 53 (1944).
827. H. Brode, Ann. Physik **37**, 344 (1940).
828. P. H. Brodersen, Z. Physik **79**, 613 (1932).
829. ——— and H. J. Schumacher, Z. Naturforsch. **2a**, 358 (1947).
830. H. H. Brons, Physica **1**, 739 (1934).
831. ———, Proc. Amst. **37**, 793 (1934).

832. W. G. Brown, Physic. Rev. **38**, 1187 (1931).
833. A. Budó and I. Kovács, Z. Physik **116**, 693 (1940); **117**, 612 (1941).
834. ―――― ――――, Physik. Z. **45**, 122 (1944).
834a. ―――― ――――, Hung. Acta Physica **1**, No. 3 (1948).
835. A. Burris and C. D. Hause, J. Chem. Phys. **11**, 442 (1943).
836. K. Butkow and A. Terenin, Z. Physik **49**, 865 (1928).
837. J. Cabannes and J. Dufay, Ann. de Géophys. **1**, 1 (1944).
838. ―――― ――――, Ann. de Géophys. **2**, 290 (1946).
839. W. H. B. Cameron and A. Elliott, Proc. Roy. Soc. London **169**, 463 (1939).
840. R. J. Cashman, J. Opt. Soc. **36**, 356 (1946).
841. H. B. G. Casimir, Physica **2**, 719 (1935), Archives du Musée Teyler (III), **8**, 201 (1936).
842. ―――― and D. Polder, Physic. Rev. **73**, 360 (1948).
842a. A. D. Caunt and R. F. Barrow, Nature **164**, 753 (1949).
843. B. K. Chakraborti, Ind. J. Phys. **10**, 155 (1935).
844. D. Chalonge, Ann. de Phys. **1**, 123 (1933).
845. S. Chapman, Rep. Progress in Physics **9**, 92 (1943).
846. S. Y. Ch'en, Physic. Rev. **58**, 884 (1940).
847. W. H. J. Childs, Proc. Roy. Soc. London **137**, 641 (1932).
848. A. Christy, Astrophys. J. **70**, 1 (1929).
849. A. Ciccone, Ricerca Sci. **2**, 3 (1935).
850. C. H. D. Clark, Trans. Faraday Soc. **36**, 370 (1940).
851. ―――― and K. R. Webb, Trans. Faraday Soc. **37**, 293 (1941).
852. F. Coheur-Dehalu, Bull. Acad. Roy. Belg. **23**, 604 (1937).
852a. ―――― and B. Rosen, Mem. Soc. Roy. Sci. Liége **1941**, 405.
853. E. H. Coleman and A. G. Gaydon, Discuss. Faraday Soc. No. 2, 166 (1947).
854. ―――― ―――― and W. M. Vaidya, Nature **162**, 108 (1948).
855. D. K. Coles and W. E. Good, Physic. Rev. **70**, 979 (1946).
856. F. C. Connelly, Proc. Phys. Soc. London **45**, 780 (1933).
856a. ――――, Proc. Phys. Soc. London **46**, 790 (1934).
857. J. P. Cooley and J. H. Rohrbaugh, Physic. Rev. **67**, 296 (1945).
858. A. S. Coolidge, Physic. Rev. **65**, 236 (1944).
859. ―――― and H. M. James, J. Chem. Phys. **6**, 730 (1938).
860. H. Cordes, Z. Physik **97**, 603 (1935).
861. ――――, Z. Physik **105**, 251 (1937).
862. ―――― and P. W. Schenk, Z. Elektrochem. **39**, 594 (1933).
863. ―――― and H. Sponer, Z. Physik **63**, 334 (1930).
863a. R. E. Cornish and E. D. Eastman, J. Am. Chem. Soc. **50**, 627 (1928).
864. D. Coster and F. Brons, Z. Physik **97**, 570 (1935).
865. ――――, H. H. Brons and H. Bulthuis, Z. Physik **79**, 787 (1932).
866. ――――, E. W. van Dijk and A. J. Lameris, Physica **2**, 267 (1935).
867. C. A. Coulson, Proc. Roy. Soc. Edinburgh **61**, 20 (1941).
868. ―――― and R. P. Bell, Trans. Faraday Soc. **41**, 141 (1945).
869. ―――― and C. M. Gillam, Proc. Roy. Soc. Edinburgh **62A**, 360 (1948).
870. N. H. Coy and H. Sponer, Physic. Rev. **58**, 709 (1940).
871. W. Cram, Physic. Rev. **46**, 205 (1934).
872. W. O. Crane and A. Christy, Physic. Rev. **36**, 421 (1930).
873. F. H. Crawford and P. M. Tsai, Proc. Am. Acad. Ark. Sci. **69**, 407 (1935).
873a. M. F. Crawford, H. L. Welsh and J. L. Locke, Physic. Rev. **75**, 1607 (1949); **76**, 580 (1949).
874. W. E. Curtis and S. F. Evans, Proc. Roy. Soc. London **141**, 603 (1933).
875. ―――― and A. Harvey, Proc. Roy. Soc. London **121**, 381 (1928).
876. ―――― ――――, Proc. Roy. Soc. London **125**, 484 (1929).
877. ―――― and R. G. Long, Proc. Roy. Soc. London **108**, 513 (1925).
878. M. Czerny and P. Mollet, Z. Physik **108**, 85 (1937).
879. B. P. Dailey, R. L. Kyhl, M. W. P. Strandberg, J. H. Van Vleck and E. B. Wilson, Jr., Physic. Rev. **70**, 984 (1946).
879a. T. W. Dakin, W. E. Good and D. K. Coles, Physic. Rev. **70**, 560 (1946).

880. O. Darbyshire, Physic. Rev. **40**, 366 (1932).
881. A. Daudin and C. Fehrenbach, C. R. Paris **222**, 1083 (1946).
882. ————, J. de phys. **9**, 163 (1948).
883. C. O. Davis and H. L. Johnston, J. Am. Chem. Soc. **56**, 1045 (1934).
884. D. N. Davis, Astrophys. J. **106**, 28 (1947).
885. O. Deile, Z. Physik **106**, 405 (1937).
886. A. Delsemme and B. Rosen, Bull. Soc. Roy. Sci. Liége **1945**, 70.
887. D. M. Dennison, Physic. Rev. **28**, 318 (1926).
888. M. Désirant and A. Minne, C. R. Paris **202**, 1272 (1936).
889. G. H. Dieke, Proc. Amst. **28**, 174 (1925).
890. ————, Z. Physik **57**, 71 (1929).
891. ————, Physic. Rev. **48**, 608, 610 (1935).
892. ————, Physic. Rev. **50**, 797 (1936).
893. ————, Physic. Rev. **54**, 439 (1938).
894. ————, Physic. Rev. **60**, 523 (1941).
894a. ————, Physic. Rev. **76**, 50 (1949).
895. ———— and R. W. Blue, Physic. Rev. **47**, 261 (1935).
895a. ———— and H. M. Crosswhite, Bumblebee Report No. 87 Johns Hopkins University, 1948.
896. ————, S. Imanishi and T. Takamine, Z. Physik **54**, 826; **57**, 305 (1929).
897. ————, T. Takamine and T. Suga, Z. Physik **49**, 637 (1928).
897a. ———— and F. S. Tomkins, Physic. Rev. **76**, 283 (1949).
898. P. P. Dobronravin, C. R. U.R.S.S. **18**, 399 (1937).
899. A. E. Douglas, Can. J. Res. A **19**, 27 (1941).
900. ———— and G. Herzberg, Can. J. Res. A **18**, 165 (1940).
901. ———— ————, Can. J. Res. A **18**, 179 (1940).
902. ———— ————, Astrophys. J. **94**, 381 (1941); Can. J. Res. A **20**, 71 (1942).
902a. ———— ————, Physic. Rev. **76**, 1529 (1949).
903. A. R. Downie and R. F. Barrow, Nature **160**, 198 (1947).
904. J. Dufay, C. R. Paris **206**, 1948 (1938).
905. ————, Astrophys. J. **91**, 91 (1940).
906. ————, C. R. Paris **215**, 284 (1941).
907. ————, Ann. d'Astrophys. **5**, 93 (1942).
908. ————, C. R. Paris **223**, 783 (1946).
909. ————, Ann. de Géophys. **3**, 1 (1947).
909a. ———— and M. Bloch, Ann. d'Astrophys. **11**, 58, 107 (1948).
910. ———— and G. Déjardin, Ann. de Géophys. **2**, 249 (1946).
911. ———— and T. Mao-Lin, Ann. de Géophys. **3**, 282 (1947).
912. O. S. Duffendack and J. M. La Rue, J. Opt. Soc. **31**, 146 (1941).
913. R. B. Dull, Physic. Rev. **47**, 458 (1935).
914. J. W. M. DuMond and E. R. Cohen, Rev. Mod. Phys. **20**, 82 (1948).
915. J. L. Dunham, Physic. Rev. **35**, 1347 (1930).
916. T. Dunham, Jr., Pub. Am. Astron. Soc. **10**, 123 (1941).
917. ————, in G. P. Kuiper's "The Atmospheres of the Earth and Planets" (Chicago, 1949) p. 286.
918. ———— and W. S. Adams, Publ. Astron. Soc. Pac. **49**, 26 (1937).
918a. R. A. Durie and T. Iredale, Trans. Faraday Soc. **44**, 806 (1948).
919. F. Duschinsky, E. Hirschlaff and P. Pringsheim, Physica **2**, 439 (1935).
920. R. J. Dwyer, J. Chem. Phys. **7**, 40 (1939).
921. ————, Astrophys. J. **100**, 300 (1944).
922. ———— and O. Oldenberg, J. Chem. Phys. **12**, 351 (1944).
922a. B. Eisler and R. F. Barrow, Proc. Phys. Soc. London **62A**, 740 (1949)
923. A. Elliott, Proc. Amst. **33**, 644 (1930).
924. ————, Proc. Roy. Soc. London **174**, 273 (1940).
924a. A. Eucken and G. Hoffmann, Z. physik. Chem. B **5**, 442 (1929).
925. E. Fagerholm, Ark. Mat. Astr. Fys. **27A**, No. 19 (1940).
926. K. Faltings, W. Groth and P. Harteck, Z. physik. Chem. B **41**, 15 (1938)
926a. U. Fano, Bur. Stand. J. Res. **40**, 215 (1948).
927. A. Farkas, Z. physik. Chem. B **5**, 467 (1929).
928. ————, Z. physik. Chem. B **10**, 419 (1930).

929. L. Farkas and H. Sachsse, Z. physik. Chem. B **23**, 1, 19 (1933).
930. ——— and L. Sandler, J. Chem. Phys. **8**, 248 (1940).
931. M. W. Feast, Nature **162**, 214 (1948).
931a. ———, Proc. Phys. Soc. A **62**, 121 (1949).
931b. ———, to be published.
932. B. T. Feld, Physic. Rev. **72**, 1116 (1947).
933. ——— and W. E. Lamb, Jr., Physic. Rev. **67**, 15 (1945).
934. W. F. C. Ferguson and I. Hudes, Physic. Rev. **57**, 705 (1940).
935. W. Finkelnburg and H. Hess, Physik. Z. **39**, 666 (1938).
936. ——— and W. Weizel, Z. Physik **68**, 577 (1931).
937. P. J. Flory and H. L. Johnston, J. Chem. Phys. **14**, 212 (1946).
938. H. M. Foley, Physic. Rev. **69**, 616 (1946).
939. ———, Physic. Rev. **71**, 747 (1947).
940. E. W. Foster and O. Richardson, Proc. Roy. Soc. London **189**, 149, 175 (1947).
941. C. A. Fowler, Jr., Physic. Rev. **59**, 645 (1941).
942. ———, Physic. Rev. **62**, 141 (1942).
943. W. R. Frederickson, M. E. Hogan, Jr., and W. W. Watson, Physic. Rev. **48**, 602 (1935).
944. ——— and C. R. Stannard, Physic. Rev. **44**, 632 (1933).
945. R. Frisch and O. Stern, Z. Physik **85**, 4 (1933).
946. H. M. Froslie and J. G. Winans, Physic. Rev. **72**, 481 (1947).
947. Y. Fujioka, Z. Physik **119**, 182 (1942).
948. G. W. Funke, Z. Physik **101**, 104 (1936).
949. J. Funke and C. F. E. Simons, Proc. Amst. **38**, 142 (1935).
949a. N. Fuson, J. Opt. Soc. **38**, 845 (1948).
949b. S. N. Garg, Ind. J. Phys. **23**, 161 (1949).
950. A. Gatterer, Ricerche spettroscop. **1**, 139 (1942).
951. ———, Ricerche spettroscop. **1**, 153 (1942).
952. ———, G. Piccardi and F. Vincenzi, Ricerche spettroscop. **1**, 181 (1942).
953. A. G. Gaydon, Proc. Roy. Soc. London **178**, 61 (1941).
954. ———, Proc. Roy. Soc. London **182**, 286 (1944).
955. ———, Proc. Phys. Soc. London **56**, 85 (1944).
956. ———, Proc. Phys. Soc. London **56**, 95, 160 (1944).
957. ———, Proc. Phys. Soc. London **58**, 525 (1946).
958. ———, Nature **161**, 731 (1948).
959. ——— and R. W. B. Pearse, Nature **141**, 370 (1938).
960. ——— ———, Proc. Roy. Soc. London **173**, 28 (1939).
961. ——— ———, Proc. Roy. Soc. London **173**, 37 (1939).
962. ——— and W. G. Penney, Proc. Roy. Soc. London **183**, 374 (1945).
963. ——— and G. Whittingham, Proc. Roy. Soc. London **189**, 313 (1947).
964. ——— and H. G. Wolfhard, Proc. Roy. Soc. London **194**, 169 (1948).
965. ——— and R. E. Worley, Nature **153**, 747 (1944).
966. L. Gerö, Z. Physik **95**, 747 (1935).
967. ———, Z. Physik **96**, 669 (1935).
968. ———, Z. Physik **109**, 204 (1938).
969. ———, Z. Physik **109**, 210 (1938).
970. ———, Z. Physik **109**, 216 (1938).
971. ———, Ann. Physik **35**, 597 (1939).
972. ———, Z. Physik **116**, 379 (1940).
973. ———, Z. Physik **117**, 709 (1941).
974. ———, Z. Physik **118**, 27 (1941).
975. ——— and K. Lörinczi, Z. Physik **113**, 449 (1939).
976. ——— and R. Schmid, Z. Physik **116**, 598 (1940).
977. ——— ———, Physic. Rev. **60**, 363 (1941).
978. ——— ———, Z. Physik **118**, 210 (1941).
978a. ——— ———, Z. Physik **118**, 250 (1941).
979. ——— ———, Z. Physik **121**, 459 (1943).
980. ——— ———, Proc. Phys. Soc. London **60**, 533 (1948).
981. ——— ——— and K. F. von Szily, Physica **11**, 144 (1944).

982. W. F. Giauque, J. Am. Chem. Soc. **52**, 4816 (1930).
983. ——— and R. Overstreet, J. Am. Chem. Soc. **54**, 1731 (1932).
984. ——— and T. M. Powell, J. Am. Chem. Soc. **61**, 1970 (1939).
985. G. E. Gibson and A. Macfarlane, Physic. Rev. **46**, 1058 (1934).
985a. D. A. Gilbert, A. Roberts and P. A. Griswold, Physic. Rev. **76**, 1723 (1949).
986. R. H. Gillette and E. H. Eyster, Physic. Rev. **56**, 1113 (1939).
986a. N. S. Gingrich, Rev. Mod. Phys. **15**, 90 (1943).
987. N. Ginsburg and G. H. Dieke, Physic. Rev. **59**, 632 (1941).
988. G. Glockler and G. E. Evans, J. Chem. Phys. **10**, 606 (1942).
989. W. E. Good, Physic. Rev. **70**, 213 (1946).
990. C. F. Goodeve and B. A. Stephens, Trans. Faraday Soc. **32**, 1517 (1936).
991. A. R. Gordon, J. Chem. Phys. **5**, 350 (1937).
992. ——— and C. Barnes, J. Chem. Phys. **1**, 692 (1933).
993. W. Gordy, J. Chem. Phys. **14**, 305 (1946).
994. ———, J. Chem. Phys. **15**, 305 (1947).
994a. ———, Rev. Mod. Phys. **20**, 668 (1948).
994b. ———, H. Ring and A. B. Burg, Physic. Rev. **74**, 1191 (1948); **75**, 208 (1949).
995. ———, J. W. Simmons and A. G. Smith, Physic. Rev. **72**, 344 (1947).
996. B. Grabe and E. Hulthén, Z. Physik **114**, 470 (1939).
997. L. Grabner and V. Hughes, to be published.
998. C. Gregory, Physic. Rev. **61**, 465 (1942).
999. W. Grether, Ann. Physik **26**, 1 (1936).
1000. B. Grundström, Z. Physik **75**, 302 (1932).
1001. ———, Z. Physik **97**, 171 (1935).
1002. ———, Z. Physik **98**, 128 (1935).
1003. ———, Z. Physik **113**, 721 (1939).
1004. ———, Z. Physik **115**, 120 (1940).
1005. ——— and A. Lagerqvist, Ark. Mat. Astr. Fys. **28B**, No. 8 (1942).
1006. K. M. Guggenheimer, Proc. Phys. Soc. London **58**, 456 (1946).
1007. R. Guillien, C. R. Paris **198**, 1223, 1486 (1934); **202**, 1373, 1423 (1936).
1008. A. Guntsch, Z. Physik **107**, 420 (1937).
1009. ———, Z. Physik **110**, 549 (1938).
1010. ———, Dissertation (Stockholm, 1939).
1011. ———, Ark. Mat. Astr. Fys. **31A**, No. 22 (1945).
1012. ———, Ark. Mat. Astr. Fys. **33A**, No. 2 (1945).
1013. ———, Nature **157**, 662 (1946).
1013a. D. ter Haar, Bull. Astron. Inst. Netherlands **10**, 1 (1943).
1014. D. ter Haar, Physic. Rev. **70**, 222 (1946).
1015. H. D. Hagstrum, Physic. Rev. **72**, 947 (1947).
1016. ———, J. Chem. Phys. **16**, 848 (1948).
1017. ——— and J. T. Tate, Physic. Rev. **59**, 354 (1941).
1018. ——— ———, Physic. Rev. **59**, 509 (1941).
1018a. H. Hamada, Phil. Mag. **12**, 50 (1931).
1019. G. E. Hansche, Physic. Rev. **57**, 289 (1940).
1020. R. E. Harrington, Dissertation (University of California, 1942).
1021. A. Harvey and H. Bell, Proc. Phys. Soc. London **47**, 415 (1935).
1022. A. Heimer, Z. Physik **105**, 56 (1937).
1023. T. Heimer, Naturwiss. **24**, 521 (1936).
1024. W. Heisenberg, Z. Physik **31**, 617 (1926).
1024a. W. Heitler, Physik. Z. **31**, 185 (1930).
1025. V. Henri and F. Wolff, J. de phys. (6) **10**, 81 (1929).
1026. L. R. Henrich, Astrophys. J. **99**, 59 (1944).
1026a. G. H. Herbig, Astrophys. J. **109**, 109 (1949).
1027. L. Herman and R. Herman, C. R. Paris **201**, 714 (1935).
1028. R. Herman, C. R. Paris **217**, 141 (1943).
1029. ———, Ann. de Phys. (11) **20**, 241 (1945).
1030. ———, Nature **157**, 843 (1946).
1031. ———, C. R. Paris **222**, 492 (1946).

1032. R. Herman and A. G. Gaydon, J. de phys. (8) **7**, 121 (1946).
1033. ———— ————, Proc. Phys. Soc. London **58**, 292 (1946).
1033a. ———— and L. Herman, J. de phys. (8) **7**, 203 (1946).
1034. ———— and L. Herman, J. de phys. (8) **9**, 160 (1948).
1035. ———— ———— and J. Gauzit, Nature **156**, 114 (1945).
1035a. R. C. Herman, H. S. Hopfield, G. A. Hornbeck and S. Silverman, J. Chem. Phys. **17**, 220 (1949).
1036. G. Herzberg, Z. Physik **49**, 512 (1928).
1037. ————, Z. Physik **52**, 815 (1929).
1038. ————, J. Chem. Phys. **10**, 306 (1942).
1039. ————, Rev. Mod. Phys. **14**, 195 (1942).
1040. ————, Astrophys. J. **96**, 314 (1942).
1041. ————, Physic. Rev. **69**, 362 (1946).
1042. ————, Physic. Rev. **70**, 762 (1946).
1043. ————, in Ann. Rep. Dir. Yerkes Obs., Astronom. J. **52**, 146 (1947).
1044. ————, in G. P. Kuiper's "The Atmospheres of the Earth and Planets" (Chicago, 1949) p. 346.
1045. ————, Nature **163**, 170 (1949).
1046. ————, Can. J. Res. A **28**, 144 (1950).
1047. ———— and L. Herzberg, Nature **161**, 283 (1948).
1048. ———— ———— and G. G. Milne, Can. J. Res. A **18**, 139 (1940).
1049. ———— and W. Hushley, Can. J. Res. A **19**, 127 (1941).
1050. ———— and L. G. Mundie, J. Chem. Phys. **8**, 263 (1940).
1051. ———— and J. G. Phillips, Astrophys. J. **108**, 163 (1948).
1051a. ———— and K. N. Rao, J. Chem. Phys. **17**, 1099 (1949).
1052. ———— and R. B. Sutton, Can. J. Res. A **18**, 74 (1940).
1053. L. Herzberg and G. Herzberg, Astrophys. J. **105**, 353 (1947).
1054. C. N. Hinshelwood, Proc. Roy. Soc. London **188**, 1 (1946).
1055. J. A. Hipple, R. E. Fox and E. U. Condon, Physic. Rev. **69**, 347 (1946).
1056. E. Hirschlaff, Z. Physik **75**, 315 (1932).
1057. W. Holst, Z. Physik **86**, 338 (1933).
1058. ————, Z. Physik **89**, 40 (1934).
1059. ————, Thesis (Stockholm, 1935).
1060. J. J. Hopfield, Physic. Rev. **31**, 1131 (1928).
1061. ————, Astrophys. J. **72**, 133 (1930).
1062. ———— and H. E. Clearman, Jr., Physic. Rev. **73**, 876 (1948).
1063. T. Hori, Z. Physik **61**, 481 (1930).
1064. ————, Z. Physik **62**, 352 (1930).
1065. ————, Mem. Ryojun Coll. Eng. **6**, 115 (1933).
1066. ———— and Y. Endô, Proc. Phys. Math. Soc. Jap. **23**, 834 (1941).
1067. T. Horie, Proc. Phys. Math. Soc. Japan **21**, 143 (1939).
1068. H. G. Howell, Proc. Roy. Soc. London **182**, 95 (1943).
1069. ————, Proc. Phys. Soc. London **57**, 32 (1945).
1070. ————, Proc. Phys. Soc. London **57**, 37 (1945).
1071. ————, Proc. Phys. Soc. London **59**, 107 (1947).
1072. ———— and N. Coulson, Proc. Roy. Soc. London **166**, 238 (1938).
1073. ———— ————, Proc. Phys. Soc. London **53**, 706 (1941).
1074. M. L. Huggins, J. Chem. Phys. **3**, 473 (1935); **4**, 308 (1936).
1075. H. K. Hughes, Physic. Rev. **70**, 570 (1946).
1076. ————, Physic. Rev. **72**, 614 (1947).
1077. Y. Hukumoto, Sci. Rep. Tohoku Univ. **18**, 581, 585, 597 (1929); **19**, 301 (1930).
1078. H. M. Hulburt and J. O. Hirschfelder, J. Chem. Phys. **9**, 61 (1941).
1079. E. Hulthén, Z. Physik **50**, 319 (1928).
1080. ————, Z. Physik **113**, 126 (1939).
1081. J. Hunaerts, Ann. d'Astrophys. **10**, 237 (1947).
1082. E. Hutchisson, Physic. Rev. **36**, 410 (1930).
1083. S. Imanishi, Sci. Pap. Inst Physic. Chem. Res. (Tokyo) **10**, 193, 237 (1929).
1084. ————, Sci. Pap. Inst. Physic. Chem. Res. (Tokyo) **11**, 139 (1929).

1084a. S. Imanishi, Sci. Pap. Inst. Physic. Chem. Res. (Tokyo) **39**, 45 (1941).
1085. M. Ishaq, Proc. Phys. Soc. London **53**, 355 (1941).
1086. ———, Ind. J. Phys. **18**, 52 (1944).
1087. ——— and R. W. B. Pearse, Proc. Roy. Soc. London **173**, 265 (1939).
1088. A. Jabloński, Acta Phys. Pol. **6**, 350 (1937).
1089. ———, Physica **7**, 541 (1940).
1090. ———, Physic. Rev. **68**, 78 (1945).
1091. D. Jack, Proc. Roy. Soc. London **120**, 222 (1928).
1092. C. V. Jackson, Proc. Roy. Soc. London **126**, 373 (1930).
1093. A. V. Jakovleva, Z. Physik **69**, 548 (1931).
1094. ———, J. Exper. and Theor. Phys. U.S.S.R. **9**, 302 (1939).
1095. ——— and V. Kondratjew, Acta Physicochim. U.S.S.R. **13**, 241 (1940).
1096. J. Janin, Ann. de Phys. (12) **1**, 538 (1946).
1097. C. K. Jen, Physic. Rev. **74**, 1396 (1948).
1098. F. A. Jenkins, Physic. Rev. **72**, 169 (1947).
1099. ———, Physic. Rev. **74**, 355 (1948).
1100. C. G. Jennergren, Nature **161**, 315 (1948); Ark. Mat. Astr. Fys. **35A**, No. 22 (1948).
1101. C. R. Jeppesen, Physic. Rev. **54**, 68 (1938).
1102. W. Jevons, Proc. Roy. Soc. London **106**, 174 (1924).
1103. ———, Proc. Roy. Soc. London **122**, 211 (1929).
1104. ———, Proc. Phys. Soc. London **41**, 520 (1929).
1104a. J. Joffe, Physic. Rev. **45**, 468 (1934).
1105. R. C. Johnson and H. G. Jenkins, Proc. Roy. Soc. London **116**, 327 (1927).
1106. ——— and N. R. Tawde, Phil. Mag. **13**, 501 (1932).
1107. V. A. Johnson, Physic. Rev. **60**, 373 (1941).
1108. H. L. Johnston, Physic. Rev. **45**, 79 (1934).
1109. ——— and A. T. Chapman, J. Am. Chem. Soc. **55**, 153 (1933).
1110. ——— and D. H. Dawson, J. Am. Chem. Soc. **55**, 2744 (1933).
1111. ——— ——— and M. K. Walker, Physic. Rev. **43**, 473, 980 (1933).
1112. ——— and E. A. Long, J. Chem. Phys. **2**, 389, 710 (1934).
1113. ——— and M. K. Walker, J. Am. Chem. Soc. **55**, 172 (1933); **57**, 682 (1935).
1113a. W. M. Jones, J. Chem. Phys. **16**, 1077 (1948).
1113b. A. V. Jones, to be published.
1114. J. Kaplan, Physic. Rev. **31**, 997 (1928).
1115. ———, Physic. Rev. **35**, 1298 (1930); **36**, 784, 788 (1930).
1116. ———, Physic. Rev. **46**, 534, 631 (1934).
1117. ———, Physic. Rev. **47**, 193 (1935).
1118. ———, Physic. Rev. **47**, 259 (1935).
1119. ———, Physic. Rev. **71**, 274 (1947).
1120. A. Kastler, Ann. de Géophys. **2**, 315 (1946).
1121. P. C. Keenan, Astrophys. J. **96**, 101 (1942).
1122. ———, Astrophys. J. **107**, 420 (1948).
1123. ——— and W. W. Morgan, Astrophys. J. **94**, 501 (1941).
1124. W. H. Keesom and K. W. Taconis, Physica **3**, 237 (1936).
1125. J. M. B. Kellogg and S. Millman, Rev. Mod. Phys. **18**, 323 (1946).
1126. ———, I. I. Rabi, N. F. Ramsey, Jr., and J. R. Zacharias, Physic. Rev. **56**, 728 (1939).
1127. ——— ——— ——— ——— ———, Physic. Rev. **57**, 677 (1940).
1127a. C. C. Kiess and E. Z. Stowell, Bur. Stand. J. Res. **12**, 459 (1934).
1128. M. Kimura and M. Miyanishi, Sci. Pap. Inst. Physic. Chem. Res. (Tokyo) **10**, 33 (1929).
1129. ——— and Y. Uchida, Sci. Pap. Inst. Physic. Chem. Res. (Tokyo) **18**, 109 (1932).
1130. A. S. King, Astrophys. J. **70**, 105 (1929).
1131. ———, Publ. Astron. Soc. Pac. **54**, 157 (1942).
1132. ——— and P. Swings, Astrophys. J. **101**, 6 (1945).
1133. G. W. King and J. H. Van Vleck, Physic. Rev. **55**, 1165 (1939).
1134. G. B. Kistiakowsky and H. Gershinowitz, J. Chem. Phys. **1**, 432 (1933).

1135. B. Kleman and E. Lindholm, Ark. Mat. Astr. Fys. **32B**, No. 10 (1945).
1136. J. K. Knipp, Physic. Rev. **53**, 734 (1938).
1136a. E. Kondratieva and V. Kondratiev, J. Phys. Chem. U.S.S.R. **14**, 1528 (1940).
1137. V. Kondratiev, C. R. U.S.S.R. **20**, 547 (1938).
1138. ——— and M. Ziskin, Acta Phys. Chim. **6**, 307 (1937).
1139. I. Kovács and A. Budó, J. Chem. Phys. **15**, 166 (1947).
1139a. ——— ———, Hung. Acta Physica **1**, No. 4 (1949).
1139b. ——— and S. Singer, Physik. Z. **43**, 362 (1942).
1140. H. A. Kramers and D. ter Haar, Bull. Astron. Inst. Netherlands **10**, 137 (1946).
1141. A. Kratzer, Ann. de Phys. **71**, 72 (1923).
1142. N. V. Kremenewsky, C. R. U.S.S.R. **3**, 251 (1935).
1143. M. K. Krogdahl, Astrophys. J. **100**, 311, 333 (1944).
1144. R. de L. Kronig, Z. Physik **75**, 468 (1932).
1145. ———and I. I. Rabi, Physic. Rev. **29**, 262 (1927).
1146. G. P. Kuiper, Astrophys. J. **100**, 378 (1944).
1147. ———, Astrophys. J. **106**, 251 (1947).
1148. ———, The Atmospheres of the Earth and Planets (Chicago, 1949) p. 304.
1148a. ———, Astrophys. J. **109**, 540 (1949).
1149. ———, W. Wilson and R. J. Cashman, Astrophys. J. **106**, 243 (1947).
1150. P. Kusch and F. W. Loomis, Physic. Rev. **49**, 217 (1936).
1151. R. Ladenburg and C. C. Van Voorhis, Physic. Rev. **43**, 315 (1933).
1152. R. T. Lageman, A. H. Nielsen and F. P. Dickey, Physic. Rev. **72**, 284 (1947).
1153. A. Lagerqvist, Ark. Mat. Astr. Fys. **29A**, No. 25 (1943).
1154. ———, Ark. Mat. Astr. Fys. **33A**, No. 8 (1946).
1155. ———, Ark. Mat. Astr. Fys. **34B**, No. 23 (1947).
1156. ———, Dissertation (Stockholm, 1948).
1158. ——— and U. Uhler, Arkiv Fysik, **1**, 459 (1949).
1159. ——— and R. Westöö, Ark. Mat. Astr. Fys. **31A**, No. 21 (1945).
1160. ——— ———, Ark. Mat. Astr. Fys. **32A**, No. 10 (1945).
1161. L. Lal, Physic. Rev. **73**, 255 (1948).
1162. ———, Nature **161**, 477 (1948).
1163. H. R. L. Lamont, Proc. Phys. Soc. London **61**, 562 (1948).
1164. O. G. Landsverk, Physic. Rev. **56**, 769 (1939).
1165. J. M. Lejeune, Bull. Soc. Roy. Sci. Liége **14**, 318 (1945).
1166. ——— and B. Rosen, Bull. Soc. Roy. Sci. Liége **14**, 81 (1945).
1167. ——— ———, Bull. Soc. Roy. Sci. Liége **14**, 322 (1945).
1168. L. C. Lewis, Z. Physik **69**, 786 (1931).
1169. A. Lindh and Å. Nilsson, Ark. Mat. Astr. Fys. **29A**, No. 27 (1943).
1170. E. Lindholm, Naturwiss. **27**, 470 (1939).
1171. ———, Ark. Mat. Astr. Fys. **28B**, No. 3 (1941).
1172. ———, Dissertation (Uppsala, 1942).
1173. ———, Ark. Mat. Astr. Fys. **29B**, No. 15 (1943).
1174. ———, Ark. Mat. Astr. Fys. **32A**, No. 17 (1945).
1175. J. W. Linnett, Trans. Faraday Soc. **36**, 1123 (1940).
1176. ———, Trans. Faraday Soc. **38**, 1 (1942).
1177. L. H. Long, J. Chem. Phys. **16**, 1087 (1948).
1178. F. W. Loomis and A. J. Allen, Physic. Rev. **33**, 639 (1929).
1179. ——— and T. F. Watson, Physic. Rev. **45**, 805 (1934).
1179a. W. Low and C. H. Townes, Physic. Rev. **75**, 529 (1949).
1180. F. Lowater, Proc. Phys. Soc. London **41**, 557 (1929).
1181. R. W. Lunt, R. W. B. Pearse and E. C. W. Smith, Proc. Roy. Soc. London **151**, 602 (1935).
1182. R. H. Lyddane, F. T. Rogers, Jr., and F. E. Roach, Physic. Rev. **60**, 281 (1941).
1183. D. Maeder, Helv. Phys. Acta **16**, 503 (1943).
1184. R. Maeder, Helv. Phys. Acta **21**, 411 (1948).
1185. ——— and E. Miescher, Nature **161**, 393 (1948).

1186. P. C. Mahanti, Nature **127**, 557 (1931).
1187. ———, Z. Physik **68**, 114 (1931).
1188. ———, Z. Physik **88**, 550 (1934).
1189. ———, Ind. J. Phys. **9**, 369 (1935).
1190. ———, Ind. J. Phys. **9**, 455 (1935).
1191. ——— and A. K. Sen Gupta, Ind. J. Phys. **13**, 331 (1939).
1192. L. Malet and B. Rosen, Bull. Soc. Roy. Sci. Liége **14**, 382 (1945).
1193. ——— ———, Bull. Inst. Roy. Col. Belge **1945**, 377.
1194. M. M. Mann, A. Hustrulid and J. T. Tate, Physic. Rev. **58**, 340 (1940).
1195. E. J. Marais, Physic. Rev. **70**, 499 (1946).
1196. H. Margenau, Rev. Mod. Phys. **11**, 1 (1939).
1197. ———, Physic. Rev. **56**, 1000 (1939).
1198. L. S. Mathur, Proc. Roy. Soc. London **162**, 83 (1937).
1198a. ———, Ind. J. Phys. **11**, 177 (1937).
1199. L. R. Maxwell, J. Opt. Soc. **30**, 374 (1940).
1200. ——— and V. M. Mosley, Physic. Rev. **57**, 21 (1940).
1201. J. E. Mayer and L. Helmholz, Z. Physik **75**, 19 (1932).
1202. K. J. McCallum and E. Leifer, J. Chem. Phys. **8**, 505 (1940).
1203 F. C. McDonald, Physic. Rev. **29**, 212 (1927).
1204. F. K. McGowan and J. G. Winans, Physic. Rev. **65**, 349A (1945).
1205. A. McKellar, Publ. Astron. Soc. Pac. **52**, 307 (1940).
1206. ———, Publ. Dominion Astrophys. Observ. **7**, 251 (1941).
1207. ———, Rev. Mod. Phys. **14**, 179 (1942).
1208. ———, Astrophys. J. **98**, 1 (1943).
1209. ———, Astrophys. J. **99**, 162 (1944).
1210. ———, Astrophys. J. **100**, 69 (1944).
1211. ———, J. Roy. Astron. Soc. Can. **41**, 147 (1947).
1212. ———, Astrophys. J. **108**, 453 (1948).
1213. ———, Publ. Astron. Soc. Pac. **59**, 186 (1947); Publ. Dom. Astrophys.
    Obs. **7**, 395 (1949).
1213a. ———, Publ. Astron. Soc. Pac. **61**, 34 (1949).
1214. ——— and W. Buscombe, Publ. Dom. Astrophys. Observ. **7**, 361 (1948).
1215. ——— and F. A. Jenkins, Publ. Dom. Astrophys. Observ. **7**, 155 (1939).
1216. ——— and P. Swings, Astrophys. J. **108**, 458 (1948).
1217. R. R. McMath, O. C. Mohler and L. Goldberg, Physic. Rev. **73**, 1203
    (1948); **74**, 623 (1948).
1218. R. Mecke, Ann. Physik **71**, 104 (1923).
1219. W. F. Meggers, Bur. Stand. J. Res. **10**, 669 (1933).
1220. ——— and G. H. Dieke, Bur. Stand. J. Res. **9**, 121 (1932).
1221. ——— and J. A. Wheeler, Bur. Stand. J. Res. **6**, 239 (1931).
1222. ——— ———, Bur. Stand. J. Res. **9**, 268 (1932).
1222a. A. B. Meinel, Astrophys. J. **111** (1950).
1223. P. W. Merrill, Astrophys. J. **83**, 126 (1936).
1224. ——— and O. C. Wilson, Astrophys. J. **87**, 9 (1938).
1225. P. Mesnage, Ann. de Phys. (11) **12**, 5 (1939).
1226. N. Metropolis, Physic. Rev. **55**, 636 (1939).
1227. ——— and H. Beutler, Physic. Rev. **55**, 1113 (1939).
1228. K. Mie, Z. Physik **91**, 475 (1934).
1229. E. Miescher, Helv. Phys. Acta **8**, 279 (1935).
1230. ———, Helv. Phys. Acta **8**, 587 (1935).
1231. ———, Helv. Phys. Acta **9**, 693 (1936).
1232. ———, Helv. Phys. Acta **14**, 148 (1941).
1233. ——— and M. Chrétien, Nature **163**, 996 (1949); Helv. Phys. Acta **22**,
    588 (1949).
1233a. ——— ———, (unpublished).
1234. M. V. Migeotte, Physic. Rev. **73**, 519 (1948); Astrophys. J. **107**, 400
    (1948).
1235. ———, Physic. Rev. **74**, 112 (1948).
1235a. ———, in G. P. Kuiper's "The Atmospheres of the Earth and Planets"
    (Chicago, 1949) p. 284.
1235b. ———, Physic. Rev. **75**, 1108 (1949).

1235c. M. V. Migeotte and R. M. Chapman, Physic. Rev. **75**, 1611 (1949).
1236. R. Migeotte, Bull. Soc. Roy. Sci. Liége **12**, 658 (1941).
1237. ———, Bull. Soc. Roy. Sci. Liége **1942**, 48.
1238. ———, Mem. Soc. Roy. Sci. Liége (4) **5**, 3 (1942).
1239. ——— and B. Rosen, Bull. Soc. Roy. Sci. Liége **1944**, 248.
1240. C. E. Miller, J. Chem. Phys. **6**, 902 (1938).
1240a. R. M. Milton, Chem. Rev. **39**, 419 (1946).
1240b. O. C. Mohler, L. Goldberg and R. R. McMath, Physic. Rev. **74**, 352
        (1948).
1241. A. Monfils and B. Rosen, Nature **164**, 713 (1949).
1242. H. Moureu, B. Rosen and G. Wetroff, C. R. Paris **209**, 207 (1939).
1243. S. Mrozowski, Z. Physik **62**, 314 (1930).
1244. ———, Z. Physik **95**, 524 (1935); **99**, 236 (1936).
1245. ———, Acta Phys. Pol. **4**, 405 (1935).
1246. ———, Acta Phys. Pol. **5**, 85 (1936).
1247. ———, Bull. Acad. Pol. **1937A**, 295.
1248. ———, Acta Phys. Pol. **7**, 45 (1938).
1249. ———, Physic. Rev. **58**, 332 (1940).
1250. ———, Physic. Rev. **58**, 597 (1940).
1251. ———, Physic. Rev. **63**, 63 (1943).
1252. E. F. Mueller and F. D. Rossini, Am. J. Phys. **12**, 1 (1944).
1253. L. A. Müller, Ann. Physik **82**, 39 (1927).
1254. W. Müller, Helv. Phys. Acta **16**, 3 (1943).
1255. R. S. Mulliken, Physic. Rev. **26**, 1 (1925).
1256. ———, Physic. Rev. **26**, 319 (1925).
1257. ———, Physic. Rev. **32**, 388 (1928).
1258. ———, Physic. Rev. **38**, 836 (1931).
1259. ———, J. Chem. Phys. **3**, 573 (1935).
1260. ———, Physic. Rev. **56**, 778 (1939).
1261. ———, J. Chem. Phys. **7**, 14, 20 (1939).
1262. ———, Astrophys. J. **89**, 283 (1939).
1263. ———, Physic. Rev. **57**, 500 (1940).
1264. ———, J. Chem. Phys. **8**, 234 (1940).
1265. ———, J. Chem. Phys. **8**, 382 (1940).
1266. ———, Physic. Rev. **61**, 277 (1942).
1266a. ———, J. Chim. Phys. **46**, 497 (1949).
1267. ——— and G. S. Monk, Physic. Rev. **34**, 1530 (1929).
1268. ——— and C. A. Rieke, Rep. Progress in Physics **8**, 231 (1941).
1269. G. M. Murphy and J. E. Vance, J. Chem. Phys. **7**, 806 (1939).
1270. G. Nakamura and T. Shidei, Jap. J. Phys. **10**, 5 (1934).
1270a. J. J. Nassau, G. B. van Albada and P. C. Keenan, Astrophys. J. **109**,
        333 (1949).
1271. L. Natanson, Acta Phys. Pol. **7**, 275 (1938).
1272. S. M. Naudé, South African J. Sci. **41**, 128 (1945).
1273. ———, Nature **155**, 426 (1945).
1273a. ——— and H. Verleger, Proc. Phys. Soc. London (in press).
1274. H. Neuimin and A. Terenin, Acta Phys. Chim. **5**, 465 (1936).
1275. T. E. Nevin, Proc. Phys. Soc. London **43**, 554 (1931).
1276. ———, Phil. Mag. **20**, 347 (1935).
1277. ———, Proc. Roy. Soc. London **174**, 371 (1940).
1278. ———, Proc. Roy. Irish Acad. A **48**, 1 (1942); **50**, 123 (1945).
1279. ——— and T. Murphy, Proc. Roy. Irish Acad. **46A**, 169 (1941).
1279a. R. Newburgh and K. Wieland, Helv. Phys. Acta **22**, 590 (1949) and
        private communication.
1280. J. L. Nickerson, Physic. Rev. **47**, 707 (1935).
1281. M. Nicolet, Mem. Soc. Roy. Sci. Liége (4) **2**, 89 (1937).
1282. ———, Z. Astrophysik **15**, 145 (1938).
1282a. W. A. Nierenberg, I. I. Rabi and M. Slotnick, Physic. Rev. **73**, 1430
        (1948).
1283. W. A. Nierenberg and N. F. Ramsey, Physic. Rev. **72**, 1075 (1947).

1284. B. E. Nilsson, Ark. Mat. Astr. Fys. **35A**, No. 19 (1948).
1284a. ———, Ark. Mat. Astr. Fys. **31**, No. 4 (1944).
1285. ———, Dissertation (Stockholm, 1948).
1285a. P. Nolan and F. A. Jenkins, Physic. Rev. **50**, 943 (1936).
1286. B. O'Brien, J. Opt. Soc. **36**, 369 (1946).
1287. E. Oeser, Z. Physik **95**, 699 (1935).
1288. Y. Öhman, Stockholm Obs. Ann. **13**, No. 11 (1941).
1289. T. Okuda, Nature **138**, 168 (1936).
1290. O. Oldenberg, J. Chem. Phys. **3**, 266 (1935).
1291. ——— and A. A. Frost, Chem. Rev. **20**, 99 (1937).
1292. ———, J. E. Morris, C. J. Morrow, E. G. Schneider and H. S. Sommers, Jr., J. Chem. Phys. **14**, 16 (1946).
1293. ——— and H. S. Sommers, Jr., J. Chem. Phys. **8**, 468 (1940); **9**, 114, 432, 573 (1941).
1294. C. M. Olmsted, Z. wiss. Phot. **4**, 293 (1906).
1295. ———, Z. wiss. Phot. **7**, 300 (1906).
1296. E. Olsson, C. R. Paris **204**, 1182 (1937).
1297. ———, Ark. Mat. Astr. Fys. **26B**, No. 9 (1938).
1298. ———, Ark. Mat. Astr. Fys. **26B**, No. 10 (1938).
1299. L. S. Ornstein and H. Brinkman, Proc. Amst. **34**, 33 (1931).
1300. ——— ———, Proc. Amst. **34**, 498 (1931).
1300a. H. Oura and M. Ninomiya, Proc. Phys. Math. Soc. Japan **25**, 335 (1943).
1300b. M. L. Oxholm and D. Williams, Physic. Rev. **76**, 151 (1949).
1301. R. C. Pankhurst, Proc. Phys. Soc. (London) **62A**, 191 (1949).
1302. G. Pannetier and A. G. Gaydon, Nature **161**, 242 (1948).
1303. A. E. Parker, Physic. Rev. **41**, 274 (1932).
1304. ———, Physic. Rev. **44**, 90, 914 (1933).
1305. ———, Physic. Rev. **45**, 752A (1934).
1305a. T. F. Parkinson, E. A. Jones and A. H. Nielsen, Physic. Rev. **76**, 199 (1949).
1306. L. Pauling and J. Y. Beach, Physic. Rev. **47**, 686 (1935).
1307. R. W. B. Pearse, Proc. Roy. Soc. London **125**, 157 (1929).
1307a. ——— and M. W. Feast, Nature **163**, 686 (1949).
1308. ——— and A. G. Gaydon, Proc. Phys. Soc. London **50**, 711 (1938).
1309. ——— and S. P. Sinha, Nature **160**, 159 (1947).
1310. C. L. Pekeris, Physic. Rev. **45**, 98 (1934).
1311. M. L. Perlman and G. K. Rollefson, J. Chem. Phys. **9**, 362 (1941).
1312. H. Petersen, Z. Physik **98**, 569 (1936).
1313. A. Petrikaln, Z. Physik **51**, 395 (1928).
1314. J. G. Phillips, Astrophys. J. **107**, 389 (1948).
1315. ———, Astrophys. J. **108**, 434 (1948).
1316. ———, Astrophys. J. **110**, 73 (1949).
1316a. ———, Astrophys. J., to be published.
1317. G. Piccardi, Gazz. chim. Ital. **63**, 127 (1933).
1317a. ———, Atti Accad. Lincei **17**, 654 (1933).
1318. ———, Atti Accad. Lincei **21**, 584 (1935).
1318a. ———, Atti Accad. Lincei **21**, 589 (1935); **25**, 86 (1937).
1318b. M. E. Pillow, Proc. Phys. Soc. London **62A**, 237 (1949).
1319. G. Placzek and E. Teller, Z. Physik **81**, 209 (1933).
1320. D. Porrett, Proc. Roy. Soc. London **162**, 414 (1937).
1321. W. M. Preston, Nature **145**, 623 (1940); Physic. Rev. **57**, 887 (1940).
1322. P. Pringsheim, Physica **5**, 489 (1938).
1323. ——— and B. Rosen, Z. Physik **50**, 1 (1928).
1324. E. M. Purcell, R. V. Pound and N. Bloembergen, Physic. Rev. **70**, 986 (1946).
1325. ———, H. C. Torrey and R. V. Pound, Physic. Rev. **69**, 37 (1946).
1326. J. Querbach, Z. Physik **60**, 109 (1930).
1327. R. Ramanadham, G. V. S. R. Rao and C. Ramasastry, Ind. J. Phys. **20**, 161 (1946).
1328. C. Ramasastry, Ind. J. Phys. **22**, 95 (1948).

1329. C. Ramasastry, Ind. J. Phys. **22,** 119 (1948).
1329a. ———, Ind. J. Phys. **23,** 35 (1949).
1330. ——— and K. R. Rao, Ind. J. Phys. **20,** 100 (1946).
1331. ——— ———, Ind. J. Phys. **21,** 143 (1947).
1332. N. F. Ramsey, Jr., Physic. Rev. **58,** 226 (1940).
1333. ———, Physic. Rev. **74,** 286 (1948).
1334. D. H. Rank, J. Opt. Soc. **36,** 239 (1946).
1335. K. N. Rao, Astrophys. J. **110,** 304 (1949).
1335a. ———, Astrophys. J. **111** (1950).
1336. K. R. Rao and G. V. S. R. Rao, Ind. J. Phys. **18,** 281 (1944).
1337. ——— ———, Current Science **13,** 279 (1944).
1338. ———, M. G. Sastry and V. G. Krishnamurty, Ind. J. Phys. **18,** 323 (1944).
1339. P. T. Rao and K. R. Rao, Ind. J. Phys. **20,** 49 (1946).
1339a. ——— ———, Ind. J. Phys. **23,** 185 (1949).
1340. F. Rasetti, Proc. Nat. Acad. Am. **15,** 411 (1929).
1341. ———, Z. Physik **66,** 646 (1930).
1342. S. K. Ray, Ind. J. Phys. **16,** 35 (1942).
1343. A. L. G. Rees, Proc. Phys. Soc. London **59,** 998 (1947).
1344. ———, Proc. Phys. Soc. London **59,** 1008 (1947).
1345. F. Reiche and H. Rademacher, Z. Physik **39,** 444 (1926); **41,** 453 (1927).
1346. O. K. Rice, Physic. Rev. **35,** 1538, 1551 (1930).
1347. ———, J. Chem. Phys. **9,** 258 (1941).
1348. O. W. Richardson, Nuovo cimento **15,** 232 (1938).
1350. A. Roberts, Rev. Sci. Instruments **18,** 845 (1947).
1351. G. D. Rochester, Proc. Roy. Soc. London **150,** 668 (1935).
1352. ———, Physic. Rev. **56,** 305 (1939).
1353. ——— and E. Olsson, Z. Physik **114,** 495 (1939).
1354. M. T. Rogers, V. Schomaker and D. P. Stevenson, J. Am. Chem. Soc. **63,** 2610 (1941).
1355. B. Rosen, Z. Physik **43,** 69 (1927).
1356. ———, Bull. Soc. Roy. Sci. Liége **1944,** 176.
1357. ———, Physic. Rev. **68,** 124 (1945).
1358. ———, Nature **156,** 570 (1945).
1359. ——— and F. Bouffieux, Bull. Acad. Roy. Belg. **1936,** 885.
1360. ——— and M. Désirant, Bull. Soc. Roy. Sci. Liége **1935,** 233.
1361. ——— ——— and J. Duchesne, Physic. Rev. **48,** 916 (1935).
1362. ——— and F. Monfort, Physica **3,** 257 (1936).
1363. E. J. Rosenbaum, J. Chem. Phys. **6,** 16 (1938).
1364. E. Rosenthaler, Helv. Phys. Acta **13,** 355 (1940).
1365. D. Roy, Ind. J. Phys. **13,** 231 (1939).
1366. R. Rydberg, Z. Physik **73,** 74 (1931).
1367. B. Saha, Ind. J. Phys. **14,** 123 (1940).
1368. R. Samuel, J. Chem. Phys. **12,** 167, 180, 380, 521 (1944).
1369. I. Sandeman, Proc. Roy. Soc. Edinburgh **59,** 1, 130 (1938).
1370. ———, Proc. Roy. Soc. Edinburgh **60,** 210 (1940).
1371. R. F. Sanford, Publ. Astron. Soc. Pac. **38,** 177 (1926).
1372. ———, Publ. Astron. Soc. Pac. **41,** 271 (1929); **44,** 246 (1932).
1373. ———, Publ. Astron. Soc. Pac. **52,** 203 (1940).
1374. C. R. Sastry and K. R. Rao, Ind. J. Phys. **19,** 136 (1945).
1375. M. G. Sastry, Current Science **10,** 197 (1941).
1376. ———, Ind. J. Phys. **15,** 95, 455 (1941); **16,** 27, 169, 343 (1942).
1377. ——— and K. R. Rao, Ind. J. Phys. **15,** 27 (1941).
1378. S. Satoh, Sci. Pap. Inst. Physic. Chem. Res. (Tokyo) **29,** 53 (1936).
1379. P. W. Schenk, Z. Physik **106,** 271 (1937).
1380. ———, Z. physik. Chem. B **51,** 113 (1942).
1381. R. Schlapp, Physic. Rev. **39,** 806 (1932).
1382. R. Schmid, Z. Physik **64,** 279 (1930).
1383. ———, A. Budó and J. Zemplén, Z. Physik **103,** 250 (1936).
1384. ——— and L. Gerö, Z. Physik **86,** 297 (1933).

1385. R. Schmid and L. Gerö, Z. Physik **96**, 198 (1935).
1386. ——— ———, Z. Physik **96**, 546 (1935).
1387. ——— ———, Z. Physik **106**, 205 (1937).
1388. ——— ———, Ann. Physik **33**, 70 (1938).
1389. ——— ———, Z. Physik **115**, 47 (1940).
1390. ——— ———, Proc. Phys. Soc. London **58**, 701 (1946).
1391. H. D. Schmidt-Ott, Z. Physik **69**, 724 (1931).
1392. H. Schmitz and H. J. Schumacher, Z. Naturforsch. **2a**, 359 (1947).
1393. ——— ———, Z. Naturforsch. **2a**, 362 (1947).
1393a. H. Schüler, private comm.
1394. H. Schüler, H. Gollnow and E. Fechner, Ann. Physik **36**, 328 (1939).
1395. ——— ——— and H. Haber, Z. Physik **111**, 484 (1939).
1395a. ——— and A. Woeldike, Physik. Z. **44**, 335 (1943).
1396. ——— and A. Woeldike, Physik. Z. **45**, 171 (1944).
1397. M. K. Sen, Ind. J. Phys. **10**, 429 (1936).
1398. ———, Ind. J. Phys. **11**, 251 (1937).
1399. A. K. Sen Gupta, Ind. J. Phys. **13**, 145 (1939).
1400. ———, Ind. J. Phys. **17**, 216 (1943).
1401. ———, Ind. J. Phys. **18**, 182 (1944).
1401a. P. K. Sen-Gupta, Proc. Roy. Soc. London **143**, 438 (1933).
1402. R. B. Setlow, Physic. Rev. **74**, 153 (1948).
1404. G. A. Shajn, Bull. Abastumani Astrophys. Obs. No. 6 (1942).
1405. D. Sharma, Proc. Nat. Acad. India **14A**, 37 (1944).
1406. ———, Proc. Nat. Acad. India **14A**, 133 (1944).
1407. ———, Proc. Nat. Acad. India **14A**, 217 (1945).
1408. ———, Proc. Nat. Acad. India **14A**, 224, 228 (1945).
1409. ———, Proc. Nat. Acad. India **14A**, 232 (1945).
1410. ———, Nature **157**, 663 (1946).
1411. C. H. Shaw, Physic. Rev. **57**, 877, 881 (1940).
1412. J. H. Shaw, G. B. B. M. Sutherland and T. W. Wormell, Physic. Rev. **74**, 978 (1948).
1413. R. W. Shaw and H. C. Ketcham, Physic. Rev. **45**, 753A (1934).
1414. E. N. Shawhan and F. Morgan, Physic. Rev. **47**, 377 (1935).
1415. A. G. Shenstone, Physic. Rev. **72**, 411 (1947).
1416. C. Shin-Piaw, Ann. de Phys. (11) **10**, 173 (1938).
1417. E. F. Shrader, Physic. Rev. **64**, 57 (1943).
1418. N. L. Singh, Current Science **11**, 276 (1942).
1419. ———, Proc. Ind. Acad. **25A**, 1 (1946).
1420. ——— and L. Lal, Science and Culture **9**, 89 (1943).
1421. S. P. Sinha, Proc. Phys. Soc. London **59**, 610 (1947).
1422. ———, Proc. Phys. Soc. London **60**, 436 (1948).
1423. ———, Proc. Phys. Soc. London **60**, 443 (1948).
1424. ———, Proc. Phys. Soc. London **60**, 447 (1948).
1424a. ———, Ind. J. Phys. **22**, 401 (1948).
1424b. ———, Current Science **17**, 208 (1948).
1424c. ———, Proc. Phys. Soc. London **62A**, 124 (1949).
1425. J. C. Slater, Physic. Rev. **32**, 349 (1928).
1426. J. A. Smit, Physica **12**, 683 (1946).
1426a. D. F. Smith, M. Tidwell and D. V. P. Williams, Physic. Rev. **77**, 420 (1950).
1427. H. D. Smith and N. Barton, Trans. Roy. Soc. Canada **39** (III), 25 (1945).
1428. L. G. Smith, Rev. Sci. Instruments **13**, 54 (1942).
1429. N. D. Smith, Physic. Rev. **49**, 345 (1936).
1430. M. M. Smit-Miessen and J. L. Spier, Physica **9**, 193, 422 (1942).
1431. L. Sosnowski, J. Starkiewicz and O. Simpson, Nature **159**, 818 (1947).
1432. J. L. Spier and J. A. Smit, Physica **9**, 597 (1942).
1433. J. W. T. Spinks, Can. J. Res. A **20**, 1 (1942).
1433a. L. Spitzer, Jr., in G. P. Kuiper's "The Atmospheres of the Earth and Planets" (Chicago, 1949) p. 213.
1434. H. Sponer and W. W. Watson, Z. Physik **56**, 184 (1929).

1434a. J. Stebbins, A. E. Whitford and P. Swings, Astrophys. J. **101**, 39 (1945).
1435. G. Stenvinkel, Z. Physik **114**, 602 (1939).
1436. ———, E. Svensson and E. Olsson, Ark. Mat. Astr. Fys. **26A**, No. 10 (1938).
1437. B. Stepanov, J. Phys. U.S.S.R. **9**, 317 (1945).
1437a. M. W. P. Strandberg, C. Y. Meng and J. G. Ingersoll, Physic. Rev. **75**, 1524 (1949).
1437b. ———, T. Wentink, Jr., and A. G. Hill, Physic. Rev. **75**, 827 (1949).
1438. B. Strömgren, Astrophys. J. **108**, 242 (1948).
1439. J. Strong, Physic. Rev. **45**, 877 (1934).
1440. E. C. G. Stueckelberg, Physic. Rev. **42**, 518 (1932).
1441. T. S. Subbaraya, B. N. Rao and N. A. N. Rao, Proc. Ind. Acad. **5A**, 365 (1937).
1442. ———, N. A. N. Rao and B. N. Rao, Proc. Ind. Acad. **5A**, 372 (1937).
1443. G. B. B. M. Sutherland, J. Chem. Phys. **8**, 161 (1940).
1444. ———, D. E. Blackwell and P. B. Fellgett, Nature **158**, 874 (1946).
1445. ——— and G. S. Callendar, Rep. Progress in Physics **9**, 18 (1943).
1446. P. Swings, Astrophys. J. **95**, 270 (1941).
1447. ———, Lick Obs. Bull. **19**, 131 (1941).
1448. ———, Rev. Mod. Phys. **14**, 190 (1942).
1449. ———, Monthly Notices Roy. Astron. Soc. **103**, 86 (1943).
1450. ———, Astrophys. J. **97**, 72 (1943).
1451. ———, in G. P. Kuiper's "The Atmospheres of the Earth and Planets" (Chicago, 1949) p. 159.
1452. ———, C. T. Elvey and H. W. Babcock, Astrophys. J. **94**, 320 (1941); **95**, 218 (1942).
1453. ———, A. McKellar and R. Minkowski, Astrophys. J. **98**, 142 (1943).
1454. ——— and L. Rosenfeld, Astrophys. J. **86**, 483 (1937).
1455. T. Takamine, T. Suga and Y. Tanaka, Sci. Pap. Inst. Physic. Chem. Res. (Tokyo) **36**, 437 (1939).
1455a. ———, Y. Tanaka and M. Iwata, Sci. Pap. Inst. Physic. Chem. Res. (Tokyo) **40**, 371 (1943).
1455b. R. M. Talley, H. M. Kaylor and A. H. Nielsen, Physic. Rev. **77**, 529 (1950).
1456. T. Tanaka and T. Horie, Proc. Phys. Math. Soc. Japan **23**, 464 (1941).
1456a. Y. Tanaka, Sci. Pap. Inst. Physic. Chem. Res. (Tokyo) **39**, 447 (1942).
1456b. ———, Sci. Pap. Inst. Physic. Chem. Res. (Tokyo) **39**, 456 (1942).
1456c. ———, Sci. Pap. Inst. Physic. Chem. Res. (Tokyo) **39**, 465 (1942).
1456d. ———, Sci. Pap. Inst. Physic. Chem. Res. (Tokyo) **42**, 49 (1944).
1456e. ———, Sci. Pap. Inst. Physic. Chem. Res. (Tokyo) **43**, 28 (1948).
1457. Y. Tanaka and T. Takamine, Physic. Rev. **59**, 613 (1941); Sci. Pap. Inst. Physic. Chem. Res. (Tokyo) **39**, 427 (1942).
1458. ——— ———, Physic. Rev. **59**, 771 (1941); Sci. Pap. Inst. Physic. Chem. Res. (Tokyo) **39**, 437 (1942).
1459. N. R. Tawde and S. A. Trivedi, Proc. Phys. Soc. London **51**, 733 (1939).
1460 A. Terenin, Physica **10**, 209 (1930).
1461. ———, Physic. Rev. **36**, 147 (1930).
1462. J. Terrien, Ann. de Phys. (11) **9**, 477 (1938).
1463. S. Toh, Proc. Phys. Math. Soc. Japan **22**, 119 (1940).
1464. R. C. Tolman and R. M. Badger, J. Am. Chem. Soc. **45**, 2277 (1923).
1465. A. Tournaire and E. Vassy, C. R. Paris **201**, 957 (1935); **202**, 562 (1936).
1466. C. H. Townes, Physic. Rev. **71**, 909 (1947).
1466a. ——— and S. Geschwind, Physic. Rev. **74**, 626 (1948).
1467. ———, A. N. Holden, J. Bardeen and F. R. Merritt, Physic. Rev. **71**, 644 (1947).
1468. ——— ——— and F. R. Merritt, Physic. Rev. **72**, 513 (1947); **74**, 1113 (1948).
1469. ———, F. R. Merritt and B. D. Wright, Physic. Rev. **73**, 1334 (1948).
1470. ——— and W. R. Smythe, Physic. Rev. **56**, 850, 1210 (1939).
1471. J. W. Trischka, Physic. Rev. **74**, 718 (1948); **76**, 1365 (1949).

1472. W. M. Tschulanowsky, Bull. Acad. Sci. U.R.S.S. **1**, 1313 (1935).
1473. N. Tsi-Ze and C. Shang-Yi, J. de phys. **9**, 169 (1938).
1474. Y. Uchida, Sci. Pap. Inst. Physic. Chem. Res. (Tokyo) **30**, 71 (1936).
1475. —— and Y. Ota, Jap. J. Phys. **5**, 59 (1928).
1476. H. C. Urey, J. Am. Chem. Soc. **45**, 1445 (1923).
1477. E. E. Vago and R. F. Barrow, Proc. Phys. Soc. London **58**, 538 (1946).
1478. —— ——, Proc. Phys. Soc. London **58**, 707 (1946).
1479. —— ——, Proc. Phys. Soc. London **59**, 449 (1947).
1480. —— ——, Victor Henri commemorative vol. **1948**, 201.
1481. —— ——, private comm.
1482. W. M. Vaidya, Proc. Ind. Acad. **6A**, 122 (1937).
1483. ——, Proc. Ind. Acad. **7A**, 321 (1938).
1484. ——, Proc. Nat. Inst. Sci. India **7**, 89 (1941).
1485. J. G. Valatin, Proc. Phys. Soc. London **58**, 695 (1946).
1486. J. E. Vance and J. R. Huffman, Physic. Rev. **47**, 215 (1935).
1487. H. C. Van de Hulst, Ann. d'Astrophys. **8**, 12 (1945).
1488. J. H. Van Vleck, Physic. Rev. **71**, 413 (1947).
1489. L. Vegard, Physica **12**, 606 (1946).
1490. ——, Nature **162**, 300 (1948).
1491. —— and G. Kvifte, Geophys. Publ. **16**, No. 7 (1945).
1492. —— and E. Tønsberg, Geophys. Publ. **13**, No. 1 (1940); **16**, No. 2 (1944).
1493. P. Venkateswarlu, Proc. Ind. Acad. **24**, 473 (1946).
1494. ——, Proc. Ind. Acad. **24**, 480 (1946).
1495. ——, Proc. Ind. Acad. **25**, 119, 133 (1947).
1496. ——, Proc. Ind. Acad. **25**, 138 (1947).
1497. ——, Proc. Ind. Acad. **26**, 22 (1947).
1498. ——, Ind. J. Phys. **21**, 43 (1947).
1499. ——, private comm.
1499a. ——, Physic. Rev. **77**, 79 (1950).
1500. J. Verhaeghe, Wis-Natuurk. Tijdschr. **7**, 224 (1935).
1501. B. Vodar, C. R. Paris **204**, 1467 (1937).
1502. ——, J. de phys. **9**, 166 (1948).
1503. H. H. Voge, J. Chem. Phys. **16**, 984 (1948).
1504. P. G. Voorhoeve, Dissertation (Utrecht, 1946).
1505. A. T. Wager, Physic. Rev. **64**, 18 (1943).
1506. D. D. Wagman, J. E. Kilpatrick, W. J. Taylor, K. S. Pitzer and F. D. Rossini, Bur. Stand. J. Res. **34**, 143 (1945).
1507. A. L. Wahrhaftig, J. Chem. Phys. **10**, 248 (1942).
1508. R. Wajnkranc, Z. Physik **104**, 122 (1937); **105**, 516 (1937).
1509. J. M. Walter and S. Barratt, Proc. Roy. Soc. London **119**, 257 (1928).
1510. ——, Proc. Roy. Soc. London **122**, 201 (1929).
1511. O. H. Walters and S. Barratt, Proc. Roy. Soc. London **118**, 120 (1928).
1511a. R. K. Waring, Physic. Rev. **32**, 435 (1928).
1512. J. Waser and K. Wieland, Nature **160**, 643 (1947).
1513. W. W. Watson, Nature **117**, 157 (1926).
1514. ——, Physic. Rev. **47**, 213 (1935).
1515. —— and A. E. Parker, Physic. Rev. **37**, 167 (1931).
1516. —— and R. Simon, Physic. Rev. **57**, 708 (1940).
1517. —— and R. L. Weber, Physic. Rev. **48**, 732 (1935).
1518. M. Wehrli, Naturwiss. **22**, 289 (1934).
1519. ——, Helv. Phys. Acta **7**, 611 (1934).
1520. —— and W. Hälg, Helv. Phys. Acta **15**, 315 (1942); **16**, 371 (1943).
1521. R. T. Weidner, Physic. Rev. **72**, 1268 (1947); **73**, 254 (1948).
1522. W. Weizel, Z. Physik **51**, 328 (1928).
1523. ——, Z. Physik **52**, 175 (1928).
1524. —— and C. Füchtbauer, Z. Physik **44**, 431 (1927).
1525. ——, H. W. Wolff and H. E. Binkele, Z. physik. Chem. B **10**, 459 (1930).
1526. W. Wenk, Helv. Phys. Acta **14**, 355 (1941).

1527. G. Wentzel, Z. Physik **43**, 524 (1927); Physik. Z. **29**, 321 (1928).
1528. W. West, J. Chem. Phys. **7**, 795 (1939).
1529. G. W. Wheland, J. Chem. Phys. **13**, 239 (1945).
1530. J. U. White, J. Chem. Phys. **8**, 79, 459 (1940).
1531. G. C. Wick, Z. Physik **85**, 25 (1933); Nuovo cimento **10**, 118 (1933)
1532. ――――, Physic. Rev. **73**, 51 (1948).
1533. K. Wieland, Helv. Phys. Acta **12**, 295 (1939).
1534. ――――, Helv. Phys. Acta **14**, 420 (1941).
1535. ――――, Helv. Phys. Acta **19**, 408 (1946).
1536. ――――, J. Chim. Phys. **45**, 3 (1948).
1536a. ――――, V. Henri Volume, p. 229 (Liége, 1948).
1537. ――――, private comm.
1538. ―――― and A. Herczog, Helv. Chim. Acta **29**, 1702 (1946).
1538a. ―――― ――――, Helv. Chim. Acta **32**, 889 (1949).
1539. E. Wigner, Z. physik. Chem. B **23**, 28 (1933).
1540. ――――, J. Chem. Phys. **5**, 720 (1937); **7**, 646 (1939).
1541. R. Wildt, Astrophys. J. **93**, 502 (1941).
1542. H. A. Wilhelm, Iowa State Coll. J. Sci. **6**, 475 (1932).
1543. E. B. Wilson, Jr., Chem. Rev. **27**, 17 (1940).
1544. F. J. Wilson, M. A. thesis, Saskatoon, 1941 (unpublished).
1544a. O. C. Wilson, Publ. Astron. Soc. Pac. **60**, 198 (1948).
1545. J. G. Winans, Physic. Rev. **42**, 800 (1932).
1546. ――――, Physic. Rev. **60**, 169A (1941).
1547. ―――― and M. P. Heitz, Physic. Rev. **65**, 65 (1944).
1548. ―――― and W. J. Pearse, Physic. Rev. **74**, 1262A (1948).
1549. H. G. Wolfhard, Z. Physik **112**, 107 (1939).
1550. R. W. Wood, Phil. Mag. (6) **12**, 499 (1906).
1551. ――――, Phil. Mag. (6) **16**, 184 (1908); Physik. Z. **9**, 590 (1908); **12**, 1204 (1911).
1552. ―――― and G. H. Dieke, J. Chem. Phys. **8**, 351 (1940).
1553. ―――― and F. E. Hackett, Astrophys. J. **30**, 339 (1909).
1554. L. H. Woods, Physic. Rev. **63**, 426 (1943).
1555. ――――, Physic. Rev. **63**, 431 (1943).
1556. ――――, Physic. Rev. **64**, 259 (1943).
1557. H. W. Woolley, J. Chem. Phys. **9**, 470 (1941).
1558. ――――, Bur. Stand. J. Res. **40**, 163 (1948).
1558a. ――――, R. B. Scott and F. G. Brickwedde, Bur. Stand. J. Res. **41**, 379 (1948).
1559. R. E. Worley, Physic. Rev. **64**, 207 (1943).
1560. ――――, Physic. Rev. **65**, 249 (1944).
1561. K. O. Wright, Astrophys. J. **99**, 249 (1944).
1562. C. K. Wu and S. C. Chao, Physic. Rev. **71**, 118 (1947).
1563. O. R. Wulf and L. S. Deming, Terr. Mag. **41**, 299, 375 (1936); **42**, 195 (1937).
1564. ―――― and E. H. Melvin, Physic. Rev. **55**, 687 (1939).
1565. K. Wurm, Z. Physik **79**, 736 (1932).
1566. ――――, Vierteljahresschr. d. Astr. Ges. **78**, 18 (1943).
1567. ―――― and H. J. Meister, Z. Astrophysik **13**, 199 (1937).
1568. H. Yoshinaga, Proc. Phys. Math. Soc. Jap. **19**, 847, 1073 (1937).
1569. T. Yuasa, Sci. Rep. Tokyo Bun. Dai. **3**, 195 (1937).
1570. ――――, Sci. Rep. Tokyo Bun. Dai. **3**, 239 (1938).
1571. ――――, Proc. Phys. Math. Soc. Japan **21**, 498 (1939).
1572. C. Zener, Proc. Roy. Soc. London **140**, 660 (1933).
1573. A. van der Ziel, Physica **1**, 513 (1934).
1574. ――――, Physica **13**, 240 (1947).

AUTHOR INDEX

SUBJECT INDEX

This index serves also as an index of the more frequently used symbols. These symbols are listed at the beginning of the section devoted to the corresponding letter. The Greek letters are arranged under the letter with which they begin when they are written in English (for example, π and Ψ are listed under P). Symbols to which a word is joined are arranged under the corresponding symbol (for example, P *branch* is under P, not under Pb).

Italicized page numbers refer to more detailed discussions of the subjects than ordinary page numbers, or to definitions; boldface page numbers refer to illustrations.

Individual molecules and atoms are listed only under their chemical symbols. Under this symbol is also given all that concerns the molecule or atom it represents (for example, the heat of dissociation of hydrogen is listed as H_2, *heat of dissociation*).

The alphabeting of subjects has been done first on the basis of the part before the comma except in combinations of symbols and words (see above). In alphabeting the second part, after the comma, prepositions and articles at the beginning have been disregarded.

H₂ molecule (*Cont.*)
Lyman bands, 159, 202, 385
magnetic resonance spectrum, **60f.**, *300*, **312f.**
a mixture of two modifications (ortho and para H₂), *139f.*
molecular constants of all electronic states, *530–533*
in planetary atmospheres, *488*
predissociation, 411, 442
pre-ionization, *414*, *419*, 459, 531
probability density distribution of the electrons, *351*, **352**
quadrupole spectrum, *279*, 488
quantized rotation in liquid state, *464*
radio-frequency spectrum, **60f.**, *300*
Raman spectrum, 62f., 65, 115
recombination, 444
repulsive state ³Σ_u⁺, 340, 350f., 352f., 378f., *403*
rotational levels, statistical weight and nuclear spin orientation, **134f.**
rotation-vibration spectrum, *279f.*, 488
spectrum, 34, **36**, 158f., 247, 255f., 273, 307
in stellar atmospheres, 492, 494
thermodynamic functions, 471f.
van der Waals interaction for excited states, **380f.**
Werner bands, 255
H₂⁺ molecule, 326f., 343, *359*, **360f.**, *384*, 402, *534f.*
Half-width of a level, *407*
Halogen hydrides, 374, 383, 446
continuous absorption spectra, 34, 386, 389, 441
infrared and Raman spectra, 53f., 61f., 81f., 105 (*see also individual hydrides*)
Halogens:
absorption bands in the visible region, *273f.*, *385f.*, 438, 445, 465 (*see also individual halogens*)
emission bands, 436f.
recombination spectra, *402*
valency of, *356f.*
Hamiltonian function (operator), *11f.*, 17
Harmonic oscillator, *73–82*
according to the old quantum theory, 76, 163
classical motion, *73f.*
eigenfunctions, *76*, **77f.**
energy levels, term values, **75f.**
heat capacity (content), *468f.*
infrared spectrum, **75**, *79f.*
isotope effect, *141*
potential curve, *74*, **75**
probability density distribution, *76*, **77f.**
probability distribution of momenta (velocities), *78*, **79**
Raman spectrum, *84–88*

Harmonic oscillator (*Cont.*)
selection rules for, *79*, *86*
vibrational frequency, ν_{osc.}, *74f.*, **78f.**, *82*
Harmonics, 92, *94f.*
HBr, 53f., 62, 341f., 386, 393, 441, 464, 472, *534f.*
HBr⁺, 341, *534f.*
HCl:
an atomic molecule, *374*
electronic structure, 341f.
far infrared spectrum, **57f.**, 64, *81f.*, 105
infrared and Raman spectra in liquid, solid and dissolved state, *464*
isotope effect, *141f.*
molecular constants, **95f.**, 98, 105, *113f.*, *534f.*
near infrared spectrum (bands), **53f.**, 62, *81f.*, *94f.*, 127
fine structure of, **55f.**, **57**, **126**
potential curve, **91**, 440
Raman spectrum, *61*, **62f.**, 65
thermal distribution in the ground state, *123*, **124f.**
thermodynamic functions, 472
H²Cl (DCl), *143*, *534f.*
HCl⁺, 264, 341, *534f.*
HD, 61, 141, 202, 300, 472, *532f.* (*see also* H₂)
HDO, 483
He³, He⁴ nuclei, 135f., 460f.
He₂ (He + He), 136, 229f., 287, 348, 459
continuum, *404*
electronic structure, *343f.*, 355, *364f.*, 535f.
molecular constants, 535f.
potential energy, *378f.*
spectrum, 34, 132, 136, 247, 256f., 273, 304, 322, 327
He₂⁺, 343, 364, 536f.
Head:
of a band, 30, *47f.*, 113, *171f.*, 205f., 306
anomalous (extra heads), **174f.**
distance from the zero line, *171f.*, 193
of a comet, 488, *490f.*
of heads, *160f.*, 174f., 198
of the Q branch, *172*
Headless bands, 30, **39**, 50, *171f.*, *174*
Heat:
of combustion, 481
of dissociation, *see* Dissociation energy
of formation, atomic, *480f.*
of reaction, *474f.*, 480f.
of sublimation, 481
Heat capacity, molar, *467f.*
difference for ortho and para modifications, 140, **470f.**
Heat content of a gas, *466f.*
contribution of vibration and rotation, 123, *468*
Heavy hydrogen, *see* D₂, T₂, HD, HT
HeH, HeH⁺, *354f.*

Life (*Cont.*)
  of symmetrical (antisymmetrical) states, *139*
Light scattering, *61*, *82f.*
Light, velocity of, 2
LiH, 181, 183, 192, *341f.*, **376f.**, *546f.*
LiI, 546f.
LiK, 546f.
Limit of a band (line) series, *see* Convergence limit
Limiting curve of dissociation, **428f.**, *432*
Linear combinations, 214
  of atomic orbitals, *330*, *362*, 368, 384
Line broadening:
  due to radiationless decomposition, *406f.*, *410f.*, 418, *429*, 432
  by pressure, *397f.*, *405*
Line strength, *127f.*, *208f.*, *250*, 258, 271, 465
Liquid state, nature of, *464f.*
LiRb, 547
London dispersion forces, *378f.*
Long-wave-length limit:
  of a dissociation continuum, *390*, *394*, *441f.*
  of excitation of atomic fluorescence, *443*
  of an ionization continuum, *388*
  of photodissociation, 443
Loomis-Wood procedure for picking out branches, **191f.**
Lorentz-Weisskopf broadening, 398
(**L,S**) coupling, *26*, 216, 237
LuO, *547*
Luminosity of stars, 492
Lyman bands, *see* H_2
Lyman-Birge-Hopfield bands, *see* N_2

M

m, electron mass, 2
m, running number, *42f.*, *56f.*, 63, *111f.*, 169f., 174
m_l, component of *l* in the field direction, *17*, 22, *324f.*, 331f.
m_s, component of *s* in the field direction, 22, 331f.
M, M_x, M_y, M_z, dipole moment, *19*, 80, 96, 130f., *199f.*, 382
M, M_J, quantum number of the component of **J**, *28*, *69*, 118, *298*, 311
  selection rule for, 29, *73*, 243, *299f.*, 313
M shell, 22
M-type stars, 492f.
M_I, quantum number of the component of **I**, 311, 313
M_K, quantum number of the component of **K**, 303
M_L, quantum number of the component of **L**, 28, *212f.*, *316f.*, 322
  selection rule for, 29, 241

M_S, quantum number of the component of **S**, 28, 214, 303
  selection rule for, 29, 242, 304f.
M_T, quantum number of the component of **T**, *133*, 137, *311*, *313*
μ, electric dipole moment of the molecule, *307f.*, 377
μ, reduced mass, *67*, 75, 141
$\bar{\mu}$, mean magnetic moment, 462
μ_A, reduced mass in atomic-weight units, 77f., 98, 180, 501
$\bar{\mu}_E$, mean component of electric moment in the direction of *E*, *307f.*
$\bar{\mu}_H$, mean component of magnetic moment in the direction of *H*, *298f.*, *303*
μ_I, $\bar{\mu}_I$, nuclear magnetic moment, *311f.*
μ_J, $\bar{\mu}_J$, component of μ in the direction of **J**, *299f.*, *303*, 309
μ_Λ, μ_Ω, μ_S, magnetic moment of Λ, Ω or **S**, *300f.*, *303*
μ_0, Bohr magneton, *299*
μ_{0n}, nuclear magneton, *299*
Magnetic dipole radiation, 19f., 131, 275, *277f.*, *300*
  selection rules for, 130, *277*, 300
Magnetic field:
  influence on bands, *298–307*
  producing forbidden predissociations, *419*
Magnetic moment:
  of the atomic nucleus, 139, *309f.*, *460f.*
  of the molecule, 222f., *298f.*, *462*
Magnetic quantum number, *see* m_l, *M*
Magnetic resonance spectra, **60f.**, *300*, 311, **312f.**
Magnetic rotation spectra, *306f.*
Magneton, *299f.*
Main branches, 244, 249, 251, 262, 270
Manifold:
  of electronic terms, *315–348*
    determination:
      from the electron configuration, 315, *322–348*
      from the terms of the separated atoms, *315–322*
      from the terms of the united atom, 315, *322*
  of states of an electron, *324*
Many-line spectrum, 34, **36**, 49, 340
Mars atmosphere, *487*
Mass of atoms, determination from band spectra, 145, 168
Matrix elements:
  of the dipole moment, *19f.*, *71f.*, *79f.*, *130*, *199f.*, *242*, *382*
  of the perturbation function, *14*, 93, *283f.*, 286, 288, 379, 408, 418, *421f.*
  of the scattering moment (polarizability), *86f.*, *90*
Maxwell-Boltzmann distribution, 122